PAVEMENT, ROADWAY, AND BRIDGE LIFE CYCLE ASSESSMENT 2020

PROCEEDINGS OF THE INTERNATIONAL SYMPOSIUM ON PAVEMENT. ROADWAY, AND BRIDGE LIFE CYCLE ASSESSMENT 2020 (LCA 2020), SACRAMENTO, CA, USA, 3-6 JUNE 2020

Pavement, Roadway, and Bridge Life Cycle Assessment 2020

Editors

John Harvey
University of California, Davis, USA

Imad L. Al-Qadi
University of Illinois at Urbana-Champaign, USA

Hasan Ozer
Arizona State University, Tempe, USA

Gerardo Flintsch
Virginia Polytechnic Institute and State University, Blacksburg, USA

CRC Press
Taylor & Francis Group
Boca Raton London New York

CRC Press is an imprint of the
Taylor & Francis Group, an **informa** business

A BALKEMA BOOK

Published by:
CRC Press/Balkema
Schipholweg 107C, 2316 XC Leiden, The Netherlands

© 2020 by Taylor & Francis Group, LLC
CRC Press/Balkema is an imprint of Taylor & Francis Group, an informa business

No claim to original U.S. Government works

ISBN-13: 978-0-367-55166-7 (hbk)
ISBN-13: 978-1-003-09227-8 (eBook)

DOI: 10.1201/9781003092278
https://doi.org/10.1201/9781003092278

Typeset by Integra Software Services Pvt. Ltd., Pondicherry, India

Visit the Taylor & Francis Web site at
http://www.taylorandfrancis.com

and the CRC Press Web site at
http://www.crcpress.com

Library of Congress Cataloging-in-Publication Data

Table of contents

Forword

An increasing number of agencies, academic institutes, and governmental and industrial bodies are embracing the principles of sustainability in managing their activities. Life Cycle Assessment (LCA) is an approach developed to provide decision support regarding the environmental impact of industrial processes and products. LCA, which is undergoing continued improvement, is being implemented world-wide, particularly in the areas of pavement, roadways and bridges. This includes standardization of practice, better alignment with international norms in other fields, resolution of gaps in data and technical approaches, and greater understanding of LCA advantages and challenges in assist decision-makers.

The International Symposium on Pavement, Roadway, and Bridge Life Cycle Assessment 2020 provided a forum for sharing and discussing experiences and results, assessing status, plans for implementation, challenges, and identification of the extent of consensus on current issues. This symposium was organized by the University of California Pavement Research Center and the National Center for Sustainable Transportation. The symposium is a follow-on to the 2010 Pavement LCA Workshop in Davis, California; 2012 RILEM Symposium on LCA for Construction Materials in Nantes, France; 2014 Pavement LCA Symposium in Davis, California, and Pavement Life-Cycle Symposium 2017 in Champaign, Illinois.

The conference brought together academic, government, and industrial leaders from around the world. Selected papers are included in these proceedings. Each paper has been peer-reviewed by at least three professionals in this field. Based on the reviewers' recommendations, the papers that suited the conference goals and objectives were included in the proceedings. The proceedings papers cover various research and practical issues related to pavement, roadway and bridge LCA, including data and tools, asset management, environmental product declarations, procurement, planning, vehicle interaction, and impact of materials, structure, and construction.

The technical program of the symposium consisted of keynote speeches, short invited presentations on key topics, approximately 40 oral presentations of papers, and panel discussions. The symposium was developed in consultation with government and industry advisors and a scientific committee of over 60 members. The organizers of the symposium acknowledge the efforts of all members of the scientific committee whose help has vastly contributed to the success of the symposium. We are thankful for all those who volunteered their time to thoroughly review the submitted papers and offer constructive comments to authors.

John Harvey
University of California, Davis

Imad L. Al-Qadi
University of Illinois at Urbana-Champaign

Hasan Ozer
Arizona State University

Gerardo Flintsch
Virginia Polytechnic Institute and State University

Symposium organization

Chairmen

John Harvey
University of California, Davis

Imad L. Al-Qadi
University of Illinois at Urbana-Champaign

Hasan Ozer
Arizona State University

Gerardo Flintsch
Virginia Polytechnic Institute and State University

On-Site Organizing Committee

Natalie Ruiz, *National Center for Sustainable Transportation, UC Davis*
Leon Szeto, *National Center for Sustainable Transportation, UC Davis*
Jon Lea, *UC Pavement Research Center, UC Davis*
Ali Butt, *UC Pavement Research Center, UC Davis*
Arash Saboori, *UC Pavement Research Center, UC Davis*
David Miller, *UC Pavement Research Center, UC Davis*
Laura Melendy, *Institute of Transportation Studies, UC Berkeley*
Nick Burmas, *California Department of Transportation*

Government, Academic, and Industry Supporters

National Center for Sustainable Transportation (UC Davis), Transportation Research Board, Federal Highway Administration, National Asphalt Pavement Association, National Ready Mix Concrete Association, Interlocking Concrete Pavement Institute, Tongji University, American Concrete Pavement Association, Asphalt Recycling & Reclaiming Association, Asphalt Institute, International Society for Concrete Pavement, European Asphalt Technology Association, International Society for Asphalt Pavement/Asphalt Pavement and the Environment, International Grooving and Grinding Association, Foundation for Pavement Preservation, Southwest Concrete Pavement Association, California Nevada Cement Association, California Asphalt Pavement Association, California Construction and Industrial Materials Association, University of Illinois at Urbana-Champaign, Arizona State University, Virginia Polytechnic Institute and State University.

Scientific Committee

Gordon Airey, *University of Nottingham, UK*
Warda Ashraf, *University of Maine, USA*
Danilo Balzarini, *ARA, USA*

Andrew Braham, *University of Arkansas, USA*
Ali Butt, *University of California, Davis, USA*
Francesco Canestrari, *Università Politecnica delle Marche, Italy*
Silvia Caro, *Universidad de los Andes, Colombia*
Alan Carter, *Université du Québec, Canada*
Dingxin Cheng, *California State University, Chico, USA*
Ghassan Chehab, *American University Beirut, Lebanon*
Bouzid Choubane, *Florida Department of Transportation, USA*
Erdem Coleri, *Oregon State University, USA*
Steve Cross, *Oklahoma State University, USA*
Eshan Dave, *University of New Hampshire, USA*
Anne de Bortoli, Université de Paris-Est, France
Rodrigo Delgadillo, *Universidad Técnica Federico Santa María, Chile*
Samer Dessouky, *University of Texas, San Antonio, USA*
Heather Dylla, *Federal Highway Administration, USA*
Ahmed Faheem, *Temple University, USA*
Navneet Garg, *Federal Aviation Administration, USA*
Daba Gedafa, *University of North Dakota, USA*
Jeremy Gregory, *Massachusetts Institute of Technology, USA*
Elie Hajj, *University of Nevada, Reno, USA*
Noe Hernandez, *Universidad Nacional Autónoma de México, Mexico*
Bernhard Hofko, *Technische Universität Wien, Austria*
Joe Holland, *California Department of Transportation, USA*
Arpad Horvath, *University of California, Berkeley, USA*
Sallie Houston, *VSS Emultech, USA*
Xiaoming Huang, *Southeast University, China*
Erdin Ibraim, *University of Bristol, UK*
Björn Kalman, *Swedish National Road and Transport Research Institute, Sweden*
Sampat Kedarisetty, *University of California, Davis, USA*
Jon Lea, *University of California, Davis, USA*
Jeremy Lea, *University of California, Davis, USA*
Zhen Leng, *Hong Kong Polytechnic University, China*
Eyal Levenberg, *Technical University of Denmark, Denmark*
Hui Li, *Tongji University, China*
Yanlong Liang, *University of California, Davis, USA*
Jeffrey Lidicker, *California Air Resources Board, USA*
Asa Lindgren, *Swedish Transport Administration, Sweden*
Mark Lozano, *University of California, Davis, USA*
Qing Lu, *University of South Florida, USA*
James Mack, *CEMEX, USA*
Mehran Mazari, *California State University, Los Angeles, USA*
Sabbie Miller, *University of California, Davis, USA*
Amlan Mukherjee, *Michigan Technological University, USA*
Maryam Ostovar, *University of California, Davis, USA*
Tony Parry, *University of Nottingham, UK*
Matteo Pettinari, *Danish Road Directorate, Denmark*
Prashant Ram, *Applied Pavement Technology, USA*
Maria Carmen Rubio, *University of Granada, Spain*
Shadi Saadeh, *California State University, Long Beach, USA*
Arash Saboori, *University of California, Davis, USA*
Shoshanna Saxe, *University of Toronto, Canada*
Joseph Shacat, *National Asphalt Pavement Association, USA*
Jo Sias, *University of New Hampshire, USA*

Nadarajah Sivaneswaran, *Federal Highway Administration, USA*
Wynand Steyn, *University of Pretoria, South Africa*
Gabriele Tebaldi, *Università degli Studi di Parma, Italy*
Susan Tighe, *University of Waterloo, Canada*
Shane Underwood, *North Carolina State University, USA*
Hao Wang, *Rutgers University, USA*

Mapping of unit product system/processes for pavement life-cycle assessment

Chaitanya G. Bhat
Graduate Student, Michigan Technological University, Houghton, USA

Amlan Mukherjee
Associate Professor, Michigan Technological University, Houghton, USA

Joep Meijer
The Right Environment, Austin, USA

ABSTRACT: An impediment to the application of Life-Cycle Assessment (LCA) in supporting decision-making is the consistent definition of goal and scope and use of reliable and high-quality datasets. Building on the FHWA pavement LCA framework, this paper presents an ontology that reflects a relational grammar to represent the unit and product system processes required for pavement LCAs. Methodologically, the paper builds upon previous research that defined lineage and process ontologies for LCA using Resource Description Frameworks (RDF) to represent the data semantics for pavement LCA. RDF relates LCA flows and processes in a subject-predicate-object relationship. The resulting ontology provides a foundational structure for libraries that can be used to ensure the transparency and consistency of any software tool that is developed to compute pavement LCAs. An ongoing research effort is mapping the product systems developed through this ontology onto publicly available background databases for consistent use in conducting pavement LCA. This is carried out in collaboration with the Federal LCA Commons to ensure the use of public datasets that are already being collected across different agencies. In addition, the ongoing effort aims to characterize data quality, and review background data as 'sufficient and appropriate' by developing a pavement-specific pedigree matrix. As LCA software tools are developed to support decision-making for departments of transportation, this paper along with the ongoing research provides a common set of protocols and LCA inventories that can support reliable and easy to use decision support applications.

1 INTRODUCTION

An impediment to the application of Life-Cycle Assessment (LCA) in supporting decision-making is ensuring the consistent definition of goal and scope and use of reliable and high-quality datasets. As LCA begins to inform public procurement for highway construction (Buy Clean California Act, 2017), this is becoming a challenge in informing decision-making in Departments of Transportation (DOT). In addition, it is critical to ensure that the software tools being developed to support DOT decision-making, meet standards of quality and reliability through compliance with the pavement LCA system boundaries, and by using agreed-upon life cycle inventories that meet specific data quality standards. This need can be met by developing a standard pavement construction specific mapping of the relationships between flows, processes and product sub-systems within an agreed-upon system boundary. Hence, the objective of this paper is to develop an ontology for the pavement LCA system that can be used to map unit processes within the pavement LCA system boundary as developed by the Federal Highways Administration (FHWA) Sustainable Pavements Program (fhwa.dot.gov, 2019).

Specifically, the paper presents an ontology that reflects a relational grammar and a classification of the information on pavement LCA flows and processes, to represent the unit and product system processes required for pavement LCAs.

Methodologically, the research builds upon the previous research (Ingwersen et al., (2015), Cashman et al., (2016), Edelen et al., (2017)) that defines lineage and process ontologies for LCA using Resource Description Frameworks (RDF) to represent the data semantics for pavement LCA. RDF relates the LCA flows and processes in a subject-predicate-object relationship.

The work presented in this paper is part of an ongoing research effort that is also mapping the product systems developed through this ontology onto publicly available background databases for consistent use in conducting pavement LCA. The ontology provides a relational grammar that can be used to represent interactions between flows and processes as defined within the pavement LCA framework. This effort is happening in collaboration with the Federal LCA Commons (Lcacommons.gov, 2019) to ensure the use of public datasets that are already being collected across different agencies. In addition, the ongoing effort aims to characterize data quality and review background data as 'sufficient and appropriate' by developing a pavement-specific pedigree matrix based on the updated pedigree matrix document from the United States Environmental Protection Agency (Edelen and Ingwersen, 2016). The scope of this paper is limited to presenting the ontology development and discusses its applications to software tool development for pavement LCA. It is expected that this ontology will contribute to a foundational structure for libraries that can support any software tool that is developed to compute pavement LCAs. The contribution of this paper along with the ongoing research is that it provides a common set of protocols and LCA inventories that can support reliable and easy to use decision support applications.

2 BACKGROUND

The state of academic inquiry and practice in the field of pavement LCA has come a long way since the inception of the FHWA's Sustainable Pavement Program and the associated Sustainable Pavements: The Technical Working Group (SPTWG) in 2011 (fhwa.dot.gov, 2019). The outcomes of these efforts are the FHWA's pavement sustainability reference manual (Van Dam et al., 2015) and the pavement LCA framework (Harvey et al., 2016). These are useful resources for state and federal agencies to develop a better understanding of pavement LCA and its relationship to building sustainable pavements.

Meanwhile, the pavement construction materials industry has embraced LCA and both the concrete (carbonleadershipforum.org, 2019) and asphalt (asphaltpavement.org, 2019) industries have developed ISO 14025 (ISO, 2006) and EN 15804 (CEN, 2012) compliant Environmental Product Declaration (EPD) programs. This has laid the foundations for the development of Product Category Rules (PCRs) for pavement construction materials, empowering industry organizations to step in as program operators and develop industry-specific PCRs, and tools that encourage the use of EPDs in communicating the impacts of cradle-to-gate LCAs. These efforts have been supported by and conducted in collaboration with the ongoing academic inquiry (Mukherjee, 2016), particularly in furthering models of the use phase of pavement LCA, with emphasis on topics such as Pavement-Vehicle Interaction (PVI) (Zaabar and Chatti, (2010) cshub.mit.edu, (2019), Harvey et al., (2016)) and heat island effect (Sen and Roesler, (2016), Li, (2012)).

A recent outcome of this collaborative work has been the development of the pyramid that provides perspective on the organization of the pavement LCA framework and provides direction for the road ahead. Figure 5 (present in Supporting Information) illustrates the pyramid. It provides perspective on the organization of the LCA framework and identifies subsequent steps that move towards the implementation of pavement LCAs, through data organization, management, and eventual tool development (Ram et al., 2017).

At the very base of the pyramid is the development of LCA framework that has been completed. The top of the pyramid represents the software tools and the specifications and policies for the implementation of pavement LCAs. In between are the crucial levels of organization that require consideration of the decision-making contexts in which LCAs are to be applied as well as the consistent and structured use of the different kinds of data that lie at the heart of a successful LCA.

The organizational context of the application of LCAs lies in the area of pavement construction procurement, and while it is relevant to the work presented through this paper, it is beyond the scope of this paper. The paper focuses on developing an ontology that maps the processes in the pavement LCA system boundary and the ongoing research is mapping available life cycle inventories to the processes. This completes the work in the third and fourth levels (from the bottom) of the pyramid.

The paper builds on previous work on defining ontologies for LCA. Specifically, different ways of developing an ontology and collection methods in other LCA application areas were reviewed. This provided a methodological foundation as well as the points of departure for this paper. The relevant literature surveyed were:

- Ingwersen et al., (2015) initiated the development of an LCA Harmonization Tool (LCA-HT), using RDF to define a new data architecture for chemicals. The purpose of this study was to enhance the data interoperability by automatically combining, storing, and annotating LCA data.
- Janowicz et al., (2015) presented an ontology pattern to specify the notions and properties of flows, activities, agents, and products with an intention to reduce the inefficiencies in LCA data collection and management for commonly used chemicals.
- Cashman et al., (2016) identified challenges in the current data management practices and developed a data mining method based on lineage and process ontology to automate data inventory modeling. This study encouraged the use of publicly available data at the United States Environmental Protection Agency in the field of chemical manufacturing.
- Edelen et al., (2017) identified the gaps in the existing definition of elementary flows, the essential components of data used for LCA, and, proposed an approach to benchmark the collection, definition, and evaluation of data through typology, use of unique identifiers and standard nomenclature.
- Zhang et al., (2015) developed flow and process ontologies using the Web Ontology Language, followed by a semantic representation model to present the relationships between flows and processes as an RDF graph. However, the literal meaning of terminology "process ontology" used in Zhang et al., (2015) are different to the one defined in Cashman et al., (2016) as the former refers to the literal definition of "flow" and "process".

The following representational constructs adopted from the literature provide the foundation for the ontology proposed in this paper.

- The use of RDF for relating entities such as flows and processes in a subject-predicate-object relationship within the context of pavement LCAs as per Ingwersen et al., (2015).
- The use of a set parameter typology for each entity that is being mapped as per Edelen et al., (2017). This helps to specify the parameters that are provided as input by pavement engineers and other stakeholders in conducting LCA.
- The use of lineage and process ontologies, as used by Cashman et al., (2016) in organizing and relating the generic relationships between LCA entities such as flows and processes, and the specific relationships between materials, energy and processes used that are specific to pavement LCAs. This provides the underlying structure for the ontology that can reflect the RDF relationships.
- Adoption of an object-oriented approach to associating the parameter typology for pavement LCA to the underlying ontology framework reflecting the pavement product system. As the OpenLCA application programming interface (API) (github, 2019) already has a set of class definitions in place for critical entities representing flows, processes, and product systems; a decision was made to inherit their class definitions. This will also allow easy interoperability with the OpenLCA platform in the future. Each of these components is described in the next section.

A critical difference between the discussed literature and this paper lies in the underlying motivation for developing the ontology. For example, Zhang et al., (2015) aided in deductive reasoning for automating the modeling for LCA in general and they implemented the model for a case study of ball bearings. Similarly, Cashman et al., (2016) used data mining to infer a framework of relationships that reflect inventory models in the chemical manufacturing industry. In comparison, the pavement LCA framework already provides us with a framework of product and process relationships specific to pavements. Hence, the discovery of the underlying framework using inductive reasoning is not necessary for this paper.

Instead, the paper uses the existing pavement LCA framework to define "legal" product systems to ensure the consistency of any software tool that is developed to compute pavement LCAs. Recently, a similar mapping effort was conducted by Bernstein et al., (2019) to provide process-level parametric relationships for unit manufacturing process (UMP) models as defined by ASTM E3012, (2016). Their study first generated a life-cycle inventory (LCI) model for a vertical milling process and then linked the results obtained with an existing LCI database. Further, they compared the LCA results from their mapping effort against LCA results of a similar milling unit process model from a commercial database. The work presented in this paper is a parallel effort for the pavement LCA domain.

The ontology presented in this paper was not used to conduct automated reasoning, but instead to develop the representation for pavement LCA information that can in future support automated reasoning for LCA software tools. These tools can support decision-making for transportation agencies. The common ontology presented in this paper and the mapping to public data sources in the ongoing research effort will ensure transparency and reliability of these tools and create trust among industry stakeholders.

3 METHODOLOGY

This section discusses the building blocks of the proposed ontology. It is expected that any software that is built to support pavement LCA will follow this ontology to ensure that the underlying processes and product systems are consistently represented. To that extent, the building blocks of the ontology consist of:

1. *Process Ontology*: A relational grammar using RDF – this is a pavement LCA specific representation that builds on the underlying LCA graphical network. It is used to define the relationships between flows and processes in LCA.
2. *Lineage Ontology*: This is a classification of all the different flows and processes specific to pavement LCAs.
3. *Product System Diagrams*: Using the Process Ontology and the Lineage Ontology, the Product System diagrams create an exhaustive mapping of all possible flows and processes in the pavement LCA framework. For a given product system and a reference output flow, the input flows and output flows are designated by a set of "legal" combinations that are deemed consistent with the pavement LCA framework. Anything inconsistent with the framework, are not part of the product system diagrams and cannot be represented using this ontology.
4. *Class Definitions*: This is an object-oriented programming approach to developing class definitions for all the processes and flows in pavement LCA and extending the OpenLCA API to enable the development of a specification for a software that can support pavement LCA.
5. *Data Types*: The data used in pavement LCA comes in different forms through different stakeholders. Some of them are parameters provided from design, material production, and construction processes, while others are upstream life cycle inventories. Classification of these data types is necessary, including a definition of which stakeholder the data is sourced from.

The rest of this section discusses each of these components in detail.

3.1 Definition of flows, processes and product systems

A *product system* is a collection of unit processes with elementary and product flows, performing one or more defined functions, and which models the life cycle of a product (ISO, 2006).

A *unit process* smallest element considered in the life cycle inventory analysis for which input and output data are quantified (ISO, 2006). This paper will further refer to this as *process* for simplification.

A *flow* is an input or output to a process representing raw materials, energy used in a process or emissions of pollutants into the environment (ISO, 2006). This definition of *flow* is representative of *elementary flows, product flows,* and *waste flows.*

In this mapping, a product system is represented as a one-to-one mapping of flows to processes and represented using the graph-theoretic notation: where, P is a set of processes represented as nodes, and F is a set of flows represented as directed edges. The graph π (P, F) is a directed graph denoting the product system. Figure 1, indicates a product system π (P, F), that can be represented as follows:

$$P = \{P_1, P_2, P_3, P_4, P_5, P_6\};$$
$$F = \{F_1, F_2, F_3, F_4, F_5, F_6\}$$
$$\pi (P, F) = \{(P_1, P_3), (P_2, P_3), (P_3, P_5), (P_4, P_5), (P_5, P_7), (P_6, P_7)\}$$

The *flow, process* and *product system* definitions are consistent with the model graph in Open-LCA and form the basis of this ontology. As illustrated in Figure 1, a product system π (P, F), can be defined as a collection of *processes* and *flows* using a graphical representation. This is the fundamental construct for representing the relationships between the *flows* and *processes* in LCA, in general.

3.2 Definition of process ontology

Building on the network constructs defined in the previous section, this section discusses the relationships between different *flows, processes* and *product systems* using the RDF methodology as it applies to pavement LCA. The RDF methods defined in this paper are:

1. *IsProducedBy*: Relates a flow object to a process object that it is an output of.
2. *IsConsumedBy*: Relates a flow object to a process object that it is an input to.
3. *IsMovedBy*: Relates flow object to a specific transportation process object that is used to move it between locations.

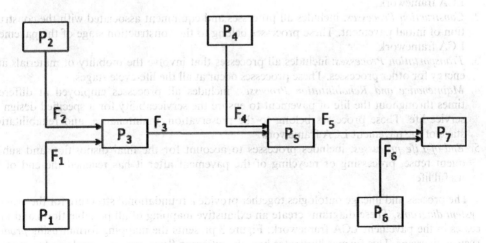

Figure 1. Definition of a product system.

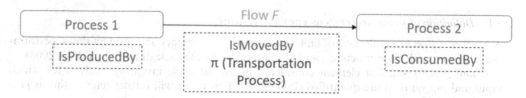

Figure 2. Basic resource description framework schema for pavement LCA.

These methods are already implemented in the OpenLCA API, excepting for the *IsMovedBy* construct that is defined specifically to indicate the role of transportation processes in pavement LCA. Figure 2 illustrates the fundamental relationships between *flows* and *processes* in a general format. The flow *F* is an output of *Process 1*, denoted by the method *IsProducedBy* and it is an input to *Process 2*, denoted by *IsConsumedBy*. In between the two *processes*, it is moved by the method *IsMovedBy* by a *transportation process*, that is represented by a product system. The transportation process product system - π (*transportation process*) can in turn be comprised of a set of flows and processes describing the outcomes of a specific mode of transportation. For example, if *F* is asphalt binder, *Process 1*, is a process called *Refining of Crude Oil*, *Process 2* is a process called *Production of Asphalt Mixture*, and the *transportation process* involved is *Transportation by Train*, then the RDF triple will represent the life cycle exchange of: Asphalt binder *IsProducedBy Refining of Crude Oil, IsMovedBy Transportation by Train* and *IsConsumedBy Production of Asphalt Mixture*. A series of these RDF triples connected using a network as expressed in the previous section will be used to construct the narrative of the pavement LCA system boundary.

3.3 *Definition of lineage ontology*

The lineage ontology identifies the different kinds of *flows* and *processes* and classifies them in an ordered fashion. Generically, all *flows* can be classified as either *material flows* or *energy flows*. Further to reflect the context of pavement LCAs, as per the definition of the FHWA framework, the *material*, and *energy* flows are subdivided into materials and energy of interest as shown in the lineage tree in This tree classifies all the different flows in the pavement LCA based on whether they are *materials* or *energy*. Similarly, it also classifies all the *processes* as follows:

1. *Production Processes*: includes all processes used in the acquisition and processing of pavement materials. These processes belong to the material production stage of the pavement LCA framework.
2. *Construction Processes*: includes all processes and equipment associated with the construction of initial pavement. These processes belong to the construction stage of the pavement LCA framework.
3. *Transportation Processes*: includes all processes that involve the mobility of materials and energy for other processes. These processes occur at all the life-cycle stages.
4. *Maintenance and Rehabilitation Processes*: includes all processes employed at different times throughout the life of pavement to ensure the serviceability for a specified design or service life. These processes belong to the preservation, maintenance, and rehabilitation stage of the pavement LCA framework.
5. *End-of-Life processes*: includes processes to account for the final disposition and subsequent reuse, processing or recycling of the pavement after it has reached the end of its useful life.

The process and lineage ontologies together provide a foundational structure for the *product system diagrams*. These diagrams create an exhaustive mapping of all possible flows and processes in the pavement LCA framework. Figure 3 presents the mapping format using *product system diagrams*. This format illustrates how the different *flows*, *processes* and *product systems*

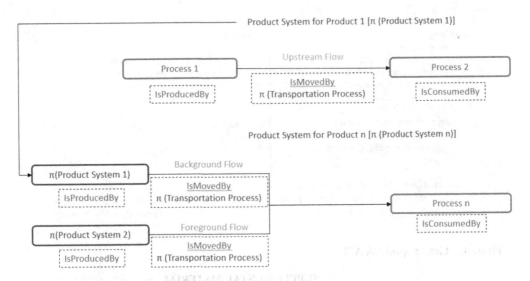

Figure 3. Mapping format using product system diagrams.

within the system boundary relate to each other, along with the data types necessary to relate them to each other. Hence, each product system diagram is a statement of the allowable lists of processes, flows and transportation processes that can be combined in a network to produce the reference product.

3.4 *Definition of class hierarchy*

This section details an object-oriented programming approach to develop class definitions for all the *processes* and *flows* in pavement LCA and extend the OpenLCA API to enable the development of a specification for a software that can support pavement LCA. The class definition mentioned in this section is consistent and will inherit the class definitions for flows and processes as per the Java™ API for OpenLCA. Each of the specific *material* and *energy* flows, and the associated processes defined in the pavement LCA are defined as child classes of the classes defined in the previous section. Hence, a material such as an asphalt mixture will be defined by an *AsphaltMixture* class that is a child class of the abstract class *Material* which in turn is a child of the *Flow* class from the OpenLCA API. Similarly, all *material* and *energy* flows used in pavement LCA are defined as sub-classes of the abstract classes *Material* and *Energy* which in turn inherit the OpenLCA *Flow* class. For processes, the OpenLCA *Process* class is parent to the abstract classes for *production processes, transportation processes, construction processes, maintenance and rehabilitation processes,* and *end-of-life processes*. This class hierarchy is illustrated in Figure 6 (present in the Supporting Information).

Figure 4 represents the glimpse of the software architecture through a pseudo code. This code is for the definition of the abstract class materials that inherited the flow class from the OpenLCA API, and in turn, has *AsphaltMixture* as a child class. Similarly, the abstract class *ProductionProcess* that inherits the process class from the OpenLCA API, and in turn has *AsphaltProduction* as a child class. This will help easy interoperability within the OpenLCA platform, of any future pavement LCA software that can directly take advantage of the OpenLCA API, and all the LCA computational methods that OpenLCA already provides.

Another advantage of this method is that when defining class properties and methods, this effort can build on existing definitions of identifier and location information for products and processes already defined in data inventories available in OpenLCA's *.zolca* format. These fields include Universally Unique Identifier (UUID), and Chemical Abstracts Service number (CAS).

7

```
org.openlca.core.model                              org.openlca.core.model
Class Flow                                          Class Process
    java.lang.Object                                    java.lang.Object
        org.openlca.core.model.AbstractEntity               org.openlca.core.model.AbstractEntity
        org.openlca.core.model.RootEntity                   org.openlca.core.model.RootEntity
            org.openlca.core.model.CategorizedEntity            org.openlca.core.model.CategorizedEntity
            org.openlca.core.model.Flow                         org.openlca.core.model.Process


public abstract class Materials extends Flow {      public abstract class ProductionProcess extends Process {
    // declare fields                                   // declare fields
    // declare nonabstract methods                      // declare nonabstract methods
}                                                   }


Public class AsphaltMixture extends Materials {     Public class AsphaltProduction extends Process {
    // declare asphalt mixture specific fields          // declare asphalt production specific fields
    // declare asphalt mixture methods                  // declare asphalt production specific methods
}                                                   }
```

Figure 4.   Link to openLCA API.

SUPPLEMENTAL MATERIAL

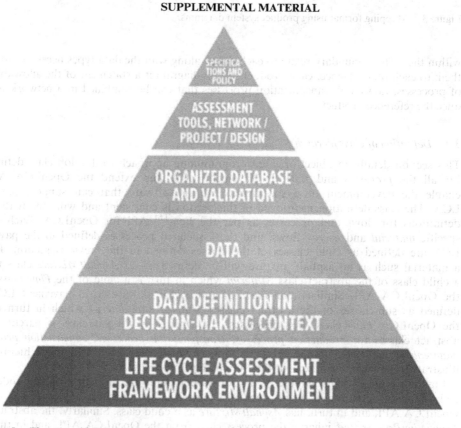

Figure 5.   The pyramid (Ram et al., 2017).

The newly defined classes also provide the opportunity to include methods that are specific to the pavement LCA framework. These methods will supplement the methods already inherited from OpenLCA for conducting standard LCA calculations such as *.ProductSystem( )* to create a product system and *.ImpactMethod( )* to apply impact methods such as TRACI (Bare, 2012). These new methods are currently being developed in an

8

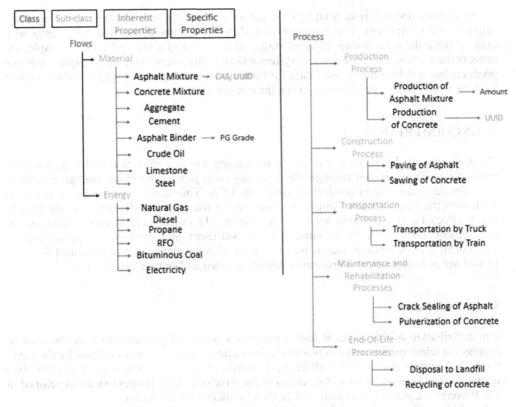

Figure 6. Lineage ontology.

ongoing research effort, considering pavement LCA specific functions such as comparing pavement LCA outcomes using different kinds of mixtures or having different perform-ance expectations over a defined lifetime. The discussion of these methods is currently beyond the scope of this paper.

3.5 Data types

To account for granularity, the paper defines three different kinds of data to characterize flows, processes and product systems, as follows:

1. Upstream: This is a flow, process or product system that is characterized entirely by generic public inventories. The person conducting a pavement LCA has no influence on it and does not collect it. For example, Crude Oil is an upstream flow that *IsProducedBy* the process Extraction of Crude Oil from Onshore and Offshore Wells.
2. Background: This is a flow that is characterized by generic public inventories, though the person conducting a pavement LCA has to input specific parameters based on national and regional averages defining it. However, there is no way to control these average parameters through pavement design or construction. For example, percentages defining the electricity mix in a region denoted by Electricity at Grid is a background flow that *IsProducedBy* the product system Electricity at Grid.
3. Foreground: This is the data that is specific to a pavement LCA and the person conducting the LCA has direct control over it through modification of design and construction strat-egies. For example, the percentage of recycled content in pavement, or the quantity of diesel used by equipment during construction, or the distance traveled through transporta-tion are foreground flows.

The ongoing research is associating different data types with stakeholders such as agencies, suppliers and contractors. This is important as different stakeholders interact differently with each of these datasets during different stages of a pavement's life cycle. For example, the impacts from *AsphaltMixture* product system will be a *foreground* data for an asphalt mixture producer but will be a *background* data for both contractors procuring and installing asphalt mixture as well as for agencies supervising the construction process.

4 ONGOING RESEARCH

The work presented in this paper is part of an ongoing research effort that is also mapping the product systems developed through this ontology onto publicly available background databases for consistent use in conducting pavement LCA. This effort is happening in collaboration with the Federal LCA Commons to ensure the use of public datasets that are already being collected across different agencies. In addition, the ongoing effort aims to characterize data quality and review background data as 'sufficient and appropriate' by developing a pavement-specific pedigree matrix based on the updated pedigree matrix document from the United States Environmental Protection Agency (Edelen and Ingwersen, 2016).

5 CONCLUSION

The contribution of this paper is that it presents a relational grammar that can be used to express the relationships between *processes*, *flows* and *product systems* as defined by the pavement LCA system boundary for all life cycle stages, excluding the use-stage. The exhaustive mapping of all possible *flows* and *processes* in the pavement LCA framework are developed in MS PowerPoint, and the document will be made available upon request.

In addition, the ontology definitions using object-oriented programming is presented that can directly inherit the OpenLCA architecture. This ontology is extensible and provides a software architecture for any stakeholder who wants to build a tool. The extensive class definitions along with the association of different data types to stakeholders are developed in MS Excel, and the document will be made available upon request.

REFERENCES

Asphaltpavement.org. (2019). *NAPA EPD Program*. [online] Available at: http://www.asphaltpavement. org/EPD [Accessed 10 Apr. 2019].
ASTM E3012 - 16 Standard Guide for Characterizing Environmental Aspects of Manufacturing Processes. [online] Astm.org. Available at: https://www.astm.org/Standards/E3012.htm [Accessed 13 Sep. 2019].
Bare, J. C. (2012). *Tool for the Reduction and Assessment of Chemical and Other Environmental Impacts (TRACI), Version 2.1 - User's Manual*, United States Environmental Protection Agency. Available at: https://www.epa.gov/chemical-research/tool-reduction-and-assessment-chemicals-and-other-environ mental-impacts-traci
Bernstein, W., Tamayo, C., Lechevalier, D. and Brundage, M. (2019). Incorporating unit manufacturing process models into life cycle assessment workflows. *Procedia CIRP*, 80, pp. 364–369.
Buy Clean California Act.AB-262.
Carbonleadershipforum.org. (2019). *Concrete Product Category Rule | Carbon Leadership Forum*. [online] Available at: http://www.carbonleadershipforum.org/2017/01/03/concrete-pcr/ [Accessed 6 Apr. 2019].
Cashman, S., Meyer, D., Edelen, A., Ingwersen, W., Abraham, J., Barrett, W., Gonzalez, M., Randall, P., Ruiz-Mercado, G. and Smith, R. (2016). Mining Available Data from the United States Environmental Protection Agency to Support Rapid Life Cycle Inventory Modeling of Chemical Manufacturing. *Environmental Science & Technology*, 50(17), pp. 9013–9025.
Cshub.mit.edu. (2019). *Pavement Vehicle Interaction (PVI) | Concrete Sustainability Hub*. [online] Available at: http://cshub.mit.edu/pavements/pvi [Accessed 6 Apr. 2019].

Edelen, A. and Ingwersen, W. (2016). *Guidance on Data Quality Assessment for Life Cycle Inventory Data.* Life Cycle Assessment Research Center, National Risk Management Research Laboratory, United States Environmental Protection Agency.

Edelen, A., Ingwersen, W., Rodríguez, C., Alvarenga, R., de Almeida, A. and Wernet, G. (2017). Critical review of elementary flows in LCA data. *The International Journal of Life Cycle Assessment,* 23(6), pp.1261–1273.

Fhwa.dot.gov. (2019). *Sustainable Pavement Program - Sustainability - Pavements - Federal Highway Administration.* [online] Available at: https://www.fhwa.dot.gov/pavement/sustainability/ [Accessed 5 Apr. 2019].

GitHub. (2019). *GreenDelta/olca-modules.* [online] Available at: https://github.com/GreenDelta/olca-modules/tree/master/olca-core/src/main/java/org/openlca/core/model [Accessed 6 Apr. 2019].

Harvey, J., Meijer, J., Ozer, H., Al-Qadi, I., Saboori, A. and Kendall, A. (2016). *Pavement Life-Cycle Assessment Framework.* No. FHWA-HIF-16-014. Washington, D.C.: Federal Highway Administration.

Harvey, J., Lea, J., Kim, C., Coleri, E., Zaabar, I., Louhghalam, A., Chatti, K., Buscheck, J. and Butt, A., 2016. *Simulation of cumulative annual impact of pavement structural response on vehicle fuel economy for California test sections* (No. UCPRC-RR–2015–05).

Ingwersen, W., Hawkins, T., Transue, T., Meyer, D., Moore, G., Kahn, E., Arbuckle, P., Paulsen, H. and Norris, G. (2015). A new data architecture for advancing life cycle assessment. *The International Journal of Life Cycle Assessment,* 20(4), pp. 520–526.

ISO (2006). 14025: *Environmental labels and declarations—Type III environmental declarations—Principles and procedures.* International Organization for Standardization: Geneva, Switzerland. Available at: https://www.iso.org/standard/38131.html

Janowicz, K., Krisnadhi, A., Hu, Y., Suh, S., Weidema, B., Rivela, B., Tivander, J., Meyer, D., Berg-Cross, G., Hitzler, P., Ingwersen, W., Kuczenski, B., Vardeman, C., Ju, Y. and Cheatham, M. (2015). A Minimal Ontology Pattern for Life Cycle Assessment Data. In: *Workshop on Ontology and Semantic Web Patterns (6th edition).*

Lcacommons.gov. (2019). *LCA Collaboration Server.* [online] Available at: https://www.lcacommons.gov/lca-collaboration/ [Accessed 6 Apr. 2019].

Li, H. (2012). *Evaluation of Cool Pavement Strategies for Heat Island Mitigation.* Doctorate. University of California, Davis.

Mukherjee, A. (2016). *Life Cycle Assessment of Asphalt Mixtures in Support of an Environmental Product Declaration.* [online] Lanham, MD: National Asphalt Pavement Association. Available at: https://www.asphaltpavement.org/PDFs/EPD_Program/LCA_final.pdf [Accessed 6 Apr. 2019].

Openlca.org. (2019). [online] Available at: http://www.openlca.org/download/ [Accessed 6 Apr. 2019].

Ram, P., Harvey, J., Muench, S., Al-Qadi, I., Flintsch, G., Meijer, J., Ozer, H., Van Dam, T., Snyder, M. and Smith, K. (2017). *Sustainable Pavements Program Road Map Draft Document.* FHWA-HIF-17-029. Federal Highway Administration.

Sen, S. and Roesler, J. (2016). Contextual heat island assessment for pavement preservation. *International Journal of Pavement Engineering,* 19(10), pp.865–873.

CEN (2012). Sustainability of Construction Works-Environmental Product Declarations-Core Rules for the Product Category of Construction Products. EN, 15804. The European Committee for Standardization (CEN: Comité Européen de Normalisation), Brussels, Belgium.

Van Dam, T., Harvey, J., Muench, S., Smith, K., Snyder, M., Al-Qadi, I., Ozer, H., Meijer, J., Ram, P., Roesler, J. and Kendall, A. (2015). *Towards Sustainable Pavement Systems: A Reference Document.* FHWA-HIF-15-002. Washington, D.C.: Federal Highway Administration.

Zaabar, I. and Chatti, K. (2010). Calibration of HDM-4 Models for Estimating the Effect of Pavement Roughness on Fuel Consumption for U.S. Conditions. *Transportation Research Record: Journal of the Transportation Research Board,* 2155(1), pp.105–116.

Zhang, Y., Luo, X., Buis, J. and Sutherland, J. (2015). LCA-oriented semantic representation for the product life cycle. *Journal of Cleaner Production,* 86, pp.146–162.

11

Asphalt pavement resurfacing: A review toward a better selection and representativeness of LCI

A. de Bortoli

LVMT, Ecole des Ponts ParisTech, University of Paris-East, Marne la Vallée, France

ABSTRACT: Road maintenance, and especially resurfacing, has become a major concern for the environmental performance of the road industry. Life cycle assessment (LCA) is used to provide science-based recommendations to enhance this performance. But LCA results are widely acknowledged to be context-dependent, and their robustness largely depends on the quality of the input data. This study provides knowledge on how to prioritize efforts to contextualize resurfacing life cycle inventories (LCI), by analyzing the impact contribution of the different life cycle stages and sub-stages of resurfacing operations, from cradle-to-laid. Then, it gives an overview of existing LCIs for asphalt resurfacing: it reviews and qualitatively details the selection and development of these LCIs, published in both academic and institutional literature. It highlights an overall need for better quality data and easier access to information on this quality. A collaborative "LCI observatory" is suggested to help the infrastructure LCA community to gather and scientifically select LCIs according to the specificity of each study, to produce more robust recommendations for decision-makers.

1 INTRODUCTION

1.1 *Context and objective*

In most developed countries, road networks are considered to be fully developed, but ageing. Roads are the infrastructure supporting the most intense ground transportation modes: their maintenance should now be an important element of mobility planning. With the rise of environmental issues and their threat to human way-of-life, studies are carried out to alleviate the environmental burden that road maintenance brings. Road construction activities have received a special attention from Life Cycle Assessment (LCA) practitioners to decrease their environmental impacts, by changing materials, their industrial modes of production, the construction processes, the material End-of-Life (EoL), or even their albedo. Nevertheless, the robustness of these assessments depends partly on the quality of input data, i.e. their reliability, completeness, temporal, geographical, and further technological correlations (Weidema et al., 2013), as well as the number of elementary flows considered. These data, called Life Cycle Inventories (LCIs), are a quantified list of input and output flows of energy and chemical substances related to a unitary quantity of industrial process/product/activity at the smallest scale (or process-scale), while it is a quantified list of input and output manufactured materials, energy and other flows at the biggest scale (or product-system scale). Insuring robust LCA recommendations requires an adequate selection of LCIs as well as a potential adaptation or development of new LCIs. But prioritizing the effort provided to LCI data quality implies further knowledge on process main contributors to the impacts and existing LCIs.

The objectives of this study are to provide elements to apprehend the variability in road resurfacing LCAs, to give first insights about contextualization priorities for pavement resurfacing LCIs and to offer an overview on existing dedicated LCIs. First, we propose to analyze the impact contribution of the different life cycle stages of resurfacing operations from cradle-to-laid – namely material production, asphalt mix manufacturing, transportation and construction. Then, we review and detail the representativeness of existing LCIs for the different elements in these stages, published in both academic and institutional literatures.

1.2 Pavement asphalt resurfacing

1.2.1 Definitions
Two families of asphalt resurfacing will be studied: thick asphalt overlay and surface treatment. A "thick asphalt overlay" is a pavement superficial layer made of asphalt mix, with a thickness after compaction between 15 and 80 mm. A "surface treatment" consists in spraying a binder on the road before recovering it with sand or aggregates. This operation can be doubled to make a "double surface dressing", and can also start with covering the pre-existing pavement with aggregates before spraying the first binder layer for "double prechipped surface treatment".

1.2.2 Supply chain description
The production system for road resurfacing is presented on Figure 1 (Supplemental Material). It includes the production chains of aggregates and binders, their manufacture as asphalt mixes and/or their spreading on the road work zone.

For the two families of resurfacing operation, the materials used are quite similar, and composed of a matrix of aggregates, bounded together with a petroleum-based binder called bitumen, and potentially stabilized or optimized with additive agents. These materials are individually produced and transported, then mixed together in asphalt plants before laying for asphalt overlays or direct spreading on roads in the case of treatment using building machines, after the potential removal of the former superficial layer of the road and at least a preparation of the pavement surface. Aggregates are extracted from quarries: they can be hard-rock or soft-rock aggregates, which requires different technologies of extraction (explosives, loaders, etc.). Extracted rocks are then sent to the asphalt plant to be dried, screened and mixed with other asphalt mix materials, or processed separately and sent directly on surface treatment work sites.

SUPPLEMENTAL MATERIAL

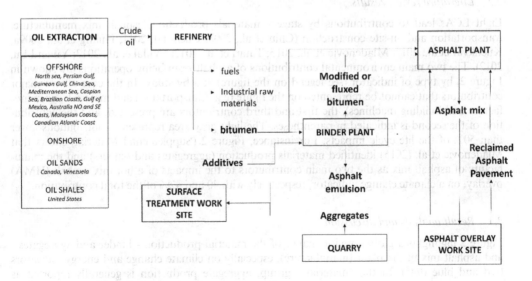

Figure 1. Road resurfacing production system.

Bitumen is a residue from crude oil distillation. Depending on the crude oil nature, i.e. where it is extracted from, distillation of crude oil produces different byproducts. Only the heaviest crude oils contain bitumen. Crude oil can be extracted offshore, onshore, or obtained from oil sands or oil shales. It is then refined to obtain different products depending on volatility. Bitumen is one of the less volatile part, remaining in the distillation column at the end of the process. It is sent to asphalt plant, sometimes after a chemical modification at the binder plant. In this plant, additive agents can be added to the bitumen to enhance its physicochemical properties, or the bitumen can be processed with water and other components to produce asphalt emulsions for surface treatment operations. Binder additive agents are mostly polymers and oils. Polymers are very common in French binders: they enhance resistance to oxidation.

Resurfacing building machines can be diverse, but they generally consume low-sulfured fuel.

At the EoL, the surface can be overlaid or removed. Reclaimed Asphalt Pavement (RAP) are removed asphalt materials that can be added to virgin asphalt mixes in proportions ranging from 0 to more recently close to 100% even for the wearing course, in average at a rate of 20% in France.

2 ENVIRONMENTAL HOTSPOTS OF PAVEMENT ASPHALT RESURFACING: A REVIEW

To spot environmental key contributing stages of pavement asphalt resurfacing, we will review LCA results of pavement asphalt resurfacing. After presenting succinctly a method of review to get the environmental hotspots of pavement asphalt resurfacing operations, we analyze existing LCA studies and present their results.

2.1 Method for the literature review

ISO standard 14040 sets out the different steps for performing an LCA. The results of the LCA show the potential environmental impacts of the system from cradle-to-grave, for a given functional unit. We will only review the LCA studies that present results for the different life cycle stages of road pavement resurfacing, using asphalt mixes. We will then present the contribution of each life cycle stage of the resurfacing operation. This literature review has been mainly conducted using ScienceDirect (exhaustive review), in fall 2017.

2.2 Literature review results

Eight LCAs lead to contributions by stage – materials production, asphalt mix manufacture, transportation and on-site construction (Chiu et al., 2008; Cuenoud, 2011; Huang et al., 2009a; Kucukvar et al., 2014; Mladenovič et al., 2015; Tatari et al., 2012a; Vidal et al., 2013; Yu and Lu, 2012). The two main environmental contributions of asphalt resurfacing operations are shown in Figure 2, by type of indicator (see legend on the figure) and by study. In the case of two major contributors that cannot be represented on the figure (e.g. transportation and materials, or manufacture and building machines), the first and third contributors are presented, and the contribution of the second is indicated in parentheses. The light gray area represents contributions lower than 50% of the life cycle impacts. For instance, Figure 2 (Supplemental Material) shows that Mladenovic et al. (2015) identified materials production (aggregates and binders) and the manufacture of asphalt mix as the two main contributors to the impacts of a hot mix asphalt (HMA) overlay, on a climate change indicator, respectively with 40 and 42% of the total contribution.

2.3 Result analysis and discussion

Figure 2 highlights a global predominance of the material production - binder and aggregates - and asphalt mix in the plant (manufacture), especially on climate change and energy indicators (red and blue dots). In the "materials" group, aggregate production is generally reported as

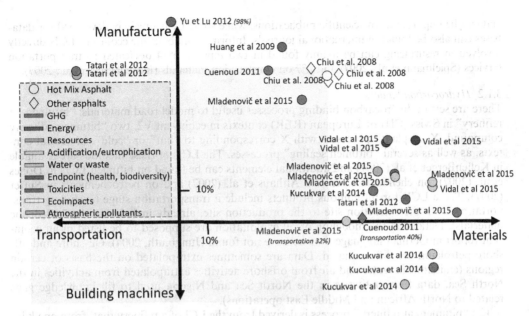

Figure 2. Representation of the two main stage contributors to the environmental impacts of resurfacing by type of indicator, type of asphalt, and LCA study.

having a low impact contribution, above 15% (Chiu et al., 2008; Cuenoud, 2011; Mladenovič et al., 2015), although this is not systematic, since it might concentrate up to 60% of the impacts in human toxicity (Mladenovič et al., 2015). Jullien et al. (2012) report that this impact varies widely depending on the site and the operating process (e.g. type of explosive). The construction stage (building machines) rarely appears in the two major contributions except for the study by Kucukvar et al. (2014), just like the transportation stage that appears in the top 2 for only four indicators in three studies (Kucukvar et al., 2014; Mladenovič et al., 2015; Tatari et al., 2012a). These exceptions can be explained by the distance of supply for aggregates, an important parameter of the total fossil fuel consumption of resurfacing techniques (Thenoux et al., 2007).

Although the main contributions are not consensual, as they depend on various parameters – e.g. material extraction or manufacturing techniques, transportation distances, plant emission performances, overlay thickness and techniques in general, but also the type of indicator considered and probably the characterization and allocation methods - this systematic review pushes toward considering in priority the contextualization of the LCIs for primary material manufacturing and more especially of binders, as well as those for the asphalt plants.

3 LIFE CYCLE INVENTORIES REVIEW

We propose here to thoroughly review and detail existing LCIs involved in road asphalt resurfacing. We start by discussing LCIs in ecoinvent V2 and V3, before presenting academic and syndicate's LCIs that brought more recent data or differently spatialized LCIs. Again, the academic literature review has been conducted using ScienceDirect, in fall 2017.

3.1 The ecoinvent LCIs

3.1.1 Database and ecoinvent documentation

Among the different LCI databases existing across the world, e.g. Gabi Database, ecoinvent, US LCI or Chinese Core Life Cycle Database, we chose to review ecoinvent, said to be one of the most complete LCI database. LCIs are not systematically developed for a specific database and can be imported from scientific publications, while some LCIs precisely developed for data-bases can also be found in international journals. Information about the ecoinvent LCIs directly involved in resurfacing can be mainly found in two reports: n°14 dedicated to transportation services (Spielmann et al., 2007) and n°7 on construction materials (Kellenberger et al., 2007).

3.1.2 *Hydrocarbon binders*

There are several hydrocarbon binding processes useful to model road materials: "bitumen, at refinery" in Swiss (CH) or European (RER) contexts in ecoinvent V2, two "bitumen adhesive compound, X, at plant" processes with X corresponding to "hot" or "cold", in several con-texts, as well as several "bitumen sealing" processes. The LCIs of these methods are available in KellenBerger et al. (2007) and additional elements can be found on bitumen LCIs in Dones et al. (2007), on chemical solvents in Althaus et al. (2007) and on petrochemicals in Sutter (2007). These LCIs for bituminous products include a transportation stage (excluding trans-portation from the extraction site to the production site, already included in the LCI of the materials). Details on the modeling of oil exploitation are supposed to be listed in an ecoin-vent report in German language that we have not found (Jungbluth, 2004). On-shore and off-shore petroleum are differentiated. Data are sometimes extrapolated on the basis of certain regions (emissions in water and air from offshore activities extrapolated from activities in the North Sea, data from activities in the North Sea and Nigeria used to fill knowledge gaps related to North African and Middle East operations).

The "bitumen, at refinery" process is derived from the LCI of a refinery that, from one kilo-gram of crude oil, produces various co-products: fuels, other specialty products including bitu-men, and industrial raw materials (naphtha and some gases for petrochemicals). These LCIs come from two Swiss refineries – in Collombey and Cressier - studied in 2000 and whose data were supplemented by older literature data. The European process is extrapolated from the Swiss process, from a sample of 1 to 5 refineries supposedly representative of the European refinery stock. The adaptation is notable for some flows consumed in the technosphere - crude oil extraction location, electricity mix, transportation to the refinery - while the outflows in the natural environment are globally similar in quality and quantity.

"Bitumen adhesive compound, cold, at plant" is the bitumen used in the construction sector to waterproof roadways (dressing courses) and roofs. Originally, it was representative of an emulsion – a cold mixture with more than 30 % bitumen by weight, water (50-60%), solvents (<8%) and emulsifiers (<3%) - in 2000 (Kellenberger et al., 2007, p. 756). A hot process (nat-ural gas burner, 140° C) is also available, originally developed in 1994 (spatial perimeter: RER or "rest of the world" (=RoW), i.e. global process (=GLO) excluding Switzerland (CH)). The "bitumen sealing" processes, "V60" and others, is also used for waterproofing of structures and infrastructures. According to the description, they nevertheless correspond more to a method for adhering sand, talc or aggregates to a structure, i.e. an emulsion used for build-ing surface treatments.

3.1.3 *Aggregates*

The calculation of the aggregate LCI called "mine, gravel/sand" in ecoinvent V2.2 is detailed in the report n°7 (Kellenberger et al., 2007). It is based on the study of 4 Swiss quarries con-sidered representative of the country between 1997 and 2001, and on operating assumptions (quarry surface, operating time, annual production, machinery (silos, sieves and crushers, with lifespans of 25 years)). An allocation then makes it possible to differentiate the sand from the aggregates ("sand at mine" on one side, "gravel, crushed, at mine" on the other): for 1 metric ton of crushed rock, 35% of sand and 65% of aggregates are produced. Other sources do not consider any allocation between sand and aggregates from a quarry production (UNPG, 2011a, 2011b, 2011c).

3.1.4 *Other ecoinvent LCIs*

An "asphalt mastic" process corresponds more to the material used for crack-filing or side-walks. The pavement EoL was modeled with Swiss statistics: 1.5 Mt of RAP and 1.3 Mt of gravels per year (based on the document "Bundesamt für Konjunkturfragen" published in

1991 that we did not find on the internet), of which 20% recycled and re-used at its maximum level of performance (bituminous concrete used in wearing course or base course according to its original layer (thus its binder rate/quality)) and almost all of the remainder going to the sub-base layer, when a negligible percentage is stocked or sent to landfill. Ecoinvent V3 provides recycling processes for "waste asphalt" or "waste bitumen" for different geographical contexts (CH, GLO, RoW).

3.1.5 Conclusion
Ecoinvent LCIs are used worldwide to conduct LCAs. But in the road sector, they generally date back to the 1990s or the 2000s. They have been updated, often by uninformed "extrapolation". These updates make difficult to judge the quality of the data in a specific context. For instance, ecoinvent website does not give access to LCI data quality indicators directly on process cards, despite such data may exist in the versions 3 of the database (Weidema et al., 2013). Simultaneously, how to make these LCIs project-specific is difficult to figure out and may bring to non-adequate practices and unreliable results, despite a steady interest in this subject and especially about spatialization (Patouillard et al., 2018).

3.2 Review of other academic and industrial LCIs
Some studies use software dedicated to pavement LCA to evaluate resurfacing operations, such as PaLate (Cross et al., 2011) or ECORCE (Jullien et al., 2014). Their LCIs are not exploitable because too aggregated and/or inaccessible. All pavement LCAs potentially contain useful data, however, they are often insufficiently detailed to be re-usable. A meta-analysis presents the LCIs choices in road LCAs, conducted between 2010 and 2015 and most cited in the Web of Science search engine (AzariJafari et al., 2016). In this section, we only report the studies including disclosed disaggregated LCIs for resurfacing (complete operation and/or materials).

3.2.1 LCI selection and development in the academic literature
Among the articles dealing with road maintenance LCA, many do not address the questions of geographical, technological or temporal adequation of the LCIs selected to their study. Zhang et al. (2010) use various LCIs from ageing studies or specifically adapted to the US context - data from Portland Cement Association, Athena Institute for Bitumen or US databases (Franklin US LCI Database linked to Simapro). Jullien et al. (2014) also use now somehow dated LCIs as the ECORCE software used as been developed at the end of the 2000s. Some others only inventory few flows compared to what is done in more exhaustive LCIs (e.g. in ecoinvent). Yu and Lu (2012) compare resurfacing (including the use stage) using LCIs with a restricted number of flows from literature published between 2001 and 2007. Mladeno-vic et al. (2015) compare two resurfacing techniques using normal aggregates and slag aggregates from foundries.

Most of the LCIs come from the GaBi database and date back to the 2000s. Asphalt emissions during the compaction and cooling stages are neglected due to a lack of data, which is justified by the fact that in a comparative approach of two mixes, these emissions are relatively identical. Huang et al. (2009b) model the environmental impacts of a resurfacing project, including inconvenience to the user due to road capacity loss in the road works zone. Emissions from the combustion of fossil products, including those occurring in asphalt plants, are arbitrarily set at the levels of European emission standards, in order to circumvent the difficulty of accessing supposedly available but practically rather confidential data.

Weiland and Muench (2010) compare three resurfacing techniques on a Washington DC interstate case: the LCIs used come from the US EPA, the GREET model on transportation energy emissions and consumptions, and Stripple's study for bitumen (2001). Jullien et al. (2012) present partial aggregates LCIs from 3 French quarries: a massive rock quarry and two soft rock quarries. The study finds significant differences between the environmental impacts related to these LCIs and those of ecoinvent. But these LCIs are unfortunately not reusable as are only disclosed the energy consumption and some main flows (emissions to air), then the final impacts (acidification, eutrophication, etc.) related to the quarries. Moretti et al. (2017) propose inventories relating to asphalt mixes produced by two Italian plants, with details on the elementary flows. Giani et al. (2015) publish data on plant energy consumptions, but also equipment and material quantities needed for resurfacing, with consideration of RAP. Several studies also consider the impacts of adding recycled or alternative materials to virgin asphalt mixes (Aurangzeb et al., 2014; Balaguera et al., 2018; Chowdhury et al., 2010; Farina et al., 2017; Huang et al., 2007; Wang, 2016).

Nonetheless, there were no published LCIs in 2010 for additives - adhesives or interface additives, emulsifying agents, fillers, curing agents, resins, fibers - and concrete admixtures (Sayagh et al., 2010). Vidal et al. (2013) have since then developed an inventory for one additive used in warm mix asphalt. This article compares HMA resurfacing operations with and without RAP. The authors adapt ecoinvent's LCIs: "gravel" and "sand" respectively for aggregates and fines, and "bitumen" for binder. The fillers - cements and lime - which stiffen the mix and strengthen against ageing, are also modeled from the ecoinvent LCIs. The modeled additive for warm asphalt is synthetic zeolite, whose LCI comes from detergents (Fawer et al., 1998). The authors also created an inventory for asphalt plants (infrastructure) and a complete process for producing asphalt mixes; they offer LCIs for construction equipment, truck transport and leaching of pavements according to the RAP rate.

Other authors have performed LCAs of warm mix asphalt processes (Cheng et al., 2010; Tatari et al., 2012b) - sometimes without reusable data (data aggregation or non-disclosure) (Mazumder et al., 2016) - and cold-mix asphalt (Giani et al., 2015). Santos (2015) proposes a compilation of pavement LCA studies in the Portuguese context, under the following LCI choices: Eurobitume (Blomberg et al., 2011) for hydrocarbon binders, Jullien et al. (2012) or Stripple (2001) for aggregates depending on the manuscript chapter, Althaus et al. (2007) for tap water, US EPA (2004) for the production of asphalt mixes in the plant and the transport of materials, EEA (2013) for construction equipment, road operation, electricity, coal, and crude oil, and Dones et al. (2007) among others for fuels.

Yang has developed regional models of hydrocarbon binders adapted to 5 regions of the USA, and whose procedure is very well detailed (2014). The environmental impacts related to binder production are significant for a number of LCA impact categories (Figure 2), and Yang shows that sources of crude oil and refinery consumption can vary significantly within the USA. This also recalls that sources of crude oil supply can also vary greatly depending on the geopolitical context and the market, and therefore fluctuate over time. Yang also includes an energy conversion loss between crude oil and hydrocarbon binder of 6%. She estimates that leaching during the use stage due to RAP is negligible.

Transportation stages are obviously variable. French studies mainly model transportation by road over relatively short distances: studies by Hoang (2005), Chappat and Bilal (2003) and CimBéton (2011) consider bitumen transportation over 300 km between refineries and asphalt plants (over 354 km directly on resurfacing site (Hoang, 2005)), aggregates transportation from 39 to 100 km, and transportation of asphalt mixes and materials at their EoL over around 20km. The distances considered in other countries can be even lower: for example in the US, transportation as short as 30 to 50 km between quarry and asphalt plants, and 80 to 100 km between refinery and asphalt plants (Ozer et al., 2017).

3.2.2 LCIs from syndicates

LCIs carried out by industrial unions are an important source of data but are also sensitive - access to basic data is sometimes difficult or even completely confidential – and require to be critically reviewed by a neutral organism to assess their quality.

3.2.2.1 HYDROCARBON BINDER LCIS FROM EUROBITUME

The seminal study by Blomberg et al. (2011) proposes LCIs for several types of binders: bitumen, polymer modified bitumen, and bitumen emulsion. These kinds of bitumen comply with EN 12591 and correspond to penetrations of 20 to 220 (1/10mm). LCIs are available in ecoinvent V2.2 format as sub-processes (report) or elementary flows (complementary Excel spreadsheet). Attention must be paid to the functional unit which is one metric ton for pure bitumen processes but 1.52 metric ton for emulsion (corresponding to the transformation of one metric ton of bitumen). The refinery impacts allocation is made according to economic considerations, using standard values of manufactured products, according to the recommendations of ISO TR 14049. This allocation choice necessarily impacts the LCIs. The 2007 data on oil extraction come from the International Association of Oil and Gas Producers (OGP): they cover 32% of global oil production. Refinery emissions data come from reports from CONCAWE, the association for air and clean water conservation in Europe. Regarding the distribution of crude oils and energy consumption of refineries, anonymous questionnaires were distributed to members of Eurobitume.

For pipeline and ship transportation modes, transportation company data were used and supplemented with data from the US EPA. The other data comes from ecoinvent 2.2. The quality of the data is considered representative of the European context by the critical review, with one third of the worldwide extractions represented but also 20% of the European bitumen production (via 7 refineries). The most important environmental flows according to the report are the consumption of crude oil and natural gas, the air emissions of CO_2, SO_2, NOx, CH_4 and NMVOC. The data quality of these flows is considered particularly good, while those of other flows is only considered as satisfactory. The OGP report gives energy consumption of oil extraction within a range of 490 to 1580 MJ/t depending on the type of oil and its origin. The distribution of energy consumptions for oil exploitation, between gas and diesel, is provided by ecoinvent (Dones et al., 2007).

Nevertheless, the energy return rate relating to oil extraction continues to fall (Hall et al., 2014), a trend that must not reverse: it should be considered in LCIs of bituminous products. The bitumen modified by addition of polymers is modeled according to an original case study conducted by Eurobitume with the most widely used polymer in Europe, styrene butadiene styrene (SBS) in its granular form, at 3.5% by weight of the final product. But many different polymers exist on the market. The bitumen emulsion modeled corresponds to a cationic emulsion formulation with an amine type emulsifier and a hydrochloric acid, the most popular emulsion in Europe. The emissions relating to gas and heavy fuel oil burned in power plants are those of ecoinvent "Refinery gas, burned in furnace/RER" and "Heavy fuel oil, burned in refinery furnace/RER" with the exception of SO_2 emissions related to fuel oil which are taken from data from oil refineries in Europe in 2006 (1.33% of sulfur in fuel and 26.6 gSO_2/kg of fuel oil burned).

3.2.2.2 AGGREGATE LCIS FROM THE RELATED FRENCH SYNDICATE

UNPG has carried out three studies relating to the main types of aggregates used in the construction sector in France: aggregates from massive rocks (UNPG, 2011a) and soft rocks (UNPG, 2011b), as well as recycled aggregates (UNPG, 2011c). In the first two cases, the analysis considers all the activities carried out on the production site: clearing and exploitation of the site, processing and marketing of the finished products, then final site rearrangement after shutdown. It omits certain elements judged as negligible by the former French standard NF P 01-010. The system boundaries for recycled aggregates - coming from concrete blocks, RAP, ballast, hydraulically bounded gravel, natural gravels or earthworks - include the treatment of materials from the demolition, planning or earthmoving site: sorting and/or crushing and/or screening, carried out in fixed or mobile installations. It does not include the demolition itself, nor the transportation or energy consumption related to the possible aggregate drying. The massive rock aggregate LCI is based on the study of 8 sites in 2007, including 4 eruptive rock sites and 4 calcareous sites, considered representative for the distribution of the national production. For the LCI of loose rock aggregates, the site measures have also been conducted in 2007 on 8 different sites, including 5 sites in water, 2 sites out of water, and a mixed site (60% in water and 40% out). Finally, the recycled aggregates LCI comes from 7 recycling facilities including 3 fixed and 4 mobile installations, also considered as national representative conditions. LCIs of these three types of aggregates are provided in an exploitable format and have been reviewed according to NF EN 15804+A1 standard for LCA of the built environment in Europe and France (AFNOR, 2014).

3.2.2.3 CONCLUSION

Pavement LCAs, regardless of the scale they consider – road section, wearing course, materials - use a variety of LCIs. Some LCIs are largely used: sometimes for their geographical suitability, but perhaps sometimes by conformism, as they come from seminal studies and thus became the most popular LCIs. Other LCIs has been developed for a particular article, and are more or less complete in terms of flows considered. For instance, about the air emissions included, often restricted to regulated emissions. Finally, if the LCI selection must be important in the final LCA results, the selection process is rarely justified, probably because of the work overload it requires on top of an already highly time-consuming activity.

4 CONCLUSION

First, a literature review shows that "primary" materials – especially bitumen - and final material manufacture are the two most frequent and major contributors to road resurfacing environmental impacts, from cradle-to-laid. The contribution weight depends on the indicator analyzed, the context-specificity, as well as characterization and allocation methods. Then, existing inventories for road resurfacing LCA are reviewed, especially focusing on their quality characteristics. The quality of a LCI depends on reliability, completeness, temporal geographical, and further technological correlations (Weidema et al., 2013), as well as the number of elementary flows considered. LCI quality and uncertainty analysis attract a growing interest in publications without being systematic. But understanding the quality of an LCI is time-consuming, and our literature exploration shows the quality heterogeneity in the LCIs chosen in specific contexts, possibly due to a lack of overview on existing LCIs. Even when allocating time to it, traceability of modeling assumptions can be difficult. Some updates or adaptations are conducted on background datasets with questionable transparency on the process. As a result, little energy is generally dedicated to choosing adapted LCIs. The resulting robustness issue may reduce the significance of recommendations from LCA studies to decision-makers. Considering the reasonable number of important process contributors to pavement LCA results, an « LCI observatory », i.e. a collaborative platform inventorying specific LCIs and their quality characteristics, may be a solution to help the community in producing more robust recommendations.

REFERENCES

Abdo, J., 2011. Analyse du cycle de vie de structures routières (No. T89), Collection Technique CimBéton - BÉTON ET DÉVELOPPEMENT DURABLE. CIMBETON.

AFNOR, 2014. NF EN 15804+A1 Avril 2014 - Contribution des ouvrages de construction au développement durable - Déclarations environnementales sur les produits - Règles régissant les catégories de produits de construction.

Althaus, H.-J., Hischier, R., Osses, M., Primas, A., Jungbluth, N., 2007. Life Cycle Inventories of Chemicals - Data v2.9 (2007) (EcoInvent Report No. n°8).

Aurangzeb, Q., Al-Qadi, I.L., Ozer, H., Yang, R., 2014. Hybrid life cycle assessment for asphalt mixtures with high RAP content. Resour. Conserv. Recycl. 83, 77–86. https://doi.org/10.1016/j.resconrec.2013.12.004

AzariJafari, H., Yahia, A., Ben Amor, M., 2016. Life cycle assessment of pavements: reviewing research challenges and opportunities. J. Clean. Prod. 112, 2187–2197. https://doi.org/10.1016/j.jclepro.2015.09.080

Balaguera, A., Carvajal, G.I., Albertí, J., Fullana-i-Palmer, P., 2018. Life cycle assessment of road construction alternative materials: A literature review. Resour. Conserv. Recycl. 132, 37–48. https://doi.org/10.1016/j.resconrec.2018.01.003

Blomberg, T., Barnes, J., Bernard, F., Dewez, P., Le Clerc, S., Pfitzmann, M., Porot, L., Southern, M., Taylor, R., 2011. Life Cycle Inventory: Bitumen. Brussels, Belgium.

Chappat, M., Bilal, J., 2003. The Environmental Road of the Future - Life Cycle Analysis - Energy Consumption and Greenhouse Gas Emissions. Colas Group.

Cheng, L., Chen, D., Yan, G., Zheng, H., 2010. Life Cycle Assessment of Road Surface Paving with Warm Mix Asphalt (WMA) Replacing Hot Mix Asphalt (HMA). IEEE, pp. 1–5. https://doi.org/10.1109/ICEEE.2010.5660713

Chiu, C.-T., Hsu, T.-H., Yang, W.-F., 2008. Life cycle assessment on using recycled materials for rehabilitating asphalt pavements. Resour. Conserv. Recycl. 52, 545–556. https://doi.org/10.1016/j.resconrec.2007.07.001

Chowdhury, R., Apul, D., Fry, T., 2010. A life cycle based environmental impacts assessment of construction materials used in road construction. Resour. Conserv. Recycl. 54, 250–255. https://doi.org/10.1016/j.resconrec.2009.08.007

Cross, S., Chesner, W., Justus, H., Kearney, E., 2011. Life-Cycle Environmental Analysis for Evaluation of Pavement Rehabilitation Options. Transp. Res. Rec. J. Transp. Res. Board 2227, 43–52. https://doi.org/10.3141/2227-05

Cuenoud, 2011. Valorcol: asphalt mix complying with environment and sustainable development. Eur. Roads Rev. Spring.

Dones, R., Bauer, C., Bolliger, R., Burger, B., Heck, T., Röder, A., Faist Emmenegger, M., Frischknecht, R., Jungbluth, N., Tuchschmid, M., 2007. Life Cycle Inventories of Energy Systems: results for current systems in Switzerland and other UCTE countries - Data v2.0 (EcoInvent Report).

EEA, 2013. EMEP/EEA air pollutant emission inventory guidebook 2009. Technical guidance to prepare national emission inventories (No. European Environment Agency Technical report n°12/2013). European Environment Agency, Luxembourg.

Farina, A., Zanetti, M.C., Santagata, E., Blengini, G.A., 2017. Life cycle assessment applied to bituminous mixtures containing recycled materials: Crumb rubber and reclaimed asphalt pavement. Resour. Conserv. Recycl. 117, 204–212. https://doi.org/10.1016/j.resconrec.2016.10.015

Fawer, M., Postlethwaite, D., Klüppel, H.-J., 1998. Life cycle inventory for the production of zeolite a for detergents. Int. J. Life Cycle Assess. pp 71–74.

Giani, M.I., Dotelli, G., Brandini, N., Zampori, L., 2015. Comparative life cycle assessment of asphalt pavements using reclaimed asphalt, warm mix technology and cold in-place recycling. Resour. Conserv. Recycl. 104, 224–238. https://doi.org/10.1016/j.resconrec.2015.08.006

Hall, C.A.S., Lambert, J.G., Balogh, S.B., 2014. EROI of different fuels and the implications for society. Energy Policy 64, 141–152. https://doi.org/10.1016/j.enpol.2013.05.049

Hoang, 2005. Tronçons autoroutiers: une méthodologie de modélisation environnementale et économique pour différents scenarios de construction et d'entretian.

Huang, Y., Bird, R., Bell, M., 2009a. A comparative study of the emissions by road maintenance works and the disrupted traffic using life cycle assessment and micro-simulation. Transp. Res. Part Transp. Environ. 14, 197–204. https://doi.org/10.1016/j.trd.2008.12.003

Huang, Y., Bird, R., Bell, M., 2009b. A comparative study of the emissions by road maintenance works and the disrupted traffic using life cycle assessment and micro-simulation. Transp. Res. Part Transp. Environ. 14, 197–204. https://doi.org/10.1016/j.trd.2008.12.003

Huang, Y., Bird, R.N., Heidrich, O., 2007. A review of the use of recycled solid waste materials in asphalt pavements. Resour. Conserv. Recycl. 52, 58–73. https://doi.org/10.1016/j.resconrec.2007.02.002

Jullien, A., Dauvergne, M., Cerezo, V., 2014. Environmental assessment of road construction and maintenance policies using LCA. Transp. Res. Part Transp. Environ. 29, 56–65. https://doi.org/10.1016/j.trd.2014.03.006

Jullien, A., Proust, C., Martaud, T., Rayssac, E., Ropert, C., 2012. Variability in the environmental impacts of aggregate production. Resour. Conserv. Recycl. 62, 1–13. https://doi.org/10.1016/j.resconrec.2012.02.002

Jungbluth, N., 2004. Erdöl. In: Sachbilanzen von energiesystmeen: Grundlagen für den ökologischen vergleich von energisystemen und den einbezug von energisystemen in Ökobilanzen für fie Scheiz (Ed. Dones R.). EcoInvent.

Kellenberger, D., Althaus, H.-J., Künniger, T., Lehmann, M., Jungbluth, N., Thalmann, P., 2007. Life Cycle Inventories of Building Products (EcoInvent Report No. n°7). EcoInvent.

Kucukvar, M., Noori, M., Egilmez, G., Tatari, O., 2014. Stochastic decision modeling for sustainable pavement designs. Int. J. Life Cycle Assess. 19, 1185–1199. https://doi.org/10.1007/s11367-014-0723-4

Mazumder, M., Sriraman, V., Kim, H.H., Lee, S.-J., 2016. Quantifying the environmental burdens of the hot mix asphalt (HMA) pavements and the production of warm mix asphalt (WMA). Int. J. Pavement Res. Technol. 9, 190–201. https://doi.org/10.1016/j.ijprt.2016.06.001

Mladenovič, A., Turk, J., Kovač, J., Mauko, A., Cotič, Z., 2015. Environmental evaluation of two scenarios for the selection of materials for asphalt wearing courses. J. Clean. Prod. 87, 683–691. https://doi.org/10.1016/j.jclepro.2014.10.013

Moretti, L., Mandrone, V., D'Andrea, A., Caro, S., 2017. Comparative "from Cradle to Gate" Life Cycle Assessments of Hot Mix Asphalt (HMA) Materials. Sustainability 9, 400. https://doi.org/10.3390/su9030400

Ozer, H., Yang, R., Al-Qadi, I.L., 2017. Quantifying sustainable strategies for the construction of highway pavements in Illinois. Transp. Res. Part Transp. Environ. 51, 1–13. https://doi.org/10.1016/j.trd.2016.12.005

Patouillard, L., Bulle, C., Querleu, C., Maxime, D., Osset, P., Margni, M., 2018. Critical review and practical recommendations to integrate the spatial dimension into life cycle assessment. J. Clean. Prod. 177, 398–412. https://doi.org/10.1016/j.jclepro.2017.12.192

Santos, J., 2015. A comprehensive life cycle approach for managing pavement systems. Universidade de Coimbra, Portugal.

Sayagh, S., Ventura, A., Hoang, T., François, D., Jullien, A., 2010. Sensitivity of the LCA allocation procedure for BFS recycled into pavement structures. Resour. Conserv. Recycl. 54, 348–358. https://doi.org/10.1016/j.resconrec.2009.08.011

Spielmann, M., Bauer, C., Dones, R., 2007. Transport services: Ecoinvent report no. 14, EcoInvent report. Swiss Centre for Life Cycle Inventories, Dübendorf.

Stripple, H., 2001. Life cycle assessment of road. A Pilot Study for Inventory Analysis (No. 2nd revised Edition). Report from the IVL Swedish EnvironmentalResearch Institute.

Sutter, J., 2007. Life Cycle inventories of Petrochemical solvents - Data v2.0 (2007) (EcoInvent Report No. n°22).

Tatari, O., Nazzal, M., Kucukvar, M., 2012a. Comparative sustainability assessment of warm-mix asphalts: A thermodynamic based hybrid life cycle analysis. Resour. Conserv. Recycl. 58, 18–24. https://doi.org/10.1016/j.resconrec.2011.07.005

Tatari, O., Nazzal, M., Kucukvar, M., 2012b. Comparative sustainability assessment of warm-mix asphalts: A thermodynamic based hybrid life cycle analysis. Resour. Conserv. Recycl. 58, 18–24. https://doi.org/10.1016/j.resconrec.2011.07.005

Thenoux, G., González, Á., Dowling, R., 2007. Energy consumption comparison for different asphalt pavements rehabilitation techniques used in Chile. Resour. Conserv. Recycl. 49, 325–339. https://doi.org/10.1016/j.resconrec.2006.02.005

UNPG, 2011a. Module d'informations environnmentales de la production de granulats issus de roches meubles - données sous format FDES conforme à la norme NF 10–01010. Union Nationale des Producteurs de Granulats.

UNPG, 2011b. Module d'informations environnmentales de la production de granulats issus de roches massives - données sous format FDES conforme à la norme NF 10–01010. Union Nationale des Producteurs de Granulats.

UNPG, 2011c. Module d'informations environnmentales de la production de granulats recyclés - données sous format FDES conforme à la norme NF 10–01010. Union Nationale des Producteurs de Granulats.

US EPA, 2004. AP-42: compilation of air pollutant emission factors (Volume 1: Stationary point and area sources, Chapter 11: Mineral products industry, 11.1). United States Environmental Protection Agency.

Vidal, R., Moliner, E., Martínez, G., Rubio, M.C., 2013. Life cycle assessment of hot mix asphalt and zeolite-based warm mix asphalt with reclaimed asphalt pavement. Resour. Conserv. Recycl. 74, 101–114. https://doi.org/10.1016/j.resconrec.2013.02.018

Wang, Y., 2016. The effects of using reclaimed asphalt pavements (RAP) on the long-term performance of asphalt concrete overlays. Constr. Build. Mater. 120, 335–348. https://doi.org/10.1016/j.conbuildmat.2016.05.115

Weidema, B.P., Hischier, R., Mutel, C., Nemecek, T., Reinhard, J., Vadenbo, C.O., Wernet, G., 2013. Overview and methodology - Data quality guideline for the ecoinvent database version 3 (ecoinvent report No. No. 1). St. Gallen.

Weiland, C., Muench, S., 2010. Life-Cycle Assessment of Reconstruction Options for Interstate Highway Pavement in Seattle, Washington. Transp. Res. Rec. J. Transp. Res. Board 2170, 18–27. https://doi.org/10.3141/2170-03

Yang, R., 2014. Development of a pavement life cycle assessment tool utilizing regional data and introducing an asphalt binder model. University of Illinois at Urbana-Champaign, Urbana, Illinois.

Yu, B., Lu, Q., 2012. Life cycle assessment of pavement: Methodology and case study. Transp. Res. Part Transp. Environ. 17, 380–388. https://doi.org/10.1016/j.trd.2012.03.004

Zhang, H., Lepech, M.D., Keoleian, G.A., Qian, S., Li, V.C., 2010. Dynamic Life-Cycle Modeling of Pavement Overlay Systems: Capturing the Impacts of Users, Construction, and Roadway Deterioration. J. Infrastruct. Syst. 16, 299–309. https://doi.org/10.1061/(ASCE)IS.1943-555X.0000017

Pavement, Roadway, and Bridge Life Cycle Assessment 2020 – Harvey et al (eds)
© 2020 Taylor & Francis Group, London, ISBN 978-0-367-55166-7

Towards more sustainable airfield pavements using life-cycle assessment of design alternatives

K. Mantalovas, R. Roberts, G. Giancontieri, L. Inzerillo & G. Di Mino
University of Palermo, Palermo, Italy

ABSTRACT: Airports are critical infrastructures and their success is paramount to development through tourism, trade and connectivity. Within the airport, the runway pavement must always be in a pristine condition. Consequently, pavement design and maintenance decisions are vital. Authorities must make decisions concerning preferred materials for design and maintenance. Decisions should be balanced by both economic and environmental factors. This paper considers a case study at Falcone Borsellino Airport (PMO), Palermo, Italy, where air traffic has been steadily increasing. Different pavements are proposed: flexible Asphaltic concrete designs with both traditional Asphaltic concrete and using reclaimed asphalt during maintenance, and a rigid concrete pavement. Environmental and economic impacts are assessed utilizing frameworks of life cycle assessment and life cycle costing analysis on the alternatives. The results highlight impacts material design changes have on the environment for airfield pavements and offer insights for management to help establish sustainable strategies moving forward.

1 INTRODUCTION

1.1 *Increasing air traffic and related needs for runway expansions and maintenance*

With increases in travel across the world, there is a constant need for maintenance and upgrades to airport pavements. This is especially in the case of cities, which are considered 'Tourist destinations' such as Palermo, Italy where there has been a steady increase in visits to the Sicilian region. The main airport in Palermo is the Falcone Borsellino Airport (PMO) – Punta Raisi, which was established in 1953. Upgrades and construction work have been carried out on the runway as recent as 2001 and it has a runway length of approximately 3km. This airport has seen continuous increases in air traffic over the last few years (Scianni 2017). As a result, it is essential for the local airport authorities to make decisions that will help reduce economic and environmental burdens to the state given possible upgrades to this airport and the runway to cater for increased tourism and thus additional flights. To this end, the Airport authorities need to make decisions on the best and most sustainable materials for any upgrades and maintenance activities.

1.2 *The sustainability of the choice of design for airfield pavements*

For any runway pavement design, there are typically two main design approaches; the use of a flexible pavement or a rigid one (there is also the possibility of a combination, but this study will focus on the two general designs). Each of these choices has a different construction methodology and its own maintenance pipeline. Traditionally design and maintenance approaches for airfield pavements are based on economic and social factors associated with materials and design schemes. Additionally, site-specific factors such as climatic and geographic conditions associated with the site featuring in design and maintenance decisions.

However, as a result of recent political and social interest, there is now a need to ensure any chosen methods are also done in a sustainable manner to reduce environmental damages with authorities in some countries making it mandatory to have Life Cycle and Costing Analysis (LCCA) studies implemented within the designs (Federal Aviation Administration 2009). Whilst this is a good approach and helps reduce environmental damages, the use of these types of restrictions is not worldwide as yet. To this end, this research was done to bridge this gap using a case study in Palermo, Italy wherein an LCA was done on alternative designs for the Airport runway to determine differences in environmental impacts on the construction and maintenance of different designs. The costs of alternating designs were also considered by carrying out a brief LCCA considering similar parameters. The case study is described in Section 2.

2 METHODOLOGY

2.1 Design of pavement alternatives

For the design of the Pavements for the case study, the FAARFIELD software was utilized (Federal Aviation Administration 2017). This software was released by the Federal Aviation Administration in conjunction with their updated guidelines for Airport pavement designs (Federal Aviation Administration 2009). This software makes use of the Cumulative Damage Factor (CDF), which considers mixed traffic. Within its algorithms, it utilizes layered elastic analysis (LEAF) for flexible pavement designs and 3D finite element analysis (NIKE3D) for rigid pavement designs. It also contains a wide aircraft library, which is used to input the traffic for the pavement to be designed. This is important as this method allows the input of the entire expected mix of traffic and not just a single equivalent load. For this to be done, the traffic at the PMO Airport was developed based on previous traffic counts at the airport (Scianni 2017). This data was compiled within the software and an annual growth rate of two percent was applied for expected growth in air traffic. To this end, two pavements were designed based on the traffic and local conditions at the airport. This resulted in the layer designs depicted in Table 1 for the Portland Cement Concrete (PCC) Rigid Pavement design and the Asphalt Concrete (AC) Flexible Pavement.

As the study considers the life cycle of the pavements the maintenance plans with an alternate plan utilizing Reclaimed Asphalt (RA) were also developed and this is further examined in the subsequent section.

2.2 The use of Life Cycle Assessment – goal and scope

Life Cycle Assessment is a tool that has been globally standardized according to the international standards of ISO 14040 and ISO 14044 (International Organization for standardization 2006a,b). It is seen as a tool which offers a look at the impacts inflicted onto the environment by a given process, product or possibly even service. Within this study, the assessment was done on two different pavement design alternatives for the Airport Runway utilizing the related maintenance

Table 1. Specifications of layers for each pavement design.

Pavement	Layer material	Thickness (mm)	Modulus or R (MPa)
PCC	PCC Surface	401.5	4.48
	P-401/P-403 St (flex)	130.0	2,757.90
	P-209 Cr Ag	160.0	267.03
	Subgrade (k=44.1)		95.87
AC	P-401/P-403 HMA Surface	130.0	1,378.95
	P-401/P-403 St (flex)	127.0	2,757.90
	P-209 Cr Ag	160.0	267.03
	Subgrade (CBR = 9.3)		95.87

strategies for each design. The end goal of this was to measure and compare the environmental impacts imposed by these two designs at the same location and facing the same air traffic. The scope of this LCA exercise was to quantify and compare the environmental impacts of two different design alternatives within a cradle to laid + maintenance approach, as depicted by the defined system boundaries The LCA was carried out in line with recommendations utilized in recently published guidelines on the use of LCAs for airport structures (Butt et al. 2019). The LCA carried out followed a comparative and process-based approach. The environmental impacts imposed as a result of constructing and maintaining an Asphaltic Concrete Flexible pavement are compared to equivalent impacts coming from the construction and maintenance of a Portland Cement Concrete Rigid Pavement over the same time period and at the same location. The construction was based on requirements defined by the FAA(Federal Aviation Administration 2018) and the maintenance was done based on input from local experts and previous research (Wang et al. 2016). For the maintenance pipelines, different activities were considered for each pavement design. For the PCC design, concrete repairs were utilized every 8 years within the life cycle which involved patching and depth repair to the PCC. For the AC design, the maintenance consisted of milling of 50mm and overlaying of 75mm at similar time intervals as the PCC so as to make the two pipelines comparable. The study also considered a maintenance pipeline for the AC design in which Reclaimed Asphalt was utilized for the maintenance activities. Reclaimed asphalt has been tested and used in several airports for runways and for construction and maintenance activities within the US (Hajj et al., 2010). Other studies have also recommended the inclusion of RAP into FAA guidelines (Guercio and McCarthy, 2015).

Reclaimed Asphalt is generally defined as asphaltic pavement materials that were previously used in pavement construction but were removed during the typical resurfacing, rehabilitation and reconstruction activities of a pavement and were then subsequently processed (Federal Highway Administration 2011). The use of RA within general pavement construction has steadily increased with a percentage of ten to thirty percent being applied in mixtures with legislative and technical limitations of its use (Mantalovas & Di Mino 2019). As a recycled product, its use offers both financial and environmental benefits. There have also been several case studies of the uses of the RA in airfields for resurfacing activities and maintenance. (Hajj et al. 2010, White 2018, 2019). RAP can offer savings of up to 27% when utilized in pavements so its use is expected to increase (White 2018) as the technology is more proven. Based on these previous works, this justified considering an additional maintenance pipeline for the AC pavement including the use of 30% of RA. This thus created three alternatives for the LCA utilizing the two pavement designs.

2.3 *System boundaries and functional unit*

For this case study, the processes which were included in the system boundaries of the product system were the initial production of the materials, the construction of the pavement itself and the maintenance pipelines for both pavement designs. It was done over an analysis period of 30 years with the use of a function unit of $1m^2$ of the pavement surface including the underlaying layers until the subgrade (BRE 2013, European Asphalt Pavement Association 2016, The Norwegian EPD Foundation 2009). These system boundaries are depicted in Figure 1 for the two design alternatives and they were defined and modelled within the Gabi ts software.

It should be noted further that both the construction and maintenance phases of the LCA considered the required raw material extraction and transportation to the mixing plant and the runway site. To carry out this LCA, Gabi ts, (Thinkstep (a sphera company) 2019) was used with the Gabi Professional database. No primary data was collected for this study but instead, previous studies and reputable sources of literature, reports, standards, Product Category Rules (PCRs) and Environmental Product Declarations (EPDs) were exploited. (BRE 2013, European Asphalt Pavement Association 2016, European Union Commission 2013, European Bitumen Association 2012, Giani et al. 2015, International Organization for Standardization 2006a,b, Mohammadafzali et al. 2017, National Asphalt Pavement Association 2017, The Norwegian EPD Foundation 2009). As a result, no cut-offs were applied. Table 2 depicts specific sources of data that were utilized for the inventory.

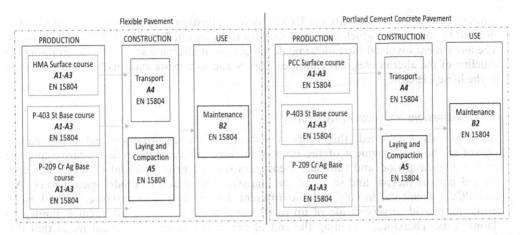

Figure 1. System boundaries of the product system of the two design alternatives analysed.

Table 2. Data sources for LCA components.

LCA component	Data Source
Portland cement	Gabi ts Professional Database (Thinkstep (a sphera company) 2019)
Coarse Aggregates Fine Aggregates Filler Bitumen	Allback2pave Mix Design and Workshops (Wellner et al. 2015)
Natural Gas Fuel Electricity Water	(Anthonissen et al. 2016; Aurangzeb et al. 2014; Gillespie 2012; Huang 2007)
Transport distances Asphalt Paver Asphalt compactor Asphalt milling	Allback2pave Mix Design and Workshops (Wellner et al. 2015)

For the impact assessment, ReCiPe 2016, Hierarchist(H) (Huijbregts, M.A.J. et al. 2016) was used for Mid and Endpoint impact category indicators. The midpoint indicators are a standardized set of impact category indicators such as Human toxicity. The endpoint indicators are narrowed and aggregated areas expressing damages to human health, ecosystems and resource availability. By considering both indicators, approaches the results yielded provide a thorough and exhaustive look at how the product affects the environment. Furthermore, as required by definitions of ISO 14040 and 14044 standards, the steps followed in this study were as follows: Definition of goal and scope of the study, Life Cycle Inventory (LCI) Analysis, Life Cycle Impact Assessment (LCIA) and finally Life Cycle Interpretation.

2.4 Consideration of Life Cycle Costing (LCCA) of alternatives

Whilst it is critical for the pavement design to be environmentally sustainable, it is also very important to choose alternatives that are cost-effective. To this end, a brief LCCA was carried out on pavement alternatives utilized in the study. An LCCA is carried out to provide guidance to an agency on the most cost-effective alternative over its life cycle from a list of many. (Anik Das et al. 2015). The Federal Highway Administration (FHWA) (Walls & Smith 1998) produced a guide on how this should be carried out for pavement design which is used as the foundation for subsequent software tools and applications in the field. For the purpose of this study, the AirCost Excel-based application (Federal Aviation Administration 2011) was utilized which was developed for carrying out LCCAs on Airfield pavements. The assessment involved the following: development of alternate strategies, determining the timeline of the alternatives, estimation of agency and user costs and finally a determination of the life-cycle cost.

2.5 Assumptions and focus of LCA study

As previously mentioned, this study was done on the construction and maintenances phases of the same segment of the runway in three different life cycle scenarios and the over a 30-year period and thus has a comparative end goal. The end of life is not considered part of thegoal and scope of the analysis. This was a methodological choice. AC and PCC pavements, usually, exhibit different service lives. For instance, a rigid pavement can provide a service life of up to 80 years whilst a flexible pavement is limited to 45-60 years. Therefore, including the end of life in the assessment would mean that different analysis periods would have to be considered for the two alternatives making the comparison biased. Furthermore, in the recent guidelines produced by the FAA the analysis period needs to be at least to the first major rehabilitation ((Butt et al., 2019)). Moreover, including the end of life of only one of the alternatives, the AC option for example, since it would be earlier, again, would have made the analysis and comparison biased. This goes against ISO 14044(The International Standards Organisation, 2006), which states that the scope of the study needs to be defined in such a way that the systems can be compared. This is why it was decided to omit the end of life stages of both the alternatives. As a result, the omission of stages C1-C4, according to the European Asphalt Pavement Association (Huijbregts, M.A.J et al. 2016) is justified. Additionally, for the study, for the AC pavement, the distances of raw materials to mixing plant was assumed to be 10km with the exception being the bitumen, which is assumed to be transported from Southern Sicily to the Runway site. For the mixing plant, the distances to the construction site were considered to be 35km. Concerning the PCC pavement, the distance of a ready-mix concrete supplier to the site was established as 10km. These distances were based on actual distances of plants within the area in Sicily to the airstrip in Palermo.

3 RESULTS AND DISCUSSION

3.1 Analysis of midpoint indicators

With respect to the interpretation of the LCA, the midpoint indicators as described by ReCiPe 2016 (Huijbregts, M.A.J et al. 2016) were attained and are depicted in Figure 2. In the figure, impacts are similar for the pavement designs with similar responses made for the AC pavement options with and without the RA.

Figure 4 Illustrates that at the midpoint level, the values of most of the indicators for all the alternatives seem to be quite similar. Noteworthy exceptions can be observed for the indicators: climate change, fossil depletion, and metal depletion. Specifically, PCC and specifically cement seem to be more impactful for the climate change indicators as cement is energy-intensive and emissive compared to asphalt. Fossil and metal depletion indicators' values for the AC option are elevated compared to PCC, due to the fact that higher quantities of materials are utilized over the period of the 30-year analysis, mainly required for the maintenance of the AC pavement.

Based on these results, the research team wanted to distinguish the impacts occurring during the construction and maintenance phases of the two pavement designs. Further analysis was subsequently done to separate these impacts and determine how much each individual phase contributes to the total environmental damage. This was carried out as a brief hotspot analysis and this is described in section 3.2

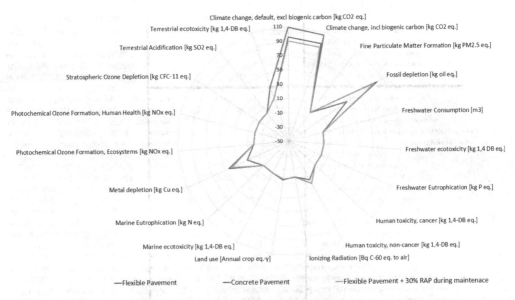

—Flexible Pavement   —Concrete Pavement   —Flexible Pavement + 30% RAP during maintenace

Figure 2.   Relative percentage contribution of the Midpoint indicators for the three analyzed scenarios.

3.2   *Hotspot analysis*

The impacts given by the Midpoint indicators showcased the overall differences between the options. However, it does not distinguish when these impacts are occurring during the different phases of the life cycle. Consequently, a hotspot analysis was done on the midpoint analysis to distinguish the impacts occurring during the construction and maintenance phases of the two general pavement types. This analysis separated these impacts to determine how much each individual phase contributes to the total environmental damage. This is depicted in Figure 3.

Figure 3 Sheds light upon the most environmentally critical stages of the analysed alternatives. For the flexible pavement, it can be clearly seen that a hot spot is exhibited at the maintenance phase of the pavement. Significant amounts of raw materials are required for the maintenance of the AC pavement, compared to its construction and thus, the impacts of these stages are greater.

For the rigid pavement, the situation is different. The construction stage seems to be the most impactful in terms of environmental impacts. Higher amounts of materials are used during its construction compared to its maintenance. PCC maintenance includes patching and depth repairs that are significantly less impactful for the environment compared to the maintenance strategy of the AC option.

Therefore, it can be surmised that for the Rigid Pavement, most of the environmental damage occurs at the construction phase and thus whilst the overall impacts over the 30-year period are close, the Rigid pavement is providing a one-time significant blow to the environment with its pipeline. For the Flexible Pavement, the impacts are more distributed throughout the pavement's life cycle. This is a significant point to make notwithstanding the economic costs.

3.3   *Analysis of endpoint indicators*

With respect to the interpretation of the LCA, the endpoint indicators as described by ReCiPe 2016 (Huijbregts, M.A.J. et al. 2016) were attained and are depicted in Figure 4. Within the context of the research parameters, the endpoint indicators indicate that AC seems to be more environmentally friendly when compared to the PCC option for the specific study, and the specific system boundaries and analysis period and can offer a guide as which material is a more sustainable choice for this type of work and materials observed. Indeed, a PCC option would provide a longer service life compared to an AC pavement. Thus, the environmental benefits of the PCC option would start being exhibited in the long run, since the service life of

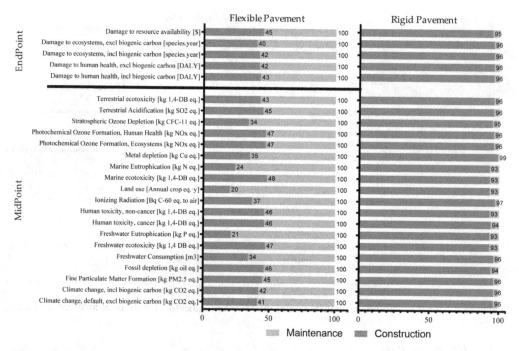

Figure 3. Relative percentage variation of the midpoint and the endpoint indicators over the two main phases of the LCA for the two analysed scenarios.

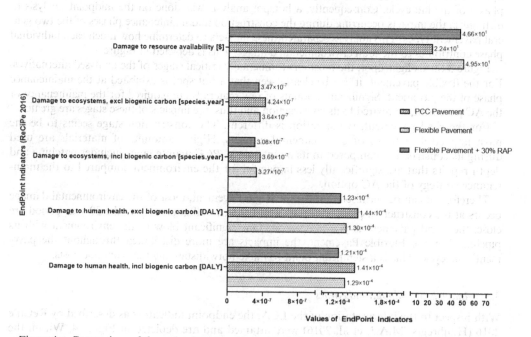

Figure 4. Comparison of the end point impact category indicators values for the two scenarios.

a PCC could reach up to 60-80 years. On the contrary, the usual service life of an AC based pavement can be up to 45years. When the RA is incorporated into the maintenance program of the AC pavement the impacts are further reduced highlighting the value of utilizing the RA material. Further use of the RA in the design and construction phases would also likely further reduce the overall environmental impacts but the addition of the material during this phase would need to be regulated over time given the importance of the pavement in the airport and the related safety concerns.

3.4 LCCA of alternatives

Whilst the LCA of the study highlighted the environmental impacts and concerns of the different materials, it was also considered necessary to have an understanding of the costs involved in the alternatives. As a result, the LCCA was carried out to bridge this gap. This was done using the AirCost Excel Application (Federal Aviation Administration 2011).

Within the software, the three alternatives were inserted through the use of the FAA PAVEAIR web-based application (Federal Aviation Administration 2019) which was developed in order to fulfil requirements of the Airport Pavement Management system (Federal Aviation Administration 2009). For this case study, the LCCA was done on the runway pavement with a discount rate of 4% was applied and a probabilistic assessment approach was undertaken for the costs with 2000 iterations carried out for the assessment. Indirect user costs were not included as this assessment was done to have a simple understanding of the alternatives and further work will include these costs. It should, however, be noted that the conclusions drawn from this assessment could be revised when and if the indirect costs are factored into the analysis.

For the timeline of assessment, two assessments were made. The first of these was over a period of 30 years. However, in this simulation, it was shown that the cost of the PCC option was approximately five times that of the flexible pavement options. This was expected given the fact that the general life expectancy of concrete pavement is 60 to 80 years whereas with the flexible pavement it is usually designed for a 30-year period. As a result of this, the initial costs associated with the PCC is much more than that of AC and as such, it would take much more time to see the financial benefits of using the PCC. Based on this a second simulation was run with a timescale of 50 years. The differences between the simulated net present value costs are shown in Figure 5 below. Within the LCCA, the cheapest alternative was the AC pavement with the maintenance pipeline that utilized RA over the 50-year period, and as a result figure 8 depicts the cost differences related to this option. The most expensive option over the timeframe was the normal AC pavement. This again demonstrates the value of utilizing the RA even only when applied during the maintenance phase. It also demonstrates that the PCC option becomes more cost-effective when the costing analysis is done over a larger timeframe which is expected based on typical costs of these materials.

4 CONCLUSIONS AND RECOMMENDATIONS

The results of the study provide a clear guide into the cause of the impacts of alternative designs of an airfield pavement runway. The study showcases that overall an AC surface for a runway is more environmentally friendly, when compared to that of a PCC pavement, for an analysis period of 30 years and for the assumptions and considerations made for this study. This also shows that this is due to substantial initial impact by the PCC pavement. This can be attributed to the substantial amount of concrete utilized in this design and the materials involved in this type of construction. The analysis further shows that concrete pavements are generally able to project their economic and environmental benefits over a longer period as compared to the asphaltic materials. This is directly related to the project service life of the concrete which is usually 60 to 80 years whereas with asphalt which is 30 to 40 years. This is significant as if the materials used in the PCC construction can be altered to be more environmentally friendly then this design approach could then be superior to the AC design given the minimal impacts of the maintenance of the PCC pavement. Moreover, it became evident that the most impactful stage of the AC surface product system, as defined in this study, is its maintenance, while the most environmentally impactful stage for the PCC pavement is its

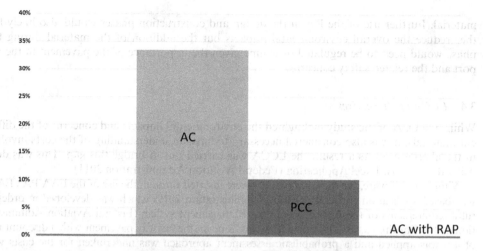

Figure 5. Percentage differences of simulated alternatives during the LCCA.

construction. Thus, it can be recommended to the interested stakeholders, that their main actions focusing on the environmental improvement of AC and PCC types of pavements, should be targeting their maintenance and construction phases, respectively.

The study also demonstrated the value in utilizing RA in the approach as this enables both cost and environmental benefits. The alternative produced using RA displayed the best economic and environmental results of the options under the study. For future work altering concrete materials and Asphaltic materials can be considered to establish more environmentally friendly design and maintenance pipelines.

ACKNOWLEDGEMENTS

The research presented in this paper was carried out as part of the H2020-MSCA-ETN-2016. This project has received funding from the European Union's H2020 Programme for research, technological development and demonstration under grant agreement number 721493.

REFERENCES

Das, A., Barua.S., Khan, N., Rahman, M. 2015. An Evaluation of Life Cycle Cost Analysis of Airport Pavement. International Journal of Engineering Research and Technology. 4(8): 352–356. https://doi.org/10.17577/ijertv4is080396

Anthonissen, J., Van den bergh, W., Braet, J. 2016. Review and environmental impact assessment of green technologies for base courses in bituminous pavements. Environmental Impact Assessment Review. 60: 139–147. http://dx.doi.org/10.1016/j.eiar.2016.04.005

Aurangzeb, Q., Al-Qadi, I.L., Ozer, H., Yang, R., 2014. Hybrid life cycle assessment for asphalt mixtures with high RAP content. Resources, Conservation and Recycling. 83: 77–86.

BRE. 2013. Product Category Rules for Type III environmental product declaration of construction products to EN 15804:2012. Watford: BRE Group.

Butt, A.., Harvey, J.T., Reger, D., Saboori, A., Ostovar, M., Bejarano, M. 2019. Life-Cycle Assessment of Airfield Pavements and Other Airside Features: Framework, Guidelines, and Case Studies Report No. DOT/FAA/TC-19/2. Washington, DC.

Das A., Barua S., Khan N., Rahman M, 2015. An Evaluation of Life Cycle Cost Analysis of Airport International Journal of Engineering & Technology. 4(8): 352–356. https://doi.org/10.17577/ijertv4is080396

European Asphalt Pavement Association. 2016. Guidance Document for preparing Product Category Rules (PCR) and Environmental Product Declarations (EPD) for Asphalt Mixtures by the European Asphalt Pavement Association. Brussels: European Asphalt Pavement Association.

European Bitumen Association, 2012. Life cycle inventory: Bitumen. Brussels: European Bitumen Association.

European Union Commission. 2013. PEF recommendations. Official Journal of the European Union. 56 https://doi.org/doi:10.3000/19770677.L_2013.124.eng

Federal Aviation Administration, 2009. Airport Pavement Design and Evaluation. Advisory Circular 150/5320-6E. Washington, DC: Federal Aviation Administration.

Federal Aviation Administration, 2011. AirCost. v1.0.0

Federal Aviation Administration, 2017. FAARFIELD v1.42.

Federal Aviation Administration, 2019. FAA PAVEAIR v3.3.

Federal Highway Administration, 2011. Reclaimed Asphalt Pavement in Asphalt Mixtures: State of the Practice. Rep. No. FHWA-HRT-11-021. McLean, Virginia.

Giani, M.I., Dotelli, G., Brandini, N., Zampori, L. 2015. Comparative life cycle assessment of asphalt pavements using reclaimed asphalt, warm mix technology and cold in-place recycling. Resources, Conservation and Recycling. 104(Part A): 224–238. https://doi.org/10.1016/j.resconrec.2015.08.006

Gillespie, I., 2012. Quantifying the Energy Used in an Asphalt Coating Plant. University of Strathclyde.

Guercio, M.C., McCarthy, L.M., 2015. Quantifying the performance of warm-Mix asphalt and reclaimed asphalt pavement in flexible airfield pavements. Transportation Research Record 2471: 33–39.

Hajj, E.Y., Sebaaly, P.E., Kandiah, P., 2010. Evaluation of the use of reclaimed asphalt pavement in airfield HMA pavements. Journal of Transport Engineering. 136(3): 181–189. https://doi.org/10.1061/(ASCE)TE.1943-5436.0000090

Huang, Y., 2007. Life Cycle Assessment of Use of Recycled Materials in Asphalt Pavements Thesis submitted to the Newcastle University for the Degree of Doctor of Philosophy. Newcastle University.

Huijbregts, M.A.J et al. 2016. ReCiPe 2016: A harmonized life cycle impact assessment method at midpoint and endpoint level Report I: Characterization. Bilthoven: National Institute for Public Health and the Environment.

International Organization for Standardization. 2006a. ISO 14040: 2006 Environmental management - Life Cycle Assessment - Principles and Framework.

International Organization for Standardization. 2006b. ISO 14044: 2006 Environmental Management - Life Cycle Assessment - Requirements and Guidelines.

Mantalovas, K. & Di Mino, G. 2019. The sustainability of reclaimed asphalt as a resource for road pavement management through a circular economic model. Sustainability. 11(8): 2234. https://doi.org/10.3390/su11082234

Mohammadafzali, M., Koohifar, F., Ali, H., Baqersad, M., Massahi, A. 2017. Investigation of pavement raveling performance using smartphone. International Journal of Pavement Research and Technology. 11(6): 553–563. https://doi.org/10.1016/j.ijprt.2017.11.007

National Asphalt Pavement Association. 2017. Product Category Rules (PCR) For Asphalt Mixtures. Lanham, Maryland: National Asphalt Pavement Association.

Scianni, L. 2017. Il metodo empirico -meccanicistico applicato alle pavimentazioni flessibili aeroportuali. Analisi degli elementi di input e output. Tesi di Laurea. Palermo: Universita Degli Studi di Palermo.

The Norwegian EPD Foundation. 2009. Product-Category Rules (PCR) for preparing an Environmental Product Declaration (EPD) for Product Group Asphalt and crushed stone. Oslo: The Norweigan EPD Foundation.

Thinkstep (a sphera company), 2019. Gabi ts. Software-System and Databases for Life Cycle Engineering, Stuttgart.

Walls, J. & Smith M.R. 1998. Life-Cycle Cost Analysis in Pavement Design, FHWA-SA-98-079. Washington, DC.

Wang, H., Thakkar, C., Chen, X., Murrel, S. 2016. Life-cycle assessment of airport pavement design alternatives for energy and environmental impacts. Journal of Cleaner Production. 133(1): 163–171. https://doi.org/10.1016/j.jclepro.2016.05.090

Wellner, F., Canon Falla, G., Milow, R., Blasl, A., Di Mino, G., Di Liberto, C.M., Noto, S., Lo Presti, D., Jimenez Del Barcon Carrion, A., Airey, G., 2015. AllBack2Pave Project Results [WWW Document]. 2. URL https://www.cedr.eu/strategic-plan-tasks/research/cedr-call-2012/call-2012-recycling-road-construction-post-fossil-fuel-society/allback2pave/ (accessed 11.15.19).

White, G. 2018. State of the art: Asphalt for airport pavement surfacing. International Journal of Pavement Research and Technology. 11(1): 77–98. https://doi.org/10.1016/j.ijprt.2017.07.008

White, G. 2019. Incorporating RAP into Airport Asphalt Resurfacing. Proc. 7th International Conference on Bituminous Mixtures and Pavements 12–14 June 2019: 543–550 Thessaloniki, Greece. https://doi.org/10.1201/9781351063265-73

Recommendations for airfield life cycle assessment tool development

A.A. Butt, J.T. Harvey, J. Lea & A. Saboori
University of California Pavement Research Center, Davis, CA, USA

N. Garg
Federal Aviation Administration, USA

ABSTRACT: In a recent (2019) Federal Aviation Administration funded project, guidelines and a framework for conducting life cycle assessment (LCA) studies for airfields was developed. Four comprehensive example case studies were prepared, with data from recently completed projects and participation from four U.S. airports and/or their consultants, and were presented in the report. The scope of the study was limited to airside civil infrastructure. The framework provides a starting point for the development of an LCA tool for airfields, which can be adapted from roadway pavement LCA tools. This paper reviews the data, models and interface changes needed to create an airfield LCA tool. Different data/unit processes and a different graphical user interface (GUI) will be needed, however, no changes are required for the LCA engine. An understanding of the performance models for the use stage for different types of airfield features that include runways, taxiways, aprons, gate pavements, access roadways, the landscaping and fencing, lighting, drainage, de-icing handling, will also be studied. Some of the steps that would be required to develop an airfield LCA tool are:

- Finalize the scope of features and operations to be considered in the tool
- Define the events in the airfield life cycle for each feature and operation
- Define the materials and equipment and transports in relation to airfields
- Locate and localize life cycle inventory data
- Build unit processes and models (special attention to materials that are specific to airfields, and the use stage models)
- Working with practicing airfield engineers, prototype the GUI, reports, graphics, etc.
- Implement, test, document
- Online help system
- Support, bug and enhancement tracking

1 INTRODUCTION

An increasing number of agencies, companies, organizations, institutes, and governing bodies are embracing principles of sustainability in managing their activities and conducting business. These principles focus on the overarching goal of proactively bringing key environmental, social, and economic factors into the decision-making process. Sustainability considerations are not new, as they were often considered indirectly or informally in the past. However, recent years have seen increased efforts to quantify sustainability effects as they pertain to pavements and other transportation infrastructure, systematically incorporating them into decision making in a more organized fashion (Van Dam et al. 2015).

The airports in the United States (US) have a growing need to be able to quantify their environmental impacts from airfield infrastructure and operations, and to consider these impacts in airfield management, conceptual design, design, materials selection, and construction project delivery decisions. LCA can also be used to evaluate the life cycle environmental impacts as part of policy and standards development. All of these tasks can be performed using life cycle assessment (LCA), although there are different constraints and requirements with respect to the scope of the LCA and the data available for each of these different applications.

In a recent FAA funded project, guidelines and a framework for conducting LCA studies for airfields was developed in order to support the airports to be able to quantify the environmental impacts in a life cycle approach (Butt et al. 2019). Furthermore, four comprehensive example case studies were also performed and presented. Four USA airports (including John F. Kennedy International Airport, Chicago O'Hare International Airport, Boston Logan International Airport and Nashville International Airport) actively participated and shared data of recently completed projects at their respected airports. This was achieved with the help of Port Authorities (Port Authority of New York and New Jersey and Massachusetts Port Authority) and consultant/organizations (Bowman, Barrett and Associates Inc. and Atkins). The scope of the study was limited to the runways, taxiways, shoulders, aprons, fences, and other airside civil infrastructures.

There may be tools, other than a few available online (Kulikowski et al. 2016, Yang and Al-Qadi 2017), used by the US airports (in-house) to estimate GHG emissions for different life cycle stages, and with different systems boundaries and functional units for a variety of purposes. However, these tools may not be freely available and no reviews of such tool are available online.

To be relevant to decision-making, an LCA tool must model the details of the construction and maintenance life cycle of an airfield infrastructure project when a user needs more detailed environmental impact results and has the additional input data required for a more detailed analysis. In addition, there is a need for a project-level LCA tool that uses life cycle inventories (LCIs) specific to the materials and equipment typically used at the airports; environmental life cycle assessment of airfields (eLCAa) has potential to fill this need.

2 PROJECT OVERVIEW

One of the goals of the FAA is to provide guidance and assistance to the airports in order to be able to incorporate innovation in their processes. An indicated earlier, the FAA has recently provided US airports with guidelines and an airside pavement LCA framework that can be used to evaluate the environmental part of the sustainability matrix. The FAA is further interested in a tool that can be accessed easily and simplifies the LCA modeling and analyses for the users (in this case, the US airports), and has relevant data for US airports.

The goal of this project is to outline the development process of a web-based airfield LCA software which is similar to a web-based pavement LCA tool (eLCAP) being developed for the California Department of Transportation (Caltrans; Lea et al. 2019). The web-based environmental LCA for airfield (eLCAa) software will be a project-level LCA tool that uses US averaged airfield-specific life cycle inventories and processes. eLCAa is expected to perform a formal mass-balancing procedure for an airfield LCA project model, and then compute 18 different impact categories (including Human Health Particulate Air, Acidification, Primary Renewable Energy, etc). A detailed Excel™ report will be generated to download display graphs and tables of results.

Construction-type events require user input specifications for materials (e.g., hot mix asphalt, portland cement concrete, aggregate base) and their associated quantities, transports and their associated distances, and construction equipment (e.g., asphalt paver and roller) and their associated times of operation. Similar to eLCAP, eLCAa will have built-in library versions for these processes based on the practices in the US. These library-based processes will allow a user to analyze a specific airside project or create user-defined processes based on library versions, and then customize the amounts and sources of inputs that go into that user-defined process. For example, the library process for Electricity Grid Mix uses 43.4 percent from Natural Gas, but a user can create a user-defined Electricity Process, based on the

Electricity Grid Mix library process, which instead the locally relevant percentage from Natural Gas. Further, any custom, user-defined process set up—either by using the "Manage User Processes" page or within a project—becomes available globally to that user for any project.

For airfield pavements, an examination of the use stage includes the pavement effects on aircraft, such as damage and increased fuel consumption, and the pavement effect on other processes, such as air conditioning use, storm water treatment, and lighting. Currently, there are no models available that can analyze the roughness, macrotexture or structural response of the airside pavements. There is research needed in this field to provide data for the future tool.

Users will interact with eLCAa via a web browser that accesses its user interface (UI). The main UI web page will contain the controls necessary to define the life cycle of a pavement project: Construction, Maintenance/Rehab, Materials, Transport and Equipment. Data for a pavement project is grouped into a project trial; there can be an unlimited number of project trials for a project, and a user can have an unlimited number of projects. All user data are stored in a database, currently an SQL Server. Additionally, the tool will have capability to export the data for a project trial to a local hard disk in a "json"-formatted file. These downloaded files can act as a backup to the user database or as project documentation; they can also be uploaded to eLCAa for further processing.

3 METHODOLOGY

3.1 *Steps involved in the development of the tool*

The project will start with the existing tool (eLCAP) which will be modified based on the scope of the airfield LCA tool. The steps that will be essential include;

- Finalize scope of the systems and operations to be considered
- Develop airfield airside LCA inventories
- Update and develop new user input capabilities
- Results presenting, reporting and exporting
- Convert research grade database
- Update documentation, online help and manual, online user support
- Develop training materials for the airports
- Arrange a pilot workshop to go over most common cases

3.2 *The body of the tool*

The main function of an LCA tool is to simulate (i.e., to model) the life span of a project in order to compute the environmental effects of construction and maintenance, and of traffic on the section (Use Stage). This ability is embedded in eLCAP and the experience can easily be used to apply it to a tool specific to the airfields (as being proposed in this project). Such a tool (eLCAa) can help users make informed decisions on the best course of action to pursue to minimize harmful environment effects and maximize pavement performance over the long term. This is important because sometimes what sounds like a good idea turns out not to be so good when all the "upstream" activities (processes) that go into the extraction and production of materials, construction equipment, and traffic over the life of a pavement project are considered. The balancing, process data, flow data, models and assessment are few other important building blocks of a tool development process and are discussed in brief in later sub-sections.

3.2.1 *Balancing*
A smallest process within a life cycle stage is called a unit process. In a unit process, there are input and output flows that are required to be mass balanced in order for the unit process to

36

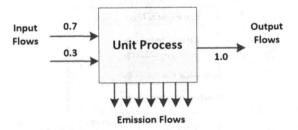

Figure 1. A unit process (Lea et al. 2019).

be valid. Input flows include energy and resources whereas output flows consist of product, waste and emissions as presented in Figure 1 below.

A typical modeled pavement project may include several hundred unit processes for one construction-type event, and there may be many construction-type events in the overall life cycle. These collected unit processes form the LCA balance model for the construction-type event.

3.2.2 *Process and flow data*

As mentioned earlier, the building block of any LCA is a unit process and high-quality representative process-level data are key to obtaining representative results. Since LCA is basically an accounting activity, that is, data items (flows) are simply multiplied by factors and then added up. If process-level data is representative of what is being modeled (e.g., a pavement project), then the LCA should result in a good estimate of the environmental impact of what was included in the LCA. But if the process-level data is not representative, if perhaps some of the data are for industries from a different country because "local" data are unavailable, then the LCA will not result in realistic assessments of what is being modeled.

Intimately associated with the unit process discussed above are flows. Flow objects are used to model the flows of materials and emissions. Flows connect the input items of one unit process to the outputs of another unit process.

3.2.3 *Models*

A model may consist of series of input and output flows to produce a product. A model may include upstream effects (i.e. unit processes that have been scaled as input to the model, e.g. crude oil (agg) is upstream effect as shown in Figure 2) along with input flows of itself. Figure 2 shows a simple representation of a refinery model however, a model can be combination of several models making up a complex project model as shown in Figure 3.

3.2.4 *Other important considerations*

There are several other important aspects to be covered for the tool development process which include;

- User interface
  - ◦ Authentication and authorization
  - ◦ Project trials management
  - ◦ Life cycle definition
  - ◦ User defined processes
  - ◦ Analyses

Result generation

- Architecture
- Data Access

These all will be covered in the eLCAa documentation.

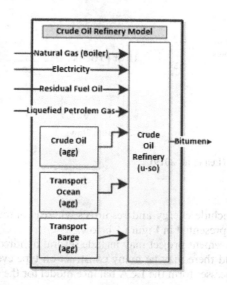

Figure 2. An example of a crude oil refinery model from eLCAP (Lea et al. 2019).

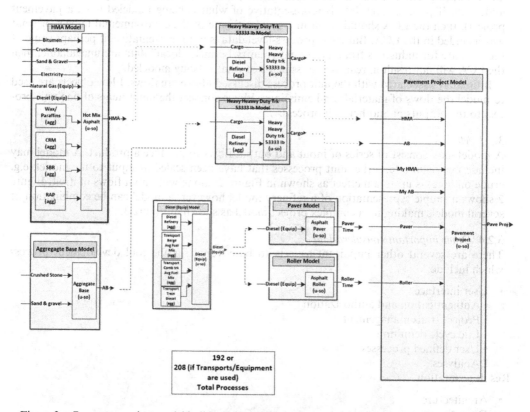

Figure 3. Pavement project model buildup of several upstream models from eLCAP as an example (Lea et al. 2019).

4 FUTURE DIRECTIONS

The intended use of the tool is to be able to assess design standards, specifications, and policies, and project-specific design alternatives, for a wide range of environmental impacts over the life cycle stages of materials production, construction, maintenance and rehabilitation, use, and end-of-life.

Currently, eLCAP has 58 specific LCIs (exported from GaBi) and 43 user-addressable processes (grouped into 21 types of models, such as hot mix asphalt, portland cement concrete, electricity, paver, grinder, etc.) for construction-type events (materials and equipment). Most of these LCIs and processes are relevant to airfields or can be easily adjusted to reflect differences between roadway and airfield materials and construction for the development of eLCAa library. Existing literature and some primary data collection will be needed to develop inventories for airfield lighting, landscaping, drainage and fencing. New models will need to be developed for the interactions of pavement conditions and aircraft. Existing models can be used for most airside land vehicles but will need to be updated where possible for specialized landside vehicles. Following are likely new information that will be required for the development of eLCAa:

- Airfield specific materials
- Specialized airport materials transportation and construction procedures and equipment
- Airfield lighting systems
- Airfield landscape systems
- Airfield fencing systems
- Airfield drainage and water treatment systems
- Use Stage
  ○ Lighting
  ○ Drainage
  ○ Effects of pavement condition on aircraft fuel use while taxiing
  ○ Effects of pavement condition on aircraft life
  ○ Effects of pavement on specialized landside vehicles

- Allow users to compare one Project Trial to another Project Trial

REFERENCES

Butt, A.A., Harvey, J.T., Reger, D., Saboori, A., Ostovar, M. and Bejarano, M. 2019. Life-Cycle Assessment of Airfield Pavements and Other Airside Features: Framework, Guidelines, and Case Studies (No. DOT/FAA/TC-19/2).

Kulikowski, J., Sawalha, M., Sladek, M., and Roesler, J. 2016. Development of LCA-AIR – An Airport Pavement Life Cycle Assessment Tool, International Conference on Concrete Pavements, San Antonio, Texas.

Lea, J., Harvey, J.T. and Saboori, A. (expected 2019). eLCAP: A Web Application for Environmental Life Cycle Assessment of Pavements. A report for California Department of Transportation developed by University of California Pavement Research Center.

Yang, R. and Al-Qadi, I.L. 2017. Development of a life-cycle assessment tool to quantify the environmental impacts of airport pavement construction. Transportation Research Record, 2603(1),pp.89–97.

Van Dam, T., Harvey, J.T., Muench, S., Smith, K., Snyder, M., Al-Qadi, I., Ozer, H., Meijer, J., Ram, P., Roesler, J. and Kendall. A. 2015. Towards sustainable pavement systems: a reference document. Report FHWA-HIF-15-002. https://www.fhwa.dot.gov/pavement/sustainability/hif15002/hif15002.pdf

Incorporating the impacts of climate change into a life cycle assessment of a slab-on-girder highway bridge

G. Guest, J. Zhang, B. Kadhom & J. Singh
National Research Council of Canada, Ottawa, Canada

ABSTRACT: Climate change is expected to impact both the operational and structural performance of infrastructure such as buildings, roads and bridges. Decision making in climate change adaptation of core public infrastructure without integrating life cycle assessment (LCA) can potentially create unforeseen problems shifting into the future. However, most past life cycle assessment studies do not consider how the structural and operational performance of infrastructure will be affected by a changing climate. The goal of this research is to understand how a changing climate affects the environmental life cycle performance of several bridge deck designs for a reinforced concrete bridge case study in Ottawa, Canada. With focus on chloride-induced corrosion as the primary decision metric for bridge deck replacement, the variation in deck design was focused on choice of bridge deck steel reinforcement, where carbon steel (CS), galvanized steel (GS) and stainless steel (SS) were considered. The results indicate a high level of sensitivity towards assumed distribution of surface chloride concentration (Co). Even under light Co conditions, the CS rebar deck performs worse (in terms of climate change potential [CCP]) than the GS- and SS- rebar decks by 22% and 13%, respectively, and this finding becomes more apparent under the higher Co sensitivity scenarios. On the other hand, the deck with GS-rebar out-performed the other decks for light-to-moderate Co conditions, whereas the SS-rebar deck had the lowest life cycle CCP under extreme Co conditions.

1 INTRODUCTION

Climate change is expected to have considerable widespread impacts on our infrastructure systems as a consequence of both extreme and chronic changes in precipitation, temperature, wind and sea level rise (Boyle et al. 2013; Neumann et al. 2015; NRC 2008). Climate plays an important role in terms of the direct and indirect impacts on both structural and operational performance of various infrastructures (Hunt and Watkiss 2011). Climate change will further exacerbate the existing state of corrosion of reinforced concrete (RC) infrastructure in the world, which is affected by four critical factors: temperature, humidity, rate of de-icing salt application and CO_2 concentration for concrete carbonation (Wang et al. 2010a; Wang et al. 2010b). The temporal sensitivity of how these factors change the initiation and propagation of corrosion is especially important in a changing climate. In Canada, corrosion of reinforcing steel bars (rebar) in concrete bridge decks is mainly due to the use of de-icing salts in wintertime, and therefore, chloride-induced corrosion is the most important bridge deck deterioration mechanism to explore.

The current practice of bridge environmental life cycle assessment (LCA) typically assumes a static service life of key components like superstructure, beams and deck. Utilizing LCA that integrates the impacts on the system due to a changing climate is relatively new to the field and requires an understanding of how infrastructure degradation, service life and maintenance schedules change as a function of forecasted climate predictions. There are a number of bridge LCA studies, but none that consider how climate change impacts on bridge performance (Kendall et al. 2008; Kuik et al. 2008; Sharrard et al. 2008; Strauss and Bergmeister 2008;

Hammervold et al. 2013; Du and Karoumi 2014; Pang et al. 2015; Zhang et al. 2016; Born 2018; Navarro et al. 2018; Xie et al. 2018).

On the other hand, Navarro et al. (2018) undertook a bridge deck LCA that considered several corrosion-prevention alternatives for bridge decks exposed to chloride laden (marine – Galicia, Spain) surroundings. They applied the Tuutti model (Tuutti 1982) (time to corrosion initiation assuming a Fickian process of chloride ingress) across 15 designs to determine the maintenance schedule and corresponding impacts. They found that such preventative designs as improved concrete properties by adding silica fume and reducing water-to-cement ratio, or applying surface sealants can lead up to 30-40% reduced life cycle impacts due to maintenance operations and material requirements.

The goal of this study was to apply a recently developed framework (Guest et al. 2019) for incorporating the effects of climate change impacts on infrastructure service life to a single-span reinforced concrete bridge. Several reinforced steel choices were considered, where a modified bridge deck deterioration model (Guest et al., accepted; Nickless and Atadero 2018) was utilized to capture the performance of each scenario, and performance distresses were used as inputs into a bridge LCA model that considered construction and maintenance/rehabilitation materials and activities, end-of-life (EOL) across the service life of the bridge. By applying the proposed framework, the results from this case study provide insight into how climate change affects the LCA results of RC concrete bridge systems and indicate whether adaptive measures are required to improve bridge system performance.

2 METHODS

2.1 Case study

2.1.1 Goal and scope

A comparative attributional LCA was undertaken where several scenarios of RC highway bridge designs were considered assuming a 2-lane (12.45 meter width, one-way) and 25.6 meter span. The bridge was assumed to consist of a cast-in-place deck/diaphragms on precast, pre-stressed concrete I-girders (commonly known as a slab-on-girder bridge). This bridge design was chosen because it represents the most common type of pre-stressed concrete bridge constructed in North America. The main objective was to see if the performance of several bridge deck designs would significantly change between a recent historic climate versus a high (RCP8.5, see section 2.6) climate change scenario. The bridge, highway traffic and climate data utilized are representative of the city of Ottawa, Canada. Selecting an appropriate functional unit (FU) remains a challenge in bridge LCAs, and therefore, rather than select a single FU, results are presented based on several: the total impact of the bridge over its 75 year service life; FUs considering only area or distance: 1 m²-year and 1 lane-m-year; FUs considering both distance and utilization: 1 ESAL-m and 1 person-m (ESAL = Equivalent Single Axle Load).

Figure 1 provides an overview of the bridge system from a life cycle perspective with key activities listed. Five ReCiPe (NIPHE 2016) (hierarchical) midpoint impact categories were considered: climate change potential (CCP), water depletion potential (WDP), fossil depletion potential (FDP), metal depletion potential (MDP) and particulate matter formation potential (PMFP). For the purposes of vehicle fleet dynamics, this prospective LCA assumes new bridge construction in the year 2020 where a seventy-five year service life to the year 2094 was assumed. The bridge substructure (abutments) and concrete girders were assumed to be replaced upon the prescribed service life of 75 years whereas the necessity for bridge deck replacement depended on results from a chloride-induced corrosion model.

Following the framework on including climate impacts in LCA infrastructure system (Guest et al. 2019), the primary goal of this study was to consider how climate change would affect the rate of chloride-induced bridge deck corrosion and corresponding intervention schedule of the bridge system over time. A comparison between the bridge designs using historic versus forecasted climate data from a high climate change scenario was undertaken to

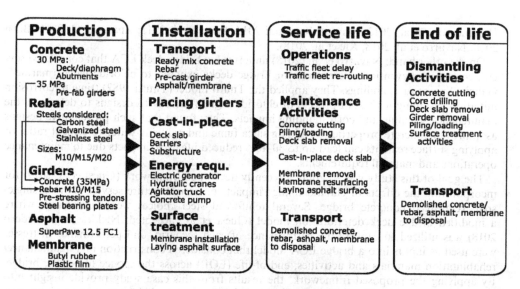

Production	Installation	Service life	End of life
Concrete	**Transport**	**Operations**	
30 MPa:	Ready mix concrete	Traffic fleet delay	**Dismantling**
Deck/diaphragm	Rebar	Traffic fleet re-routing	**Activities**
Abutments	Pre-cast girder		
35 MPa	Asphalt/membrane		Concrete cutting
Pre-fab girders	**Placing girders**	**Maintenance**	Core drilling
Rebar		**Activities**	Deck slab removal
Steels considered:	**Cast-in-place**	Concrete cutting	Girder removal
Carbon steel	Deck slab	Piling/loading	Piling/loading
Galvanized steel	Barriers	Deck slab removal	Surface treatment
Stainless steel	Substructure		activities
Sizes:	**Energy requ.**	Cast-in-place deck slab	
M10/M15/M20	Electric generator		
Girders	Hydraulic cranes	Membrane removal	
→Concrete (35MPa)	Agitator truck	Membrane resurfacing	
→Rebar M10/M15	Concrete pump	Laying asphalt surface	**Transport**
Pre-stressing tendons			Demolished concrete/
Steel bearing plates	**Surface**		rebar, asphalt, membrane
Asphalt	**treatment**	**Transport**	to disposal
SuperPave 12.5 FC1	Membrane installation	Demolished concrete,	
Membrane	Laying asphalt surface	rebar, ashpalt, membrane	
Butyl rubber		to disposal	
Plastic film			

Figure 1. Simplified life cycle flow diagram of the reinforced concrete bridge systems considered in this study.

illustrate the degree of importance that climate plays on bridge deck deterioration and to determine if one of the considered adaptive designs would be sensible from an environmental life cycle impact perspective.

2.2 *Bridge deck deterioration modeling*

Results were utilized from a chloride-induced corrosion model proposed by Nickless and Atadero (2018), and with modifications (Guest et al. *accepted*), which was created for predicting deterioration of modern RC bridge decks with protective systems such as waterproofing membranes and epoxy coated rebar. The model calculates deterioration of individual cells within a deck with consideration of three stages: T1 = time to corrosion initiation; T2 = time to crack initiation; and, T3 = time to cell failure (i.e. surface failure crack width > 0.3mm [AASHTO 2002]). The bridge deck (12.45x25.6m) was divided into 30.48cmx30.48cm cells (total of 3078 cells), where the cell deterioration model was largely based on the work of Hu et al. (2013) which in turn was based on a number of existing models from literature. Nickless and Atadero (2018) updated the model in Hu et al. (2013) with other models for determining the rate of non-uniform corrosion (Keßler et al. 2015), cracking pressure (Jang and Oh 2010; Šavija et al. 2013) and linear crack-width (Chen and Leung 2015).

The model proposed by Nickless and Atadero (2018) also provided preliminary ways to model the impact of asphalt overlays, waterproofing membranes and epoxy disbondment from rebar, if considered in the design. The individual cell deterioration model was also linked to a global deck deterioration model to allow for Monte Carlo Simulation with probabilistic inputs of key parameters and to include the effect that expansion joints and their deterioration have on corrosion of nearby cells to reflect real variations that exist in deck deterioration processes. A bridge deck was assumed to be replaced when 30% of the 3078 cells reached a failure state which was assumed to be a surface crack width greater than 0.3 mm. Table 1 summarizes the LCA scenarios considered alongside key characteristics of each bridge deck design.

Table 1. Bridge deck design and climate scenarios considered.

#	Scenario code	Scenario name	State of practice	Surface seal	deck cover (mm) min	max	Rebar	Climate[1]	C_{th} mean	cov	Dc-ref.	E-D(KJ/mol)	# bridge decks for each Co_{mean} (kg/m3) 1.8	3.5	5.3	7.4
1	OPC-NS-CS-HC	OPC, Carbon Steel, no seal, recent climate	uncommon	no	0.043	0.058	CS	HIST	1.21	0.2	8.0E-12	35	2	3	3	4
2	OPC-NS-GS-HC	OPC, Carbon Steel, no seal, recent climate	uncommon	no	0.043	0.058	GS	HIST	2.42	0.2	8.0E-12	35	1	2	2	3
3	OPC-S-CS-HC	OPC, Carbon Steel, with seal, recent climate	common	yes	0.043	0.058	CS	HIST	1.21	0.2	8.0E-12	35	2	2	2	4
4	OPC-S-GS-HC	OPC, Galvanized steel rebar, recent climate	common	yes	0.043	0.058	GS	HIST	2.42	0.2	8.0E-12	35	1	2	2	2
5	OPC-NS-SS-HC	OPC, Stainless steel rebar, recent climate	new	no	0.043	0.058	SS	HIST	12.1	0.2	8.0E-12	35	1	1	1	1
6	OPC-NS-CS-CCC	OPC, Carbon Steel, no seal, RCP 8.5	uncommon	no	0.043	0.058	CS	RCP8.5	1.21	0.2	8.0E-12	35	2	3	4	4
7	OPC-NS-GS-CC	OPC, Carbon Steel, no seal, RCP 8.5	uncommon	no	0.043	0.058	GS	RCP8.5	2.42	0.2	8.0E-12	35	1	2	2	3
8	OPC-S-CS-CC	OPC, Carbon Steel, with seal, RCP 8.5	common	yes	0.043	0.058	CS	RCP8.5	1.21	0.2	8.0E-12	35	1	2	2	4
9	OPC-S-GS-CC	OPC, Galvanized steel rebar, RCP 8.5	common	yes	0.043	0.058	GS	RCP8.5	2.42	0.2	8.0E-12	35	1	1	2	2
10	OPC-NS-SS-CC	OPC, Stainless steel rebar, RCP 8.5	new	no	0.043	0.058	SS	RCP8.5	12.1	0.2	8.0E-12	35	1	1	1	1

1. HIST = historical climate change assuming a 1981-2010, 30 year climate normal; RCP8.5 = climate change scenario, representative concentration pathway with global average +8.5 W/m2 by 2100.
Acronyms: CC = climate change; COV = coefficient of variation; Co_{mean} = mean surface chloride concentration (kg Cl-/m3); C_{th} = steel corrosion threshold; Dc-ref = reference chloride diffusion coefficient; GS = galvanized steel; HC = historic climate (1981-2010); OPC = ordinary Portland concrete; SS = stainless steel.

2.3 Bridge design details

A slab on precast concrete I-girder bridge was assumed as it has been most widely used in North America for nearly sixty years (Fu and Wang 2015). This type of bridge consists of a number of longitudinal prestressed concrete I-beams (girders) connected either compositely or non-compositely across the tops by a continuous cast-in-place reinforced concrete slab. The main design differences resided in the decks. As indicated in Table 1, carbon steel (CS), galvanized steel (GS) and stainless steel (SS) were considered for their significantly different chloride (probabilistic) thresholds for corrosion initiation. A combination of traffic loading and environmental effects leads to deck, pavement and membrane deterioration, and over time, they can fall to a critical level where safety becomes a concern and rehabilitation is required. In this study, the pavement and waterproof membrane layers were assumed to be either replaced every 20 years based on best practice or when the bridge deck failure year occurred.

2.4 Traffic conditions and operational performance modelling

Traffic survey data (MTO 2015) was utilized to predict the fleet composition into the future where vehicle counts were taken for highway 417 at Panmure Road which is east of Ottawa, Ontario, Canada. Based on trends on annual average daily traffic (AADT, includes all vehicle types) from 1998 to 2010, an average annual growth rate in AADT of 3% was assumed with linear growth from 11,650 in 2016 to 25,500 in 2055 and assumed to remain steady thereafter. A constant relative hourly fleet mix was used and based on a traffic data survey (MTO, *personal communication*) which provided an average hourly weekday-specific breakdown in terms of number of vehicles of a particular range in length. The light duty vehicle mix was based on province-wide statistics and a medium electric vehicle penetration scenario (IESO 2016). Total ESALs and commuters during the 75-year service life was estimated to be 98.2 Million ESALs and 908 Million commuters (i.e. number of estimated people to cross the bridge for commuting purposes).

The operational performance of the pavement was based on the fuel efficiency of the vehicle. A physics based transport modelling approach as explained in Maness et al. (2015) and based on the Ross model was utilized (Ross 1997). Nine vehicle typologies were considered: mid-sized passenger car (gasoline, diesel and EV), lightweight truck[1] (gasoline, diesel and EV), bus (diesel), mid-duty truck (diesel), and heavy-duty truck with trailer (diesel). The average hourly weekday traffic count as explained above (MTO 2015b) was used in a model that relates free-flow speed and traffic volume (vehicles-hr^{-1}-lane^{-1}) to average fleet speed (Brilon and Lohoff 2011), which was utilized in the Ross model to determine vehicle fuel consumption of vehicles passing over the bridge section every hour of the service life.

2.5 Climate data utilized

The Canadian Regional Climate Model (CanRCM4) developed by the Canadian Centre for Climate Modelling and Analysis (Scinocca et al. 2016) was used in this study to project the required climatic data at daily resolution. As this study aims to consider likely worst case impacts due to climate change on the bridge deck performance, only the output of CanRCM4 under the RCP8.5 scenario was used. To study the climate change impact on bridge deck performance, daily climatic parameters (temperature and relative humidity) were used considering two climate periods: a historical period (1981-2010) and a future climate period (2020-2094) considering a changing climate under the RCP8.5 scenario. These time periods were selected to illustrate the extent to which each bridge deck design performs in light of a changing climate.

---

1. Light weight truck includes minivans, light trucks, sports utility vehicles, and vans.

2.6 Life cycle inventory data and modelling assumptions

The ecoinvent v3.4 database (default cutoff) (ecoinvent 2018) was utilized as the background system while unit processes for the foreground system that were integrated with ecoinvent v3.4 are described below. The two concrete mixes (specified to 30MPa and 35Mpa) assumed for the various bridge components were based on an ecoinvent Quebec-specific process (currently there is no Ontario specific process in ecoinvent, where the designed bridge is located) where concrete mix inputs were updated according to bridge design specs, and the electricity mix was updated to the Ontario grid mix. The unit processes representing reinforcing steel (carbon steel, stainless steel and galvanized) for various rebar dimensions were based on global or 'rest of world' average datasets from ecoinvent. Galvanized rebar was assumed to be carbon steel bar dipped in zinc coating with an assumed thickness of 86 microns or 610 grams-m^{-2} (AGA 2011) for which an existing ecoinvent process for hot dip zinc coating of generic metals pieces was updated to better represent zinc coating of carbon steel rebar.

The five pre-stressed, pre-cast concrete girders were fabricated offsite with a 35 MPa concrete mix, reinforcing carbon steel and pre-stressed/high-strength steel tendons. The utility/capital requirements for the girder fabrication plant were assumed to be similar (m^3 product flow basis) to an existing ecoinvent process for concrete sole-plate and foundation manufacturing. The waterproofing membrane was based on a roll-type membrane of 1.65 mm (65 mils) thickness and composed of nominally polymeric membrane on a shrink-resistant, heavy-duty, polypropylene woven carrier fabric (Meadows 2018). Two ecoinvent processes representing synthetic rubber production and plastic film extrusion were used assuming mass requirements equivalent to the membrane material thicknesses and coverage for the entire bridge deck area.

The bridge abutment footing, back wall and stem were assumed to be cast-in-place. In turn, the five pre-fabricated girders were transported via flatbed truck assuming 100 km from plant to construction site, and set in place with a truck crane. The bridge deck and diaphragms were cast-in-place where the pre-mixed concrete was assumed to be transported 50 km from mixing plant and the rebar transported 100 km to site from regional storage. On-site energy/machinery requirements were estimated for hydraulic truck cranes of three capacities (11t, 16t and 22t), agitator truck, truck-mounted concrete pump and diesel generators. The on-site machinery energy requirements were based on a study developing LCI for concrete construction which considered requirements for a similar RC bridge case study (Kawai et al. 2005). The asphalt surface layer (50 mm of SuperPave 12.5 FC2) was assumed to be replaced every 20 years along with the degraded waterproof membrane in the case of the bridge deck designs using CS or GS rebar. A routine maintenance schedule of spot repairs and routing/sealing was based on specifications from Sharif (2013). The on-site energy/machinery requirements for demolition of the various concrete elements was based on LCI results from Kawai et al. (2005), while all waste material was assumed to be sent to a landfill 50 km from the site.

Pavement material and activity and electricity inputs were regionalized to the province of Ontario, where pavement material/activity libraries from Athena's Pavement LCA tool (Athena Sustainbale Materials Institute 2018) were utilized and augmented with additional inputs stemming from ecoinvent.

2.7 Sensitivity analysis

Surface chloride concentration (Co) is an important parameter that governs chloride-induced corrosion of bridge decks. It is generally quite dynamic and site specific due to factors such as concrete design, proximity to sea water, and the rate at which de-icing salts are applied. Due to a lack of site-specific or regional measurements, each bridge deck design was simulated assuming four Co logarithmic distributions with mean surface chloride concentration (Co$_{mean}$) values assuming a coefficient of variation (COV) of 0.5. Weyers et al. (1993) classified the corrosive environments of bridge decks into four categories in terms of surface chloride concentration: light, moderate, high and severe exposures and the suggested Co$_{mean}$ values (units: kg Cl$^-$ m^{-3}) were 1.8 (low), 3.5 (moderate), 5.3 (high) and 7.4 (severe). While some Co distributions are likely not realistic for the given site, we present the results for sensitivity analysis purposes.

3 RESULTS

3.1 *Bridge deck deterioration*

The bridge deck deterioration curves (i.e. cumulative cells reaching time of failure, T3) resulting from the chloride-induced corrosion modelling are presented in Figure 2 for each bridge design/climate/Co_{mean} scenario, where 30% cell failure (i.e. surface crack > 0.3mm) is assumed to trigger a decision to replace the bridge deck.

In all cases, the SS-rebar deck design does not require replacement even in extreme Co conditions with a Co_{mean} of 7.4 kg Cl$^-$ m^{-3} (Figure 2) and this was due to the high corrosion threshold of SS ($C_{th,mean}$ = 12.1 kg-m^{-3}). In comparing decks constructed with CS- and GS-rebar with and without membrane seals, the seal was found to play an influential role in extending service life for light-to-high and all Co_{mean} scenarios, respectively. For example, under moderate Co_{mean} conditions (3.5 kg Cl$^-$ m^{-3}), the deck fail years were estimated to be year 30 and 60 for the OPC-CS-NS-HC scenario and only year 39 for the OPC-CS-S-HC scenario.

3.2 *Life cycle impact assessment results*

Figure 3 presents the life-cycle CCP results breakdown between major system areas across each bridge design scenario along with sensitivity results as it pertains to the assumed distribution of surface chloride concentration with mean (Co_{mean}) values (units: kg Cl$^-$m^{-3}) of 1.8 (light), 3.5 (moderate), 5.3 (high) and 7.4 (extreme).

The best performing bridge deck design depended, to some degree, on the level of Co concentration assumed and how it corresponds to whether or not one or several deck failures occur. For instance, under a severe Co (Co_{mean} = 7.5 kg m^{-3}) and assuming CC, the SS-deck scenario had a CCP that was 11% and 35% lower than the GS-deck and CS-deck designs (with seal), respectively. While the SS-deck performed better in terms of CCP for the high and severe Co conditions, the GS-deck (regardless of using a seal) performed significantly better under light Co conditions. The main reason was because the improved chloride threshold (Cl_{th}) of GS-rebar compared to CS-rebar (Co_{mean}: 1.21 vs. 2.42 kg m^{-3}) permitted no deck

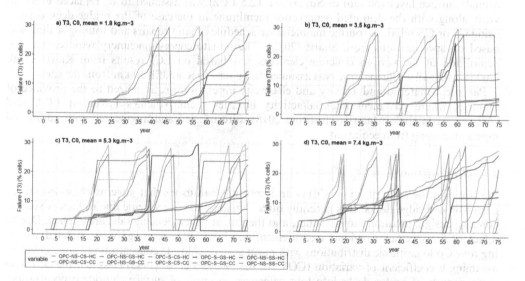

Figure 2.   Bridge deck deterioration curves for each deck design considering both a historic climate (HC [1981-2010]) and a changing climate (CC [2020-2094]) where each sub-plot depicts results for a given surface chloride distribution with Co_{mean} values stated. When/if failure returns to zero it indicates a deck replacement. See Table 1 for scenario acronym explanations.

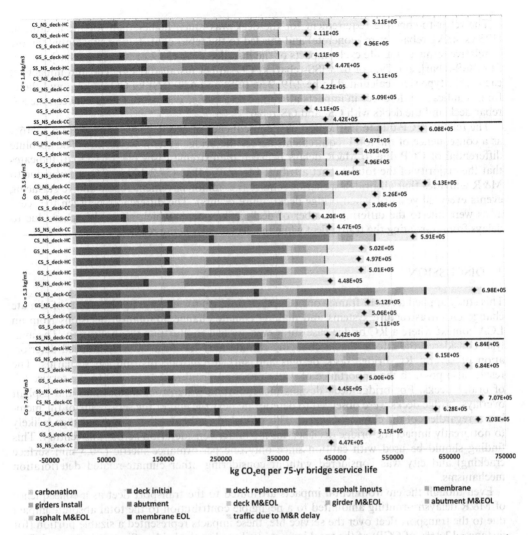

Figure 3. LCIA results (climate change potential in kg CO_2eq) across each scenario divided between the main contribution areas; absolute results reflect a functional unit of one RC bridge (12.45x256m) operating for its assumed 75 year service life; net impacts are indicated with black diamonds with values labelled. M&EOL = maintenance and end-of-life; M&R = maintenance and rehabilitation.

replacement for the GS-deck scenarios which led to significantly reduced impacts. In the case of the CS-rebar deck scenarios, the seal plays an important role in reducing impacts under scenarios with moderate-to-high Co_{mean} (i.e. 3.5 and 5.3 kg Cl⁻ m⁻³) where a deck replacement can be avoided.

3.2.1 *Hotspot analysis*

For the purposes of hotspot analysis in terms of CCP, the scenarios under moderate Co (i.e. Co_{mean} = 3.5 kg m⁻³) concentration are used for explaining relative. The top four contributing sub-systems were due to the abutment, deck (initial+replacement), fleet delays due to M&R and the girders. The impact due to the abutments, assumed to require 14.2 tonnes of CS-rebar and 285 m³ of concrete (30MPa), were the greatest contributor (30-33% or 136 t CO_2eq for installed abutments), if only one bridge deck was required; otherwise, the impacts due to the initial bridge deck and subsequent replacement(s) led to the highest contribution.

The rebar-to-concrete requirement for the deck was much higher than for the abutments (195 vs. 48 kg rebar per m^3 concrete) and this led to the rebar contributing substantially more in relative terms to the life cycle impacts of the installed deck (38.7 or 50% [CS-rebar], 41.0 or 51% [GS-rebar], and 79.3 or 67% [SS-rebar] kt CO_2eq per installed deck). The unit impacts of each rebar type were calculated to be 2.19, 2.36 (GS) and 4.89 (SS) kg CO_2eq kg^{-1}, and therefore, significant differences in impacts attributed to a bridge deck were found between the SS-rebar deck and the decks with CS- and GS-rebar.

The relative CCP due to traffic fleet as a consequence of M&R delays ranged from 12-16% as a consequence of the total requirements across the bridge system service life. The absolute differential of CCP due to M&R delays between the extreme cases (44.5 t CO_2eq) indicates that the majority of the total impact attributed to traffic fleet (57.2-102 kt CO_2eq) was due to M&R activities that all bridge scenarios require (i.e. minor pavement repair and re-surfacing events every 20 years or when bridge deck failure occurs). The differences between the scenarios were due to the differing number of deck replacements which affected impacts due to delays from re-routing the traffic fleet while the deck(s) was being replaced.

4 DISCUSSION

This study applied a recent framework (Guest et al. 2019) for integrating the effects of climate change on infrastructure systems and making climate change adaptation decisions from an LCA context where a RC bridge case study was applied to illustrate its application. The analysis considered the effects that a changing climate may have on chloride-induced deterioration of several RC bridge deck designs and surface chloride distribution scenarios. The results also point to the importance of water-proofing membrane longevity to the service life of bridge decks. For bridge deck designs incorporating waterproof membranes and asphalt overlays, bridge decks were observed to deteriorate near the end of the service life of the membrane regardless of the climate scenario. This case study indicates that climate change is likely to not greatly impact the bridge deck service life under consideration across this century. This finding should be used with caution since only one performance metric (>0.3 mm surface cracking) and city was considered without considering other climate-related deterioration mechanisms.

Even though the environmental impacts attributed to the transport fleet as a consequence of M&R delays/re-routing amounted to a negligible contribution to the total absolute impact due to the transport fleet over the service life, these impacts represented a sizable portion (for instance, 12-16% of CCP) of the total impact attributed to the bridge. Since the transport fleet was found to be a major hotspot in the system, it is essential to capture the differences in traffic fleet delays/speed as it relates to fuel efficiency of the fleet mix. By attributing these impacts to the bridge system, it provides a fair way to compare various bridges by capturing the fleet impact trade-offs between bridges with significantly different deck replacement requirements and intervention strategies.

REFERENCES

AASHTO (American Association of State Highway and Transportation Officials) (2002) Standard specification for highway bridges. Washington, DC.

AGA (American Galvanizer's Association) (2011) Hot-dip galvanized reinforcing steel - A specifier's guide.

Athena Institute (2018) Athena Pavement LCA tool. https://calculatelca.com/software/pavement-lca/. Accessed 8 Aug 2017.

Basheer PA., Chidiac SE and Long AE (1996) Predictive models for deterioration of concrete structures. Constr Build Mater 10:11.

Bhardwaj A, Misra V, Mishra A, Wootten A, Boyles R, Bowden JH and Terando AJ (2018) Downscaling future climate change projections over Puerto Rico using a non-hydrostatic atmospheric model. Clim Change 1–15.

Born RO (2018) Life cycle assessment of large scale timber bridges : A case study from the world' s long-est timber bridge design in Norway. Transp Res Part D 59:301–312. doi: 10.1016/j.trd.2018.01.018

Boyle J, Cunningham M and Dekens J (2013) Climate change adaptation and water resource management: a review of the literature. Clim Chang Adapt Can Infrastruct. doi: 10.1016/j.eneco.2013.09.005

Brilon W and Lohoff J (2011) Speed-flow models for freeways. Procedia - Soc Behav Sci 16:26–36. doi: 10.1016/j.sbspro.2011.04.426

Chen E and Leung CKY (2015) Finite element modeling of concrete cover cracking due to non-uniform steel corrosion. Eng Fract Mech 134:61–78. doi: 10.1016/j.engfracmech.2014.12.011

Du G and Karoumi R (2014) Life cycle assessment framework for railway bridges: Literature survey and critical issues. Struct Infrastruct Eng 10:277–294. doi: 10.1080/15732479.2012.749289

Ecoinvent (2018) Introduction to ecoinvent Version 3. https://www.ecoinvent.org/database/introduction-to-ecoinvent-3/introduction-to-ecoinvent-version-3.html. Accessed 15 May 2018.

Fu CC and Wang S (2015) Computational Analysis and Design of Bridge Structures. Taylor & Francis Group, Florida.

Government of Canada (GoC) (2018) Ottawa Climate Normals. http://climate.weather.gc.ca/climate_normals/results_1981_2010_e.html?stnID=4337. Accessed 6 Feb 2019.

Guest G, Zhang J, Atadero R and Shirkhani H Incorporating the effects of climate change into bridge deterioration modelling: the case of slab-on-girder highway bridge deck designs across Canada. Accepted. to J. Mater. Civ. Eng..

Guest G, Zhang J, Maadani O and Shirkhani H (2019) Incorporating the impacts of climate change into infrastructure life cycle assessments: A case study of pavement service life performance. J Ind Ecol. doi: https://doi.org/10.1111/jiec.12915

Hammervold J, Reenaas M and Brattebø H (2013) Environmental Life Cycle Assessment of Bridges. Bridg Eng 18:153–161. doi: 10.1061/(ASCE)BE.1943-5592.0000328.

Hu N, Haider SW and Burgueno R (2013) Development and Validation of Deterioration Moels fo Concree Bridge Decks - Phase 2: Mechanics-based Degradation Models. 1–131. doi: 10.1016/S0008-8846(98)00259-2

Hunt A and Watkiss P (2011) Opus : University of Bath Online Publication Store Climate Change Impacts and Adaptation in Cities : A Review of the Literature. Online 104:13–49. doi: http://dx.doi.org/10.1007/s10584-010-9975-6

IESO (Independent Electricity System Operator) (2016) Module 2: Demand outlook.

Jang BS and Oh BH (2010) Effects of non-uniform corrosion on the cracking and service life of reinforced concrete structures. Cem Concr Res 40:1441–1450. doi: 10.1016/j.cemconres.2010.03.018

Kawai K, Sugiyama T, Kobayashi K and Sano S (2005) Inventory Data and Case Studies for Environmental Performance Evaluation of Concrete Structure Construction. 3:435–456.

Kendall A, Keoleian G a. and Helfand GE (2008) Integrated Life-Cycle Assessment and Life-Cycle Cost Analysis Model for Concrete Bridge Deck Applications. J Infrastruct Syst 14: 214–222. doi: 10.1061/(ASCE)1076-0342(2008)14:3(214)

Keßler S, Angst U, Zintel M and Gehlen C (2015) Defects in epoxy-coated reinforcement and their impact on the service life of a concrete structure: A study of critical chloride content and macro-cell corrosion. Struct Concr 16:398–405. doi: 10.1002/suco.201400085

Kuik O, Brander L, Nikitina N, Navrud S, Magnussen K and Fall E. (2008) Report on the monetary valuation of energy related impacts on land use changes, acidificaiton, eutrophication, visual intrusion and climate change. CASES project, WP3.

Maness HL, Thurlow ME, McDonald BC and Harley RA (2015) Estimates of CO2 traffic emissions from mobile concentration measurements. J Geophys Res Atmos 120:2087–2102. doi: 10.1002/2014JD022876

Meadows WR (2018) MEL-DEK Waterproofing System. https://www.wrmeadows.com/mel-dek-deck-waterproofing-system/#ds. Accessed 12 Dec 2018.

MTO (Ministry of Transport Ontario) (2015) Traffic volume characteristics on Provincial highways. https://www.ontario.ca/data/traffic-volume.

Navarro IJ, Yepes V, Martí JV and Gonzalez-Vidosa F (2018) Life cycle impact assessment of corrosion preventive designs applied to prestressed concrete bridge decks. J Clean Prod 196:698–713. doi: 10.1016/j.jclepro.2018.06.110

Neumann JE, Price J, Chinowsky P, Wright L, Ludwig L, Streeter R, Jones R, Smith JB, Perkins W, Jantarasami L et al. (2015) Climate change risks to US infrastructure: impacts on roads, bridges, coastal development, and urban drainage. Clim Change 131:97–109. doi: 10.1007/s10584-013-1037-4

Nickless K and Atadero RA (2018) Mechanistic Deterioration Modeling for Bridge Design and Management. J Bridg Eng 23:04018018. doi: 10.1061/(ASCE)BE.1943-5592.0001223

NIPHE (National Institute for Public Health and the Environment) (2016) ReCiPe 2016: A harmonized life cycle impact assessment method at midpoing and endpoint level. https://www.rivm.nl/en/Topics/L/Life_Cycle_Assessment_LCA/Downloads/Documents_ReCiPe2017/ReCiPe2016_CFs_v1_1_20170929. Accessed 5 May 2018.

NRC (National Research Council of the National Acadamies) (2008) Potential Impacts of Climate Change on U.S. Transportation. doi: 10.17226/12179

Pang B, Yang P, Wang Y, Kendall A, Xie H and Zhang Y (2015) Life cycle environmental impact assessment of a bridge with different strengthening schemes. Int J Life Cycle Assess 20:1300–1311. doi: 10.1007/s11367-015-0936-1

Peters GP, Andrew RM, Boden T, Canadell JG, Ciais P, Quéré LC and Wilson C (2012) The challenge to keep global warming below 2 C. Nat Clim Chang 3:4.

Riahi K, Rao S, Krey V, Cho C, Chirkov V, Fischer G and Rafaj P (2011) RCP 8.5—A scenario of comparatively high greenhouse gas emissions. Clim Change 109: 33.

Ross M (1997) Fuel efficiency and the physics of automobiles. Contemp Phys 38:381–394. doi: 10.1080/001075197182199

Safiuddin M and Soudki KA (2011) Sealer and coating systems for the protection of concrete bridge structures. Int J Phys Sci 6:8188–8199. doi: 10.5897/IJPSX11.005

Šavija B, Luković M, Pacheco J and Schlangen E (2013) Cracking of the concrete cover due to reinforcement corrosion: A two-dimensional lattice model study. Constr Build Mater 44:626–638. doi: 10.1016/j.conbuildmat.2013.03.063

Schlaepfer DR, Bradford JB, Lauenroth WK, Munson SM, Tietjen B, Hall SA and Lkhagva A (2017) Climate change reduces extent of temperate drylands and intensifies drought in deep soils. Nat. Commun. 8.

Scinocca JF, Kharin V V., Jiao Y, Qian MW, Lazare M, Solheim L, Flato GM, Biner S, Desgagne M and Dugas B (2016) Coordinated global and regional climate modeling. J Clim 29:17–35. doi: 10.1175/JCLI-D-15-0161.1

Sharif MO (2013) LIFE CYCLE COSTING ANALYSIS USING THE MECHANISTIC-EMPIRICAL PAVEMENT DESIGN GUIDE FOR FLEXIBLE PAVEMENTS. Ryerson University.

Sharrard AL, Matthews HS, Asce AM and Ries RJ (2008) Using an Input-Output-Based Hybrid Life-Cycle Assessment Model. J Infrastruct Syst 14:327–336. doi: 10.1061/_ASCE1076-0342_200814:4_327

Strauss A and Bergmeister K (2008) Advanced life-cycle analysis of existing concrete bridges. J Mater. 20:9–19.

Trenouth WR, Gharabaghi B and Perera N (2015) Road salt application planning tool for winter de-icing operations. J Hydrol. 524:401–410. doi: 10.1016/j.jhydrol.2015.03.004

Tuutti K (1982) Corrosion of steel in concrete. CBI Research Report 4:82,. Stockholm.

Wang X, Nguyen M, Stewart MG, Syme M and Leitch A (2010a) Analysis of climate change impacts on the deterioration of concrete infrastructure part 3: case studies of concrete deterioration and adaptation. doi: 10.13140/RG.2.1.4115.6244

Wang X, Nguyen M, Stewart MG, Syme M and Leitch A (2010b) Analysis of climate change impacts on the deterioration of concrete infrastructure Part 1: Mechanisms, Practices, Modelling and Simulations – A Review. doi: 10.13140/RG.2.1.4115.6244

Weyers RE, Prowell BD and Sprinkel MM (1993) Concrete bridge protection, repair and rehabilitation relative to reinforcement corrossion: a methods application manual.

Xie H, Wu W and Wang Y (2018) Life-time reliability based optimization of bridge maintenance strategy considering LCA and LCC. J Clean Prod. 176:36–45. doi: 10.1016/j.jclepro.2017.12.123

Zhang Y-R, Wu W-J and Wang Y-F (2016) Bridge life cycle assessment with data uncertainty. Int J Life Cycle Assess 21:569–576. doi: 10.1007/s11367-016-1035-7

Life cycle environmental impact considerations for structural concrete in transportation infrastructure

S.A. Miller
University of California Davis, Davis, USA

ABSTRACT: Growing demand for infrastructure and its maintenance is resulting in a spike in concrete demand. Concrete is composed of several constituents, including granular rocks (aggregates), water, and cement. Conventional cement contains ground clinker, a kilned and quenched material with notable energy-derived and process-derived greenhouse gas (GHG) emissions. Many mitigation strategies have been proposed to reduce these GHG emissions; among which is improved efficiency in the use of cement in concrete and/or efficiency in the use of concrete in infrastructure systems. In this work, the environmental impacts of several efficiency-related mitigation strategies are presented including increased use of mineral admixtures to reduce cement demand, the effects of concrete strength for axial and flexural load bearing applications, and the tradeoffs in using steel reinforcement in beams and columns. Results show that depending on several factors, different mitigation methods to improve efficiency of cement or concrete result in greater reductions of GHG emissions.

1 INTRODUCTION

Rapid urbanization leading to the need for new infrastructure as well as infrastructure replacement is resulting in a spike in associated materials demand. Of these infrastructure materials, concrete is experiencing the steepest rise in demand (Monteiro et al., 2017). Concrete typically contains 7 to 15% cement, 14 to 21% water, and 60 to 75% aggregates by volume as well as up to 8% air (PCA, 2018). Concrete production has grown faster than other infrastructure materials in the past few decades, leading to it being the single greatest man-made material consumed annually (Monteiro et al., 2017). The production of cement currently exceeds 4 billion metric tones annually (van Oss, 2018) and that consumption is accompanied by notable consumption of aggregates and water. Because aggregates and water are typically locally sourced for the production of concrete, this spike in demand has led to potential resource scarcity issues associated with these constituents (Miller et al., 2018a, Ioannidou et al., 2017).

Coupled with this growing material consumption to meet societal demand, there have been notable environmental impacts from the production of concrete and its constituents. Among the most discussed impacts are the high greenhouse gas (GHG) emissions from the production of cement. Conventional cement is composed of finely ground clinker, a kilned and quenched material, and mineral admixtures. The production of clinker requires high temperature heating, at ~1400°C, and results in calcination of limestone (the conversion of $CaCO_3 \rightarrow CaO + CO_2$) to form a reactive material. The high levels of cement production and the GHG emissions from both energy-derived and process-derived (i.e., those from calcination) sources contribute to concrete being responsible for over 8% of anthropogenic GHG emissions (Miller et al., 2016a). As a result of both the energy resources used in the cement kilns and the raw materials in the kilns, there has also been growing concern related to air pollutant production from cement manufacture, such as SO_X emissions (USEPA, 2016a). Additionally, the production of particulate matter (PM) has been noted for almost every stage of raw material acquisition through concrete production (USEPA, 1994, USEPA, 1995, USEPA, 2006). Much of the PM emissions from energy resources and cement manufacture have controls to reduce the amount of particulates that enter that atmosphere; however, fugitive emissions from sources such as handling, leakages, and transportation can still contribute notable PM emissions.

Efforts to mitigate the environmental impacts from cement and concrete production have gained global attention. In a recent roadmap developed by the International Energy Agency (IEA) in collaboration with the World Business Council for Sustainable Developments' Cement Sustainability Initiative, a goal of mitigating 24% of global CO_2 emissions from cement production below current levels was articulated (IEA, 2018). While not part of the roadmap, the IEA's report mentions the potentially critical nature of efficient use of cement and concrete materials. The efficient use of cement and concrete is a still an evolving area of research. Considerable research has been conducted assessing environmental benefits of reducing cement content in concrete through use of mineral admixtures (e.g., (Celik et al., 2015)). More recently examination of the efficient use of concrete mixture proportioning, steel reinforcement, and maintenance strategies have been explored for their ability to contribute to more environmentally sustainable concrete infrastructure (Miller et al., 2018b, Kourehpaz and Miller, 2019, Fantilli et al., 2019, Lepech et al., 2014).

This work explores the potential effects of cement and concrete efficiency methods on three environmental impacts: GHG emissions, acidification, and respiratory effects. Three primary means of improving efficiency are considered: cement efficiency through partial replacement, efficiency through selection of concrete based on properties desired, and efficiency through concurrent design with steel reinforcement.

2 MATERIALS AND METHODS

2.1 Materials

To exemplify the effects of the different material efficiency measures considered in this work, 8 concrete mixtures were assessed. These mixtures contain CEM I Portland cement (PC) and some mixtures contain a partial replacement of cement with ground granulated blast furnace slag (GBS). Mixture proportions and strength data are from (Oner and Akyuz, 2007). These 8 concrete mixtures and their compressive strengths at 3 different curing ages are presented in Table 1. These mixtures were selected because they represent two 28-strength groupings, namely ~27 MPa and ~40 MPa, and three partial replacement levels of cement, namely GBS to PC ratios of 0:1, 0.71:1, 1:1, and 1.3:1.

2.2 Environmental impact assessment

The scope of the environmental impact assessment was for cradle-to-gate production of concrete, incorporating material, energy, and emissions flows from raw material acquisition through concrete batching (see process flow diagram, Figure 1). To perform an impact

Table 1. Concrete mixture proportions and compressive strength.

	Mixture Proportions (kg/m^3)					Compressive Strength (MPa)		
	PC	GBS	Fine Agg.	Coarse Agg.	Water	28 days	63 days	180 days
Mix 400-0	400	0	659	999	239	40.4	41.5	44
Mix 245-175	245	175	647	982	239	41.4	47.7	54
Mix 245-245	245	245	609	924	250	42.3	48.5	56
Mix 245-315	245	315	569	864	263	41.5	48.1	56.3
Mix 280-0	280	0	716	1087	224	27.5	28.3	29.8
Mix 175-125	175	125	707	1073	223	27	31.7	36.2
Mix 175-175	175	175	681	1033	230	27.8	33.1	38.2
Mix 175-225	175	225	654	991	238	27.2	32.9	38

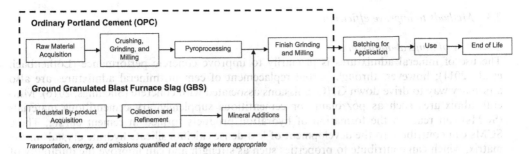

Figure 1. Process flow diagram of scope of environmental impact assessment. Dashed line indicates components considered within this study.

assessment for the production of concrete, it was assumed that cement production and batching would occur just outside Sacramento, California. The raw materials for PC manufacture (e.g., limestone and clay) as well as the aggregates were assumed to be locally sourced. The cement kiln was modeled based on the thermal energy mix reported for dry kilns by the United States Geological Survey (USGS) (van Oss, 2018); the kiln efficiency was determined through data from the USGS and fuel LHV from Argonne National Laboratory (GREET, 2010). CO_2 emissions from calcination were approximated as 0.48 kg of CO_2 emissions per kg of cement based on a 95% clinker content with 65% CaO composition. The electricity mix for processes in California were modeled using the State's average data from 2016 (CEC, 2018). For cement production, the electricity demand was modeled based on the requirements for a preheater/precalciner kiln reported by (Marceau et al., 2006). The electricity requirements to acquire aggregates, both fine and coarse, were based on the University of California's Green-Concrete Tool (Gursel and Horvath, 2012) as was the electricity requirement for batching. In addition to the electricity demand for cement, aggregates, and batching, fugitive particulate matter emissions were incorporated based on data reported by the United States Environmental Protection Agency (USEPA, 1995, USEPA, 2006). Because GBS is not produced in California, it was modeled as being acquired from Pennsylvania; the electricity demand associated with collecting this slag was based on (Gursel and Horvath, 2012) and the electricity grid was modeled based on (USEIA, 2018). For the GBS, only collection, processing, and transportation were incorporated in the assessment of impacts; no impacts from the production of the primary good were allocated to this by-product. For the materials sourced locally, transportation was modeled as by truck, using data from (NREL, 2012), over a distance of 50 km. For the GBS, transportation was modeled as by train, using data from (NREL, 2012), over a distance of 3000 km.

While there are a variety of environmental impacts that are of interest in assessing concrete production, recent attention has been given to greenhouse gas (GHG) emissions (e.g., (IEA, 2018)), air pollutant emissions such as sulfur oxides (e.g., (Lei et al., 2011)), and particulate matter emissions (e.g., (Abdul-Wahab, 2006)). For all three of these environmental impact categories, impacts were derived from energy and raw material processing, herein referred to as energy-derived impacts and process-derived impacts, respectively. To assess each of these three environmental impact categories, the United States EPA's TRACI method (USEPA, 2016b) was applied, namely examining its global warming (in kg CO_2-eq), acidification (in SO_2-eq), and respiratory effects (in $PM_{2.5}$-eq) impacts. While not quantitatively addressed in this work, assumptions made in the environmental impact assessment modeling could lead to uncertainty in environmental impacts assessed. Particularly noteworthy is the anticipated higher uncertainty for impact categories such as respiratory effects relative to global warming impacts. The high uncertainty from impacts, such as respiratory effects, stems from several factors. These can include variability in intake fraction and the influence of current air pollutant concentrations in a region.

2.3 Methods to improve efficiency

2.3.1 Use of mineral admixtures

The use of mineral admixtures is primarily to improve concrete performance (Lothenbach et al., 2011); however, through partial replacement of cement, mineral admixtures are also a primary way to drive down GHG emissions associated with concrete (Snellings, 2016). Mineral admixtures such as pozzolanic or cementitious supplementary cementitious materials (SCMs) can react in the formation of hydration products formed in cement curing. These SCMs can contribute to the development of less-dense calcium silicate hydrate crystals in the matrix, which can contribute to properties such as strength and can improve the durability of concrete members. In doing so, the partial replacement of cement with SCMs can contribute to more efficient use of cement: less cement can be necessary for a given volume of concrete while providing similar or improved material properties. At the same time, it has been shown that excessive use of SCMs could lead to a loss in properties, especially at early ages (Miller, 2018), so quantities should be used appropriately.

To investigate the effects of efficient use of cement through use of mineral admixtures, the difference in environmental impacts that could be achieved through the partial replacement of PC with GBS while maintaining the same strength was examined for a cubic meter of concrete.

2.3.2 Concrete strength and applications

While compressive strength is among the most common metrics used for specifying concrete, it is not necessarily indicative of performance, especially in transportation infrastructure. To exemplify the effects of different design requirements in a simplified manner, the use of compressive strength and flexural strength were highlighted. Two functional units of comparison were considered, ignoring the effects of steel reinforcement, which is addressed in the subsequent section. The first was the quantity of concrete to hold up an axial force for a constant member length, assuming it is designed such that no other loadings affect behavior. Past work has shown minimization of environmental impacts for such an application would occur at the lowest ratio of the environmental impact per cubic meter, i, divided by the compressive strength, f_c, based on (Miller et al., 2016b). The second functional unit of comparison was a flexural member designed for initial cracking on its tensile surface carrying a uniform load, assuming a simply supported constant span-length member, no shear effects, and rupture strength is equivalent to $0.62\sqrt{f_c'}$ based on (ACI, 2011). In this case, past work has shown that minimization of environmental impacts would occur at the lowest ratio of the environmental impact per cubic meter, i, divided by the compressive strength to the power of $\frac{1}{4}$, $f_c^{\frac{1}{4}}$, based on (Miller et al., 2016b). Each of the eight concrete mixtures discussed were compared using these ratios as a means of addressing strength effects for specific applications and environmental impacts concurrently.

2.3.3 Effects of steel reinforcement

In structural applications, such as bridges, concrete is typically reinforced with steel rebar to help bear tensile stresses. Steel, not unlike concrete, has notable environmental impacts associated with production, in large part a reflection of the energy intensity of its production (Davis et al., 2018). As such, weighing between using more concrete in a member or using a higher steel reinforcement ratio becomes a tradeoff. To examine this tradeoff, two case studies were examined for the effects of steel reinforcement on cumulative environmental impacts. For this assessment, the environmental impacts of steel were modeled using the European Commissions ELCD model (European Commission, 2017). The two applications assessed were: a reinforced concrete column and a reinforced concrete beam designed for yield in the moment-curvature relationship. The column was assumed to be 3 m tall, supporting a 1500 kN load, having a reinforcement cover depth of 75 mm, having a fixed width of 30 cm, and containing an area of steel of $0.001\ m^2$; the steel was considered to have a yield strength of 420 MPa. The beam was assumed to be 7 m in length, simply supported, carrying a uniform load of 30 kN/m, having a fixed width of 20 cm, having a cover depth of 75 mm, containing an area of steel of $0.001\ m^2$; the steel was considered to have a yield strength of 420 MPa. For both designs, the concrete mixtures were modeled to capture differences in strength: the depth of the member varies such that stronger concrete required a smaller cross-sectional area. Equations to relate environmental impacts of concrete, considering variation in mixture proportions and rebar, from Kourehpaz and Miller (Kourehpaz and Miller, 2019) were used for each of these two case studies.

3 RESULTS

3.1 *Use of mineral admixtures*

As anticipated, GHG emissions for the concrete mixtures assessed were shown to be predominantly driven by PC content. The PC resulted in 59 to 88% of the total GHG impacts for each mixture (see Figure 2). These GHG emissions were a function of the clinker production: a result both of the energy-derived emissions associated with kilning the clinker at ~1400°C and the process-derived CO_2 emissions from calcination. The GBS and aggregates each contributed limited GHG emissions, ranging from 0 to 21% and 2 to 4%, respectively. Batching contributed negligibly to GHG emissions. Transportation impacts were similar to those from GBS, approximately 9 to 18% of the total GHG impacts.

A similar trend to GHG emissions was present for acidification because acidification impacts are primarily driven by energy demand for concrete. For this environmental impact category, PC resulted in 42 to 79% of the total acidification impacts for each mixture. Transportation was the next largest contributor to impacts at 18 to 35% of the total acidification impacts, followed by GBS at 0 to 22%. The remaining constituents and processes accounted for negligible impacts up to 4% of the total acidification impacts.

Respiratory effects were dissimilar from the other two environmental impact categories considered. For respiratory effects, the greatest contributor to emissions was the acquisition of aggregates, leading to 56 to 68% of the total impacts. PC contributed the next highest impacts, but unlike the other environmental impact categories for which batching resulted in negligible contributions, for respiratory effects, batching resulted in 11 to 13% of the total impact. This change in relative contributions was primarily driven by the fugitive emissions modeled in the acquisition of raw materials and batching. While cement kilns produce a cement kiln dust that would be expected to exceed these fugitive emissions (Marceau et al., 2006), leading to the PC being the greatest contributor to respiratory effects, the controls in place on cement kilns severely limit the escape of this dust.

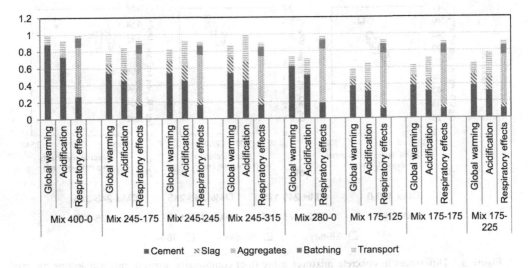

Figure 2. Environmental impacts for concrete mixtures. Impacts are presented by main processes or constituents and are normalized to the greatest impact mixture (Mix 400-0 for global warming and respiratory effects; Mix 245-315 for acidification).

The interest in using GBS as a partial replacement in cement in this work was to assess the potential for improving the efficient use of PC. For the 40 MPa concrete mixtures, a 22% reduction in GHG emissions, a 9% reduction in acidification, and a 10% reduction in respiratory effects were achieved. For the 27 MPa concrete mixtures, the use of GBS allowed for up to a 20% reduction in GHG emissions, a 7% reduction in acidification, and a 7% reduction in respiratory effects. However, due to the long transportation distances for the GBS and the impacts of transportation on acidification, the highest GBS content mixtures (Mix 245-315 and Mix 175-225) actually exceeded the acidification impacts of the PC-only mixtures (Mix 400-0 and Mix 280-0). These findings suggest that while efficient use of cement could drive down certain environmental impacts, the replacements for cement must be selected carefully to not result in unintended consequences, increasing other environmental impacts.

3.2 Concrete strength and applications

When utilizing comparison ratios that incorporate concrete strength, in this case the axial and flexural ratios assessed at 28-day, 63-day, and 180-day strengths, the mechanical properties of the concrete start to influence which would be the lowest environmental impact mixture (see Figure 3). For the axial ratio at 28 days, Mix 245-175 offered the best combination of strength

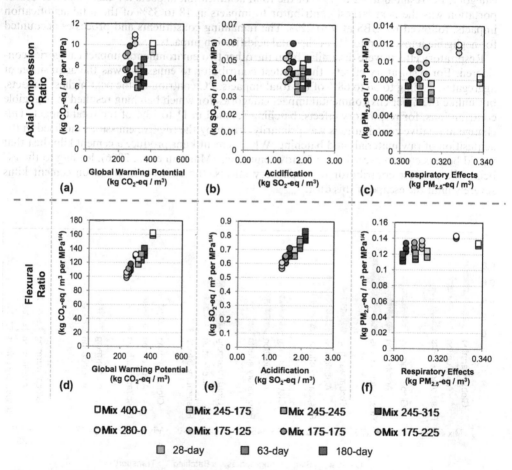

Figure 3. Differences in concrete mixtures using (a-c) compressive strength ratios reflecting an axial load and (d-f) flexural design ratios reflecting a uniformly loaded member. (Note: different shades and shapes refer to different concrete mixtures; different colors reflect different curing ages).

56

and both GHG emissions per cubic meter as well as acidification per cubic meter. However, due to the lower aggregate content, and hence lower associated fugitive emissions, Mix 245-245 offered the best ratio of respiratory effects and compressive strength. For the flexural ratio at 28 days, because strength is less significant, Mix 175-125 provided the best combination of strength and both GHG emissions per cubic meter as well as acidification per cubic meter. Mix 245-315 provides the lowest, and therefore best, ratio of respiratory effects and strength.

As noted previously, because of the less significant role strength plays in the flexural ratio, the evolution of strength with higher curing times did not change which mixtures would result in the best combination of strength and environmental impacts, but strength gain did affect the axial ratios. At 63 days of curing and at 180 days of curing, Mix 245-175 continued to provide the best combination of GHG emissions and acidification with strength, but Mix 245-315 became favorable over Mix 245-245 for respiratory effects.

The highest impact mixture for the axial ratio was typically Mix 280-0, with the exception of 28-day acidification. However, for the flexural ratio, there was greater variability in which mixture would lead to the highest impacts. Mix 400-0 led to the highest GHG emissions to strength; Mix 245-315 lead to the highest acidification; Mix 280-0 lead to the highest respiratory effects. This variability suggests high subjectivity that could arise based on which environmental impact categories assessed. Further, the mixture that led to the highest ratio of strength and acidification was the same mixture that minimized strength and respiratory effects, suggesting potential unintended consequences if one environmental impact is mitigated without assessing others.

3.3 Effects of steel reinforcement

For each of the concrete mixtures, the same area of steel reinforcement was used in each the column and the beam design, respectively. However, because the mixtures had different strengths, the cross-section of concrete necessary to bear the load varied, hence the reinforcement ratio varied, and the environmental impacts of each concrete mixture varied. As such, the relative contribution of the steel rebar to the cumulative impacts of the designed members differed for each mixture and each design (see Figure 4). For the column, the rebar contributed 40 to 49% of the total GHG emissions, 27 to 34% of the total acidification impacts, and 16 to 23% of the respiratory effects. For the beam, the steel rebar contributed less to the cumulative impacts of the designed members, namely 15 to 22% of the total GHG emissions, 9 to 13% of the total acidification impacts, and 6 to 7% of the respiratory effects.

For the two designed members, different mixtures resulted in the lowest cumulative impacts. For the column, compressive strength of concrete has a strong correlation to a reduction in size of the member (a nearly 1:1 ratio). As a result, the 40 MPa concrete mixtures consistently had lower environmental impacts than the 27 MPa mixtures for each of the three environmental impact categories. With its high strength and relatively low binder content (i.e., sum of PC and GBS per cubic meter), Mix 245-175 led to the lowest GHG emissions and acidification impacts. However, for respiratory effects, there was a tradeoff between reducing aggregate content with a higher level of GBS and the additional impacts from transporting GBS. This tradeoff led to Mix 245-245 and Mix 245-315 having the same, and lowest, respiratory effects for the column. For the beam, in addition to strength, the second moment of area plays a large role in the volume of concrete needed to withstand member bending. As a result, there was a lower correlation to increased strength and member volume than with the column. In this case, the 27 MPa concrete mixtures led to lower GHG emissions and acidification impacts than the 40 MPa concrete mixtures; however, due to the higher aggregate content, and associated fugitive particulate emissions, the 27 MPa concrete mixtures had higher respiratory effects than the 40 MPa concrete mixtures. For the concrete beam, the lowest GHG emissions were for Mix 175-125; this mixture results in 7% higher GHG emissions than the minimum mixture for the concrete column. For acidification impacts for the designed beams, the lowest impact mixture was again Mix 175-125, which has 11% higher impact than the lowest acidification impact column mixture. For respiratory effects, the lowest impact mixture was the Mix 245-315 for the beam, which was one of the two mixtures that offered the lowest respiratory effects for the designed column.

57

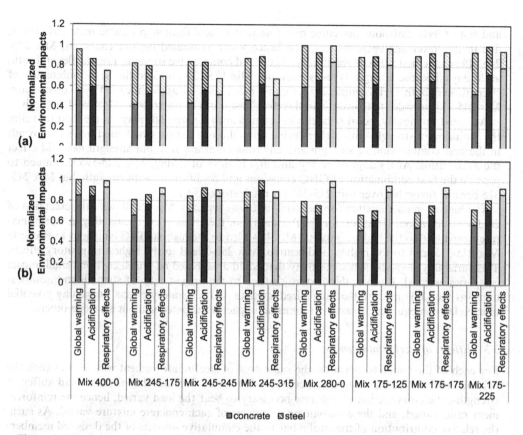

Figure 4. Environmental impacts of designed reinforced concrete members normalized to the highest environmental impact member where (a) represented a reinforced concrete column design and (b) represents a reinforced concrete beam design.

Beyond the important finding that different mixtures would minimize impacts depending on application, it is also important to note the differences between these mixtures relative to the volume comparisons drawn in Section 3.1. For the column, the lowest environmental impact mixtures resulted in a reduction of 18% for GHG emissions, 21% for acidification, and 33% for respiratory effects (comparisons drawn between the lowest impact concrete mixture and the highest impact concrete mixture). Had those same mixtures been selected based on volume comparisons relative to the highest impact mixture, these mixtures would have resulted in a 22% reduction of GHG emissions, 16% reduction in acidification, and a 10% reduction in respiratory effects. While these are all lower than the highest impact concrete mixtures, if decisions had been made based on the volume comparisons, the lowest impact mixtures would have led to what would have appeared to be a 41% reduction in GHG emissions, a 34% reduction in acidification, and a 10% reduction in respiratory effects. With the exception of the respiratory effects, the mixtures that would be selected to minimize environmental impacts based on volume comparisons would not have lead to the lowest environmental impact reinforced concrete member possible and the reductions noted from the volume comparisons would not inform the potential reductions in the designed member. Because the reinforced concrete beam is more closely linked to the volume of concrete used to achieve a high second moment of area, these differences between the volume comparisons and the member comparisons are less pronounced, but the issue remains: mitigations strategies, emissions cap policies, and advancement of environmentally sustainable cement-based materials engineering must include consideration for design and performance; they cannot be limited to environmental impact assessments conducted with a volume or mass functional unit.

4 CONCLUSIONS

By assessing different means to efficiently use cement and concrete, this work derived critical findings in how comparisons in concrete are drawn to use cement and concrete most effectively. In this work, the influence of efficient cement use through partial replacement with a common supplementary cementitious material, ground granulated blast furnace slag (GBS) was explored for three environmental impact categories: greenhouse gas (GHG) emissions, acidification, and respiratory effects. Three means of understanding cement and concrete efficiency were explored: volumetric comparisons of mixtures, comparisons based on axial and flexural loading, and comparisons based on designed reinforced concrete members. Some significant findings from this work included:

- The partial replacement of cement can drive down GHG emissions and energy-derived environmental impacts; however, there is a less strong tie between reductions in cement content and fugitive-emissions derived impacts;
- The incorporation of mechanical properties can change which mixtures appear to reduce environmental impacts and the application selected can further alter desirability of mixture proportions;
- Steel reinforcement can contribute notably to the environmental impacts of reinforced concrete members and selection of a mixture based solely on volume comparisons could lead to unintended consequences.

The findings of this work will act as a strong foundation for further study into the efficient use of cement and concrete. In future work, examination will incorporate other environmental impact categories, other types of concrete mixtures, other properties – particularly durability-related properties – and other applications.

REFERENCES

ABDUL-WAHAB, S. A. 2006. Impact of fugitive dust emissions from cement plants on nearby communities. *Ecological Modelling*, 195, 338–348.

ACI 2011. 318-11: Building Code Requirements for Structural Concrete. Farmington Hills, MI: American Concrete Institute.

CEC. 2018. *Total System Electric Generation* [Online]. California Energy Commission, State of California. Available: http://www.energy.ca.gov/almanac/electricity_data/total_system_power.html [Accessed May 24 2018].

CELIK, K., MERAL, C., GURSEL, A. P., MEHTA, P. K., HORVATH, A. & MONTEIRO, P. J. M. 2015. Mechanical properties, durability, and life-cycle assessment of self-consolidating concrete mixtures made with blended portland cements containing fly ash and limestone powder. *Cement and Concrete Composites*, 56, 59–72.

DAVIS, S. J., et al. 2018. Net-zero emissions energy systems. *Science*, 360, eaas9793.

EUROPEAN COMMISSION. 2017. *Joint Research Centre* [Online]. European Commission. Available: http://eplca.jrc.ec.europa.eu/ELCD3/ [Accessed April 1 2018].

FANTILLI, A. P., TONDOLO, F., CHIAIA, B. & HABERT, G. 2019. Designing Reinforced Concrete Beams Containing Supplementary Cementitious Materials. 12, 1248.

GREET 2010. The Greenhouse Gases, Regulated Emissions, and Energy Use In Transportation Model, GREET 1.8d.1. Argonne, IL.

GURSEL, A. P. & HORVATH, A. 2012. *GreenConcrete LCA Webtool* [Online]. Berkeley, CA: University of California, Berkeley. Available: http://greenconcrete.berkeley.edu/concretewebtool.html [Accessed November 13 2014].

IEA 2018. Technology Roadmap: Low-Carbon Transition in the Cement Industry. Paris, France.

IOANNIDOU, D., MEYLAN, G., SONNEMANN, G. & HABERT, G. 2017. Is gravel becoming scarce? Evaluating the local criticality of construction aggregates. *Resources, Conservation and Recycling*, 126, 25–33.

KOUREHPAZ, P. & MILLER, S. A. 2019. Eco-efficient design indices for reinforced concrete members. *Materials Structures*, 52, 96–109.

LEI, Y., ZHANG, Q., NIELSEN, C. & HE, K. 2011. An inventory of primary air pollutants and CO2 emissions from cement production in China, 1990–2020. *Atmospheric Environment*, 45, 147–154.

LEPECH, M. D., GEIKER, M. & STANG, H. 2014. Probabilistic design and management of environmentally sustainable repair and rehabilitation of reinforced concrete structures. *Cement and Concrete Composites*, 47, 19–31.

LOTHENBACH, B., SCRIVENER, K. & HOOTON, R. D. 2011. Supplementary cementitious materials. *Cement and Concrete Research*, 41, 1244–1256.

MARCEAU, M. L., NISBET, M. A. & VANGEEM, M. G. 2006. Life cycle inventory of Portland cement manufacture. Skokie, Illinois: Portland Cement Association.

MILLER, S. A. 2018. Supplementary cementitious materials to mitigate greenhouse gas emissions from concrete: can there be too much of a good thing? *Journal of Cleaner Production*, 178, 587–598.

MILLER, S. A., HORVATH, A. & MONTEIRO, P. J. M. 2016a. Readily implementable techniques can cut annual CO2 emissions from the production of concrete by over 20%. *Environmental Research Letters*, 11, 074029.

MILLER, S. A., HORVATH, A. & MONTEIRO, P. J. M. 2018a. Impacts of booming concrete production on water resources worldwide. *Nature Sustainability*, 1, 69–76.

MILLER, S. A., JOHN, V. M., PACCA, S. A. & HORVATH, A. 2018b. Carbon dioxide reduction potential in the global cement industry by 2050. *Cement and Concrete Research*, 114, 115–124.

MILLER, S. A., MONTEIRO, P. J. M., OSTERTAG, C. P. & HORVATH, A. 2016b. Comparison indices for design and proportioning of concrete mixtures taking environmental impacts into account. *Cement and Concrete Composites*, 68, 131–143.

MONTEIRO, P. J. M., MILLER, S. A. & HORVATH, A. 2017. Towards sustainable concrete. *Nature Materials*, 16, 698–699.

NREL. 2012. *U.S. Life Cycle Inventory Database* [Online]. U.S. Department of Energy. Available: https://www.lcacommons.gov/nrel/search [Accessed November, 19 2012].

ONER, A. & AKYUZ, S. 2007. An experimental study on optimum usage of GGBS for the compressive strength of concrete. *Cement and Concrete Composites*, 29, 505–514.

PCA. 2018. *How Concrete is Made* [Online]. Portland Cement Association. Available: https://www.cement.org/cement-concrete-applications/how-concrete-is-made [Accessed September 24 2018].

SNELLINGS, R. 2016. Assessing, Understanding and Unlocking Supplementary Cementitious Materials. *RILEM Technical Letters*, 1, 50–55.

USEIA. 2018. *Pennsylvania: State Profile and Energy Estimates* [Online]. United States Energy Information Administration. Available: https://www.eia.gov/state/?sid=PA#tabs-4 [Accessed August 12 2018].

USEPA 1994. Emission Factor Documentation for AP-42, Section 11.6: Portland Cement Manufacturing.

USEPA 1995. AP 42, Fifth Edition, Volume I Chapter 11: Minerals Products Industry. Research Triangle Park, NC.

USEPA 2006. AP 42, Fifth Edition, Volume I Chapter 11: Minerals Products Industry: Concrete Batching. Research Triangle Park, NC.

USEPA. 2016a. *Air Emissions Inventories: Air Emissions Sources* [Online]. U.S. Environmental Protection Agency. Available: https://www.epa.gov/air-emissions-inventories/air-emissions-sources/ [Accessed October 7 2016].

USEPA. 2016b. *Tool for Reduction and Assessment of Chemicals and Other Environmental Impacts (TRACI)* [Online]. Available: https://19january2017snapshot.epa.gov/chemical-research/tool-reduction-and-assessment-chemicals-and-other-environmental-impacts-traci_.html [Accessed December 31 2018].

VAN OSS, H. G. 2018. Minerals yearbook: cement 2015. United States Geological Survey.

Life cycle assessment of ultra-high performance concrete bridge deck overlays

M. Rangelov
Research Engineer, NRC Post-Doctoral Fellow, Federal Highway Administration, Washington D.C.

R.P. Spragg
Research Materials Engineer, Federal Highway Administration, McLean, VA

Z.B. Haber
Research Structural Engineer, Federal Highway Administration, McLean, VA

H. Dylla
Sustainability Program Manager, Federal Highway Administration, Washington D.C.

ABSTRACT: There is an urgent need for effective, long-lasting rehabilitation solutions for deteriorated concrete bridge decks. Traditionally, deteriorated bridge decks are repaired using localized patches or thin, bonded overlays composed of conventional concrete (CC) materials. An alternative solution that is gaining popularity is ultra-high performance concrete (UHPC) overlays, in part due to superior mechanical and durability properties compared to CC. Bridge owners commonly ask about life cycle costs and environmental impacts of UHPC overlays. As such, this study evaluates and compares the environmental impacts of different bridge deck overlay technologies: CC, latex-modified concrete (LMC), and UHPC using life cycle assessment (LCA). The scope of the LCA is cradle-to-grave for a service life of 50 years. The analysis is conducted assuming the overlay technologies are deployed on a hypothetical, yet representative long-span, signature highway bridge structure. The major rehabilitation schedules are estimated based on the literature and communication with stakeholders. LCA results for cradle-to-built scope indicate that the UHPC overlay demonstrates comparable impacts to CC and LMC options in most impact categories. For the 50-year life cycle, UHPC overlay is the lowest-impact option, because of lower overlay thickness, less impacts of the existing bridge deck demolition and improved durability. The initial benefits of CC and LMC overlays are offset by the impacts of reconstruction. Unless a full 50-year service life is achieved by using CC and LMC overlays with no need for reconstruction, UHPC remains a favorable option from a life-cycle perspective.

1 INTRODUCTION

Maintenance and rehabilitation of highway bridge decks is a continual challenge for bridge owners. Traditionally, conventional cement- or asphalt-based overlays have been economical and constructible options to repair bridge decks suffering from freeze-thaw damage, spalling of cover concrete, and/or cracking. However, many transportation agencies recognize that traditional rehabilitation methods and materials provide limited service life extension as noted in the report by Krauss et al. (2009). Thus, there is a need for resilient and durable overlay solutions for aging reinforced concrete bridge decks.

Ultra-high performance concrete (UHPC) is emerging as an innovative solution to a variety of bridge design, construction and repair challenges. UHPC-class materials typically exhibit compressive strengths beyond 150 MPa, tensile strengths above 5 MPa, have sustained post-cracking ductility. UHPC has low water-to-cementitious ratio and a disconnected microstructure. As a result, UHPC has a substantially lower permeability compared to conventional concrete. As such, UHPC demonstrates neglectable freeze-thaw degradation and overall outstanding durability performance. Additional material property information can be found in literature (Graybeal et al. 2018; Haber et al. 2018). These properties make UHPC an attractive candidate material for use as a bridge deck overlay. As an overlay, UHPC can provide structural strengthening (Habel et al. 2007) and protection from chloride and water ingress (Charron et al. 2007; Graybeal and Tanesi, 2007), which can be achieved with a 25-mm to 50-mm thick layer of UHPC. The concept and use of UHPC overlays was pioneered in Europe over ten years ago, and is gaining popularity in the U.S.; the first UHPC bridge deck overlay in the U.S. was installed in 2016 (Haber et al. 2018). In the U.S. market, UHPC overlays appear likely to be economically competitive for use on signature, long-span bridge structures when replacement of the bridge deck is the primary alternative, and when conventional concrete overlay solutions are unlikely to extend the service life of the structure to meet the owner's requirements (FHWA, 2017). The study described herein was prompted by feedback from industry stakeholders and bridge owners who have inquired about UHPC overlays. This study was developed through a direct collaboration with industry stakeholders in response to their inquiries and to identify representative information essential for LCA.

2 OBJECTIVE AND SCOPE

The objective of this research was to conduct a preliminary life cycle assessment (LCA) to quantify and compare the resource-use and environmental impacts of three different bridge deck overlay systems: conventional concrete (CC) overlays, latex-modified concrete (LMC) overlays, and UHPC overlays. The scope of the LCA is cradle-to-grave for a service life extension of 50 years. The analysis conducted assumes the overlay systems are installed on a previously built, in-service hypothetical, yet representative long-span, signature, highway bridge structure located in the north-eastern part of the U.S. The study specifically investigates the environmental impacts of material type, the impact of the initial overlay installation (termed cradle-to-gate), and finally a cradle-to-grave analysis for a period of 50 years following initial overlay installation, which assumes that conventional solutions, CC and LMC, will require removal and re-installation.

3 PREVIOUS LCA-UHPC RESEARCH

LCA is an analytical method to quantify used resources and environmental impacts of product or processes over the life cycle (ISO, 1997). Since UHPC usage in infrastructure is relatively new, the corresponding number of LCA studies in literature is limited. Some studies provided direct comparisons of impacts of a volumetric unit of CC and UHPC, which misrepresents the functionality of the two materials (Müller et al. 2014; Randl et al. 2014). Randl et al. (2014) demonstrated the reduction of environmental impacts of UHPC relative to CC. In this study, impacts of a volumetric unit of UHPC were reduced using the coefficients from the literature to account for improved mechanical and durability properties compared to CC (Randl et al. 2014). (Müller et al. 2014) normalized environmental impacts of CC and UHPC by the respective water absorption coefficients, and concluded that the normalized impacts of UHPC were markedly lower. (Stengel and Schießl, 2014) compared the environmental impacts of producing different bridge structures and structural components (bridge girders, complete bridge structures, columns) with UHPC or CC. The results demonstrated that, depending on the type of structure, UHPC options can be favorable in terms of environmental impacts (Stengel and Schießl, 2014). Adding life-cycle stages beyond production would account for the durability of UHPC and likely result in lower life cycle impacts of UHPC. (Bizjak et al. 2016) compared the environmental impact of different rehabilitation solutions for a concrete railway bridge structure. The authors showed that rehabilitation of the old steel structure with a new UHPC deck had environmental benefits compared with a new bridge construction. (Sameer et al., 2019) provided a comprehensive LCA's of conventional concrete bridge in comparison with UHPC-strengthened alternative over the full 90-year life cycle. This study provided a thorough list of underlying assumptions and data sources; however, the assessment was limited to water, material usage, and carbon footprint (Sameer et al. 2019). In the domain of bridge deck overlays, (Habert et al. 2013) conducted a case study with different rehabilitation scenarios for a bridge in Slovenia and concluded that a UHPC overlay is beneficial compared to the conventional concrete option. However, this study solely included the assessment of greenhouse gas emissions, which may be associated with unintended tradeoffs. In summary, the literature on LCA of UHPC is limited, particularly in terms of impact analysis throughout the life cycle and assessment of multiple environmental indicators. Additionally, many studies are specific to Europe, while comparable efforts in the United States are yet to be initiated.

3.1 *Hypothetical Bridge Structure (HBS) and overlay details*

Figure 1 Presents the overview of the hypothetical bridge structure (HBS), used in the analysis. As seen in Figure 1-a, the structure consists of a 670-m long main suspension span and two 213-m approach spans. The cross-section of the HBS, shown in Figure 1-b, consists of four 3.7-m wide travel lanes with no shoulders. Figure 2-a presents a cross-section view of the bridge deck prior to the overlay construction. The original deck thickness was assumed to 30 cm, and contained two mats of steel reinforcement. Figures 2-b and -c present the schematics of deck section rehabilitated with CC/LMC and UHPC overlays, respectively. Note, there is a slight thickness difference between the overlay solutions. The thicknesses shown are representative of those which might be applied to a bridge like the HBS. Also note that in both cases the deck surfaces have been prepared using hydromilling (sometimes referred to as hydrodemolition). Lastly, the HBS is assumed to be located in the north-eastern part of the U.S.

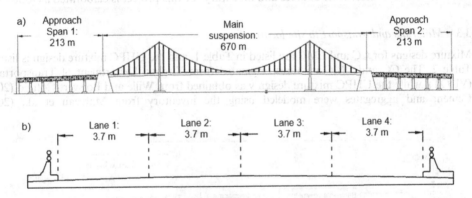

Figure 1. Hypothetical bridge structure (HBS) details: a) elevation view of bridge structure and b) cross-section view of the structure's deck and travel lanes.

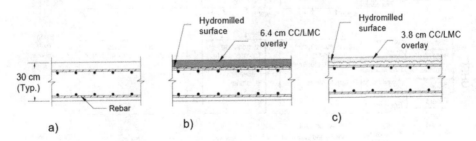

Figure 2. HBS deck and overlay details: a) Pre-overlay, b) CC or LMC overlay c) UHPC overlay.

3.2 LCA methodology and data sources

The goal of this study was to compare conventional bridge deck overlay solutions (CC or LMC) to a UHPC overlay solution considering that these solutions are applied to the same HBS (shown in Figure 1). The functional unit is the overlay applied to the full deck area of the HBS over a service life of fifty years. The scope of the analysis includes the following life cycle phases: materials extraction and production (A1), transportation to the plant, mixing, transport to the site (A2), initial construction of the overlay systems (A3), and reconstruction of the conventional overlay systems as a part of the use phase (B). End-of-life (EOL) phase (C) was excluded from the analysis assuming that the impacts of EOL treatment would be similar among the studied scenarios. Moreover, since UHPC is relatively new from a deployment standpoint, the methods for its EOL treatment require further research. Analyzed product systems are schematically represented in Figure 3. LCA was performed in the OpenLCA software and the impact assessment was done using TRACI 2.1 impact assessment methodology (Bare, 2011, 2012). The analysis included a total of five environmental impact categories: acidification potential (AP), eutrophication potential (EP), global warming potential (GWP), ozone depletion potential (ODP), and smog creation potential (SCP). As discussed in the subsequent sections, environmental product declarations (EPDs) were used as the data sources and five included impact categories are reported on all utilized EPDs. Details of studied scenarios, product system, LCA methodological choices, and implemented data sources for each phase is elaborated as follows.

The data used in the different life cycle stages must be sourced from different entities. For instance, material producers would control of material production data, while the contractors typically have the data pertinent to construction and maintenance. Hence, the data inventory to support LCA of a transportation asset should be identified through collaboration with these key stakeholders. Data sources and inventory for this project is elaborated as follows.

3.3 Materials and transport to site location

Mixture designs for CC and LMC are listed in Table 1, and the UHPC mixture design is listed in Table 2. The CC and LMC mixtures originated from Virginia Department of Transportation (VDOT), while the UHPC mixture design was obtained from Wille and Boisvert-Cotulio (2013). Cement and aggregates were modeled using the inventory from Marceau et al., (2007).

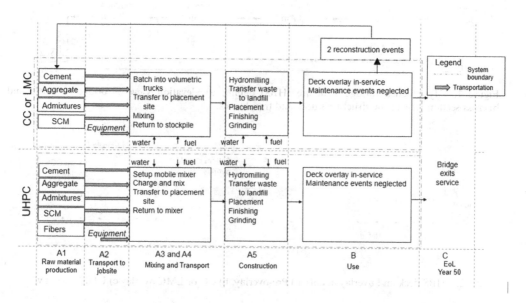

Figure 3. Scheme of LMC, CC and UHPC bridge deck overlays product systems.

Table 1. Mixture designs, material transportation distances and life cycle inventory (LCI) data sources for CC and LMC cradle-to-gate analysis.

Constituent	Quantity [kg/m³] CC (LMC)	Transportation distance to site [km]	LCI data source
Cement	249 (390)	119	(Marceau et al. 2007)
SCM	132 (0)	137	Only upstream transportation (NREL, 2012)
Water	163 (81)	Local	(National Renewable Energy Laboratory, 2012) (NREL, 2012)
Fine Aggregate	676 (932)	40	(Marceau et al. 2007)
Coarse Aggregate	1080 (732)	40	(Marceau et al. 2007)
Air Entraining Admixture	0.24 (0)	805	(EFCA, 2015a)
Water Reducing Admixture	0 (0.76)	805	(EFCA, 2015b)
Latex Admixture	0 (122)	805	(EFCA, 2015c)

Table 2. Mixture designs, material transportation distances and life cycle inventory (LCI) data sources for UHPC cradle-to-gate analysis.

Constituent	Quantity [kg/m³]	Transportation distance to site [km]	LCI data source
Cement	790	145	(Marceau et al. 2007)
Silica Fume	198	483	Only upstream transportation, (NREL, 2012)
SCM	192	470	Only upstream transportation, (NREL, 2012)
Water	146	Local	(NREL, 2012)
Fine Aggregate	1141	362	(Marceau et al. 2007)
Steel Fibers	118	84	(Stengel and Schießl, 2014)
Water Reducer	28	1101	(EFCA, 2015b)

Supplementary cementitious materials (SCM) were considered industrial waste, and only their upstream transportation was considered in LCA, which is known as cutoff allocation approach, prescribed in both Product Category Rules documents for concrete (Carbon Leadership Forum, 2013; NSF International, 2019). For modeling of chemical admixtures the study used publicly available European EPDs from the European Federation of Concrete Admixtures Association (EFCA) ((EFCA), 2015a, b, and c). Steel fibers were modeled using data reported in the study by (Stengel and Schießl, 2014). All utilized EPDs were based on cradle-to-gate scope of the analysis (phases A1-A3) reflective of material manufacturing. Transportation of all constituents was modeled using truck transportation from US LCI database (NREL, 2012). Transportation distances for mixture constituents from the conventional materials were determined by an analysis from the nearest production point, e.g., the nearest cement plant or aggregate quarry. For the UHPC materials, the aforementioned report details the production location of each constituent material (Wille and Boisvert-Cotulio, 2013), because the constituents are specifically detailed to result in the desired engineering properties.

3.4 *Mixing, transport, and construction*

The analysis segmented the overlay installation sequence into mixing, transport and construction. In this analysis, the sequences are discussed as separate events, but in reality, they would be conducted simultaneously. To obtain realistic data, the authors worked closely with a major bridge rehabilitation contractor as a key stakeholder to identify construction sequences, types of equipment, and duration of construction activities.

The mixing and transport processes were different for CC and LMC, as opposed to UHPC. It was assumed that CC and LMC were batched in volumetric mixer trucks on-site, which were driven to the site of placement on the bridge deck, mixed and placed simultaneously. For UHPC, it was assumed that mixing was performed in portable on-site plant at one end of the bridge and that UHPC was transported by a construction vehicle to the placement site.

The construction process was similar for all three cases. The bridge deck was prepared using hydromilling, which is an important step to ensure the overlay is properly bonded to the existing deck and delivers satisfactory long-term performance. The amount of concrete removed by hydromilling was replaced in the construction phase with an overlay, as shown in Figure 2-b and 2-c. The concrete debris removed from the deck and the water used in hydro-milling process were collected using the vacuum truck and transported to the nearest landfill for disposal. The additional construction activities correspond to placement and consolidation of the mixtures, finishing the placed materials, and grinding and grooving the surface after placement to prepare the final ride surface. It was assumed that the same equipment and pro-cedures would be used for both conventional and UHPC overlays, per the authors' discussions with the aforementioned bridge contractor. For each activity, contractors have identified the types of equipment likely to be used, productivity, use rate, and horsepower. This information was used to calculate fuel usage for each piece of equipment and for the entire overlay. The correlation between horsepower and hourly fuel consumption from the reference (Wang *et al.*, 2012) and modeled using the flow "Diesel combusted in industrial equipment" from US LCI database (NREL, 2012). For surfacing operations and use of crane for the assembly on on-site plant, the emissions from the Environmental Protection Agency's software MOVES was used (EPA, 2012). Lastly, the transportation of all equipment to the site was estimated using truck transportation from US LCI database (NREL, 2012).

3.5   *Element use phase*

Herein, the element use phase consisted of the in-service period after the installation of the overlay; recall, this was 50 years. In traditional applications, there would be periodic routine maintenance activities associated with the overlay surface maintenance, which are consider as equal among the three scenarios. More importantly, the major component of the use phase is the reconstruction of CC and LMC overlays. CC and LMC concrete can have a wide range of usable lives, studies have proposed 22-26 years and 14-29 years (Weyers et al., 1993; Krauss et al. 2009; Balakumaran et al. 2017). This means that the CC and LMC overlays would require two or more reconstruction cycles for an analysis period of 50 years. UHPC indicates a time to corrosion initiation exceeding 50 years (Spragg et al. 2019). As such, it was assumed that the UHPC overlay would not require reconstruction. Herein, reconstruction was defined as the need to completely remove and replace the bridge deck overlay. Reconstruction was modeled by re-analyzing the product system starting back at material production phase (A1) for CC and LMC (shown in Figure 3). The analysis assumed *two* reconstruction events for CC and LMC.

4   RESULTS AND DISCUSSION

Table 3 summarizes total environmental impacts of CC, LMC, and UHPC for different life cycle phases. Impacts for material production are reported for 1 m^3 of material (Balakumaran et al. 2017). Impacts of CC are somewhat lower than that of LMC, which can be the effect of a lower cement content in the CC mixture. The environmental impacts of UHPC were for the most part similar to those reported by (Stengel and Schießl, 2014), and the GWP per unit volume of UHPC matched the finding of (Sameer et al., 2019). As seen in Table 3, the impacts of UHPC are higher than that of the same volume of CC or LMC, which is an expected trend.

Table 3. Environmental impacts of CC, LMC and UHPC for different life cycle phases.

Life cycle phase		Material production (A1-A3)			Life cycle (A1-C)		
Declared or functional Unit		1 m³ of CC	1 m³ of LMC	1 m³ of UHPC	CC overlay*	LMC overlay*	UHPC overlay
Environmental impacts	AP [kg SO₂ eq.]	1.8	3.1	8.5	24,700	28,400	2,630
	EP [kg N eq.]	0.06	0.2	0.7	1,420	1,810	564
	GWP [kg CO₂ eq.]	434	745	1,930	2,800,000	3,710,000	651,000
	ODP [kg CFC-11 eq.]	2.4 E-6	3.8 E-6	30 E-6	7.1 E-3	11.1 E-3	6.8 E-3
	SCP [kg O₃ eq.]	27	38	101	764,000	796,000	41,700

* Based on two reconstruction cycles as discussed in Section 3.5

However, the conventional concretes and UHPC have substantially different functionality, therefore direct comparison of the two material types in terms of unit volume is not justified. The functional unit, namely a bridge deck overlay with the specified geometry and analysis period of fifty years, provides for the comparison between three options. For this analysis, this required two additional reconstruction events for the CC and LMC materials. Environmental impacts for the 50-year life cycle listed in Table 3 indicate that the UHPC overlay is outperforming the CC and LMC option. Phase-by-phase impacts are elaborated in the following sections.

4.1 *Materials production*

Figure 4 presents the contribution of different unit processes to the environmental impacts of production of CC, LMC and UHPC. Portland cement is the largest contributor to the overall impacts of CC, accounting for 64 to 99 % of the environmental footprint. Upstream transportation of raw materials contributes up to 27 % of impacts, while aggregate contributes to up to 9.5 % of environmental impacts of CC. For LMC, in addition to Portland cement, the effects of admixtures (latex modifier, in particular) appears to be prominent (up to 58% in EP). However, it is noteworthy that the production of latex modifier was approximated using water resisting admixture model EPD (EFCA, 2015), which serves the same purpose as latex modifier but might differ slightly in terms of chemical composition. Therefore, the conclusions regarding the effect of this admixture to overall impacts cannot be drawn with high certainty.

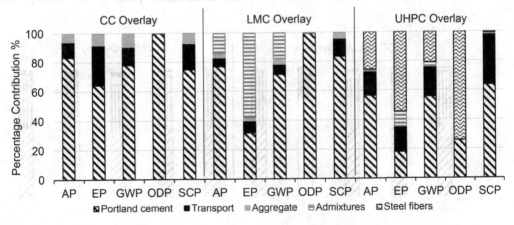

Figure 4. Environmental impacts of unit processes for life-cycle stages associated with material production (A1-A3).

In the case of UHPC, Portland cement makes the largest contribution in terms of AP, GWP, and SCP (33-69%), while for EP and ODP, steel fibers contribute the most (54 and 74 %, respectively). Impact of the upstream transportation for UHPC adds from 12 to 35% of environmental impacts, while admixtures constitute up to 5 % of environmental impacts of UHPC. Contribution of steel fibers to overall UHPC impact in correspond to literature findings (Stengel and Schießl, 2014; Sameer et al., 2019). Lowering the amounts of Portland cement and steel fibers without compromising performance characteristics may be an effective way to improve the footprint of UHPC. Additionally, implementing alternative material constituents, such as synthetic fibers, may result in environmental benefits. Transportation of material constituents can be another process worth considering when aiming for UHPC environmental impact reduction.

4.2   Construction

Figure 5 presents the environmental impact of different construction processes during overall construction. Hydromilling and slurry containment are the processes associated with the highest impacts in all categories, except for ODP. Paving, surfacing, and transportation account for up to 13 % of the construction impacts. In terms of ODP, main contributors are waste and equipment transportation, however, the magnitude of these impacts is relatively low, when normalized by the annual U.S. production from 2008, which is the updated normalization dataset matching TRACI 2.1 (Ryberg et al., 2014).

4.3   Cradle-to-built impacts

Figure 6-a presents comparison of the environmental impacts of CC, LMC, and UHPC overlays for the cradle-to-built scope of the analysis. For ease of interpretation, impacts were normalized using the corresponding values from UHPC. As seen in Figure 6-a, the UHPC overlay shows lower AP and SCP compared to CC and LMC. In terms of EP and GWP, CC shows the lowest and LMC the highest impacts. However, impacts of all three overlay types are relatively similar. The only exception is ODP. In this category, the UHPC overlay presents the highest impacts for cradle-to-built analysis, but which represents negligible effects when normalized by annual U.S. production. The relatively low impacts of UHPC overlay can be attributed to several reasons. First, the higher performance of the UHPC means a thinner overlay can be used, which means the influence of the relatively intense materials production phase were limited. The second implication of a thinner overlay is the lower amount of hydromilling and slurry containment, which, as seen in Figure 5, constitute a major portion of the construction impacts. Lastly, the way in which construction processes are set up, the

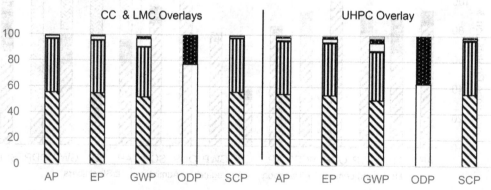

Figure 5.   Environmental impacts of different construction processes.

68

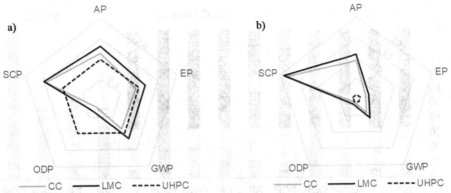

Figure 6. Comparison of bridge deck overlay for: a) cradle-to-built and b) life cycle, with values normalized by the UHPC production values.

construction equipment is generally used for its intended purpose (e.g., a mobile plant is used for mixing, instead of mixing trucks in the case of CC and LMC), which also has an effect of increasing mixing efficiency and reducing emissions.

4.4 *Life cycle impacts of the overlay*

Comparison of life-cycle impacts of CC, LMC, and UHPC overlays are presented in Figure 6-b. Once again, these results are normalized to be relative to UHPC. The results demonstrate that the UHPC overlay is associated with the lowest environmental burdens when considering the life-cycle. Here, two reconstruction events were assumed for CC and LMC cases, which correspond to total life-cycle impacts three times higher than the cradle-to-built impacts seen in Figure 6-a. These results indicate that from a life-cycle perspective, CC and LMC overlays would likely be less environmentally intense than UHPC overlays if they could endure the full 50-year service life. However, literature findings indicate that this is rarely the case, particularly in the environments with chloride ingress (marine environments or in the presence of deicing chemicals), where reinforcement corrosion is a common driver of deck distress. UHPC has a dense microstructure, low porosity, and discontinuous pore system, which impedes the ingress of water, gasses, and ions, as a result, prevents reinforcement corrosion. Accordingly, the UHPC overlay is characterized by better durability, eliminating the need for reconstruction over a 50-year life cycle and consequentially delivering lower life-cycle impacts.

Relative contributions of different processes to environmental impacts over the life cycle are shown in Figure 7. As already stated, for CC and LMC, reconstruction contributes to two thirds of life cycle impacts, since the same construction process takes place three times. When cradle-to-built impacts are broken down for CC and LMC, construction is the most environmentally intensive process in categories AP, EP, and SCP. In terms of GWP, impacts of construction and materials production are similar for the CC and LMC overlays, while in terms of ODP, materials production is the most intense. Mixing process constitutes up to 5 percent of CC and LMC overlay impacts. For UHPC, materials production is the most intensive process in all analyzed impacts categories (45-99%), except for SCP. UHPC construction is responsible for 75% of SCP over the life cycle, while the production of the material contributes approximately 23% of SCP. Impacts of mixing are limited to 3% of the life cycle impacts for UHPC overlay. Because of the limited study scope, the assessment of data quality and uncertainty was not performed. However, this assessment can be utilized in the future to confirm the conclusions of this study.

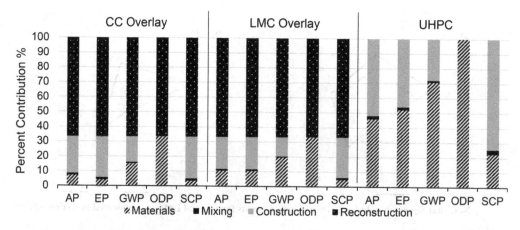

Figure 7.   Life cycle environmental impacts of different construction processes.

5   CONCLUSIONS

This study compared the environmental impacts of conventional concrete (CC), latex modified concrete (LMC), and ultra-high performance concrete (UHPC) bridge deck overlays for a signature bridge structure in the Northeastern U.S. A 50-year life cycle was assumed. Furthermore, the analysis assumed that the CC and LMC overlays would need to undergo two overlay reconstruction events over the analysis period; it was assumed that the UHPC overlay would not require reconstruction given its superior durability characteristics. Results of life cycle assessment (LCA) for cradle-to-built analysis scope indicate that the UHPC overlay demonstrates comparable impacts to CC and LMC options in most impact categories. For the 50-year life cycle, UHPC overlay is the lowest-impact option, since the initial benefits of CC and LMC overlays are offset by the impacts of reconstruction. Portland cement and steel fibers constitute the majority of the environmental impacts of UHPC. The reduction of the quantities of these constituents or replacement with less environmentally aggressive alternatives with lower environmental footprint (without compromising performance), may be an efficient way to improve the environmental profile of UHPC. In terms of construction, preparatory works, namely hydromilling and slurry containment, contribute to the most of the environmental impacts. The number of reconstruction events is the decisive factor in life-cycle impacts.

ACKNOWLEDGMENT

This research was performed while the author, M. Rangelov, held an NRC Research Associateship award at Federal Highway Administration (FHWA).

REFERENCES

Balakumaran S. et al. 2017. *Performance of Bridge Deck Overlays in Virginia: Phase I: State of Overlays.* VTRC 17-R17. Charlottesville, VA.

Bare, J. (2011) TRACI 2.0: the tool for the reduction and assessment of chemical and other environmental impacts 2.0, *Clean Technologies and Environmental Policy*. Springer, 13(5), pp. 687–696.

Bare, J. (2012) Tool for the Reduction and Assessment of Chemical and Other Environmental Impacts (TRACI), Version 2.1 - User's Manual; EPA/600/R-12/554.

Bizjak K. et al. 2016. Environmental life cycle assessment of railway bridge materials using UHPFRC. *Materials and Geoenvironment*. De Gruyter Open, 63(4), pp. 183–198.

Carbon Leadership Forum (2013) *Product Category Rules for Concrete.*

Charron, J.-P., Denarié, E. and Brühwiler, E. (2007) 'Permeability of ultra high performance fiber reinforced concretes (UHPFRC) under high stresses', *Materials and structures*. Springer, 40(3), 269–277.

European Federation of Concrete Admixtures Associations (EFCA). 2015c. *Environmental Product Declaration as per ISO 14025 and EN 15804. Concrete Admixtures- Water resisting admixture. Declaration number: EPD-EFC-20150090-IAG2-EN.*

European Federation of Concrete Admixtures Associations (EFCA). 2015a. *Environmental Product Declaration as per ISO 14025 and EN 15804. Concrete Admixtures- Air Entrainers. Declaration number: EPD-EFC-20150086-IAG1-EN.*

European Federation of Concrete Admixtures Associations (EFCA). 2015b. *Environmental Product Declaration as per ISO 14025 and EN 15804. Concrete Admixtures- Plasticizers and Superplasticizers. Declaration number: EPD-EFC-20150091-IAG1-EN.*

Environmental Protection Agency (EPA). 2012. *Motor Vehicle Emissions Simulator: User Guide for MOVES 2010 b. Vol 37, EPA-420-B-12-001b.*

Graybeal, B. et al. 2018. Accelerated Construction of Robust Bridges through Material and Detailing Innovations. In *Proceedings of the 9th International Conference on Bridge Maintenance, Safety, and Management*. Melbourne, Australia: International Association for Bridge Maintenance and Safety.

Graybeal, B. and Tanesi, J. 2007. Durability of an ultrahigh-performance concrete. *Journal of materials in civil engineering*. American Society of Civil Engineers, 19(10), 848–854.

Habel et al. 2007. Experimental investigation of composite ultra-high-performance fiber-reinforced concrete and conventional concrete members. *ACI Structural Journal*. American Concrete Institute, 104 (1), 93.

Haber et al. 2018. *Properties and behavior of UHPC-class materials*. United States. Federal Highway Administration. Office of Infrastructure.

Habert et al. 2013. Lowering the global warming impact of bridge rehabilitations by using Ultra High Performance Fibre Reinforced Concretes. *Cement and concrete composites*. Elsevier, 38, 1–11.

International Standard Organization (ISO). 1997. *ISO 14040: Environmental management-Life cycle assessment-Principles and framework.*

Krauss et al. 2009. *Guideline for Selection of Bridge Deck Overlays, Sealers and Treatments*. NCHRP 20-07/Task 234. Washington, DC.

Marceau et al. 2007. *Life cycle inventory of portland cement concrete*. Portland Cement Association.

Müller et al. 2014. Assessment of the sustainability potential of concrete and concrete structures considering their environmental impact, performance and lifetime. *Construction and Building Materials*. Elsevier, 67, 321–337.

National Renewable Energy Laboratory (NREL). 2012. *U.S. Life Cycle Inventory Database*. Available at: https://www.lcacommons.gov/nrel/search.

NSF International (2019) *Product Category Rule for Environmental Product Declarations for Concrete.*

Randl et al. 2014. Development of UHPC mixtures from an ecological point of view. *Construction and Building Materials*. Elsevier, 67, 373–378.

Ryberg, M. *et al.* (2014) Updated US and Canadian normalization factors for TRACI 2.1, *Clean Technologies and Environmental Policy*. Springer, 16(2), 329–339.

Sameer *et al.* 2019. Environmental Assessment of Ultra-High-Performance Concrete Using Carbon, Material, and Water Footprint. *Materials*. Multidisciplinary Digital Publishing Institute, 12(6), 851.

Spragg et al. 2019. Using Formation Factor to Define the Durability of Ultra-High Performance Concrete. In *2nd International Interactive Symposium on Ultra-High Perfomrance Concrete*. Albany, NY: Iowa State University.

Stengel, T. and Schießl, P. 2014. Life cycle assessment (LCA) of ultra high performance concrete (UHPC) structures. in *Eco-efficient Construction and Building Materials*. Elsevier, 528–56.

Federal Highway Administration (FHWA) (2017) *Ultra-High Performance Concrete for Bridge Deck Overlays FHWA-HRT-17-097.*

Wang et al. 2012. UCPRC Life Cycle Assessment Methodology and Initial Case Studies for Energy Consumption and GHG Emissions for Pavement. *Univ. of California Pavement Research Center, Davis, CA.*

Weyers, R. 1993. *Concrete bridge protection, repair, and rehabilitation relative to reinforcement corrosion: A methods application manual*. SHRP-S-360. Washington, DC.

Wille, K. and Boisvert-Cotulio, C. 2013. *Development of Non-Proprietary Ultra-High Performance Concrete for Use in the Highway Bridge Sector (FHWA-HRT-13-100)*. FHWA-HRT-13-100. Storrs, CT.

How do funds allocation and maintenance intervention affect the deterioration process of urban bridges in Shanghai

Yu Fang & Lijun Sun*

Key Laboratory of Road and Traffic Engineering, Ministry of Education, Tongji University, China

ABSTRACT: The performance of urban bridges gradually deteriorates with the growth of service time because of environmental action and traffic loads. The bridge deterioration prediction model has a significant impact on bridge maintenance effectiveness and saving repair costs. In this paper, a Weibull distribution based semi-Markov prediction model was developed by using the historical bridge inspection records to investigate how do funds allocation and maintenance intervention affect the deterioration process of urban bridges in shanghai. The deterioration behavior of three major bridge parts was compared, and the effect of bridge location was also investigated. The prediction result shows the decay rates of the deck system are the fastest. Moreover, the service life expectancy of bridges in the suburban area is shorter than the downtown area. Therefore, the overall deterioration trends for urban bridges with or without the medium and major repair were predicted based on the investigated actual repair rates. The result shows that the medium and major repair can significantly decrease the proportion of bridges with condition rating 4 and 5, especially for bridges in the suburban area. But the conventional maintenance intervention has a limited overall effect on network-level urban bridges. There is a noticeable trend of bridge degradation from condition rating 1 and 2 to condition rating 2 and 3. Therefore, to reduce the maintenance cost of the entire life cycle of urban bridges in Shanghai, it is necessary to adopt a more reasonable preventive maintenance strategy.

1   INTRODUCTION

Urban bridges have complex structural forms and diverse material types, playing a pivotal role in road networks. As the service time continues to increase, the bridge performance will gradually decay at different rates due to the external environment and traffic loads (Mishalani et al. 2002). The difference in the allocation of maintenance funds and the decision of maintenance measures may also significantly affect the deterioration process of urban bridges. However, during the life cycle of urban bridges (Itoh et al. 2003), accurate prediction of the future bridge performance can ensure the service performance and operation safety of urban bridges to the maximum extent under the limited financial budget (Du et al. 2013).

Bridge management systems (BMSs) has been applied to help the management of bridges since the 1990s (Morcous 2006). In Shanghai, BMS has been applied to small or moderately sized bridges since 2004 (Fang et al. 2019). Now, more than two thousand urban bridges across the city are inspected annually in the BMS database. In which, reinforced concrete and prestressed concrete bridges account for nearly 80% of the total bridge numbers. (Fang et al. 2017). Figure 1 shows the age distribution of concrete beam bridges in Shanghai. The graph reveals that more and more bridges will reach their 50-year milestone in the next few years, as well, the deterioration trend of urban bridges will become a severe problem.

Shanghai is the largest city in mainland China with 16 districts under its jurisdiction. Among them, there are seven districts in the downtown area of Shanghai and nine districts in

---

*Corresponding author: ljsun@tongji.edu.cn

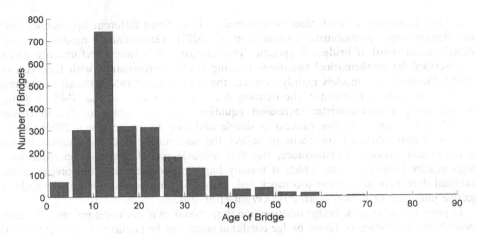

Figure 1. Age distribution of urban bridges in Shanghai.

the suburban area. Due to location conditions and historical reasons, the suburbs have led to relatively lagging development in economic and infrastructure. There is a big gap between the essential public services, the convenience of transportation infrastructure, and the finance funding allocation when compared with the downtown area (Horvath 2009). The number of urban bridges in the downtown area and the suburban area is shown in Table 1. Moreover, funds allocation and maintenance intervention are quite different for urban bridges in different regions, which may lead to significant differences in the deterioration process and service life expectancy of urban bridges.

In this paper, to investigate how do funds allocation and maintenance intervention affect the deterioration process of urban bridges in shanghai, a Weibull distribution based semi-Markov process model has been applied in this study. In which the Weibull distribution was used to characterize the service-life behavior of bridge deterioration within each CR, and the semi-Markov process was used to evaluate the transition probabilities of bridge deterioration process between different CRs. Besides, this paper also analyzed the long-term performance deterioration trend of network-level urban bridges affected by the medium and major repair intervention under different budgets.

2 LITERATURE REVIEW

The existing prediction model of bridge deterioration in the BMS of Shanghai is still using a deterministic regression method, which is hard to reflect the uncertainty and randomness of bridge deterioration process. Moreover, the assumptions (i.e., memoryless and homogeneous) of commonly used Markov chain method do not conform to the actual bridge deterioration process. Future state which only based on the current state not associated with the history is unrealistic, and the transition probability from one state to another also does not always remain constant. Hence, it is essential to devise an efficient way to forecast the decay process and the service life of the existing/new urban bridges.

Table 1. Number of urban bridges in different regions.

Location region	District name	Bridge number
Downtown area	Huangpu, Xuhui, Changning, Yangpu, Hongkou, Putuo, and Jingan	1000
Suburban area	Minhang, Baoshan, Jiading, Qingpu, Songjiang, Jinshan, Fengxian, Pudong New Area, and Chongming	1610

Bridge deterioration prediction models mainly have three different approaches, including deterministic, probabilistic (Agrawal et al. 2010). Deterministic models assume the deterioration trend of bridges is specific. The relationship between performance and time is described by mathematical equations relating bridge performance with time (Lu et al. 2015). Deterministic models mainly estimate the bridge decay rate through the regression analysis method. For instance, the existing deterioration model in the BMS of Shanghai is still using a deterministic regression equation. The advantage of the deterministic models is that the modeling process is simple and easy to revise (Chen 2005). Nonetheless, it is quite difficult for them to reflect the uncertainty and randomness of bridge deterioration process. Furthermore, the deterministic model requires the extensive and high-quality historical data, which is usually hard to provide; also, the preprocessing of original data may also cause too much risk since the preprocessing is mostly under subjective judgment (Zambon et al. 2017)(Winn et al. 2013).

In probabilistic models, bridge deterioration is supposed as a stochastic process, of which the probability distribution of future bridge condition states can be predicted to indicate the bridge performance. These models can reflect the uncertainty of bridge deterioration caused by external factors such as traffic loads, materials, environment, and maintenance (Su et al. 2015). Moreover, these stochastic models can be further subdivided into state-based and time-based models (Madanat et al. 1995). The state-based models such as Markov chains are modeled through a transition probability matrix from one condition state to another in discrete time, and the distribution of the future condition is not dependent on the past. Bridge management systems in the U.S like Pontis (Golabi et al. 1997) and BRIDGIT (Hawk et al. 1998) have adopted the Markov chain theory to calculate the bridge deterioration rates (Reardon et al. 2016). Meanwhile, the time-based models such as Weibull distribution, assume the duration time of a bridge maintains a conditional rating as a random variable (Ng et al. 1996). Probabilistic models perform well in versatility and simplicity. They can clearly describe the service-life behaviors of bridge deterioration (Medjoudj et al. 2009). The limitations of probabilistic models lie in the need of a large number of accurate observational data and empirical model assumptions (Reardon et al. 2016).

3 METHODOLOGY

By considering the time-dependent reliability at each CR in the process of bridge deterioration, this paper proposes a semi-Markov process prediction model based on the Weibull distribution. The basic assumption in reliability theory is the existence of a probabilities relationship between the observed CRs of discrete-state and the unobserved continuous deterioration process (Mishalani et al. 2002). The CRs of bridge performance are treated as a response variable subject to time and other exponential variables and continues to decay with the increase of time (Agrawal et al. 2010). And once the unobserved continuous deterioration reaches or exceeds the threshold of boundaries between different CRs, the time $[0, t]$ will be recorded as the duration T_i of a certain $CR = i$.

As shown in Formula (1), the cumulative distribution function $F_i(t)$ of duration T_i in $CR = i$ describes the probability that the bridge will transition out of $CR = i$ by the time t.

$$F_i(t) = prob(T_i \leq t) = \int_0^t f_i(x)dx \qquad (1)$$

where T_i is the duration of $CR = i$,
$f_i(t)$ is the probability distribution function of T_i.

And the survivor function $S(T_i)$ is the complement of $F_i(t)$ which means the probability that the bridge element would still be in $CR = i$ by time t, is defined as Formula (2).

$$S_i(t) = prob(T_i > t) = 1 - F_i(t) \qquad (2)$$

where T_i is the duration of $CR = i$,
$F_i(t)$ is the cumulative distribution function of duration T_i.

Then the transition probability $p_{ij}(t, \Delta)$ out of $CR = i$ into a worse condition state $CR = j$ within period $.\Delta.$ after time t can be defined as Formula (3)

$$p_{ij}(t, \Delta) = prob(t < T_i < t + \Delta | T_i > t) = \frac{F_i(t + \Delta) - F_i(t)}{S_i(t)} = 1 - \frac{S_i(t + \Delta)}{S_i(t)} \qquad (3)$$

where T_i is the duration of $CR = i$,
$F_i(t)$ is the cumulative distribution function of duration T_i,
$S_i(t)$ is the survivor function of duration T_i.

The Weibull distribution has been widely used in the analysis of time-dependent reliability and service life behavior due to its flexibility in fitting different types of service life data. And with different values of the shape parameter β, the Weibull distribution can be related to several other probability distributions such as the normal distribution, the exponential distribution, and even the Rayleigh distribution. In this paper, the Weibull distribution method is used to modeling the observed duration T_i which represents the time of the bridge staying at a particular $CR = i$. The Weibull distribution is mathematically defined by its pdf equation, and the two-parameter pdf expression of Weibull distribution is defined as follows.

$$f(T_i) = \frac{\beta_i}{\eta_i} \left(\frac{T_i}{\eta_i}\right)^{\beta_i - 1} e^{-\left(\frac{T_i}{\eta_i}\right)^{\beta_i}} \qquad (4)$$

where $T_i 0$ is the duration, $\beta_i > 0$ is the shape parameter, $\eta_i > 0$ is the scale parameter.

The transition process between different CRs of bridge deterioration can be described as a semi-Markov process Y_n with m feasible states. The associated random variables (X_n, TS_n) is the successive states and times of the n th transition, and the length of a sojourn time interval (TS_n, TS_{n+1}) is same as the random variable duration T_i of a certain condition state $X_n = i$ described in the time-dependent reliability theory. The distribution of duration T_i depends on both the state X_n being visited and the state X_{n+1} to be visited next (Ng et al. 1998).

For any states $i, j = 1, 2, \cdots, m$ and time $t0$, the associated matrix of transition probabilities Q_{ij}, formally called the semi-Markov kernel can be defined as Formula (5).

$$Q_{ij}(t) = P(X_{n+1} = j, TS_{n+1} - TS_n \leq t | (X_0, TS_0), (X_1, TS_1), \cdots (X_n, TS_n)) = P(X_{n+1} = j,$$
$$TS_{n+1} - TS_n \leq t | X_n = i) \qquad (5)$$

where X_n is the successive states visited of the nth transitions,
TS_n is the successive times of the nth transitions.

Then the cumulative distribution function $F_i(T)$ same as described in formula (1) which represents the duration T_i in the state i here, can be given as Formula (6).

$$F_i(T) = P(TS_{n-1} - TS_n \leq t | X_n = i) = \sum_{j=1}^{m} Q_{ij}(t) \qquad (6)$$

where $Q_{ij}(t)$ is the semi-Markov kernel,
X_n is the successive states visited of the n th transitions.

If we assume a semi-Markov process having spent an initial time t_0 in the state i before the time $t = 0$ (i.e. the age of bridge when the observation starts), with a destination state j and the system could go through an intermediate state l. Let τ be the time measured for the stay in state i from time $t = 0$ to the time of transition to state k. Then the transition probabilities $\phi_{ij}(t)$ of the semi-Markov process can be calculated as following (Corradi et al. 2004):

$$p_{ij}(t) = \delta_{ij}(1 - F_i(t)) + \sum_{l=1}^{m} \int_0^t p_{lj}(t - \tau)dQ_{il}(\tau) = \delta_{ij}(1 - F_i(t)) + \sum_{l=1}^{m} \int_0^t Q'_{il}(\tau)p_{lj}(t - \tau)d\tau \quad (7)$$

where δ_{ij} is the Kronecker δ (i.e. if $i = j$, $\delta_{ij} = 1$ or if $i{\neq}j$, $\delta_{ij} = 0$),
$F_i(t)$ is the cumulative distribution function,
$Q_{il}(\tau)$ is the semi-Markov process kernel.

4   MODEL DEVELOPMENT

According to the Chinese technical standard of maintenance for city bridge (CJJ 99-2017), the bridge condition index (BCI) was implemented to assess the condition of small or moderately sized bridges. Through a hierarchy-weight evaluation method based on bridge components, the bridge is decomposed in the forms of "bridge-part-component" (China 2017). The score of BCI represents the overall condition of the whole bridge by comprehensively weighted the damage of three major parts (i.e., the deck system, the superstructure, and the substructure) and other more detailed component elements (i.e., the deck pavement, the horizontal linkage, the beam component, pier and etc.). The BCI scores from 0 to 100 corresponding to bridge condition ratings (CRs) from A to E, indicating different bridge performance from intact to dangerous as listed in Table 1 (Li et al. 2016). Generally, only routine maintenance or minor repair is applied to bridges with CR from A to C. As for bridges with CR D or E, medium repair, major repair or rehabilitation may be implemented in the next year depending on the finance budget. For ease of model calculation, the condition ratings are simplified from a qualitative rating [A, B, C, D, E] to an ordinal system [1, 2, 3, 4, 5] in this paper.

Table 2.   Chinese bridge condition ratings (CJJ 99-2017).

Rating	Score	Condition
A	[90,100]	Intact
B	[80,89]	Good
C	[66,79]	Qualified
D	[50,65]	Bad
E	[0,49]	Dangerous

The demand for bridge maintenance and repair is gradually increasing in Shanghai, while the amount of financial allocations remains limited. Therefore, for many urban bridges with CR 4 or 5, it is impossible to immediately implement medium or major repair for all these bridges. In bridge management practice, according to the number of annual maintenance funds, only part of bridges with CR 4 or 5 are selected for a medium or major repair. Through the investigation of the medium and major repair implemented over the past decade, the actual repair rates for urban bridges with CR 4 and 5 in the downtown area and suburban area were obtained (Li et al. 2016). Table 3 reveals the rates of medium and major repair for bridges with CR 4 or 5 in different regions. The repair rates of bridges in the downtown area were much higher than the suburban area, which indicates that the bridge maintenance intensity and financial allocations in the downtown area are more significant than that the suburban area.

Table 3. Repair rates of CR 4 or 5 bridges in different location.

Location region	CR	Repair rates
Downtown area	4	24.4%
	5	100%
Suburban area	4	4.8%
	5	16.7%

4.1 Inspection data processing

Although there are a variety of different structural forms and material categories of urban bridges, the most common bridge types are concrete beam bridges. Besides, the basic information of concrete beam bridges is more accurate than others. Therefore, this study aims to set up the performance forecasting model to investigate the concrete beam bridge deterioration in the city of Shanghai. The original modeling data used in this study is derived from a total of 2610 concrete beam bridges and 14 years of historical inspection records stored in the BMS database. A reliable process of data preparation is crucial to develop the bridge deterioration prediction model, which can guarantee the data consistent with the deterioration phenomenon and avoid the interference of the individual deviation and the repair activity.

In this paper, five main steps of the data preparation process are as following:

- Filtering missing or abnormal records;
- Removing records with CR decline over 2 between two adjacent years;
- Resetting the age of A rating bridges upgraded from D and E to 0;
- Set reasonable upper and lower thresholds of CR for bridge ages;
- Correcting the deviation caused by the subjective manual inspection process.

When using the probabilistic forecasting method, the presence of censored observation should be taken into account. The influence of the complete, the right-censored, and the left-censored observation are different (Ng et al. 1998). In this research, compared with the left-censored observation, the sample number of right-censored observation is relatively large and can be more efficiently used to parameter estimation (Fang et al. 2018). So, while the records of left-censored observations are eliminated, the right-censored observation records are used in the estimation of the Weibull-distribution parameters together with the complete observation records after the tag processing.

4.2 Service-life of bridges with different funds allocation

Once the Weibull distribution parameters are obtained by fitting the service-life data, the Weibull probability density function can be used to model a variety of bridge service-life behaviors within different CRs (including the survival function, the hazard function, the mean value, and the quantile statistics). The shape parameter β_i and scale parameter η_i of Weibull distribution can be both obtained by fitting the observed durations of the whole bridge, the deck system, the superstructure, and the substructure at different CRs. The estimated results of the Weibull distribution parameters for bridges with different funds allocation are as shown in Table 4.

By using the mean value of durations at each CR calculated distribution parameters, the service life expectancy of bridges with different funds allocation can be predicted, as shown in Figure 2. The decay rates of the deck system within different CRs are the fastest when compared with the superstructure and the substructure. The life expectancy of urban bridges in the downtown area and the suburban area are significantly different. Taking the bridge deck system as an example, the service life expectancy of the bridge deck system for bridges in the downtown area is almost 60 years, while that for bridges in the suburban area is only 50 years. Therefore, it is necessary to increase the maintenance investment for the bridges in the suburbs while maintaining the performance of bridges in the downtown area.

Table 4. Estimated parameters of urban bridges with different funds allocation.

| | Estimated Weibull distribution parameters | | | |
| | Downtown area | | Suburban area | |
CR	Shape β_i	Scale η_i	Shape β_i	Scale η_i
1	1.537	20.647	1.438	18.415
2	1.664	22.000	1.772	19.575
3	1.372	23.747	1.719	25.614
4	0.968	17.808	1.477	22.983

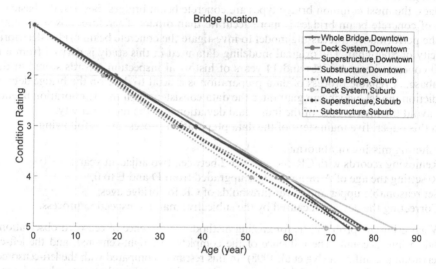

Figure 2. Deterioration prediction for bridges in different regions.

5 DETERIORATION PREDICTION WITH REPAIR INTERVENTION

5.1 *Deterioration with/without repair intervention*

In this paper, it is assumed that a bridge can degrade no more than two CRs in the interval of two adjacent years and substituting Formula (3) in which set $t = 0$ and $\Delta = t - x$ (Sobanjo 2011). Then the calculation equation of the semi-Markov transition probability can be rewritten as Formula (8).

$$p_{ij}(t) = \sum_k \sum_{x=1}^t f_{ik}(x)\left[\frac{F_{kj}(t-x) - F_{kj}(0)}{1 - F_{kj}(0)}\right], i \neq j = \sum_k \sum_{x=1}^t f_{ik}(x)F_{kj}(t-x) \qquad (8)$$

where $f_{ik}(x)$ is the probability density function of duration T_i from $CR = i$ to $CR = k$, $F_{kj}(t - x)$ is the cumulative density function of duration T_k from $CR = k$ to $CR = j$.

To predict the bridge condition for any time t, the initial condition $CR(0)$ at time $t = 0$ is required in the form of a state vector as shown in Formula (9).

$$CR(0) = [p_1 \quad p_2 \quad p_3 \quad p_4 \quad p_5] \qquad (9)$$

where p_i is the proportion of the whole bridge inventory in $CR = i$ at time $t = 0$.

If we assume at the beginning the proportion of the whole bridge, the deck system, the superstructure, and the substructure with different CRs in 2004 as the initial proportion of the bridge inventory. Then the predicted $CR(t)$ of the semi-Markov process would be a product of the initial condition vector and the transition probability matrix as shown below.

$$CR(t) = CR(0) \times P_{ij}(t) \tag{10}$$

where $P_{ij}(t)$ is the transition probability matrix of the semi-Markov process at time t.

Therefore, the network-level deterioration process of urban bridges with or without the medium and major repair intervention can be predicted by using different transition matrix $P_{ij}(t)$. The transition matrix $P_{ij}(t)$ of the bridges between different CRs without the medium and major repair is shown in Formula (11). Which just like the assumptions when processing bridge inspection records, only the routine maintenance and minor repair are applied to urban bridges during the deterioration process.

$$P_{ij}(t) = \begin{bmatrix} p_{11}(t) & p_{12}(t) & p_{13}(t) & 0 & 0 \\ 0 & p_{22}(t) & p_{23}(t) & p_{24}(t) & 0 \\ 0 & 0 & p_{33}(t) & p_{34}(t) & p_{35}(t) \\ 0 & 0 & 0 & p_{44}(t) & p_{45}(t) \\ 0 & 0 & 0 & 0 & 1 \end{bmatrix} \tag{11}$$

The transition matrix $P_{ij}(t)$ of the bridges between different CRs with the medium and major repair is shown in Formula (12). In which a proportion of bridges with CR 4 and 5 can upgrade to CR 1 after the medium and major repair intervention. The upgrade probabilities from CR 4 and 5 to CR 1 depend on the actual repair rates in different regions.

$$P_{ij}(t) = \begin{bmatrix} p_{11}(t) & p_{12}(t) & p_{13}(t) & 0 & 0 \\ 0 & p_{22}(t) & p_{23}(t) & p_{24}(t) & 0 \\ 0 & 0 & p_{33}(t) & p_{34}(t) & p_{35}(t) \\ p_{41}(t) & 0 & 0 & (1 - p_{41}(t))p_{44}(t) & (1 - p_{41}(t))p_{45}(t) \\ p_{51}(t) & 0 & 0 & 0 & (1 - p_{51}(t)) \end{bmatrix} \tag{12}$$

where $p_{41}(t), p_{51}(t)$ is the transition probability of bridges from CR 4 and 5 upgrading to CR 1 which is calculated based on the actual repair rates of the medium and major repair.

As shown in Figures 3 and 4, the prediction results of the overall bridge deterioration trend in the downtown area and the suburban area from 2004 to 2024 without the external intervention. While the green dotted lines reveal the impact of the medium and major repair on network-level bridge deterioration process. The overall condition status of bridges in downtown performs better than bridges in the suburbs. Moreover, the deck system is the most sensitive to external repair interventions. The intervention of the medium and major repair can significantly decrease the proportion of bridges with CR 4 and 5, especially for bridges in the suburban area. However, the external intervention has a limited overall effect on large-scale bridge networks because the percentage of bridges with CR 4 and 5 is relatively small.

There is a noticeable trend of bridges degradation from CR 1 and 2 to CR 2 and 3, which is in line with the actual situation in bridge management of Shanghai in recent years. It is successful in reducing the number of bridges with CR 4 and 5 through the implementation of the medium and major repair. But the conventional reaction maintenance method cannot adequately avoid the degradation of bridges with CR 1 and 2. Overall, a more appropriate preventive maintenance method are needed to reduce the maintenance cost of the entire life cycle of urban bridges in Shanghai.

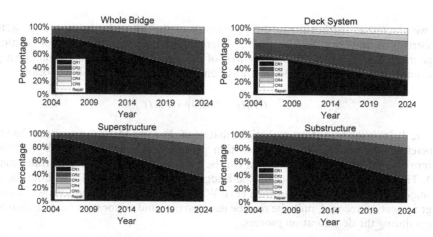

Figure 3.   Prediction of the bridge deterioration with/without repair intervention in downtown area.

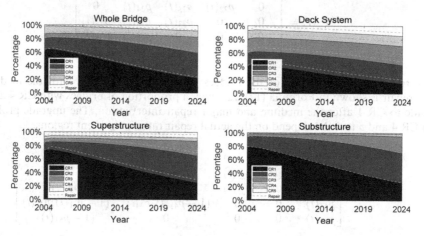

Figure 4.   Prediction of the bridge deterioration with/without repair intervention in suburban area.

6   CONCLUSION

Influenced by the external environment and traffic loads, the performance of bridges will gradually deteriorate with the growth of service time. However, the existing deterioration model of the BMS in Shanghai is still using the deterministic regression method, which is hard to reflect the uncertainty and randomness of the bridge deterioration process. By considering the time-dependent reliability in the process of bridge deterioration, a Weibull distribution based semi-Markov process prediction model was proposed in this paper. And the deterioration behavior of the three major parts was predicted, and the result shows that the decay rates of the deck system are the fastest. The impact of urban bridges in different regions with different funds allocation is analyzed, and the result reveals that the service life expectancy of bridges in the suburban area is shorter than the downtown area. Bridge deterioration trends with or without maintenance intervention were also predicted based on the investigated actual repair rates. It shows that the medium and major repair can significantly decrease the proportion of bridges with CR 4 and 5, especially for bridges in the suburban area. But the existing conservation maintenance strategies have a limited overall effect on network-level urban bridges, and there is a noticeable trend of bridges degradation from CR 1 and 2 to CR 2 and 3. Therefore, to reduce the maintenance cost of the entire life cycle of urban bridges in Shanghai, it is necessary to adopt a more reasonable preventive maintenance method.

ACKNOWLEDGEMENTS

This study was supported by the National Key R&D Program of China (No. 2018YFB1600100). The authors would like to acknowledge this financial support.

REFERENCES

Agrawal, A.K., Kawaguchi, A., and Chen, Z. (2010). "Deterioration rates of typical bridge elements in New York." *J. Bridge Eng.*, 10.1061/(ASCE)BE.1943-5592.0000123,(ASCE) 15(4), 419–429.

Chen, Z. (2005). "Research on technology structure of transportation infrastructure management system." Ph.D.

China, M.O.C.O. (2017). "Technical standard of maintenance for city bridge.".

Corradi, G., Janssen, J., and Manca, R. (2004). "Numerical treatment of homogeneous semi-Markov processes in transient case–a straightforward approach." *Methodol. Comput. Appl.*, 6(2), 233–246.

Du, G., and Karoumi, R. (2013). "Life cycle assessment of a railway bridge: comparison of two super-structure designs." *Struct. Infrastruct. E.*, 9(11), 1149–1160.

Fang, Y., Li, L., Chen, Z., and Sun, L. (2017). "Prediction Model of Concrete Girder Bridge Deterior-ation in Shanghai Using Weibull-Distribution Method.".

Fang, Y., and Sun, L. (2018). "A Weibull Distribution Based Semi-Markov Process Model for Urban Bridge Deterioration Prediction.".

Fang, Y., and Sun, L. (2019). "{Developing A Semi-Markov Process Model for Bridge Deterioration Prediction in Shanghai}." {*SUSTAINABILITY*}, {10.3390/su11195524}}, Article-Number = {{5524}, (ASCE) {11}({19}).

Golabi, K., and Shepard, R. (1997). "Pontis: A system for maintenance optimization and improvement of US bridge networks." *Interfaces*, 27(1), 71–88.

Hawk, H., and Small, E.P. (1998). "The BRIDGIT bridge management system." *Struct. Eng. Int.*, 8(4), 309–314.

Horvath, A. (2009). "Principles of using life-cycle assessment in bridge analysis." 21–22.

Itoh, Y., and Kitagawa, T. (2003). "Using CO2 emission quantities in bridge lifecycle analysis." *Eng. Struct.*, 25(5), 565–577.

Li, L., Li, F., Chen, Z., and Sun, L. (2016). "Bridge Deterioration Prediction Using Markov-Chain Model Based on the Actual Repair Status in Shanghai.".

Lu, P., Pei, S., Tolliver, D., and Jin, Z. (2015). "Data-based Evaluation of Regression Models for Bridge Component Deterioration.".

Madanat, S., Mishalani, R., and Ibrahim, W.H.W. (1995). "Estimation of infrastructure transition prob-abilities from condition rating data." *J. Infrastruct. Syst.*, 1(2), 120–125.

Medjoudj, R., Aissani, D., Boubakeur, A., and Haim, K.D. (2009). "Interruption modelling in electrical power distribution systems using the Weibull—Markov model." *Proceedings of the Institution of Mechanical Engineers, Part O: Journal of Risk and Reliability*, 223(2), 145–157.

Mishalani, R.G., and Madanat, S.M. (2002). "Computation of infrastructure transition probabilities using stochastic duration models." *J. Infrastruct. Syst.*, 8(4), 139–148.

Morcous, G. (2006). "Performance prediction of bridge deck systems using Markov chains." *J. Perform. Constr. Fac.*, 20(2), 146–155.

Ng, S., and Moses, F. (1998). "Bridge deterioration modeling using semi-Markov theory." *A. A. Balkema Uitgevers B. V, Structural Safety and Reliability.*, 1 113–120.

Ng, S.K., and Moses, F. (1996). "Prediction of bridge service life using time-dependent reliability analysis." *Bridge management*, 3 26–32.

Reardon, M.F., and Chase, S.B. (2016). "Migration of Element-Level Inspection Data for Bridge Man-agement System.".

Sobanjo, J.O. (2011). "State transition probabilities in bridge deterioration based on Weibull sojourn times." *Struct. Infrastruct. E.*, 7(10), 747–764.

Su, D., Nassif, H., and Hwang, E. (2015). "Probabilistic Approach for Forecasting Long Term Perform-ance of Girder Bridges.".

Winn, E., Burgueno, R., and Haider, S.W. (2013). "Project- and Network-Level Bridge Deck Degrad-ation Models via Neural Networks Trained on Empirical Data.".

Zambon, I., Vidovic, A., Strauss, A., Matos, J., and Amado, J. (2017). "Comparison of stochastic predic-tion models based on visual inspections of bridge decks." *J. Civ. Eng. Manag.*, 23(5), 553–561.

Assessment of asphalt concrete EPDs in Scandinavia and the United States

L. Strömberg & S. Hintze
Nordic Construction Company, (NCC), Solna, Sweden & Department of Civil and Architectural Engineering, KTH Royal Institute of Technology, Stockholm, Sweden

I.L. Al-Qadi & E. Okte
Illinois Center of Transportation, Department of Civil and Environmental Engineering, University of Illinois at Urbana–Champaign, Urbana, IL, USA

ABSTRACT: Scandinavian and the US construction industry have made noticeable progress with implementation of the environmental product declarations (EPDs) as a certificate of the environmental performance of asphalt concrete (AC). The information in an EPD can be used as a reliable measure of environmental performance and can be accepted for use in reporting, marketing, procurement and product development. However, the existing standards for the development of EPDs needs improvement. Moreover, preparation of the EPDs in different countries are performed in using various tools and methods. This may lead to incomparability of EPDs developed by manufactures. A global EPD-calculation framework would make it possible to overcome these barriers and improve communication between contractors and clients in the US and Europe. This paper compares current EPDs practices in the US asphalt industry to those of Europe, specifically the Norwegian and Swedish asphalt industry. The study has evaluated two digital EPD tools, from Norway and the US, and one LCA tool from Sweden that may be used as a Swedish EPD tool for AC. This paper presents tools used by the aforementioned countries and identifies the strengths and limitations of each system. It also highlights the difficulty in developing a universal system for analysis. This study is a first step towards a global standard relevant to the US and Europe.

1 INTRODUCTION

Life Cycle Assessment (LCA) is an established methodology for quantifying the environmental performance of any product. However, LCA has the challenge of coordinating assumptions by various products and industries. System boundaries, inventory analyses, and allocation rules may differ from one analysis to another. Environmental product declaration (EPD) is the result of LCA analysis using consistent product category rules (PCR). The aim of the EPDs is to allow comparison of the environmental impacts of similar products by different producers.

A PCR is produced by an expert group and must be approved by an EPD program operator. In Scandinavia, such operators include EPD Norway (2019) or EPD International (2019). In addition, the general program instructions (GPI) must be followed when a PCR is being developed. ISO 14025 (2006) regulates the program operator, i.e. the organization that ensures the verification and publication of EPDs in accordance with the PCR that applies to the product in question.

Within the EPD development, there are two types of PCRs: general PCRs, e.g. PCRs for all construction products and product-specific PCRs, e.g. PCRs for the asphalt concrete (AC). Both EPD Norway and EPD International have general PCRs for construction products

(EPD International, 2012; EPD Norge, 2017a) and PCRs for AC (EPD International, 2018; EPD Norway, 2017b).

There are many applications for EPDs in the industry. Internally, an EPD may be used to reduce the environmental impact during the manufacturing processes and/or to assist in an investment decision of an equipment based on efficiency. Externally, EPDs are used to communicate the environmental impact of a product transparently and credibly. An EPD can be generic or a product-specific. A generic or industry-average EPD for a particular product may be developed, which is common in the Nordic market. Another type of EPD, a product-specific EPD, is often produced by a material manufacturer to identify potential optimization measures.

The EPD development process is usually accompanied by a program operator that defines GPI and supervises the EPD process (Bovea et al., 2014). In the Nordic market, the number of EPDs for building materials has increased significantly in recent years. The Swedish-based operator, the EPD International (EPD International, 2019; Welling, 2018) has published 525 EPDs for building materials, building products or components and 11 EPDs for buildings and engineering works. The Norwegian operator, EPD Norway has published 437 EPDs for construction products (EPD Norway, 2019; Hauan, 2018). Scandinavian asphalt manufacturers are free to select the program operator for publishing their EPDs.

In the US, many sustainability and green construction-rating systems require EPDs, including Greenroads® and LEED. In the AC industry, one of the main program operators is the National Asphalt Pavement Association (NAPA) (National Asphalt Pavement Association, 2017). NAPA has published PCR guidelines specifically for asphalt mixtures, which are valid from January 31, 2017 until January 31, 2022. As with other product PCRs, only the production stage of AC is included in the LCA system boundary. For the asphalt industry in the US, there have been challenges in using EPDs to communicate LCA results (Mukherjee et al, 2017). Some of the challenges include, but are not limited to defining AC system boundaries, availability of accurate and transparent data, missing inventory data, and allocating impact of recycled materials.

In the European Union and Sweden, there is a strong political drive to develop standard and regulations to reduce the climate impact. The effect of these standards on reducing the climate impact for pavements is unknown and has not been verified with laboratory and field tests. A product-specific EPD can be used to calculate and verify the actual environmental impact of an AC mix, but it requires using product-specific data on raw materials, energy use, and emissions by the AC manufacturer. Percentage of recycled material in AC and transportation to a construction site are examples of input data for an EPD. However, these data are usually not collected by manufactures nor shared when available.

The number of AC product-specific EPDs is still limited in Scandinavia and the US. The Swedish-based EPD International has published seven EPDs for AC mixes (EPD International 2020). The EPD Norway has published six EPDs for most common AC mixes (EPD Norway 2020). In the US, NAPA has published nine EPDs for four plants in four states.

The unavailability of information limits the use of EPDs as a certificate of environmental performance of infrastructure projects in Sweden (Strömberg, 2016; Strömberg 2017a & 2017b; Strömberg et al, 2020). Hence, product-specific EPDs, in Scandinavia and the US, are currently limited to the assessment of environmental impact during the AC production stage. A more comprehensive EPD that includes other life cycle stages, such as the use and end-of-life stages, requires more specific data on the future maintenance and repair scenarios. Furthermore, the preparation of a product-specific EPD is costly, resource-intensive task, and requires LCA-expert knowledge. There is a growing trend in developing a cost effective computer-based tools to produce EPDs. However, EPDs developed by different manufactures are incompatible because of the variation in EPD development and the use of various LCA tools.

A global EPD-calculation framework would improve communication on environmental performance between material suppliers, contractors and clients in the US and Europe. In this paper, two digital EPD tools, from Norway and the US, and one LCA tool from Sweden were evaluated as potential candidates as a Swedish EPD tool for AC mixes. The paper presents the result of reviewing these tools and points out the strengths and limitations of each system. It also highlights the difficulty in developing a universal system for analysis.

The adaptation of the EN 15804 and ISO 21930 regarding calculation rules for the assessment of AC environmental impact is being considered in Sweden, Norway and the US. To have consistent system boundaries across the industry, PCR are required. The PCR are developed using the general guidelines in ISO 14025 (2006) and are meant to clarify the boundaries of LCA methodology (Ingwersen, 2014). Therefore, the development of an EPD is based on a set of standards and requirements, from basic standards on environmental management systems, ISO 14001 (2015) and the standard series for life cycle analysis ISO 14040 (2006) and ISO 14044 (2006), to more specific EPD standards EN 15804 (2012), ISO 21930 (2017) and a PCR. Figure 1 summarizes relevant standards for the development of an EPD.

The European standard, EN 15804, specifies an industry-approved method for the development of an EPD for a construction product. The US has adopted the EN 15804 with some modifications as the international standard, ISO 21930. Both EN 15804 and ISO 21930 lay the foundation for all PCRs and EPDs for building products and establishes principles regarding system boundaries, reported environmental impact categories, etc. According to these standards, there is a minimum requirement for an EPD to include the environmental impact from raw material extraction up to the factory gate ("cradle-to-gate"), which is referred to in the standard as modules A1 to A3, see Figure 2.

Another type of EPD, "cradle-to-gate with options" is based on modules A1 to A3 plus other selected optional modules, e.g. end-of-life modules C1 to C4. Module D may be included in this EPD. If an EPD covers material and production, construction process, use and maintenance, replacements, demolition, waste processing for re-use, recovery, recycling and disposal, is called "cradle-to-grave." In this EPD, the module D may be included.

To summarize some recommendations for the Swedish AC industry for using EPDs as a uniform requirement for calculation and reporting of the AC impact on environment, the scope of this study is limited to following tasks:

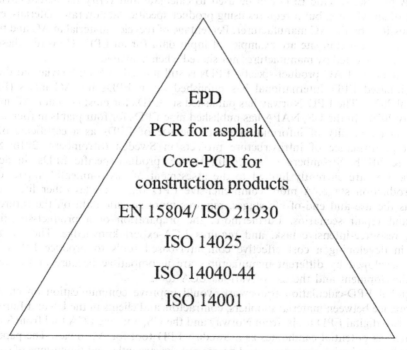

Figure 1.   Standards for the evaluation and communication of environmental impact for building products.

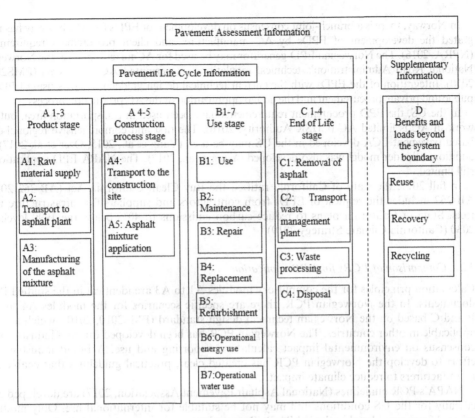

Figure 2. The environmental impact of AC pavement is divided into several life cycle stages or modules in accordance with EN 15804/ISO 21930.

1. An analysis of current practice with EPD procurement requirements for infrastructure projects in Sweden, Norway and the US.
2. A review of three PCRs for developing EPDs for AC: a PCR from EPD Norway (2017b), a PCR from EPD International (2018) and the PCR being developed by NAPA (2017).
3. A review of three national LCA tools for AC: the Swedish Transport Administration's EKA tool (Lundberg, 2015), Norwegian EPD-generator (EPD-generator for products, 2018) and NAPA's Eco-Label EPD tool. Additionally, some LCA tools commonly used in the US are reviewed to determine whether they can be used to generate EPDs (Al-Qadi et al, 2015; Yang et al., 2018, Yang et al. 2014).

3 RESULTS

3.1 *Analysis of current client briefs with EPDs*

In the Swedish part, the large number of project briefs describing the procurement requirements regarding EPDs and LCAs have been analyzed (Strömberg et al., 2020). The analysis has shown that Swedish Transport Administration (STA) (2019), Stockholm County Council (2016; 2018) and Swedish municipalities (Bjäresten, 2018; General Environmental Requirements for Contractors, 2018; Uppsala Municipality, 2016) have no uniform requirements regarding calculation, reporting, optimization and follow-up of the actual environmental impact of AC pavements with EPDs. This makes the transition to a climate-neutral AC manufacturing in Sweden a challenge. In addition, the availability of accurate and transparent data, missing inventory data, and allocating impact of recycled materials would affect EPDs.

In Norway, there is a branch-joint consensus on development of EPDs for AC. Norway has integrated the development of EPDs by AC manufactures and client procurement requirements (NPRA, 2016). The Norwegian EPD-generator may be used for AC mixes that meet the Norwegian National Road Administration's technical regulations on permitted AC pavements (PMS2010, 2010). Integration of the EPDs with the current technical design guidelines allows using EPDs in public and private procurement and the design optimization of the AC production process.

In the US, the EPD procurement requirements have been used in transportation area; but the work in AC is limited AC (CHSR Authority, 2016). The first and comprehensive regional-data based pavement LCA developed in the US may be used (Yang et al., 2015, Ozer et al. 2017). In addition, a binder model was also developed (Yang et al., 2017). The NAPA EPD application is still limited.

In fall 2017, the state of California ratified the Buy Clean California Act (AB-262 2017). AB 262 includes the request of EPDs from contractors and suppliers for infrastructure projects. State of California plans to reduce carbon emissions to 40% below the 1990 levels by 2030 (California Climate Strategy 2019).

3.2   Comparison of PCRs for EPD application

Calculation principles for the compulsory modules A1 to A3 are identical in the studied PCR documents. In the Norwegian PCR, there are specific scenarios for the modules A4 to A5, B and C based on the Norwegian technical design standard (PMS2010, 2010), which are not applicable in other countries. The Norwegian PCR has been developed through industry-wide consensus on environmental impact calculation, reporting and use. A broad industry-wide effort to develop the Norwegian PCR has created clear, practical guidelines that enable AC manufacturers to reduce climate impact.

NAPA's PCR guidelines (National Asphalt Pavement Association, 2017) are developed specifically for the US conditions and may not be suitable for international use. Only modules A1 to A3 are considered in the NAPA PCR procedures.

The PCR by EPD International has an international character and is not adapted to any specific national conditions. This PCR does not consider the Swedish national design standard for AC pavements (Swedish Transport Administration, 2011). There is a greater degree of freedom in the EPD International PCR when applied to the development of a product-specific EPD for a certain country.

The PCRs in this study do not consider modules B5 to B7, as they are relevant for AC pavements. The EPDs development based on the aforementioned PCRs are comparable only for modules A1 to A3, the production stage of AC. In the new version of the European EPD standard, EN 15804, which will be implemented around all countries in the European Unit as of March 2020, reporting on the environmental impact of modules A1-A3, C1-C4 and D will be mandatory.

It is expected that the new standard developed by European technical committee CEN/TC350 on Sustainability of Construction Works, working group TC 227/WG6, will create further comparability for AC pavement life cycle stages (Swedish Institute of Standards, 2019). Before that, clear conditions must be developed that allow comparability of EPDs with the system boundaries in addition to modules A1 to A3. A more detailed comparison of EPD International and EPD Norway PCRs is given in the Swedish study (Strömberg et al., 2020).

3.3   Summary of EPD and LCA tools in Scandinavia and the US

In Sweden, the Excel-based LCA tool Energy and Carbon Dioxide in Asphalt Production, EKA, which is developed and managed by STA is used (Lundberg, 2015). The EKA tool enables LCA calculations of the entire chain in AC manufacture—from the input material to the completed pavement construction. A calculation model has been developed for estimating carbon dioxide emissions and energy consumption from components, equipment, and manufacturing processes for common AC mixtures in Sweden. Calculation results in EKA provide an overview of factors that affect energy consumption and emissions of CO_2-eq and thus make it possible to identify areas of improvement. The units for comparison are CO_2-eq and kWh per ton or per m^2 of AC pavement.

The current version of EKA may not be verified as an EPD tool for AC because it does not meet the requirements of EN 15804 or any of the analyzed PCRs. Calculation in the EPD format requires that the environmental impact categories, such as acidification and eutrophication, are calculated in accordance with EN 15804. In addition, the documentation and quality of the emission factors as well as the LCA method used do not meet the verification requirements.

Product-specific EPDs are costly, resource-intensive, and requires LCA-expert knowledge. Hence, there is a growing trend in developing computer-based tools to produce EPDs. In Norway, a web-based digital EPD tool has been developed to support the generation of EPDs regardless of AC manufacturers' access to EPD expertise in Norway (Aakre, 2018). The tool allows the development of EPDs for AC plant and AC pavement utilizing various functional units. The mandatory modules are A1 to A5. The various calculation scenarios for modules B and C have been developed based on the life expectancy of common Norwegian AC pavements. Hence, Norwegian AC pavement product-specific EPDs for modules A to C may be developed. Three industry-wide EPDs for common AC pavements have been generated and are "cradle-to-gate" or "cradle to gate with options." Project-unique EPDs may be produced by AC manufacturers based on the parent EPDs.

In the United States, NAPA has developed its own Eco-Label EPD tool (Emerald, 2017). Similar to the Norwegian EPD-generator, these declarations are "cradle-to-gate" EPD for modules A1 to A3. Users can generate product specific EPDs using their production plants data. If there is a lack of plant data, industry average data are used to generate EPDs. Currently, there are nine EPDs published in four states by four plants. These data do not encompass all available EPD data, because each plant has the right not publish the results of their EPDs. There are also private companies that provide EPDs for contractors and AC plants. However, EPDs produced by different parties—even if they were built using the same PCR—may not be comparable, especially if they are not built using the same PCR process.

In addition to EPD tools, LCA tools may be used to create EPDs. However, any EPD created independently must be verified by a third party. There are several LCA tools that can perform cradle-to-gate analysis (Al-Qadi et al, 2015).

Several tools are built specifically for conducting pavement LCA. PaLATE 2.0, developed by the University of California, Berkeley, is an open source and freely available LCA tool. The tool uses generic data and may require changes to its LCI database based on location and process. PaLATE V2.2 was used recently to complete project requirement life cycle inventory (LCI) for the Greenroads® Rating System (dos Santos et al, 2017).

Athena Pavement LCA is another freely available LCA tool developed for pavement LCA. It is a web-based application that covers roadway construction stages in Canada. There is a desktop version for the United States that covers US Regional Data (Athena, 2018). However, it has an LCI database that makes it only applicable to selected regions from which the data was collected.

The LCA tool GaBi, which was developed by PE International in collaboration with the University of Stuttgart (dos Santos et al, 2017), may be used for a "cradle-to-gate." The tool has several commercial databases, e.g. Ecoinvent and US LCI. However, GaBi uses average data that may not be applicable to many EPDs. SimaPro, developed by the National Renewable Energy Laboratory (NREL), is another LCA tool that has a US LCI database. SimaPro provides "gate-to-gate", "cradle-to-gate" and "cradle-to-grave" accounting of the energy and material flows associated with producing a material, component or assembly in the United States (Deru et al, 2009). When industry average data are used, the accuracy of EPDs would be decreased. To address this limitation, the Illinois Center for Transportation (ICT) introduced an LCA that utilizes regional data collected from contractors, which could be used to generate EPDs (Al-Qadi et al, 2015).

4 SUMMARY

There is an ongoing effort to develop and implement policies for using EPDs to reduce the impact of transportation infrastructure projects on the environment. The use of any EPD tool must be harmonized with adopted PCR for AC and verified by an EPD operator. The lack of comparable and reliable industry wide-used EPDs for various AC is a serious barrier.

Currently, Europe and the US AC pavement industry has no requirement regarding calculating, reporting, and optimizing actual impact on the environment. This paper compares current EPD practices by the US asphalt industry to those of Europe, specifically the Norwegian and Swedish asphalt industry. Several digital EPD and LCA, that may be used to generate EPDs, tools have been discussed. The advantages and limitations of each system are presented. In addition, difficulty to develop a universal system for analysis is emphasized.

There is a real opportunity to develop a global industry-wide EPD-calculation to compare various AC mixes regardless of the country of origin. This study represents the first step in developing a universal system for AC EPDs.

ACKNOWLEDGEMENTS

The financial support by The Swedish Transport Administration and the Swedish Construction Industry Development Fund (SBUF) is acknowledged. This paper is part of the ongoing projects Digitalization of EPDs for Asphalt Pavements (project ID: 13472) and Adaptation of Road Design to LCC and LCCA (project ID: 13722). The financial support and effort of the participating companies, Skanska, NCC, Nynas and Peab Asfalt are appreciated. The authors also acknowledge the companies and organizations participated in the interviews and the support provided by the Department of Civil and Architectural Engineering at KTH Royal Institute of Technology.

REFERENCES

Aakre, A. 2018. Norwegian Contractors Association (EBA). Personal interview on May 5th, 2018.

Al-Qadi, I.L., Yang, R., Kang, S., Ozer, H., Ferrebee, E., Roesler, J.R., Salinas, A., Meijer, J., Vavrik, W.R., & Gillen, S.L. 2015. Scenarios developed for improved sustainability of Illinois Tollway: Life-cycle assessment approach. Transportation Research Record: Journal of the Transportation Research Board 2523 (1): 11–18.

Athena. 2018. User manual and transparency document for pavement LCA v3.1, Athena Sustainable Materials Institute Ottawa, Canada.

Bjäresten, E. 2018. Linköping municipality's requirements with EPDs (in Swedish). Presentation at workshop Digitalization of EPDs for asphalt pavements on Mars 27th, 2018.

Bovea, M.D., Ibáñez-Forés, V., & Agustí-Juan, I. 2014. Environmental product declaration (EPD) labelling of construction and building materials. In Eco-efficient Construction and Building Materials: 125–150. Sawston, UK.

CHSR Authority 2016. California High-Speed Rail Big Picture. Sacramento, CA: CHSR Authority.

Deru, M.P. 2009. US life cycle inventory database roadmap, National Renewable Energy Laboratory, Golden, CO.

dos Santos, J.M.O., Thyagarajan, S., Keijzer, E., Flores, R.F., & Flintsch, G. 2017. Comparison of life-cycle assessment tools for road pavement infrastructure. Transportation Research Record 2646 (1): 28–38.

Emerald 2017. Emerald eco-label EPD tool instructions. EN 15804:2012 Sustainability of construction works — Environmental product declarations — Core rules for the product category of construction products.

EPD-generator for products 2018, LCA, Eng. Version.

EPD International 2018. Product Category Rules for Construction Products and Construction Services. 2012:01, version 2.3.

EPD International 2018. Asphalt Mixtures Product category classification: UN CPC 1533 & 3794, 2017:01 Version 1.0.

EPD International, 2019. www.environdec.com, Retrieved on August 2019.

EPD Norge 2017a. NPCR Construction products and services – Part A.

EPD Norge 2017b. NPCR 025:2017 Version 1.1 Part B for Asphalt.

EPD Norway, 2019. www.epd.no, Retrieved on August 2019.

General environmental requirements for contractors 2018. (in Swedish), Malmo- and Gothenburg municipalities and Swedish Transport Administration, Stockholm, Sweden.

Hauan, H. 2018. EPD Norge. Personal interview on May 2nd, 2018.

Ingwersen, W.W., & Vairavan, S. 2014. Guidance for product category rule development: Process, outcome, and next steps. The International Journal of Life Cycle Assessment 19 (3): 532–537.

Kang, S., Yang, R., Ozer, H., and I. L. Al-Qadi, "Life-cycle greenhouse gases and energy consumption for the material and construction phases of pavement with traffic delay," Transportation Research Record, No. 2428, Washington D.C., 2014, pp. 27–34.

ISO 14001:2015 Environmental management systems – Requirements with guidance for use.

ISO 14025:2006, Environmental labels and declarations – Type III environmental declarations – Principles and procedures.

ISO 14040:2006, Environmental management – Life cycle assessment – Principles and framework.

ISO 14044:2006, Environmental management – Life cycle assessment – Requirements and guidelines.

ISO 21930:2017 Sustainability in buildings and civil engineering works - Core rules for environmental product declarations of construction products and services.

Iversen, O. 2018. Personal interviews on April 25th, 2018 and May 2nd, 2018.

Lundberg, R. 2015. Energy efficient asphalt pavements: LCA tool EKA (in Swedish), Swedish Transport Administration.

Mukherjee, A. & Dylla, H. 2017. Challenges to using environmental product declarations in communicating life-cycle assessment results: Case of the asphalt industry. Transportation Research Record 2639 (1): 84–92.

National Asphalt Pavement Association 2017. Product category rules (PCR) for asphalt mixtures. Lanham, MD: National Asphalt Pavement Association.

NPRA 2016. CO_2 emissions of the road project "E6 Column-Moelv" (in Norwegian). Publisher: Norwegian Public Road Administration.

Ozer, H., R. Yang, and I. L. Al-Qadi, "Quantifying sustainable strategies for the construction of highway pavements in Illinois," Transportation Research Part D, Vol. 51, 2017, pp. 1–13 (DOI information: 10.1016/j.trd.2016.12.005).

PMS2010 2010. Guidance document for road contractors (in Norwegian), Norwegian Public Road Administration.

Strömberg, L. 2016. Verified climate calculation of contractors' design. Proceedings of International High- Performance Built Environment Conference – A Sustainable Built Environment Conference 2016 Series (SBE16), Sydney, Australia, pp. 1–9, 1877-7058.

Strömberg, L. 2017a. Current difficulties with creation of standardized digital climate calculations for infrastructure projects, Proceedings of the Pavement Life-Cycle Assessment Symposium 2017, Champaign, Illinois, US, April 12-13, pp. 23–29, ISBN: 978-1-315-15932-4 (eBook).

Strömberg, L. 2017b. Conceptual Framework for Calculation of Climate Performance with Pre-verified LCA-Tools, Journal of Civil Engineering and Architecture, Vol. 11, pp. 29–37, 2017, DOI:10.17265/1934-7359.

Strömberg, L., Wendel M., Berglund M. & Lindgren Å. 2020. Digitalization of EPDs for asphalt – experience from Sweden and input from Norway. Accepted for publishing in proceeding of 7th Eurasphalt & Eurobitume Congress, Madrid 12-14 May 2020.

Stockholm County Council 2016. Environmental program 2017-2021 (in Swedish), Stockholm County Council.

Stockholm County Council 2018. Project management requirements of project "Akalla - Barkarby 4712 Work tunnels Robothöjden and Landingsbanan" (in Swedish), Stockholm County Council.

Swedish Institute of Standards 2019. Swedish Transport Administration 2011. Swedish Transport Administration's Technical Requirements for Road Construction (in Swedish), Swedish Transport Administration, TRV 2011:072, TDOK 2011:264.

Swedish Transport Administration 2019. TDOK 2015:0480: Climate requirements in planning, construction, maintenance phase and on technically approved railway equipment (in Swedish), Swedish Transport Administration, version 3.

Uppsala municipality 2016. Procurement requirements for project "Rosendal Stage 2B" (in Swedish), Uppsala municipality.

Welling, S. 2018. EPD International. Personal interview on May 5th, 2018.

Yang, R., S. Kang, H. Ozer, and I. L. Al-Qadi, "Environmental and economic analyses of recycled asphalt concrete mixtures based on material production and potential Performance," Resources, Conservation & Recycling Journal, Vol. 104, 2015, pp. 141–151.

Yang, R., H. Ozer, and I. L. Al-Qadi, "Regional upstream life-cycle impacts of petroleum products in the United States, Journal of Cleaner Production, Vol. 139, 2017, pp. 1138–1149.

Yang, R., I. L. Al-Qadi, and H. Ozer, "Effect of methodological choices on pavement life-cycle assessment," The Journal of Transportation Research Record, Vol. 2672, No. 40, 2018, pp. 78–87.

Life cycle assessment in public procurement of transport infrastructure

S. Toller
Swedish Transport Administration, Stockholm, Sweden

S. Miliutenko & M. Erlandsson
IVL Swedish Environmental Research Institute, Stockholm, Sweden

S. Nilsson
WSP Sweden, Stockholm, Sweden

M. Larsson
IVL Swedish Environmental Research Institute, Stockholm, Sweden

ABSTRACT: By 2045, Sweden has the goal of zero net emissions of greenhouse gases (GHG). The Swedish Transport Administration (STA), aims to reduce the life cycle climate impact of constructing, operating, and maintaining transport infrastructure by 30 % in 2025 and by 50 % in 2030 compared to 2015. To enable this, the STA has implemented LCA-based climate requirements in public procurements of road and rail infrastructure. In this paper, the results from three different evaluations of the STA climate requirements have been summarized. The development and implementation of LCA-based climate requirements would not have been possible without both methodological maturity and organisational learning. Recent work evaluating the climate requirements has resulted in an increased understanding of the procurement related mechanisms that favour effective climate mitigation measures.

1 INTRODUCTION

1.1 *Background*

The Swedish parliament has adopted the goal that by 2045, Sweden should have zero net emissions of greenhouse gases (GHG) (Government Offices of Sweden 2018). The transport sector has been identified as an important contributor to GHG emissions (EC 2016) and the EU has recognised the need to consider the climate change challenge during the planning of transport infrastructure (COM 2011). Previous works have concluded that transport infrastructure may account for a significant share of transport system impacts (e.g. Chester & Horvath 2009; Rahman et al. 2014). Therefore, emission reduction measures need to consider not only emissions from traffic, but also those related to construction, operation, maintenance, and reinvestment of the transport infrastructure.

In Liljenström et al. (2019), a network life cycle assessment (LCA) based approach was recently presented for assessing yearly emissions from transport infrastructure taking place in Sweden. They estimated the annual GHG from Swedish transport infrastructure as 2.8 million tons of carbon dioxide equivalents during the year 2015. That corresponds to about a third of the overall impact from the built environment and impact related to the construction works (Erlandsson 2019). Moreover, it corresponds to about 5% of the total GHG emissions in Sweden and 17% of the GHG emissions from industry in Sweden (where production of steel, cement, refineries and chemicals have the largest share) (Swedish Environmental Protection Agency, 2019). Management of the infrastructure stock with focus on reinvestment of road and rail infrastructure and manufacturing of high-volume materials (like asphalt, steel, and concrete) were identified by Liljenström et al. (2019) as important areas to address in order to contribute the climate mitigations.

1.2 *Climate calculations at the STA*

The Swedish Transport Administration (STA) is responsible for long term planning of the transport system as well as the construction, operation and maintenance of public roads and railways. It has established following targets for reducing the life cycle climate impact from transport infrastructure:30 % below 2015 levels by 2025, 50 % below 2015 levels by 2030, and zero emissions in 2045. To enable this, the STA has implemented LCA-based climate requirements in public procurements of new road and rail infrastructure. To facilitate the requirements, the STA has developed a climate calculation model 'Klimatkalkyl', that provides a pre-defined methodological setting (Toller & Larsson 2017). The model applies the basic principles of LCA in accordance with ISO 14040 and ISO 14044 and further specified calculation rules for all construction products EN 15804 (CEN 2013). However, GWP and primary energy use are the only impact categories included. The climate calculation model 'Klimatkalkyl' is used to streamline and to determine the resulting GWP and primary energy for transport infrastructure projects. Production (including raw material extraction, transport and refining) and maintenance (including operation, maintenance and reinvestments) are included in the calculations. Demolition may be added if relevant. The traffic using the infrastructure is not accounted for.

The model includes a generic database with background data in terms of resource templates for building components and maintenance and operation activities, together with LCA data for the material and fuels used. The resource templates and the LCA data simplifies for the user to rapidly generate an LCA. The LCA result can in brief be described as achieved by multiplying resource use (material and fuels, i.e. bill of resources) and generic background LCA data. The LCA data are normally conservative and geographical representative values. The user of 'Klimatkalkyl' can replace the generic resource templates for the building components or activities with project specific input data and the generic LCA data can be substituted by supplier specific LCA data in order to generate a more precise result. The user's possibility to replace the resource templates for maintenance and operation is limited due to verification problems (future resource needs for maintenance and operation cannot be validated).

1.3 *Climate requirements in public procurement*

In 2016, quantitative climate requirements were implemented for procurement of larger projects by the STA. The requirements are based on the expected emissions from a project in a life cycle perspective (a baseline), calculated in the STA climate calculation model with generic background data. A certain percentage decrease in GHG emissions is required, compared to the baseline, depending on the type of project. Compliance with the requirements should be verified by a climate declaration in the same model at hand-over stage. Thus, the climate requirements could be expressed as a type of contract clause rather than qualification criteria. The climate calculation model and the first version of requirements are described more in detail in Toller & Larsson (2017). Climate requirements can be met either by using less material (e.g. by choosing a more efficient technology) or by choosing material suppliers with better climate performance. In the latter case, product specific LCA data that is verified through an environmental product declaration (EPD) needs to be presented. Bonuses can be awarded if the decrease in GHG emissions is larger than requested, up to a certain limit and based on total project cost.

In 2018 the LCA-based climate requirements for the entire construction projects were complemented with requirements on climate performance for some specific construction materials and fuels in order to encourage the development of materials and fuels with lower GHG emissions. These new requirements were implemented in smaller projects and maintenance contracts for which LCA calculations in the climate calculation model are not required. Based on the results by Liljenström et al. (2019), the first version of the material requirements focuses on steel reinforcement, construction steel, and concrete and cement, as they contribute significantly to the global warming potential (GWP). Climate performance is verified through Environmental product declarations (EPDs), that needs to be in accordance with the EPD standard published by the European Committee for Standardization (CEN 2013).

There is limited published material available regarding the use of LCA-based climate requirements in public procurement, only a few examples were found. Hochschorner & Finnveden (2006) examined procurement of defence industry material and recommended the use of LCA as a tool for learning about environmental impacts, fulfilling requirements and choosing between alternatives in material acquisition. Parikka-Alhola & Nissanen (2009) report on a case study where LCA was applied as part of tender evaluation within public procurement of contractors for a major road in Finland. In that case, the LCA requirements affected the selection of contractor. Blechenberg et al. (2013) point out successful examples of LCA-based climate requirements for transport infrastructure in The Netherlands and in Great Britain. The Netherlands are using carbon dioxide performance ladder and the LCA-based tool Dubo-Calc in order to promote GHG reduction in construction projects through procurement. The Highways Agency in Great Britain has developed a set of methods and tools for the collection and reporting of carbon dioxide emissions in order to encourage voluntary initiatives from contractors and suppliers. In addition to the examples from the Netherlands and Great Britain, Kadefors et al. (2019) also show examples of the use of LCA-based climate requirements during planning of transport infrastructure in Australia and USA.

The development of LCA-based climate requirements on transport infrastructure by the STA may be seen as a case to gain knowledge and experience in order to achieve an increased understanding of the procurement related mechanisms that favour effective climate mitigation measures. The climate requirements are still relatively new, but different evaluations have already been performed. However, none of these has yet been scientifically published. The aim of this paper is to evaluate the LCA-based climate requirements that are implemented by the STA in public procurements of transport infrastructure in Sweden, and to provide suggestions on how to improve the use of climate requirements in order to further enhance climate mitigation.

2 EVALUATION OF PROJECT REQUIREMENTS

2.1 Evaluations performed

Three different evaluations have been performed for the LCA-based climate requirements implemented by the STA. In 2016, interviews were performed as a part of the research project 'LCA in public procurement' at IVL, the Swedish Environmental Research Institute. A deepened evaluation of the case 'Stockholm bypass' was later performed and presented in Larsson & Novakovic (2017) and Miliutenko et al. (2019). In 2018, an evaluation with larger scope was performed, called 'Control station 2018' (Nilsson et al. 2019). The evaluations can be seen as overlapping, as they to some extent focus on the same projects. However, as their scope and methods differ somewhat, they also complement each other.

Due to the lengthy lead times for construction of infrastructure, and the fact that the climate requirements have only been applied for a few years, no projects with climate requirements have yet been fully finished. This limits the possibilities for quantitative evaluations regarding the achieved decrease in emissions. The evaluations performed focus instead on qualitative information from the stakeholders involved.

There are different types of construction contracts represented in the evaluations. The two main types of construction contracts procured by STA are design-bid-build (DBB) and design-build (DB). In addition, a few early contracts involvement (ECI) have been procured (Eriksson & Westerberg 2011). The ECI contracts are a form of partnering: both the client and the contractor are involved in the planning. In the DB type of contract, the contractor is responsible for both design and project implementation. The DBB contracts, on the other hand, involve a separation between design and production, as the client and their consultants design the build documents before the contractor is procured and involved to carry out production. The advantage with DBB contracts is the ability for the client to ensure quality, since they specify the design in detail and design the build documents (Chen et al. 2009). A disadvantage with DBB contracts may be decreased innovation abilities due to strict specifications.

2.2 Methods

For the first evaluation in 2016, three types of LCA-based climate requirements were evaluated: 'Information requirement' (requirement 1), 'contract with clauses' (requirement 2) and 'contract clauses and bonus' (requirement 3) (Table 1). In all cases, the STA climate calculation model was used for the GWP assessment. In requirement type 2 and 3, a baseline for the construction was calculated, from which the contractor or designer was required to achieve a certain reduction. Reduction levels were discussed with contractors and material suppliers prior to this work, in order to find a level that would be challenging but not unrealistic. There are two alternatives for quantifying the resource use when calculating the baseline: i) based on the contractor's design documents, in requirement 2, or ii) based on the prescriptive documents from STA, in requirement 3. At the time of the investigation, only requirements of type 1 or 2 had yet been implemented in real procurements.

The evaluation aimed to investigate the involved consultants, contractors and STA stakeholders' opinions, incentives, ideas and experiences for the different requirements. Qualitative interviews were performed with semi-structured questions in order to cover the perspectives of the respondents (Kvale 1996). The selection of respondents to represent these projects followed a method metaphorically called snowball sampling, which is advocated in several papers by researchers studying perceptions and experiences (e.g. Denzin & Lincoln 2011). In total, there were four consultants, six contractors and six STA stakeholders interviewed. The interviews were recorded and the transcripts were synthesised into themes, according to thematic analysis (Vaismoradi et al. 2013).

The interview study in 2016 was followed by an in-depth case study of the requirements used in a specific project, the 'Stockholm Bypass'. This construction of a new 21 km motorway in Stockholm was the first major infrastructure project in Sweden where climate requirements were included in the tender documentation for the procurement work (Larsson & Novakovic 2017). Work on climate requirements for the 'Stockholm Bypass' began in 2011-2012, before formal guidelines for climate requirements had been developed at the STA in

Table 1. Different climate requirements applied in the Swedish Transport Administration (STA) during climate requirement development phase, 2011-2016.

Requirement type	Incentive	Calculation of baseline	Required GHG reduction	Contract types where requirement was applied
1	Learning	Prescriptive documents towards tender	None; stipulated but voluntarily defined in a dialogue process	DB & ECI
2	Contract clauses	Contractor's design documents after tender	10%	DB & DBB
3	Contract clauses and bonus	Client's prescriptive documents	11-27% depending on project. In case of DBB, reduction is shared between designer and contractor.	DB & DBB

2016. Larsson & Novakovic (2017) and Miliutenko et al. (2019) evaluated the use of climate requirements in the procurement of the Stockholm Bypass with the help of four workshops with STA stakeholders and contractors during 2017 and 2018. Both DBB and DB types of contracts were represented.

In the evaluation 'Control station 2018' the aim was to get a deepened understanding of effects and measures implemented as a consequence of the climate requirements and to suggest a progressive development of the requirements. The evaluation considered particularly the effectiveness of the STA climate requirements in relation to the national goals, possible improvements of the bonus system and the level of GHG decrease to be required. The evaluation combined semistructured interviews with a focus group approach (Kvale 2007) that included i) workshops with stakeholders from different projects ii) workshops with different stakeholders within the same project and iii) individual interviews. In total 80 stakeholders and 16 different companies were involved. The projects were selected based on STA's information about implemented requirements. Stakeholders (including STA stakeholders, contractors, consultants and material suppliers) were selected by snowball sampling. The results were analyzed qualitatively, based on grounded theory (Strauss & Corbin 1996). A quantitative analysis was also performed in order to understand the level of consensus between and within stakeholder groups.

2.3 Results from the interviews

In the interview study 2016, the overall response to mitigation of GWP from construction work was positive. The cooperation between the client and the contractor was reported as fundamental for facilitating active measures for GWP mitigation. Knowledge and decision making with an LCA perspective was seen as important for designers and purchasers to find new areas for decreased impacts and innovative design.

There were a variety of opinions about how to regulate and apply the GWP requirements. Several respondents stressed the opinion that GWP mitigation requirements from clients such as the STA are needed if changes in climate mitigation are to take place. All the contractors and consultants thought that climate requirements were preferable, but a participation process and dialogue was emphasised by several respondents as necessary for the requirements to be successfully implemented. Some of the companies had already made efforts to assess and decrease their GWP, as a part of their long-term visions or business models. Several respondents anticipated that the STA climate calculation model and the associated requirements would further contribute to the contractors' future financial or competitive benefits.

Some STA stakeholders, however, questioned if LCA-based climate requirements would be the most cost-effective method to meet climate goals and suggested that the requirements could instead be set based on environmental certification (e.g. SEEQUAL), or by requirements on specific materials. Cost savings were reported to be indirectly found as a consequence of GWP mitigation measures in some cases, but also the opposite was reported where the GWP mitigation alternatives led to increased costs. It was also argued that major mitigation measures can be done already in the early design stages, where contractors have not yet been involved, and that such measures should not be neglected.

An important problem identified was the uncertainty in the baseline calculation. In requirement type 3, that uncertainty was expected to be larger than in requirement type 2, as the STA's baseline estimations, based on cost items, were less precise than the contractors' baseline estimation based on build documents. The contractors, however, preferred the baseline to be set by the STA as that would enable the contractor to include their own design in the GHG reductions. Several respondents noted the importance of having an ability to change the baseline as the construction proceeds, due to the possibility of new information about the prerequisites becoming available (e.g. more complicated geological conditions than expected). The GWP baseline was suggested to follow the established process for cost regulations.

Other problems that were mentioned were related to the climate calculation model or the organisational maturity. The inclusion and exclusion of various processes associated with GWP in the climate calculation model were questioned by some respondents. In some cases, a lack of clearly defined environmental responsibility, and knowledge, in the STA project organisation was identified as an organisational barrier to overcome.

It was also pointed out that the GHG target levels have to follow the contractor's ability to act. In contracts where the design is limited due to strict technical requirements, the reductions would be more difficult to fulfil. It was suggested that contracts with higher freedom to make measures (DBs and ECIs) should get higher reduction requirements. However, although the DBB contractor cannot influence the design, they still have the freedom to select material suppliers.

2.4 Results from the case study 'Stockholm Bypass'

The workshops with the STA stakeholders and contractors resulted in many proposals for improving the use of climate requirements in procurement in Sweden (Miliutenko et al. 2019). The workshops identified several proposals that could be implemented within the current work of the STA as well as the ongoing project 'Stockholm Bypass'. These involve mainly organizational aspects, such as clear information about the climate requirements during all kinds of meetings with the contractors, preparations for unplanned increased costs due to climate targets in a timely manner, and the dissemination of good examples.

Several proposals that can be implemented in a longer-term perspective throughout the work of STA were also identified. These included such organizational aspects such as a more structured way of working and clear differences in the work between design-bid-build (DBB) and design-build (DB) contracts. In addition, it was proposed that STA should develop financial incentives for innovative solutions and increase the opportunities for approved climate mitigation measures by helping the contractors to identify a list of the important measures to reduce greenhouse gas emissions.

Several suggestions in terms of methodological improvements were also identified. These included further development of both the LCA climate calculation model and further development of climate criteria in the procurement process.

Both the STA stakeholders and contractors agreed that it is important to build expertise in climate work internally within their organizations. It was concluded that more research in climate mitigation measures is needed together with practical demonstrations in pilot projects.

2.5 Results from the 'Control station' evaluation

Through the workshops and interviews, it became obvious that the STA climate requirements are interpreted as a forceful signal of the importance of the climate issue to the industry. Most respondents (71 %) were positive about the requirements that were perceived to contribute to creating long-term game rules for the industry. The requirements were seen as valuable for contributing to knowledge sharing and the development of systematic approaches to climate mitigation.

The mitigation measures were often reported to imply cost reductions (100 % of the responding contractors agreed, but only 33 % of the material suppliers). Thus, all emission savings that were seen in the evaluated projects cannot fully be attributed to the climate requirements and the effectiveness of the requirements could not be fully revealed. The choice of material with better climate performance, e.g. cement, were reported to imply a potential increased cost. However, example calculations revealed that although the material is expensive, the effect on the total project cost could still be expected to be minor.

A number of possibilities to further improve the impact and effect of the requirements were identified through the evaluation. It was revealed that the climate requirements have not yet been passed down to affect the material suppliers to the extent that was expected. A reason could be that the requirements are still relatively new, and the projects concerned are still in the early planning phases. The required GHG decrease has also been limited to in average 15% so far. In order to increase the possibilities of reaching the climate goals, GHG reductions of 50% were suggested to be required. More targeted requirements were also suggested for the material climate performance as a complement to the project requirements (such requirements had not yet been implemented during the time for the evaluation). Further, the bonus model was suggested to be reformed. The bonus was suggested to be based on the amount of emissions rather than contract size, and it was suggested to be applied for up to 100% reduction, instead of being limited to a certain reduction level. Support was also requested by the contractors regarding how the emission reductions are expected to be achieved in projects, e.g. as demonstration examples or guidelines.

The uncertainty in the base line calculation was also brought forward as an important barrier for the requirement to work effectively towards climate mitigation. As a solution, it was suggested that instead of comparing the initial baseline (based on the STA estimates) with the final declaration, the baseline should be modified towards the end of the project to better capture the project conditions. Another solution would be to focus on the performed climate mitigation measures rather than the difference between baseline and final declaration.

3 DISCUSSION

3.1 *Organisational and methodological maturity for LCA implementation in procurement*

Based on the results from all three evaluations, aspects for successful implementation of LCA-based climate requirements in procurement could be categorised into two overarching aspects, methodological and organisational. High maturity of LCA in organisations is suggested to be signified by the ability to use LCA in an appropriate way in relation to the specific decision context. Based on the evaluation results, higher levels of organisational LCA maturity is suggested to depend on competence, experience, the level of integration of LCA in decisions and participation/cooperation internally and externally and understanding of the acceptable uncertainty depending on the decision context (Table 2).

Apart from organisational maturity, the implementation of LCA-based decisions depends on the maturity of the LCA model used, in methodological terms. Several methodological parameters affect the result. Some important methodological parameters that were identified in this study are: the precision of prescriptive and construction documents, the setting of system boundaries: (e.g. inclusion and exclusion of components; life cycle steps); the extent of

Table 2. Organisational maturity for implementing LCA-based requirements in public procurement.

Maturity level	LCA knowledge in the organisation	LCA-based decisions	Participation and cooperation for climate mitigation	Consistent use of LCA methodology
Lower	Limited internal competence and experience of LCA	LCAs are made, but do not affect decisions	Low participation and cooperation in and between organisations	Diversity of methods
Higher	Broad acceptance of LCA, LCA experts within the organisation	Proactive LCA management, where LCA influences decisions	Participation processes are well developed, trust is established in and between organisations	Consistent and established use in a structured manner in the organisation
	Extensive LCA competence and experience	LCA-based decisions in all projects and project phases, the method is adapted to each decision context		

compliance with standards; the number of environmental categories; quality and amount of LCA data; quality, transparency and precision of LCA assessment tools; and verification processes. As suggested by Butt et al. (2015), in order to make the relevant methodological choices for an LCA, the stage within the planning process in which the LCA results will be used should be considered. Thus, high methodological maturity implies that the method is well adapted to the context in which it will be used.

It is suggested that implementation of LCA-based requirements in public procurements require both high methodological and organisational LCA maturity, as a combined concept. Using the maturity concept may provide insights for managing and fostering LCA implementation in organisations, and in public procurement. Since the STA started with the implementation of LCA-based climate requirements, both STA and its contractors have taken steps towards increased maturity.

3.2 *Managing uncertainties in the base line for climate requirements*

The outcome from an LCA calculation can never be better than the source data used. Fundamental in LCA is therefore to have control over the so-called reference flow, or in other words, all resources used during the life cycle for the assessed construction works. Differences in life cycle inventory datasets may arise due to either variability (i.e. regional or technological differences) or uncertainties (i.e. data unavailability, data inaccuracy or data ambiguity) (Kendall et al. 2009).

Based on the evaluations described in this paper, one important problem for LCA-based climate requirements on project level is the uncertainty derived from the level of details included in the LCA. The most ambitious LCAs include many components, and will therefore also show higher environmental impact compared to a less ambitious LCAs. When a project is finished, there is more information available than before it was initiated, and, consequently, a higher level of details will therefore be available for the declaration than for the baseline calculation. As suggested in the evaluations, this problem could be handled by recalculation of the initial baseline, or by a larger focus on the performed mitigation measures.

In addition, there is also the possibility to decrease this type of uncertainty by utilising the progress that has taken place lately regarding the availability of digital data through BIM applications. Digitalization could enable full data cover and a good quality of the mapping of the resources used as input to an LCA. We regard the data gathered for the cost calculation as currently the most sufficient input for such LCA calculation for a new building before build. These data include typically between 5000 to 15 000 items that all together define the bill of resources (BoR) for the construction phase (A1-A5). Such cost calculations are made in specialized software's or other BIM applications. Digitalization is also in an EC context regarded as an essential part of the over-all roadmap for increased sustainability in the construction sector. In 2012, the European Commission published a Communication Strategy for the sustainable competitiveness of the construction sector and its enterprises, as a part of the Europe 2020 initiative.

3.3 *Current practice and development*

The fact that there is no yet any finished project with climate requirements limits the possibility to draw quantitative conclusions regarding the effectiveness of the requirements. For projects procured between Jan 2018 and March 2019, a 22 percent reduction in GHG is expected, but the actual outcome cannot be verified before they are finished. However, through the different evaluations performed, an increased understanding was obtained of how the LCA-based climate requirements are experienced by the contractors, consultants, material suppliers and different type of STA stakeholders. As a result, several improvement measures were identified and a progressive development of the require GHG reduction and bonus model was suggested. Some of the suggested measures have already been implemented. A 50% GHG

97

reduction is now required for projects that are planned to be finished 2030 or later, and bonus is possible for up to 100% reduction. Further, requirements on specific materials are implemented for the smaller STA projects and maintenance contracts that were not covered by the project climate requirements.

To enable the implementation of LCA-based climate requirements, practices at the STA had to be developed regarding reception and approval of the product specific EPD data used, as well as quality criteria that describe the representativeness of the LCA used in the calculation of the environmental performance (Erlandsson 2018). There is already an increasing number of EPDs delivered by material suppliers to the STA and a support function for internal storage and interpretation of EPD information at the STA is currently being built up. In parallel, the use of commonly agreed and openly accessible generic LCA data is also being discussed on a national level for different types of construction works, not limited to transport infrastructure.

Further improvements are under development. The different possibilities for managing uncertainties in the base line are being tested in pilot projects. A guideline for supporting contractors in finding effective mitigation measures is currently under preparation. Model development is also continuously being carried out. The STA climate calculation model is for example under revision during 2019-2020 in order to better capture material transports and correspond to the construction phase modules (A1-A5) prescribed in the EPD-standard. The possibilities for effective utilization of digitalized data are also being tested.

4 CONCLUSIONS

Since 2016, LCA-based climate requirements have been implemented in public procurements of road and rail infrastructure at the STA. Project climate requirements are based on the expected GHG emissions in a life cycle perspective (a baseline), calculated in the STA climate calculation model with background data. In the contract clause, a certain percentage decrease in GHG emissions is required from the project, compared to the baseline. In this paper, the results from three different evaluations of the STA climate requirements have been summarized.

The evaluations have resulted in an increased understanding of the procurement related mechanisms that favour effective climate mitigation measures. The development and implementation of LCA-based climate requirements would not have been possible without both methodological and organizational maturity. Management of uncertainty in LCA calculations was highlighted as an area for future development. In general, there was a positive response to the STA climate requirements, and they were considered as important in order to meet the goal of reduced GWP from the transport infrastructure.

REFERENCES

Blechingberg, M., Jalmby, M., & van Noord, M. 2013. The reduction of carbon dioxide emissions in construction projects: An international study. Esam AB. Stockholm.

Butt, A. A. Toller, S. & Birgisson, B. 2015. Life cycle assessment for the green procurement of roads: a way forward. *Journal of Cleaner Production* 90: 163–170.

CEN, 2013. Sustainability of Construction Works – Environmental Product Declarations – Core Rules for the Product Category of Construction Products. European Standard EN 15804:2012 + A1.

Chen, Y., Zhu, X. & Zhang, N. 2009. Comparison of project objectives and critical factors between DBB and DB in China. In proc. *IEEE International Conference on Industrial Engineering and Engineering Management, Hong Kong 2009*, pp. 583–587.

Chester, M. V. & Horvath, A. 2009. Environmental assessment of passenger transportation should in-clude infrastructure and supply chains. *Environmental Research Letters* 4, 024008.

COM 2011. WHITE PAPER Roadmap to a Single European Transport Area – Towards a competitive and resource efficient transport system. European Commission. [Online]. Availabe at: https://eur-lex.europa.eu/legal-content/EN/TXT/PDF/?uri=CELEX:52011DC0144&from=EN

Denzin, N.K. & Lincoln, Y.S. (eds) 2011. The Sage handbook of qualitative research. Thousand Oaks: Sage Publications.

EC 2016. A European Strategy for Low-Emission Mobility. *Communication from the commission to the European Parliament, the Council, the European Economic and Social Committee and the Committee of the Regions.* European Commission. [Online]. Availabe at: https://ec.europa.eu/clima/policies/transport_en#tab-0-0

Eriksson, P.E. & Westerberg, M. 2011. Effects of cooperative procurement procedures on construction project performance: A conceptual framework. *International Journal of Project Management* 29: 197–208.

Erlandsson M. 2018. Q metadata for EPD. Quality-assured environmental Product declarations (EPD) for healthy competition and increased transparency. *Report No C363, December 2018.* Stockholm: Smart Built Environment and IVL Swedish Environmental Research Institute.

Erlandsson, M. 2019. Modell för bedömning av svenska byggnaders klimatpåverkan – inklusive konsekvenser av befintliga åtgärder och styrmedel (in Swedish). Comissioned by Naturvårdsverket, Boverket & IVL. *Report C 433, February 2019.* Stockholm: IVL Swedish Environmental Research Institute.

Government Offices of Sweden 2018. The Swedish climate policy framework. Government Offices of Sweden, Ministry of the Environment and Energy. [Online]. Availabe at: https://www.government.se/information-material/2018/03/the-swedish-climate-policy-framework/

Kadefors, A., Uppenberg, S., Olsson, J. A., Balian, D., & Lingegård, S. 2019. Procurement Requirements for Carbon Reduction in Infrastructure Construction Projects - An International Case Study. Stockholm: KTH Royal Institute of Technology.

Kendall, A., Harvey, J. & Lee, I.-S. 2009. A critical review of Life Cycle Assessment practice for infrastructure materials, In *Proceedings of US-Japan Workshop on Life Cycle Assessment of Sustainable Infrastructure Materials. Sapporo, Japan.*

Kvale, S. 1996. InterViews: An Introduction to Qualitative Research Interviewing. Thousand Oaks: Sage Publications.

Larsson, M. & Novakovic, H. 2017. Fallstudie Hantering av Klimatkrav Förbifart Stockholm, Del 1 – Framgångsfaktorer och utmaningar med klimatkravställandet. *Report 2017/6003, Dec 2019.* Stockholm: Trafikverket.

Liljenström, C., Toller, S., Åkerman, J. & Björklund, A. 2019. Annual climate impact and primary energy use of Swedish transport infrastructure. *EJTIR* 19(2): 77–116.

Miliutenko, S., Gunnarsson K., Holmström T. & Ryding, S-O. 2019. Fallstudie Hantering av Klimatkrav Förbifart Stockholm Del 2 – Förslag på förbättringar vid användning av klimatkrav i upphandling (in Swedish). *Report 2017/6003,* June 2019. Stockholm: Trafikverket.

Rahman, H., Chin, H. C. & Haque, M. 2014. Environmental sustainability of urban road transport: an integrated analysis for life cycle emission impact. *International Journal of Environment and Sustainable Development* 13(2): 126–141.

Strauss, A. & Corbin, J. (eds) 1996. Basics of Qualitative Research. London: Sage Publications.

Swedish Environmental Protection Agency 2019. Territoriella utsläpp och upptag av växthusgaser (in Swedish). [Online]. Availabe at: http://www.naturvardsverket.se/Sa-mar-miljon/Statistik-A-O/Vaxthusgaser-territoriella-utslapp-och-upptag/

Toller, S. & Larsson, M. 2017. Implementation of life cycle thinking in planning and procurement at the Swedish Transport Administration. *In Al-Qadi, I.L., Ozer, H. & Harvey, J. (eds) Proceedings of Pavement Life-Cycle Assessment Symposium in Champaign, Illinois, USA, April 2017,* pp. 281–287. CRC Press.

Vaismoradi, M., Turunen, H. & Bondas, T. 2013. Content analysis and thematic analysis: Implications for conducting a qualitative descriptive study. *Nursing & Health Sciences* 15: 398–405.

Technical and organizational challenges to developing product category rules for asphalt pavement construction

Chaitanya G. Bhat
Michigan Technological University, Houghton, USA

Joseph Shacat
Director of Sustainable Pavements, National Asphalt Pavement Association, Lanham, USA

Amlan Mukherjee
Michigan Technological University, Houghton, USA

ABSTRACT: The objective of this paper is to assess the technical and organizational challenges to extending the current Product Category Rule (PCR) for North American asphalt mixtures to include the asphalt pavement construction stage. The extant literature on pavement construction Life Cycle Assessment (LCA) tends to focus on assessing the environmental impacts of alternative designs emphasizing the life cycle accounting. There is limited consideration of the challenges a contractor encounters when reporting LCA outcomes using International Organization of Standardization (ISO) compliant documents. The organizational context defined by different project delivery scenarios, as well as how ISO defines "construction elements", calls for a re-examination of the system boundary and other technical questions regarding cut-off and exclusions, thus determining the extent of data collection, and the relevant data quality requirements for a construction stage LCA. For instance, as per ISO Standard 21930, the asphalt pavement is considered to be a construction element, separate and independent of other construction elements and operations, and hence becomes a subset of the traditional construction stage system boundary. The paper also examines the effect of project delivery systems on construction project data reporting, the quality of available data, and its impact on the development of estimates, and their use with respect to actual reported project emissions. Scenarios in which contractors can use the PCR to improve their construction processes in addition to lowering environmental impacts are also identified. The outcomes of the analysis in this paper can inform the development of construction stage PCR for pavements while providing contractors practical guidance on reducing and reporting their environmental impacts.

1 INTRODUCTION

The National Asphalt Pavement Association (NAPA) has a cradle-to-gate Environmental Product Declaration (EPD) program called "Eco-Emerald" (Asphaltpavement.org, 2019). An International Organization of Standardization (ISO) Standard 14040 (ISO, 2006) compliant Life-Cycle Assessment (LCA) study (Mukherjee, 2016) supports the Product Category Rule (PCR) for this EPD program. The streamlined LCA studies do not reflect on the environmental burden shifted between different life-cycle and ISO standard 14025 prescribes that comparison of EPDs should only be carried out after accounting for all the life-cycle stages. With this motivation, this paper identifies the challenges for NAPA to extend the existing EPD program for construction stage as a first step towards including the complete life-cycle.

The objective of this paper is to discuss the technical and organizational challenges to extending the current PCR for North American asphalt mixtures (National Asphalt Pavement Association, 2017) to include the asphalt pavement construction stage, with NAPA continuing as the program operator. Specifically, the study addresses the following challenges:

1. Define the relevant product category for asphalt pavement construction,
2. Establish the system boundary and potential cut-off and exclusions for asphalt pavement construction,
3. Determine the extent of data collection, and the relevant data quality requirements for a construction stage LCA.
4. Identify contractors' challenges for conducting and reporting LCA outcomes within the context of different contractual scenarios.

This paper has been written using ISO Standard 21930 (ISO, 2017) as the primary compliance Standard, while also informed by ISO Standard 14025 (ISO, 2006). In addition, the Federal Highway Administration (FHWA)'s Pavement Life Cycle Assessment Framework (Harvey et al., 2016) was used as a guidance document to ensure harmony with existing PCRs and industry benchmarks.

The paper starts with an overview of the previous LCA work on pavement construction stage. Next, it establishes the technical challenges for the definition of goal and scope, data collection and data quality for the pavement construction stage. Guidance is provided to tackle these potential challenges with additional insights from the members of the PCR committee that developed the existing PCR (National Asphalt Pavement Association, 2017). Within the context of organizational challenges, the paper first establishes the role played by stakeholders in the decision-making process considering their relationships within the asphalt industry. Next, it identifies scenarios in which contractors can use the PCR to improve their construction processes, in addition to lowering environmental impacts. The paper limits its discussion on the analysis of use cases to the organizational context of a design-bid-build delivery system.

2 BACKGROUND

As per Muench, (2010) which is an extensive review paper on the sustainability impacts from roadway construction, previous literature obtained less than 5% energy and CO_2 emissions from construction activities at jobsite. In addition, Muench, (2010) summarized that the total energy consumption and CO_2 emissions varied based on pavement section, maintenance activities and LCA scope. Specifically, energy varied from 3 to 7 TJ per lane mile while the CO2 emissions varied from 200 to 600 Mg per lane mile. The FHWA's Towards Sustainable Pavement Systems: A Reference Document (Van Dam et al., 2015) and the Pavement Life Cycle Assessment (LCA) Framework (Harvey et al., 2016) are useful resources for state, local and federal agencies to learn about and adopt practices rooted in life-cycle thinking, and LCA to build sustainable pavements. Towards Sustainable Pavement Systems: A Reference Document details practices intended to reduce emissions, and energy consumption from material, and equipment during construction of pavements. This document included excavation, earth movement, material processing, and placement, and compaction/consolidation operations within the construction stage when considering an entire pavement structure as their product system.

As per ISO Standard 14025 (ISO, 2006), there needs to be an underlying process-based LCA supporting a PCR. The construction stage has been included in various pavement LCA studies (Zapata & Gambatese, 2005) that compared the LCA outcomes for both concrete and asphalt pavements using a system boundary from cradle-to-laid (laid: pavement placement). Aurangzeb et al., (2014) conducted a hybrid LCA for asphalt mixtures by including material extraction, construction, maintenance, and rehabilitation life-cycle stages. However, a review of previous LCA efforts that quantify environmental impacts during the construction stage (Muench, 2010) shows that there is no significant consensus regarding a system boundary definition for the construction stage, i.e. what different operations must be included while conducting an LCA study for the construction stage.

There are several studies focusing primarily on the pavement construction stage that are informed by the FHWA Pavement LCA Technical Working Group efforts. Mukherjee et al., (2013) considered a project-based framework that included the products and the processes that were associated with a pavement construction project to define the system boundary of a construction project. They illustrated the approach by collecting material, equipment and transportation data from 14 construction, maintenance, and rehabilitation projects across the state of Michigan and used them to develop benchmark estimates of carbon emissions for typical pavement construction and rehabilitation projects. Along the same lines, recent work at the Illinois Tollway has furthered this paradigm, conducting an in-depth, and detailed LCA of Illinois Tollways' design and construction processes and tying them to contractual pay items (Ozer & Al-Qadi, 2017). While this approach has the implementation advantage of tying LCA components directly to project pay items, the underlying perspective is that of an agency, and the focus is on a transaction between the agency and a general contractor.

There are also currently a set of tools that can be used to estimate environmental impacts due to construction activities. Horvath, (2004) developed PaLATE, an MS Excel tool that uses an LCA based approach to quantify environmental impacts from road construction and maintenance. This work has been further modified to PaLATE II correcting the initial inconsistencies of data present in PaLATE (rmrc.wisc.edu, 2019). Additionally, methods have been developed to determine the impact of engine idling on CO2 emissions (Lewis, Leming and Rasdorf, 2012). Tang et al., (2013) Also discussed the impact of schedule delays on construction greenhouse gas emissions. The MOVES (Motor Vehicle Emission Simulator) developed to calculate traffic delay emissions (US EPA, 2019), and NONROAD (both tools developed by USEPA) to estimate emission factors from non-road vehicles (US EPA, 2019) have also been used across multiple studies to estimate the impacts of traffic delays during construction (Cass and Mukherjee, 2011). FHWA's recent attempt, the Infrastructure Carbon Estimator (ICE) tool similarly attempts to tie project-level construction stage impacts to design decisions made at the network level.

When it comes to developing a process-based LCA to support a compliant construction stage EPD, much of this work does not meet the requirements because of the following reasons:

1. System boundary definitions organized using the project definition within a contractual scope and tied to pay-items are useful to agencies, but cannot be used to inform the system boundary of a specific pavement material related construction element during the construction stage as required by an EPD for asphalt pavement construction.
2. Most of the available tools focus on estimating "greenhouse gas emissions" often using "emission factors" gathered from the literature, rather than using process Life-Cycle Impact Assessment (LCIA) methods to estimate mid-point indicators such as Global Warming Potential (GWP).
3. The most critical gap is the nature of construction project data reporting, the quality of available data, and how this informs the development of estimates, and their use with respect to actual reported project emissions.

The rest of this paper addresses these gaps, providing recommendations for a potential PCR committee in addressing these challenges if the current "cradle-to-gate" EPD programs are to be extended to include the construction stage.

3 TECHNICAL CHALLENGES

3.1 *Goal and scope definition*

The processes of defining the product category, establishing the system boundary and the functional/declared unit, and determining the extent of data collection are all interlinked for an LCA study. Definition of the product category needs to be compliant with the Standards and documents stated in the previous section. In addition, the definition of the product category should fall within the scope of the program operator. The multi-level compliance requirements for a construction stage PCR are illustrated in X.

The asphalt pavement construction PCR is required to use the ISO Standard 21930 (ISO, 2017) as the core PCR. ISO Standard 21930 (ISO, 2006) complements the previously existing ISO Standard 14025 (ISO, 2006). ISO Standard 21930 (ISO, 2017) provides core rules for EPDs of construction products and processes. In addition, the goal and scope of the PCR should ideally use the Pavement Life Cycle Assessment (LCA) Framework (Harvey et al., 2016) as a guideline. Finally, the PCR should be compliant and in harmony with the existing asphalt mixture PCR (National Asphalt Pavement Association, 2017). The multi-level compliance requirements for a construction stage PCR are illustrated in Figure 3 (present in the Supporting Information).

The following paragraphs set the context for identifying challenges by defining some of the relevant terms followed by a discussion on a typical pavement section, and on the system boundary for construction stage requirements as per Harvey et al., (2016).

3.1.1 *Definitions*

Product Category Rule (PCR): a set of specific rules, requirements, and guidelines for developing EPDs for one or more product categories (ISO, 2006).

Product category: a group of construction products, construction elements or integrated technical systems that can fulfill equivalent functions (ISO, 2006).

Construction works: everything that is constructed or results from construction operations (ISO, 2006).

Construction product: an item manufactured or processed for incorporation in construction works (ISO, 2006).

Construction element: refers to a part of a construction containing a defined combination of construction products (ISO, 2006).

Sub-category PCR: a set of specific rules, requirements, and guidelines, which provide additional, consistent requirements to the core PCR for developing EPDs for sub-categories of the overall product category of construction products (ISO, 2006).

Scenario: a collection of assumptions and information relevant to possible future events (ISO, 2006).

Owner: An owner or agency (can be either public or private) is the party instigating the construction of a pavement. Public owners include agencies of the federal government, the state department of transportation, county and municipal entities (Clough et al., 2015).

Designer: A pavement designer is the party, organization or firm that designs the project. Public agencies may house a design professional within their firm or employ a private individual or a firm to design a pavement project (Clough et al., 2015).

Contractor: An individual entity or a firm that is in contact with the owner for construction of the project, either in its entirety or for some specialized portion thereof. There can exist a single prime contractor and multiple sub-contractors who contract with the prime contractor (Clough et al., 2015).

Supplier: A person or a firm producing and supplying a material (Clough et al., 2015).

For asphalt pavements, the term construction works refers to a complete pavement consisting of surface, base and sub-base courses as indicated by Figure 1, construction product refers to the amount of Hot-Mix Asphalt (HMA) delivered to a construction site, and construction element refers to layers of asphalt surface course with tack coat between these layers. A typical pavement section with an asphalt surface course contains a base course, an optional sub-base course, and a subgrade as presented in Figure 1. There will be a tack coat placed between the different asphalt layers and a prime coat between the base and asphalt layers. The base may be made up of course aggregates while the sub-base may consist of sand or gravel.

3.2 *Definition of product category and system boundary*

A questionnaire was used to consult with an advisory committee (consisted of members from previous PCR committee for cradle to gate PCR), on choosing the product category,

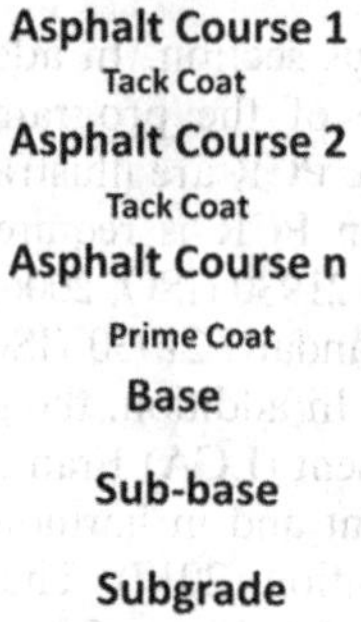

Asphalt Course 1
Tack Coat
Asphalt Course 2
Tack Coat
Asphalt Course n
Prime Coat
Base
Sub-base
Subgrade

Figure 1. Conventional pavement section with HMA surface course.

and the system boundary for the potential construction stage PCR. The committee's suggestion was to consider layers of HMA with tack coat placed between the layers as the product category and to include placement and milling operations within the system boundary for the construction stage. This would be a construction product or a construction element as per the definition in ISO Standard 21930 (ISO, 2017) i.e. a construction product if prime coat and tack coat are not present within the system boundary otherwise a construction element. Hence, the potential PCR would address United Nations Standards Products and Services Code (UNSPSC) 721410: Highway and Road Construction Services. Specifically, the different types of Highway and Road Construction services relevant to asphalt pavement will be:

1. Placing a new Hot-Mix Asphalt pavement falls under UNSPSC 72141103: Highway and road paving service (bidsync.com, 2019) and,
2. Placing overlay over the existing concrete pavement falls under UNSPSC 72141104: Highway and resurfacing service (bidsync.com, 2019)

The system boundary will be different for each of these services as the underlying processes involved are different.

3.3 *Definition of declared/functional unit*

Mukherjee and Cass, (2012) developed a functional unit definition for the construction stage and stated that the selection of a functional unit depends on the decisions being considered from both the project and network-level perspectives. The point of departures from their study on the definition of the functional unit are listed below:

1. Lane-mile was used as the functional unit in previous LCA studies. However, lane-mile is not an appropriate functional unit, as the emissions do not scale in any uniform fashion as the number of lane-miles increases. This is because the length of the shoulder does not increase at the same rate as the number of lanes increase per mile.
2. There is an impact of statistical smoothening as the number of lane-miles increases, resulting in larger projects having relatively lower impacts.

Also, the definition of the declared or functional unit for the construction stage depends on the definition of scenarios. In the context of the United States, it is not possible to holistically define a single functional unit across all states as there are different performance measures/Standards in each state. Hence, it is suggested that a declared unit be used for the construction stage. All the activities mentioned in the system boundary can be reduced to and expressed per short ton of HMA. For some materials within the construction element, such as tack coat or prime coat, the amount of material and the level of effort needed to place the material is typically specified as a function of area and will vary as a function of pavement thickness or short tons of HMA. This potential conflict can be resolved through the development and definition of scenarios. Hence, the recommendation is to use a short ton of HMA as the declared unit when conducting the LCA supporting a construction stage PCR.

3.4 *Cut-offs and allocation*

Construction stage includes operations such as earthwork, drainage, sub-base and base placements apart from placing the asphalt pavement. Additionally, the construction of an asphalt pavement utilizes a paver, roller, and other construction equipment. The pre-placement operations, as well as the manufacture and maintenance of this equipment, fall beyond the scope of the program operator's sphere of influence, and hence, are not included within the system boundary. However, the environmental impacts from these excluded activities need to be allocated to respective sub-contractors as part of the production process.

Potential challenges may arise when allocating impacts for the construction of a thin asphalt overlay over the existing concrete pavement. This construction activity would include removal of joints, edge and joint repair, and cleaning of cracks in the concrete pavement prior to the placement of an asphalt overlay. Preferably, the allocation of environmental impacts in such scenarios can be done based on pay-items. This paper does not engage in actual data collection. Hence, detailed explanations regarding cut-offs and allocation of different processes within the construction stage system boundary need to be made once their environmental significance is quantified.

3.5 *Data collection*

Next step in establishing the potential PCR will be to determine the extent of data collection for the LCA. The following classification for the types of data to be collected can be used to inform the granularity and the level of specificity they need to be collected at:

1. Upstream Data: Data sets for flows, processes or product systems that are characterized entirely by generic public inventories. The person conducting a pavement LCA has no influence on it and does not collect it. For example, Crude Oil, at refinery is an upstream flow. The quality of the data used will be a function of the generic database selected. Transparent, publicly available datasets should be preferred.
2. Background Data: Data sets for flows that are characterized by generic public inventories, though the person conducting a pavement LCA has to provide specific parameters based on national, and regional averages to customize the data to reflect regional/context-specific needs. There is no way to control these average parameters through pavement design, or construction, however, they cannot be generalized to all contexts either. For example, percentages defining the electricity mix in a region defining Electricity at Grid is a background flow.
3. Foreground Data: Data sets that are specifically characterized input parameters to a pavement LCA. The person conducting the LCA has direct control over this data through modification of design and construction strategies. For example, the percentage of recycled content in the pavement, or the quantity of diesel used by equipment during construction, or the distance traveled through transportation.

A data collection instrument has been prepared in a Microsoft Excel spreadsheet to collect the parametric data for material and energy used for both the new HMA pavement construction, and for an overlay construction. Drop-down lists are provided to capture the list of equipment and respective units for materials and energy consumed. Also, for construction and transportation processes, data is differentiated into foreground, background and upstream data types and defined from the perspective of three stakeholders: agency, contractor and supplier. The level of granularity for data collection will be clear while mapping background and upstream processes during the life-cycle inventory stage of the LCA. Level of granularity refers to the scope of data collection (product-specific, region-specific or industry average).

The data quality of the life-cycle inventory can be ensured by using the data quality guidance as defined in the updated pedigree matrix (Edelen and Ingwersen, 2016). The quality indicators range from a score of 1 to 5 with 1 indicating the highest quality data and 5 representing the data of the lowest quality. Specifically, these data quality indicators identify the flow representativeness with respect to the temporal, geographical, and, technological correlations along with examining data collection methods, the extent of process review and completeness.

4 ORGANIZATIONAL CHALLENGES

A significant gap in the extant literature on LCA is a lack of consideration of the challenges a contractor encounters when reporting LCA outcomes using ISO compliant documents. LCA outcomes do not reflect the pragmatic challenges encountered by a pavement contractor during construction. This section discusses the organizational challenges from the context of different project delivery scenarios and provides some insights on overcoming the same.

4.1 Definition of scenario

ISO 21930 Standard (ISO, 2006) mandates defining scenarios for all the other life-cycle stages apart from material extraction and production stage. As per the definition of the scenario mentioned in the earlier section, there is a need to account for future actions as well. Hence, scenario definitions can be viewed from two different perspectives for the pavement construction stage, as follows:

1. LCA practitioner's perspective: motivated by LCA parameters, and the choice of declared unit
2. Business/stakeholder perspective: motivated by who produces the EPDs and how it is required

In the first scenario, the declared unit for the supporting LCA should ideally be associated with a performance metric such as distress or design life with the defined product category. However, in the context of the United States, a scenario cannot be linked to a performance metric or pavement design life due to the many different Standards and regulations across different states. Besides, while construction does influence pavement performance, it is one of many other context-specific factors that can confound the association. Hence, scenarios are best defined using a declared unit of one short ton of HMA. Single, or multiple scenarios can be defined by associating the declared unit to different construction activities.

For example, a scenario could be defined as "short tons of HMA laid per hour in a paving operation." This scenario will include the impacts of the materials placed as included in the construction element, and the impacts of the equipment required including the trucks, spreader, and roller. The cumulative impacts can be expressed as a function of the total tonnage of HMA used. An advantage of this scenario is that it can be seamlessly integrated with the current PCR for asphalt mixtures. In addition, a quality metric, such as the required density of pavement after compaction, can be associated with it to communicate performance at construction without associating it with pavement performance. This scenario reflects the contractor responsibilities at construction and can be aligned with existing pay-items.

The second scenario considers stakeholders in the following cases:

1. An asphalt mixture producer who is also the contractor responsible for performing the asphalt pavement construction
2. A pavement contractor who is responsible for performing the asphalt pavement construction, but procures the asphalt material from a separate producer

These cases can be considered under different combinations of contracting mechanisms and project delivery methods. Typical DOT projects tend to be delivered as design-bid-build (DBB) projects using contracts that are either lump sum or unit price. Other project delivery methods that are more prevalent in large complex projects could be versions of design-build (DB) delivery, and the construction manager/general contractor (CM/GC) delivery. Each of these methods could have a guaranteed maximum price, and/or incentive/disincentive-based contracts. The choice of project delivery and type of contract used will decide the stakeholders involved in requiring and/or producing the EPD for the construction phase. A listing of these possible contractual scenarios is presented in Figure 2. Figure 2 suggests that the lump-sum contract is most favorable to accommodate the inclusion of EPDs across different scenarios.

The following describes the sub-scenarios for various actors along with the scope of EPD they need to produce.

1. Owner: For public works, the owner will typically be a state or local Department of Transportation (DOT). They wouldn't produce EPDs however, will require other actors to submit EPDs based on the scope of the project. Depending on the state, the requirement could be due to legislative mandates or (e.g., Buy Clean California Act, (2017) in California) or, for private owners be motivated by efforts at getting certification under a green rating system such as LEED v4 (new.usgbc.org, 2019). More on the owner's dilemma is described in the paragraphs following this section.

2. Material producer: This actor will typically be an asphalt mix producer and/or asphalt plant owner. They will be required to produce a cradle-to-gate EPD irrespective of the project delivery method. This can be done using the current NAPA "Emerald Eco-Label" program (1) if the scope is limited to cradle to gate. If the scope is expanded to include the construction stage, the use of a short ton of HMA as a unit can seamlessly build upon a cradle-to-gate EPD.

3. Pavement contractor: This actor is a contractor or sub-contractor who will procure the material, and rent/own equipment and is responsible for performing the asphalt paving. They will be required to produce a gate-to-gate EPD with options when the owner employs DBB or CM/GC project delivery methods. In such a scenario, the use of a short

Project Delivery Methods \ Actors	Owner	Material Producer	Pavement Contractor	Pavement Contractor + Material Producer	Pavement Designer + Pavement Contractor	Material Producer + Pavement Designer + Pavement Contractor	Construction Manager
Design-Bid-Build	X	**Cradle to Gate**	**Gate to Laid** *(Lump-Sum)*	**Cradle to Laid** *(Lump-Sum)*	X	X	X
CM/GC	X	**Cradle to Gate**	**Gate to Laid**	**Cradle to Laid**	X	X	X *(Lump-Sum)*
Design-Build	X	**Cradle to Gate**	X	X	**Gate to Laid** *(Lump-Sum)*	**Cradle to Laid** *(Lump-Sum)*	X

X= Not applicable
Bold= EPDs to be submitted by respective actors
Italics= Corresponding Contracting Methods

Figure 2. Proposed definition of contractual scenarios.

ton of HMA as a declared unit that can seamlessly build upon a cradle-to-gate EPD, as described in the last case is ideal.

4. Pavement contractor plus material producer: This actor is typically a contractor that owns an asphalt plant. They would be required to produce a cradle to gate EPD with options when the owner employs DBB, or CM/GC project delivery methods. They are subject to the same procurement requirements as a contractor, with the advantage that they also have ownership over the manufacturing of materials along with the raw material procurement.

5. Pavement designer plus pavement contractor: This actor is a contractor that houses a designer within the firm or on a contractual basis. They would be required to produce a cradle to gate EPD with options when the owner engages a DB project delivery method. They are subject to the same procurement requirements as a contractor, with the advantage that they also have ownership over the pavement design component. However, they do not have control over the manufacturing of materials along with raw material procurement.

6. Material producer plus pavement designer plus pavement contractor: This will typically refer to a contractor that owns an asphalt plant and houses a pavement designer within the firm as well. They would be required to produce a cradle-to-gate EPD with options when the owner employs a DB project delivery method. They are subject to the same procurement requirements as a contractor, with the advantage that they also have ownership over the pavement design, the manufacturing of materials, and the raw materials procurement components.

7. Construction manager: These actors come into the picture when an owner subordinates the construction activities to a construction management firm. They wouldn't be required to produce any EPDs themselves but would oversee and collect EPDs from the suppliers, contractors, and subcontractors as appropriate.

The contractual scenarios mentioned in Figure 3 will emerge only when environmental impacts become a part of procurement decisions through Best-Value Procurement (BVP) methods as described in (Scott III et al., 2019), or when legislative mandates such as the Buy Clean Act (2017) are implemented. The Buy Clean Act requires EPDs to be provided for informational purposes only for successful bids at the time of installation, for a list of eligible materials. Currently, this list is short and does not include asphalt mixtures. Also, such requirements, though part of the procurement process, are not being used to inform the bidding process, or to support design decision-making. However, in the near future, it is likely that once California has a record of EPD information, the list of eligible materials may expand to asphalt mixtures, and the EPD information will become part of the decision-making process. This inclusion can either be from a design perspective or from a procurement perspective.

A significant confounding factor within this regulatory environment is: at what point in the procurement process would a cradle-to-gate EPD with options be required. For instance, the cradle-to-gate EPDs can be required at the point of procurement when the life-cycle stages of the product from A1 to A3 as mentioned in ISO Standard 14025 (ISO, 2006) have already been completed and impacts recorded where relevant as foreground data. However, the cradle-to-gate EPD with options can be required either at the point of project procurement or after the construction process is over.

In the first case, similar to project cost, the contractor can provide only an estimate of what the construction impacts are likely to be. Depending on the contractual relationship, owner's requirements, and PCR rules, it's possible that the construction stage impacts could be determined by the manufacturer, regardless of who the actual contractor is and what means and methods they employ.

In the second case, the contractor can provide the actual or recorded impacts during the construction process using observed foreground data such as fuel usage. This presents a dilemma for the owner: either use an estimate for discriminating between contractors at procurement or use an actual impact for informational purposes. Alternatively, both the estimate

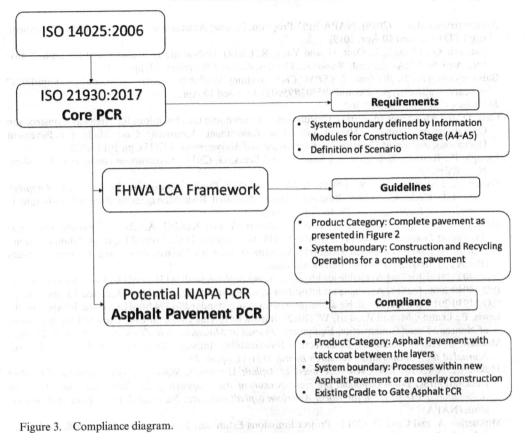

Figure 3. Compliance diagram.

and the actual impacts could be used to create an incentive where the contractor strives to improve their actual impacts compared to their declared estimates (similar to cost reduction incentives). This will require new contractual language and discernment on behalf of the owner, on how such information can be used for improving project outcomes.

Private owners can require EPDs to fulfill LEED v4 (new.usgbc.org, 2019) requirements. In such cases, the owner requiring the EPD will likely prefer producers who can show impacts less than the industry average to achieve maximum LEED points. However, LEED v4 does not require EPDs to include the construction stage.

5 CONCLUSION

This paper provides a blueprint for the development of construction stage PCRs and the corresponding challenges in extending the current asphalt mixture PCR (4) to cover the construction stage. The primary challenge lies in establishing a set of rules that can cover all the aspects of the construction element, for all the different stakeholder engagement scenarios as defined by the project delivery system and contracting mechanism in use. Developing an extension to the existing PCR will require a significant investment on behalf of the program operator - NAPA - in a regulatory regime that is currently focusing primarily on requiring cradle-to-gate EPDs. In addition, as discussed, there is ambiguity regarding what process should be best used to integrate cradle-to-gate EPDs with options into the procurement process.

REFERENCES

Asphaltpavement.org. (2019). NAPA EPD Program. [online] Available at: http://www.asphaltpavement. org/EPD [Accessed 10 Apr. 2019].

Aurangzeb, Q., Al-Qadi, I., Ozer, H. and Yang, R. (2014). Hybrid life cycle assessment for asphalt mixtures with high RAP content. *Resources, Conservation and Recycling*, 83, pp.77–86.

Bidsync.com. (2019). *BidSync: UNSPSC Code*. [online] Available at: https://www.bidsync.com/DPX? ac=catviewall&cattype=2&catid=95902#95902 [Accessed 10 Apr. 2019].

Buy Clean California Act.AB-262.

Cass, D. and Mukherjee, A. (2011). Calculation of Greenhouse Gas Emissions for Highway Construction Operations by Using a Hybrid Life-Cycle Assessment Approach: Case Study for Pavement Operations. *Journal of Construction Engineering and Management*, 137(11), pp.1015–1025.

Clough, R., Rounds, J., Segner, R., Sears, S. and Sears, G. (2015). *Construction contracting*. Hoboken, N.J.: Wiley.

Edelen, A. and Ingwersen, W. (2016). *Guidance on Data Quality Assessment for Life Cycle Inventory Data*. Life Cycle Assessment Research Center, National Risk Management Research Laboratory, United states Environmental Protection Agency.

Harvey, J., Meijer, J., Ozer, H., Al-Qadi, I., Saboori, A. and Kendall, A. (2016). *Pavement Life-Cycle Assessment Framework*. No. FHWA-HIF-16-014. Washington, D.C.: Federal Highway Administration.

Horvarth, A. (2004). Pavement Life-cycle Assessment Tool for Environmental and Economic Effects (PaLATE). In: *IEEE Personal Communications*.

ISO 14025:2006. [online] Available at: https://www.iso.org/standard/38131.html [Accessed 10 Apr. 2019].

ISO 14040:2006. [online] Available at: https://www.iso.org/standard/37456.html [Accessed 10 Apr. 2019].

ISO 21930:2017. [online] Available at: https://www.iso.org/standard/61694.html [Accessed 10 Apr. 2019].

Lewis, P., Leming, M. and Rasdorf, W. (2012). Impact of Engine Idling on Fuel Use and CO2 Emissions of Nonroad Diesel Construction Equipment. *Journal of Management in Engineering*, 28(1), pp.31–38.

Muench, S. (2010). Roadway Construction Sustainability Impacts. *Transportation Research Record: Journal of the Transportation Research Board*, 2151(1), pp.36–45.

Mukherjee, A. (2016). *Life Cycle Assessment of Asphalt Mixtures in Support of an Environmental Product Declaration ISO 14040 compliant life cycle assessment study supporting the National Asphalt Pavement Association, EPD program for North American asphalt mixtures*. National Asphalt Pavement Association (NAPA).

Mukherjee, A. and Cass, D. (2012). Project Emissions Estimator. *Transportation Research Record: Journal of the Transportation Research Board*, 2282(1), pp.91–99.

Mukherjee, A., Stawowy, B. and Cass, D. (2013). Project Emission Estimator. *Transportation Research Record: Journal of the Transportation Research Board*, 2366(1), pp.3–12.

National Asphalt Pavement Association (2017). *Product Category Rules (PCR) for Asphalt Mixtures*. Lanham, MD.

New.usgbc.org. (2019). LEED v4 | USGBC. [online] Available at: https://new.usgbc.org/leed-v4 [Accessed 10 Apr. 2019].

Ozer, H. and Al-Qadi, I. (2017). Regional LCA Tool Development and Applications. In: *Pavement LCA Symposium*.

Rmrc.wisc.edu. (2019). *PaLATE | Recycled Materials Resource Center*. [online] Available at: https:// rmrc.wisc.edu/palate/ [Accessed 10 Apr. 2019].

Scott III, S., Molenaar, K., Gransberg, D. and Smith, N. (2019). *NCHRP Report 561: Best-Value Procurement Methods for Highway Construction Projects*. Transportation Research Board.

Tang, P., Cass, D. and Mukherjee, A. (2013). Investigating the effect of construction management strategies on project greenhouse gas emissions using interactive simulation. *Journal of Cleaner Production*, 54, pp.78–88.

US EPA. (2019). *Latest Version of MOtor Vehicle Emission Simulator (MOVES) | US EPA*. [online] Available at: https://www.epa.gov/moves/latest-version-motor-vehicle-emission-simulator-moves [Accessed 10 Apr. 2019].

US EPA. (2019). *NONROAD Model (Nonroad Engines, Equipment, and Vehicles) | US EPA*. [online] Available at: https://www.epa.gov/moves/nonroad-model-nonroad-engines-equipment-and-vehicles [Accessed 10 Apr. 2019].

Van Dam, T., Harvey, J., Muench, S., Smith, K., Snyder, M., Al-Qadi, I., Ozer, H., Meijer, J., Ram, P., Roesler, J. and Kendall, A. (2015). *Towards Sustainable Pavement Systems: A Reference Document*. FHWA-HIF-15-002. Washington, D.C.: Federal Highway Administration.

Zapata, P. and Gambatese, J. (2005). Energy Consumption of Asphalt and Reinforced Concrete Pavement Materials and Construction. *Journal of Infrastructure Systems*, 11(1), pp.9–20.

Using environmental product declarations to support pavement green public procurement

M. Rangelov
National Academy of Sciences/Federal Highway Administration, Washington, USA

H. Dylla & N. Sivaneswaran
Federal Highway Administration, Washington, USA

ABSTRACT: Green public procurement (GPP) represents the inclusion of environmental considerations into procurement decisions. GPP is gaining more popularity, but its adoption is hindered by the lack of knowledge, clear implementation frameworks, and sound comparison criteria. Quantified and verifiable environmental performance of products or goods must serve as a basis for the fair comparison between bidding options. This study investigates the potential use of environmental product declarations (EPDs) to support GPP implementation in the pavement infrastructure sector. EPDs report quantified environmental impacts of different materials, providing meaningful comparisons and fair bidding. This study reviews current EPD programs of pavement materials (asphalt, concrete, cement, aggregate), their suitability for GPP, as well as the consistency of EPDs and underlying product category rules (PCRs) among different materials. The applicability of EPDs in construction material procurement, as well as the pavement design process supported by life-cycle assessment (LCA) was investigated. Findings indicate that the technically sound implementation of EPDs in GPP is contingent on: 1) harmonization among materials' PCRs, 2) availability of reliable public background datasets, and 3) stakeholders' education. Addressing these needs is expected to provide for transparency, technically sound and fair comparison of bids, and cut back on costs of environmental impact assessment, all of which are essential for GPP uptake. Development of systems and databases that streamline the data transfer among different stakeholders is also expected to facilitate these efforts.

1 INTRODUCTION

1.1 Green public procurement

Public procurement (PP) is the primary economic activity of the government, contributing to between 10 and 19 percent of gross domestic product (GDP) in developed countries (Zhu et al 2013; Chiappinelli & Zipperer 2017). Purchasing decisions of government have significant repercussions; therefore, PP has been recognized as an instrument that supports the government's pursuit of various strategic objectives, such as sustainability (Chiappinelli & Zipperer 2017). PP practices and underlying policies also have the potential to influence the remainder of the market and stakeholders from the private sector (Simcoe & Toffel 2011).

Green public procurement (GPP) is a practice of selecting products and services with lower life-cycle environmental impacts compared to the typically procured products and services (European Comission 2008). GPP can therefore incentivize improvement of environmental performance and increase the knowledge of sustainable production and consumption. Interest in GPP is growing worldwide. European Procurement Directives have defined a PP framework that explicitly allows for the incorporation of environmental and social considerations in procurement (European Parliament 2014). This regulation, however, does not mandate the use of GPP. European Union (EU) states can choose to what extent GPP will be deployed. In China, interest in GPP has been on the rise since 2004 (Zhu et al. 2013). Various GPP initiatives have also been investigated in Africa, Southeast Asia, and Australia (Akenroye et al. 2013; Bohari et al. 2017; Adham & Siwar 2012; Sanchez et al. 2013; Sparks 2018).

Based on the literature, environmental impacts are typically introduced into bidding in one of two ways (Chiappinelli & Zipperer 2017). First, environmental criteria that bids are expected to satisfy are included in technical requirements in call for tenders. Alternatively, environmental performance is treated as one contributor to the award criteria, either through monetization [e.g. (Ministerie van Verkeer and Waterstaat. 2015)] or weighting of different environmental performance metrics [e.g. (Baron 2016)]. From a technical standpoint, the method of environmental impact quantification in support of GPP varies from case to case. Ideally, a technically sound methodology, such as life-cycle assessment (LCA) should be used, however, that is rarely the case (Igarashi et al. 2015; Parikka-Alhola 2008; Testa et al. 2016b). An environmental impact analysis accompanying GPP is typically simplified because of a lack of personnel with LCA knowledge, and because of the technical and economic burden (Parikka-Alhola & Nissinen 2012). Environmental criteria commonly used in GPP include energy use and savings, water efficiency, chemical content, and use of renewable resources (Testa et al. 2016b). Some GPP efforts are focused solely on carbon footprint reduction, which is the practice known as low-carbon PP [e.g. (Antón & Díaz 2014; Correia et al. 2013)].

1.2 *GPP of pavements*

In the domain of pavement infrastructure, GPP efforts were undertaken in the Netherlands and Belgium. In the Netherlands, the LCA tool—Dubocalc—is used to quantify environmental impacts and calculate the environmental cost indicator (ECI) value, which is used as the basis for the reduction of the bidding prices (Ministerie van Verkeer and Waterstaat 2015). In a pilot project in Belgium, 11 social and environmental indicators were used as a basis for the reduction of bidding price, however, the full pavement LCA was not performed (Maeck & Redant 2018). In addition to the price reduction based on the proposed design, credits were also earned based on a follow-up evaluation of the agreement between designed and as-built performance (Maeck & Redant 2018). Contractors demanded low administrative burden to comply with tender requirements, as well as transparent and objective evaluation (Maeck & Redant 2018). In the United States, recent legislative acts and initiatives confirmed that environmental impacts of construction materials are becoming a topic of interest. Specifically, in 2017, the State of California legislated the Buy Clean act, which requires reporting of global warming potential (GWP) for select building products through environmental product declarations (EPDs) beginning from 2020 (California Legislative Organziation 2017). Collected EPDs will be used to develop material-specific maximum GWP values as benchmarks. Beginning in 2021, procured materials should demonstrate performance below the specified benchmark. Steps toward similar initiatives have been undertaken in Oregon, Washington, and Minnesota (OCAPA and Oregon DEQ 2016; Carbon Leadership Forum 2019; Minnesota Legislature 2019b & 2019a). Technical and administrative frameworks that will support the implementation of these initiatives are in development.

1.3 *GPP challenges*

The interest in GPP is rising; however, its implementation is still limited and relatively immature. In Europe, more than a decade after GPP framework was initiated (European Parliament 2014), only a handful of sectors enacted the corresponding legislation (information technology equipment, road vehicles, and buildings) (Antón & Díaz 2014). Challenges with wider GPP uptake referenced in the literature include a lack of technical and legal expertise, as well as a perceived cost increase (Chiappinelli & Zipperer 2017). Testa et al. 2016a indicated that training and knowledge increased the implementation of GPP, while the impact of public authority size was not found significant. Varnäs et al. (2009) highlighted that the lack of

knowledge on how to establish specific, verifiable, and quantifiable environmental performance is hindering GPP adoption. Lack of resources and clear stakeholders' responsibilities
were identified as primary obstacles to GPP uptake in China (Zhu et al. 2013).

2 RESEARCH OBJECTIVES

Despite the increasing interest in GPP, its implementation is still limited and hindered by
a lack of knowledge, a well-defined implementation framework, and objective comparison criteria. In the field of infrastructure, GPP efforts were made in material procurement [e.g.(California Legislative Organziation 2017)], and for assisting pavement design selection [e.g.
(Maeck & Redant 2018; Ministerie van Verkeer and Waterstaat 2015)]. This study investigates
the potential ways to address the identified GPP challenges in the context of materials and
pavements. The focus of this study is to investigate the role of material EPDs for supporting
the GPP of pavement infrastructure, identify challenges, and propose future steps.

3 METHODOLOGY

To achieve the objectives of the study, researchers used a simple product system for pavement
LCA to evaluate the potential fit of EPDs and determine the key requirements that EPDs
must satisfy for use as a building block of pavement LCA. Subsequently, currently available
product category rules (PCRs) for pavement materials were identified and investigated. The
aim was to evaluate if PCRs provide for consistency and applicability of resulting EPDs in
GPP, both for materials and pavement-procurement. Inconsistencies and potential challenges
were identified and solutions that can facilitate GPP implementation in the future were
proposed.

4 EPDS AND PCRS

An emerging interest to improve the environmental performance of various products has
incentivized the development of standard protocols for environmental labels. The International Organization for Standardization (ISO) has standardized the EPD development
framework through ISO 14025 standard (ISO 2006b), congruent with ISO 14040 (ISO 1997)
and ISO 14044 (ISO 2006a). EPDs are Type III eco-labels that report the environmental
impacts of a product, which are primarily intended for business-to-business communication.
EPDs are produced according to a rigorous LCA procedure established through PCRs and
third-party verification (Simonen 2014). EPDs, therefore, have the potential to enable meaningful comparison between the materials and to incentivize production improvement (Simonen & Haselbach 2012). From the manufacturers' perspective, driving forces for EPD
development are an efficient response to customer requirements, ease of marketing, communication, and potential use in bidding (Strömberg 2017). ISO 14025 advocates for the use of
EPDs to encourage purchasers to include environmental performance considerations into
decision-making (ISO 2006b). Additionally, because EPDs typically contain the impacts associated with cradle-to-gate materials production, EPDs have the potential to be used as a data
source in LCA analyses beyond cradle-to-gate.

4.1 *Use of EPDs as a data source in LCA analysis with a broader scope*

LCA is a data-intensive analysis methodology. The lack of reliable, affordable, and regionally
specific input data has been identified as a major obstacle for pavement LCA (U.S. Department of Transportation Federal Highway Administration (FHWA) 2017). A common
approach to conducting LCA, to use commercially available proprietary databases. An

advantage of using commercially available proprietary databases is that the data are up to date and satisfies certain quality requirements. A disadvantage of using commercially available proprietary databases is the lack of transparency and costs. As concrete and asphalt mixtures are engineered materials, the material data from the commercial databases likely do not represent the nuances of the local material production and specific mixture designs. In the context of pavement PP, considerations of cost, transparency, material, and local specificity may be crucial. It is, therefore, worthwhile to seek alternative reliable data sources, and, in terms of materials, EPDs present a promising solution.

Figure 1 shows a simplified schematic of the asphalt pavement product system. Dashed lines in Figure 1 show the system boundaries for producing each material that would potentially be used to develop corresponding EPDs. As seen in Figure 1, production of aggregate, asphalt binder, and admixtures can be represented by corresponding EPDs. Additionally, the production of these constituents feed into the production of asphalt mixture, which can also be represented by an EPD. In the case of concrete pavement, EPDs for aggregate, cement, and admixtures can be used in the product system and/or feed into an EPD for a concrete mixture. LCA analyses with relatively narrow scope (cradle-to-gate) used in the creation of EPDs are implemented as components of LCA with broader scope (cradle-to-grave, in the example in Figure 1). For the technically sound implementation of this approach, all combined LCA analyses must be consistent from a methodological standpoint and in terms of underlying data. As Figure 1 indicates, flows such as transportation (represented by thick black arrows in Figure 1) and energy are common for multiple life-cycle phases and the production of various constituents. To provide for the consistency of LCA, data sources for these flows must be harmonized among all life-cycle phases and EPDs used in the analysis. Assumptions and methodological choices specified in PCRs must be mutually consistent and in accordance with pavement LCA. If EPDs are used solely for the material procurement, any type of benchmarking or comparison among EPDs can only be performed if methodological and data consistency with EPDs is confirmed. Methodological consistencies include the definition of key elements of LCA, such as the goal and scope of the analysis, impact assessment method, and methodological choices such as allocation, choice of inventory, and impact categories, etc. Data consistency encompasses uniformity of primary and secondary data sources and treatment of proxies. This study further investigates the methodological and data consistency among PCR and EPD programs of different materials used jointly on pavement projects.

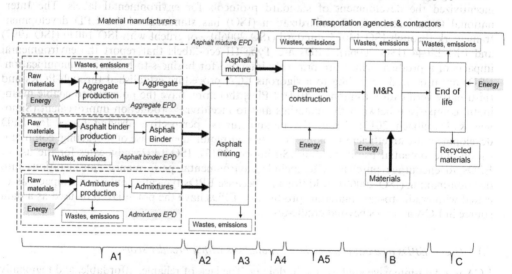

Figure 1. An example of asphalt pavement product system for LCA for cradle-to-grave scope. Thick black arrows indicate transportation. Designation of life-cycle phases: A1- materials extraction and upstream production, A2- transport to the production facility, A3- manufacturing, A4- transport to site, A5-installation, B-use, C-end of life (ISO 2017).

The scheme presented in Figure 1 also indicates that material manufacturers control and own the data in phases A1-3, while phases A4-C fall under the responsibility of transportation agencies and contractors executing construction activities. Production of EPDs has created initiatives to improve the footprint of the upstream material production. Life-cycle stages beyond material production are primarily in the control of transportation agencies. Accordingly, the systems that enable data transfer between different stakeholders should be implemented to facilitate the development of pavement LCA. Building Information Model (BIM) is a system that can help collect, store, and efficiently provide access to the environmental performance data of materials and construction activities (in addition to parameters such as cost, material quantities, etc.). For instance, for every material used in a project, EPDs can be uploaded in BIM, while BIM can be used to track project-specific equipment usage for the construction equipment. As a result, data stored in BIM can be used to facilitate life-cycle inventory for pavement LCA.

The use of EPDs for material procurement is limited to cradle-to-gate impacts of material production. To evaluate pavement LCA, life-cycle stages following material production (A4, A5, B and C in Figure 1), which are in control of contractors and pavement agencies, should be additionally quantified. Due to the limited space, the scope of this article could not address technical and organizational challenges of pavement LCA in detail. The analysis of EPDs was selected because of the urgency brought about by the legislation such as Buy Clean California Act. EPDs are already being produced by manufacturers and therefore represent a suitable starting point for impact reduction. Material selection is a relatively short-term decision over which contractors and agencies have control, which strengthens the case for the inclusion of EPDs as GPP aid.

4.2 Available PCRs and EPDs for pavement materials

This section summarizes current efforts in the development of PCRs and EPDs of the following pavement materials in the United States: asphalt mixtures, concrete mixtures, cement, and aggregate. These pavement materials contribute to the significant portion of the pavement material footprint. PCRs of other building materials (flat glass, mineral wool, structural steel, precast concrete elements, carbon steel reinforcing bars) are also available but were out of the scope of this study.

PCR for asphalt mixtures was created in 2016 by the National Asphalt Pavement Association (NAPA) and is valid from 2017 to 2022 (NAPA 2016). Stakeholders from the asphalt industry and public and private owners of transportation assets were involved in the development of this document (Mukherjee & Dylla 2017). PCR is accompanied by an online tool, Emerald Eco Label, which facilitates EPD production.

PCR for portland cement concrete was initially published in 2013 under Carbon Leadership Forum (CLF) and with the National Ready Mixed Concrete Association (NRMCA) as a program operator and stakeholders from concrete industry and academia (Carbon Leadership Forum 2013). With an extension, PCR lasted until March 2019 and was complemented by several tools to generate EPDs, listed on the NRMCA website. Industry average EPDs for concrete were developed for the United States (NRMCA 2016) and Canada (CRMCA 2017). Upon the expiration, concrete PCR was replaced with a new version in March 2019, with NSF International as a program operator (NSF International 2019). The new concrete PCR is compliant to ISO 21930 (ISO 2017) and has a 5-year validity period. This document was created through the collaboration of stakeholders from the building industry (manufacturers, organizations), users, Oregon Department of Environmental Quality (DEQ), and LCA experts.

PCR for construction aggregates was issued in 2016 under the American Society of Testing and Materials (ASTM) as a program operator (ASTM International 2017). The category of construction aggregates encompasses sand, gravel, crushed stone, crushed concrete, iron slag, steel slag, or any combination thereof. Interested parties involved in PCR development were producers and suppliers of construction materials (aggregates, primarily), industry associations, and academia.

ASTM also served as a program operator for two PCRs for cementitious products, issued in 2014, with a 5-year validity period. One PCR pertains to portland, blended hydraulic, masonry, mortar, and plastic cements (ASTM International 2014a), while the other PCR covers ground granulated blast furnace slag (slag cement) (ASTM International 2014b). Industry average EPDs for portland and slag cement were created by Portland Cement Association and Slag Cement Association, respectively (PCA 2016; Slag Cement Association 2015).

Additives, admixtures, and other chemicals have been recognized as an important data gap (NAPA 2016). In Europe, for instance, a PCR for concrete admixtures has been created by IBU (Institut für Bauen and Unwelt- German Institute for Building and Environment) and European Federation of Concrete Admixtures (EFCA) (IBU 2014). Model European EPDs for concrete admixtures can be found on the EFCA website. It is noteworthy that these EPDs present the conservative estimations of environmental impacts of admixtures production (IBU 2014). The comparable efforts in the United States are yet to be undertaken. However, because concrete, asphalt, and their constituents are typically produced locally, efforts to put forth PCRs for these materials are likely prioritized over admixtures.

4.3 *Challenges*

As evident from the previous section, PCRs for different products were created by different entities and with different stakeholders involved. No overarching authority was overseeing and harmonizing these efforts (Simonen & Haselbach 2012). The mutual consistency between the existing EPDs may be questionable. Some inconsistencies in methodologies and in background data sources were identified by Mukherjee et al. (2018) (Amlan Mukherjee, Chaitanya Bhat 2018). Part of the current EPD production started with producers voluntarily reporting the environmental performance of their products for marketing purposes (Mukherjee and Dylla 2017). For building products such as concrete, EPDs yield credits through Green Building Standards and Certification Systems (e.g. LEED v4) (Braune, Kreißig, and Sedlbauer 2007). The corresponding PCRs were created by stakeholders from the building industry; however, the domain of infrastructure and public works differs from the private sector in terms of funding, contracting mechanisms, procurement, and stakeholders involved. These differences might cause the existing PCRs for the building products to not fully accommodate the needs and concerns related to GPP and infrastructure. Moreover, one LEED credit is awarded for the submission of 20 EPDs. One more LEED credit is obtained if the products that constitute at least 50% of cost demonstrate impacts below industry average in at least three categories. The consistency and comparability of EPDs, therefore, was likely considered as a secondary concern.

5 COMPATIBILITY OF DIFFERENT MATERIALS PCRS

5.1 *Methodological compatibility*

In the most general terms, the reviewed PCRs follow the same cradle-to-gate scope and include raw materials acquisition, transportation, and manufacturing (phases A1, A2, and A3). Cutoff criteria, impact assessment methodology, and mandatory reported impact categories closely correspond among the PCRs. Methodological differences among PCRs are discussed as follows.

All reviewed PCRs have a 5-year validity period. Cement, aggregate, and concrete EPDs developed under these PCRs also last for 5 years from the published date. This is advantageous for the material manufacturers because EPDs are valid for 5 years regardless of the production date. Conversely, asphalt EPDs are valid until the end of a PCRs validity. The advantage of using this approach is to establish consistency between the EPDs and PCRs that are valid concurrently. No overlap between EPDs produced under different PCRs provides for the consistency necessary for comparing EPDs and the development of benchmarks.

Another methodological difference among PCRs is in the declared unit. Specifically, the declared unit for concrete is volumetric (1 m^3), while the declared unit for an asphalt mixture is mass-based (a short ton). The difference in the declared unit likely stems from the units in which the material is typically purchased. If used in pavement LCA, these EPDs may have to be complemented with density to ensure mass balance. Declared units for aggregate and cement are mass-based.

Table 1 shows the summary of the impact categories specified in different PCRs . The required environmental impacts are largely the same. In addition to mandatory impact categories, concrete PCR from 2013 specifies several optional impact categories, while the PCR from 2019 specifies carbon sequestration (the details on how this impact should be evaluated are not outlined). These optional categories are not listed in PCRs for aggregate and cement. Accordingly, if aggregate and cement EPDs are used as data sources for concrete EPD or pavement LCA, optional impact categories should be assessed separately. Differences exist in reporting of energy and resource use among PCRs. For instance, the asphalt PCR does not require waste-generation reporting because of the assumption that asphalt production is a closed-loop process with no waste output. In other PCRs, waste production is either reported as hazardous and nonhazardous waste or total waste, as shown in Table 1. The asphalt PCR does not prescribe reporting of renewable and nonrenewable material resources,

Table 1. Impact categories and inventory terms specified in different PCRs.

PCR		Asphalt	Concrete PCR 2013	Concrete PCR 2019	Aggregate	Portland cement
Impact categories	Global warming potential	R	R	R	R	R
	Acidification potential	R	R	R	R	R
	Eutrophication potential	R	R	R	R	R
	Smog creation potential	R	R	R	R	R
	Ozone depletion potential	R	R	R	R	R
Energy consumption inventory	Nonrenewable fossil fuel energy	R*	R	R	R	R
	Nonrenewable nuclear energy		R		R	R
	Renewable energy (solar, wind, hydro, geothermal)	R*	R	N	R	R
	Renewable energy (biomass)		R		R	R
Resource consumption inventory	Nonrenewable material resources	N	R	R	R	R
	Renewable material resources	N	R	N	R	R
	Fresh water consumption	R	R	R	R	R
	Nonhazardous waste	N	R	N	R	R
	Hazardous waste	N	R	N	R	R
Optional impact categories	Carbon emissions from biofuel consumption	N	O	N	N	N
	Energy from waste recovery	N	O	N	N	N
	Carbon sequestered in product	N	O	O	N	N
	Total waste disposed/recycled/reused	N	O	R	N	N
	Content declaration/chemical of concern	N	O	N	N	N
	Particulate matter emission	N	O	N	N	N

Note: R- requested, O- optional, N- not specified; *- reported both as material and as energy.

while this assessment is the part of aggregate, cement, and concrete PCR from 2013. Renewable and nonrenewable energy consumption is mandated in all PCRs, either in separated categories shown in Table 1, or aggregated. The exception is the new concrete PCR, which requires only reporting of nonrenewable energy and material consumption.

Another methodological inconsistency among the PCRs is the allocation of slag (Mukherjee et al. 2018). In aggregate PCR, steel slag aggregate is considered with the economic allocation, while in all other PCRs, slag is treated as a byproduct with no economic value; it is represented as a raw material for which only upstream transport impacts are accounted. PCR for steel products defines slag as a co-product and recommends a system expansion approach to account for its impacts (SCS Global Services 2015). Slag is an example of a product that crosses boundaries of different industries that produce slag or use slag in their manufacturing process and, as such, should be treated with caution in LCA to avoid omission or double-counting. This example illustrates the necessity of harmonizing PCRs across the industry sectors that share product system boundaries to allow for consistent use of EPDs in pavement LCA.

5.2 Compatibility of recommended background data sources

LCA is a data-intensive process. The data quality and selection of background data sources from PCRs directly translate into the reliability of EPD (Mukherjee et al. 2018). Based on all PCRs, users are required to report the assessment of background data quality of their EPDs in the categories of technological, temporal, geographical representativeness, completeness, and reliability. In aggregate and cement PCRs, default background data sources are not specified. The background data sources for both concrete PCRs and asphalt PCR are provided in Table 2. As seen in Table 2, there is a consensus on background data for portland cement and

Table 2. Background data sources in asphalt and concrete PCRs.

	Asphalt PCR (2016) (NAPA 2016)	Concrete PCR (Carbon Leadership Forum 2013)	Concrete PCR (NSF International 2019)
Prescribed vs. recommended	Prescribed	Recommended	Prescribed
Portland cement	PCA 2016	NREL 2012	NRMCA 2016; CRMCA 2017
Fly ash	Not applicable	Confirm no additional processing	Considered as recovered material
Slag cement	Cutoff rule	LCI of Slag Cement, USA, 2003	Slag Cement Association (2015)
Natural aggregate	Marceau et al. (2007)	ecoinvent or GaBi	ecoinvent 3.4. World 2001
Crushed aggregate	Marceau et al. (2007)	ecoinvent or GaBi	ecoinvent 3.4. World 2001
Admixtures	No data at this time	EFCA2005	EFCA 2015 (IBU 2014)
Water	Not applicable	Site specific data	ecoinvent 3.4.
Crude oil at refinery	NREL 2012	Not specified	Not specified
Natural gas at refinery	NREL 2012	Not specified	Not specified
Transportation	NREL 2012	NREL 2012/EPA 2003	NREL 2012
Electricity generation	Greet 2015	US EPA eGrid	ecoinvent 3.4. USA (2015)
Electricity emissions	Not specified	NREL, U.S. LCI (Laboratory 2012)	Not applicable
Site generated electricity	Not specified	NREL 2012	NREL 2012
Electricity loss	Greet 2013 (6.5%)	Not specified	Not applicable
Recycled asphalt shingles	Mukherjee (2016)	Not applicable	Not applicable

transportation, but other background data sources differ. Mukherjee et al. (2018) argued that while asphalt PCR prescribes public data sources, thereby prioritizing data availability and transparency, concrete PCR includes proprietary background data sources, which indicates the prioritization of data quality. Concrete PCR from 2013 recommends background data sources, while the PCR from 2019 prescribes the background data sources, as does the asphalt PCR. Additionally, in asphalt PCR, flows that should be covered by primary (foreground) versus secondary (background) data are differentiated, which is not the case with either of the concrete PCRs.

6 CONCLUSIONS

To date, EPD programs for pavement materials were initiated, which sets the stage for the use of EPDs in GPP. EPDs can be supplied as a part of tender documentation without further investigating environmental impacts or making comparisons. Using EPDs to benchmark the production, however, incentivizes improvements in production efficiency and assists agencies in pursuing their environmental goals, but further harmonization efforts are necessary. In terms of EPDs validity period, expiration of EPDs with the expiration of the corresponding PCR is appropriate for GPP. In that case, only EPDs produced by the same methodology will be valid concurrently, which enables a technically justified comparison. In terms of methodological consistencies, harmonization should be made in the quantification and reporting of materials and energy inventories. Inconsistencies in the allocation of slag and materials from other industries should be addressed by PCRs harmonization. For data sources, a disparity in the prescribed background data sources is an important obstacle for using EPDs in GPP when pavement designs are compared. Reliable public background databases can offer a potential solution in GPP context by providing for transparency and keeping the cost of EPDs low and available to the practitioners. The literature on GPP clearly indicated that transparency, fairness, and low burdens for costs and administrative efforts are key concerns of stakeholders (Chiappinelli and Zipperer 2017; J. Maeck 2018). Tools for pavement LCA that include public background data sets, provide for regular updates, and enable consistent reporting can be key resources to facilitate GPP uptake in the domain of pavements. Education on environmental assessment, roles of stakeholders, state-of-the-art, and challenges is an important prerequisite for collaborative work on PCR harmonization. Development of databases through BIM and similar systems that would streamline the communication between stakeholders and facilitate LCA is another relevant effort in this area. Conclusively, EPDs of building materials have the potential to facilitate GPP; however, harmonization of PCR programs, identification of reliable public background datasets, and the development of tools and databases that streamline the analysis and reporting are challenges yet to be addressed.

ACKNOWLEDGMENT

This research was performed while the author, M. Rangelov, held an NRC Research Associateship award at Federal Highway Administration (FHWA).

REFERENCES

Adham et al. 2012. Empirical Investigation of Government Green Procurement (GGP) Practices in Malaysia. *OIDA International Journal of Sustainable Development* 4 (4): 77–88.

Akenroye et al. 2013. Development of a Framework for the Implementation of Green Public Procurement in Nigeria. *International Journal of Procurement Management* 6 (1): 1–23.

Antón et al. 2014. Integration of LCA and BIM for Sustainable Construction. *International Journal of Social, Behavioral, Educational, Economic, Business and Industrial Engineering* 8 (5): 1378–82.

ASTM International. 2014a. Product Category Rules for Preparing Environmental Product Declaration for Portland, Blended Hydraulic, Masonry, Mortar, and Plastic (Stucco) Cements.

ASTM International. 2014b. Product Category Rules for Slag Cement UN CPC 374.

ASTM International. 2017. Product Category Rules for Preparing an Environmental Product Declaration for Construction Aggregates: Natural Aggregate, Crushed Concrete, and Iron/Steel Furnace Slag.

Baron, R. 2016. The Role of Public Procurement in Low-Carbon Innovation. *Round Table on Sustainable Development*, 12–13.

Bohari et al. 2017. Green Oriented Procurement for Building Projects: Preliminary Findings from Malaysia. *Journal of Cleaner Production* 148: 690–700.

Braune et al. 2007. The Use of EPDs in Building Assessment–towards the Complete Picture. *Sustainable Construction–Materials and Practices*, 299–304.

California Legislative Organziation. 2017. *Buy Clean California Act [3500-3505]*.

CRMCA 2017. CRMCA Member Industry-Wide EPD for Canadian Ready-Mixed Concrete.

Carbon Leadership Forum. 2013. Product Category Rules for Concrete.

Chiappinelli, O. & Zipperer V. 2017. Using Public Procurement as a Decarbonisation Policy: A Look at Germany. *DIW Economic Bulletin* 7 (49): 523–32.

Correia et al. 2013. Low Carbon Procurement: An Emerging Agenda. *Journal of Purchasing and Supply Management* 19 (1): 58–64.

European Commission. 2008. Public Procurement for a Better Environment. s.l. : Communication from the Commission to the European Parliament, the Council, the European Economic and Social Committee and the Committee of the Regions.

European Parliament and The Council of European Union. 2014. Directive 2014/24/EU of the European Parliament and of the Council on Public Procurement and Repealing Directive 2001/18/EC.

FHWA 2017. Sustainable Pavements Program Road Map. FHWA-HIF-17-029. Washington, DC.

Carbon Leadership Forum. 2019. Buy Clean Washington: Study Overview.

Igarashi et al. 2015. "Investigating the Anatomy of Supplier Selection in Green Public Procurement." *Journal of Cleaner Production* 108: 442–50.

Institut Bauen and Umwelt (IBU). 2014. Product Category Rules for Construction Products Part B Requirements on the EPD for Concrete Admixtures.

ISO 1997. *ISO 14040: Environmental Management-Life Cycle Assessment-Principles and Framework*.

ISO 2006a. ISO 14044, Life Cycle Assessment - Requirements and Guidelines.

ISO 2017. ISO 21930: 2017- Sustainability in Buildings and Civil Engineering Works- Core Rules for Environemntal Product Declarations of Construction Products and Services.

ISO 2006b. ISO 14025: Environmental Labels and Declarations—Type III Environmental Declarations—Principles and Procedures.

Maeck J. & Redant K. 2018. Moving toward Green Public Procurement in Belgium. In *Proceedings of 7th Transport Research Arena (TRA)*. Vienna, Austria.

Marceau et al. 2007. *Life Cycle Inventory of Portland Cement Concrete*. Portland Cement Association.

Ministerie van Verkeer and Waterstaat. 2015. Green Public Procurement. The Rijkswaterstaat Approach. http://primes-eu.net/media/8772517/6_presentation-riga-blue-version-pp.pdf.

Minnesota Legislature. 2019a. *Minesotta Bill HF2204: Maximum Acceptable Global Warming Potential*.

Minnesota Legislature. 2019b. *Minnesota Bill HF2203: Buy Clean Minnesota Act*.

Mukherjee, A. 2016. Life Cycle Assessment of Asphalt Mixtures in Support of an Environmental Product Declaration. *National Asphalt Pavement Institute: Lanham, MD, USA*.

Mukherjee, A. & Dylla H. 2017. Challenges to Using Environmental Product Declarations in Communicating Life-Cycle Assessment Results: Case of the Asphalt Industry. *Transportation Research Record* 2639 (1): 84–92.

Mukherjee al. 2018. Challenges in Meeting Data Needs for Use of Environmental Product Declarations in Pavement Design and Construction: State of Practice and Future Scope. Final Report in Preparation, FHWA.

National Asphalt Pavement Association (NAPA). 2016. Product Category Rules for Asphalt Mixtures.

National Renewable Energy Laboratory. 2012. U.S. Life Cycle Inventory Database. https://www.lcacommons.gov/nrel/search.

NRMCA. 2016. NRMCA Member Industry-Wide EPD for Ready Mixed Concrete.

NSF International. 2019. Product Category Rule for Environmental Product Declarations for Concrete.

OCAPA and Oregon DEQ. 2016. Oregon Concrete Environmental Product Declaration (EPD) Program.

Parikka-Alhola, K. 2008. Promoting Environmentally Sound Furniture by Green Public Procurement. *Ecological Economics* 68 (1–2): 472–85.

Parikka-Alhola, K., Nissinen, A. 2012. Environmental Impacts and the Most Economically Advantageous Tender in Public Procurement. *Journal of Public Procurement* 12 (1): 43–80.

PCA. 2016. Environmental Product Declaration (EPD) for Portland Cements (per ASTM C150, ASTM C1157, AASHTO M85 or CSA A3001).

Sanchez et al. 2013. Sustainable Infrastructure Procurement in Australia: Standard vs. Project Practices. *Sustainable Infrastructure Procurement in Australia: Standard vs. Project Practices.*

SCS Global Services. 2015. North American Product Category Rule for Designated Steel Construction Products.

Simcoe et al. 2011. LEED Adopters: Public Procurement and Private Certification. *Boston University School of Management and Harvard Business School.*

Simonen, K. 2014. *Life Cycle Assessment.* Routledge.

Simonen, K. & Haselbach, L. 2012. Environmental Product Declarations for Building Materials and Products: US Policy and Market Drivers. In *Proceedings of the International Symposium on Life Cycle Assessment and Construction, IFSTTAR.*

Slag Cement Association. 2015. Industry Average EPD for Slag Cement.

Sparks, D. 2018. Exploring Public Procurement as a Mechanism for Transitioning to Low-Carbon Buildings. Queensland University of Technology.

Strömberg, L. 2017. Conceptual Framework for Calculation of Climate Performance with Pre-Verified LCA-Tools. *Journal of Civil Engineering and Architecture* 11: 29–37.

Testa et al. 2016b. Drawbacks and Opportunities of Green Public Procurement: An Effective Tool for Sustainable Production. *Journal of Cleaner Production* 112: 1893–1900.

Testa et al. 2016a. Examining Green Public Procurement Using Content Analysis: Existing Difficulties for Procurers and Useful Recommendations. *Environment, Development and Sustainability* 18 (1): 197–219.

Varnäs et al. 2009. Environmental Consideration in Procurement of Construction Contracts: Current Practice, Problems and Opportunities in Green Procurement in the Swedish Construction Industry. *Journal of Cleaner Production* 17 (13): 1214–22.

Zhu et al. 2013. Motivating Green Public Procurement in China: An Individual Level Perspective. *Journal of Environmental Management* 126: 85–95.

A feasibility study to assess the use of EPDs in public procurement

L.K. Miller & B.T. Ciavola
Trisight, Houghton, USA

ABSTRACT: As the science of life cycle assessment (LCA) has advanced so has the public's understanding of its usefulness in supporting decision-making. This has led to pressure on regulatory bodies in considering the use of LCA and Environmental Product Declarations (EPDs) to inform public procurement. Legislatively, the Buy Clean California Act, 2017, (Assembly Bill 262), is requiring successful bids on public construction projects to produce EPDs for a list of eligible materials, at the point of installation. Similar bills have been introduced in Washington and Minnesota. In this context, it is imperative for state Departments of Transportation (DOTs) to proactively identify the best strategies for integrating EPDs into the procurement process. This paper discusses the outcomes of a feasibility study that was conducted for Minnesota DOT to identify the challenges and barriers faced by DOT and industry stakeholders in adopting the use of EPD in the procurement process. The study included two workshops for MnDOT and industry stakeholders to educate them on the current state of LCAs and EPDs and to identify the barriers and challenges based on available resources and capabilities. A roadmap was developed for MnDOT that lays out a recommendation of activities that could be adopted to successfully implement the adoption of EPDs in public procurement. The workforce development components of this work will also support the pavement construction industry create the capacity to adapt to new practices from an informed position, rather than being passively responsive to external mandates.

1   INTRODUCTION

The field of Life Cycle Assessment (LCA) is growing, and its use in business and public domains is expanding. Environmental Product Declarations (EPDs) have been targeted as an effective method of communicating environmental impacts, and their provision is being rewarded or required in a growing number of projects, especially in building and infrastructure construction. Within the pavement sector, the Federal Highway Administration's (FHWA) Sustainable Pavements Program along with the associated Sustainable Pavements Technical Working Group (TWG) has led the development of guidelines outlined in FHWA's documents *Towards Sustainable Pavement Systems: A Reference Document* (Van Dam et al. 2015) and *Pavement Life Cycle Assessment Framework* (Harvey et al. 2016) to support state, local and federal agencies in developing a better understanding of LCA and its relationship to building sustainable pavements.

2   BACKGROUND AND DRIVERS

The Buy Clean California Act, 2017, Assembly Bill 262 signed October 15, 2017, was the first legislation to mandate the use of EPDs. Specifically, it required provision of EPDs for a short list of common building used in state funded construction projects. Pavement materials were not present on the list of building materials in the finalized bill. The Buy Clean California Act presented a number of implementation challenges, particularly in the mandated timeline and requirements for industry averages. The bill required creation of maximum acceptable Global

Warming Potential (GWP) benchmarks for each of the building material types by January 1, 2019, despite most effected industries not having an industry average EPD available as a robust basis for comparison. As a result, the California Department of General Services (DGS), which has been charged with the implementation of the Buy Clean Act, now has a deadline of January 1, 2021 to finalize the GWP benchmarks.

Caltrans, the California state agency responsible for the state's highway and public transportation systems, is currently piloting projects that will require EPDs for a wider range of construction materials in parallel with implementing AB 262. The Caltrans effort is aimed towards gathering up-to-date and regionally applicable environmental impact information for use in LCA for pavement design, pavement management and the development of specifications and other policies to reduce environmental impacts, including global warming potential and others. Interest in EPDs in pavement procurement is burgeoning: a recent webinar on LCA in Procurement offered by the Transportation Research Board (TRB), with presenters from CalTrans, MnDOT, and FHWA attracted 190 attendees.

Based on California's AB262, similar 'Buy Clean' bills were introduced in Minnesota and Washington. This legislative effort gave energy to efforts within MnDOT to proactively evaluate and determine an appropriate path forward for the use of EPDs in pavement procurement without requiring legislative action. MnDOT recognized that it did not have the internal infrastructure, institutional awareness of LCA and EPDs, and industry buy-in to use EPDs in procurement. They wished to avoid the difficulties experienced in California with implementation and create a potential path forward informed by the agencies and industries affected. It was under these circumstances that MnDOT proposed a study to assess the preparedness of stakeholders and the feasibility of using EPDs to inform procurement in Minnesota.

The study consisted of two workshops to educate and gather input from MnDOT employees and industry stakeholders, while creating a roadmap for incorporating EPDs into MnDOT's procurement process. The following sections outline the topics of discussion and outcomes of the workshops.

3 METHODOLOGY

Initial discussions identified two main groups of stakeholders in the potential use of EPDs in procurement. The first group, referred to as Tier 1 stakeholders, are the engineers and administrators internal to MnDOT who would be responsible for the long term deployment and management of the system, including pavement design and construction engineers and managers of the procurement process and environmental decision-making. The second group, referred to as Tier 2 stakeholders, are the external stakeholders, including contractors and construction materials industry representatives. Two workshops were planned to gather input from both tiers of stakeholders, as well as provide educational resources and an introduction to the fundamentals of LCA and the use of EPDs in the pavement industry.

3.1 Workshop one

Workshop One focused on Tier 1 stakeholders (from within MnDOT). Representatives from FHWA Minnesota Division Office and Office of Infrastructure were also present. Once the attendees were presented with the basics of LCA and current state of EPDs in pavement procurement, an interactive activity was conducted. It included a guided discussion to identify stakeholder opportunities and challenges, and creation of a draft roadmap for EPD usage in pavement procurement in Minnesota.

The following questions were used to seed brainstorming and discussion:

1. How do you envision the use of EPDs in design/procurement/construction (or your area of expertise)?
2. What are the current barriers to using EPDs within MnDOT?
3. What are the immediate opportunities that will facilitate the use of EPDs within MnDOT?

4. What are the long-term opportunities that will facilitate the use of EPDs within MnDOT?

The subsequent discussion identified and recorded common themes from the breakout groups and Tier 2 external stakeholders were identified for invitation to Workshop Two.

3.2 *Workshop two*

Workshop Two focused on engaging the Tier 2 stakeholders with the process begun in Workshop One. The workshop began with a brief introduction to LCA and EPDs, followed by a peer experience exchange activity was conducted for the stakeholders to learn from the experiences of other agencies. Representatives from Caltrans and the Oregon Department of Environmental Quality (DEQ) presented recent experiences in integrating EPDs with their respective procurement processes. Breakout groups were formed for discussion with the following questions as the topics of interest:

1. At what point in the construction process do you see EPDs being required?
2. How should EPDs be included in contracts?
3. What would be appropriate between facility-specific EPDs participating in an industry average EPDs?
4. What aspects should be highlighted when communicating the use of EPD in procurement to potential participants?

The outcomes of the discussions were recorded and used to inform the further development of the roadmap for EPD usage in pavement procurement in Minnesota.

4 OUTCOMES

4.1 *Workshop one*

Workshop One was held in September of 2018 with 15 participants attending. Discussion topics included how EPDs may be used in MnDOT's operations, current barriers to their use, and near and long term opportunities to facilitate the use of EPDs. Additional topics that came up were the role of MnDOT in developing a plan for EPD use, and piloting the use of EPDs.

Workshop attendees discussed how EPDs could be used within their areas of expertise. It was agreed that EPD use should be piloted simply as a data collection tool, without driving any decisions, and once past the initial phase, EPDs should not drive pavement selection between different industries, but could be used to drive specific mix selection within a given pavement type. This concept of data collection without value judgements must be very clearly communicated to industry stakeholders.

A need for leadership and an education plan were identified as current barriers to using EPDs within MnDOT. Within MnDOT, a committee would need to be formed to provide leadership and drive towards long term acceptance of any EPD program. The Approved Product List (APL) initiative was identified as a model of an internally led initiative. A similar approach to committee creation and championing was recommended for EPD use.

Both an internal and external education plan would need to be developed. The decentralized nature of MnDOT will require an internal education plan to inform and include city and county engineers and administrators. Additionally, education is a crucial need to attain buy in from industry partners. Since EPDs are not currently required, industry awareness in some areas is low, and any use of EPDs will need to clearly identify the benefits to industry, with special emphasis on data collection without value judgements as previously noted.

Immediate and long term opportunities for facilitating the use of EPDs within MnDOT were identified. Best value procurement (BVP) was discussed as a way to include EPDs but ultimately identified as being too early stage for inclusion in BVP. Finally, the ICE (Infrastructure carbon estimator) tool from FHWA was discussed as a potential complementary tool to any EPD collection process. The ICE tool is currently being used by Minnesota to provide annual updates to emissions estimates. EPDs could be used to support these estimates; this would be a topic of study in an EPD pilot study.

Several additional themes emerged from the discussion. First, there was wide consensus that MnDOT and the pavement industries have been highly involved in EPD development and are therefore well-positioned assess the potential use of EPDs at the State level. MnDOT personnel have participated closely in the FHWA Sustainable Pavements Technical Working Group, while the national organizations for both concrete and asphalt producers have supported manufacturers in EPD creation. Additionally, MnDOT's Office of Sustainability has related experience in implementing a mandate for greenhouse gas reduction in Minnesota. Attendees agreed that given this background, MnDOT and industry stakeholders should drive the development of a potential plan for EPDs in procurement. Legislation (if any) should be informed by these efforts, rather than pushing them.

Second, attendees suggested an EPD pilot program would be necessary to determine the most appropriate step in the procurement process to collect EPDs, to identify the appropriate database resources and administrative support necessary to collect and store the documentation, to integrate EPD data into the ICE tool, and to estimate the costs to both MnDOT and industry to support such a requirement. This was identified as a topic needing input from the Tier II (external) stakeholders, which would be gathered in Workshop Two.

4.2 *Workshop two*

Workshop Two included both Tier 1 (internal) and Tier 2 (external) stakeholders. It was held in March of 2019 with 25 participants. Discussion topics included where in the procurement process to include EPDs, how to incorporate EPDs into contracts, and how to address facility-specific EPDs vs. industry average EPDs. Additional program training needs were identified, as well as areas for further study. Finally, workshop participants gave input in the development of a roadmap for EPD use in pavement procurement in Minnesota.

Discussion began by asking what step in the procurement process would be best for requiring EPDs. The following points in the process were suggested for pilot:

1. Collected up front as estimate and have an "as-built", with mix design, during the pre-qualification phase,
2. Before construction starts but not at bidding, or
3. Post construction (as used in California to meet the Buy Clean requirements).

The following ways to include EPDs in contracts were proposed:

1. As incentives and/or through alternative contracting methods,
2. To add points to a technical score in design/bid, or
3. For informational purposes only.

Requiring plant specific or industry average EPDs was discussed, with the consensus falling on using plant specific EPDs as not all industries do not have an average yet. It would be necessary consider that small operations may find EPD creation difficult, time consuming, or costly. Developing partnerships between contractors and DOT (as in the case of Oregon DEQ) was also considered.

Needs for internal and external workforce training about EPDs in procurement were discussed. The workshop participants felt the following topics should be highlighted:

1. Identify goals of program
2. Provide case studies for contractors
3. Ensure that all the information is uniformly available to all participants at the same time
4. Avoid explicit or implicit comparisons between pavement types
5. Industry and MnDOT must be aware and prepared for the potential for EPDs to become mandatory, and work together on messaging and communication.

Over the course of Workshop Two, additional questions were raised, including:

1. If a Special Provision is used to integrate EPDs into contracts, how would it be drafted?
2. How can contractors be meaningfully included in implementing EPDs in procurement?
3. What is the appropriate benchmarking values if EPD benchmarks are used?
4. What unintended consequences, such as changes in the supply chain, can impact this effort?
5. How to develop appropriate balance and tradeoffs between reported impact categories, if EPDs are used in decision-making?

Based on these discussions, the investigators and MnDOT co-developed a roadmap for studying EPDs in pavement procurement in Minnesota, as seen in Figure 1.

4.3 *Roadmap and next steps*

The roadmap developed from the workshop outcomes highlights background knowledge and drivers and the technical, organizational, and industry integration actions necessary to fully evaluate the use of EPDs in procurement. Technical roadblocks like benchmarking and data quality assessments are being addressed through an FHWA study. Organizational and industrial integration are proposed to be the subject of a pilot study of EPD usage. The outline of the proposed pilot was informed by the outputs of the workshops and consists of three steps:

1. Implementing the plan for piloting EPD use within MnDOT,
2. Identifying potential for impacts on external stakeholders, and
3. Establishing how EPDs will be used in future.

The core function of the proposed study is to create a baseline by piloting EPDs with small number of contractors. Step 1 and Step 2 would occur concurrently, with Step 1 focusing on the stakeholders that are internal to MnDOT and Step 2 focusing on those external to the agency.

4.3.1 *Step 1: Implement pilot plan for EPD use within MnDOT*
The first part of the Pilot study involves studying policies and procedures for requesting, managing, evaluating, and deploying EPD documents in the procurement process. Mechanisms to include EPDs in contracts and specifications and administrative infrastructure for the

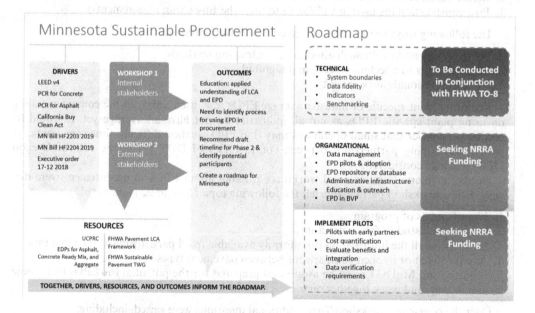

Figure 1. Minnesota sustainable procurement roadmap.

program would be identified. An initial EPD database would be created while studying options for data management and process flow. A method to include and educate distributed MnDOT jurisdictions, districts (counties and cities), and contractors on EPD use would be developed. Finally, methods for the inclusion of EPDs in specifications, Approved Product List (APL), and Best Value Procurement (BVP) processes would be evaluated.

4.3.2 *Step 2: identify potential for impact on external business contractor practices by Piloting EPD use*

In parallel with Step 1's evaluation of MnDOT-internal processes, Step 2 would explore policies and procedures that directly affect private industry partners, such as pavement producers and contractors. External stakeholders such as mix producers and contractors would be engaged to evaluate the process of creating EPDs. Early partners for pilot study would be identified and educated on EPD best practices. The cost and benefits for a contractor to create an EPD would be quantified, and disclosure rules established.

4.3.3 *Step 3: Establish how EPDs will be used in future*

Steps 1 and 2 would then be integrated to establish guidelines for a complete program for the use of EPDs in procurement. All outcomes of the pilot study would be codified, and the potential for integration within MnDOT and other state agencies evaluated. Integration with cradle-to-gate MnDOT systems would be balanced with other procurement processes. Additional tasks are to determine program vs. project level interactions and verify that data requirements meet the Code of Federal Regulations. A baseline of environmental impacts may be created, and if impacts are evaluated, categories should be selected for evaluation. Finally, additional training and documentation needs would be identified.

Future work is looking to implement the proposed pilot study.

5 CONCLUSIONS

The need for proactive planning for EPDs in procurement is clear. While industry and state agencies recognize the necessity for using EPDs to measure, communicate and reduce environmental impacts of construction materials, there is a strong motivation for state agencies, contractors, and LCA experts to drive the design of business processes and approaches that best support the use of sustainable design and construction options. Driving the integration of EPDs to pavement procurement from within MnDOT and with external stakeholders can take advantage of existing procurement processes, create appropriate administrative infrastructure and institutional knowledge, and increase industry buy-in.

Limiting factors for the use of EPDs in procurement have become less technical and more organizational and educational. While the LCA community is familiar with EPDs and their usage, the discussions at the MnDOT workshops clearly showed that fundamental knowledge in the topic is still not commonplace, even among affected stakeholders. For instance, many stakeholders expressed concerns that EPDs could be used to make value judgements or comparisons of one pavement material to another, though one of the fundamental concepts of EPDs is that they may not be used for the comparison of dissimilar items.

Workforce training will be essential in any practical implementation of EPD use. Stakeholders will need the fundamentals of LCA/EPD and the state of the art and practice in life cycle thinking. This training will need to be internal to agencies requesting EPDs, as well as the industries providing them.

Agencies and industry partners will also need to develop resources, like case studies, webinars, and reference sheets, that can support contractors in developing and effectively participating in LCA/EPD informed procurement.

Lastly, a pilot program will be essential to develop the institutional mechanisms to include EPDs in procurement, assess the impacts on internal and external stakeholders, and develop appropriate workforce training.

ACKNOWLEDGEMENTS

The work reported in this paper was supported by the Minnesota Department of Transportation Contract No. 1003321, with Amlan Mukherjee, Michigan Tech, as Principal Investigator (PI) and the first author, Lianna Miller, Trisight, as Co-PI. Hearty acknowledgement is also due to Curt Turgeon of MnDOT for facilitating this project.

REFERENCES

Buy Clean California Act 2017. Legislative Counsel Bureau, State of California. "Assembly Bill No. 262 Chapter 816."
Buy Clean Minnesota Bill 2019. HF 2204. 91st Legislature, House of Representatives, State of Minnesota.
Creating the Buy Clean Washington Act Bill 2018. HB 2412. 65th Legislature, House of Representatives, State of Washington.
Harvey, J.T., Meijer, J., Ozer, H., Al-Qadi, I.L., Saboori, A. & Kendall, A. 2016. Pavement Life-Cycle Assessment Framework (No. FHWA-HIF-16-014).
Van Dam, T.J., Harvey, J.T., Muench, S.T., Smith, K.D, Snyder, M.B., Al-Qadi, I.L., & Ozer, H. et al. 2015. Towards Sustainable Pavement Systems: A Reference Document.

Integrated evaluation method of life cycle economic and environmental impact for permeable pavement: Model development and case study

Jiawen Liu, Hui Li* & Yu Wang
College of Transportation Engineering, Tongji University, Shanghai, China

ABSTRACT: Multi-scheme comparison is of great significance for sustainable pavement. The existing methods including Life Cycle Cost Analysis (LCCA) and Life Cycle Assessment (LCA) are often isolated from each other, which cannot achieve the comprehensive evaluation of whole life cycle of the pavement. This paper proposes an integrated evaluation method of economic cost and environmental impact based on life cycle analysis framework. In view of the permeable pavement, the life cycle cost and environmental impacts of typical permeable asphalt pavement (PA) and dense-graded asphalt pavement (DA) are studied. The results show that the economic cost of PA is higher than that of DA, but the environmental impact is much lower. The gaps of economic cost and environmental impact will expand with the increase of traffic volume. This study could provide some useful insights for the decision-making in pavement engineering.

1 INTRODUCTION

Permeable pavement is of great significance for the reconstruction of urban water circulation (Pratt *et al.*, 1988). With the increasing diversification of pervious materials, pavement construction and maintenance technologies, the comparison of permeable pavement alternatives has become an urgent research direction in pavement engineering. So far, research methods widely used include life cycle cost analysis (LCCA) and life cycle assessment (LCA) (Harvey *et al.*, 2014).

The concept of LCCA was first proposed by the US military in the 1960s, now it has become a necessary part of road planning in the United States (Chan *et al.*, 2008). The existing evaluation process of LCCA mainly considers the construction and maintenance costs of the owner and the economic, time and safety costs of the user throughout the life cycle, and converts them into a unified economic indicator through a uniform discount rate (Walls & Smith, 1998). At the same time, the LCCA software "Realcost" developed by FHWA has gradually been extended from the United States to the world, to help solve the life cycle cost problems of pavement schemes.

LCA, precisely, process-based LCA (PLCA), derived from the study of Coca-Cola bottles by the Midwest Institute of the United States in the late 1960s. Over the past decades, the framework of life cycle environmental assessment has been relatively complete. According to ISO standards, the evaluation process can be divided into four steps: goal and scope, life cycle inventory (LCI), life cycle impact assessment (LCIA) and interpretation. As for the application of LCA method in road industry, research mostly focused on the environmental impact of various pavement materials, construction or maintenance techniques at specific pavement life cycle stages (Thenoux *et al.*, 2007b; Giustozzi *et al.*, 2012b). In 2011, Nicholas J. Santero (Santero *et al.*, 2011) pointed out that in the current pavement LCA studies, traffic delay, rolling resistance, pavement albedo and end of life allocation were not effectively incorporated into the LCA framework, and these factors would affect the accuracy of the conclusions.

However, the application of LCCA and LCA methods are often isolated from each other, which cannot fully discuss the comprehensive impact of pavement alternatives. In 2012, Giustozzi et al. (Giustozzi et al., 2012) linked preventive maintenance measures to performance and cost over the life cycle of pavement through a multi-attribute life cycle cost, performance and environment analysis. In 2013, Faisal Hameed (Hameed, 2013) firstly integrated LCA and LCCA, which he called the ILCA2 approach, to estimate the environmental impacts and costs of stormwater runoff, and pointed out that low-impact development (LID) had better economic benefits than traditional drainage practices. In 2017, Batouli et al. (Batouli & Mostafavi, 2017) evaluated the life cycle cost and environmental impact of different pavement design schemes, and the results showed that the initial cost of flexible pavement was lower, however it would bring higher long-term cost and greater environmental impact. The same year, Adil Umer (Umer et al., 2017) proposed an integrated LCCA and LCA framework based on an economic and environmental trade-off analysis, in which environmental impact and cost were normalized into sustainability index to evaluate different pavement technologies.

The objective of this paper is to combine LCCA with LCA to establish an integrated life cycle evaluation method for the comprehensive impact of permeable pavement. Meanwhile, corresponding calculation models are adopted to predict the pavement characteristics and performance. Combined with case analysis and data calculation, this paper provides some reference for the multi-scheme selection in pavement engineering.

2 INTEGRATED EVALUATION METHOD OF PAVEMENT LIFE CYCLE ECONOMIC COST AND ENVIRONMENTAL IMPACT

2.1 *Integration of LCCA and LCA*

The calculation process of LCCA divides the total cost into two categories according to the undertaker: owner or agency cost and user cost (Walls & Smith, 1998). The pavement life cycle inventory analysis (LCI), in order of time and space, calculates the environmental impact of the whole life cycle, including raw material acquisition stage, construction stage, use stage, maintenance stage and end of life. Further discussion of each part will find a lot of similarities between the objects evaluated by the two (Figure 1). Compared with LCCA's classification method based on the undertaking subject, the spatio-temporal sequence of LCI is relatively easier to understand and develop, and LCA's framework is also more extensive and logical. Therefore, it is possible to integrate LCCA's evaluation goals into the process of LCI to realize synchronous analysis of economic cost and environmental impact.

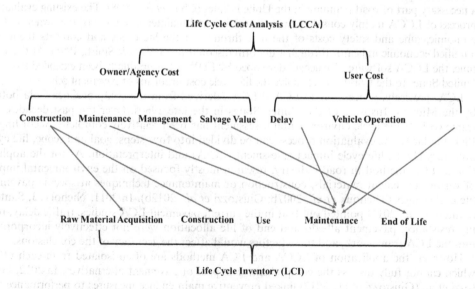

Figure 1. Similarities and differences between LCCA and LCI.

It should be noted that some life cycle costs need to be calculated as independent economic costs, such as: labor cost and direct monetary input; user cost in the maintenance stage, which can be further divided into vehicle operation cost ($/km), delay cost ($/h) and safety cost ($/unit accident) (FHWA, 1998); and net present value (NPV) in Equation 1.

$$NPV = \text{Initial construction cost} + \sum_{K=1}^{N} K_{\text{Expected future cost}} \left[\frac{1}{(1 + Discount\ rate)^{Number\ of\ years}} \right]$$

(1)

2.2 Life cycle assessment models for permeable pavement

The environmental impact of each pavement life cycle stage is usually listed in the inventory analysis, including data collection and data calculation.

2.2.1 Raw material acquisition stage

This stage calculates the environmental impact of pavement material production processes, including not only the environmental impact of the production process of materials such as asphalt, cement, aggregates, but also the transportation and mixing processes of these materials (Cai, 2013). The overall environmental impact is calculated via the product of the amount of material and equipment used and the corresponding environmental impact per unit amount.

2.2.2 Construction stage

This stage calculates the environmental impacts of pavement levelling, spreading, and rolling. Also, the transportation of raw materials should be classified. The specific energy consumption can be calculated according to the quota and consumption of the corresponding construction code (Zhu et al., 2016), and can also be analogized according to similar projects.

2.2.3 Use stage

This stage calculates the environmental impact caused by the interaction of pavement surface with vehicles and the environment. Among these impacts, the research on pavement rolling resistance and albedo is especially numerous.

1. Pavement rolling resistance model

Rolling resistance is one of the main factors affecting vehicle loss during use stage. Many existing models are used to evaluate the impact of rolling resistance on vehicle fuel consumption, including the HDM-4 model released by the World Bank (Zaabar & Chatti, 2010), which considers the change of rolling resistance while keeping the speed constant, and the MOVES model released by the USEPA (USEPA, 2010), which considers both the change of rolling resistance and vehicle speed. Based on Wang's research (Wang et al., 2010) on environmental impact assessment of rolling resistance, this paper adopted a simplified method shown in Equation 2, along with the decay formula and maintenance formula of IRI in Equation 3-4, continuous pavement parameters can be obtained within a period of time.

Additional Fuel Consumption of Gasoline/Diesel Vehicles $= (IRI - Initial\ IRI) \times 0.0313 \times$

$Standard\ Gasoline/Diesel\ Fuel\ Consumption \times Traffic\ Volume \times Road\ Length^2$

(2)

$$\sqrt{IRI} = -0.174 + 9.66 \times 10^{-5} \times \sqrt{Cumulative\ ESAL} + 1.15 \times \sqrt{Initial\ IRI}$$ (3)

$$\Delta IRI = -0.6839 + 0.6197 \times IRI(\text{Before Maintenance})$$ (4)

Where IRI =the international roughness index of pavement at any time, m/km; Cumulative ESAL =accumulative number of equivalent standard axle load after maintenance, times; Initial IRI =the IRI after maintenance, m/km.

2. Pavement albedo model

In order to evaluate the impact of pavement albedo on environment, radiation forcing is used as a measurement. Radiation forcing is a measure of the extent to which a factor changes the energy balance of the earth-atmosphere system. In this paper, the radiation forcing method considering time variation is adopted (Yu & Lu, 2014), as Equation 5 shows. 0.01α represents the change in pavement albedo per unit area, and RF refers to the radiation forcing change, whose value is between 1.12 to 2.14 W/m^2.

$$0.01\alpha = \frac{1.087 \times RF \times t}{0.217 \times t - 44.78e^{-t/172.9} - 6.26e^{-t/18.51} - 0.22e^{-t/1.186} + 51.26} [kgCO_2] \quad (5)$$

3. Permeability model of permeable pavement

According to the related study of Wang et.al (Wang et al., 2018), the environmental impacts of permeability will be evaluated from three categories: urban flooding, water recycling and water purification.

① Urban flooding: urban flooding usually causes vehicles to wade or detour, the corresponding environmental impacts are shown in Equation 6. Cf (kg/km) is the fuel consumption caused by urban flooding; ni (days) is the flooding duration; AADT (pcu/d) is the annual average daily traffic; β (%) is the vehicles detour rate; d1,i (km) is the vehicle's detour distance; γ1 (kg/km) is the average vehicle fuel consumption on dry pavement; d2,i (km) is the length of flooded pavement; γ2 (kg/km) is the average vehicle fuel consumption on flooded pavement.

$$C_f = \sum n_i \times AADT \times [\beta \times d_{1,i} \times \gamma_1 + (1 - \beta) \times d_{2,i} \times (\gamma_2 - \gamma_1)] \quad (6)$$

② Water recycling: water recycling of PA is evaluated from the prospect of alleviating the irrigation water demand (IWD), and it is considered that the reduced IWD is equal to the smaller value of the water recycling amount (V) in Equation 7 and the IWD (W) in Equation 8.

$$V = P \times (1 - \varphi) \times A_P \times \sigma \quad (7)$$

$V(m^3)$ is the average annual water recycling amount of PA; P (mm) is the local annual rainfall; φ is the runoff coefficient; A_p (m^2) is the pavement area;σis the coordinate parameter, 0.001 is selected here.

$$W = (ET_0 \times K - \gamma P) \times S \times \beta \quad (8)$$

$W(m^3)$ is the average annual IWD; ET_0 (mm/year) is the evapo-transpiration; K is the adjustment coefficient, whose value is between 0.5~1.0; γ is the effective coefficient of irrigation, varies from 0.45 to 0.75; S (m^2) is the vegetation area; β is the coordinate parameter,0.001. The saved energy consumption of PA is determined by Equation 9.

$$E = n \times d \times \alpha \times \min\{P \times (1 - \varphi) \times A_P \times \sigma, (ET_0 \times K - \gamma P) \times S \times \beta\} \quad (9)$$

E (kW) is the the saved energy consumption; n (year) is the analysis period; d (km) is the transport distance; α (kWh/t•km) is the transportation efficiency, since electricity is the main consumption, α is usually between 0.025 kWh/t•km to 0.1kWh/t•km.

③ Water purification: water purification counts on the porous structure of PA. This paper will directly adapt the corresponding fitting results of the pollutant removal rate of permeable pavement (Table 1).

Table 1. Pollutants removal rate of permeable pavement.

Pollutants	Maximum (mm)	Removal Rate	Regression Coefficient R^2
TP	63	$R_{TP\ 63}=0.110h_{63}+19.75$	0.996
	40	$R_{TP\ 40}=0.085h_{40}+26.16$	0.964
	12	$R_{TP\ 12}=0.120h_{12}+23.73$	0.973
TN	63	$R_{TN\ 63}=0.005h_{63}+1.577$	0.936
	40	$R_{TN\ 40}=0.004h_{40}+1.202$	0.981
	12	$R_{TN\ 12}=0.007h_{12}+0.599$	0.926
Pb		$R_{Pb}=0.630\gamma+28.485$	0.858
Zn		$R_{Zn}=0.738\gamma+23.224$	0.992

2.2.4 *Maintenance stage*

The main environmental impacts at this stage are divided into direct and indirect impacts. Direct impacts include the environmental impacts of material production and maintenance construction required for maintenance activities, which are similar to the material acquisition and construction stages. Indirect impacts refer to traffic delays caused by maintenance activities, which create an additional environmental burden.

Since few studies focused on the evaluation model of end of life, this stage is not included in this paper. this stage is not included in this paper. The EOL mainly calculates the environmental impact caused by different treatment methods when the pavement life is over. The main disposal methods are divided into two categories: landfill and recycling. Due to the diversity of recycling methods, research focusing on end-of-life assessment models is limited, and their relationship with pavement types is not obvious, and their effects are often overlooked.

3 CASE STUDY ON PERMEABLE PAVEMENT LIFE CYCLE ECONOMIC COST AND ENVIRONMENTAL IMPACT ASSESSMENT

3.1 *Goal and scope*

Assuming a secondary road in Shanghai as the object, and taking PA and DA as comparison. The basic pavement parameters are as follows: road length 10km; pavement width 7m; subgrade width 10m; design life PA: 9 years, DA:12 years; design speed 60km/h; proportion of heavy-duty diesel vehicle (HDDV): 5%; traffic growth rate 5%. Other detailed design parameters are listed in Table 2. Meanwhile, in order to consider the impact of traffic volume on economic cost and environmental burden of pavement, the control group with the initial AADT of 2000 pcu and 10000 pcu was set. Taking 36 years as the analysis period for evaluation. The pavement structures are listed in Figure 2 as well.

The evaluation scope and system boundary include the raw material acquisition stage, construction stage, use stage and maintenance stage. It is worth noting that due to the particularity of permeable pavement, the common overlay technology is not applicable. The specific maintenance schedule for DA and PA can be referred to *Drainage Asphalt Pavement Technical Specification in Shanghai*.

3.2 *Life cycle inventory*

3.2.1 *Raw material acquisition*

During the analysis period, PA and DA need to be repaired four times and three times respectively. According to the pavement structure and the material cost and environmental impact per unit, the economic cost and environmental impact can be calculated.

Table 2. Detailed design parameters.

Design Parameters		DA	PA	Data Source
Initial annual average daily ESAL (Equivalent Single Axle Loads)		150/300/750		Pavement Engineering(Sun, 2012)
Discount		8		The Theoretical and Methodological Basis and Measurement of the Social Discount Rate in Different Regions of China(Tan et al., 2009)
Local annual rainfall (mm)		1140		National Meteorological Information Center(National Meteorological Information Center)
Time value ($/h·per car)		14		Life-Cycle Cost Analysis in Pavement Design(FHWA, 1998b)
Vehicle operating value ($/km·per car)		0.14		
Speed (km/h)		40		
Urban flooding	average depth (m)	0.4	0.2	Long-Term Stormwater Quantity and Quality Performance of Permeable Pave-ment Systems(Brattebo & Booth, 2003) Analytical Equation for Estimating the Stormwater Capture Efficiency of Perme-able Pavement Systems(Zhang & Guo, 2014)
	average duration (d)	4	2	
	frequency (times/year)	2	1	
	flooded length (km)	1		
Water recycling	detour distance (km)	20	0.3	Assumption
	runoff coefficient	1		Analytical Equation for Estimating the Stormwater Capture Efficiency of Per-meable Pavement
	transport distance (km)	30		Assumption
Pollutant con-centration (mg/L)	TP	5		A Comparative Study of the Emissions by Road Maintenance Works and The Disrupted Traffic Using Life Cycle Assessment and Micro-simulation (Huang et al., 2009) Purification Effect-iveness of Runoff Pollution by the Per-meable Asphalt Pavement(Song et al., 2009) Study on the Removal Mechanism of Porous Asphalt Pavement to Heavy Metals from Road Runoff(Zhao, 2014)
	TN	3		
	Pb	0.1		
	Zn	1		

3.2.2 Construction

Based on the assumption of the transportation distance is 20 km, 10 km is assumed as the distance from mixing plant to the construction cite, and all mixing machines' power are within 320 t/h. The sum of mechanical cost and labor cost is the economic cost in the construction process, and the energy consumption of pavement can be further calculated for its environmental impact.

3.2.3 Use

This stage synthesizes the three models mentioned before, and compares the difference of economic cost and environmental impact between PA and DA.

1. Effect of permeability:
   1) Urban flooding: Wang et.al (Wang et al., 2018) found that when the water depth was 0.4m, 100% vehicles chose to detour, while 62.6% of the vehicles chose to detour when the water depth was 0.2m. Assume that the fuel consumption of the wading section is 100% more than normal driving.
   2) Water recycling: According to the assumptions, the average annual water recycling amount of PA is 7.7×10^4 m^3. The mean plant evapotranspiration in summer in

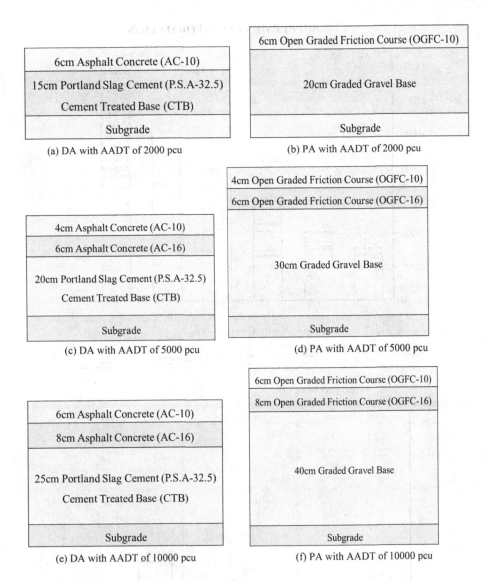

| 6cm Asphalt Concrete (AC-10) |
| 15cm Portland Slag Cement (P.S.A-32.5) Cement Treated Base (CTB) |
| Subgrade |

(a) DA with AADT of 2000 pcu

| 6cm Open Graded Friction Course (OGFC-10) |
| 20cm Graded Gravel Base |
| Subgrade |

(b) PA with AADT of 2000 pcu

| 4cm Asphalt Concrete (AC-10) |
| 6cm Asphalt Concrete (AC-16) |
| 20cm Portland Slag Cement (P.S.A-32.5) Cement Treated Base (CTB) |
| Subgrade |

(c) DA with AADT of 5000 pcu

| 4cm Open Graded Friction Course (OGFC-10) |
| 6cm Open Graded Friction Course (OGFC-16) |
| 30cm Graded Gravel Base |
| Subgrade |

(d) PA with AADT of 5000 pcu

| 6cm Asphalt Concrete (AC-10) |
| 8cm Asphalt Concrete (AC-16) |
| 25cm Portland Slag Cement (P.S.A-32.5) Cement Treated Base (CTB) |
| Subgrade |

(e) DA with AADT of 10000 pcu

| 6cm Open Graded Friction Course (OGFC-10) |
| 8cm Open Graded Friction Course (OGFC-16) |
| 40cm Graded Gravel Base |
| Subgrade |

(f) PA with AADT of 10000 pcu

Figure 2. Different pavement structure alternatives.

Shanghai is 6.4 mm/day, and the vegetation area affected is 0.2 km^2. After the coefficients in Equation 8 are all averaged, the average annual IWD is 2.15×10^5 m^3. Finally, the transport distance is 10 km, corresponding cost is 0.06 kwh/(t·km), and the saved power within the time range evaluated is 1.73 million kWh. This energy saving has nothing to do with traffic volume but only with the type of pavement.

3) Water purification: The removal of pollutants from PA is not related to traffic volume but to pavement type. It should be noted that, due to the insufficient data of the impact factors of TP and TN, these two pollutants are only part of the model calculation.

2. Effect of rolling resistance: Assuming that the initial IRI is 2m/km, it can be calculated according to the maintenance schedule and the decay formula.

3. Effect of albedo: RF is set at 1.5 W/m^2, according to the report of Lawrence Laboratory, the albedo of PA is 0.12, that of DA is 0.10, and that of the control group is 0.10, therefore the equivalent increment of carbon dioxide emission is 553 metric tons.

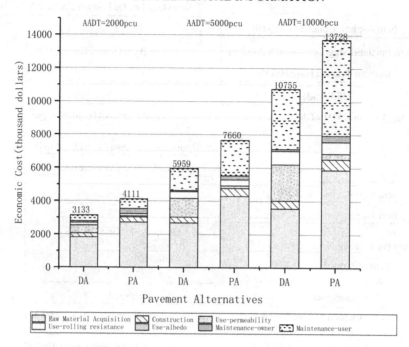

(a) Economic Cost

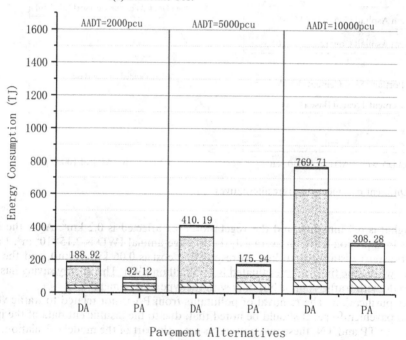

(b) Energy Consumption

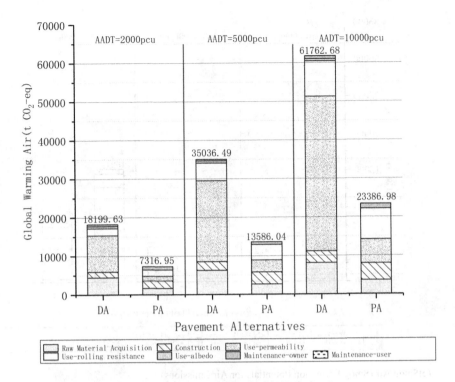

(c) Global Warming Air (Global Warming Potentials for Air Emissions)

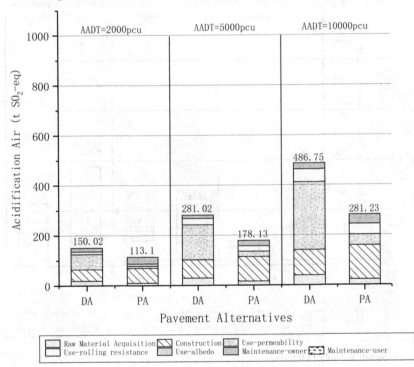

(d) Acidification Air (Acidification Potentials for Air Emissions)

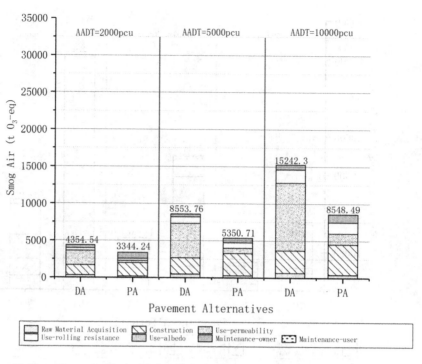

(e)Smog Air (Smog Formation Potentials for Air Emissions)

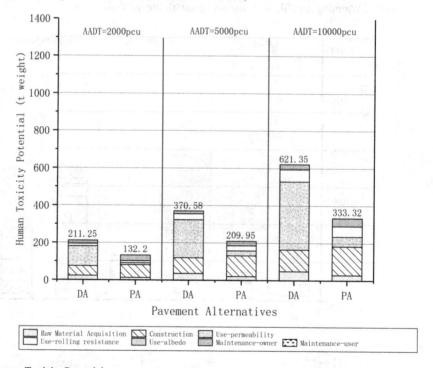

(f) Human Toxicity Potential

Figure 3.   Economic cost and environmental impact assessment results.

3.2.4 *Maintenance*

The purpose of this paper's life cycle cost analysis is to evaluate the cost differences due to various pavement structures, materials, construction or maintenance schemes during the life cycle. The user-related part of these costs mainly focuses on the impact of construction and maintenance on traffic operation. Therefore, the calculation of user costs only considers the costs caused by the impact of the construction work on traffic and does not include the user's fuel costs.

4 RESULTS AND DISCUSSIONS

Based on the results of inventory analysis, the performance of PA and DA in different environmental impact categories is evaluated. The evaluation method adopts the standard steps of classification, distribution, characterization of the USEPA (quantification is not included). The categories of impact assessment, characterization parameters and impact factors are selected and distributed according to the TRACI method. The results can be further visualized to evaluate the advantages and disadvantages of each alternative in different impact categories (Figure 3 in Supplemental Materials).

From the view of life cycle economic cost:

1. Considering the whole life cycle, the initial construction cost of PA is 47%, 57% and 62% higher than that of DA. It can be inferred that in order to maintain the porous structure and the mechanical properties of PA, high viscosity modified asphalt is commonly used, and its particularity also requires higher maintenance frequency due to relatively short service life.
2. Considering the cost of use and maintenance stages, the economic cost of PA is 31%, 29% and 28% higher than DA under different traffic volume. When the analysis period is extended to the life cycle, the disadvantage of PA will be significantly reduced, which can be regarded as the advantage brought by the benefits of maintenance and use stage. When the traffic volume rises, the cost gap between the two types of pavement narrows slightly, this is because PA with high traffic volume will also produce certain economic benefits during the use phase.

From the view of life cycle environmental impact:

1. Take the two pavement schemes with AADT of 2000pcu as the example, the energy consumption, global warming air, acidification air, smog air and human toxicity of PA decreased by 51%, 60%, 25%, 23% and 37% respectively. The reduction of energy consumption and emissions is larger, while the reduction of acidification air, smog air and human toxicity is smaller, which is caused by the high energy consumption and emissions of cement for semi-rigid base.
2. With the increase of traffic volume, the environmental benefit advantage of PA tends to become more obvious, due to the environmental benefit of high traffic pavement in the use stage is more significant than the additional environmental burden during the construction stage.

5 CONCLUSIONS

This paper proposes a comprehensive evaluation method of pavement life cycle economic cost and environmental impact in an innovative way, and takes permeable pavement as the example to carry out a case study, verifying the feasibility of this method. The main conclusions are as follows:

1. Through integrating LCCA and LCA methods and considering the economic cost and environmental impact of the pavement, the merits of different pavement alternatives can be

evaluated in a more comprehensive and complete perspective, thus providing the qualified support for decision-making.

2. The life cycle economic cost of PA is much higher than that of DA. This difference is mainly caused by the high-viscosity asphalt required for PA. Therefore, from the perspective of economic cost optimization, DA has advantages over PA. On the other hand, the life cycle environmental impact of PA is much lower than that of DA. This difference is primarily caused by the environmental benefits of PA during use stage. Therefore, from the perspective of environmental protection, PA is the better choice.

3. From the perspective of overall evaluation, in most cases, due to the environmental benefits of it, PA can be used as an optimal solution for pavement design; whereas when the strength of PA does not meet the requirements or its environmental benefits can not be reflected, DA can be the better solution.

4. Regardless of PA or DA, as to the economic cost or environmental impact, the use stage occupies a non-negligible part of the life cycle. However, the traditional LCCA method does not consider the change of economic cost caused by human-vehicle interaction during the long-term use phase; the model of use phase in LCA method is still in the research stage. Therefore, the model and method of use phase in LCCA and LCA still need more attention and research.

REFERENCES

Batouli, M. & Mostafavi, A. 2017. Service and performance adjusted life cycle assessment: a methodology for dynamic assessment of environmental impacts in infrastructure systems, 1–19.

Brattebo, B. O. & Booth, D. B. 2003. Long-term stormwater quantity and quality performance of permeable pavement systems, *Water Research*, Vol. 37 No. 18, pp. 4369–4376.

Cai, R. 2013. Research On Quantitative Analysing System On Energy Consumption and Carbon Emission of Asphalt Mixtures. Xi'an, Chang'an University.

Chan, A., Keoleian, G. & Gabler, E. 2008. Evaluation of Life-Cycle Cost Analysis Practices Used by the Michigan Department of Transportation, *Journal of Transportation Engineering*, Vol. 134 No. 6, pp. 236–245.

FHWA. 1998a. LIFE-CYCLE COST ANALYSIS IN PAVEMENT DESIGN: EXECUTIVE SESSION. DEMONSTRATION PROJECT NO. 115. Washington, DC United States.

FHWA. 1998b. Life-Cycle Cost Analysis in Pavement Design: Executive Session. Demonstration Project No. 115.

Giustozzi, F., Crispino, M. & Flintsch, G. 2012a. Multi-attribute life cycle assessment of preventive maintenance treatments on road pavements for achieving environmental sustainability, *International Journal of Life Cycle Assessment*, Vol. 17 No. 4, pp. 409–419.

Giustozzi, F., Crispino, M. & Flintsch, G. 2012b. Multi-attribute life cycle assessment of preventive maintenance treatments on road pavements for achieving environmental sustainability, *International Journal of Life Cycle Assessment*, Vol. 17 No. 4, pp. 409–419.

Hameed, F. (2013) Integrated Life Cycle Analysis Approach (ILCA2) for transportation project and program development, ProQuest Dissertations and Thesis.

Harvey, J., Meijer, J. & Kendall, A. 2014. Life Cycle Assessment of Pavements, *Techbrief*.

Huang, Y., Bird, R. & Bell, M. 2009. A comparative study of the emissions by road maintenance works and the disrupted traffic using life cycle assessment and micro-simulation, *Transportation Research Part D Transport & Environment*, Vol. 14 No. 3, pp. 197–204.

National Meteorological Information Center (no date). Available at: http://data.cma.cn/dataService/cdcindex/datacode/A.0029.0005/show_value/normal.html. 2018/4/15.

Pratt, C. J., Mantle, J. D. G. & Schofield, P. A. 1988. Urban stormwater reduction and quality improvement through the use of permeable pavements, *Water Pollution Research & Control Brighton*, Vol. 21 No. 8-9, pp. 769–778.

Santero, N. J., Masanet, E. & Horvath, A. 2011. Life-cycle assessment of pavements. Part I: Critical review, *Resources, Conservation and Recycling*, Vol. 55 No. 9-10, pp. 801–809.

Song, Q., Yongpeng, X. U. & Yong, E. 2009. Purification effectiveness of runoff pollution by the permeable asphalt pavement, *Journal of Northeast Agricultural University*.

Sun, L. 2012. Pavement Engineering., Tongji University Press.

Tan, Y., Li, D. & Wang, F. 2009. The Theoretical and Methodological Basis and Measurement of the Social Discount Rate in Different Regions of China., *Journal of Industrial Technological Economics*, Vol. 28: 66–69.

Thenoux, G., González, Á. & Dowling, R. 2007. Energy consumption comparison for different asphalt pavements rehabilitation techniques used in Chile, *Resources Conservation & Recycling*, Vol. 49 No. 4, pp. 325–339.

Umer, A., Hewage, K., Haider, H. & Sadiq, R. 2017. Sustainability evaluation framework for pavement technologies: An integrated life cycle economic and environmental trade-off analysis, *Transportation Research Part D Transport & Environment*, Vol. 5388-101.

USEPA. 2010. MOVES (Motor Vehicle Emission Simulator), *United States Environmental Protection Agency*.

Walls, J. & Smith, M. R. 1998. Life-cycle cost analysis in pavement design: in search of better investment decisions, *Operating Costs*.

Wang, T., Harvey, J. T. & Jones, D. 2010. A Framework for Life-Cycle Cost Analyses and Environmental Life-Cycle Assessments for Permeable Pavements, *University of California Pavement Research Center Technical Memorandum*.

Wang, Y., Li, H. & Behzad, G. 2018. Initial evaluation methodology and case studies for life cycle impact of permeability of permeable pavements, *International Journal of Transportation Science and Technology*.

Yu, B. & Lu, Q. 2014. Estimation of albedo effect in pavement life cycle assessment, *Journal of Cleaner Production*, Vol. 64 No. 2, pp. 306–309.

Zaabar, I. & Chatti, K. 2010. Calibration of HDM-4 Models for Estimating the Effect of Pavement Roughness on Fuel Consumption for U. S. Conditions, *Transportation Research Record Journal of the Transportation Research Board*, Vol. 2155 No. -1, pp. 105–116.

Zhang, S. & Guo, Y. 2014. Analytical Equation for Estimating the Stormwater Capture Efficiency of Permeable Pavement Systems, *Journal of Irrigation & Drainage Engineering*, No. 141:6014004, pp.

Zhao, Y. 2014. Study On the Removal Mechanism of Porous Asphalt Pavement to Heavy Metals From Road Runoff., Nanjing Forestry University.

Zhu, H., Cai, H. & Yan, J. 2016. Development of Life-Cycle Inventory of Emerging Asphalt Pavement Technologies and Life-Cycle Evaluation System in China. *Transportation Research Board 95th Annual Meeting*. Washington DC, United States.

Pavement life cycle management: Towards a sustainability assessment framework in Europe

A. Jiménez del Barco Carrión
University of Granada, Spain

Tony Parry
University of Nottingham, UK

Elisabeth Keijzer
TNO, The Netherlands

Björn Kalman
VTI, Sweden

Konstantinos Mantalovas
University of Palermo, Italy

Ali A. Butt & John T. Harvey
UC Davis, USA

Davide Lo Presti
University of Palermo / University of Nottingham, Italy / UK

ABSTRACT: Pavement Life Cycle Management is a 2-year international project aiming at supporting European National Road Authorities (NRAs) to introduce sustainability in their practices by providing training on Life Cycle techniques and a user-friendly package to support their widespread implementation. The first task in Pavement Life Cycle Management (PavementLCM) project is the creation of a Sustainability Assessment (SA) framework that complies with EN15643-5 and consequently include the three pillars of sustainability, use a life cycle approach and use quantifiable sustainability performance indicators. This paper presents the first steps towards the creation of the framework which includes the following steps: 1) review of the available Product Category Rules (PCRs) related to asphalt mixtures and pavement activities; 2) definition of the object of the assessment; 3) review and survey of NRAs practices of the main research efforts in Europe towards the definition of sustainability performance indicators.

1 INTRODUCTION

1.1 *Background on sustainability assessment*

Sustainability is defined as "state in which components of the ecosystem and their functions are maintained for the present and future generations" (EN 15643-5, 2017). For National Road Authorities (NRAs), this is an aspect to consider when carrying out their asset management and project planning. However, there is no single way to assess the sustainability of assets. There are many different methods, tools and databases claiming to support sustainability assessment, but there is not one method defined as the most appropriate for national road authorities.

Amongst all available methods there is a core principle, namely that the impacts of a project or product should be analysed over its whole life cycle. This principle is generally referred to as Life Cycle Management (LCM) and comprises, amongst others, environmental Life Cycle Assessment (LCA), Life Cycle Costing analysis (LCC, sometimes referred to as LCCA) and Social-Life Cycle Assessment (S-LCA). For these methodologies, general approaches exist, for example the ISO 14040 (2006) and ISO 14044 (2006) which describes the standard procedures for LCA. More specific guidelines for construction works exist as well, such as the EN 15804 (2012) standard for LCAs of building products, the ISO 15392 (2008) and ISO 15686-5 (2017) standards for LCC of buildings and constructed assets respectively. When the three pillars of sustainability are considered in the evaluation of a product or system, the exercise can be referred as Sustainability Assessment (SA), which is part of LCM and can be used for different purposes (i.e. reporting sustainability performance of a pavement or help decision-making for the use of an innovative asphalt mixture). In this respect, EN 15643-5 (2017) was recently published to set the general principles and requirements of SA for civil engineering works.

Despite these techniques are now widely used and despite the existence of ostensibly clear standards, the route towards daily practice is not unambiguously. First of all, this is due to the procedure followed by these life cycle approaches: the first step in every life cycle analysis is to define the goal and scope of the analysis and thereby determining the functional units, system boundaries and type of indicators. For example, whilst an asphalt producer might be most interested in the carbon footprint of the production process, a road asset manager might be more interested in the life time expectations and a wider variety of environmental impacts (e.g. water pollution, toxic emissions to air, etc.). These scoping differences imply that two different studies are, despite the clear standards they are based on, generally hard to compare. As Santero, Masanet and Horvath (2011) concluded, there is a need for standardization of functional units, system boundary expansion, improved data quality and reliability and broadening of study scopes.

The subsequent problem is that over the past years many different approaches have been followed, datasets created, and tools developed for SA of pavements. This increase in available tooling has however not lead to clear and comparable information, but merely to an overload of slightly or hugely differing information sources. These sources have been inventoried in past CEDR and FP7 projects (for example SUNRA (Sowerby et al., 2014)), integral frameworks for multi-criteria analysis have been developed (for example AllBack2Pave (Lo Presti and D'Angelo, 2015)), Product Category Rules (PCRs) have been published, and attempts have been made to harmonize indicators' use (for example, the CWA 17089 (CEN, 2016) developed in LCE4ROADS). Some countries, like The Netherlands, have gone beyond EN 15804 (2012) and specified the system boundaries for product assessment in infrastructure by means of more specific guidelines and tools (SBK Bepalingsmethode, DuboCalc tool and VBW asphalt model). These projects marked the first steps towards international availability of reliable and comparable sustainability data, but practical implementation is still lacking in most countries.

As a result, there is an imminent need of harmonization of how to perform Sustainability Assessment (SA) exercises in pavement engineering. The next steps towards simpler and better sustainability assessment for NRAs, are:

1. to set clear and practical guidelines for LCM practices for road authorities,
2. to inventory and characterize the available data sources and sketch the path towards harmonization, and
3. to generate reliable durability data for green asphalt.

1.2 *Pavement life cycle management (PavementLCM) project*

The CEDR Transnational Research Programme was launched by the Conference of European Directors of Roads (CEDR). CEDR is the Road Directors' platform for cooperation and promotion of improvements to the road system and its infrastructure, as an integral part of a sustainable transport system in Europe. Its members represent their respective National

Road Authorities (NRA) or equivalents and provide support and advice on decisions concerning the road transport system that are taken at national or international level. Within this CEDR Call 2017, PavementLCM is a 2-year international project aiming at supporting European National Road Authorities to introduce sustainability at the core of their practices by providing training on Life Cycle Management techniques and a user-friendly package to support their widespread implementation.

One of the first tasks of PavementLCM project is the creation of PavementLCM Sustainability Assessment Framework within work package 2 "Transfer of Knowledge" of PavementLCM project. The aim of PavementLCM Sustainability Assessment Framework is to provide recommendations and guide NRAs to perform the SA of asphalt mixtures and pavement activities.

1.3 *Aim of the paper*

Within the context described above, the aim of this paper is to introduce the first steps undertaken towards the creation of PavementLCM Sustainability Assessment Framework for NRAs. In this regard, the framework has to be built adapting the requirements of EN 15643-5 (2017) "Sustainability of construction works – Sustainability assessment of buildings and civil engineering works Part 5 Framework on specific principles and requirements for civil engineering works" to the pavement industry. For this purpose, firstly, the results of the review conducted on the latest available Product Category Rules (PCRs) regarding the definition of the object of assessment, and of sets of sustainability performance indicators for asphalt mixtures and pavement activities are presented. As part of it, the results of a series of interviews carried out with European road authorities as well as with the California Department of Transportation (Caltrans) related to the use of indicators are presented. Finally, the conclusions and basis on which PavementLCM SA Framework will be built are established.

2 METHODOLOGY

2.1 *Development of PavementLCM sustainability assessment framework*

PavementLCM Sustainability Assessment Framework should meet the requirements of the standard EN 15643-5 (2017), for which the main highlights can be summarized as:

- Environmental, social and economic performance must be assessed.
- The assessment should use a life cycle approach.
- The assessment should use quantifiable indicators measured without value judgements.

In this regard, the most important points to start defining an SA framework is to identify the systems to analyze, which will be in the scope of the framework, using a life cycle approach (named objects of assessment from now on as per EN 15643-5) and the indicators to measure the sustainability performance of such objects of assessment.

The scope of PavementLCM SA Framework includes two different objects of assessment which are considered crucial for NRAs: "pavement materials" and "road pavement activities" (Figure 1). Within the pavement materials category, only asphalt mixtures are covered due to its importance in Europe, while pavement activities category includes any new construction, maintenance and rehabilitation operations in asphalt pavements. It is important to highlight the importance of the differentiation of these two objects of assessment because:

1. The stakeholders involved in the SA of pavement materials and pavement activities are different. While contractors and material designers are more interested in the SA of materials, i.e. to declare the sustainability performance of their products, the SA of activities is more interesting for NRAs to help their decision-making process.
2. The life cycle of each object of assessment is different. Asphalt mixtures are materials and therefore their life cycle only covers cradle-to-gate, while the assessment of pavement

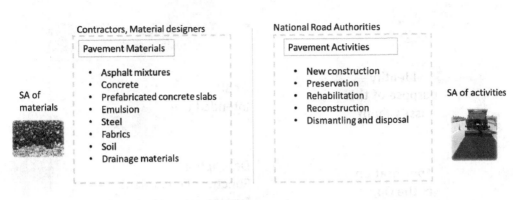

Figure 1. Pavement materials vs. Pavement activities in PavementLCM SA Framework.

activities may include any other life cycle stage of interest for the practitioner (i.e. cradle-to -laid, cradle-to-grave).

3. Different indicators are needed to perform the SA of each object of assessment.

In order to carry out and complete the SA of pavement materials and activities, the steps illustrated in Figure 2 shall be followed. which represent the core of PavementLCM SA Framework. These steps have been defined on the basis of the series of standards for calculation methods for buildings (EN 15978, 2011; EN 16309, 2014; EN 16627, 2015). The aim of PavementLCM SA Framework is to provide guidelines for NRAs to perform each of such steps. In this regard, the process to develop the framework started with the review and survey with NRAs of their SA practices focusing on defining (1) the objects of assessment within the scope of the framework and (2) sets of sustainability performance indicators for each of them, which are the results presented in this paper.

2.2 Review of the definition of objects of assessment

The definition of an object of assessment in a SA (EN 15643-5) should include/be accompanied by:

- A description of the object.
- Life cycle stages included.
- Functional/declared unit. The functional unit defines the way in which the identified functions or performance characteristics of the product are quantified. The primary purpose of the functional unit is to provide a reference by which material flows (input and output data) and any other information are normalized to produce data expressed on a common basis. The declared unit is used instead of the functional unit when the precise function of the object of assessment is not stated or is unknown.
- Analysis period (if applicable): period over which the time-dependent characteristics of the object of assessment are analysed.

The main efforts focused on the harmonization of the definition of objects of assessment for a SA are those made for LCA in Product Category Rules (PCRs). In this regard, the latest PCRs for asphalt mixtures and pavement activities available were reviewed and compared.

2.3 Review and survey about sustainability performance indicators

The main latest efforts dedicated to defining sets of sustainability performance indicators were identified and reviewed for comparison. Next, a survey of NRAs practices with interviews with 11 NRAs: Each NRA was contacted separately and at least one member, expert in "Sustainability", was interviewed for an average time of two hours. In most cases the NRA's

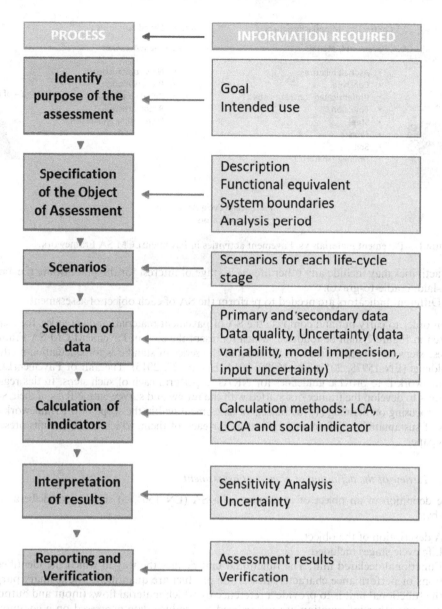

PROCESS		INFORMATION REQUIRED
Identify purpose of the assessment	←	Goal Intended use
Specification of the Object of Assessment	←	Description Functional equivalent System boundaries Analysis period
Scenarios	←	Scenarios for each life-cycle stage
Selection of data	←	Primary and secondary data Data quality, Uncertainty (data variability, model imprecision, input uncertainty)
Calculation of indicators	←	Calculation methods: LCA, LCCA and social indicator
Interpretation of results	←	Sensitivity Analysis Uncertainty
Reporting and Verification	←	Assessment results Verification

Figure 2. Steps to set up a SA.

representatives were more than one. In the case of Caltrans, 5 experts from different departments agreed to provide us the requested information. The full questionnaire can be requested to the authors. In this paper, only the results related to the awareness and use of sustainability performance indicators are discussed.

3 RESULTS AND DISCUSSION

3.1 *Definition of objects of assessment*

Table 1 summarizes the features of the PCRs reviewed for the definition of the objects of assessment of Pavement SA Framework. The nomenclature in Table 1 for the life cycle stages

Table 1. PCR's object of assessment definition ("NA" = Not Applicable; "-"= not defined).

PCR	Object	Life Cycle stages	Functional/Declared Unit	Analysis period
NAPA (2017)	Asphalt mixture	Cradle-to-gate (A1 to A3)	Declared unit: 1 short tonne of asphalt mixture	NA
EAPA (2017)	Asphalt mixture	Cradle-to-gate (A1 to A3)	Declared unit: 1 metric tonne of asphalt mixture	NA
The International EPD System (2018a)	Asphalt mixture	Cradle-to-gate (A1 to A3)	Declared unit: 1 metric tonne of manufactured asphalt mixture	NA
	Asphalt mixture	Cradle-to-gate with options (A1 to A4)	Declared unit: 1 metric tonne of manufactured asphalt mixture delivered to the construction site	NA
	Pavement activities	Cradle-to-gate with options (A1 to A5)	Functional Unit: A paved surface of $1m^2$, which fulfils the specified quality criteria during the Reference Service Life	NA
	Pavement activities	Cradle-to-gate with options (A1 to A5 + B1 and B4 minimum)	Functional Unit: A paved surface of $1m^2$, which fulfils the specified quality criteria during the Reference Service Life of the construction (default value of 40 years)	-
	Pavement activities	Cradle-to-grave (A1-A3, A4-A5, B1-B4, C1-C4)	Functional Unit: A paved surface of $1m^2$, which fulfils the specified quality criteria during the Reference Service Life of the construction (default value of 40 years)	-
The Norwegian EPD Foundation (2017)	Asphalt mixture	Cradle-to-gate with options (A1 to A3)	Declared Unit: 1 tonne of manufactured asphalt mixture	NA
	Pavement activities	Cradle-to-gate with options (A1 to A5)	Functional Unit: $1 m^2$ surface covered with asphalt, which fulfils the specified quality criteria during the service life of asphalt surfacing.	NA
	Pavement activities	Cradle-to-gate with options (A1 to any beyond A5)	Functional Unit: $1 m^2$ surface covered with asphalt, which fulfils the specified quality criteria during the Estimated Service Life of a construction work	-
	Pavement activities	Cradle-to-grave (A1-A3, A4-A5, B1-B4, C1-C4)	Functional Unit: $1 m^2$ surface covered with asphalt, which fulfils the specified quality criteria during the Estimated Service Life of a construction work	-
The International EPD System (2018b)	Highways (except elevated highways), streets and roads	Cradle-to-gate with options (A1-A3 mandatory)	Declared unit: $1 m^2$ of road	NA
		Cradle-to-grave (A1-A3, A4-A5, B1-B7, C1-C4)	Functional unit: $1 m^2$ of road with a specific intended use	-

(A1-A5, B1-B7, C1-C4) refers to that defined in EN 15643-5 (2017) for civil engineering works and used in the PCRs for declared information modules and is further specified here. According to Table 1, the most common declared unit for asphalt mixtures is 1 metric tonne of asphalt mixtures. Most PCRs define the life cycle stages of this material as cradle-to-gate which includes the following information modules: A1 Raw materials acquisition, A2 Transport to manufacturing site and A3 Manufacturing; while only the International Environmental Product Declaration (EPD) System (2018a) gives the option of including A4 Transport to construction site. There is no analysis period for this type of analysis. PavementLCM SA Framework will consider modules A1-A3 for asphalt mixtures.

In the case of pavement activities, all PCRs, except the International EPD System (2018b), present the same functional unit as 1 m^2 of paved surface meeting the quality requirements during its reference service life, which will be the same in PavementLCM SA Framework. the International EPD System (2018b) define the functional unit as 1 m^2 of road with a specific extended use because it covers a wider object of assessment (e.g. whole infrastructure of highways, streets and roads) than the rest, which are dedicated to pavement activities related to asphalt layers. Since PavementLCM SA Framework is focused on asphalt mixtures and pavement activities, the International EPD System (2018b) PCR is considered only for information.

In terms of life cycle stages, the International EPD System (2018a) and the Norwegian EPD Foundation (2017) define a particular system as "cradle-to-gate with options" including, in addition to A1-A3, the modules: A4 Transport to construction site and A5 Construction. Regarding A5, the construction process of asphalt mixtures consists of the installation of the material in the road, e.g. laying and compaction. These options can therefore be referred as "cradle-to-laid" (Butt et al., 2019). More systems are defined in these PCRs for pavement activities in which the practitioner can perform a "cradle-to-gate" assessment plus any other module of interest depending on the goal of the study and are referred as "cradle-to-gate with options". On the other hand, if a "cradle-to-grave" analysis is to be performed, the International EPD System (2018a) and the Norwegian EPD Foundation (2017) define as mandatory the following modules: B1 Use, B2 Maintenance, B3 Repair, B4 Replacement, C1 Removal, C2 Transport to waste management plant, C3 Waste processing and C4 Disposal. Only the International EPD System (2018b) includes modules B5 Refurbishment, B6 Operational energy use and B7 Operational water use. This difference is again due to the fact the International EPD System (2018b) is dedicated to a wider object of assessment than the rest. It is therefore stated that PavementLCM SA Framework should include information modules A1-A3, B1-B4 and C1-C4 for pavement activities.

Finally, there is no specification of analysis period in any of the PCRs reviewed. This means that the analysis period is open to the decision of the practitioner of the LCA. In the case of performing a cradle-to-laid analysis, the definition of an analysis period does not apply, while it is needed and of crucial importance for cradle-to-gate with options and cradle-to-grave analysis. In this regard, Harvey et al. (2016) gave a recommendation to use the same principle as in LCCA establishing an analysis period which covers at least one major rehabilitation operation. For pavement activities, 40 years is considered a reasonable value. PavementLCM SA Framework follows this recommendation.

3.2 *Sustainability performance indicators*

EN 15643-5 (2017) highlights the need of using quantifiable indicators to measure the sustainability performance of civil engineering works. Table 2 shows the different sets of indicators proposed by several European standardization and research efforts.

EDGAR (De Visscher et al., 2016) is a CEDR project funded in the call for Energy Efficiency 2013 in which one of the objectives was to select appropriate sustainability criteria for asphalt mixtures. The basket of indicators proposed was the result of a literature review followed by a screening process. The screening consisted of the normalization of the environmental indicators declared in four asphalt mixtures EPDs in order to identify the least relevant for asphalt mixtures. Finally, the basket of indicators was approved by an advisory board and project executive board. A total of 11 indicators are finally proposed.

CWA 17089 (CEN, 2016) was the result of the FP7 project LCE4ROADS. This set of sustainability performance indicators was selected to consider as environmental criteria the

Table 2. Sets of sustainability performance indicators.

Sustainability pillar	Pavement Activities		Asphalt Mixture
	CWA 17089 (2016)	SUP&R ITN (2017)	EDGAR (2015)
Environmental	Global warming potential (GWP)	Global warming potential (GWP)	Global warming potential (GWP)
	Formation potential of tropospheric ozone (POCP)		Air pollution
	Depletion potential of the stratospheric ozone layer (ODP)	Depletion potential of the stratospheric ozone layer (ODP)	
	Acidification potential of soil and water (AP)	Acidification potential of soil and water (AP)	Leaching potential
	Eutrophication potential (EP)	Eutrophication potential (EP)	
	Abiotic depletion potential for non-fossil resources (ADP-elements)		
	Abiotic depletion potential for fossil resources (ADP-fossil fuels)		
	Human toxicity potential (HTP)		
	Ecotoxicity potential (ETP)		
	Energy use	Energy demand	
	Secondary materials	Secondary materials used	Depletion of resources & waste management
	Primary materials consumption (including water)	Water consumption	
	Waste		
		Particulate matter	
	Materials or components to be reused or recycled, and exported energy	Materials or components to be reused or recycled	Recyclability
Economic	Whole life cost	Life cycle Agency Cost	Financial cost
		Life cycle User Cost	
Social	Comfort index	User comfort	Skid resistance
	Safety audits and safety inspections	Safety audits and safety inspections	
	Adaptation to climate change		
	Tire-pavement noise	Noise reduction	Noise
	Responsible sourcing		Responsible sourcing
	Traffic congestion due to maintenance activities	Traffic congestion	Traffic congestion
			Durability

materials to be used, environmental impact associated to the whole infrastructure of roads following EN 15804 and extending it to the construction process, maintenance rehabilitation and use phases. In terms of economic criteria, it is based on ISO 15686-5 (2017), while the social aspect covers criteria for comfort, safety (EU harmonized safety audit and safety inspection (Directive 2008/96EC) applied to the TEN-T: Trans European Transport Network) and noise. In the CWA, 20 European institutions participated including NRAs, research centers and private companies, who agreed on the adequacy of the indicators proposed. Although lower than an EN, this CWA provides already a first level of standardization and consensus for the use of sustainability performance indicators for roads. A total of 21 indicators are proposed.

SUP&R ITN (2017) is a FP7 project which aimed at developing a decision support system that could help pavement engineers at the design stage to rank different alternatives in terms of sustainability. As part of this effort, a set of sustainability performance indicators was selected to be incorporated in the system as the multi-criteria for the decision-making. To select these indicators, firstly, an extensive literature review was conducted to identify the available indicators. Those collected were then screened according to two sets of principles: (1) measurability, unique and globally accepted definition and recurrence; and (2) sensitivity, updatable data, available data, and non-corruptibility. Next, a score based on a three-point scale (i.e. 0, 1 and 2) was assigned to each indicator and those with zero points were already discarded. The remaining indicators were organized in terms of recurrence and any indicator below the 75th percentile was excluded from the list. This list was subjected to critical judgement that would determine their inclusion in the final set of sustainability performance indicators. A total of 15 indicators are finally proposed (Santos *et al.*, 2019).

In Table 2, the indicators have been organized in a comparative basis to those of the CWA 17089 for having this the higher consensual level. It can be observed that CWA 17089 include most of the sustainability performance indicators proposed in SUP&R ITN and EDGAR, except particulate matter and durability, showing therefore a significant agreement between the different research efforts performed towards the harmonization of sustainability indicators.

PavementLCM SA Framework is specifically tailored for NRAs, and therefore, in order to understand the use of indicators in their practices, the set of 21 indicators proposed in the CWA 17089 were presented in the interviews and four options were given to the interviewees to select for each indicator: 1) Not aware; 2) Aware but not using (3) Aware and using (4) Critical.

In order to analyze the results, the score in brackets in the list above was given to each answer. Next, a total score was assigned to each indicator as the sum of all the replies. Next, a threshold of 20 points was taken to consider an indicator as relevant, taking into account that the maximum score could be 44 and the minimum 11. As a result, the relevant indicators in sustainability assessment practices for the NRAs interviewed are:

1. Primary materials consumption
2. Secondary materials used
3. Energy use
4. Waste
5. Global Warming Potential
6. Whole life cots
7. Comfort index
8. Safety audits and inspections
9. Tyre-pavement noise
10. Traffic congestion due to maintenance operations

From this list, five indicators are environmental, one indicator is economic (the only one proposed in CWA 17089) and four indicators are social. In this regard, comparing the number of relevant environmental and social indicators, and considering that "safety audits and inspections" and "tire-pavement noise" are those with the highest score, which means that, opposite to what can be thought, the social indicators in CWA 17089 are in fact very relevant to NRAs.

During the interviews with NRAs, one of the main findings was the different levels of implementation of SA that exists among the different NRAs in Europe (i.e. some NRAs do not know how to use SA while other are already using it for green procurement). For those not using SA, the level of complexity and number of indicators that the presented efforts have proposed is an obstacle to start using it. PavementLCM SA Framework aims at helping NRAs introduce SA on their practices, and therefore, on the basis of this result, the list of the most relevant 10 indicators will be used as an initial point to further screen and select indictors for PavementLCM SA Framework according to NRAs' priorities for the performance of SA case studies.

4 CONCLUSIONS

This paper presents the structure and first steps carried towards the creation of a Sustainability Assessment Framework for the road industry according to EN 15643-5. In this regard, a review of the available PCRs related to asphalt mixtures and pavement activities is presented focusing on the definition of these objects of assessment for life cycle techniques. Next, the main research efforts in Europe towards the definition of sustainability performance indicators for asphalt mixtures and pavement activities are shown, followed by a first screening based on interviews conducted with eleven National Road Authorities.

PavementLCM SA Framework will clearly differentiate between two objects of assessment: pavement materials (focusing on asphalt mixtures) and pavement activities, due to the different implications that the assessment of each of them have in terms of stakeholders' interest, definition of object of assessment and use of indicators. In this regard, after the review of PCRs and use of indicators carried out, it can be concluded that:

- PavementLCM SA Framework will define the object of assessment for pavement materials with a declared unit of 1 tonne of manufactured material and including the life cycle stages to cradle-to-gate (A1-A3) or cradle-to-gate with transport (A1-A4).
- For pavement activities, the functional unit is defined as 1 m^2 of paved surface meeting the quality requirement during the reference service life. The options of cradle-to-laid (A1-A5), cradle-to-gate with options (A1-A4 plus any other) and cradle-to-grave (A1-A5, B1-B4, C-1-C4) are considered, excluding modules B5, B6 and B7. The analysis period should cover at least one major rehabilitation considering 40 years as a reasonable value.
- The selected set of sustainability performance indicators are: primary materials consumption, secondary materials use, energy use, waste, global warming potential, whole life costs, comfort index, safety audits and inspection, tire-pavement noise and traffic congestion due to maintenance operations. PavementLCM SA Framework will consider this list to further define a set of 5-6 indicators in order to facilitate the introduction of SA exercise in NRAs.

PavementLCM project will provide NRAs with a complete package that will include the SA framework together with a "SA compass" to guide NRAs for the selection of SA data and tools, datasets of SA and durability assessment, recommendations for uncertainty analysis, recommendations for decision-making, a roadmap for the harmonization of SA exercises in Europe and recommendations to introduce circular economy principles into the road industry.

ACKNOWLEDGMENTS

The research in this paper is part of PavementLCM (www.pavementlcm.eu) project funded by the CEDR Transnational Research Programme launched by the Conference of European Directors of Roads (CEDR) in the Call 2017 New Materials. The participating NRAs are: New Materials are Austria, Belgium-Flanders, Denmark, Germany, Netherlands, Norway, Slovenia, Sweden and the United Kingdom.

REFERENCES

Butt, A.A., Harvey, J.T., Reger, D., Saboori, A., Ostovar, M. and Bejarano, M., (2019). *Life-Cycle Assessment of Airfield Pavements and Other Airside Features: Framework, Guidelines, and Case Studies* (No. DOT/FAA/TC-19/2). Federal Aviation Administration.

CEN (2016) *CEN Workshop Agreement 17089 Indicators for the sustainability assessment of roads.*

EAPA (2017). *GUIDANCE DOCUMENT FOR PREPARING PRODUCT CATEGORY RULES (PCR) AND ENVIRONMENTAL PRODUCT DECLARATIONS (EPD) FOR ASPHALT MIXTURES.*

EN 15643–5 (2017) *Sustainability of construction works – Sustainability assessment of buildings and civil engineering works. Part 5: Framework on specific principles and requirement for civil engineering work s.*

EN 15804 (2012) *EN 15804:2012 - Standards Publication Sustainability of construction works — Environmental product declarations — Core rules for the product category of construction products, International Standard.*

EN ISO 14044 (2006) *Environmental management — Life cycle assessment — Requirements and guidelines.*

EN15978 (2011) *Sustainability of construction works - Assessment of environmental performance of buildings - Calculation method, International Standard.*

EN16309 (2014) *Sustainability of construction works - Assessment of social performance of buildings - Calculation methodology, International Standard.*

151

EN16627 (2015) *Sustainability of construction works - Assessment of economic performance of buildings - Calculation methods, International Standard.*

Harvey, J., Meijer, J., Ozer, H., Al-Qadi, I.L., Saboori, A. and Kendall, A., (2016). Pavement Life Cycle Assessment Framework (No. FHWA-HIF-16-014). United States. Federal Highway Administration.

ISO 14040 (2006) *Environmental Management - Life Cycle Assessment - Principles and Framework.* doi: 10.1002/jtr.

ISO 15392 (2008) *Sustainability in building construction: General principles.* doi: 10.5594/J09750.

ISO 15686–5 (2017) *Buildings and constructed assets — Service life planing. Part 5: Life-cycle costing.*

NAPA (2017) 'Product Category Rules (PCR) for Asphalt Mixtures', *Environmental Product Declaration*, pp. 1–15.

Lo Presti, D. and D'Angelo, G. (2015) *AllBack2Pave. Sustainability assessment of the AllBack2Pave technologies.*

Santero, N. J., Masanet, E. and Horvath, A. (2011) 'Life-cycle assessment of pavements. Part I: Critical review', *Resources, Conservation and Recycling.* Elsevier B.V., 55(9–10), pp. 801–809. doi: 10.1016/j.resconrec.2011.03.010.

Santos, J., Bressi, S., Cerezo, V. and Presti, D.L., (2019) 'SUP&R DSS: A sustainability-based decision support system for road pavements', *Journal of Cleaner Production*, 206, pp. 524–540. doi: 10.1016/j.jclepro.2018.08.308.

Sowerby, C. *et al.* (2014) *SUNRA. Sustainability – National Road Administrations - Project Framework for a Sustainability Rating System for Roads. Organisational Level User Guide.*

The International EPD System (2018a) *Product Category Rules: ASPHALT MIXTURES.*

The International EPD System (2018b) *Product Category Rules: Highways, streets and roads.*

The Norwegian EPD Foundation (2017) 'Product Category Rules for Asphalt', pp. 1–25.

De Visscher, J. *et al.* (2016) *Evaluation and Decision Process for Greener Asphalt Roads. Final Report. 2016. CEDR Call 2013: Energy Efficiency.*

Impacts of climate-change and realistic traffic conditions on asphalt pavement and rehabilitation decisions using life cycle assessment

K. Haslett, E. Dave & W. Mo
Department of Civil and Environmental Engineering, University of New Hampshire, Durham, USA

ABSTRACT: Typical pavement Life Cycle Assessment (LCA) are performed using historical climate data to evaluate pavement performance and provide recommendations for budgeting and planning of M&R strategies in the future. However, due to climate change, this assumption may not be appropriate as flexible pavements' performance is influenced by climate stressors. This study explores the impacts of future climate data and realistic traffic data (RTD) in the pavement M&R evaluation process. A 26-km stretch of Interstate-495 was used to evaluate costs and environmental impacts with varying M&R scenarios and pavement structures. Predicted performance using historical and future projected climate data in combination with RTD is used for life cycle cost and global warming potential estimation. Results show that incorporating future project climate data and RTD can lead to a substantial increase in agency LCA impacts (up to 20% for the presented case-study), the increase is function of pavement structure and M&R alternative.

1 INTRODUCTION AND BACKGROUND

Since the late 19th century, climate change has consisted of a global temperature rise (0.9 degrees Celsius), global sea level rise (203 mm) and an increase in extreme weather events, among others (Climate Change: Vital Signs of the Planet, 2019). While these changes alone are of concern, the implications to human life, infrastructure systems and the economy is grave. Climate change poses a serious threat to both natural and built systems including transportation infrastructure systems (i.e. bridges, rail, road networks, airports etc.). In context of the road network, future increases in very hot days and heat waves pose a concern with pavement integrity and permanent deformation (Gudipudi et al. 2017). Accelerated sea level rise and increased extreme precipitation events will cause changes in subgrade moisture level, water table depth and flooding susceptibility; alter bearing capacity of the pavement system; and in turn, degrade the performance of the road infrastructure system (Daniel et al. 2014, Knott et al. 2017 and Knott et al. 2019).

This may result in serious implications to freight movement, which is multimodal and moves approximately 50 million tons of freight across the US every day. For example, truck freight, which relies on the efficient and safe transportation of goods via the road network reported a movement of 11.5 billion tons of freight in 2015 and is expected to increase by 44% to move 16.5 billion tons by 2045 (USDOT, 2015). Meanwhile, traffic congestion, road closures and delays affect citizens daily where on average a road-user spends 42 hours waiting in traffic each year, and the annual cost of truck congestion is 28 billion dollars (USDOT, 2015).

The use of life cycle assessment (LCA) and life cycle cost analysis (LCCA) are increasing in pavement management due to the growing need to consider sustainable, cost and environmentally effective maintenance and rehabilitation (M&R) plans. Future planning incorporating the effects of a changing climate and realistic traffic conditions is critical. Typically, pavement LCA are performed using historical climate data to plan pavement life expectancy and inform M&R plans. However, pavements systems are constantly exposed to the natural environment and impacts of climate change, therefore it may not be applicable to use historical climate data to inform decisions about future pavement performance.

In recent years, there have been several studies conducted that explore the impacts on pavement performance using future projected climate data in the form of temperature (Meagher et al., 2012), precipitation and the combination of both (Heitzman et al. 2011 and Mndawe et al. 2015). In 2017, Gudipudi et al. conducted a study with the primary object to predict performance of freeway sections in different climatic regions across the US using different climate models. Performance predicted using historical climate data compared to incorporating projected climate data was performed focusing on various pavement distresses such as fatigue cracking, asphalt concrete (AC) rutting and total rutting. However, an LCA approach was not used to make comparisons among various M&R treatments and the timing of those treatments using historical versus projected climate data in combination with realistic traffic data (RTD).

The study herein explores the use of incorporating future projected climate data and RTD into a pavement LCA analysis. Building upon a LCA framework that includes real time traffic data and considers both user and agency costs (Haslett et al. 2019), the addition of future climate data (temperature and precipitation) in the pavement performance analysis is demonstrated in this paper using a case study for two flexible pavement cross sections and three M&R alternatives. The primary study objective of this study was to use an LCA framework to investigate impacts in terms of global warming potential (GWP) and life cycle cost (LCC) when using historical climate data compared to projected future climate data in combination with realistic traffic conditions.

2 METHODOLOGY

2.1 Life cycle assessment framework

A pavement LCA framework that included raw materials and excavation, material transportation, operational and maintenance impacts and end-of-life was used in this study. A 30-year analysis period was considered and all impacts (GWP and LCC) were quantified in terms of agency, user and total impact. LCA software programs including Sima-Pro 8.3 and the Pavement-Life-Cycle Assessment Tool (PaLATE2.0) were used to collect unit impact information on raw materials, transportation and construction impacts. Pavement performance curve information was generated using the American Association of State Highway and Transportation Officials (AASHTO) PavementME design software and then used to determine fuel consumption and carbon emission factors for different vehicle classes under various International Roughness Index (IRI) and speeds. Scenarios were calculated using a combination of Google Maps®, the U.S. Environmental Protection Agency (EPA) Motor Vehicle Emission Simulator (MOVES), the SHRP2 Naturalistic Driving Study, and Massachusetts Department of Transportation (MassDOT) Data Management System. Further detail on the incorporation of RTD into an LCA framework is discussed in a prior study (Haslett et al. 2019). Lastly, end-of-life was account for by discounting salvage value (remaining life) of each pavement M&R scenario at the end of the 30-year analysis period.

2.2 Case study location and pavement cross sections

This case study utilized a 26-km section of Interstate 495 (I-495) in Massachusetts, from Chelmsford to Methuen. This highway is comprised of 3 lanes in each direction with a distributional factor of 50% (of 24 hr. peak volume). Traffic volume information for this section was collected from MassDOT's traffic data management system. To be conservative, all simulations assumed no annual change in total traffic volume, i.e. annual growth rate of 0%.

Two different pavement cross sections were simulated, a "thick pavement" and "thin pavement" section. Simulating two different cross section allows for comparisons to be drawn on the role of pavement structure on performance under future climate conditions with respect to the different M&R strategies explored in this study. Table 1 summarizes the pavement cross section information used in this study for three different M&R scenarios; do nothing and reconstruct (DNR), mill and overlay (MO) and cold-in-place recycling (CIR).

2.3 Maintenance and rehabilitation scenarios

For both the thick and thin pavement cross section, three different M&R strategies (DNR, MO and CIR) were consider using historical climate data and future projected climate data following the RCP 8.5 emission pathway resulting in a total of 12 scenarios. Predicted initial IRI values used in development of performance curves for each cross section were determined using a 90% reliability factor in PavementME design software. The thick pavement cross section required an initial IRI value of 1.55 m/km, meanwhile the thin pavement section required an predicted initial IRI of 1.66 m/km. While the initial IRI values are similar when using either historical or future climate projection data, the performance curve slopes vary due to the rate of accumulated pavement distresses. Figure 1 provides an example of the predicted performance curves from PavementME design software for both the thick and thin pavement section. The analysis period of 30 years from 2020-2050 was held constant among all scenarios. The dashed lines represent the respective M&R scenario performance curves assuming historical climate data, while the solid lines represent the same M&R scenarios but assuming future projected climate data.

The DNR scenario assumes that no M&R activity is conducted over the analysis period except for reconstruction when a terminal IRI value of 2.71 m/km is reached. The MO scenario begins with the same initial IRI values as the DNR scenario, however when the pavement degrades to an IRI of 2.21 m/km a MO treatment is triggered (the milling depth and overlay

Table 1.   Pavement cross-sections and AASHTO PavementME design software input parameters used for given case study location.

Input Type	Variable	DNR[b]	MO[d]	CIR[f]
Thick Cross Section	Layer 1	50.8 mm AC[c]	50.8 mm OL[e]	50.8 mm AC
	Layer 2	101.6 mm AC	101.6 mm AC	101.6 mm CIR
	Layer 3	203.2 mm Crushed Stone Base		
	Layer 4	609.6 mm Prepared Subgrade (A-2-6)		
	Layer 5	Natural Subgrade (A-2-6) Semi-infinite		
Thin Cross Section	Layer 1	50.8 mm AC	50.8 mm OL	25.4 mm OL
	Layer 2	50.8 mm AC	50.8 mm AC	76.2 mm CIR
	Layer 3	101.6 mm Crushed Stone Base		
	Layer 4	Prepared Subgrade (A-2-6) 609.6 mm		
	Layer 5	Natural Subgrade (A-2-6) Semi-infinite		
Traffic	AADTT[a]	12,175		
	Lanes	3		
	Speed (km/hr)	105		
Climate	Elevation (m)	5.8		
	Location	Boston, MA		

[a]AADTT = Two way annual average daily truck traffic.
[b]DNR = Do nothing and reconstruct
[c]AC= Flexible Asphalt concrete
[d]MO = Mill and Overlay
[e]OL = Flexible Overlay
[f]CIR = Cold-In-Place Recycling

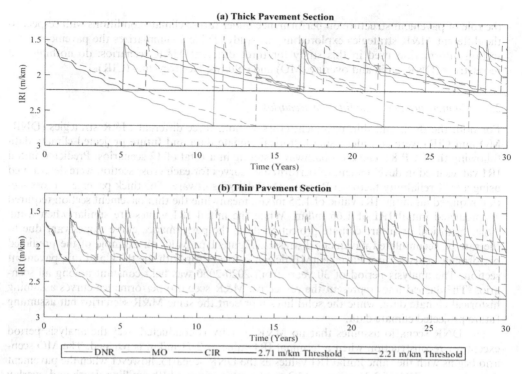

Figure 1. Performance curves for (a) thick pavement section and (b) thin pavement section comparing maintenance and rehabilitation timing for historical climate data (dashed line) and future climate data (solid line) scenarios.

thickness are equal and assumed to be constant in each MO application). The MO treatment is repeated until the end of the 30-year analysis period. Similarly, the CIR scenario begins with the same initial IRI curve as the DNR scenario but when an IRI threshold of 2.21 m/km is reached it triggers the CIR treatment, at which point the CIR performance curve is repeated until the end of the analysis period. Please note that as with typical CIR practice, the CIR layer is surfaces with an asphalt concrete overlay.

2.4 *Climate data integration*

Climate data averaged from 21 models assuming RCP 8.5 (highest emission pathway) and RCP 4.5 (intermediate emission pathway) was integrated into AASHTO's PavementME design system to evaluate pavement performance using pavement IRI. This paper presents only the comparison of historical climate data to RCP 8.5 future climate projection data for brevity and due to early findings in the study that there was not a significant difference in performance curves using average RCP 4.5 or RCP 8.5 emission scenarios, therefore the timing of M&R treatments being triggered would be comparable. It is also acknowledged that future research efforts should undertake a probabilistic approach with the use of 21 different global circulation models as opposed to taking the average of them.

Climate data was procured from Coupled Model Intercomparison Project Phase 5 (CMIP5) for a 12 km square grid near the case study location (latitude 42° 30′ 58.68"N and longitude 71° 44′ 44.88"W). Daily precipitation and daily maximum and minimum temperature were obtained for the years 2020-2050. To incorporate climate date in to PavementME, further processing is required to convert temperature and precipitation to hourly climatic data. Daily precipitation values were divided into 24 equal increments and spaced over the course of a given day. Converting daily maximum and minimum temperatures into hourly data was slightly more computationally intensive. A method adapted by Valle et al. in 2017 was followed in this study where the minimum daily temperature (t_{min}) occurs at sunrise and the maximum temperature (t_{max}) occurs at 2 p.m. Equations 1-4 are used to calculate intervening temperatures:

$$\text{for } 0:00 < h < \text{rise and } 14:00 < h < 24:00, T(h) = t_{ave} + amp(\cos(\pi \times h')/(10 + \text{rise})) \quad (1)$$

$$\text{for rise} < h < 14:00, T(h) = t_{ave} - amp(\cos(\pi(h' - \text{rise})/14))) \quad (2)$$

where

$$h' = h + 10 \text{ if } h < \text{rise} \quad (3)$$

$$h' = 14 \text{ if } h > 14 \quad (4)$$

where rise = time of sunrise in hours; $T(h)$ = temperature at any hour; h = time in hours, $h' = h + 10$ if $h < $ rise, $h' = 14:00$ if $h > 14:00$; $t_{ave} = (t_{min} + t_{max})/2$; $amp = (t_{max} - t_{min})/2$

In addition to hourly temperature and precipitation data, PavementME also requires wind speed, percent sunshine, percent humidity and water table depth as climatic input factors. For wind, sunshine and humidity inputs the last year of historical climate data available was repeated for subsequent years until the end of the analysis period. This input assumption is deemed appropriate as the effect on pavement performance (in terms of IRI) is not considered dominate, therefore effects are negligible (Qiao et al. 2013). For water table depth, it was assumed to remain at a constant depth of 3.05 m. Change in groundwater level due to climate change is of concern and may have considerable implications on pavement performance and service life depending on the location as demonstrated by Knott et al. in 2017 and 2019. However, due to groundwater projection not being available in the CMIP5 dataset researchers verified groundwater level trends provided by the United States Geological Survey (USGS) and determined that groundwater level may not be of major concern for the current case study location, therefore it was not considered in this analysis.

2.5 Life cycle cost determination

To make fair comparisons among all M&R scenarios over the 30 year analysis period, all life cycle costs for both users and agencies were converted to net present value (NPV) using Equation 5:

$$NPV = \frac{Future\ Cost}{(1 + i)^n} \quad (5)$$

where i = discount rate (percent); n = number of years from initial construction.

If a pavement alternative had any remaining life determined from its respective IRI performance curve, its salvage was accounted for in the calculation of LCC. This was accomplished by taking the percentage of remaining months in service and discounting that percent of remaining life to present value. For example, if a given M&R alternative had 15 months of service life remaining based on the its IRI performance curve and it was expected to have 60 months of total service life, then 25% of the M&R treatment cost was discounted back to present value.

3 RESULTS AND DISCUSSION

LCA impacts for all scenarios explored in this case were quantified in terms of GWP (CO_2 eq) and LCC (dollars). Comparisons are drawn on the impacts of using historical climate data or future climate projection data in the LCA framework for operations (user impact), construction and maintenance (agency impact) and total impact (combination of user and agency).

3.1 Global Warming Potential (GWP)

The GWP of a greenhouse gas (GHG) indicates the amount of warming a gas can cause over a given period (typically 100 years). GWP is an index, where CO_2 has an index value of 1 and for all other GHG (methane, nitrous oxides, hydrofluorocarbons etc.) the index value is the number of times more warming they cause compared to CO_2 (Brander et al. 2012). For this study, carbon dioxide equivalent (CO_2 eq) was used, which signifies the amount of CO_2 that would have the equivalent global warming impact. Table 2 summarizes the GWP impact for all scenarios using historical, future projected climate data and the percent difference with respect to using historical climate data. Results are tabulated for the thick pavement section and for the thin pavement section separately.

It can be observed that there is a larger difference in GWP impact when using future climate data as compared to historical due to construction and maintenance (C/M) activities as oppose to operational impacts from roadway users, regardless of cross section type. For the thick pavement cross section, the total difference in impact is highest for the DNR scenario (6.03%) followed by the MO (3.69%) and CIR treatment (0.17%). The same trend is observed for the thin pavement cross section. The direct implication for pavement designers and policymakers is the potential underestimation of GWP impact over the lifetime of pavement system. The magnitude of the underestimation is dependent on pavement structure, as well as the type of M&R undertaken over the pavement service life (i.e. no maintenance, MO only or use of CIR). One reason why the percent difference is higher for thicker pavements compared to thin pavements while holding all traffic, material characteristics and climate scenarios constant is the critical role of having a sufficient pavement structure to withstand the current traffic level. From Figure 1a and Figure 1b the increase in frequency of M&R treatments is clearly shown for two different pavement structures. There is a higher frequency of M&R activity for the thin pavement section under the given traffic loading over the 30-year analysis period. This trend holds true when comparing performance curves generated with historical climate data and future projected climate data.

Table 2 . Global warming potential impact summary table for all maintenance and rehabilitation scenarios during historical and future projected climate data.

Global Warming Potential (Gg CO_2eq)								
		Thick Pavement Section			Thin Pavement Section			
M&R Scenario	Impact	Historical	8.5 RCP[f]	%Diff[g]	Impact	Historical	8.5 RCP	%Diff
DNR[a]	C/M[d]	4.57	5.71	20.00	C/M	12.76	14.48	11.86
	Ops[e]	13.22	13.22	0.00	Ops	13.23	13.23	0.00
	Total	17.79	18.93	6.03	Total	25.99	27.71	6.20
MO[b]	C/M	3.92	4.57	14.34	C/M	7.88	8.14	3.23
	Ops	13.21	13.21	0.00	Ops	13.21	13.21	0.01
	Total	17.13	17.78	3.69	Total	21.09	21.35	1.24
CIR[c]	C/M	2.90	2.93	0.95	C/M	2.58	2.65	2.84
	Ops	13.205	13.204	-0.01	Ops	13.21	13.21	0.00
	Total	16.11	16.13	0.17	Total	15.78	15.86	0.48

[a] DNR = Do nothing and reconstruct
[b] MO = Mill and overlay
[c] CIR = Cold-in-place recycling
[d] C/M = Construction and maintenance (agency impact)
[e] Ops = Operations (user impact)
[g] RCP = Representative concentration pathway
[f] %Diff = Percent difference with respect to impacts using historical climate data

The operational impact (Ops) compared for each scenario includes the impact from all vehicles traveling on the 26 km stretch of interstate for 30 years (entire analysis period). Operational impact considers pavement roughness as well as realistic traffic condition (daily traffic congestion) based on average travel patterns from Google Maps® to quantify the user impact over the analysis period.

Another interesting observation is when comparing all M&R scenarios (historical and future climate data together), the maximum percent difference with respect to the highest impact scenario in terms of CO_2 eq for agency impact was 49.2% and 82.2% for the thick and thin pavement respectively. Similarly, the maximum percent difference from a road user perspective was 0.13% and 0.19% for the thick and thin pavement section. While the difference in impact is higher for agencies due to construction and maintenance costs, it is important to note that the 0.13% and 0.19% is based on the current traffic volume with assumption of no traffic growth over the 30-year analysis period. It is recommended that further analysis be performed to verify the user impact while assuming varying traffic growth percentages. Other variables that may influence operational impact include vehicle fuel efficiency and future fuel costs (Haslett et al. 2019). In practice, when selecting a M&R treatment plan, it is recommended that varying traffic growth levels, vehicle efficiencies and fuel costs be assessed using either a probabilistic approach or sensitivity analysis to ensure uncertainty associated with user impacts can be accounted for.

3.2 Life Cycle Cost Analysis (LCCA)

All life cycle costs (LCC) were calculated in terms of NPV in billions of dollars, while taking into account salvage value at the end of the analysis period. Table 3 summarizes the LCC impacts for all M&R scenarios using historical and future projected climate data and the percent difference with respect to historical climate data.

The percent difference by incorporating future project climate data into the LCA framework resulted in a total percent increase of 1.97% and 1.41% in LCC for the DNR scenario for thick and thin pavement section respectively. Once again, a larger difference is observed for agency costs as compared to user costs and most notably for the DNR scenario followed by MO and CIR. The maximum percent difference in LCC for agencies (C/M impact) was

Table 3. Life cycle cost impact summary table for all maintenance and rehabilitation scenarios during historical and future projected climate data.

Life Cycle Cost (Billions of Dollars)								
	Thick Pavement Section				Thin Pavement Section			
M&R Scenario	Impact	Historical	8.5 RCP	(%diff)	Impact	Historical	8.5 RCP	(%diff)
DNR	C/M	0.26	0.31	16.55	C/M	0.62	0.70	11.62
	Ops	2.33	2.33	0.01	Ops	2.33	2.29	-1.69
	Total	2.59	2.64	1.97	Total	2.94	2.99	1.41
MO	C/M	0.16	0.16	0.74	C/M	0.17	0.17	0.27
	Ops	2.32	2.32	0.01	Ops	2.29	2.29	0.01
	Total	2.48	2.49	0.05	Total	2.45	2.45	0.03
CIR	C/M	0.16	0.16	0.22	C/M	0.16	0.16	0.38
	Ops	2.32	2.32	0.00	Ops	2.29	2.29	0.00
	Total	2.48	2.49	0.01	Total	2.45	2.45	0.03

a DNR = Do nothing and reconstruct
b MO = Mill and overlay
c CIR = Cold-in-place recycling
d C/M = Construction and maintenance (agency impact)
e Ops = Operations (user impact)
g RCP = Representative concentration pathway
f %Diff = Percent difference with respect to impacts using historical climate data

48.8% for the thick pavement cross section and 76.8% for the thin pavement cross section. While results will depend substantially on the highway section of choice, location and other case specific inputs, the difference in impacts (both GWP and LCC) when using future projected climate data can result in a potential for significant savings over the service life of a roadway.

Meanwhile, the users (ops impact) maximum percent difference in LCC for thick and thin pavement was 0.13% and 1.86% respectively. The slight increase in operational impact when the total pavement structure thickness decrease emphasizes the need to consider operational impact for varying pavement sections as the rate of deterioration (quicker drop in IRI) may impact road user's fuel consumption and in turn increase GWP.

3.3 Impact of incorporating future climate data

The assumption of using historical climate data to design and predict pavement performance in the future where the climate is changing may lead to under designed pavements and lack of budgeting for M&R over the service life of the road. Pavements are constantly exposed to the natural environment and design to perform under given temperature ranges and environmental conditions. However, if those design criteria do not consider an accurate representation of what the pavement will be exposed to over the course of its service life, it can lead to increased user and agency life-cycle impacts.

Incorporation of future climate data projections within PavementME design software showed that for this evaluated case-study location and all M&R scenarios, pavements will experience a higher distress accumulation and early failure. As a result, the frequency of M&R activity increased for all scenario regardless of pavement structure. There is a need to consider future climate projections when conducting a pavement LCA to ensure that appropriate user and agency impacts are correctly accounted. Results from this case study are in agreement with literature that have incorporated future projected climate (temperature and or precipitation data) showing the increase impacts on pavement distresses (Mills et al. 2009, Heitzman et al. 2011, Daniel et al. 2014, Gudipudi et al. 2017, Mallick et al. 2018, Stoner et al. 2019).

In this study, agency GWP impacts increase by as high as 20% and as low as 0.97% depending on the M&R scenario and pavement structure. Similarly, agency LCC impacts may increase by as much as 16.6% and as low as 0.22%. While this is a fairly high range of increase in agency impact, it reiterates the need to properly predict pavement performance in a changing climate as an opportunity to optimize M&R treatments and long-term budgeting. From a user perspective, the difference in GWP and LCC impacts were not as significant (less than a percent) across all M&R scenarios and pavement section when comparing historical and future climate data. It is important to note the constraints of the case study presented and the possible causation it has on agency and user impacts. By holding the M&R treatment IRI trigger values constant for both the thick and thin pavement section over the entire analysis period, what is observed is an increase in M&R activity as pavement structure decreases and with the incorporation of future projected climate data. The pavement roughness is allowed only to reach an IRI value of 2.71 or 2.21 m/km depending on the M&R scenario before a given M&R treatment is applied and pavement condition restored. However, if M&R activity was held constant (i.e. every 5 or 10 years) and pavement condition allowed to continue to degrade (i.e. IRI continues to increase) it is suspected that operational impact would increase with the incorporation of future projected climate data.

4 SUMMARY

A primary objective of this study was to first illustrate the need to consider future climate data when performing a pavement LCA, this driven by the changing climatic conditions and its impacts on performance and longevity of pavement infrastructure. A method to incorporate future projected climate data an LCA framework that considers realistic traffic conditions was presented along with results from a case study consisting of two different pavement structures and three M&R alternatives. Results from this study show that in general LCA impacts in terms of GWP and LCC increase when using future projected climate data. A higher percent difference in GWP and LCC from use of historical to future projected climate data is observed for agencies due to increased number of construction activities in the future climate conditions. Whereas, since this study limited the pavement performance within a close range of roughness (in terms of IRI), the road user's operational impacts were consistent between historic and future climate. Agency GWP impacts may increase by as much as 20% and as little as 0.97% depending on pavement structure and the M&R scenario while LCC impacts increase by as much as 16.6% and as little as 0.22%. It should be clear that for this study M&R timing was based on set IRI trigger values, however if M&R timing was held to a consistent schedule (i.e. every 5 or 10 years) and pavement roughness allowed to continue to degrade it is expected that operational impact would increase with the incorporation of future projected climate data. Additionally, instead of suing a single treatment repeatedly over the analysis period, a combination of maintenance and rehabilitation treatments would be performed. Further analysis is required to verify the assumption regarding the constraint of M&R timing activity and the effect on operational impacts.

Another recommendation for future work or extension of this study would be to perform a probabilistic analysis with the use of 21 different global circulation models available from CMIP5 rather than taking the average of them. However, as demonstrated in this study by simply incorporating future projected climate data into the pavement LCA framework there is a substantial opportunity to improve reliability of the planning and budgeting process for pavement management over the infrastructure life, while minimizing the environmental impact from agencies and users. RCP emission pathways 4.5 (intermediate) and 8.5 (high) were evaluated initially, however it was determined that there was minimal difference on the timing of M&R activities, therefore results from only RCP 8.5 were presented. The implication for agencies is that there is a difference in LCA impacts when using historical compared to future projected climate data regardless of which concentration pathway is assumed. Therefore, it is important to consider future rather than historical climate data when performing a pavement LCA to accurately capture the timing of M&R activities and change in operational impacts due to the higher rate of pavement distress accumulation.

REFERENCES

Brander, M., Davis, G. (2012) Greenhouse Gases, CO2, CO2e, and Carbon: What Do All These Terms Mean? Ecometrica. [online] Available at: https://ecometrica.com/assets/GHGs-CO2-CO2e-and-Carbon-What-Do-These-Mean-v2.1.pdf. [Accessed 11 Sep. 2019].

Climate Change: Vital Signs of the Planet. (2019). *Climate Change Evidence: How Do We Know?* National Aeronautics and Space Administration (NASA). [online] Available at: https://climate.nasa.gov/evidence/[Accessed 11 Sep. 2019].

Daniel, Jo Sias, et al. "Impact of climate change on pavement performance: Preliminary les- sons learned through the infrastructure and climate network (ICNet)." *Climatic Effects on Pavement and Geotechnical Infrastructure*. 2014. 1–9.

Gudipudi, Padmini P., B. Shane Underwood, and Ali Zalghout. "Impact of climate change on pavement structural performance in the United States." Transportation Research Part D: Transport and Environment 57 (2017): 172–184.

Haslett, Katie E., Eshan V. Dave, and Weiwei Mo. "Realistic Traffic Condition Informed Life Cycle Assessment: Interstate 495 Maintenance and Rehabilitation Case Study." Sustainability 11.12 (2019): 3245.

Heitzman, Michael, et al. Developing MEPDG Climate Data Input Files for Mississippi. No. FHWA/MS-DOT-RD-11-232. 2011.

Knott, Jayne F., et al. "Assessing the effects of rising groundwater from sea level rise on the service life of pavements in coastal road infrastructure." Transportation Research Record 2639.1 (2017): 1–10. https://doi.org/10.3141/2639-01

Knott, Jayne F., et al., "A Framework for Introducing Climate-Change Adaptation for Pavement Management," Sustainability, 11 (16),p. 4382, 2019. https://doi.org/10.3390/su11164382

Mallick, Rajib B., et al. "Understanding the impact of climate change on pavements with CMIP5, system dynamics and simulation." International Journal of Pavement Engineering 19.8 (2018): 697–705.

Meagher, William, et al. "Method for evaluating implications of climate change for design and performance of flexible pavements." Transportation Research Record 2305.1 (2012): 111–120.

Mills, Brian N., et al. "Climate change implications for flexible pavement design and performance in southern Canada." Journal of Transportation Engineering 135.10 (2009): 773–782.

Mndawe, M. B., et al. "Assessment of the effects of climate change on the performance of pavement subgrade." African Journal of Science, Technology, Innovation and Development 7.2 (2015): 111–115.

Qiao, Y., Flintsch, G. W., Dawson, A. R., & Parry, T. (2013). Examining Effects of Climatic Factors on Flexible Pavement Performance and Service Life. Transportation Research Record, 2349(1),100–107. https://doi.org/10.3141/2349-12

Stoner, A. M. K., Daniel, J. S., Jacobs, J. M., Hayhoe, K., & Scott-Fleming, I. (2019). Quantifying the Impact of Climate Change on Flexible Pavement Performance and Lifetime in the United States. Transportation Research Record, 2673(1),110–122. https://doi.org/10.1177/0361198118821877

US Department of Transportation, 2015. Beyond Traffic 2045: Trends and Choices. DOT, US.

Pavement life cycle assessment of state highway network with Caltrans PaveM system

I.A. Basheer
Office of Pavement Management, Pavement Program, Caltrans, Sacramento, California, USA

ABSTRACT: The California state highway system is comprised of over 50,000 lane-miles of pavements serving a daily vehicle-miles-travelled (VMT) of 1.6 billion. The State highway system continues to age; requiring frequent maintenance and rehabilitation (M&R) to improve ride quality and safety and extend pavement life. In response to the California State Assembly Bill AB 32, the California Department of Transportation (Caltrans) has been promoting sustainable solutions to control greenhouse gases (GHG) emissions from all pavement-related activities including pavement M&R and vehicles operation. There has been no in-depth study utilizing network-level data to accurately evaluate GHG emissions due to M&R needs and ride quality deterioration. In this study, the Caltrans pavement management system (PaveM) was used to quantify GHG emissions attributed to M&R activities (Materials & Construction Stage) and vehicles operation affected by pavement roughness/smoothness (Use Stage) over a 30-year analysis period. Several scenarios employing different performance-based optimization/prioritization schemes commonly used in project selection simulated under variousfunding plans were analyzed in addition to freefall (do-nothing) scenarios. The estimated pavement repair budgets, condition improvement, GHG emissions reductions, and fuel cost savings were compared. The analysis demonstrated the effectiveness of the Caltrans' PaveM system in evaluating the pavement carbon footprints of the state highway network, and identifying M&R funds allocation methods that minimize both the repair cost and negative impact on environment.

1 INTRODUCTION

The California Department of Transportation (Caltrans) manages approximately 50,400 lane-miles of highway pavements comprised of approximately 13,100 lane-miles of concrete-surfaced pavements (26%) and 37,300 lane-miles of asphalt-surfaced pavements (74%). This state highway network is used by a total of 1.6 billion VMT each day. This network continues to age and deteriorate, and the State responds with M&R repairs to extend the pavement life and improve ride quality and safety. Pavement M&R activities that take place on the network and the normal daily traffic contribute to increasing GHG emissions through utilization of fuel used in materials production and placement, and for vehicles' operation. In response to the CA State Assembly Bill AB 32 (Global Warming Solutions Act of 2006), Caltrans has been promoting sustainable solutions relevant to pavement construction and repair to control GHG emissions from sources attributed to active highway construction and vehicles operation. The 2016 and 2017 California State transportation sector's share of the statewide annual total GHG emissions was estimated at 170 MMTCO2e (million metric tons of CO_2 equivalent) constituting 41% of the statewide total from all sources (CARB 2018). In 2016, the all-sectors statewide GHG emissions totaled about 430 MMTCO2e and the "on-road" transportation share was 155 MMTCO2e (CARB 2018).

In this study, the Caltrans data-rich pavement management system PaveM was enhanced to quantify the statewide GHG emissions from all the relevant sources attributed to the pavements of the state highway network. As with all PMS's, the PaveM optimization tool enables the use of several optimization/prioritization schemes and funding scenarios in all Caltrans' funding programs for project selection (prioritization) and budget allocation. This study evaluates and compares effectiveness of optimization schemes and funding levels for addressing the state highway network's repair needs with the lowest budget possible while achieving the lowest possible GHG emissions over an analysis period of 30 years. The study also compares benefits to pavement agencies, highway users, and the environment.

2 APPROACH

In relation to pavements, GHG emissions are produced in two stages. The first stage is the Materials and Construction (M&C) Stage; which is concerned with all the GHG emissions attributed to M&R activities that are primarily driven by funded projects. The second stage is the Use Stage in which GHG emissions are attributed to on-road vehicles' operation following pavement construction. The GHG amount models for the two stages are discussed in greater detail in Wang (2013). Using these models, it is possible to estimate, for a given project the GHG emissions associated with the M&C Stage of an M&R treatment and then evaluate the additional GHG emissions due to vehicles' operation during the Use Stage following construction. The GHG emissions in these two stages differ from treatment to treatment and from pavement to pavement. To quantify the two main sources of emissions during the useful life of a pavement, it is important to analyze the benefits of applying M&R treatments during these two stages. Keeping pavements smoother (with lower IRI), reduces vehicle's rolling resistance; thus, improves on-road vehicle fuel economy, reduces vehicle's fuel consumption, and as a result reduces GHG emissions during the Use Stage. But to keep the pavement smooth, M&R treatments are required over the life of the pavement; demanding additional resources and energy and producing GHG emissions.

On a pavement life timeline, these have opposing effects on GHG emissions. The net benefit that can be achieved by applying GHG-producing M&R treatments to improve pavement condition and reduce vehicle operation-attributed GHG is quantified with a life-cycle assessment (LCA) framework. When an M&R treatment is undertaken, the M&C Stage represents a GHG source in the pavement life cycle, but the pavement becomes smoother; which reduces GHG emissions related to vehicles' operation. Doing nothing (no treatment) contributes to zero construction-related GHG; however, it allows the pavement condition (e.g., IRI) to deteriorate which increases GHG emissions in the Use Stage. Working on a timeline of pavement M&R events, it is possible to conduct a comprehensive "carbon accounting" to determine benefits and costs.

The GHG accounting procedure is illustrated in the example of a newly constructed pavement shown in Figure 1, which tracks performance (Figure 1-a) and GHG accumulation (Figure 1-b) over time. Right after initial construction (at time zero), the IRI is lowest, and GHG is equal to the M&C Stage's GHG. The IRI continues to increase as pavement smoothness deteriorates and the Use Stage GHG accumulates until the pavement IRI reaches a threshold value; at which time an M&R treatment is due. The "appropriately" selected M&R treatment resets the IRI to a brand new like condition and produces a smoother pavement that saves on both vehicles fuel consumption and GHG emissions. Because of M&R treatment application, a new M&C Stage's GHG is added, and the Use Stage's GHG starts to accumulate; however, at a much slower rate than would have been if a treatment was not applied. The repeated application of an M&R treatment over a number of cycles (green curve) over the analysis period brings the IRI down to below the threshold level, as shown in Figure 1-a. This is to be compared with the red curve which represents the IRI without the treatment (i.e., Do-Nothing or Freefall). It is possible to compare two possible benefits from applying an M&R treatment: (i) benefit of making the pavement smoother; quantified by the magnitude of the enclosed area between the M&R deterioration curve (green) and the freefall curve (red) shown in Figure 1-a; and (ii) benefit of producing lesser GHG emissions quantified by the numerical difference in accumulated GHG emissions represented by numerical difference in y-value of the freefall and M&R curves in Figure 1-b. To quantify these two benefits, IRI

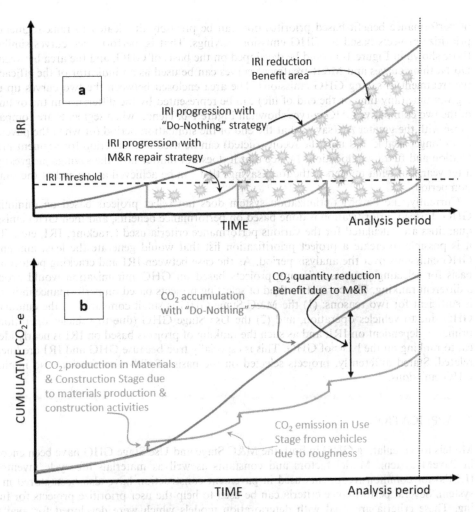

Figure 1. (a) Benefit area in terms of IRI improvement, and (b) Savings in GHG emission quantity between applying successive M&R treatments and do-nothing.

deterioration models representing the IRI curves shown in Figure 1-a for both the original pavement (red) and the M&R treatment (green) must be available along with IRI resetting rules. Additionally, the GHG quantification models presented earlier must be utilized to produce the red and green curves in Figure 1-b.

In the Caltrans PaveM system, the state highway network is divided into pavement management segments using fine segmentation. Based on condition improvement (e.g., using IRI) the management segments are compared relative to the magnitude of benefit area (enclosed between the red and green performance curves shown in Figure 1-a above). Segments with greater performance benefit area are prioritized higher for receiving repair funds than those with smaller area. Project prioritization in PaveM is carried out based on either ride quality (IRI) or cracking improvement; both weighted for traffic. The appropriate performance function for the existing pavement is selected based on the last treatment. With known conditions and preselected performance thresholds (triggers), an M&R treatment (with known performance model) is selected using applicable decision trees.

The benefit equation representing the benefit area (the numerical integration of the difference between the two performance models) is calculated. This procedure is run on all the segments considered, and then the segments are ranked based on the benefit value. This method of performance benefit-based prioritization can be precisely duplicated to rank segments or prioritize projects based on GHG emissions savings. That is, performance curves similar to those shown in Figure 1-a could be developed on the basis of GHG, and the area between the Do-Nothing curves and M&R treatment curves can be used as an indicator of the efficacy of the treatment in saving GHG emissions. The area enclosed between the two curves up until a given time (any time or the end of life) can be represented by the difference in the ordinates of the two cumulative GHG curves shown in Figure 1-b. Hence, when segments are compared, those with the greater Δy (savings) at the end of the simulation period (or when the pavement is no longer usable and must be reconstructed) can be used as a criterion for segment prioritization and funding allocation. Notice that it does not matter how the savings progress year after year; but what matters is the total savings that can be achieved at the end of the simulation period.

Currently, the PaveM optimization system does not select projects based on minimizing GHG. Instead, prioritization is done based on performance benefits, and then GHG emission quantities are calculated for the various performance criteria used (cracking, IRI, etc.). Then it is possible to create a project prioritization list that would generate the least amount of GHG emissions over the analysis period. As the case between IRI and cracking performance basis for selecting projects, selecting projects based on GHG minimization would produce a different ranking. The current method of selecting projects based on performance indicators is sufficient for two reasons: (1) the M&C Stage GHG is small compared to the cumulative GHG due to vehicles operation, and (2) the Use Stage GHG (due to vehicles operation) is primarily dependent on IRI; and as such the ranking of projects based on IRI is nearly identical to ranking on the basis of GHG. This is especially true because GHG and IRI are linearly related. Stated differently, projects selected on the basis of IRI minimization also minimize GHG emissions.

3 APPLICATION

Models for calculating GHG in both the M&C Stage and Use Stage GHG have been encoded in PaveM system. Model factors and constants as well as materials life cycle inventories (LCI's) for common treatments used in pavement construction have also been stored in the system. Several performance criteria can be used to help the user prioritize projects for funding. These criteria are used with deterioration models which were developed for available treatments for evaluating the benefits achieved by applying these treatments. From the numerous criteria available in the PaveM system, the most common and effective ones are the cracking benefits and IRI benefits; both weighted for traffic.

The approach developed in this study has been used to evaluate several scenarios that vary by (i) performance criteria selected for segment prioritization, and (ii) funds availability. Based on performance benefits, two criteria were examined: (a) cracking, and (b) IRI. For funding availability, two extreme funding levels were tested; as well as the current funding level. The extreme levels were the zero-funding representing a freefall (do nothing), and the unlimited funding which allows any management segment to get enough funding just to address its condition based on treatment decision trees. The freefall scenario allows assessment of pavement condition as it deteriorates over the analysis period without any M&R interventions. While unrealistic, these scenarios define the upper and lower boundaries that GHG improvement could fall within relative to the amount of funding spent. Figure 2 is an illustration of the effect of funding level on GHG emissions accumulation.

The freefall scenario, representing the undesirable extreme, results in the maximum deterioration of the pavements; hence it accumulates the largest amount of GHG emissions over the analysis period. This level represents the upper limit of GHG emissions (worst case scenario). The unlimited funding for upgrading every pavement on the highway network as triggered by the decision trees produces a practically minimum GHG level. In comparison, unlimited spending for producing a virtually zero-IRI level on each pavement surface (that could be maintained over the analysis period) in the network, is the desirable extreme (although

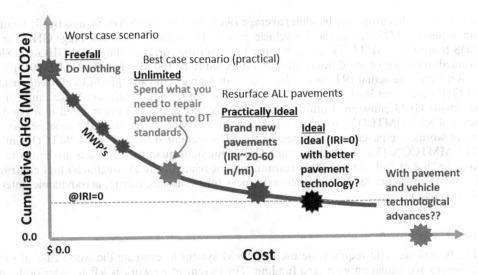

Figure 2.   Spending level versus cumulative GHG emissions level.

unattainable with current construction technologies), producing the least amount of GHG emissions accumulations with the available vehicle engine technology. This represents the lower limit of GHG emissions as shown in Figure 2. Converting every vehicle on the highway system to electric would produce a much lower GHG emissions limit than fossil fuel powered vehicles. Various actual funding levels (labelled as MWP's), including the current Caltrans funding level, would produce GHG accumulation levels over the analysis period that fall on the curve shown in Figure 2 somewhere between these extremes.

This study also quantified the current (baseline) GHG emissions level of the system. The baseline level was assumed to correspond to year 2016; the year in which the APCS was conducted. No construction related GHG was assumed for the baseline level; hence, all GHG emissions are attributed to vehicles operation (Use Stage). The baseline GHG helps evaluate the effectiveness of funding plans and optimization schemes used in projects selection.

Results of the evaluation scenarios run for 30-year analysis period, are presented and discussed next. Assumptions made with regard to this time span. For example, vehicle fuel economy remains constant, despite the current trend of improving vehicle fuel economy which can reduce the amount of GHG savings due to smoother pavements. Similarly, the percentage of gasoline and diesel engine vehicle remain constant with only small fraction of zero-emission vehicles. The current trend of cleaner vehicles will also reduce the GHG savings from smoother pavements. Also, VMT's remain constant, and vehicle class percentages unchanged. Determining traffic growth factors is rather complex and is pinned to predicting future technological improvements; which depend on (or will influence) governmental policies. The higher the number of VMTs, the greater the savings in GHG emissions due to smoother pavements. Finally, effect of pavement smoothness on vehicle's maintenance and repair costs was ignored. Accounting for the reduced costs, extended vehicle life, reduced accidents, etc. can increase the on-road vehicle user's savings besides the cost of fuel reduced with smoother pavements.

5   RESULTS

5.1   Baseline and ideal (minimum) GHG

The state highway system (SHS) was finely segmented 69,667 segments using the fine segmentation process embedded in the PaveM tool. Other needed data for this evaluation included segment length, current (or baseline) average IRI level (from 2016 APCS), and traffic volumes (in terms of AADT) for all the five vehicle groups. The annual (2016) Use Stage GHG for the SHS totaled 88.1 MMTCO2e; which more than half the corresponding total GHG emissions from all statewide on-road transportation estimated to be 150 MMTCO2e (CARB 2018).

Replacing the actual IRI values of all segments with a *zero* value produces an *ideal* scenario of GHG emissions level for year 2016; i.e., the lowest (hypothetical) Use Stage limit for the statewide GHG emissions from the state highway network. This case resulted in a one-year total of 85.3 MMTCO2e; which is a reduction of 2.8 MMTCO2e from the baseline level. In other words, the pavement roughness of the SHS accounted for only 2.8 MMTCO2e of the 88.1 MMTCO2e (3.2%). The remaining GHG emissions quantify (96.8%) is attributed to the vehicles' use of fuel to resist other resistances (not related to road roughness) that contribute to rolling resistance as well as other resistances (gravitational, inertia, aerodynamic, internal friction, etc.).

5.2 Freefall (do-nothing) scenario

The freefall scenario requires the use of PaveM system to evaluate the worst case of GHG emissions accumulation with zero funding. The pavement network is left to deteriorate over the 30-year analysis period without intervention (i.e., M&C Stage GHG=0). This scenario provides a useful benchmark for evaluating other reasonable scenarios. The scenario results indicated that over 30-year period the SHS resulted in a total of 2,688 MMTCO2e when left to deteriorate. This level represents the maximum GHG emissions level over 30-year period resulting from the Use Stage.

5.3 Unlimited (as needed) funding scenario

As the freefall scenario represents an upper limit on GHG accumulation over time, the unlimited (as needed) funding scenario represents the highest level of savings in GHG emissions that could be achieved if every pavement on the state highway system is upgraded to meet the decision trees' standards. No pavement segment is allowed to deteriorate worse than the thresholds. The aging (deterioration) of each pavement and how long would it take to deteriorate to a performance threshold value so that it qualifies for receiving repair funds depend on the treatment's performance model that is used. Although funds are not constrained, the PaveM optimization program calculates benefit area (or value related to area) associated with applying the treatment needed. The benefit can be defined in different ways; however, most commonly, based on either improving IRI or Cracking condition. Depending on the treatment selected and current pavement condition, it is possible that the benefit value be positive (indicating treatment efficacy), zero, or negative, and that depends on how the benefit would be expressed. As an example, consider concrete slab replacement and grinding as two treatments. If benefit has been expressed in terms of cracking, "slab replacement" treatment will show a positive benefit, whereas "grinding" treatment will show a zero benefit. Alternatively, if benefit is expressed in terms of IRI improvement, then "slab replacement" would probably show zero benefit, and "grinding" will show positive benefit. Numerically, benefits can also fall to negative values. PaveM has been set up to calculate benefit in many ways, but in this study, the IRI and Cracking benefits have been selected such that treatments will be applied on the pavement only if the benefit values are determined to be positive. That is, a treatment will not be selected if benefit value is zero or negative; even when the decision trees call for that treatment based on condition. Therefore, depending on the performance benefit area value (on the basis or either IRI or Cracking), the results in terms of selected projects, amount of funding to be spent and when to spend it can be quite different. For this reason, the required cost, overall network performance (e.g., ride quality), and GHG accumulation over time can differ depending on the selected performance criterion; even though funds are not constrained.

Figure 3 shows the results of this type of analysis for the state highway network on the basis of cracking (Figure 3-a) and IRI (Figure 3-b) performance. The 30-year cumulative GHG dropped from 2,688 MMTCO2e without any funding (freefall) down to 2,650 with cracking as the performance criterion (a difference of 38 MMTCO2e), then to 2,642 MMTCO2e with IRI as the performance criterion (a difference of 46 MMTCO2e). This emphasizes the importance of selecting projects on the basis of IRI performance because it also results in an additional reduction in GHG accumulation (8 MMTCO2e) compared to using cracking performance. The additional reduction is to be expected because roughness directly affects GHG in the Use Stage. It can also be noticed, and as expected, how the network-weighted average IRI is lower in the IRI-based simulation than in the cracking-based simulation.

Figure 4 shows the year-to-year spending and related GHG emissions quantity for the unlimited funding IRI-based scenario. Also shown in Figure 4 is the 2016 baseline GHG level (solid circle), and the ideal (theoretical minimum) GHG level corresponding to zero-IRI. A total of $3.5B will be needed in the first year to upgrade the entire highway network to meet the asphalt and concrete pavement decision trees' standards. Afterwards, lesser funds will be needed to maintain the pavement conditions to within the decision trees' thresholds. With the biggest improvement in the first year, a large increase in GHG over the 2016 level is attributed to M&C Stage for the entire network. This is followed by the reduced Use Stage GHG (below the 2016 baseline) that lasts for 4 years returning to the 2016 level. Once construction has been completed, IRI starts to deteriorate and GHG levels increase until the pavements are repaired again with funds being

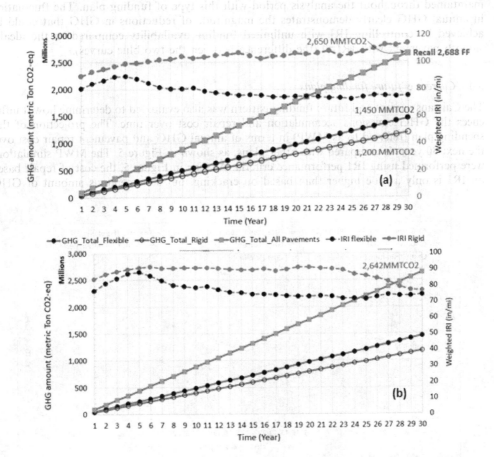

Figure 3.   GHG accumulation and network-weighted IRI for unlimited funding scenarios: (a) cracking performance, and (b) IRI performance.

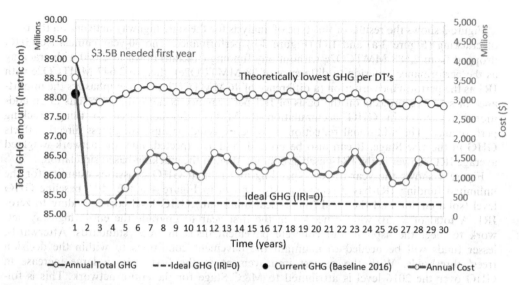

Figure 4. Annual GHG emissions versus cost for IRI-unlimited funding scenario.

expended annually; and so on. This trough and peak pattern in both GHG and cost is maintained throughout the analysis period with this type of funding plan. The fluctuation in annual GHG clearly demonstrates the magnitude of reductions in GHG that could be achieved by controlling IRI with unlimited funding availability compared to the ideally smooth pavement with IRI=0 (the difference between the two blue curves).

5.4 Current & future funding plans

The Caltrans current and future funding pattern was also evaluated to determine how it influences the GHG emissions accumulation and repair cost over time. The projection of this spending plan (referred to as MWP) in terms of annual GHG and pavement repair cost over the next 30 years is compared with other plans, as shown in Figure 5. The MWP simulations were performed using IRI performance criteria. As seen in Figure 5, the cost of repair based on IRI is only a little higher than based on cracking, but a tremendous amount of GHG

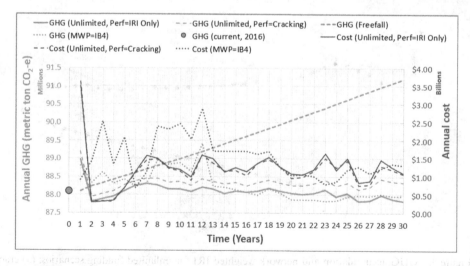

Figure 5. Annual GHG emissions and cost of all spending plans and optimization schemes.

emissions could be saved. The freefall annual GHG is higher than when funds are expended (on as needed basis) based on either cracking or IRI as the performance criterion. The simulated current funding plan could result in more GHG than even the freefall in some years, because in part it was not designed entirely with optimization algorithm to reduce IRI, and the GHG from the M&C Stage is higher than that saved from the Use Stage compared to the freefall scenario. Inclusion of programmed projects when they may not be needed while delaying more urgent projects may have been another reason for this observed trend. The current spending plan, if implemented, can result in great GHG savings with the additional funding of \$1.7B a year starting on the 15th year. Finally, large amount of spending is done with the unlimited funding scenarios in the first year to repair the network. Obviously, this is not always practically possible. As can be seen in Figure 5, the more practical MWP scenario needs more money in the first 15 years to match up with the unlimited funding cases in terms of GHG savings. Optimized spending of smaller funds when they are critically needed can have a much more positive influence on both the pavement performance and GHG savings over the analysis period than a much greater amount of funds expended in a non-optimized fashion. The benefits of using an optimization system for project selection to achieve the best results in terms of ride quality and GHG savings is clearly noticeable from Figure 5.

Table 1 summarizes the differences betwee the various spending plans in relation to their effect on repair cost and GHG emission totals. Clearly, the unlimited IRI-based optimization produces both the lowest GHG over the 30-year analysis period and the greatest savings compared to the freefall case. Interestingly, the MWP spending plan costs more than the unlimited funding plans but produces more GHG over the analysis period; mainly due to non-optimized project selection for the first 15 years as currently planned. The unlimited plan based on IRI-performance spends \$10B less and saves an additional 5 MMTCO2e over 30-year period than IRI-performance based MWP plan. Also shown in Table 1 is the cost of GHG reduction relative to the freefall case. It is seen that on the SHS, the cost of GHG reduction with pavement repair can be tremendous; but it is the lowest for the unlimited (spend-as-needed) plan, and highest for the MWP plan. The high cost per unit GHG is due to the tremendous funds Caltrans programs for preventive maintenance which tend to keep smooth pavements smooth. That is, the additional large cost of preventive maintenance is not significantly offset by a comparable reduction in GHG emissions due to smoother pavements, but it adds more tons of GHG emissions due to materials production and construction. It is important to keep in mind that the goal of pavement repair is not primarily to reduce GHG, but to improve transportation and safety; which can be achieved while GHG reduction is a side benefit.

Fuel cost and vehicle maintenance savings can also be achieved with smooth pavements. The fuel cost saving was calculated from established correlations with GHG. Table 1 includes the estimated savings in fuel cost. Vehicle maintenance saving was not calculated. The net cost

Table 1. 30-year cumulative GHG, construction cost, fuel cost saving, GHG unit cost, and net cost for several spending plans.

Spending plan	Cumu. GHG (MMTCO2e)	Cumulative construction cost (\$ Billion)	GHG reduction relative to Freefall (MMTCO2e)	Unit cost of reduction (\$/TON CO2e)	Cumulative fuel cost saving (\$ Billion)	Net cost (\$ Billion)
MWP-IRI	2647.7	45.67	40.37	1,131	19.5	26.1
Unlimited-Cr.	2649.8	35.36	38.23	924	18.1	17.2
Unlimited-IRI	2642.5	36.69	45.61	804	21.4	15.3
Freefall	2688.1	0.0	0.0	N.A.	N.A.	0.0
Ideal (IRI=0)	2559.6	Not calc	128.47	Not calc.	61.0	N.A.

is direct agency cost less the fuel cost saving for road users; which was found to be the lowest for the IRI optimization scheme. The optimization based on improving IRI resulted in the greatest fuel savings; with an additional $3.3B in savings over 30-year period compared to cracking performance.

6 CONCLUDING REMARKS

This paper presented findings from a first large-scale study to evaluate pavement-related GHG emissions using the Caltrans' PaveM system. The PaveM system was enhanced to enable Caltrans to quantify more accurately, GHG emissions related to pavements including baseline GHG level, and evaluate and compare effectiveness of various funding plans and project prioritization schemes in optimizing long-term pavement performance and reducing GHG emissions. The following was concluded: (1) Optimization on the basis of IRI performance results in the lowest pavement carbon footprint compared to other performance criteria. Project selection using IRI as the performance criterion results in greater savings in GHG emissions, agency cost, and cost of fuel over 30-year year analysis period compared to other criteria. It was found that the greater the deviation from PaveM-optimized selection the greater the GHG emissions and the higher the cost. It was also found that on the network level, M&C Stage's contribution to total GHG emissions quantities was very small (~0.5%) compared to the use stage GHG over the 30-year simulation period. Even though M&C Stage-related GHG is relatively small, Caltrans should continue to consider utilizing sustainable products on all construction projects, mandating Environmental Product Declarations (EPD's) for pavement construction materials, and promoting cleaner construction technologies and practices. Because of the high Use Stage-related GHG, it is important that Caltrans continues promoting and constructing long-life pavements that are constructed smoother (i.e., born smoother) and can stay smooth longer, and adopting aggressive proactive preservation program to prolong quality pavement performance. Finally, a Caltrans' long-range goal should Caltrans should consider expanding PaveM system capabilities (long range goal) to integrate both life cycle cost analysis (LCCA) and life cycle assessment (LCA) within its framework to capture the interaction between cost reduction and environmental impact benefits in project selection and M&R treatments.

DISCLAIMER

The contents of this paper reflect the views of the author who is responsible for the facts and the accuracy of the data presented. The paper's contents do not necessarily reflect the official views, polices, or standards of the California Department of Transportation.

REFERENCES

CARB (2018). *Annual State-Wide Greenhouse Gas Inventory*. California Air Resources Board 2018. Available online: https://www.arb.ca.gov/cc/inventory/inventory.htm
Wang T. 2013. Reducing Greenhouse Gas Emissions and Energy Consumption Using Pavement Maintenance and Rehabilitation: Refinement and Application of a Life Cycle Assessment Approach. Ph.D. Dissertation, University of California Davis. 374 p.

Pavement, Roadway, and Bridge Life Cycle Assessment 2020 – Harvey et al (eds)
© 2020 Taylor & Francis Group, London, ISBN 978-0-367-55166-7

Exploring the cost benefit value and relative emissions of pavement preservation treatments using RoadResource.org

S. Casillas & A. Braham
University of Arkansas, Fayetteville, AR, USA

ABSTRACT: Pavement preservation is important for the life cycle of roads, as proper pavement preservation lowers long-term costs while extending the life of roads. Pavement preservation treatments also have lower emissions than more conventional pavement treatments. These benefits are clearly showcased in a new online resource called RoadResource.org. This resource has many tools, but this paper will focus on two calculators: the Cost Benefit Value (CBV) calculator and the Sustainability Calculator. The CBV provides a single number that is based on the traffic level, a constraint factor, the life extension of a treatment, the unit cost of a treatment, and the existing roadway's Pavement Condition Index (PCI). Using data from the Arkansas Department of Transportation, the highest CBV achieved was 1933.53 for a rejuvenating fog seal placed on a freeway with a PCI F, while the lowest CBV achieved was a 3.48 for a base stabilization + 4.0" HMA placed on a two-lane highway with a PCI A. It is obvious a rejuvenating fog seal would not be appropriate for a PCI of F, which led to the "culling" of treatments based on their proper application. Therefore, a second analysis was performed that was able to provide guidance on the appropriate treatment for each roadway type and roadway condition, combining the power of the CBV calculator with smart inputs to identify optimal life cycle cost savings. This analysis not only optimized life cycle cost savings but also reduced greenhouse gas emissions by the equivalent of 137,000 passenger cars being removed from the road per year according to the Sustainability Calculator.

1 BACKGROUND

A Life Cycle Analysis (LCa) contains an economic and environmental analysis over the lifespan of a product or process. An economic analysis is often called a Life Cycle Cost Analysis (LCCA). An LCCA combines present and future costs into a single number, allowing for a comparison between different pavement material types (Walls and Smith, 1998). An environmental analysis is often called a Life Cycle Assessment (LCA). Instead of quantifying economically, an LCA quantifies emissions over the life span of the pavements in question (Braham, 2017). These two techniques can also be applied to pavement preservation, maintenance, or rehabilitation treatments.

The website RoadResource.org has two calculators that bring future costs into today's dollars: the Life Cycle Cost (LCC) calculator and the Equivalent Annualized Cost (EAC) calculator. The LCC calculator calculates the cost of a series of pavement treatments over the life of a pavement design, whereas the EAC calculator calculates the current cost over the life of a single treatment. A robust examination of these two calculators can be found elsewhere (Kiihnl and Braham, 2019; Casillas and Braham, 2019). These studies showed significant cost savings if proper pavement preservation is followed, upwards to $155 million per year in Arkansas. From an environmental perspective, there is also a "Sustainability Calculator" that compares the greenhouse gas emissions (GHG) from "conventional" treatments (such as a mill and fill or remove and replace) to what are referred to as "preservation and recycling" treatments (such as fog seals, chip seals, micro surfacing, or Cold In-place Recycling, to name a few). In addition to these three calculators, another calculator of interest is an economic calculator of sorts, but also includes traffic, life extension, and existing pavement condition index. This calculator is called the Cost Benefit Value (CBV) calculator. This paper will first examine the CBV calculator. Second, it will demonstrate the potential environmental savings using preservation and recycling treatments using the Sustainability Calculator.

2 GETTING STARTED

The Cost Benefit Value, or CBV, is a function of characteristics of an existing roadway and the type of treatment considered for application to the roadway. Equation 1 shows how the CBV is calculated.

$$CBV = \frac{(Traffic/Constraint\ Factor) \times (Life\ Extension)}{(Unit\ Cost) \times (Pavement\ Condition\ Index)} \tag{1}$$

In Equation 1, the traffic is provided as the annual average daily traffic (AADT). The constraint factor is intended to prevent the AADT from disproportionately influencing CBV comparison. If one road within a roadway network has a significantly higher level of traffic than all other roads, the CBV value for the one road would always be quite high. Therefore, the constraint factor, typically varying from 4-10, can be used to provide a more level playing field across an entire network. The last characteristic of the existing road is the Pavement Condition Index, or PCI. The PCI is a scale to indicate general condition of pavement where 100 is a perfect pavement and 0 is the worst rating possible. More details on calculating PCI can be found in ASTM D6433.

In addition to characteristics of the existing roadway, the type of treatment chosen to place on the road is also accounted for in the CBV. Two values are considered, life extension and unit cost. Life extension is the time for pavement to return to the same condition as prior to applying the treatment. Unit cost is the cost of applying the treatment. It is worth noting that default values based on nation-wide averages are available for both life extension and unit cost on RoadResource.org (the default values were used in this paper), but all numbers can be changed if desired to meet local conditions. In this study, the CBV was explored using data from 2018 on the Arkansas Department of Transportation (ArDOT) highway network.

3 ARDOT HIGHWAY NETWORK DATA

The data used for the CBV calculator and Sustainability Calculator was from ArDOT, and included traffic level, length of each highway segment, budget, and condition of the current highway network. In order to more easily use RoadResource.org, the seven existing types of highway were condensed into three types: freeways, multilane highways, and two lane highways. The traffic (in AADT and Vehicle Miles Traveled, or VMT) and length (in lane-miles) of each of these highway segments is in Table 1.

In order to maintain the highway network, ArDOT's budget for resurfacing and rehabilitation alone was $236,585,135 in 2018. This budget is for every type of highway in the network.

Table 1. Simplified characteristics of Arkansas' highway network.

Highway Segment	Average AADT[1]	Annual VMT[2] (millions)	Lane Miles
Freeways	60,455	9,922	3,702
Multilane highways	15,214	5,141	4,587
Two lane highways	3,793	9,844	28,943

1 Average AADT (Annual Average Daily Traffic), measured in vehicles per day (vpd)
2 VMT (Vehicle Miles Traveled)

174

However, in order to distribute the money across each type of highway segment, it was assumed the budget was allocated proportionally by VMT. While this is not necessarily ArDOT's current practice, it provided a sound method for assuring the funds were being fairly distributed across each highway segment type. Table 2 shows ArDOT's budget as a percentage of VMT.

Finally, the last piece of information necessary for this analysis was the existing pavement condition. For many years, ArDOT utilized the Pavement Condition Index (PCI) to quantify the condition of their roads. The most recent data available with PCI was from June 30, 2017. Table 3 provides the percentage of each highway segment in terms of PCI grade.

With characteristics of the highway network and budget established, the CBV calculator on RoadResource.org was explored.

4 EXPLORING THE COST BENEFIT VALUE CALCULATOR

In order to introduce students to the concepts of pavement preservation, pavement maintenance, and pavement rehabilitation, a five part semester long project was developed for multiple courses. The first course, a required senior level transportation course, placed a single treatment on the entire ArDOT network. The treatments are shown in Table 4.

While it is not recommended to put one single treatment on an entire pavement network, the exercise helped students in two ways. First, it allowed them to see the difference in applying a maintenance product versus a rehabilitation product. Rehabilitation products are more expensive, but they also last longer, so students were able to directly see the influence of time on the cost of a treatment. Second, since each group explored one maintenance and one rehabilitation treatment, when the class was shown the entire set of results, they were able to understand how their two treatments compared to other treatments. Again, it is not recommended to put one single treatment on an entire pavement network, but this method of analyzing the data can actually produce some very misleading results. A preliminary discussion of the CBV results and more detailed analysis of the assignments can be found elsewhere (Kiihnl and Braham, 2019).

To begin the more in-depth discussion on CBV, Figure 1 shows results from this first student project looking at all twenty-one treatments when placed on the freeway highway segment with a PCI grade of F.

Table 2. Simplified characteristics of Arkansas' highway network.

Highway Segment	Annual VMT (millions)	2018 Budget as % VMT
Freeways	9,922	$94,247,372
Multilane highways	5,141	$48,833,492
Two lane highways	9,844	$93,504,271
Total		$236,585,135

Table 3. ArDOT budget as a percentage of Vehicle Miles Traveled (VMT).

PCI Grade	Freeways (%)	Multilane Highways (%)	Two Lane Highways (%)
A (86-100)	31.5	3.4	0.7
B (72-85)	38.0	24.8	7.3
C (55-71)	18.7	35.1	24.4
D (41-54)	8.8	24.7	37.1
F (0-40)	3.0	12.0	30.5

Table 4. Maintenance and rehabilitation treatments explored in the study (HMA = Hot Mix Asphalt).

Maintenance Plan	Rehabilitation Plan
Rejuvenating fog seal	CIR + 2 chip seals
Crack seal	Minor mill & fill*
Chip seal	CIR + 1.5" HMA
Fog seal	HIR + chip seal
Slurry seal	HIR + 1.5" HMA
Scrub seal	Major mill & fill*
Micro surfacing (single)	FDR + 4.0" HMA
Micro surfacing (double)	Full depth R&R*
Cape seal	Base stabilization + 4.0" HMA
Thin lift HMA*	
Bonded wearing course*	
Ultra thin lift HMA*	

* Starred treatments are also referred to as "conventional" treatments on the website

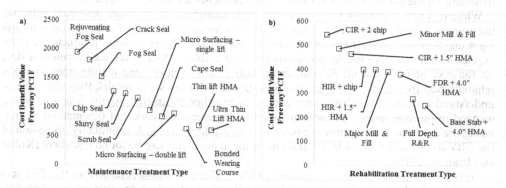

Figure 1. The Cost Benefit Value (CBV) of twelve maintenance treatments (a) and nine rehabilitation treatments (b) on ArDOT's PCI F freeways.

In theory, higher CBV values give a higher priority for a project. According to the data in Figure 1, however, the CBV calculator's highest priority treatment would be a rejuvenating fog seal on a freeway with a PCI F. This is not reasonable, as a rejuvenating fog seal is not appropriate for any road with a condition of PCI F. Another interesting observation is, in general, all of the maintenance treatments have a higher CBV than the rehabilitation treatments. When a roadway has a condition of PCI F, it most likely has some sort of structural problem, which means it would need to be rehabilitated or even reconstructed. However, by looking strictly at CBV, this is not the case. It is even more interesting that the sequence of treatments shown in Figure 1 (rejuvenating fog seal, followed by crack seal, down to Full Depth R&R and Base Stabilization with 4.0" HMA) is repeated identically for all three highway segments in all five PCI conditions. For example, Figure 2 shows the trends for two lane highways with a condition of PCI B.

The same trends are shown in Figure 2 as Figure 1, but shifted downward. In general, the highest CBV values are for freeways with a PCI F, followed by freeways with a PCI D, C, B, then A. Some multi lane highways with a PCI F are mixed in with the freeways, but most of the CBV values are lower than the freeways (and decrease as the PCI increases), with a similar relationship between multi lane highways and two lane highways. These very consistent trends indicate the traffic level, AADT, is a very heavy driver in determining the CBV value. This is one of the reasons the constraint factor is incorporated into the website. However, if

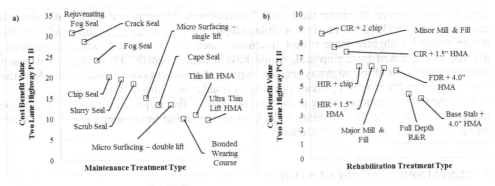

Figure 2. The Cost Benefit Value (CBV) of twelve maintenance treatments (a) and nine rehabilitation treatments (b) on ArDOT's PCI B two lane highways.

a different constraint factor is used for all of the analysis, it would only shift the CBV values up or down. Therefore, it appears that if the constraint factor is used, it would need to be changed for each highway segment and pavement condition. This would be a very cumbersome process but should be explored in more detail.

In summary, using only the CBV values of single treatments over the entire network does not produce consistent results. Therefore, a second semester long project was developed for a second course. This course, a senior level technical elective, used a combination of CBV and "engineering judgement" to determine which treatments would be placed on each highway segment in ArDOT's network. The treatments are shown in Table 5.

In Table 5, it is immediately apparent that more than just CBV values were utilized. If this were not the case, for freeways, the full depth remove and replace (full depth R&R) would have been recommended on PCI B, PCI C, and PCI D roads along with PCI F. However, full depth R&R is not an appropriate treatment for a PCI B or PCI C road. While there could be a discussion that full depth R&R is appropriate for a PCI D road depending on the distresses shown in the pavement, engineering judgement was used in order to determine that most roads with a PCI D rating could be treated using CIR + 1.5" HMA. Cold In-place Recycling can address almost all surface distresses and can often help treat structural distresses. With the addition of 1.5" of HMA, the structure is often improved enough to handle unforeseen loads applied on the pavement structure. While this discussion would be required for each road in the network and would be dependent on current/future traffic along with other considerations, this type of engineering judgement is necessary when deciding which treatments to utilize.

In addition to using both CBV and engineering judgement to choose specific treatments for each pavement condition index group, this combination was also used to choose which treatment to use on each type of highway segment. For example, for multilane highways, the CBV

Table 5. Pavement treatments utilizing CBV and engineering judgement.

Highway Segment	PCI A	PCI B	PCI C	PCI D	PCI F
Freeway	Rejuvenating fog seal (CBV = 416)	Micro surfacing: single lift (CBV = 237)	Minor mill & fill (CBV = 154)	CIR+ 1.5" HMA (CBV = 193)	Full depth R&R (CBV = 277)
Multilane	Rejuvenating fog seal (CBV = 105)	Scrub seal (CBV = 73)	CIR+2 chip (CBV = 43)	CIR+ 1.5" HMA (CBV = 49)	FDR+ 4.0" HMA (CBV = 95)
Two-lane	Rejuvenating fog sea (CBV = 26)l	Chip sea (CBV = 20)	CIR+2 chip (CBV = 11)	CIR+ 1.5" HMA (CBV = 12)	FDR+ 4.0" HMA (CBV = 24)

for a chip seal is actually higher than that of a scrub seal. The chip seal CBV = 80, while the scrub seal CBV = 73. However, since multilane highways tend to have a higher level of traffic, it was decided a slightly more robust treatment, such as a scrub seal, which is essentially a crack seal followed by a chip seal, would have a higher benefit. These types of decisions were made across multiple highway segment types and pavement conditions, but do show that only following CBV may not provide the most optimal combination of treatments.

With the CBV calculator aiding in the final decision of the proper treatments for the proper highway segments, the Sustainability Calculator was explored to examine the potential environmental benefits of alternate treatments.

5 SUSTAINBILITY CALCULATOR

The Sustainability Calculator on RoadResource.org is relatively straightforward. The calculator allows you to compare the "conventional" treatments (thin lift HMA, bonded wearing course, ultra thin lift HMA, minor mill & fill, major mill & fill, and full depth R&R) to the "preservation and recycling approach" (all of the non-conventional treatments listed in Table 4). The Sustainability Calculator calculates the percent Greenhouse Gas (GHG) change and equates the GHG savings into an emissions unit that is defined as "equivalent passenger cars removed from US roadways per year" when comparing treatments.

In short, the Sustainability Calculator estimates the amount of carbon dioxide emissions from the construction of the conventional and preservation/recycling treatments. The construction includes the raw material extraction and processing, the manufacturing, and the construction of the treatments. The carbon dioxide emissions from each treatment, referred to as GHG, were extracted from three different resources (Chappat and Bilal, 2003; Chehovits and Galehouse, 2010; Uhlman et al., 2010). Once the GHG were established for each treatment, the percent difference between the treatments between the conventional treatments and the preservation/recycling approach treatments could be calculated. This is referred to as GHG Savings and has a unit of percent. In addition, this percent savings was converted into a unit that was equivalent to removing passenger cars from the road. A document from the Environmental Protection Agency (EPA, 2018) estimated that a typical passenger vehicle has tailpipe carbon dioxide emissions (in this paper, equivalent to GHG) of approximately 4.6 metric tons. Therefore, the GHG saving could be converted into "equivalent passenger cars removed per year." While some people have trouble visualizing 4.6 metric tons of carbon dioxide, it is very easy to visualize removing one car from the road in the United States.

ArDOT commonly utilizes chip seals, mill and fill, and remove and replace for their treatments. Therefore, the treatments explored in Table 5 were compared to these three treatments using the GHG calculator. In order to provide a higher level of detail, the three types of highway segments were broken down into the five conditions for a total of fifteen comparisons. Therefore, the condition of the pavement itself does not influence the environmental savings, but the condition of the pavement does influence which two treatments were compared. During this comparison, two assumptions were made. First, it was assumed that ArDOT utilized chip seals on PCI A and B roads, minor mill & fill on PCI C roads, major mill & fill on PCI D roads, and full depth R&R on PCI F roads. However, it is rare that ArDOT treats PCI A or B roads at all, and they do not always apply the same treatment on the road with the same PCI rating. Second, chip seal treatments are not available under the conventional treatment menu on the GHG calculator page, therefore, both treatments were compared to a bonded wearing course and the difference between the two comparisons are shown. Overall, the results from the GHG calculator are presented in Table 6.

In Table 6, it is obvious that the proposed treatments reduce emissions compared to the conventional treatments. With the exception of the multilane highways with a PCI B, all of the comparisons are either equal to or provide lower emissions for the proposed treatments. Note that the existing pavement condition does not influence the GHG savings or equivalent passenger cars removed/year, but the three highway segments are divided into the five pavement condition levels in order to map the proposed treatments in

Table 6. Proposed versus conventional treatment environmental savings.

Highway Segment	Proposed Treatments	Conventional Treatments	GHG Savings (%)	Equivalent Passenger Cars Removed/Year
Freeway A	Rejuvenating fog seal	Chip seal	7	818
Freeway B	Micro: single	Chip seal	3	400
Freeway C	Minor M&F	Minor M&F	0	0
Freeway D	CIR + 1.5" HMA	Major M&F	38	1,801
Freeway F	Full Depth R&R	Full Depth R&R	0	0
Multilane A	Rejuvenating fog seal	Chip seal	17	109
Multilane B	Scrub seal	Chip seal	-2	
Multilane C	CIR + 2 chip	Minor M&F	63	8,049
Multilane D	CIR + 1.5" HMA	Major M&F	38	6,263
Multilane F	FDR + 4.0" HMA	Full Depth R&R	13	1,478
Two Lane A	Rejuvenating fog seal	Chip seal	17	142
Two Lane B	Chip seal	Chip seal	0	0
Two Lane C	CIR + 2 chip	Minor M&F	63	35,305
Two Lane D	CIR + 1.5" HMA	Major M&F	38	59,360
Two Lane F	FDR + 4.0" HMA	Full Depth R&R	13	23,704
		SUM		137,291

Table 5 to the conventional ArDOT treatments. The scrub seal actually produced slightly higher GHG emissions which lead to less passenger cars being removed from the road. However, for all other treatments, the proposed treatments go from a minimum of 3% greenhouse gas emissions savings (freeway with a PCI B) to 63% (multilane and two lane with a PCI C). In addition, the equivalent total number of passenger cars removed per year are just over 137,000. While the Sustainability Calculator does not claim to provide a comprehensive emission analysis, it does provide a glimpse into the potential savings of using preservation and recycling treatments (the proposed treatments) versus conventional treatments in the state of Arkansas.

6 CONCLUSIONS

Proper pavement preservation can not only extend the life of deteriorating pavements but also lower life cycle costs of the road. An online resource called RoadResource.org has been developed to showcase the benefits of pavement preservation treatments and provide users with the tools to quantify both economic and environmental costs. Previous research has shown that optimizing pavement treatments could save Arkansas up to $155 million/year. The Cost Benefit Calculator (CBV) calculator uses inputs of traffic level, a traffic constraint factor, life extension of a treatment, unit cost of a treatment, and existing roadway pavement condition to calculate a single number indicating priority of utilizing this treatment on the given existing roadway. The Sustainability Calculator provides the percent reduction in greenhouse gas emissions and estimates the equivalence of passenger cars removed from the roadway. In this research, an in-depth analysis of the CBV calculator was conducted using results obtained from two semester long projects in two different courses, which utilized data from the ArDOT highway network. This analysis provided a recommended treatment plan for ArDOT using CBV and engineering judgement, which was then analyzed for potential environmental benefits. The major findings from this exercise include the following:

- Traffic level, represented as AADT, is highly influential in determining the CBV value.
- If the traffic constraint factor is used, it should be changed for each highway segment and pavement condition to adequately eliminate disproportionate influence due to traffic level.

- Using the CBV of single treatments over the entire pavement network does not produce consistent results or yield reasonable recommendations.
- The use of engineering judgment to select smart inputs is necessary to utilize the CBV for treatment selection for each pavement condition.
- Combining engineering judgment with CBV results also allows for treatment selection for different highway segments.
- Using the proposed treatment plan would reduce emissions equal to removing over 137,000 passenger cars from our roadway system per year.

REFERENCES

Braham, A. Fundamentals of Sustainability in Civil Engineering, CRC Press, 1st Edition, ISBN: 978–1498775120, April, 2017.

Casillas, S., and Braham, A. Defining and Quantifying Pavement Preservation in Arkansas, accepted and in revision with Journal of Transportation Engineering, Part B: Pavements, January 2020.

Chappat, M., and Bilal, J. Sustainable Development: The Environmental Road of the Future, Life Cycle Analysis. Colas, 2003.

Chehovits, J., and Galehouse, L. Energy Usage and Greenhouse Gas Emissions of Pavement Preservation Processes for Asphalt Concrete Pavements. Okemos, Michigan: National Center for Pavement Preservation, 2010.

EPA. Greenhouse Gas Emissions from a Typical Passenger Vehicle, United States Environmental Agency (EPA), Office of Transportation and Air Quality, EPA-420-F-18-008, March 2018.

Kiihnl, L, and Braham, A. Exploring the Influence of Pavement Preservation, Maintenance, and Rehabilitation on Arkansas' Highway Network: An Educational Case Study, International Journal of Pavement Engineering, accepted and published online June 2019.

Uhlman, B., Andrews, J., Kardmas, A., Egan, L., and Harrawood, T. Micro Surfacing Eco-efficiency Analysis, Final Report – July 2010. BASF Corporation, 2010.

Walls, J., and M. Smith. Life-Cycle Cost Analysis in Pavement Design - Interim Technical Bulletin. Report FHWA-SA-98-0079. FHWA, U.S. Department of Transportation, September 1998.

A case study of using life cycle cost analysis in pavement management system

D. Cheng & K. Joslin
California Pavement Preservation Center, California State University, Chico, USA

ABSTRACT: Riverside County is one of the fifty-eight counties in California. According to the 2010 census data, the population of the County was more than 2 million, which makes it the 4th-most populous county in California and the 11th-most populous in the United States. A major challenge for Riverside County is to preserve and rehabilitate the roadway pavement in a cost-effective way. A detailed evaluation of the pavement management system, including the decision tree, is expected to help provide the public with quality roadways with limited funding and assist the County to make decisions that will lead to the best pavement management options that are cost effective and environmentally sustainable. Different funding scenarios and Pavement Condition Index (PCI) goals were evaluated to help dictate the allocation options for pavement preservation and rehabilitation for the County's future needs.

This paper summarized the results for short-term 5-year funding strategies, mid-range 10-year funding strategies, and long-range 20-year funding strategies. Over 30 different budget scenarios were analyzed, the total amount of asphalt and aggregates for each scenario was calculated, the Green House Gas (GHG) emissions were estimated, and the most beneficial budget scenarios were presented in this paper. A properly balanced preservation and fixing the worst strategy will give the County the best solution based on this study.

1 INTRODUCTION

1.1 *Background*

The Riverside County is one of the fifty-eight counties in California. According to the 2010 census data, the population of the County was more than 2 million, which makes it the 4th-most populous county in California and the 11th-most populous in the United States. The county covers 7,303 square miles in Southern California, spanning from the greater Los Angeles area to the Arizona border (County of Riverside 2018).

The mission of the County's transportation department is to provide the citizens of the County with increasingly more courteous, efficient, and cost-effective services dedicated to improving the quality of life and orderly economic development by the provision and management of a safe, efficient and convenient transportation system, enhancing the mobility of people, goods and services within the integrated agency activities (RCTLMA 2018). The County's pavement management system (StreetSaver) is crucial to determine the needs and priorities for the County. StreetSaver also aides in the decision-making on critical pavement life situations and implementing cost-effective strategies through preventative maintenance on the County roads.

A major challenge for the County is to preserve and rehabilitate the roadway pavements in a cost-effective way. The evaluation of the pavement management system will help provide the public with quality roadways with limited funding. Different scenarios of funding and Pavement Condition Index (PCI) goals help dictate the budget allocation options for the Counties future needs. Reviewing the County's decision tree within their current pavement management system will help the county make decisions that will lead to the best pavement management options in a cost-effective way.

1.2 Study objective

The objectives of this study are to evaluate the current pavement management program and treatment types of Riverside County; provide recommendations and cost performance measures for improving the County's pavement management system; and improve the overall pavement condition throughout the County Road System.

2 STUDY APPROACH

To achieve the objectives of the study, the following tasks are designed and executed:

Task 1   Collect information related to Riverside County pavement management program.
Task 2   Determine cost-effective Maintenance and Rehabilitation (M&R) strategies for the county.
Task 3   Evaluate the County's pavement management system.
Task 4   Evaluate different funding and project selection scenarios.
Task 5   Recommend best funding scenarios with consideration of material's usage and GHG emissions.

To evaluate the effectiveness of the existing pavement management program, task 1 was to collect information and data from Riverside County. The California Pavement Preservation Center (CP2 Center) developed a questionnaire for Riverside County that was focused on clarifying the county's current decision tree, practices, and prices of treatments. The information was used to analyze the current pavement management system and adjust it based on the CP2 Center's research.

The answers to the questionnaire made it possible for the CP2 Center to complete task 2 - determine cost-effective strategies for maintenance and rehabilitation. For task 3, the CP2 Center used the given treatments as well as the research to develop a customized decision tree that is proposed to the County. CP2 Center also modified the PCI breakpoints, which now follow the Metropolitan Transportation Commission (MTC) standards (MTC 2016).

As part of task 4 the Center analyzed different funding allocation scenarios and studied their effectiveness. A focus of the study was on funding scenarios and project selection that is needed to attain the goal on average network PCI and keep the roads in good condition. The budget scenarios were evaluated using discount rate of 0% (2018's dollar value), where discount rate reflects both inflation and interest, and the 2018-dollar value is a baseline value so that the budget can be adjusted based on inflation and interest rate in the future. In task 5, the optimal funding scenarios were identified, and the paving material usages, such as asphalt and aggregates, as well as GHG due to materials and construction were estimated.

3 NEW DECISISON TREE BASED ON LIFE CYCLE COSTS

A decision tree is a vital part of the pavement management program including maintenance and rehabilitation process. The decision tree is a product of tasks 2 and 3 and is expected to help guide the decision-making process of the County to ensure the most cost-effective solutions and to maximize the benefits.

The CP2 Center included all the County's current treatment options, as well as a few more viable options, to create a more thoroughly detailed decision tree. Using the research on treatment and rehabilitation options, the Center developed a new and improved decision tree based on Estimated Annual Cost (EAC) and lifespan of the treatments. The steps followed in the development of the new decision tree included the following:

• A review of the county's current decision tree, maintenance techniques and maintenance costs.
• Discussions with the County regarding potential changes.
• Research of a variety of maintenance strategies used in other California Counties.

- Adding a treatment lifespan as well as years between surface seals column to show the average expected life of each new and old treatment option.
- Using a Life Cycle Cost Analysis (LCCA) to select the best treatment for each pavement condition category and functional class. The equivalent annual cost, or EAC, methodology allows one to compare the cost-effectiveness of various treatments that have unequal lifespans the EAC, can be calculated using unit cost, x $/yd^2, and expected treatment life, n years, and an inflation rate of i% as shown in Equation 1.

$$EAC = \frac{x \times i}{1 - (1 + i)^{-n}}$$ (1)

4 EVALUATION OF VARIOUS FUNDING SCENARIOS

This section presents the results of the different budget scenarios projected by StreetSaver based on: various funding options, PCI goals, PCI breakpoints, percent Preventive Maintenance (PM) and discount rates. Changes in these factors create different budget scenarios where the best scenario is determined by evaluating the following important factors:

- Network average PCI,
- Deferred maintenance costs,
- Paving materials usage (sustainability – part of life cycle assessment), and
- Green House Gas emission.

The expected budget from the California Senate Bill 1 (SB 1) given to the CP2 Center is displayed in Table 1. The expected 5-year funding for Riverside County is detailed below. For the longer term, it was assumed a $60,000,000 annual budget after 2022 until 2027, where an annual budget of $30,000,000 was assumed for 2028 to 2037. These numbers were approved for analysis by the County.

Table 2 shows the different scenarios projected within three different time periods, 5-year, 10-year, and 20-year. The 5-year scenarios do not have much variation for such a short time period. The 20-year budget scenario only has one running because the county is only concerned about the SB 1 funding for that time span. These scenarios are discussed throughout this paper with the various annual budget options and preventive maintenance values.

4.1 Five-year budget alternatives

4.1.1 Five-year budget scenario 1 – SB1 funding with 20% - 40% PM

For most local agencies in California, the five-year budget scenario is important for their road planning. With the funding provided each year as shown in Table 2, the highest PCI reached with 0% discount rate is 76. The 0% discount rate scenario uses 2018's dollar value in the analysis.

Figure 1 depicts the 5-year analysis results of the PCI and amount of deferred maintenance costs. The varying preventive maintenance percentages (percent PM) depict what can be achieved

Table 1. Expected SB1 funding.

Year	Budget amount
2018	$26,000,000
2019	$46,000,000
2020	$51,000,000
2021	$57,000,000
2022	$60,000,000

183

Table 2. Budget scenarios analyzed.

Scenario	5-Year scenarios	10-Year scenarios	20-Year scenarios
1	SB1 Funding with 20% - 40% PM	SB1 Funding with 20% - 40% PM	SB1 Funding for first 10 years, then $30 million/year with 20% - 40% PM
2	$20 Million with 20% - 85% PM	$20 Million with 20% - 40% PM	

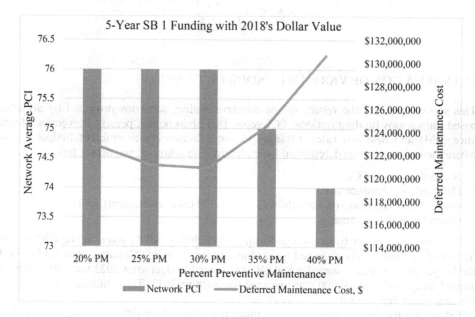

Figure 1. Five-year analysis of budget scenario 1 with 2018's dollar value.

with the given budget. Table 3 shows the estimated amount of asphalt and aggregate for each percentage of preventive maintenance. From Figure 1, the 25% and 30% PM scenarios give the highest network average PCI, and low deferred maintenance costs. However, based on the results in Table 3, the 30% PM will use 6.80×10^4 tons of asphalt and 1.16×10^6 tons of aggregate, which are lower than those of the 25% PM, 7.10×10^4 tons of asphalt and 1.24×10^6 of aggregate. In addition, the 30% PM will produce 141,163 tons of GHG $CO_{2\text{-eq}}$, which is less than 144,107 tons of GHG $CO_{2\text{-eq}}$ of the 25% PM. Therefore, the 30% PM will give the best benefits in terms of high PCI, low Deferred Maintenance Cost, relatively low material usage and low GHG emission.

The following equation was used to calculate the GHG of the PM scenarios:

$$\text{GHG Materials and Construction} = \sum_k \sum_j (g_{jX} V_j)_k \qquad (2)$$

Table 3. Paving material usage for five year budget scenario 1.

Summary	20% PM	25% PM	30% PM	35% PM	40% PM
Total Asphalt (tons)	7.26E+04	7.10E+04	6.80E+04	6.41E+04	5.93E+04
Total Aggregate (tons)	1.29E+06	1.24E+06	1.16E+06	1.07E+06	9.66E+05
GHG (tons of CO2-eq)	144,136	144,107	141,163	136,124	128,868

Where:

g_j = Estimated number of tons of $CO_{2\text{-}eq}$ per ft^3 of pavement added or removed in lane j

V_j = Pavement Thickness. Assuming Chip Seal = 0.05ft, HMA Medium Overlay = 0.2 ft base on Harvey et al. (2019).

k= sections of the pavement in the network.

GHG of the materials and construction was obtained with the following steps:

1. Divide the total funding by the cost of the treatment to obtain treated area.
2. Multiply the area by the thickness to obtain volume of paving materials. Note that thickness differs between various treatments.
3. Multiply the volume by the coefficient, g_j, in order to obtain tons of GHG emissions.

Note: The units must be ft^2 for each variable of the Equation 2.

4.1.2 Five-year budget scenario 2 – $20 million with 20% - 85% PM

Without the funding source of SB 1, the County can only have $20 million per year for roadway maintenance. The varying preventive maintenance percentages depict what can be achieved with the given budget for the scenario. Changes to the preventative maintenance percentage value will change the network PCI values, deferred maintenance costs, amount of paving material usage, as well as GHG emissions. Figure 2 depicts the 5-year analysis trend for this scenario with 0% inflation and 2018's Dollar value. Table 4 shows the estimated amount of asphalt and aggregate, and the amount of GHG emissions due to construction and materials for each percentage of preventive maintenance.

Figure 2 shows that the 25% PM through 55% PM have the same highest PCI value of 70. The 55% PM produces the lowest deferred maintenance cost for cases from the 25% PM through 55% PM. Table 4 shows that the 55% PM has the lowest materials usage, asphalt 2.57×10^4 tons and aggregate 3.85×10^5 tons, and the lowest GHG emission, 60054 tons of $CO_{2\text{-}eq}$, among the cases from the 25% PM through the 55% PM. Therefore, the 55% PM will give the best benefits with highest PCI, low deferred maintenance costs, and low paving materials usage.

Summary of 5-Year Budget Analysis

Two budget scenarios were analyzed with and without SB1 Funding with various PM percentages. With 2018's dollar value, the best scenario is the Budget Scenario 1 - SB1 Funding

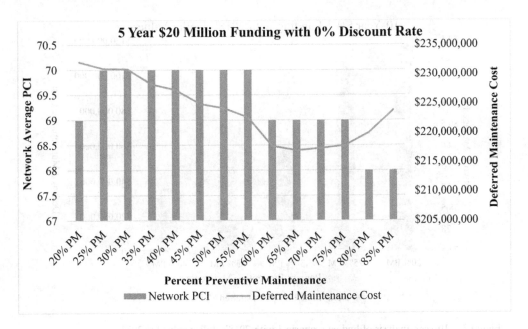

Figure 2.   Five-year analysis for budget scenario 2 with 2018's dollar value.

Table 4. Paving material usage for five year budget scenario 2.

Summary	20% PM	25% PM	30% PM	35% PM	40% PM	45% PM	50% PM
Total Asphalt (tons)	3.02E+04	2.96E+04	2.89E+04	2.83E+04	2.77E+04	2.70E+04	2.64E+04
Total Aggregate (tons)	5.37E+05	5.15E+05	4.93E+05	4.72E+05	4.50E+05	4.28E+05	4.07E+05
GHG (tons of CO2-eq)	60057	60057	60057	60056	60057	60057	60056

Summary	55% PM	60% PM	65% PM	70% PM	75% PM	80% PM	85% PM
Total Asphalt (tons)	2.57E+04	2.51E+04	2.44E+04	2.38E+04	2.29E+04	2.16E+04	2.02E+04
Total Aggregate (tons)	3.85E+05	3.63E+05	3.42E+05	3.20E+05	2.96E+05	2.65E+05	2.36E+05
GHG (tons of CO2-eq)	60,054	60,055	60,055	60,046	59,519	57,517	55,478

20% - 40% PM, where 30% PM produces the most favorable results. The PCI, deferred maintenance, total amount of paving materials, and GHG emissions are more favorable in comparison to other scenarios using 2018's dollar value. By 2022, the PCI is 76; deferred maintenance is \$120,842,614; asphalt amount is 6.80×10^4 tons and aggregate amount is 1.16×10^6 tons; and GHG emission is 141,163 tons of GHG $CO_{2\text{-eq}}$. The overall cost is \$235,048,745 with the budget provided.

4.2 Ten-year budget alternatives

4.2.1 10-year budget scenario 1 – SB1 funding with 10% - 40% PM
Figure 3 presents the 10-year analysis trend for Budget Scenario 1 with 0% discount rate in 2018's dollar value. The varying preventive maintenance percentages depict what can and will be achieved with the given budget.

Figure 3 shows that 20% PM through 30% PM have the same highest PCI value of 82. The 20% PM produces the lowest deferred maintenance cost for cases from the 20% PM through 30% PM. Table 5 shows that the 20% PM has the high materials usage, asphalt 1.63×10^5 tons and aggregate 2.89×10^6 tons, among the cases from the 20% PM through the 30% PM. Overall, the 20% PM will give the best benefits with highest PCI, lowest deferred maintenance costs, relatively

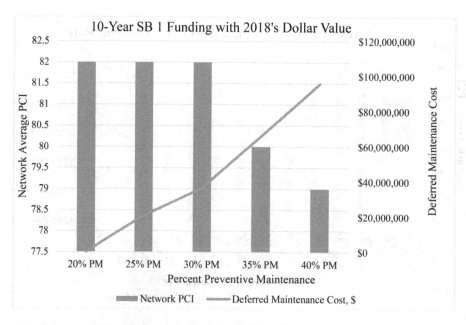

Figure 3. 10-year analysis of budget scenario 1 with 2018's dollar value for 2027.

186

Table 5. Paving material usage for ten year budget scenario 1.

Summary	20% PM	25% PM	30% PM	35% PM	40% PM
Total Asphalt (tons)	1.63E+05	1.59E+05	1.53E+05	1.46E+05	1.38E+05
Total Aggregate (tons)	2.89E+06	2.77E+06	2.61E+06	2.44E+06	2.24E+06
GHG (tons of CO2-eq)	322,911	322,854	317,905	310,105	299,547

high paving materials usage and GHG emmisions. The 20% PM produces the best results with the highest network average PCI of 82 and the lowest possible deferred maintenance cost of $0.

4.2.2 *10-year budget scenario 2 – $20 million with 20% - 40% PM*

The results show that $20 million per year budget is not enough to maintain the current average network PCI. The varying preventive maintenance percentages depict what can be achieved with the given budget for the scenario. Changes to the preventative maintenance percentage value will change the network PCI values, deferred maintenance costs, amount of paving material usage, and GHG emissions. Figure 4 depicts the 10-year analysis trend for this scenario with 0% discount rate and 2018's Dollar value. Table 6 shows the estimated amount of asphalt and aggregate, and GHG emissions for each percentage of preventive maintenance.

Figure 4 shows that the 25% PM through 40% PM have the same highest PCI value of 64. The 30% PM produces the lowest deferred maintenance cost for cases from the 25% PM through 40% PM. Table 6 shows that the 30% PM has the medium materials usage, asphalt 5.79×10^4 tons and aggregate 9.87×10^5 tons, and medium GHG emission, 120,113 tons of $CO_{2\text{-eq}}$, among the cases from the 25% PM through the 40% PM. Overall, the 30% PM will give the best benefits with highest PCI, lowest deferred maintenance costs, and medium paving materials usage and GHG emissions.

Summary of 10-Year Budget Analysis

Two budget scenarios were analyzed with and without SB1 Funding with various PM percentages. With 2018's dollar value, the best scenario is the Budget Scenario 1 - SB1 Funding 20% - 40% PM, where 20% PM produces the most favorable results. The PCI, deferred

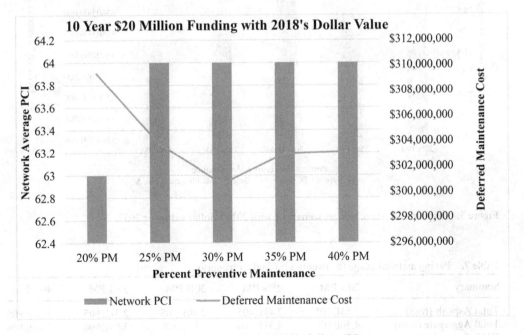

Figure 4. 10-year analysis for budget scenario 2 with 2018's dollar value.

187

Table 6.   Paving material usage for ten year budget scenario 2.

Summary	20% PM	25% PM	30% PM	35% PM	40% PM
Total Asphalt (tons)	6.05E+04	5.92E+04	5.79E+04	5.66E+04	5.53E+04
Total Aggregate (tons)	1.07E+06	1.03E+06	9.87E+05	9.43E+05	9.00E+05
GHG (tons of CO2-eq)	120,113	120,113	120,113	120,113	120,109

maintenance, and total amount of paving materials are more favorable in comparison to other scenarios using 2018's dollar value. By 2027, the PCI is 82; deferred maintenance is $0; asphalt amount is $1.63x10^5$ tons and aggregate amount is $2.89x10^6$ tons; and GHG emission is 322,911 tons of CO_{2-eq}. The overall cost is $537,674,646 with the budget provided.

4.3   Twenty-year budget scenario

Figure 5 depicts the 20-year analysis trend for this scenario with 0% discount rate and 2018's Dollar value. Table 7 shows the estimated amount of asphalt and aggregate as well as GHG emission for each percentage of preventive maintenance.

Figure 5 shows that the 20% PM has the highest PCI value of 80. The 25% PM produces the lowest deferred maintenance cost for cases from the 20% PM through 40% PM. Table 7 shows that the 25% PM has the medium materials usage, asphalt $2.48x10^5$ tons and aggregate $4.31x10^6$ tons, and medium GHG emission, 502,950 tons of CP_{2-eq}, among the cases from the

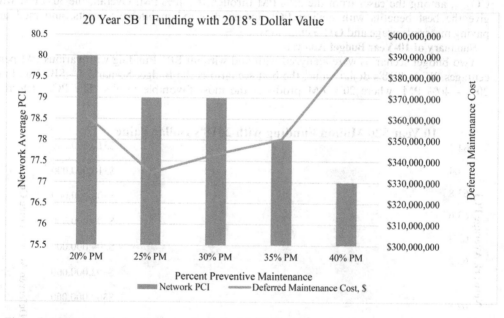

Figure 5.   20-year analysis of budget scenario 1 with 2018's dollar value for 2037.

Table 7.   Paving material usage for twenty year budget scenario.

Summary	20% PM	25% PM	30% PM	35% PM	40% PM
Total Asphalt (tons)	2.53E+05	2.48E+05	2.40E+05	2.31E+05	2.21E+05
Total Aggregate (tons)	4.50E+06	4.31E+06	4.09E+06	3.85E+06	3.60E+06
GHG (tons of CO2-eq)	503,028	502,950	498,024	490,232	480,618

20% PM through the 40% PM. Overall, the 25% PM will give the best benefits with relatively high PCI, lowest deferred maintenance costs, and medium paving materials usage and GHG emission.

5 CONCLUSIONS AND RECOMENDATIONS

5.1 *Conclusions*

The following are the conclusions of the study:

1. After reviewing the decision tree and treatments used by Riverside County, the CP2 Center worked with the County engineer to adjust decision tree, PCI breakpoints, and treatments. A new decision tree was developed based on life cycle cost.
2. Many different budget scenarios were analyzed by the CP2 Center. The most beneficial budget scenarios should result in high network average PCI, low deferred maintenance costs, low paving materials usage, and low GHG emission.
3. The $20 million per year budget is not enough to maintain the current average network PCI for the County. Based on this study, the PCI value will continue to decrease with $20 million per year of funding.
4. For the 5-year budget analysis: Budget scenario with SB 1 funding produces the best results for a 0% discount rate (2018's dollar value). This scenario includes SB1 funding with 30% PM producing the most favorable results with 0% discount rate.
5. For the 10-year budget analysis: Budget scenario with SB1 funding for the 10-year analysis gives the most favorable results. For the scenario with 0% discount rate, 20% PM is the best.
6. For the 20-Year Budget Analysis Scenarios: Budget scenario with SB 1 for the 20-year analysis gives the best results. For the scenario with 0% interest, 25% PM produces the best results.

5.2 *Recommendations*

The following are the recommendations based on this study:

1. Continually update the cost and service life for each treatment in the new decision tree.
2. A life cycle assessment should be incorporated to evaluate treatments for future alternatives. The environmental impact of the treatments should be further studied. The emission due to roadway roughness and traffic levels should be investigated.
3. A balanced budget between pavement preservation and fixing the worst roads should be utilized to improve the County's pavement conditions. Keys to achieve the desired PCI include the following:
   • Optimize percentages for preventive maintenance and rehabilitation (use lowest deferred maintenance cost).
   • Update decision tree and treatments.
   • Use innovative treatments, such as chip seal over paving mat or multiple layer system, which can improve cost effectiveness.

REFERENCES

MTC, StreetSaver Pavement Management System Program for Riverside County, 2018 Riverside County. https://countyofriverside.us/Visitors/CountyofRiversideInformation/RiversideCountyHistory. aspx, visited in December 2017.

RCTLMA, Website: http://rctlma.org/trans/, visited in January 2018.

MTC, StreetSaver PCI Distress Identification Manuals, 2016.

Harvey, J.T., Butt, A.A., Lozano, M.T., Kendall, A., Saboori, A., Lea, J. D., Kim, C., Basheer, I. Life Cycle Assessment for Transportation Infrastructure Policy Evaluation and Procurement for State and Local Governments. MDPI. November 2019.

Pavement preservation and maintenance schedule evaluation using a life-cycle assessment tool

Qingwen Zhou, Hasan Ozer & Imad L. Al-Qadi
Illinois Center for Transportation, University of Illinois at Urbana-Champaign, USA

ABSTRACT: Pavement preservation is a cost-effective and environmentally sustainable strategy for extending pavement service life, while minimizing the use of resources. Effectiveness of preservation depends on several factors, including existing pavement condition, preservation schedule, and timing of treatment application. Most of the current studies focus on the cost benefits of a single pavement preservation treatment using life-cycle cost analysis (LCCA) tools. In this study, preservation and maintenance schedules with a sequence of treatments applied at various times are designed considering the environmental impacts and energy consumption using life-cycle assessment (LCA) methodology. A user-friendly LCA tool was developed specifically for preservation of asphalt and concrete surfaced pavements by the Illinois Center for Transportation (ICT) at the University of Illinois at Urbana-Champaign (UIUC). The inventory analysis and data compiled were consistent with the ISO 14044:2006 standards. The system boundary of the LCA employed in the tool includes maintenance and use stages. Work zone is included within the construction process. The tool uses pay-items for each preservation treatment to calculate materials and construction impacts in the maintenance stage. The use-stage impact was calculated considering pavement related rolling resistance and radiative forcing. A summary of the tool development is presented along with two case studies to evaluate significance of evaluating use stage in preservation LCA as well as the importance of designing a preservation schedule as compared to a single preservation treatment.

1 INTRODUCTION

1.1 *Background*

The US roadway network is more than 4.26 million miles in total length, making it one of the largest road networks in the world. In 2017, more than 3.2 trillion of vehicle miles traveled (VMT) over this roadway network, consuming more than 187 billion gallons of fuel (FHWA, 2019). Besides, the total expenditures for over 164,000 miles national highways in the roadway network was $119.2 billion in 2017 (FHWA, 2019). As an integral part of this roadway network, pavements provide a smooth and durable all-weather traveling surface for a range of vehicles (passenger cars, trucks, buses, and motorcycles). Despite the investment needed to address deteriorated road condition, funds allocated for construction and rehabilitation have been shrinking and cost of construction has been increasing steadily. Therefore, agencies have been exploring strategies to improve pavement sustainability.

Pavement preservation is considered to be one of the sustainable strategies that can be used by state and local agencies to maintain good condition of a road in a cost-effective and environmentally friendly manner. According to the Federal Highway Administration (FHWA) Office of Asset Management, timely preservation improves safety and mobility of public transportation, reduces congestion, and may effectively extend pavement service life, while using less resources (Geiger, et al., 2005). Many studies have showed cost and environmental benefits of pavement preservation activities using life-cycle cost analysis (LCCA) and life-cycle assessment (LCA) techniques. Pittenger et al. (2011) used field test data to quantify pavement preservation service lives of asphalt and concrete pavements. Life-cycle cost (LCC) model was developed based on equivalent uniform annual cost. The research team also explored stochastic LCC model to compare pavement preservation treatment alternatives considering the risk of commodity price volatility (Pittenger et al., 2012). In addition, Wang et al. (2012) performed LCA on pavement preservation treatments to evaluate the energy consumption and greenhouse gas (GHG) emissions associated with material production, construction, and vehicle operation resulting from the progression of pavement roughness and texture.

Decision trees are commonly used to make a selection from multiple alternatives considering basic inputs like existing condition, traffic, climate, etc. (Hicks et al., 1997). Peshkin et al. (2004) proposed a framework to determine optimal timing of preventive treatment by considering various pavement condition indicators and associated costs. Overall pavement condition indicators, such as international roughness index (IRI), cracking, and rutting, were used to predict treatment service life. Data were also collected in Iowa from an existing pavement asset management system to develop a decision-making framework for local agencies (Abdelaty, et al., 2015).

However, none of those studies considered the impact of existing pavement condition on the performance of preservation treatments explicitly. Instead of scheduled activities, single treatment analysis and comparison to alternatives were the methods commonly used. There is a need to consider a more comprehensive selection criteria and preservation schedule design to provide more informed preservation guidance for agencies.

To address these gaps, a life-cycle based tool was developed exclusively for preservation. The user may select the right preservation treatment as a function of existing pavement condition and other critical variables. The tool also incorporates a list of preservation activities to design a schedule. In addition, lifetime prediction models were developed to predict the service life of different preservation treatments assigned to a schedule. The user is allowed to make the schedule design consistent with agency goals. In this study, the goal is to optimize the life-cycle energy consumption, GHG emissions, and corresponding costs

1.2 *Objective*

The main goal of this study is to conduct LCA on various pavement preservation and maintenance schedules (PPMS) for existing asphalt concrete (AC) and Portland cement concrete (PCC) pavements. A LCA tool was developed to perform the analysis on PPMS and to provide guidance to local and state transportation agencies in scheduling preservation treatments considering both service life extension and environmental benefits.

The tool has the following features:

1) Life-cycle inventories for commonly used preservation treatments on AC- and PCC-surfaced pavements.
2) Decision trees to select appropriate treatment.
3) A life-cycle estimation model to predict service life extension.
4) A preservation schedule design approach
5) Use-stage models for rolling resistance and heat island impacts.

2 METHOD

The LCA methodology integrated in the tool conforms to the International Organization for Standardization (ISO) 14044:2006 standards (ISO, 2006). The details of LCA analysis steps and tool development are discussed in the following subsections.

2.1 *Life-cycle assessment*

Life cycle assessment is a widely used approach to analyze and quantify environmental impact flows throughout the life cycle of products and systems. A product's life cycle begins with raw material acquisition and continues through production, use, end-of-life recycling or final disposal.

2.1.1 *Goal and scope definition*

The goal of the study is to quantify the environmental impacts and energy consumption of preservation and maintenance schedules for AC- and PCC-surfaced pavements. Major scope elements of this study are shown in Table 1. Table 2 lists the preservation treatments included in this study.

2.1.2 *Life-cycle inventory analysis*

Primary data were collected by gathering information about pavement preservation practices in different states. The research team distributed questionnaires to DOTs throughout the nation in 2017–18. Along with other information collected, DOTs provided their practices of

Table 1. The scope elements.

Geography	United States
System boundary	Pavement preservation is considered as part of pavement maintenance. Thus, this study only considers the materials production and construction within the maintenance stage and the interaction of the pavement with the environment in the use stage.
Functional Unit	Two functional units are used: lane miles and vehicle-miles traveled. Lane miles compute the total impacts determined by multiplying the lane numbers and section length by unit impacts within an analysis period. Vehicle-miles traveled (VMT) is used when the comparison is among alternatives with different analysis periods.
Analysis period	It is the pavement service life elapsed from the application of first preservation treatment until the reconstruction of existing pavement. It covers the length of a pavement preservation schedule. A preservation schedule may include more than one treatment.
Data assumptions	The data used are a combination of primary and secondary data from various sources, including local surveys, governmental reports and databases, industry reports, peer-reviewed papers, and commercial inventory databases.
Impact categorization	This study uses the US Environmental Protection Agency's (EPA) Tool for Reduction and Assessment of Chemicals and Other Environmental Impacts (TRACI 2.1). Four quantitative outcomes from the LCA study were analyzed: global warming potential (GWP), total energy, total energy with feedstock, and single score (SS). Single score represents a weighted score of 10 TRACI environmental impact categories and the weighting factors determined by the National Institute of Standards and Technology is specific to the United States (Bare, et al., 2006).

Table 2. Preservation and maintenance treatments considered.

Preservation Treatment for AC Surfaces	Preservation Treatment for Concrete Surfaces
Crack sealing/crack filling	Diamond grinding/grooving
Fog seal	Joint resealing
Chip seal	Dowel-bar retrofitting
Cape seal	Partial depth repair
Slurry seal	Full-depth repair
Microsurfacing	Ultra-thin bonded wearing course
Thin AC overlay	Thin AC overlay
Ultra-thin bonded wearing course (UTBWC)	Crack sealing/crack filling
Bonded-concrete overlay (BCO)	
Hot in-place recycling (HIR) and chip seal	
HIR and microsurfacing	
HIR and thin AC overlay	
Cold in-place recycling (CIR) and chip seal	
CIR and microsurfacing	
CIR and thin AC overlay	
CIR and medium overlay	

preservation schedules that produced maximum pavement performance. These practices were stored in the tool as default preservation schedule designs.

Secondary data were collected mainly from commercial life-cycle inventory (LCI) database (e.g. Ecoinvent 2.2 and 3.0 [Frischknecht, 2005]), governmental databases and reports, peer-reviewed literature, and industry reports. These inventory sources document the energy and emissions needed to produce energy sources such as fossil fuels and electricity as well as those resulting from downstream processes, such as direct combustion of fossil fuels in manufacturing of pavement materials or the use of diesel in various construction equipment.

In this study, unit processes were introduced to compile material, construction, fuel, and/or hauling processes, including upstream and downstream data and models. The impacts of a treatment can then be computed by adding the impacts of unit processes included in the system boundary. For example, the unit process of asphalt binder includes crude oil extraction, flaring, transportation, refining, and blending terminal storage. The emission resulting from the unit process was represented by summing up the upstream and downstream emissions of those subprocesses. Commercial software SimaPro (2014) was used to model the unit processes of fuel and electricity, pavement materials, materials hauling, and pavement construction.

Pay item is the common term used in the construction industry and are convenient for both agencies and contractors. A pay item represents a unit of work for which a price or environmental impact is provided. In this study, pay items were applied to compile materials, mixtures, and equipment of a unit of work by adding corresponding unit processes. To summarize, preservation activities were broken down into tasks, and each task was categorized with relevant pay items and each pay item may be further decomposed of unit processes engaged in the corresponding task.

2.1.3 *Impact assessment*

2.1.3.1 MATERIALS AND CONSTRUCTION STAGE

As discussed above, pay items were used to compile unit processes of materials production, material hauling and construction with respect to a specific activity or a task. Thus, LCA of preservation and maintenance schedule is processed by evaluating the emissions and energy consumption of pay items included in each preservation treatment. The energy consumption (EC) and environmental impacts (EI) of each pay item can be summarized as in Equations (1) and (2).

$$EC_{PItem_i} = EC_{PItem_i\_Mat} + EC_{PItem_i\_Haul} + EC_{PItem_i\_Equip} \qquad (1)$$

$$EI_{PItem_i} = EI_{PItem_i\_Mat} + EI_{PItem_i\_Haul} + EI_{PItem_i\_Equip} \qquad (2)$$

where EC_{PItem_i} and EI_{PItem_i} are the EC and EI of pay item i; and $EC_{PItem_i\_Mat}$, $EC_{PItem_i\_Haul}$, and $EC_{PItem_i\_Equip}$ are the EC of materials production, materials hauling, and construction processes included in pay item i. Similarly, $EI_{PItem_i\_Mat}$, $EI_{PItem_i\_Haul}$, and $EI_{PItem_i\_Equip}$ are the EI of materials production, hauling transportation, and construction included in pay item i.

If mix design is considered in a pay item, then the material production impacts should be formulated as the sum of material production impacts included in the mix as well as the materials that are not part of the mix design.

The impacts resulting from hauling truck operations for materials were calculated using the variable impact transportation (VIT) model developed by Kang et al. (2019). This model created a unit process, which quantifies the EC and EI of a ton of material hauled one mile at various geometric and environmental conditions. Those conditions include temperature (T), relative humidity (RH), grade (G), and hauling mode (HM).

$$EC_{PItem_i\_Equip} = \sum_j EC_{Unit\_Haul_j}(T, RH, G, HM) * Q_{PItem_i\_mat_j} * Dist_j \qquad (3)$$

$$EI_{PItem_i\_Haul} = \sum_j EI_{Unit\_Haul_j}(T, RH, G, HM) * Q_{PItem_i\_mat_j} * Dist_j \qquad (4)$$

where $EC_{Unit\_Haul_j}(T, RH, G, HM)$ and $EC_{Unit\_Haul_j}(T, RH, G, HM)$ are the EC and EI of hauling unit process for material j given T, RH, G, and HM values. $Q_{PItem_i\_mat_j}$ and $Dist_j$ are the quantity (ton) of material j to be hauled and the hauling distance (mi) of material j in pay item i, respectively.

A construction unit process computes the EC and EI of a gallon of diesel or propane used in an equipment with a specific level of horsepower during construction. As long as the equipment fuel efficiency (gal/hr), speed (mph), and number of equipment passes are known in the construction process, the EC and EI resulting from fuel consumption are calculated as follows:

$$EC_{PItem_i\_Constr} = \sum_j EC_{Unit\_Constr_j} * \frac{Fuel\ Efficiency_j}{Equipment\ Speed_j} * Num\ of\ Passes_j \qquad (5)$$

$$EI_{PItem_i\_Constr} = \sum_j EI_{Unit\_Constr_j} * \frac{Fuel\ Efficiency_j}{Equipment\ Speed_j} * Num\ of\ Passes_j \qquad (6)$$

where $EC_{Unit\_Constr_j}$ and $EI_{Unit\_Constr_j}$ are the EC and EI of construction unit process given equipment type, fuel type, and tier category.

2.1.3.2 USE STAGE

The use-stage impacts considered in this study include two parts: heat island and rolling resistance (RR). Pavement heat island impacts are resulting from thermal and reflectivity properties of pavement surface materials. Sen and Roesler (2016) introduced microscale heat island by analyzing metric pavement radiative forcing (RF) and the corresponding GWP. In another study, Sen and Roesler (2017) investigated the RF of various pavement preservation techniques considering their service lives and the geographic location. Then heat island impacts with respect to different preservation treatments in different states are calculated in terms of equivalent kg CO_2/m^2 using the corresponding RF value.

Rolling resistance-related impacts were calculated in terms of pavement roughness and texture. Pavement roughness expressed the irregularities in the pavement surface. It affects the ride quality of a vehicle and it is measured by international roughness index (IRI). A roughness speed impact model (RSI) introduced by Ziyadi et al. (2018) was used to capture extra fuel consumption due to surface roughness changes, as shown in Equation (7).

$$RSI_{t=0}^{Energy} : \hat{E}(v, IRI) = \frac{p}{v} + (k_a * IRI + d_a) + b \times v + (k_c * IRI + d_c) \times v^2 \qquad (7)$$

where $\hat{E}(v, IRI)$ is the estimated energy consumption per vehicle distance (kJ/mi) and v is the speed of the vehicle in mph. A unit of IRI is in/mi. $k_a, d_a, k_c, d_c, p,$ and b are regression coefficients.

Texture, on the other hand, corresponds to smaller wavelengths than pavement roughness, primarily controlling friction and noise. Progression of texture can be affected by traffic composition, surface materials, and mixture type. In this study, a macro-texture-related energy model developed by Zaabar and Chatti (2012) is used to account for texture effect on heavy vehicles, as shown in Equation (8).

$$\delta Etexture(\%) = 0.02 - 2.5 \times 10 - 4 \times (v - 35) \qquad (8)$$

where $\delta Etexture(\%)$ is percent increase in energy consumption per unit increase of texture in mean profile depth (MPD) and v is the speed of the vehicle in mph.

2.2   Tool development

The preservation sustainability assessment tool (PSAT) was developed using visual basic applications (VBA), an event-driven programming language in Microsoft® Excel. The motivation to develop PSAT is to provide data and necessary information to practitioners to make informed decisions on preservation schedule, considering sustainability benefits.

This tool starts with collecting basic input such as pavement structure and traffic information and then followed by preservation schedule design. A preservation schedule is a sequence of preservation treatments applied in succession. The user can schedule preservation treatments based on their experiences or using the default schedules collected from DOTs. The user may also input severity and extent of pavement distresses. Decision trees were integrated into this tool to recommend treatments based on distress input, pavement condition, and traffic volume.

Within the schedule design process, service life of each preservation treatment is estimated. The service life of a preservation treatment is defined as lifetime that elapsed from the application of a preservation treatment to the next major rehabilitation or reconstruction activity. In this project, lifetime estimation models were established by combining inputs from a questionnaire using analytical hierarchy process (AHP) method (Zhou, et al., 2019), as shown in Equation (9):

$$PCI_t = PCI_0 - m_{average} * (\alpha * PCI + \beta * \ln(AADT) + \gamma * Tr+) * t \qquad (9)$$

where PCI_0 is the initial PCI (pavement condition index, from 0-100%) after the preservation treatment is applied. PCI_t is the PCI value at year t. The average pavement condition deterioration rate after applying preservation treatment is $m_{average}$, and t is the lifetime elapsed from PCI_0 to PCI_t. α, β, γ, and ϵ are model coefficients.

The analysis period of the whole preservation schedule is determined after predicting the lifetime of each treatment included in that schedule. The user analyzes LCA stages within the preservation schedule. Pay items for each preservation treatment may be created by inputting materials, mixture, and construction information. IRI progression is also implemented using existing models to compute use stage impacts. Upon the analysis completion, LCA results are characterized for each stage and impact category and presented.

3   CASE STUDIES

3.1   Case I: Pavement preservation on high volume vs. low volume road

A treatment schedule, including two preservation treatments, was analyzed. This schedule is applied under three levels of traffic volume. The goal of this case study is to quantify the benefits of a planned preservation treatment schedule under high- and low- volume roads. General project inputs are shown in Table 3. The acceptable condition for a pavement to apply preservation should be above PCI of 65. Current pavement condition is favorable for preservation.

The preservation schedule applied in this case study is shown in Table 4. The lifetime of each treatment is calculated using Equation (9) by setting $PCI_0=100$ and $PCI_t=65$. LCA for each level of traffic was performed using PSAT. The material stage includes raw material acquisition, plant production, and transporting raw materials to the job site. The hauling distance between an asphalt plant and the project site is assumed to be 50 mi. The design parameters for thin AC overlay are 4 percent air void, 6 percent asphalt content, and maximum specific gravity of 2.500.

It is assumed that the IRI value stays at 50 in/mi right after a preservation activity; IRI is progressing linearly along the activity service life. The IRI progression rate is inputted as in Figure 1. The IRI progression rate should increase as road volume increased; and it is smaller for thin AC overlay than that of microsurfacing.

Table 3. Project information.

Basic Information	Input
Pavement Type	Conventional AC Pavement
Surface depth	4 in
Mileage	2 mi
Lane number	2 (12-ft-wide lanes)
Present PCI	70
AADT levels	3,000/15,000/30,000
Traffic growth percent	2
Speed limit	60 mph
Truck percent	10
Small-truck percent	30
Medium-truck percent	30
Large-truck percent	40

Table 4. Schedule design.

AADT	Schedule	Total Lifetime (yrs)
5,000	1) Microsurfacing (4 yrs) 2) Thin AC overlay 1 in (5 yrs)	9
15,000	1) Microsurfacing (3.5 yrs) 2) Thin AC overlay 1 in (4.5 yrs)	8
30,000	1) Microsurfacing (3 yrs) 2) Thin AC overlay 1 in (4.5 yrs)	7.5

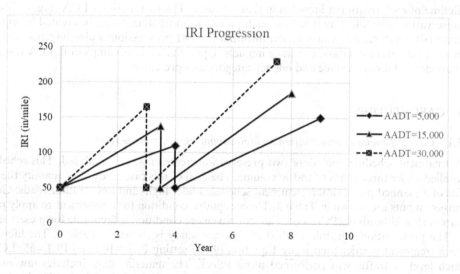

Figure 1. IRI Progression under different three traffic levels.

The LCA results of the preservation schedules are shown in Figure 2. The results are categorized into two parts: materials and construction (M&C) and use stages. Impacts resulting from traffic delay during work zone is included in materials and construction stage. And impacts in use stage are further decomposed as impacts resulting from heat island (HI), texture progression and roughness progression. Heat island impacts quantified the GWP as a result of thermal and reflectivity changes of pavement surface. Thus, the heat island does

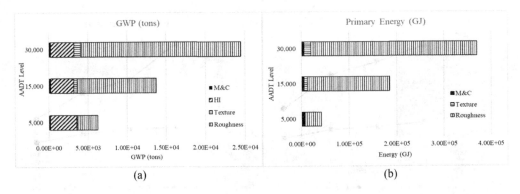

Figure 2.   (a) GWP emissions; (b) Primary energy consumption of the preservation schedule under three traffic levels.

not play a role in primary energy consumption. The overall energy consumption wasresulting roughness change.

3.2   *Case II: Sensitivity to existing pavement condition*

The goal of the second case study is to evaluate the impact of existing pavement condition in the context of a preservation schedule selected for LCA. Four existing pavement conditions and two traffic levels were considered to select a preservation treatment. The lifetime was computed using Equation (9) and IRI reduction rates were assumed based on experience, as shown in Table 5. The other information are the same as in Table 3.

As shown in Figure 3, the lower the existing PCI value, the more GWP emissions and total energy consumption it takes to preserve the pavement. However, a pavement in a good condition, only requires minor preservation, such as fog seal and crack sealing; while a pavement in a poor condition requires a costly comprehensive preservation like HIR + thin AC overlay. The overall impacts of the sequence of minor treatments may be high when the treatment is frequently applied. Therefore, this case study demonstrates that it is more effective to design a preservation schedule to evaluate the overall benefits instead of only considering a single preservation treatment.

Table 5.   Case II input.

AADT = 5,000

Existing PCI	Treatment Applied	Lifetime (yrs)	IRI Reduction Rate (in/mi/yr)
60	HIR + Thin AC overlay	5	25
70	Thin AC overlay	5	20
80	Microsurfacing	4	15
90	Fog Seal	2	10

AADT = 15,000

Existing PCI	Treatment Applied	Lifetime (yrs)	IRI Reduction Rate (in/mi/yr)
60	HIR + Thin AC overlay	4.5	30
70	Thin AC overlay	4.5	25
80	Microsurfacing	3.6	20
90	Fog Seal	2	15

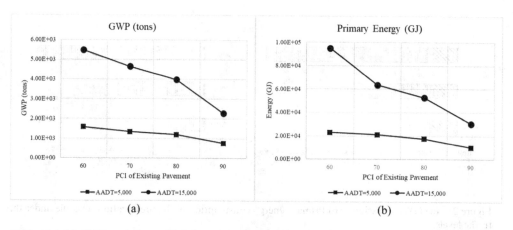

Figure 3. (a) GWP emissions; (b) Primary energy consumption of the preservation under four PCI_0 values.

4 DISCUSSION AND CONCLUSION

Life cycle assessment methodology was used to quantify sustainability impacts of preservation activities for AC- and PCC-surfaced pavements. A tool was developed and used to support project-level decision making on preservation activities. The key components of the tool development are the inventory analysis used in the LCA calculations, treatment lifetime models, and decision trees for preservation schedule design. Life-cycle inventory data were compiled using the input obtained from questionnaires, commercial databases in LCA software Sima-Pro, literature, and other publicly and commercially available databases.

The main outcome of this study is an LCA tool (PSAT), which has a user-friendly interface in the VBA platform, and pay items as a building block for ease of implementation. The tool is intended for public use to assess environmental impacts of pavement preservation and maintenance alternatives for highway pavements. The data, models, decision trees, and other relevant information were incorporated in the tool for standalone application. The tool is intended for engineers in state and local agencies, practitioners in the industry, and contractors. Two case studies are presented to evaluate the impacts of preservation schedules under various traffic volumes as well as the impacts of a preservation treatment when the initial PCI values are different.

Significance of traffic impacts were studied in the first case study. The use stage contributes as high as 93% of the total impacts; it demonstrates the importance of including use stage for preservation treatments. The second case study selected preservation treatment for a pavement at different levels of existing condition. There is a trade-off between treatment application time and application frequency. An optimal design of a schedule considering both application time and frequency is needed to yield minimum impacts.

In performing the LCA, the following assumptions were made:

1) There is no time gap between two treatments in a preservation schedule.
2) The decision tree is only based on previous experience, and it is only used for the guidance of selecting the first treatment. For the subsequent treatments in the preservation schedule, the user makes the decision.
3) The lifetime estimation model was developed for asphalt-surfaced pavements only, while the preservation lifetime for rigid pavements uses literature data as default. Lifetime prediction models should be improved with actual performance data. It is important to add weather-related parameters to the prediction models.

4) Used IRI progression models are based on historical performance data and utilized to predict the change in IRI over time. IRI was shown to be a critical parameter; therefore, prediction model for progression of IRI should be selected carefully.

REFERENCES

Abdelaty, A., H. D. Jeong, O. G. Smadi, and D. D. Gransberg. *Iowa Pavement Asset Management Decision-Making Framework*. Iowa Department of Transportation, Ames, 2015.

Bare, J., T., Gloria, and G., Norris. Development of the Method and U.S. Normalization Database for Life Cycle Impact Assessment and Sustainability Metrics. *Environmental Science and Technology*, 2006. 16:5108–5115.

Chatti, K., and I. Zaabar. *NCHRP Report 720: Estimating the Effects of Pavement Condition on Vehicle Operating Costs*. Transportation Research Board, Washington, DC, 2012.

Federal Highway Administration (FHWA). "Highway Statistics 2017." Federal Highway Administration, Washington, DC, 2019. Accessed Sep 10, 2019. https://www.fhwa.dot.gov/policyinformation/statistics/2017/.

Frischknecht, R., N. Jungbluth, H.-J. Althaus, G. Doka, R. Dones, T. Heck, S. Hellweg, R. Hischier, T. Nemecek, G. Rebitzer, and M. Spielmann. The Ecoinvent Database: Overview and Methodological Framework. *International Journal of Life Cycle Assessment*, 2005. 10: 3–9.

Geiger, D. *Memorandum: Pavement Preservation Definitions*. Federal Highway Administration, Washington, DC, 2005

Hicks, R., K. Dunn, and J. Moulthrop. Framework for Selecting Effective Preventive Maintenance Treatments for Flexible Pavements. *Transportation Research Record: Journal of the Transportation Research Board*, 1997. 1597: 1–10.

International Organization for Standardization (ISO). *Environmental Management—Life-cycle Assessment—Requirements and Guidelines*. ISO 14044:2006(E). Geneva, Switzerland, 2006.

Kang, S., M. Ziyadi, H. Ozer, and I. L. Al-Qadi. Variable Impact Transportation (VIT) Model for Energy and Environmental Impact of Hauling Truck Operation. *The International Journal of Life Cycle Assessment*, 2019. 24: 1154–1168.

Peshkin, D. G., T. E. Hoerner, and K. A. Zimmerman. *NCHRP Report 523: Optimal Timing of Pavement Preventive Maintenance Treatment Applications*. Transportation Research Board, Washington, DC, 2004. https://doi.org/10.17226/13772

Pittenger, D., D. D. Gransberg, M. Zaman, and C. Riemer. Life-cycle Cost-based Pavement Preservation Treatment Design. *Transportation Research Record*, 2011. (2235): 28–35. https//doi.org/10.3141/2235-04.

Pittenger, D., D. D. Gransberg, M. Zaman, and C. Riemer. Stochastic Life-Cycle Cost Analysis for Pavement Preservation Treatments. *Transportation Research Record*, 2012. (2292): 45–51. http://doi.org/10.3141/2292-06

Sen, S., and J. Roesler. Aging Albedo Model for Asphalt Pavement Surfaces. *Journal of Cleaner Production*, 2016. 117: 169–175.

Sen, S., and J. Roesler. Microscale Heat Island Characterization of Rigid Pavements. *Transportation Research Record: Journal of the Transportation Research Board*, 2017. 2639: 73–83.

SimaPro. Version 8.0.4. PRé Consultants [Software], Amersfoort, Netherlands. 2014.

US Census Bureau. *Annual Survey of State and Local Government Finances*, Sep. 10, 2018. www.census.gov/programs-surveys/gov-finances.html. Accessed July 20, 2019.

Wang, T., I. S. Lee, J. Harvey, A. Kendall, E. B. Lee, and C. S. Kim. *UCPRC Life Cycle Assessment Methodology and Initial Case Studies for Energy Consumption and GHG Emissions for Pavement Preservation Treatments with Different Rolling Resistance*. No. UCPRC-RR-2012-02. California Department of Transportation, Sacramento, 2012.

Ziyadi, M., H. Ozer, S. Kang, and I. L. Al-Qadi. Vehicle Energy Consumption and an Environmental Impact Calculation Model for the Transportation Infrastructure Systems. *Journal of Cleaner Production*, 2018. 174: 424–436.

Climate action plans: Review and recommendations

M.T. Lozano
Energy Systems, Energy and Efficiency Institute, University of California, Davis, USA

A.M. Kendall
Department of Civil and Environmental Engineering and Energy and Efficiency Institute, University of California, Davis, USA

A.A. Butt & J.T. Harvey
Department of Civil and Environmental Engineering, University of California Pavement Research Center, University of California, Davis, USA

ABSTRACT: Local governments in California were tasked by the 2006 Climate Change Solutions Act (Assembly Bill 32) to reduce their greenhouse gas (GHG) emissions to help achieve statewide emissions reduction goals. Many jurisdictions have developed climate action plans (CAPs) that provide context for their current rate of emissions, strategies across various sector being planned to be employed to reduce emissions, and the expected impact those strategies will have. However, not all CAPs are created equally, as certain jurisdictions develop robust CAPs that include actual emissions and economic values, list responsible parties, and identify co-benefits, while others barely meet the minimum requirements. The goal of this paper is to summarize a review of existing CAPs developed by different regions, counties and cities in the USA and determine what information makes CAPs stand out, and assess which attributes make them more actionable. The research examines strategies in the transportation sector and looks across CAPs to evaluate how actionable the strategies appear to be. For example, the CAP developed by one California county that was reviewed is exemplary in particular because it included a detailed assessment of costs—this included expected upfront costs, entities responsible for covering the upfront cost, annual net savings or costs per strategy, and entities incurring these annual savings or costs—among other key qualities. This paper also explores the role lifecycle assessment can play in the prioritization of GHG reduction strategies. Finally, the paper provides recommendations on what future iterations of CAPs should include, roadway strategies that can be considered, and introduces potential solutions to decision making.

1   INTRODUCTION

1.1   *Policy background*

Global warming caused by anthropogenic emissions of greenhouse gases (GHGs, dominated by CO_2, CH_4, and N_2O) and the resulting climate change effects of warming is perhaps the "defining" issue of our time (UN, 2019). California has been a leader in the U.S. and globally in the development of policies for reducing GHG emissions from all sectors, including the transport sector. A suite of Executive Orders (EO) and Assembly bills (AB) have codified GHG mitigation targets for the state, starting in 2005 with the Governor's EO S-3-05 which required a reduction of GHG emissions to 1990 levels by 2020, and a reduction to 80 percent below 1990 levels by 2050 (Schwarzenegger, 2005). California's 2006 Climate Change Solutions Act (AB32) made the 2020 reductions law, and tasked many government entities, including local governments and government agencies, with helping to meet those goals (CA Assembly, 2006). Since then additional policies have enhanced or expanded these targets; for example EO B-30-15 (Brown Jr., 2015) requires a reduction of 40 percent below 1990 levels by 2030, which was codified into law with Senate Bill 32 in 2016, and EO B-55-18 (Brown Jr., 2018) targets carbon neutrality for the state by 2045. Of particular relevance to California counties and cities, the Sustainable Communities and Climate Protection Act of 2008, or SB375, requires jurisdictions to develop GHG reduction targets and undertake specific actions to achieve them (ILG, 2008). In order to ensure that cities and counties are integrating GHG reduction targets in the development of regional plans, SB375 directs the California Air Resources Board to set regional targets for GHG reductions (CA Senate, 2008).

As a result, local jurisdictions in California must develop climate action plans (CAP) that identify these GHG reduction targets and the specific actions to achieve them. The transportation sector is a major contributor to GHG emissions in the US, causing 28 percent of total GHG emissions (EPA, 2018). In California, the contributions from transport are even more dominant, comprising 41% of statewide emissions (CARB, 2018). Thus, it is not a surprise that transportation is one of the key sectors identified in most CAPs and targeted for reduction. Reducing motorized travel (vehicle miles traveled or VMT) is a crucial element for most reduction targets. However, the infrastructure required for nearly all travel modes includes hardscapes, and may present an additional opportunity for GHG mitigation. Many cities and counties, and other jurisdictions such as port authorities, are responsible for managing a significant portfolio of transportation-related hardscapes including roadways, parking lots, airfields, and bike and pedestrian pathways. Today, many local governments are developing second generation CAPs which include policies, standards and specifications for transportation infrastructure and its use.

1.2 *What are CAPs*

Climate action plans are a formal arrangement of the policies and strategies that will guide jurisdictions towards their GHG reduction goals. The jurisdictions responsible for developing CAPs in California include metropolitan planning organizations (MPOs), cities, and counties. They are typically written by the jurisdictions themselves but are occasionally prepared by a third party. It is also primarily this jurisdiction's responsibility for achieving the goals outlined in the CAP. Certainly, many CAPs consider the effects of national and statewide policies on the emissions profile of the jurisdiction. However, the core purpose of CAPs is (or should be) to outline localized actions that the jurisdiction itself can take to meet emissions reduction goals.

Most CAPs follow a similar flow. They introduce the purpose of a CAP, followed by their methods in developing the CAP, including listing involved members, dates of community feedback, and any external organizations that provided quantitative data and/or consulting. This is followed by the jurisdiction's emissions inventory, which provides the annual emissions the jurisdiction is responsible for per sector—these sectors are, in varying combinations: agriculture, land use, transportation, energy, water, and waste. A forecast is provided for the expected business as usual (BAU) emissions (that is, if the jurisdiction were to refrain from implementing any GHG reduction strategies) and the establishment of emissions reduction targets. This is followed by an outline of the emissions reduction strategies being considered in each sector, with varying amounts of information provided. The CAPs typically end by outlining a timeline and/or benchmarks, sources of funding, and potential partners to assist in strategy implementation.

2 SCOPE/METHODS

2.1 *Compiling CAPs for review*

There is no single repository for CAPs, so a number of resources were used to identify CAPs and select a subset for closer review. The Institute for Local Government has a comprehensive list, which has not been updated since 2014, and internet searches were conducted for jurisdictions that were expected to have well-developed CAPs (CAILG, 2015). This list was supplemented with California local governments that have joined The Global Covenant of Mayors for Climate & Energy, an international coalition of local governments who intend on supporting and promoting actions that combat global climate change and support long-term sustainability (GCM, 2019). California local governments were selected from among all North American cities and local governments. The list was further supplemented by adding cities from any of the 12 Caltrans districts that were not yet represented. Ultimately, no CAP was found from a jurisdiction in district 9, which consists primarily of Inyo County.

2.2 *Review criteria*

When reading through CAPs, information was extracted and summarized in a spreadsheet. Reviewers noted the name of the jurisdiction (and whether it was a city or county), the year that the CAP was published, and a list of transportation strategies (in particular those that affect pavement, roadway, and bridge infrastructure elements). While reviewing strategies, it was noted whether the CAP authors provided values on expected emissions (this could be either by strategy or by sector). The same was done for cost measures. Additionally, it was noted if the CAP mentioned parties responsible for funding and/or implementing the strategies, potential co-benefits, performance metrics, or implementation timelines. Potential co-benefits include reduction in local pollutants (such a particulate matter and ozone), job production, and improved health. A portion of these findings have been summarized in Table 1.

3 FINDINGS

3.1 *Reviewed CAPs*

The preliminary list of target jurisdictions contained 40 local governments. However, not all of them had relevant CAPs, since they either: (1) had yet to release a CAP, (2) had not updated a CAP despite publicizing a past-due update (meaning they have outdated CAPs), or (3) did not have a valid CAP for other reasons [e.g. a lawsuit against San Diego's CAP resulted in a ruling late last year that required the CAP to be revised because it allowed parties to purchase out-of-region carbon credits (Jones, 2018)]. Ultimately, 36 CAPs were reviewed and included in this study. Table 1 summarizes some key points of one strategy in each reviewed Caltrans district.

3.2 *Findings on proposed strategies*

As evident in Table 1, the information provided on the strategies varied across CAPs. Further, even CAPs that were marked as having provided emissions or cost numbers may not have provided these values for all listed strategies. CAPs not marked as having included quantitative values may have provided symbolic estimates for these values (e.g. a strategy is expected to cost one out of a maximum five dollar signs, signifying it is a low cost option); however, this is not as actionable as providing a numerical estimate, and has minimal impacts on informing the expected impacts, both with regards to cost and emissions, of each strategy. For this reason, these cases were considered to not have provided quantitative information.

Below is a list of relevant strategies mentioned across the reviewed CAPs.

- **Alternative fuel vehicle fleet:** GHG emissions reductions can be achieved by switching from gasoline and diesel vehicles to renewable diesel and less carbon-intensive fuels like natural gas, electricity, and hydrogen. Cities propose changes to passenger fleets, public transit fleets, taxi fleets, and more. While alternatives are definitely cleaner during the use phase, they can have carbon intensive upstream impacts or require carbon intensive infrastructure.
- **Electric vehicle infrastructure:** As a subset of the alternative fuel vehicle movement, a particular idea is to add electric vehicle infrastructure to promote adoption. The question

Table 1. A portion of the information collected when reviewing CAPs. One example jurisdiction is included for each Caltrans district.

Jurisdiction	Year	Caltrans District	Proposed transportation strategies	Emissions numbers?	Cost numbers?	Performance measures?	Co-Benefits?
Humboldt County	2012	1	Public transportation, carsharing, telecommuting, bicycles, EV infrastructure (charging stations, encouraging adoption, etc.)	☐	☐	☐	☐
Shasta County	2012	2	Large increase in bike lanes, commute trip reduction	☒	☐	☒	☐
Woodland	2017	3	Complete streets, increased carpooling and ridesharing, public transit, walking/biking infrastructure, reduced idling, alternative fuel vehicles, EV infrastructure, V2G, electrify public fleet, autonomous	☒	☒	☒	☒
San Jose	2018	4	vehicles, ridesharing, improve public transportation infrastructure	☒	☒	☒	☐
Santa Barbara	2012	5	Alternative fuel fleet, EV infrastructure, public transit, parking policies, telecommuting	☐	☐	☐	☒
Fresno	2014	6	Alternative fuel fleet, EV charging, improved public transit, bike lanes, parking measures	☐	☐	☐	☐
Lancaster	2016	7	New bus system, electric buses, roundabouts (instead of signals), bike lanes, traffic signal synchronization, complete streets, ridesharing, autonomous vehicles	☒	☒	☒	☒
Riverside	2018	8	Preferential/improved parking, walking/bike lanes, EV charging, synchronized signals, anti-idling (for heavy-duty diesel), increased public	☒	☐	☒	☐

(Continued)

203

Table 1. (Continued)

Jurisdiction	Year	Caltrans District	Proposed transportation strategies	Emissions numbers?	Cost numbers?	Performance measures?	Co-Benefits?
Stockton	2014	10	transit. Also included improved streetlighting Parking policies, improved public transit, safe routes to school, transportation demand management	☒	☒	☒	☒
Solana Beach	2017	11	EV adoption, car-sharing, telecommuting, alternative fuel fleet, EV preferred parking, alternative work schedule, public transit, bicycling, retime some traffic signals	☒	☒	☒	☒
Santa Ana	2015	12	Signal synchronization, safe school routes and more bike/pedestrian infrastructure, parking ratios, bike sharing, alternative fuel fleet	☒	☒	☒	☐

becomes how much GHG reduction can be attributed to each additional charging station, especially when considering the emissions generated to install the charging station in the first place.

- **Electrifying public transit:** A transition from natural gas or diesel buses to battery-electric buses has the potential to decrease GHG emissions, and would certainly decrease exposure to local pollutants of those near bus routes.
- **Increase bicycle travel:** Providing infrastructure that supports bicycling can reduce VMT, but may also provide co-benefits, such as for public health.
- **LED street and traffic lights:** This strategy switches traditional incandescent lightbulbs with brighter and more efficient light-emitting diodes (LEDs). The benefits are twofold: LEDs produce a single-color light and therefore do not need any filters, and they also consume significantly less energy than their incandescent counterparts.
- **Improved road maintenance and/or increased use of recycled materials:** Improperly maintained roads cause vehicle damage while also increasing fuel consumption. There are studies that have determined that optimal road roughness at which maintenance should be made, which balances the cost of maintenance with the fuel consumption benefits achieved by a smoother road. Some CAPs also call for increased use of recycled asphalt pavement in road construction and maintenance, which decreases material use and therefore GHG emissions.

Only about half of the reviewed CAPs provided expected GHG emissions reductions, which is understandable after considering how complex it would be to arrive at these values for some of the listed strategies. However, this could also be due to limited funding to produce a CAP in the first place. Most, if not all, jurisdictions hired a consulting firm to quantify the current emissions inventory and produce baseline predictions for future emissions levels. Some jurisdictions may not have been able to get emissions and cost estimates for all considered strategies, whether it be through a consulting firm or through a research institution. This would also explain why some jurisdiction expressed expected values through symbols instead of numbers: they could not afford that level of detail.

3.3 Standout CAPs

After assessing the selection of CAPs, certain qualities were identified as those that made CAPs particularly robust. CAPs that stood out not only listed potential strategies, but also: (1) provided expected emissions reduction per strategy, (2) provided expected cost of implementation per strategy, (3) listed parties responsible for strategy implementation, (4) listed co-benefits, and/or (5) explicitly outline sources of funding. The following is a list of robust CAPs along with their outstanding qualities.

3.3.1 Cupertino (Cooke et al., 2015)
The CAP developed by Cupertino was one of the most robust reviewed. It included a detailed description of every strategy, which to varying degrees of completeness, reported the following information: implementation steps, status, parties responsible for implementation, progress indicators, expected GHG reduction potential, co-benefits, and implementation timeline—the timeline ranges used are near, medium, and long term. However, it fails to provide a cost estimate, even with simple indicators.

3.3.2 Lancaster (city of lancaster, 2016)
Lancaster's CAP stands out among other reviewed CAPs because it provides emissions reduction potentials for considered actions and does so considering different time horizons. Like many other CAPs, it also lists co-benefits and a timeline for implementation. The supporting text for each strategy also includes implementation steps, responsible parties for each step, and progress indicators. This CAP could be improved by expanding on its current cost assessment, which consists a simple indicator (a maximum three dollar signs).

3.3.3 Los angeles county (LACDRP, 2015)
Los Angeles County's CAP provided less detailed information for each action or measure compared to other robust CAPs, but still included crucial information on emissions reduction, costs, responsible parties, etc. What stood out what their robust analysis of cost per strategy, listing not only upfront cost, but also the entities responsible for covering the upfront cost, annual net savings or costs per strategy, and entities incurring these annual savings or costs.

3.3.4 Sonoma county (SCRCPA, 2016)
Sonoma County's CAP took a different approach than other county-level (as opposed to city-level) CAPs; for each strategy it provided the expected participation rate of each city within the county. This participation rate informed the reported expected emissions reductions per strategy. Co-benefits were listed per strategy, as were implementation steps, measure commitments, and progress indicators.

3.3.5 San jose (Romanow at al., 2018)
San Jose's CAP is the only one reviewed that included a marginal abatement cost curve (MACC) – which is essentially the same approach proposed in this research, and helps demonstrate the feasibility and potentially attractiveness to local governments of this approach. The provided emissions and monetary values were not calculated on a lifecycle basis; rather, they were acquired through an extended cost benefit analysis which considered direct emissions reductions and the total cost of ownership.

3.3.6 Stockton (ICF International, 2014)
While most CAPs listed in this section provided values for the expected emissions reductions, not all provided cost values. However, Stockton's CAP not only listed the expected agency

cost, but it also acknowledged uncertainty by occasionally writing "It Depends" instead of providing a concrete number; however, no lifecycle considerations were made in this CAP either. This CAP also provided great detail on both current and future sources of funding for each strategy, whether it be federal/state, city, or other. Notably, it outlines supporting strategies that would help overcome obstacles and improve the implementation efficiency of the previously listed strategies.

3.4 *Qualities that make CAPs actionable*

Identifying and outlining responsibility should certainly help in the long-term, and it makes goals clear and increases accountability. Among the CAPs that listed parties responsible for implementation, city departments were given the most responsibilities. Lancaster, for example, has multiple departments and services within the city that are responsible for researching, planning, and implementing project ideas; some of these include Development Services, Traffic Engineering, and Economic Development. Other responsible parties include the Antelope Valley Transit Authority, Lancaster Choice Energy (utility), contractors, and consultants. Similarly, Santa Cruz's CAP listed many of its own departments as responsible parties for implementation; these included Fleet and Facilities Operations, Energy Management Office, Transportation, Planning, and Public Works. Los Angeles County also indicated that the local government would incur the costs for most strategies, with the exceptions being (1) the costs of bicycle infrastructure born by businesses adding those facilities, (2) transportation signal synchronization which is eligible for grant funding, (3) car-sharing programs at least partly covered by the program operator, and (4) idling reduction goals implemented by vehicle owners.

Equally important, but less common among the reviewed CAPs, is identifying sources of funding. Many CAPs were able to determine the expected costs of the strategies, or at least determine whether they would be low or high cost, but this does not necessarily dictate which strategies are more easily implementable. Sources of funding can be at the national, state, or local level depending on the project, and identifying how each strategy will be funded can help streamline implementation down the line. Additionally, identifying sources of funding has the potential to facilitate deciding which strategies to pursue. For example, there are a variety of national-level funding sources that would go towards transit system improvements (as seen in Stockton's CAP). Therefore, even if the strategy has a higher agency cost than others, extra-local funding makes it easier to implement. Understanding what remainder of the cost would have to be covered by the city could help inform how the its limited funds should be distributed across all other strategies.

4 LIFE CYCLE ASSESSMENT TO SUPPORT DECISION MAKING

When estimating the potential for emissions reduction, it is important to not just consider the use phase of the strategy (as most studies do) but also the rest of the lifecycle. Strategies can have significant upstream (material impacts, manufacturing, and transportation) and downstream (end-of-life) impacts with significant effects on the overall change in emissions, which could alter how strategies compare to each other. Similarly, there can be significant changes to the cost beyond upfront costs of implementation. These can include additional costs for maintenance and repair, recycling, and/or disposal, as well as cost savings during the useful life of the strategy. Therefore, it is recommended that analyses of GHG reduction strategies include lifecycle assessment (LCA) to capture these impacts.

One way to visualize lifecycle emissions and lifecycle costs is through a supply curve. In a supply curve, each strategy is represented by a bar for where the emissions reduction potential is plotted on the X-axis (where a wider bar has more reduction potential), and the cost effectiveness in dollars per ton of reduction in CO2e is plotted on the Y-axis (a higher bar costs more per ton of reduction). When considering lifecycle costs, the Y-value can even be negative, meaning that the strategy produces enough money over its lifecycle (credited from energy savings) that implementing it ends up saving money per ton of reduction. These cases

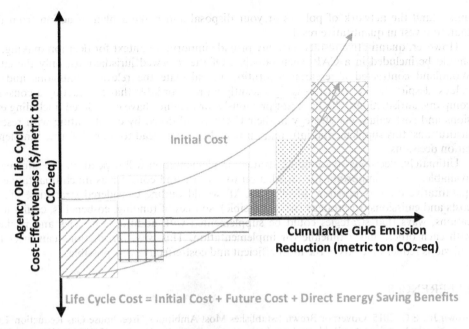

Figure 1.   An example supply curve considering agency and lifecycle costs (adapted from lutsey).

are considered "no regrets" because they both generate net profits for involved parties and reduce GHG emissions. A generic example of a supply curve, recreated from Lutsey (2008), is shown below.

A study by Harvey et al. (2019) applied this lifecycle supply curve framework to GHG reduction strategies considered by the California Department of Transportation (Caltrans). The report introduces their methodology, which accounts for the change in GHG emissions over each strategy's lifecycle, the time required for change to happen, the difficulty in bringing about the change, and both the initial and lifecycle costs of the strategies. The strategies that are analyzed in the larger study span the energy and transportation sectors, and include pavement, roadway, and bridge infrastructure elements. The listed strategies are as follows: automating bridge tolling systems, in-pavement energy harvesting using piezoelectric transducers, increased use of reclaimed asphalt pavement in road rehabilitation, alternative fuel vehicles for the Caltrans fleet, efficient maintenance of pavement roughness, and installation of solar and wind energy technologies on Caltrans assets. The final two are explored in more detail in the report to highlight the methodology and its results. Ultimately, this study confirms the feasibility of collecting the required data to analyze and compare GHG reduction strategies by their lifecycle GHG emissions reduction potential and cost-effectiveness.

5   CONCLUSIONS

From this review, it becomes apparent that not all jurisdiction can afford to include quantitative data in their CAPs, so the question becomes how a jurisdiction can produce an actionable CAP when it has limited funds to produce it. One strategy is to guide decision making through policy bundling. The approach to policy bundling is more qualitative in that it considers how different policies, or in this case strategies, interact. For example, Taeihagh et al. (2013) developed a policy bundling framework that considers synergistic, one-way facilitative, and contradictory relationships, among others. However, instead of relying on quantitative values, it considers quantitative variables like expected public approval and the perceived extent of interaction between strategies. While this is surely time consuming, it may be less expensive to understand the network of policies at your disposal and make a plan of action from there than to invest in quantitative results.

However, quantitative results certainly provide important context for decision making, and should be included in a CAP when possible. Of the reviewed jurisdictions, only the city of Woodland contracted a research institution to calculate the relevant emissions and cost values, despite the fact that this is significantly more affordable that contracting a consulting company. Jurisdictions who have not previously, or may not have considered, including emissions and cost values per strategy in their CAPs could do so by collaborating with research institutions; this supplemental information would certainly lead to more informed implementation decisions.

Ultimately, decisions surrounding strategy implementation will depend on the information available, so it is important for jurisdiction to acquire and consider as much qualitative and quantitative data as possible. Ideally, a CAP would outline considered strategies, lifecycle costs and emissions per strategy (when possible), sources of funding, co-benefits, equity implications, ways that strategies could be supplemented with additional policies and packaged with each other, and a timeline for implementation. This way, jurisdictions can meet their emissions reduction goals in the most efficient and cost-effective way possible.

REFERENCES

Brown Jr., E.G. 2015. Governor Brown Establishes Most Ambitious Greenhouse Gas Reduction Target in North America. Available online: https://www.ca.gov/archive/gov39/2015/04/29/news18938/ (accessed on September 16 2019).

Brown Jr., E.G. 2018. Executive Order B-55-18. Available online: https://www.gov.ca.gov/wp-content/uploads/2018/09/9.10.18-Executive-Order.pdf (accessed on September 16 2019).

California Air Resources Board. 2018. California Greenhouse Gas Emission Inventory - 2018 Edition. Available online: https://www.arb.ca.gov/cc/inventory/data/data.htm (accessed on September 16 2019).

CA Assembly. 2006. California Assembly Bill No. 32. Available online: http://www.leginfo.ca.gov/pub/05-06/bill/asm/ab_0001-0050/ab_32_bill_20060927_chaptered.pdf (accessed on September 16 2019).

CA Senate. 2008. Sustainable Communities and Climate Protection Act of 2008, S.B. 375. Available online: https://leginfo.legislature.ca.gov/faces/billNavClient.xhtml?bill_id=200720080SB375 (accessed on September 16 2019).

Harvey, J. T., Butt, A. A., Lozano, M. T., Kendall, A., Saboori, A., Lea, J. D., Kim, C., & Basheer, I. (2019). Life Cycle Assessment for Transportation Infrastructure Policy Evaluation and Procurement for State and Local Governments. *Sustainability*, *11*(22), 6377. https://doi.org/10.3390/su11226377

ILG. 2008. The basics of Senate Bill 375. Institute for Local Government. Available online: https://www.ca-ilg.org/post/basics-sb-375 (accessed on September 16 2019).

ICF International, 2014. City of Stockton Climate Action Plan. August. (ICF 00659.10).

Jones, H. J. (2018). County's climate action plan set aside by judge; impact on backcountry developments unclear. *The San Diego Union-Tribune.* Available online: https://www.sandiegouniontribune.com/communities/north-county/sd-no-climate-plan-ruling-20181225-story.html (accessed on September 16 2019).

Lutsey N.P. 2008. Prioritizing Climate Change Mitigation Alternatives: Comparing Transportation Technologies to Options in Other Sectors. Institute of Transportation Studies, University of California, Davis, Research Report UCD-ITS-RR-08-15.

Schwarzenegger, A. 2005. Executive Order S-3-05. Available online: http://static1.squarespace.com/static/549885d4e4b0ba0bff5dc695/t/54d7f1e0e4b0f0798cee3010/1423438304744/California+Executive+Order+S-3-05+(June+2005).pdf (accessed on September 16 2019).

Taeihagh, A., Givoni, M., & Bañares-Alcántara, R. (2013). Which policy first? A networkcentric approach for the analysis and ranking of policy measures. Environment and Planning B: Planning and Design, 40(4),595–616. doi: 10.1068/b38058

United Nations. 2019. Climate Change. Available online: https://www.un.org/en/sections/issues-depth/climate-change/(accessed on September 16 2019).

LCA consortium for a road environmental-friendly infrastructure in Mexico City

N. Hernandez-Fernández, L.P. Güereca-Hernandez, A. Ossa-Lopez & M. Flores-Guzmán
Engineering Institute, National Autonomous University of Mexico (UNAM), Mexico

C.E. Caballero-Güereca
Engineering Faculty, National Autonomous University of Mexico (UNAM), Mexico

ABSTRACT: The environmental crises in Mexico City are caused mainly by the emissions generated by the big number of vehicles. Several programs have been developed and implemented to reduce the emissions for more than three decades, lowering the concentration of some pollutants, however, pollutions contingencies are presented year to year. To contribute with the reduction of emissions the Consortium for the Life Cycle Assessment of Transport Infrastructure is created, having as a goal to propose advanced ideas and using the Life Cycle Assessment (LCA) methodology as a robust forecasting model. The initial research areas defined by the Consortium are implementing actions to reduce transport emissions, the harmonization of the LCA process in road industry, and actions to improve the condition of the primary road network of the Mexico City.

1 INTRODUCTION

Air pollution is a problem in the Mexico City Metropolitan Area (MCMA), that public authorities have tried to tackle with several programs. In an effort to successfully achieve clean air goals, the government has established air quality standards, developed an air quality monitoring network, built emission inventories, invested in a forecasting air quality model, and supported research field studies. The combined information on emissions inventory, land cover and urban morph`ology, meteorology, and atmospheric chemistry enables air quality models to be developed and to be used as a tool for forecasting potential air pollution episode, as well as evaluating past episodes and the efficiency of control measures (Molina et al. 2019).

The automatic air-quality monitoring network of the MCMA was established in the late 1980s, revealing high concentrations of all criteria pollutants measured, placing Mexico City air pollution problems among the worst in the world. Since then, the implementation of comprehensive air quality management programs has made important advances to reduce air pollution. The atmospheric concentrations of lead (Pb), sulfur dioxide (SO_2), and carbon monoxide (CO) have significantly reduced and are below the current air quality standards. However, although ozone (O_3), particulate matters less in diameter than 10 and 2.5 micrometers (PM_{10}, and $PM_{2.5}$, respectively) concentrations have also decreased substantially, they are still at levels that are above the respective air quality standards. Furthermore, no declining trends for O_3, and $PM_{2.5}$ has been observed since 2006, and several severe pollution episodes in recent years suggest that their concentrations could be increased (Molina et al. 2019).

In order to quantify the amount and type of airborne emissions in the MCMA, emission inventories have been developed since the late 1980s, in which mobile sources emissions were estimated using traffic counts while industrial emissions were estimated by voluntary survey, and since the year 1994, an automatic emissions inventory system was implemented covering four categories: point sources (industrial), area sources (services and residential), mobile sources (transportation), and natural sources (vegetation and soil) (Molina et al. 2019). The most recent pollutant emissions inventory for the MCMA performed in 2016 incorporates the addition of the criteria pollutant emissions for black carbon and greenhouse gases. Furthermore, the use of the "MOVES" model (motor vehicle emission simulator) developed by the U.S. Environmental Protection Agency, and adapted by the Eastern Research Group to the conditions of Mexico to estimate emissions coming from mobile sources, represents an improvement over previous inventories (SEDEMA 2018).

The contribution of the pollutant emitting category is shown in Figure 1. It can be observed that area sources contribute to high emissions of PM_{10}, $PM_{2.5}$, volatile organic compounds (VOCs), ammonia (NH_3), SO_2 and toxic emissions. The inventory includes emissions of 172 toxic species (gases and metals). Mobile sources produce high emissions of PM_{10}, $PM_{2.5}$, NO_X, VOC, black carbon (BC) and greenhouse gases (CO2eq). Point sources present medium to low emissions for all contaminants except CO and N_3.

The sources of mobile or transport emissions contributed to 55.5% of the total emissions shown in Figure 1 for all pollutants. Meanwhile, services and residential sources produce 29%, and industrial sources 15% of all pollutants. Key transport sources include private vehicles, taxis, microbuses, and heavy-duty diesel vehicles. According to assessments from the Secretariat of Environment for Mexico City (SEDEMA), about of 5.3 million vehicles circulate in the MCMA, of which 80% are gasoline-fueled private cars, SUVs, and motorcycles, 13% are heavy-duty freight vehicles, mostly using diesel, and 7% are public transportation vehicles (mostly diesel).

A number of complementary instruments have been developed in MCMA to control emission from vehicle fleet. The key mitigation mechanisms are a mandatory vehicle inspection program, a no-driving day, the environmental atmospheric program, the diesel vehicle self-regulation program, the comprehensive program for reducing pollutant emissions, and the school transport program. In addition, through joint action with the federal authorities, three Air Quality Programs for the MCMA (PROAIRE initiatives) have been issued, introducing measures to reduce emissions from a range of sectors. The latest program, PROAIRE 2011-2020, covers eight areas. Where Strategy 4 addresses mobility, focusing on the systematic planning of transport networks, integration of the transport systems of MCMA, and improving the fluidity of traffic flows. Specific measures promote integrated planning with an emphasis

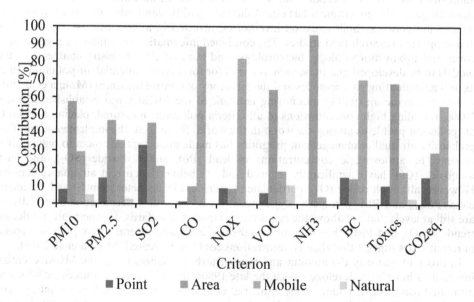

Figure 1. Sources of emission contribution in the MCMA for the 2016 emission inventory (Adapted from SEDEMA 2018).

on intermodal journeys, traffic monitoring, and management, additional express lines for public transport, promotion of cycling and rationalization of freight distribution (ITF 2017).

Notwithstanding the actions described above, in March 2016 and April 2019 pollution crisis occurred, caused in part by a large number of vehicles, the stagnant meteorological conditions, and the forest fires. For this reason, preventive actions need to be formulated with the help of forecasting modeling systems. The Consortium for the Life Cycle Assessment of Transport Infrastructure (CAIT by its acronym in Spanish) is proposed as a key stakeholder searching advance ideas and using the Life Cycle Assessment (LCA) methodology as a robust forecasting model to reduce transport emission.

2 ACTIONS TO MITIGATE POLLUTION PROBLEMS

2.1 *Background*

The following are some of the most important air quality management plans developed and implemented by the Mexican authorities to mitigate air emissions.

- Vehicle technology and fuel quality. It consisted of the modernization of the vehicle fleet, starting in 1990 and the distribution of unleaded gasoline.
- Mandatory vehicle verification program. It began in 1988 with the aim of reducing vehicle emissions through a mandatory inspection twice a year of the environmental performance of transport and ensuring its proper maintenance.
- "Hoy no circula" (no-driving day). A program that has its roots in 1987 as a citizen alternative to voluntarily participate in not using the owned vehicles once a week (Molina and Molina 2002). Nonetheless, this program became mandatory in 1989 as part of a short-term emergency protocol for the winter months and is now permanent for certain types of vehicles (the ones that do not pass the mandatory vehicle verification program).
- Program for atmospheric contingencies. It is an emergency program with the aim of warning the public during severe pollution episodes and the implementation of actions to mitigate severe levels of pollution.
- Improvement of public transport. Although air quality management programs have prioritized the expansion of public transport to improve air quality and mobility, the MCMA has not generated the transport infrastructure required for the needs of the massive population expansion. This improvement in transport systems is a long-term challenge that must include regional coordination, integrate urban planning, urban mobility and air quality management (Molina et al. 2019).
- Modernization of the asphalt mix plant in Mexico City. It consisted of the modernization of the infrastructure of Mexico City's government's asphalt plant and the transformation of most of its surface into a recreation area.
- Regulation in the production of asphalt mixtures. It was specified by the government of Mexico City that the asphalt mixture to be used in paving and pothole works, as well as for any other works carried out by the city's government should be a warm mix, i.e. produced 30°C below the conventional temperature used for its elaboration.

Nevertheless, the several air quality management programs described above, the recent air pollutants episodes in the MCMA in March 2016 and April 2019, caused in part by large number of vehicles circulating in the city and the stagnant meteorological condition, and May 2019, caused by the forest fires from surrounding areas, demonstrated that is not enough to implement contingency actions once pollution levels are increasing. As a response to this recent pollutants episodes, a new plan for enhancing urban mobility and a series of measures to further reduce O_3 and $PM_{2.5}$ pollution in the MCMA was presented by the Mexican environmental authorities in May 2019. The measures include reducing emission from liquified petroleum gas (LPG) distribution and VOCs from consumer and personal products, strengthening inspection of gas stations, eradicating practices that cause fires, reducing emissions from industries, public road infrastructure works, diesel vehicles (freight and passenger), as well providing incentives for hybrid and electrical private vehicles and improving infrastructure for public transportation and non-motorized transport (Molina et al. 2019).

However, although transport emissions are the most significative source for the air quality problem in the MCMA, to date, no actions have taken to improve the quality of the road network. The primary road network of Mexico City has a length of 6000 lane-km, and according to the Secretariat of Works and Services (SOBSE) 62% of the pavement network has a poor condition and requires major actions, meanwhile, the other 38% needs preventive maintenance.

2.2 *LCA contribution to reducing the pollution problem*

The energy use due to transportation is considerable, around 44% of the final energy use in Mexico (SENER 2017). Road transport is responsible for a larger part, with 90% (SENER 2017), and since it is mainly fossil fuels that are used, the emissions of greenhouse gases are substantial. Different actions have been taken in order to reduce the energy demand and environmental impact in this sector. Worldwide, a lot of attention has been put on developing renewable fuels and more efficient engines. However, it has become evident that other aspects of road infrastructure, such as construction and maintenance, also use a considerable amount of energy and that there can be larger saving by choosing road alignment and road surface characteristics that can be lower the fuel use of vehicles (Carlson 2011).

To describe the net environmental performance of a road and its' different phases the LCA methodology (ISO, 2006a and 2006b) can be used. The LCA methodology can be described as an advanced methodological approach to assess environmental impacts that support environmentally responsible decision-making and public policy establishment, as it allows to take into account accurate scientific information in roads, considering a cradle-to-grave approach and evaluating each of the stages that can be considered within the LCA of a given type of road, e.g.: material production, design, construction (new construction, preservation, maintenance activities, and rehabilitation), use and end-of-life stages associated with the pavement structure.

To illustrate the potential of LCA application to reduce the pollution in MCMA, some studies regarding pollutants inventory for different transport modes, energy conservation for road lifecycle, and an approach to support decision-making strategies using LCA methodology are described below.

Horvath (2006) made a study that provides a life-cycle inventory of air emissions (CO_2, NO_x, PM_{10}, and CO) associated with the transportation of goods by road, rail, and air in the U.S. It includes the manufacturing, use, maintenance, and end-of-life (EOL) of vehicles, the construction, operation, maintenance, and EOL of transportation infrastructure, as well as oil exploration, fuel refining, and fuel distribution. The study results show that the vehicle use phase is responsible for approximately 70% of the total emissions of CO_2 for all three modes. On the other hand, rail freight has the lowest associated air emissions, followed by road and air transportation. Depending on the pollutant, rail is 50-90% less polluting than the road. Air transportation is rated the least efficient in terms of air emissions, partly due to the fact that it carries low weight cargo. It is important to consider infrastructure, vehicle manufacturing, and pre-combustion processes, whose life-cycle share is likely to increase as new tailpipe emission standards are enforced.

Energy Conservation in Road Pavement Design, Maintenance, and Utilization (ECRPD) is a European Union project that was finalized in 2010 (ECRPD 2010). Within this project energy usage, on-road pavements over their lifespan were examined i.e. from the manufacture of materials, construction, maintenance, and the energy used by the vehicles on the road. The study was performed in 6 countries and different pavements, including those constructed with low energy materials, were assessed. The energy used in different maintenance regimes for each of the pavements and the impact on vehicle energy as the road deteriorates were evaluated. The main aim of the LCA-study was to compare the environmental impact of asphalt road construction and maintenance during its life cycle and determine where and in which process the major environmental damage occurs. The results showed that it is the production of asphalt mixtures and their components that are the most energy-consuming stage, around 90% of all energy needed (Carlson 2011).

In Harvey et al. (2019) the principles of consequential life cycle assessment and life cycle cost analysis are used to improve on the mitigation supply curve concept to support the evaluation and procurement decisions for transportation infrastructure. The study uses a GHG reduction "supply curve" framework to support decision-making by California Department of Transportation (Caltrans) evaluating two studies currently underway: possible changes that Caltrans can make in its operations to reduce GHG emissions and actions for transportation in climate action plans that have been developed by cities and counties in California to reduce GHG emissions. The supply curve, as is used in these studies, provides a method for selecting the most cost-effective strategies for GHG reductions taking account to net effect on the GHG emissions thorough lifecycle, the time required to make the change, process and difficulty of making the change, and calculation of the initial and lifecycle cost. The detailed methodology given in this paper for developing supply curves is a promising approach for assessing abatement strategies for GHG reduction.

3 CONSORTIUM FOR THE LIFECYCLE ASSESSMENT OF TRANSPORT INFRASTRUCTURE

3.1 Multidisciplinary group integrating the Consortium

This section describes the background, experience, and capabilities of each of the consolidated research groups that will integrate CAIT.

3.1.1 Lifecycle assessment group

The Lifecycle Assessment, Climate Change, and Sustainability group (CIVICCS by its acronym in Spanish) of the Engineering Institute, UNAM was established in 2010 with the mission of generating scientific knowledge in the discipline of LCA, climate change and sustainability, through research that invites to environmental decision-making in all areas of society. By 2020, CIVICCS is visualized as the lead LCA research group in Mexico, with national and international recognition in terms of generated scientific knowledge, support in national and international problem-solving and the formation of Human Resources for Life Cycle Assessment. Currently, CIVICCS has 19 people who carry out their master's degrees, their doctoral study, two post-doc researchers and two outsourcing. To date, CIVICCS has developed around 40 environmental footprint projects, has 26 published scientific articles, 10 book chapters, 10 articles of disclosure and more than 150 participations in different congresses; where, the work done in the areas of cement and pavement industries is noteworthy.

The CIVICCS group will be contributing to the Consortium, with its experience in developing LCA studies, as well as the access to the initial Life Cycle Inventories developed for Mexico and the specialized software to perform the LCA projects.

3.1.2 Pavements laboratory

The Pavements Laboratory (PL) began operations in 1971, initially the main objective of its research was the development of a design method for low-volume roads, with more precise consideration of Mexico's conditions in various aspects, such as local materials and construction-process, traffic characteristics, climate, and local conditions. As a result of this research, a structural design method for flexible pavements (DISPAV5) was developed and is the most widely used method in Mexico for the design of such structures. Apart from pavement design, other areas of pavement engineering have been studied in the PL, such as the characterization of new materials, construction-related aspects, and research to improve the behavior of the pavements in service.

The studies carried out by the PL in recent years have been focused on the development of pavements sustainability, promoting the use of recycled materials such as construction and demolition waste, reclaimed asphalt pavements and ground tire rubber. In addition, research carried out on the incorporation of Titanium Dioxide with the aim of reducing polluting emissions to the atmosphere has also been studied by this laboratory.

The PL group will be contributing to the Consortium bringing the knowledge in transport, specifically in pavement engineering to contribute with the ideas where the initial studies will be applied, and looking to link the projects with the Mexico City authorities.

3.1.3 *Mission & vision del CAIT*

Considering the experience of the different pre-established groups; just described, it is proposed that the CAIT's mission is to support environmental and socially viable decision-making on transport infrastructure, especially on the pavements area, considering a cradle-to-grave approach.

The vision of CAIT to 2030 is to be the most recognized research and support group for problem-solving and the establishment of public policies on pavements in Mexico.

4 CONSORTIUM WORK PLAN

The air quality conditions of MCMA offers to the CAIT a wide range of opportunities to make a substantial impact. Within its scope, some priority research needs have been identified by the working group, such are actions to reduce transport emissions, harmonization of the LCA process to road industry in Mexico City, and actions to improve the condition of the primary road network. The work plan to be followed by CAIT is described below to move towards a sustainable transport infrastructure in Mexico City.

4.1 *Actions to reduce transport emissions*

– Foment the use of photocatalytic materials on the surface layer of the pavements to capture some of the vehicle's emissions and prevent them from reaching the atmosphere.
– Quantification of the impacts generated during the construction stage, including fuel use and contribution to emissions by construction equipment and traffic congestion generated by the works, using the model MOVES-Mexico.
– Promote the road surface characteristics that can be lower the fuel use of vehicles

4.2 *Harmonization of the LCA process to road industry*

– Generation of guidelines to develop studies on the Life Cycle Assessment related to the pavement projects for Mexico City-based on the experience of the projects that will be developed by the CAIT and following the procedures described in the Standard ISO 14044 (ISO 2006b).
– Publication of guidance documents to prepare product category rules and environmental product declarations for the different types of products used in pavement construction in Mexico City.
– Definition of a decision-making approach to assessing the most cost-effective strategies for pollution reductions, using the LCA and the Life Cycle Cost Analysis approaches.

4.3 *Actions to improve the condition of the primary road network*

– Develop and implement together with the corresponding Mexico City authorities a pavement management system in order to improve the pavement condition of the road network applying a decision-making approach based on Life Cycle Assessment.
– Develop a quality control plan to improve the quality of the paving works and therefore extend its lifespan.

214

- Promote the increase of recycled materials used in the pavement layers, to reduce the exploitation of natural resources (virgin aggregates).
- Encourage the use of construction residuals coming from other industries (e.g. vehicle tires, plastics, among others) in pavement layers to avoid disposal in waste areas and reuse this material as components of the pavement layers without decreasing its performance.

5 CONCLUSIONS

Pollution in the MCMA is a severe problem that affects the health of its inhabitants throughout the year. Various factors contribute to the pollution problems in the MCMA, but one of the main problems is a large number of vehicles that circulate daily through and in the metropolitan area. Emission inventories made to quantify pollutants in the atmosphere and determine sources that generate them in the metropolitan area show that transport sources, made up of road vehicles, are the main source of pollutants.

Several programs have been developed aimed at implementing corrective and preventive actions for more than three decades. Even these programs, actions, and policies, the air quality is in poor condition, having severe episodes where the thresholds are far exceeded. For this reason and to contribute with ideas and projects for the reduction of polluting emissions, the CAIT is created by two consolidated research groups of the Institute of Engineering UNAM, each providing its expertise in its area to generate a sustainable transport infrastructure in Mexico City.

Among the plans of CAIT are implementing actions to reduce transport emissions, working in the approaches to harmonization of the LCA process to road industry, and developing actions to improve the actual poor condition of the primary road network implementing a systematic pavement management system.

REFERENCES

Carlson A. (2011). Life cycle assessment of roads and pavements - Studies made in Europe. VTI rapport 736A.

Harvey, J.T.; Butt, A.A.; Lozano, M.T.; Kendall, A.; Saboori, A.; Lea, J.D.; Kim, C.; Basheer, I. (2019). Life Cycle Assessment for Transportation Infrastructure Policy Evaluation and Procurement for State and Local Governments. Sustainability 2019, 11, 6377.

Horvath, A. (2006). Environmental Assessment of Freight Transportation in the U.S. (11 pp). Int J Life Cycle Assessment 11, 229–239. doi:10.1065/lca2006.02.244

International Organization for Standardization (2006a). Environmental management Life cycle assessment Principals and framework. ISO 14040.

International Organization for Standardization (2006b). Environmental management Life cycle assessment Requirements and guidelines. ISO 14044.

International Transportation Forum. (2017). Strategies for mitigating air pollution in Mexico City. International Transport Forum Policy Papers, 2410-8871, No 30. Paris: OECD Publishing, 2017. ITF.

Molina L. T., and Molina M. J. (2002). Air Quality Impacts: Local and Global Concern. Springer Netherlands.

Molina L. T., Velasco E., Retama A., and Zavala M. (2019). Experience from Integrated Air Quality Management in the Mexico City Metropolitan Area and Singapore. Atmosphere (Basel) 10:512. doi: 10.3390/atmos10090512

Secretaria del Medio Ambiente (2018). Inventario de Emisiones de la Ciudad de Mexico 2016. Ciudad de México. SEDEMA.

Secretaría de Energía (2017). Balance Nacional de Energia 2017. Available online: https://www.gob.mx/cms/uploads/attachment/file/414843/Balance_Nacional_de_Energ_a_2017.pdf (accessed on 20 December 2019). SENER.

Pavement, Roadway, and Bridge Life Cycle Assessment 2020 – Harvey et al (eds)
© 2020 Taylor & Francis Group, London, ISBN 978-0-367-55166-7

Lessons learned from the supply curve approach

A.A. Butt, J.T. Harvey, A. Saboori & C. Kim
University of California Pavement Research Center, Davis, USA

M.T. Lozano & A.M. Kendall
University of California, Davis, USA

ABSTRACT: This paper discusses the lessons learned from the recently completed technical memorandum titled "Life Cycle Assessment and Life Cycle Cost Analysis for Six Strategies for GHG Reduction in Caltrans Operations" which was funded by the California Department of Transportation (Caltrans). An emission reduction supply curve framework was developed that uses life cycle assessment (LCA) and life cycle cost analysis (LCCA) methodologies and ranks alternative strategies for reducing Greenhouse Gas (GHG) emissions based on benefit and cost. The report also presented the results of developing and applying the supply curve framework to six strategies for changing Caltrans operations identified by the research team which are briefly discussed in this paper. The strength of the supply curve method is that both LCA and LCCA are used to prioritize strategies for reducing environmental impacts based on cost-effectiveness. One of the major hurdles in the development of the supply curves is the availability and collection of quality data. On one hand, the research team was able to acquire primary data for some strategies while for others, they had to rely on what was available from the literature. The study only looked at GHG emissions and agency life cycle cost. This paper provides further recommendations about the supply curve methodology based on this pilot study.

1 INTRODUCTION

Supply curves are an approach that can be used for the prioritization of strategies for reducing greenhouse gas (GHG) emissions and other environmental impacts. The approach is also referred to as "marginal abatement curves," or "McKinsey curves" (named after the company that has made extensive use of them) (Creyts et al. 2007). Supply curves illustrate the economics associated with changes and policies made for climate change mitigation. In particular, the work done by Lutsey and Sperling demonstrated how alternative strategies within the transportation sector can be quantified and compared using available information, and also compared with alternatives in other sectors of the economy (Lutsey and Sperling 2009).

Using a supply curve approach provides a process for rank ordering numerous GHG reduction options based on how cost-effective they are and provides additional information for decision-making, such as the magnitude of achievable reductions. Borrowing from economic theory, the supply curve approach shows graphically the supply of a given resource (on the x-axis) that is available at a given price (on the y-axis), as can be seen in Figure 1 in the Supplemental Material. Depending on the use and derivation of the costs and cumulative emissions reduction data, the curves can more aptly be labeled as *marginal abatement, incremental cost, cost of conserved carbon, or cost-effectiveness curves*. When the individual strategies used to create the curve are shown as blocks to illustrate the effects of their discrete changes, the curves can show incremental contributions toward a goal and the decreasing cost effectiveness as additional actions are taken (Lutsey 2008).

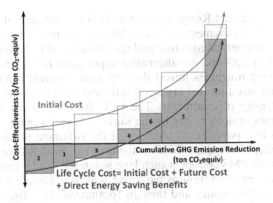

Figure 1. Example supply curve of cumulative GHG emission reduction versus cost effectiveness (adapted from Lutsey 2008).

The example shown in Figure 1 (Supplemental Material) is adapted from Lutsey's first-order assessment of alternative actions to reduce GHG emissions in the California transportation sector versus those in other sectors. The figure shows both the initial cost (transparent boxes above the x-axis) and life cycle cost (LCC; grey boxes numbered in the figure). Although all the actions have a required initial cost to make the change, only some of those changes will result in LCC savings. Those that reduce LCC, improve the efficiency of the overall economy as well as reduce GHG emissions.It is common to report the global warming effects which will happen in the future as numbers straight from the analysis i.e. damage caused by the GHG emissions after 50 years as an example will not be the same damage as that of today. The effect of GHG emissions in the future will be more detrimental to an already damaged environment. Therefore, one million metric ton of GHG emissions reduction being calculated from a strategy for a 50-year analysis period will actually be lower today. This time-adjusted warming potential (TAWP) is developed by Kendall (2012) which captures the effects of speed of change. This approach is similar to the one in LCCA where net present value (NPV) is reported. It is recommended to use TAWP along with global warming potential when reporting the results in this study.

This paper discusses the lessons learned from the recently completed technical memorandum titled "Life Cycle Assessment and Life Cycle Cost Analysis for Six Strategies for GHG Reduction in Caltrans Operations" which was funded by the California Department of Transportation (Caltrans; Harvey et al. 2019a). Details about the methodology and two of the Strategies discussed in this paper can also be found in the recently published article (Harvey et al. 2019b). An emission reduction supply curve framework was developed in the aforementioned report that uses life cycle assessment (LCA) and life cycle cost analysis (LCCA) methodologies and ranks alternative strategies for reducing GHG emissions based on benefit and cost. The supply curves for the six pilot strategy assessments can be found in the white paper (Harvey et al 2019c).

2 BRIEF DESCRIPTION OF THE STRATEGIES

The six strategies were chosen as testbeds for the framework and intentionally reflect strategies with different underlying data and technology readiness levels. Depending on the chosen strategy, the underlying data for calculating the LCCA and LCA vary from the well-documented data to first-order estimations.

2.1 Strategy 1: Pavement roughness and maintenance prioritization

Pavement condition affects the fuel economy as well as the GHG emissions of vehicles through rolling resistance (that is, energy loss due to the interaction of a vehicle and the pavement). The International Roughness Index (IRI) is a measure of roughness. Vehicle fuel use increases on rougher pavement surfaces. Caltrans and most other US states currently use a single IRI value to trigger maintenance and rehabilitation (M&R) treatment for all segments in their entire highway network. An alternative approach is to keep roads in a smoother condition (that is, keeping roughness lower) through more frequent M&R treatments where the volume of traffic and resultant fuel savings is sufficient to compensate for the GHG emissions from increased intensity of treatments. The LCC for Caltrans of keeping pavement with higher traffic volumes smoother may be the same or lower because the cost of treatment to restore smoothness to a pavement is often less if the pavement is not as badly damaged. To implement this strategy to reduce GHG emissions, the road network is divided into lane-segments (the Caltrans PMS considers each lane separately, and a lane-segment is a length of one lane with a relatively homogenous pavement structure, climate region, and traffic) based on each segment's traffic volume, and then an "optimized" IRI trigger value is identified per lane-segment that minimizes the total GHG emissions resulting from the treatment process and the smoothness-induced fuel use improvement.

2.2 Strategy 2: Energy harvesting using piezoelectric technology

Compression-based piezoelectric generation has been explored as an in-pavement energy generation source within the past decade. A popular piezoceramic is composed of lead zirconate titanate and is thus referred to as a PZT sensor. PZT sensors generate a voltage when compressed. Individual PZTs can be housed together to create a larger piezoelectric transducer. By embedding a row of PZT transducers in a highway pavement (50.8 mm [2 inches] below the pavement surface), the traffic load over the transducers generates voltage spikes that can be harvested. In-pavement piezoelectric energy generation is roughly a function of traffic load and speed: the more vehicles that pass, the heavier they are, and the faster they travel, the higher the power output will be.

2.3 Strategy 3: Automation of bridge tolling systems

All-electronic tolling (AET) technologies are available through vendors statewide. This case study examined the effects on GHG emissions resulting from implementing AET technology as a replacement for existing cash- collecting toll booths. AET uses a transponder device or license plate recognition without traffic flow interruption. At seven state-owned toll bridges, drivers currently pay their toll either with cash or via an electronic transaction with a FasTrak device. Cash-paying vehicles must stop at a tollbooth and then re-accelerate to reach to free-flow traffic speed. Although an AET system requires a reliable electronic system and real-time management, it improves traffic flow and reduces additional fuel consumption by eliminating cash tollbooth stops. Other studies have shown that a vehicle, compared to free flow, consumes more fuel and emits more pollutants when accelerating from a stop to free flow speed depending on vehicle types, traffic condition, and driving patterns (Ahn et al. 2002, Rakha and Ding 2003, Kim et al. 2018).

2.4 Strategy 4: Increased use of reclaimed asphalt pavement

A significant portion of the environmental impacts attributable to Caltrans result from projects it awards to contractors each year to maintain it's close to 80,467 lane-km (50,000 lane-miles) of California highway infrastructure. At the end of their service life, the flexible pavements in those many lane miles can be pulverized and reused as reclaimed asphalt pavement (RAP) in new hot mix asphalt (HMA). Use of RAP in HMA both reduces aggregate consumption and, more importantly, helps reduce the amount of virgin binder needed in new mixes. Caltrans currently allows contractors to use up to 15% of RAP (by weight) in HMA, which has been considered as the base scenario in this case. The goal of this examination was to calculate how much GHG emissions can be reduced by increasing the maximum RAP content in HMA mixes from 15% to 25, 40, and 50% and scaling those results to the California network.

2.5 Strategy 5: Alternative fuel technologies for the caltrans vehicle fleet

The California economic sector that contributes most to statewide emissions is transportation, and 89% of these emissions come from on-road transportation, primarily from the combustion of gasoline by light-duty vehicles and diesel by heavy-duty vehicles (CARB 2018). One statewide strategy for reducing GHG emissions is to move to a light-duty vehicle fleet that relies much more heavily on electricity than on petroleum combustion for propulsion. For heavy-duty vehicles, a second potential alternative to parallel electrification of light-duty vehicles would be to produce combustible fuels, such as biodiesel, from renewable sources. Although Caltrans vehicles make up only a very small part of the statewide vehicle fleet, the department's introduction of alternative propulsion methods could contribute to reducing the fleet's GHG emissions.

2.6 Strategy 6: Solar and wind energy production on state right-of-ways

A strategy for reducing GHG emissions from the California economy's energy sector is to increase statewide electric power generation from renewable sources, such as solar and wind, and to reduce the amount of electricity derived from nonrenewable sources, such as natural gas and coal—the primarily nonrenewable sources for in-state and out-of-state power production respectively. To date, Caltrans has implemented 74 solar projects and has proposed 14 more, but these have all been on buildings (Fox et al. 2018). And while no documentation was found online regarding solar panel installations implemented by Caltrans along highway right-of-ways or as solar canopies, these ideas were frequently found in the literature. This scenario evaluates the net GHG impacts of generating solar energy and wind energy on appropriate locations in Caltrans right-of-ways. The department owns more than 24,140 km (15,000 miles) of highway centerline, with a large but unknown amount of acreage in those right-of-ways The department can also place solar PV canopies over Caltrans-owned parking lots (Note: the solar energy generated in this strategy does not include any generated from pavements).

3 SUPPLY CURVE APPROACH DISCUSSION

The supply curve approach is briefly discussed in the next sub-section followed by the lessons learned from the process of doing LCA and LCCA for each strategy.

3.1 Supply curves

Developing supply curves to review alternative strategies provides a way to bring full system analysis, life cycle thinking, and, above all, quantification, to the decision-making environment where these considerations are often absent, and to support decision-making for prioritization that includes consideration of economics.

However, supply curves are only one tool for GHG and other pollutant reduction decision-making, and they require cautious use because they have several limitations. Specifically, past use of supply curves has at times omitted any ancillary benefits from abating GHG emissions, done a poor job of considering data uncertainty, not considered dynamic interactions over time, and lacked transparency about their assumptions. Supply curves based on single assessments of abatement measures suffer from additional shortcomings: they do not consider interactions, non-economic costs, or behavioral changes; they count benefits incorrectly; and they have inconsistent baselines (Kesicki and Akins 2012). It has been suggested that supply curves should be used more for comparing alternatives than for quantifying cumulative abatement progress (Huang et al. 2016). Further, supply curves' inability to predict future abatement has been critiqued because they fail to consider longer-term market changes driven by consumer changes, the timing of policy actions, actions taken by other market actors, and changes in future technologies (Morris et al. 2012). And even though most of these critiques have focused on national-level supply curves—rather than more granular and often less complex curves for agency- and local-level decision making—they must be kept in mind when using supply curves to support decision-making.

These critiques of past use of supply curves were addressed to the extent possible in the framework and initial case studies that are included in Harvey et al. (2019a). Furthermore, LCA and LCCA methodological approaches were used and LCA rules for documentation were intentionally followed to remedy many of the problems associated with past use of supply curves. Before supply curves and their documentation are used for decision-making, it is recommended that they be submitted to critical review by interested stakeholders, and that they be accompanied by documentation of the critiques and responses by the supply curve developers, following International Organization for Standardization LCA principles (Harvey et al. 2016).

3.2 Lessons learned (common for all strategies)

There are several important things to consider while developing supply curves:

- Analysis period
- Difficulties in setting comparable system boundaries
- Consistent functional/declared units
- Differences in maturity of the technology
- Documentation

3.2.1 Analysis period

The analysis period should be long enough to capture future consequences of current decisions. In the study for Caltrans, Harvey et al. 2019a, the analysis period was selected to be 35 years (which is a reasonable period), until 2050; a date set by the California state government to reach the GHG reduction goals to 80% below 1990 levels (Schwarzenegger 2005). For other states or governments that do not have a target date need to evaluate strategies for similar or longer analysis periods as a strategy may show GHG reductions in first few years but may not be too effective in a long term. An example can be taken of maintaining the state highway network where a cheap and an ineffective method of surface treatment can be applied today which produces least GHG emissions. However, one may find this alternative to be the worst if it requires more frequent pavement surface maintenance/treatments which sum up to greater emissions over the analysis period. Another example could be taken of replacing piezo-electric sensors in the pavement every 5 years where the sensors generate electricity due to traffic flow over them. A 5-year analysis period might show huge benefits of using such a technology however, in a long-term analysis, maintaining the pavement every 5 years may result in higher GHG emissions due to pavement maintenance compared to the benefits achieved from reduction in electricity production by the conventional methods. According to the FHWA report, "the analysis period should be long enough to capture the next rehabilitation or other major event whose timing is influenced by the current decision" (Harvey et al. 2016). The statement is applicable to non-pavement related fields as well.

3.2.2 Comparable system boundaries setting difficulties

One of the difficulties faced during this project was to set comparable system boundaries across a wide range of strategies. One example is the capital equipment impacts which are considered in strategy 5 (Section 2.5) but not in strategy 1 (Section 2.1) i.e. strategy 1 is not considering the manufacturing of the new construction equipment to be used for increased construction work. It is acknowledged that defining the scope of a strategy could be challenging due to the definition of goal and scope, unavailable models and data, and incomplete information however, consideration of all life cycle stages should be prioritized. The life cycle stages selected were not the same among all the strategies. As an example, materials stage was considered for all the strategies however end of life (EOL) stage was not scoped for strategies 1, 3 (Section 2.3) and 4 (Section 2.4). This was mainly because of the goal and scope definitions for those strategies.

3.2.3 *Consistent functional/declared units*
The strategies selected for the project were mainly under three main categories; pavement related (strategies 1 and 4), renewable energy generation related (strategies 2 and 6) and Caltrans operation related (strategies 3 and 5). It was challenging to define a functional unit which was similar among all the selected strategies. It was even difficult to define a constant functional unit within the main category such as Caltrans operation related; strategy 3 was mainly looking at all the bridge tolling operations in California whereas strategy 5 was evaluating the change in fleet of Caltrans that used renewable energy resources. Defining a functional unit that could be used for both the strategies in this case was not possible. Therefore, an appropriate functional unit was selected for each strategy. Importantly, the intended audience for all the strategies was Caltrans, and the strategies were evaluated at the state network level considering an analysis period of 35 years[1].

3.2.4 *Data*
Data collection is probably the most time-consuming process especially where data had to be collected by contacting industries/organizations. Sometimes the data are available publicly but not well advertised or straightforward to locate. It is also naïve to assume that if data/information is not available electronically, probably the data does not exist. An effort must be made to collect information/data by first determining the correct source and then contacting the source for it. Another common observation from this project was that the data required for the study were not being collected by the industry/organization because of having no need for it. As an example, a few decades ago information on the energy (electricity and fuels) used to produce a product was not important to record or store in the databases, however, since LCA has evolved, this information has become a necessity and will become essential with implementation of environmental product declarations requirement and thus, may much more widely be available in the future. There are also other general and market dependent parameters in the study, such as the discount rate for calculating the net present value in LCCA. The discount rate applied by any organization depends on both the market economic conditions in general and the internal rate of return considered by the organization. Forecasting economic conditions for 35 years in the future is challenging and adds significant uncertainties to the model output. Caltrans uses 4% discount rate and therefore, same value was used for calculating the NPV for each strategy (Caltrans 2010).

3.2.5 *Differences in maturity of the technology*
The state of readiness of the change of technology was adapted from the Technology Readiness Level (TRL) approach developed by the National Aeronautics and Space Administration (NASA 2012).

TRL 1: basic principles observed

- TRL 2: technology concept formulated
- TRL 3 and 4: experimental proof of concept/technology validated in lab
- TRL 5 and 6: technology validated and demonstrated in relevant environment at less than full scale (industrially relevant environment in the case of key enabling technologies)
- TRL 7: system prototype demonstration in operational environment (full scale)
- TRL 8: actual system completed and determined to be operational through test and demonstration
- TRL 9: actual system proven in operational environment elsewhere or less-than-full-market penetration

---

1 33 years analysis period was used for Strategies 4 and 5

Evaluation of a technology at TRL 1 will require more effort due to the unavailable information and data compared to a technology at TRL 9 where the data are readily available. The uncertainty at TRL 9 will be lower. All the strategies evaluated were at TRL 5 and 6 with an exception of strategy 1, strategy 2 and part of the strategy 6[2]. Strategy 1 (Section 2.1) is considered to be at TRL 7 as the system prototype has been demonstrated at full scale in the operational environment. Strategy 2 looked at the energy harvesting using piezoelectric technology (Section 2.2) which is relatively a new technology that has seen limited application around the world and is considered to be at TRL 5. This means that the information about this strategy is limited and thus several major assumptions are to be made in order to evaluate it. Uncertainty in the results of this strategy will be higher compared to strategy 1. Part of strategy 6 looked at electricity production using solar canopies on parking lots (Section 2.6) which is a technology well established today and is considered at TRL 9.

3.2.6 Documentation

Transparency is the key for the acceptance of any method, approach, process, or study being analyzed. Transparency gives confidence in accepting the results of a study i.e. a reader will have trust in a study if the information/data provided in the study is verifiable and reproducible. Therefore, each strategy needs to be well documented so that the resultant supply curve (from several different strategies) can help in decision support with higher level of confidence. In this project, several revisions were made for each strategy in order to properly define and describe the assumptions that were vaguely presented, perform calculations for each evaluation step, and present all the data used so that the results are reproduceable. A great deal of effort went into completing documentation of data and calculations and properly reporting them. The final structure of the documentation for evaluating a strategy was finalized:

- Description of the GHG abatement strategy,
- Study goals,
- Scope and system boundary,
- Study assumptions, limitations and gaps,
- Calculation methods used,
- Data, data sources, and data quality assessment,
- Numerical results,
- Sensitivity or uncertainty analysis,
- Appendix containing strategy related additional information and data that were used for the strategy evaluation, and
- The answer to the questionnaire that summarized each strategy's purpose and outcome.

3.3 Lessons learned (strategy specific)

A life cycle approach which includes the background and upstream processes, activities and data can provide with a more accurate picture of what a strategy can/will achieve in terms of real benefits. First order estimates can be made for such high-level studies where strategy specific data and information may be scarce. The major assumptions that were made in order to do a complete analysis for each strategy may be an under or over estimation of the results. The following section lists the challenges faced and lessons learned from each analyzed strategy (Section 2).

3.3.1 Strategy 1

PaveM was used to evaluate this strategy. The user costs equations are not included in the PaveM therefore, the strategy results only reflected the life cycle agency costs. The life cycle benefits (agency plus the user costs) were not determined that could have given a better picture of whether this strategy has positive or negative impacts. Furthermore, there are no emission

---

2 - Solar canopies over parking spaces: TRL 9
  - Wind turbines in interchanges and solar panel along right-of-ways: TRL 5 and 6

equations for construction work zone traffic congestion related in PaveM which could also affect the results.

This strategy didn't include fuel consumption and pavement damage effects that come from the alternative fuel vehicles (AFVs) currently operational on California highways. It is unknown whether these vehicles have negative pavement damage effects especially electric vehicles that have a higher gross weight compared to the conventional vehicles. No pavement damage model or results are available in the literature that could have been included in the analysis. It is acknowledged that the emissions from AFVs will be lower as compared to the fossil fueled vehicles, however, was not included in the analysis.

Driving behavior also affects fuel economy. Vehicles on smooth and safe pavements tend to be driven faster hence, consuming more fuel. The increase in fuel consumption due to vehicles moving faster on smooth pavements was not included but can be an important parameter to be considered.

3.3.2 *Strategy 2*

Energy harvesting from the pavements via traffic movement over the embedded piezoelectric devices is not a mature or fully operational/implemented technology. The electricity produced can either be used by Caltrans to store and operate electric devices (such as signals, street-lights, etc.) or can be sold back to the state utilities. The price of electricity varies throughout the state and using a single price value could result in high uncertainty. Therefore, a sensitivity analysis was performed considering cases where low and high electricity prices were used. This provides a range of profit generation within the framework, and the large discrepancy reduces the ability of the results to inform decision making. It can help to consider whether or not to continue exploring this option. Additionally, it is unknown how much excess fuel will be con-sumed by the vehicles due to the higher pavement roughness, therefore, a one percent increase in fuel use was assumed for the sensitivity analysis.

A major challenge with such a study was using standard data sources as the available litera-ture was limited and of varying scales, materials and setups. There are large discrepancies in the literature regarding in-pavement piezoelectric energy production potential. Developers claim it to be 100-200 kW/km whereas universities such as UC Berkeley and Virginia Tech have critiqued it to be around 0.01 kW/km (Hill et al. 2014). This study uses values from a model developed by Najini and Muthukumaraswamy (2017) which are close to developers reported range, with the caveat that it is likely that the results of this strategy overestimate energy production.

3.3.3 *Strategy 3*

One of the major assumptions made in this strategy was that the demolition of existing struc-tures and construction of new tollbooths is less than 5% of the total life cycle emissions of the strategy. This assumption was made as there were only seven metal toll gantries that were to be erected and the fuel consumption and emissions from the construction stage was far smaller than the environmental impacts from millions of vehicles passing the gantries annually. How-ever, for a complete system analysis, it should have been included.

Another assumption was to consider emissions from only four vehicle types as the analysis was done using the US Energy Protection Agency (EPA) program MOtor Vehicle Emission Simulator (MOVES 2015). The major reason to use *MOVES* was to include the energy and emission effects due to the drive cycles of the vehicles in the analysis as in current tolling sys-tems, stop-and-start at the tollbooths is common.

Data for the hourly toll schedules, annual traffic, and revenue for all seven bridges in Cali-fornia were supplied by the Metropolitan Transportation Commission, the transportation planning, financing and coordinating agency for the nine-county San Francisco Bay Area. Caltrans provided data for annual average daily traffic (AADT) and hourly traffic distribu-tion for state-owned toll bridges. Such data gives confidence in the results produced and helps minimizing uncertainties by avoiding making assumptions.

3.3.4 *Strategy 4*

In this strategy, there were two ways determined to calculate the quantity of materials used by Caltrans for the construction and maintenance of California highway network. One was by using PaveM which only has data of highways but not shoulders, ramps, parking lots, gore areas, or other places where Caltrans uses asphalt materials. The second method was to use Construction Cost Data Book (CCDB) which is a record of total materials and cost used annually. There were discrepancies between the two method estimates; for example, a PaveM run conducted under the current default budgeting scenario projected an expenditure of $267 million for asphalt paving materials in 2018. However, the data in the CCDB showed an expenditure of $545 million for the same items in that same year. To address this discrepancy and to calculate material consumption in each year during the analysis period, the tonnages of materials from the 2018 CCDB were multiplied in every year after 2018 by the ratio of 2018 CCDB purchases to the PaveM projections for 2018 purchases. The strategy assumed this process would account for the additional materials used outside the traveled way lanes.

It was assumed in this strategy that the performance of mixes with higher RAP content is similar to that of mixes currently in use in Caltrans projects. This assumption is currently being investigated and verified through research experiments, field studies, and pilot projects. Investigation is required because all the possible savings in the materials stage due to the use of higher percentage of RAP can be offset by potential performance reductions during the use stage because increased RAP content often results in more frequent maintenance and rehabilitation. All possible savings in the materials stage due to higher percentage of RAP use can be offset by possible reduced performance during the use stage as it results in more frequent maintenance and rehabilitation in the future. Furthermore, the quality of materials recycled again at the end of life of HMA with high RAP content is another issue not included in the scope of this study. Possible reductions in quality after multiple rounds of recycling is an issue to be considered in a more detailed study once research results in this area are available.

The study assumed that current recycling strategies will not show much improvement. Although these assumptions are considered to be highly unlikely, they are also considered to be reasonable for at least the next 5 to 10 years.

3.3.5 *Strategy 5*

The Caltrans fleet project required extensive cost and environmental data related to three separate categories: vehicle technologies, fuels, and agency data (fleet characteristics, policies, and standard practices) and for each category, two sets of data were needed, historical data and future projections; challenges and uncertainties in this specific study heavily depended on the data category and how far in the future it was needed. Some of the data were only related to calculating LCC while others were needed for both LCA and LCCA, making uncertainty in those data more critical. For example, the vehicle purchase price is only needed for LCCA. However, fuel efficiency combined with average vehicle miles traveled (VMT) results in total fuel consumption which is then either combined with fuel cost projections to calculate future fuel costs or combined with LCA of each fuel type to calculate the environmental impacts.

Regulations can also change dramatically in the future depending on political developments. For example, California implements low carbon fuel standard (LCFS) that mandates deep cuts in fuel carbon intensity averaged across all the fuel used in the state. Possible radical changes in LCFS in the future by the state legislature towards decarbonization or possible interventions by the federal government to revoke LCSF are unknown at this time and can dramatically change the study outputs. Some of the market dependent parameters that are unknown using the alternative fuel vehicles case as an example are:

- Incentives offered by local/federal government for alternative fuel technologies,
- Development of innovative materials for vehicle manufacturing,
- Future breakthroughs in alternative fuel,

- Changes in VMT due to changes in land use or possible implemented by the agency,
- Structural damage of the vehicle on transportation infrastructure,
- Congestion and changes in traffic flow due to the integration of technology into transportation
- New technologies such as autonomous vehicles and connected transportation.

3.3.6 *Strategy 6*

The largest hurdle evaluating this strategy was estimating the capacity for installation. When considering solar canopies, it was relatively easy because Caltrans annually tracks the capacity of the parking lots it owns and operates. Thus, necessary information was acquired from Caltrans for the analysis. However, it was not nearly as straightforward to acquire data needs for solar panel installation along the highway right-of-ways or wind turbines in clover leaves. For the former, the length of highway that could accommodate solar panels was assumed to be a subsection of the total miles of highway across the state of California. Subsequently, estimates were made on the average spacing between solar panels to determine the system-wide capacity. For the wind turbine case, however, installation capacity was not easily determined because there was no data on the number of clover leaves in California. Additionally, the clover leaves often differ in size. Therefore, all clover leaves were manually assessed using a web-based mapping system and were later categorized by available area to determine whether they could accommodate a wind turbine; this process was very time consuming.

Another necessary assumption was considering a single type of solar panel and wind turbine for all installations. Accounting for flexible sizing and models could lead to large differences in the final capacity and cost. In the case of wind turbines, some clover leaves could accommodate larger turbines while others would need to have the turbine size scaled down. Considering smaller turbines could also increase potential sites of installation. This would affect total energy production capacity and costs. Additionally, there is the question of how wind turbines would perform in this environment. A sensitivity analysis was performed for the case where the wind turbine produced less energy than the average statewide expected production. Additional sensitivity analysis was performed using low and high electricity sale price value, as was done in the piezo-electric case (see Strategy 2). This was necessary as the price of electricity differs from region to region within California.

33 years analysis period was used for Strategies 4 and 5
- Solar canopies over parking spaces: TRL 9
- Wind turbines in interchanges and solar panel along right-of-ways: TRL 5 and 6

ACKNOWLEDGEMENTS

The authors would like to thank Julia Biggar, Tracie Frost, and Rebecca Parker from the Caltrans Office of Smart Mobility and Climate Change Programs, and Joe Holland and Nick Burmas of the Caltrans Division of Research and Innovation, Office of Materials and Infrastructure, for their support, participation, and oversight of the project.

DISCLAIMER

REFERENCES

Ahn, K., Rakha, H., Trani, A. and Aerde, M.V. 2002. Estimating Vehicle Fuel Consumption and Emissions based on Instantaneous Speed and Acceleration Levels. ASCE Journal of Transportation Engineering, Vol. 128 (2): 182–190.

Caltrans. 2010. Life Cycle Cost Analysis Procedures Manual.

Creyts, J., Durkach, A., Nyquist, S., Ostrowski, K. and Stephenson, J. 2007. Reducing U.S. Greenhouse Gas Emissions: How Much at What Cost? McKinsey and Company for the Conference Board. U.S. Greenhouse Gas Abatement Mapping Initiative.

Fox, D., Matsuo, J. and Miner, D., 2018. Caltrans Sustainability Roadmap 2018–2019.

Harvey, J., Meijer, J., Ozer, H., Al-Qadi, I.L., Saboori, A. and Kendall, A., 2016. Pavement Life Cycle Assessment Framework. Report No. FHWA-HIF-16-014. Federal Highway Administration. United States.

Harvey, J.T., Butt, A.A., Saboori, A., Lozano, M., Kim, C. and Kendall, A. 2019a. Life Cycle Assessment and Life Cycle Cost Analysis for Six Strategies for GHG Reduction in Caltrans Operations. Technical memorandum UCPRC-TM-2019-XX (under-review by Caltrans) prepared by University of California Pavement Research Center for Caltrans.

Harvey, J.T., Butt, A.A., Lozano, M.T., Kendall, A., Saboori, A., Lea, J.D., Kim, C. and Basheer, I., 2019b. Life Cycle Assessment for Transportation Infrastructure Policy Evaluation and Procurement for State and Local Governments. Sustainability, Vol. 11 (22): 6377.

Harvey, J.T., Butt, A.A., Saboori, A., Lozano, M., Kim, C. and Kendall, A. 2019c. Alternative Strategies for Reducing Greenhouse Gas Emissions: A Life Cycle Approach using A Supply Curve. White paper (under-review by Caltrans) prepared by University of California Pavement Research Center for Caltrans.

Hill, D., Agarwal, A. and Tong, N., 2014. Assessment of Piezoelectric Materials for Roadway Energy Harvesting: Cost of Energy and Demonstration Roadmap: Final Project Report. California Energy Commission.

Huang, S., Lopin, K. and Chou, K-L. 2016. The applicability of marginal abatement cost approach: A comprehensive review. Journal of Cleaner Production, Vol. 127(20): 59–71.

Kendall, A. 2012. Time-Adjusted Global Warming Potentials for LCA and Carbon Footprints. International Journal of Life Cycle Assessment, Vol. 17(3): 1042–1049.

Kesicki, F. and Akins, P. 2012. Marginal Abatement Cost Curves: A Call for Caution. Climate Policy, Vol. 12(2): 219–236.

Kim, C., Ostovar, M., Butt, A.A. and Harvey, J.T. 2018. Fuel Consumption and Greenhouse Gas Emissions from On-Road Vehicles on Highway Construction Work Zones. In ASCE International Conference on Transportation and Development 2018 (pp. 288–298). Reston, VA.

Lutsey, N.P. 2008. Prioritizing Climate Change Mitigation Alternatives: Comparing Transportation Technologies to Options in Other Sectors. Institute of Transportation Studies, University of California, Davis, Research Report UCD-ITS-RR–08–15.

Lutsey, N. and Sperling, D. 2009. Greenhouse gas mitigation supply curve for the United States for transport versus other sectors. Transportation Research Part D: Transport and Environment, Vol. 14 (3): 222–229.

Morris, J., Patlsev, S. and Reilly, J. 2012. Marginal Abatement Costs and Marginal Welfare Costs for Greenhouse Gas Emissions Reductions: Results from the EPPA Model. Journal of Environmental Modeling and Assessment. Vol. 17(4): 325–336.

Najini, H., and Muthukumaraswamy, S. A. 2017. Piezoelectric Energy Generation from Vehicle Traffic with Technoeconomic Analysis. Journal of Renewable Energy, Vol. 2017: 1–16 (Article ID 9643858).

National Aeronautics and Space Administration. 2012. Technology Readiness Level. www.nasa.gov/directorates/heo/scan/engineering/technology/txt_accordion1.html (Accessed April 10, 2019).

Rakha, H. and Ding, Y. 2003. Impact of Stops on Vehicle Fuel Consumption and Emissions. ASCE Journal of Transportation Engineering, Vol. 129(1): 23–32.

Schwarzenegger, A. 2005. Executive Order S-3-05. California Office of the Governor.

U.S. Environmental Protection Agency (EPA). 2015. MOVES2014a Users Guide. Assessment and Standards Division, Office of Transportation and Air Quality. EPA-420-B–15–095.

226

Comparison of life cycle greenhouse gas emissions and energy consumption between complete streets vs. conventional streets

Arash Saboori*, John T. Harvey, Maryam Ostovar & Ali Azhar Butt
Department of Civil and Environmental Engineering, University of California Pavement Research Center, University of California, Davis, USA

Alissa M. Kendall
Department of Civil and Environmental Engineering, University of California, Davis, USA

ABSTRACT: Complete streets are an infrastructure-oriented intervention intended to improve social, economic, and environmental conditions of a neighborhood or corridor. However, few qualitative and quantitative approaches for objective evaluation have been developed that can help evaluate or anticipate the environmental effects of complete streets interventions. Furthermore, there is no life cycle assessment (LCA)-based tool for verifying/quantifying the full system, life cycle expected savings in impacts of implementing complete streets versus conventional ones. This study addresses this gap by providing a quantifying of designing urban streets under complete streets guidelines versus conventional ones, using LCA. The initial results indicate that application of the complete streets networks to streets where there is little negative impact on vehicle drive cycles from speed change will have the most likelihood of causing overall net reductions in environmental impacts.

1 INTRODUCTION

Complete streets are an infrastructure-oriented intervention intended to improve social, economic, and environmental conditions of a neighborhood or corridor. If successfully implemented they can provide safety benefits for active transportation (bicycle and pedestrian) users, and if they result in more active transportation, they can produce health benefits. However, few qualitative and quantitative approaches for objective evaluation have been developed that can help evaluate or anticipate the environmental impacts of complete streets interventions. Implementing complete streets in urban areas is one of the methods recommended to Metropolitan Planning Organizations (MPO) in California for meeting the Senate Bill 375 (Sustainable Communities) mandates for percent reduction in greenhouse gas (GHG) emissions due to transportation in metropolitan areas (CARB website). However, there is no LCA-based tool for verifying/quantifying the full system, life cycle expected savings in impacts of implementing complete streets interventions.

This study addresses this gap by quantifying the environmental impacts of designing urban streets under complete streets guidelines versus conventional ones, using the life cycle assessment methodology. The comparison was made for a thoroughfare type street using two designs:

- A design for the conventional street from Section Four of the Sacramento County Office of Engineering Improvement Standard (Sacramento County, 2009), currently in use for designing streets, and

- A design for the alternative complete street using the Complete Streets Manual from the Department of Urban Planning of the City of Los Angeles (City of LA, 2014).

Throughout this paper, the Sacramento County Standard is referred to as SAC-DG (Sacramento Design Guide), and the design option for each urban street type under SAC-DG is referred to as Conv-Option (Conventional [Design] Option). Similarly, the manual developed by the City of Los Angeles is referred to as LA-DG (LA Design Guide), and the design option for each street type under LA-DG is referred to as the CS-Option (Complete Street [Design] Option).

2 GOAL & SCOPE DEFINITION

Six major urban street types identified in SAC-DG are listed in Table 1. The goal of the LCAs in this section is to benchmark the selected environmental impacts, primary energy demand, and material consumption of building a thoroughfare under the two different conventional and complete street design guidelines. The main audience for this work is transportation managers and planners in metropolitan transportation agencies and local governments.

The scope of the LCA study is limited to material production, transportation of materials to the site, construction activities, future maintenance and rehabilitations, and changes in vehicle miles traveled and vehicle speed, and their effects on selected emissions from the production (well to pump) and combustion (pump to wheel) of vehicle fuel in the use stage. The assessment does not include the end-of-life of the built infrastructure, or any other effects on vehicles or the use of alternative modes of transportation in lieu of motorized vehicles.

The system boundary for each street type includes material production, transportation of raw materials from extraction site to processing plant and from there to the construction site, and construction activities; this part of the scope is referred to as a cradle-to-laid LCA. Throughout this chapter, the LCA results for these stages of the life cycle are referred to as MAC impacts (Material, transportation, And Construction).

An analysis period of 30 years was selected for conducting the LCA and calculating payback periods for offsetting the differences in environmental impacts due to CS compared with conventional designs. The functional unit is considered as one block, except where stated otherwise.

In the use stage part of the LCA, all vehicles are assumed to burn gasoline and only passenger cars and light-duty trucks (SUVs) are considered, which means that consideration of any heavier freight vehicles is excluded. It is important to note that complete streets can have other important impacts not considered in the limited scope of this study. Additional case studies for field projects can include consideration of expansion of the system boundaries for LCA, which were limited by the scope and budget of this framework development project.

Table 1. Conventional street dimensions (Sacramento County 2009).

Street Type	Minimum Conv. Asphalt Thickness (in)	Minimum Aggregate Base Thickness (in)	Pavement Width (ft)	Block Length (ft)
32-ft Minor Residential	3	10	26	300
38-ft Primary Residential	3	10	32	400
48-ft Collector	3.5	13	42	500
60-ft Major Collector	4	14	54	600
74-ft Arterial	5.5	20.5	56	720
96-ft Thoroughfare	6.5	23	78	860

3 LIFE CYCLE INVENTORY & LIFE CYCLE IMPACT ASSESSMENT

SAC-DG has detailed drawings for the cross-section of each conventional street type and other elements such as curb and gutters. SAC-DG drawings were used to determine the dimensions needed to calculate the quantity of materials (Figure 1). Minimum aggregate base (AB) and asphalt concrete (AC) thicknesses were also taken from the same reference. Figure 2 and Figure 3 show design recommendations for thoroughfares by LA-DG.

SAC-DG does not offer any recommendations for block length, but it does have requirements regarding maximum speed and minimum stopping sight distance. Due to lack of data availability and because this study is more focused on relative changes in environmental impacts of CS-Options versus Conv-Options, block length for each street type was considered as the minimum stopping sight distance multiplied by two.

SAC-DG and LA-DG use different terminologies for street types. Table 2 shows how the street types in the two guidelines are matched based on width and traffic levels.

After determining the amount of materials needed under each design guideline, LCIs and LCIA results were calculated using the UCPRC LCI Database (Saboori et al., 2019) with appropriate units (per kg of materials and mixes, per ton-km of materials transported, or per lane-km of construction activities). Full details of all the assumptions and data sources used

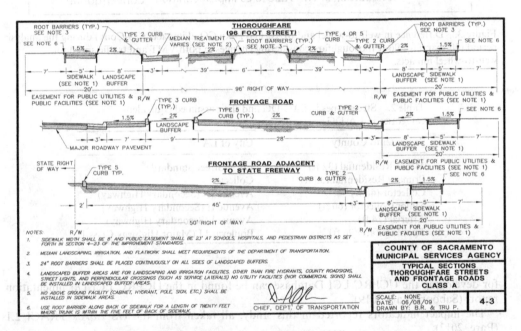

Figure 1. Cross section of a thoroughfare (Sacramento County 2009).

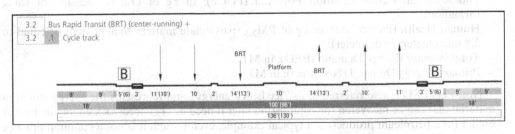

Figure 2. LA-DG recommendation for thoroughfare as a complete street (City of LA, 2014).

229

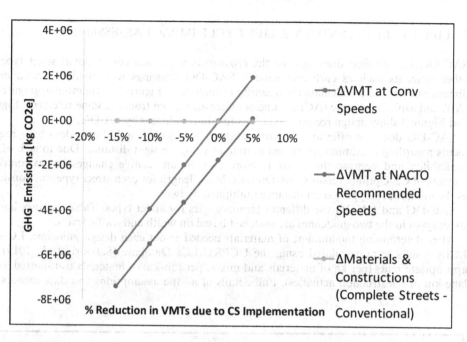

Figure 3. Difference in well-to-wheel and MAC GWP [kg CO_2e] impacts (CS-Conv) during the analysis period (30 years) for 96-ft thoroughfare considering changes in both VMT and traffic speed for well-to-wheel impacts.

Table 2. Street types in SAC-DG and their assumed equivalent categories in LA-DG.

Sacramento County	City of LA
Minor Residential (32 ft)	Local Street Standard
Primary Residential (38 ft)	Collector
Collector (48 ft)	Avenue III (Secondary Highway)
Major Collector (60 ft)	Avenue II (Secondary Highway)
Arterial (74 ft)	Avenue I (Secondary Highway)
Thoroughfare (96 ft)	Boulevard I (Major Highway Class I)

for developing the UCPRC LCI Database can be found in the UCPRC LCI Documentation report (Saboori et al., 2019).

The impact indicators reported in this study, all taken from the US EPA's TRACI 2.1 (Bare, 2012):

– Global Warming Potential (GWP): in kg of CO_2e.
– Photochemical Ozone Creation Potential (POCP): in kg of O_3e (a measure of smog formation).
– Human Health (Particulate): in kg of $PM_{2.5}$ (particulate matters smaller than or equal to 2.5 micrometers in diameter).
– Total Primary Energy Demand (PED): in MJ.
– Primary Energy Demand (Non-Fuel): in MJ.

Non-Fuel PED is also referred to as "feedstock energy" and represents the energy stored in material that can be recovered for combustion later if need be. The feedstock energy in asphalt binder (as a petroleum product) is a typical example: even though it is not a common practice to recycle binder out of pavement to combust it for energy purposes because of the cost and high emissions, the primary energy stored in the binder can theoretically be recovered for this purpose. On the contrary, the energy used in various combustion processes in the system boundary cannot be recovered. Therefore, the PED (Non-Fuel) should be reported separately (Harvey et al., 2016).

The CS elements recommended in LA-DG which are sorted into four categories:

- Intersections and Crossings
- Off-Street Non-Vehicular Treatments and Strategies
- Roadways
- Sidewalk Area

For each element (either conventional or CS) a service life was assumed and used to determine the number of times that each will be treated with a typical maintenance, rehabilitation, or reconstruction treatment during the 30-year analysis period. Although the assumptions used in this study are generally more conservative than those in actual practice, it was assumed that the entire conventional street and complete street infrastructure would be replaced at the end of their service life. If any items have remaining service life at the end of the analysis period, a linearly pro-rated salvage value was calculated and credited to the item.

4 RESULTS AND DISCUSSION

Table 3 shows MAC impacts (absolute values) under different impact categories for Conv-Option and CS-Option and the absolute change in each impact category when switching from the Conv-Option to the CS-Option. The results show that GWP impacts of complete streets elements of a thoroughfare are only 10 percent of the total MAC GWP (CS and conventional elements together).

As stated earlier, comparing different street types with each other is not part of the goal of this study, and it does not add much value since functionalities are different. Furthermore, comparing the same street type under two design methods by just looking at the absolute values of impacts is not very beneficial as these numbers are directly proportional to the length of each block. This approach would have been suitable if the goal was to minimize total emissions or reduce project-level emissions by a certain amount (where absolute values of emissions are important). However, for this specific research, because the goal is to conduct a preliminary comparison between the two designs and get a first order estimate of changes in impacts, calculating relative values seems most appropriate. The last row in Table 3 shows the relative change in MAC impacts between the two design alternatives.

As the relative change in impacts for thoroughfares shows, switching from the Conv-Option to the CS-Option results in impact reductions across all categories, ranging between 1.0 percent

Table 3. Absolute values of various impacts categories for the two design options for thoroughfares. The materials, transportation, and construction (MAC) stages: conventional (Conv) and complete streets (CS).

Design Option	GWP (kg CO_2e)	POCP (kg O_3e)	PM 2.5 (kg)	PED (Total) (MJ)	PED (Non-Fuel) (MJ)
Conv-Option	8.21E+5	1.11E+5	4.52E+2	9.81E+6	1.62E+7
CS-Option	8.13E+5	1.07E+5	4.47E+2	9.38E+6	1.42E+7
Absolute Change (CS-Conv.)	-8.10E+3	-4.41E+3	-5.92E+0	-4.30E+5	-2.02E+6
% Change in MAC (CS-Conv.)/Conv.	-1.0%	-4.0%	-1.3%	-4.4%	-12.4%

decrease in GWP, 4.0 percent decrease in POCP (smog formation) and 1.3 percent decrease in particulate matter. These changes are due to differences in the quantities used for different types of materials, primarily asphalt, concrete, and aggregate base resulting from the reduction in total pavement surface area in the complete street designs for these types. These changes in the impact indicators are nearly all less than +/- 10 percent, reflecting the fact that the conversion to a complete street involves relatively small changes in the amounts and types of materials used on a complete street versus a conventional street. PED (Non-Fuel) shows a 12.4 percent under complete street design method. This decrease is mostly because CS elements replace asphalt pavement that has high PED (Non-Fuel) values compared to other items. As mentioned, PED (Non-Fuel) has no environmental impact and is a measure of use of a non-renewable resource (oil).

5 OTHER ISSUES TO CONSIDER

There are two important factors that can significantly change the results of this analysis but were not included in this stage of the analysis. The two factors are:

- Change in vehicle-miles-traveled (VMT)
- Changes in Traffic Speed

Reducing vehicle miles traveled by facilitating active modes of transportation (biking and walking) is a major goal of CS design guideline which if they materialize can result in significant changes in the use stage impacts between the two design methods.

In addition to reducing VMT and encouraging active modes of transportation, urban designers prefer using traffic calming designs that reduce traffic speed to increase safety and make active transportation modes more attractive to the public. The impact of such measures should be quantified by considering the effects of reduced speed on vehicle fuel consumption using the lower speed limits recommended in the NACTO design guide (NACTO 2016).

Reducing traffic speed can improve safety, and potentially increase mode change from vehicles to active transportation, however, it can have negative impacts on the fuel efficiency of vehicles. However, for the case of thoroughfare, the speed limit reduction further intensifies the reduction of well-to-wheel impacts due to VMT reduction because the design speed is changed from the conventional value of 50 miles per hour (mph) to the NACTO recommended speed of 45 mph for thoroughfares and the fuel consumption of passenger cars decreases to its minimum value across all speeds. According to the MOVES model, mph is an optimal speed for fuel consumption (MOVES website).

6 SUMMARY

A preliminary set of preliminary assumptions was used to demonstrate the use of LCA to consider the full life cycle environmental impacts of conversion conventional thoroughfares to complete streets. The importance of objective and reliable models for changes in traffic volume and congestion from the implementation of complete streets and comparison with conventional streets cannot be overstated. The full system impacts of complete streets on environmental impact indicators, considering materials, construction, and traffic changes, are driven by changes in reduction in VMT and changes in the operation of the vehicles regarding speed and drive cycle changes caused by congestion, if it occurs. To avoid situations where well-intended efforts might result in greater environmental impacts, utilization of life cycle assessment should be used as a robust and objective methodology that consider the full life cycle of the alternatives. Each LCA study should use 1) high-quality data, 2) a correct definition of the system boundary, and 3) include a thorough investigation, identification, and quantification of possible significant unintended consequences. For the case of thoroughfare, the speed limit reduction further intensifies the reduction of well-to-wheel impacts due to VMT reduction because the design speed is changed from the conventional value to the optimal speed for fuel consumption. This, however, is not always true and depends on the new design speed limit.

DISCLAIMER

This study was funded, partially or entirely, by a grant from the National Center for Sustainable Transportation (NCST), supported by U.S. Department of Transportation's University Transportation Centers Program. The contents of this report reflect the views of the authors, who are responsible for the facts and the accuracy of the information presented herein. This document is disseminated in the interest of information exchange. The U.S. Government assumes no liability for the contents or use thereof. The contents of this report reflect the views of the author(s), who is/are responsible for the facts and the accuracy of the information presented. This document is disseminated in the interest of information exchange.

REFERENCES

CARB website. SB 375 Regional Plan Climate Targets. California Air Resources Board, Sacramento, CA. https://ww2.arb.ca.gov/our-work/programs/sustainable-communities-program/regional-plan-targets, accessed Sep. 1, 2019.

City of Los Angeles (LA-DG) 2014. Complete Streets Manual. *Department of City Planning, City of Los Angeles*, CA, CPC-2013.910.GPA.SP.CA.MSC, http://planning.lacity.org/Cwd/GnlPln/MobiltyElement/Text/CompStManual.pdf.

MOVES website. The United States Environmental Protection Agency, Washington, D.C., https://www.epa.gov/moves, last accessed Sep. 1, 2019.

National Association of City Transportation Officials (NACTO) and Global Designing Cities Initiative 2016. Global Street Design Guide: Street Typologies. *Global Designing Cities Initiative. Island Press; 2nd ed.* https://globaldesigningcities.org/publication/global-street-design-guide/, accessed: Nov. 27, 2017.

Saboori, A., John Harvey, Hui Li, Jon Lea, Alissa Kendall, and Ting Wang 2019. Documentation of UCPRC Life Cycle Inventory used in CARB/Caltrans LBNL Heat Island Project and other LCA Studies, *University of California Pavement Research Center, Technical Report*, Davis, CA.

Sacramento County (SAC-DG) 2009. Improvement Standards, Section 4: Streets. Office of County Engineering, *Sacramento County, Sacramento, California*, 2009. http://www.engineering.saccounty.net/Documents/Sect4StreetStds_Saccounty_%20Ver11_01_09.pdf, accessed Jun. 1, 2018.

Network-level life cycle assessment of reclaimed asphalt pavement in Washington State

M. Zokaei Ashtiani & S.T. Muench
University of Washington, Seattle, WA, USA

ABSTRACT: Virgin material savings is known as the major advantage of asphalt recycling while these benefits could be offset by plant and reclaimed asphalt pavement (RAP) processing operations. Initially in this study, an attributional life cycle assessment (LCA) model was formulated on three scenarios differing in RAP content. Furthermore, a sensitivity analysis on RAP transportation suggested that the environmental benefits of using RAP would break even at a distance of about 50 miles. To obtain a spatial distribution of RAP stockpiles, remote sensing techniques have then been implemented to locate and estimate asphalt plants' inventory of RAP in Washington State. Knowledge about the exact location and quantity of RAP stockpiles combined with LCA-driven results enables a network-level understanding of where RAP recycling remains environmentally beneficial regarding energy consumption and greenhouse gas emissions. Choropleth maps were ultimately generated to illustrate the boundaries where RAP incorporation remains environmentally friendly with regard to hauling distances.

1 INTRODUCTION

Asphalt pavements contribute to the majority of road surfaces worldwide. Its two main components are crushed aggregates and asphalt binder. These two elements are not readily available and must undergo several processes before being used. These processes in short consist of materials extraction, hot mix asphalt (HMA) production, construction operations, maintenance and rehabilitation, and disposal or recycling of the out of service product. In addition, transportation of materials is embedded into each of these stages. The use of reclaimed asphalt pavements (RAP) has also become a favorable practice among the road agencies and contractors since it is believed to satisfy what is best known as a sustainable design. Being the most recycled material by weight in the US (Williams et al. 2018) adds up to the importance of understanding how these valuable materials should be used efficiently to be considered environmentally friendly.

RAP recycling seems to be environmentally beneficial only when adverse effects of post-processing, transportation, and plant production adjustments are smaller than the eliminated impacts from virgin material production and solid waste disposal (Mukherjee 2016, Li et al. 2019). RAP is generated from milling operation of out of order asphalt pavements, full-depth pavement reclamation and demolition, and asphalt plant waste (West 2015). RAP is most commonly hauled to a processing facility (often co-located with an asphalt plant) and stockpiled in designated areas. It is usually crushed into pieces and screened as needed rather than being crushed in advance and then stockpiled. Since asphalt binder from RAP is aged and stiffened throughout the years of service, plant operation is usually affected and altered in a way to compensate for the lack of final mix consistency. Moreover, RAP moisture content is usually higher than natural aggregates and drying can incur some additional costs to contractors and thus to the environment (Chen & Wang 2018). Superheating of aggregate particles is a common practice to help with gaining high quality final HMA product containing RAP (Brock & Richmond 2007, Yang et al. 2015, Liu et al. 2017). It is also worth mentioning that researchers stressed out the concern of losing service life of HMA mixtures produced with RAP when compared to all virgin materials (Aurangzeb et al. 2014).

Life cycle assessment (LCA) is a useful method to differentiate products and processes in terms of their impacts, chiefly on the environment (Santero et al. 2011a, Balaguera et al. 2018). As described in the International Organization of Standardization (ISO) 14040 and 14044, a complete LCA study includes the i) definition of goal and scope, ii) inventory analysis, iii) impact assessment, and iv) result interpretation. Several studies have been carried out employing the approach specified in the ISO standards to evaluate and compare different applications in pavement industry. However, there seems not to be a coherent and universally agreed procedure to adopt. It can be also realized from the fact that a product category rule (PCR) has not yet been developed to treat asphalt pavements (Mukherjee & Dylla 2017). Previous studies, thus, emphasized either on a case by case basis evaluation of paving projects (e.g. Weiland & Muench 2010, Farina et al. 2016), or reviewed the existing literature to offer more insight into what has been done and what needs to be done (e.g. Santero et al. 2011a and 2011b, Balaguera et al. 2018).

For the purpose of this study, an asphalt pavement design to meet Washington State Department of Transportation (WSDOT) requirements will be proposed and an LCA is performed to compare alternative RAP use scenarios. Three scenarios of producing HMA with RAP contents of 0, 20, and 40 percent are considered. Sensitivity of total energy consumption and global warming potentials (GWP) on materials transportation distance will then be evaluated. Remote sensing techniques are used to locate RAP stockpiles and estimate their weights within the state of Washington. Having the spatial distribution of RAP stockpiles in hand along with a critical transportation distance obtained from the LCA results offers a network-level perspective into the statewide availability of RAP so that it can be recycled into HMA without sacrificing its environmental benefits. In short, within the state of Washington, where does it make environmental sense to use RAP in new asphalt pavement? The following sections elaborate on how the four fundamental LCA steps are carried out to accomplish the objectives of this research effort.

2 GOAL AND SCOPE

The Federal Highway Administration (FHWA) report, which is prepared following ISO 14040 and 14044, is pursued in this study as a framework to perform LCA (Harvey et al. 2015). The intended audience of this investigation are researchers in the LCA discipline, as well as concerned federal government, and state DOT personnel. The result of this study could also be helpful to decision/policy makers within the asphalt industry. The majority of studies done on this topic use attributional (process-based) cradle-to-grave LCA methods and so will this study (Yang et al. 2014, Liu et al. 2017). Details on the impact categories will be elaborated in subsequent sections; however, this study will investigate two broad categories of energy consumption and emissions to the environment. As the main objective of this study, sensitivity of LCA results on RAP transportation will be examined to discover a distance after which using RAP would not make environmental sense.

2.1 Functional unit

A functional unit defines the system in terms of magnitude, duration of service, and the expected quality. There seems not to be a unanimous agreement within the literature on the selection of functional unit for running LCA on pavements. Balaguera et al. (2018) provides a list of existing studies confining different aspects of LCA characteristics for researches carried out on pavements. It appears that a handful of studies used a unit of distance (kilometer, meter, mile), area (square meter, square yard), or volume (cubic meter, cubic yard) as the functional unit. In this study, definition of *a lane-kilometer of asphalt pavement constructed in 2017 which maintains its durability and smoothness during a service life of 50 years without major reconstruction activities* turns out to be the most appropriate. The location of interest is within the state of Washington, with an average traffic loading of about 3.5 million equivalent single axle load (ESAL) (Weiland & Muench 2010) over its lifetime. Following 1993 AASHTO design method, a pavement layer totaling 10 inches (25.4 cm) of asphalt in three lifts (two 3-in (7.6 cm) thick base course layers, and one 4-in (10.2 cm) thick surface course) was designed.

2.2 System boundaries

Since the goal of this study is to compare three alternatives mentioned earlier, several practices are in common when constructing a road and hence are omitted from the system boundary. Examples of such shared characteristics are subgrade stabilization, effect of construction on traffic delay, noise pollution, rolling resistance, albedo effects, lighting, and heat island effect (Aurangzeb et al. 2014). RAP is further assumed to have minimal adverse effect on pavement longevity. As a result, maintenance and rehabilitation operations were left out of the system boundary. Furthermore, this study focuses on the construction and end-of-life operations of pavement sections and ignores the impacts coming from the use phase, i.e. vehicles' fuel consumption. Transportation distances from upstream processes (e.g. in-plant transportation) are assumed to be considered within the life cycle inventory (LCI) of materials production such as asphalt and aggregates (Mukherjee 2016, Chen & Wang 2018). Life cycle of each raw material including extraction and processing, electricity and fuel consumption at each stage, asphalt mixture preparation which involves heating up aggregates and liquid asphalt, construction equipment emissions, and several transportations to/from facilities are to be accounted for. Although required per ISO, feedstock energy is left out of the system boundary.

2.3 Allocation procedures

Allocation of RAP is done using an open loop approach. It is assumed that RAP is available through the [local] asphalt plant as a free material (Mukherjee 2016). To close the loop, a unit process must be defined for removal/demolition, processing, and stockpiling RAP after the service life of the pavement section has reached. Although it is difficult to identify the responsible party for the out-of-service product, the emissions associated with RAP preparation is at least taken into account in this formulation. This approach is in line with other studies that assumed "zero [upstream] burden" for RAP (Miliutenko et al. 2013, Yang et al. 2015, Mukherjee 2016). This assumption would result in a hollow unit process to account for the open-loop recycling concept within the LCA computational structure.

2.4 Impact categories and data requirements

Classification and characterization of impacts were selected as total energy consumption, GWP, acidification, photochemical ozone formation effects on human health, and particulate matter formation. The majority of input data would be extracted from Greenhouse Gases, Regulated Emissions, and Energy Use in Transportation Model (GREET 2018), US Environmental Protection Agency (EPA) (RTI International 2004), and EPA MOVES beta model (MOVES2014b, 2014)(MOVES2014b 2014).

3 INVENTORY ANALYSIS

Following subsections describe how the life cycle inventory has been formed to adequately mimic the system boundary and achieve the goals and scopes of this study.

3.1 Reference flows

In representing the functional unit, several processes are involved within the system boundary. In general, these processes are divided into materials production, materials transportation, and construction practices. Asphalt mixtures consist mainly of stone particles (aggregates) and liquid asphalt binder (bitumen). Emulsified asphalt binder is used as a tack coat, bonding two layers of asphalt mixture together. The transportation phase typically involves the use of diesel combusting engines and considers only the use phase of these vehicles, not the whole life cycle. The construction phase also involves a series of equipment such as paver, roller, material transfer vehicle (MTV), loader, and excavator. Asphalt final density of 150 lb/ft^3 is presumed as the basis to calculate the material requirements. Asphalt binder, sand, and crushed coarse aggregates are assumed to comprise respectively 5.5, 44.5, and 50 percent of the total mass as part of a representative mix design. Three alternatives under study contain 0, 20, and 40 percent RAP, assuming that RAP has asphalt binder content of 5.5% and it consists more of finer aggregate structure. Thus, when using RAP in a mixture, it compensates for 60% of the sand and 40% of crushed coarse aggregate. 100% asphalt binder contribution from RAP is assumed for all mixtures.

On the transportation side, all materials are assumed to be carried with a conventional diesel consuming heavy-duty truck running an average speed of 45 miles per hour, each having a capacity of 20 tons. There are four major distances need to be considered for material transportation: i) asphalt plant to construction site, ii) aggregate production site to asphalt plant, iii) project location to dump site (for RAP disposal), and iv) asphalt refinery to asphalt plant.

For the construction phase, seven types of vehicles were used as paver (175 hp engine), material transfer vehicle (300 hp), breakdown roller (175 hp), finish roller (100 hp), excavator (300 hp), loader (100 hp), and HMA milling machine (600 hp). Balancing the production of asphalt (300 tons/hour) with the construction procedure for this specific pavement design will yield the operation hours of equipment to lay down pavement. A same approach as stated in Weiland & Muench (2010) has been adopted to calculate the productivity of construction activities.

The intervention (environmental) matrix is designed to accomplish the impacts associated with releases of emissions to the environment. Based on the impact categories decided to be studied in this paper, 10 emission flows have been accounted for; namely, sulfur dioxide (SO_2), volatile organic compounds (VOCs), carbon monoxide (CO), oxides of nitrogen (NO_x), particulate matters (PM_{10} and $PM_{2.5}$), oxides of sulfur (SO_x), methane (CH_4), nitrous oxide (N_2O), and carbon dioxide (CO_2).

3.2 Data collection and assumptions

A total of 28 unit-processes have been identified to satisfy the functional unit. Unit processes have been chosen with an aim to overcome multifunctionality and cut-off criteria within processes. For example, electricity generation utilizes a mix of fuels as input and thus the life cycle of each fuel must be considered in a separate unit process. Or as another example, construction practices involve several stages that use different equipment which necessitate individually independent unit processes. Where possible, data for each category of unit processes were gathered from same references as a rule of thumb for omitting discrepancies. As a result, four distinct categories were determined to describe unit processes as follows.

3.2.1 Materials production

As discussed in previous section, asphalt mixtures consist of aggregates, bitumen, and RAP. Three sources of information were used to collect data for each type of material. For bitumen (and emulsion), data were collected from four references (Stripple 2001, Weiland 2008, Eurobitume 2012, Lin 2012). In assessing their data, feedstock energy was left out of the system boundary since bitumen is hardly used as a fuel, and considering it might lead to unrealistic impact on the whole pavement life cycle (Weiland & Muench, Life-cycle assessment of reconstruction options for interstate highway pavement in Seattle, Washington, 2010). A limitation, however, is that most of the data are gathered in Europe and an assumption has been made that the input energy requirements will remain the same in the location under study. Although an electricity mix has been chosen to represent the state of Washington, other fuel types are directly in use as part of the refinery facilities. Life cycle inputs and outputs for aggregate production have been collected mainly from two recourses = (Lin, 2012) (PCA 2007, Lin 2012).

Production of hot mix asphalt in plants requires energy and releases emissions mostly to the air. A comprehensive EPA study of more than three hundred asphalt mix plants provides a reliable source for the use of this study = (RTI International 2004). Despite its age, this remains an authoritative data source. Asphalt was assumed to be produced in a conventional counter-flow drum mix plant that consumes only electricity and natural gas to provide heat and other appliances and are equipped with fabric filters.

Results drawn from the EPA study for natural gas fueled asphalt plants with fabric filters suggests that the amount of CO_2 emissions is not tightly related to plant type, and the assumption on plant type might be nontrivial. Furthermore, it can also be concluded from the same report that RAP content has no tangible effect on the CO_2 emissions (and thus energy consumption) when producing hot mix asphalt. There is also evidence from other field studies that RAP content has minimal effects on plant energy consumption such that energy required to superheat raw aggregates might be counterbalanced by the reduction in their quantity as RAP content increases (Aurangzeb et al. 2014, Yang et al. 2015). Nevertheless, studies that attempted on modeling asphalt plant operations based off thermodynamic principles found a proportional dependency between the amount of RAP added to the mixture and accumulated energy consumption to produce the same amount of asphalt mixtures without any RAP (Santos et al. 2017, Chen & Wang 2018).

To overcome the inconsistencies in the effect of RAP content on asphalt plant energy consumption, this study only considers drying of RAP particles, which is believed to be an influential factor (Chen & Wang 2018, Li et al. 2019). Moisture content of RAP was assumed to be 5% which is in line with other studies (Yang et al. 2015, Santos et al. 2017, Chen & Wang 2018). RAP is also assumed to be dried out separately before introduction to the plant (as suggested by Liu et al. (2017)) where the emission factors from EPA can be used directly without any further adjustments to consider the RAP content. Energy required to evaporate moisture from RAP was calculated having water specific and latent heat of 4.186 kJ/kg/°C and 2260 kJ/kg, respectively.

3.2.2 *Material transportation*
Since local resources are assumed to provide raw materials, transportation of goods will be done only by conventional heavy-duty diesel trucks. The GREET (2018) database was used in life cycle inventory analysis of truck fuel usage. As mentioned earlier, an assumption has been made that RAP is obtained from the same site that the paving operations are taking place. No deterministic value was assumed for the distances from asphalt plant to construction sites. Instead, a range of values from 0 to 100 miles are presumed. Performance graded (PG) asphalt binder and emulsified asphalt were assumed to be obtained from one of the three refineries within/around Washington State: US Oil & Refining Co., McCall Oil & Chemical Corporation, and Idaho Asphalt Supply Inc, with an average transportation distance of 90 miles.

3.2.3 *Initial construction and end of life demolition*
Construction equipment explained in section 3.1. are needed to construct (paving and rolling and excavating) and demolish (milling and excavating) asphalt pavements. Information on fuel consumptions and efficiencies were gathered originally from MOVES2014b (2014). The effect of construction on noise pollution, traffic delay costs, vehicle idling, and transportation of equipment to construction site (mobilization) was considered out of the scope of this study. RAP processing activities were also considered as part of the end-of-life stage. Although there is no consensus on emissions associated with RAP processing (including but not limited to crushing, screening, and stockpiling), a rather wide range of energy usage from 1.6 to 25 MJ/ton was reported to represent RAP processing (Miliutenko et al. 2013, Yang et al. 2014, Mukherjee 2016). In this study, a type of miscellaneous equipment that translates the consumption of 20 MJ per ton of RAP was selected from MOVES database to depict emissions associated with the entire processing stage.

3.2.4 *Fuel and electricity*
Washington State electricity mix consists of 69% hydropower, 10% natural gas, 9% nuclear, 7% wind, 4% coal, and 1% other resources with a grid gross loss factor of 4.4%. This electricity

mix was used to assess the point of use electricity generation input/output as well as fuel production. A list of fuels considered in this study are i) conventional and LS diesel at fueling station, ii) diesel for construction equipment, iii) natural gas as a stationary fuel, iv) natural gas for electricity generation, v) residual oil at the point of use, and vi) coal power plant. Emissions related to electricity generation were obtained from GREET (2018).

3.3 Inventory problem solution

A computational methodology described in Heijungs & Suh (2002) has been followed to solve the inventory problem. In that, two separate matrices concerning a technology matrix (**A**) and an environmental matrix (**B**) would form the skeleton of the LCI. An inventory vector (**g**) is defined in a same way as the functional unit vector (**f**). The approach to which this problem can be solved relies on the non-singularity of **A** matrix. Since the form of the technology matrix in this study was structured to use hollow processes and allocation using sub-processes, a square invertible technology matrix was resulted. As a result, $g=B \times A^{-1} \times f$ would be the LCI solution.

4 IMPACT ASSESSMENT

The next step "involves the conversion of LCI results to common units and the aggregation of the converted results within the impact categories [by mean multiplication of inventory vector by the characterization matrix,]" (Heijungs & Suh 2002) which is formally known as characterization. Characterization factors are defined to quantitatively translate emissions and resource extractions to environmental impacts. These factors are obtained from complex models which explanation is out of the scope of this study. There are several references that provide these factors, among which ReCiPe is one of the most referred resources. 18 midpoint indicators and 3 endpoint indicators are explained in ReCiPe (Huijbregts et al. 2016). Midpoint impact categories selected in this study are GWP (CO_2-eq), acidification (SO_2-eq), photochemical ozone formation (effects on human health, NO_x-eq), and particulate matter formation ($PM_{2.5}$). An individualistic perspective has been adopted wherever that mattered.

5 INTERPRETATION

This section makes an effort to shed more light on the significant issues around each RAP alternative and explains which operations during the life cycle dominate the effects on energy consumption and emissions to the environment. Thus, within the subsections that will follow, first a sensitivity analysis of energy consumption and GWP on RAP trucking distance is presented. Then, the unit processes involved in this project are classified into certain categories based on their similarities (e.g. transportation, materials production, etc.). The last subsection devotes on the application of remote sensing technique to locate RAP stockpiles within Washington State and discusses the potential of using LCA results on a network-level resource allocation viewpoint.

5.1 Sensitivity analysis

As discussed earlier, RAP transportation distance was considered as a range of values from 0 to 100 miles. Effect of varying transportation distances were then considered after life cycle impact assessment (LCIA). Figure 1 illustrates the effect of RAP transportation distance on energy consumption and GWP of the three alternatives proposed here. There seems to be a distance after which adverse environmental impacts of using RAP would outweigh its benefits. Commonly known as *critical transportation distance*, Figure 1 suggests that RAP trucking distances of over 42 and 64 miles can pose limitations on environmental advantages sourcing from recycling in terms of energy consumption and global warming potential, respectively. Li et al. (2019) adopted a same strategy and yielded a critical RAP transportation distance of 50 miles, which closely matches the finding presented here. Another study concerning the use of mixed recycled aggregate in place of

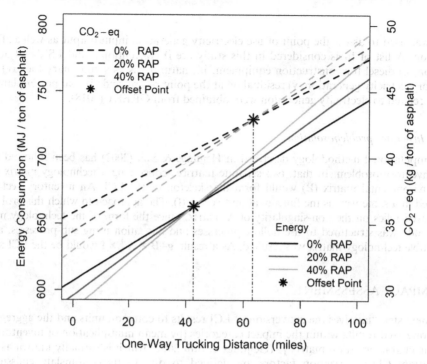

Figure 1. Critical transportation distance of RAP in view of energy consumption and CO_2-eq.

natural aggregates contented that the offset transportation distance to be environmentally sound is 25 miles (Rosado et al. 2017). Two other studies on the use of steel slag as a substitute to raw aggregates found a critical transportation distance of 90 (Anastasiou et al. 2015) and 100 (Mladenovic et al. 2015) miles in view of GWP offset point. According to the existing literature and the present study, an average distance of 50 miles seems to be a good estimate of how far RAP could be sourced in order to remain environmentally beneficial.

It must also be noted that this analysis and the consequent conclusions about the critical transport distance are dependent on the assumptions and system boundary defined here. Functional unit selected here is not universal and result interpretations are under the influence of its selection. To exemplify, traffic loadings, equipment types and their productivity, neglecting feedstock energy, inventory data sources and material properties can all in part influence this analysis. However, the decision to normalize results to a per-ton of asphalt basis is believed to eliminate some, if not all, of the presumptions. Regardless, this study is intended to showcase a practical application of LCA into real-world problems rather than reworking previous pavement LCA efforts. For this very reason, quantitative sensitivity of all variables is not analyzed elaborately.

5.2 Contribution analysis

Unit processes defined within the system boundary can be grouped into five main categories. These are: i) material production, ii) material transportation, iii) fuel consumption and electricity generation, iv) end of life demolition, and v) construction activities. Contribution analysis was carried out for the case where RAP is hauled 50 miles. At this distance, the environmental impacts of all three alternatives fall close together (according to Section 5.1.) and enables valid comparisons. In this regard, Figure 2 is produced to show the contribution of each LCA stage on impact categories. As expected, increasing RAP content would result in less environmental footprint in terms of materials production, while materials transportation and end of life operation shares tend to rise.

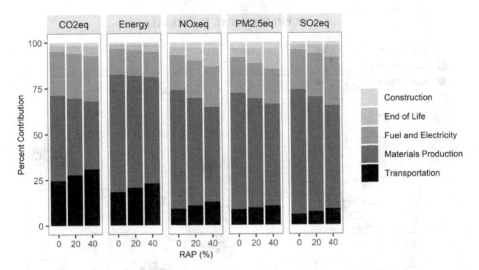

Figure 2.   Life cycle impact assessment breakdown at offset hauling distance.

5.3   *Remote sensing application*

To gain a geographically larger perspective into RAP inclusion in hot mix asphalt production, satellite imagery from Google Earth software along with some simplifying assumptions were used to locate, identify, and estimate the weights of RAP stockpiles within the state of Washington. The details on how this procedure was undertaken is expressed elsewhere (Zokaei Ashtiani et al. 2019). 1.16 million metric tons of RAP owned by 55 facilities were found to be stored in 65 locations throughout the State in 2017, which is well in line with the reported values from National Asphalt Pavement Association (NAPA) survey on RAP and HMA (Williams et al. 2018). Google Earth 3D terrain simulations were also used to identify the plant types wherever applicable. Results show that drum plants comprise 75% of asphalt production facilities within the State, while the other 25% are of batch type. Although it was argued earlier that the type of plant has no significant effect on energy consumption and hence greenhouse gas emission, information on the scatter of plant types might become trivial when more detailed plant data is available. Figure 3 shows on a map of Washington the spatial distribution, relative quantity, and the location of asphalt plants as well as their types.

5.4   *Statewide RAP availability*

Knowledge on RAP storage locations and their quantities can be used in conjunction with the critical transportation distance to find, geographically, where it is environmentally reasonable to use RAP in asphalt mixtures. To do this, 242 Washington State sub-county divisions were considered as the smallest geographic boundary of which RAP resources can be allocated to. A mathematical algorithm has been developed to assign RAP stockpiles to each region if the distance of RAP location was less than 50 miles to the division's centroid. Figure 4 (Supplemental Material) illustrates on a choropleth map of Washington the quantities of RAP that can be recycled into hot mix asphalt without sacrificing its environmental values. Such information can help inform policy makers and state DOTs to ratify regulations on the use of RAP based on availability and location. With that, some regions are better off using all virgin materials to avoid negative impacts of recycling on the environment. According to Figure 4, densely populated areas (Seattle as an example, which is located at the darkest shaded part of this map) are the first candidates to utilize RAP as most of this valuable resource is located within urban areas.

This problem may already be largely worked out using an economic surrogate metric: the cost or trucking RAP. A contractor's use of RAP on a project is most often controlled by its availability and the cost of trucking it to/from the asphalt plant. Essentially, where it may be

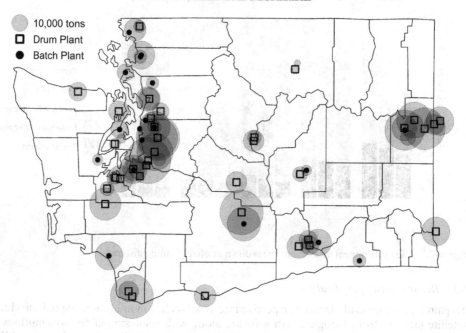

Figure 3. Spatial distribution of RAP piles and plant types across Washington State.

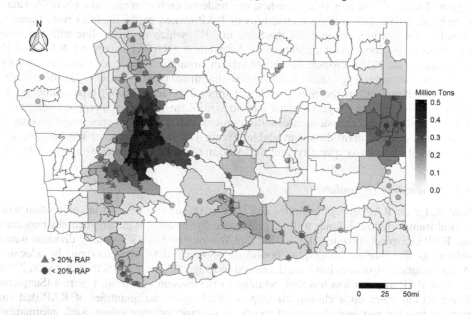

Figure 4. Spread of paving projects from 2013 to 2017 that used RAP mapped over Washington State RAP resource availability based off critical transportation distance.

environmentally unreasonable to use RAP, it is also likely economically unreasonable. A quick check of this idea is made by superimposing asphalt paving projects conducted in Washington State during 2013 thru 2017 over Figure 4. While WSDOT records to not record RAP percentages below 20% (all are listed as < 20%), they do identify RAP percentages higher than 20%. Figure 4 seems to show reasonable agreement between locations of high-RAP projects and areas where RAP use would be deemed environmentally reasonable. Furthermore, there seems to be the potential for conducting more high-RAP projects in some regions (e.g. eastern Washington) than historically performed. While these observations are tantalizing, a more thorough investigation is needed before any strong conclusions can be reached.

6 SUMMARY AND CONCLUSIONS

Reclaimed asphalt pavements are well known to be the most recycled materials by weight in the US. However, using RAP as a substitute for virgin materials may endanger the quality of final product, thus calls for some pretreatments and reconsiderations in asphalt plant operation. In addition to RAP post-processing, drying RAP particles and superheating of virgin aggregates requires surplus energy input in the form of fossil fuels and electricity which in turn would cause greenhouse gas surcharge into the environment. This study has formed an attributional cradle to grave LCA structure on three hot mix asphalt production scenarios in Washington State with 0, 20, and 40 percent RAP. Lifecycle inventory matrix was constructed based on the information gathered from open access resources including governmental reports and academic literatures.

System boundary was defined in this study to encompass four broad categories of materials production, materials transportation, construction operations, and end of life demolition. The objective of the present research aims at investigating the sensitivity of materials transportation, RAP in particular, to the overall energy consumption and greenhouse gas emissions. It was discovered that using RAP in hot mix asphalt production is environmentally beneficial only when it is located about 50 miles away from the construction site (aka critical transportation distance). This finding is valuable in the essence that using RAP seems not to be always advantageous to the environment, rather the dependency on hauling distances can impose substantial limitations.

To gain a better understanding of how the findings of this study could be expanded and applied to a much wider geographic scale, a remote sensing approach was taken to locate and estimate the quantity of RAP resources stockpiled within the state of Washington. A total of 55 asphalt plants and 65 locations that stored RAP have been identified to build a database consisting of geographic coordinates and their associated RAP weights. US census sub-county divisions were selected as the smallest geographic area. A mathematical algorithm was further developed to assign weights to regions that fall within a 50-mile radius around RAP stockpiles. As a result, a choropleth map of Washington showing each division's accessibility to RAP was developed. Such network-level representation of RAP availability can be used to inform policy makers and state DOT personnel on expected RAP use, and perhaps to adopt more comprehensive regulations that flex or restrict requirements on RAP inclusion in hot mix asphalt production.

REFERENCES

Anastasiou, E. K., Liapis, A., & Papayianni, I. 2015. Comparative life cycle assessment of concrete road pavements using industrial by-products as alternative materials. *Resources, Conservation and recycling* 101: 1–8.

Aurangzeb, Q., Al-Qadi, I. L., Ozer, H., & Yang, R. 2014. Hybrid life cycle assessment for asphalt mixtures with high RAP content. *Resources, Conservation and Recycling* 83: 77–86.

Balaguera, A., Carvajal, G. I., Albertí, J., & Palmer, P. F. 2018. Life cycle assessment of road construction alternative materials: A literature review. *Resources, Conservation & Recycling* 132: 37–48.

Brock, J. D., & Richmond, J. L. 2007. *Milling and Recycling*. Chattanooga, TN: Astec, Inc.

Chen, X., & Wang, H. 2018. Life cycle assessment of asphalt pavement recycling for greenhouse gas emission with temporal aspect. *Journal of Cleaner Production* 187: 148–157.

Eurobitume. 2012. *Life cycle inventory: Bitumen*. European Bitumen Association. Brussels, Belgium.

Farina, A., Zanetti, M. C., Santagata, E., & Blengini, G. A. 2016. Life cycle assessment applied to bituminous mixtures containing recycled materials: Crumb rubber and reclaimed asphalt pavement. *Resources, Conservation & Recycling* 117: 204–212.

GREET. 2018. *The greenhouse gases, regulated emissions, and energy use in transportation model*. Retrieved February 2018, from Argonne National Laboratory: https://greet.es.anl.gov.

Harvey, J. T., Meijer, J., Ozer, H., Al-Qadi, I. L., Saboori, A., & Kendall, A. 2015. *Pavement life-cycle assessment framework*. Federal Highway Administration (FHWA), US Department of Transportation. Washington, D.C.

Heijungs, R., & Suh, S. 2002. *The computational structure of life cycle assessment* (Vol. 11). Dordrecht: Springer.

Huijbregts, M. A., Steinmann, Z. J., Elshout, P. M., Stam, G., Verones, F., Vieira, M., Zelm, R. 2016. ReCiPe2016: a harmonised life cycle impact assessment method at midpoint and endpoint level. *International Journal of Life Cycle Assessment* 22(2): 138–147.

Li, J., Xiao, F., Zhang, L., & Amirkhanian, S. N. 2019. Life cycle assessment and life cycle cost analysis of recycled solid waste materials in highway pavement: A review. *Journal of Cleaner Production* 233: 1182–1206.

Lin, Y.-Y. 2012. *Eco-decision making for pavement construction projects*. Department of Civil and Environmental Engineering. Seattle, WA: University of Washington.

Liu, S., Shukla, A., & Nandra, T. 2017. Technological environmental and economic aspects of asphalt recycling for road construction. *Renewable and Sustainable Energy Reviews* 75: 879–893.

Miliutenko, S., Bjorklund, A., & Carlsson, A. 2013. Opportunities for environmentally improved asphalt recycling: the example of Sweden. *Journal of Cleaner Production* 43: 146–165.

Mladenovic, A., Turk, J., Kovac, J., Mauko, A., & Cotic, Z. 2015. Environmental evaluation of two scenarios for the selection of materials for asphalt wearing courses. *Journal of Cleaner Production* 87: 683–691.

MOVES2014b. 2014. *Latest Version of MOtor Vehicle Emission Simulator (MOVES)*. Retrieved September 2019, from US Environmental Protection Agency: https://www.epa.gov/moves/latest-version-motor-vehicle-emission-simulator-moves.

Mukherjee, A. 2016. *Life cycle assessment of asphalt mixtures in support of an environmental product declaration*. National Asphalt Pavement Association (NAPA), Lanham, MD.

Mukherjee, A., & Dylla, H. 2017. Challenges to using environmental product declarations in communicating life-cycle assessment results. *Transportation Research Record: Journal of the Transportation Research Board* 2639: 84–92.

PCA. 2007. *Life cycle inventory of Portland cement concrete*. Portland Cement Association, Sokie, IL.

Rosado, L., Vitale, P., & Penteado, C. G. 2017. Life cycle assessment of natural and mixed recycled aggregate production in Brazil. *Journal of Cleaner Production* 151: 634–642.

RTI International. 2004. *Emission factor documentation for AP-42 Section 11.1: Hot mix asphalt plants*. Prepared for US Environmental Protection Agency (EPA), Office of Air Quality Planning and Standards. Prepared by RTI International. Research Triangle Park, NC.

Santero, N. J., Masanet, E., & Horvath, A. 2011a. Life-cycle assessment of pavements part I: Critical review. *Resources, Conservation & Recycling* 55: 801–809.

Santero, N. J., Masanet, E., & Horvath, A. 2011b. Life-cycle assessment of pavements part II: Filling the research gaps. *Resources, Conservation & Recycling* 55: 810–818.

Santos, J., Flintsch, G., & Ferreira, A. 2017. Environmental and economic assessment of pavement construction and management practices for enhancing pavement sustainability. *Resources, Conservation and Recycling* 116: 15–31.

Stripple, H. 2001. *Life cycle assessment of road: A pilot study for inventory analysis*. Swedish Environmental Research Institute, Gothenburg, Sweden.

Weiland, C. D. 2008. *Life cycle assessment of Portland cement concrete interstate highway rehabilitation and replacement*. Department of Civil and Environmental Engineering. Seattle, WA: University of Washington.

Weiland, C. D., & Muench, S. T. 2010. Life-cycle assessment of reconstruction options for interstate highway pavement in Seattle, Washington. *Transportation Research Record: Journal of the Transportation Research Board* 2170: 18–27.

West, R. C. 2015. *Best Practices for RAP and RAS Management*. National Asphalt Pavement Association (NAPA), Lanham, MD.

Williams, B. A., Copeland, A., & Ross, T. C. 2018. *Asphalt pavement industry survey on recycled materials and warm-mix asphalt usage: 2017*. National Asphalt Pavement Association (NAPA), Lanham, MD.

Yang, R., Kang, S., Ozer, H., & Al-Qadi, I. L. 2015. Environmental and economic analyses of recycled asphalt concrete mixtures based on material production and potential performance. *Resources, Conservation and Recycling* 104: 141–151.

Yang, R., Ozer, H., Kang, S., & Al-Qadi, I. L. 2014. Environmental impacts of producing asphalt mixtures with varying degrees of recycled asphalt materials. *Pavement LCA 2014; Proc. intern. symp., Davis, 14-16 October 2014*. Davis, CA.

Zokaei Ashtiani, M., Muench, S. T., Gent, D., & Uhlmeyer, J. S. 2019. Application of satellite imagery in estimating stockpiled reclaimed asphalt pavement (RAP) inventory: a Washington State case study. *Construction and Building Materials* 217: 292–300.

Including sustainability and life cycle perspectives in decision making

R.I. Karlsson & Å. Lindgren
Swedish Transport Administration, Sweden

ABSTRACT: Agenda 2030 has 17 global development goals, and these include an economic dimension as the driving force to reaching social sustainability within the limit of the planet's boundaries. In the long term, all decisions taken must be sustainable, which means that they must be based on:

- Well-founded selection of measures
- Established working methods
- Measurable indicators for sustainability

We find tools for this in both life cycle assessments (LCA) and life cycle cost (LCC) calculations. The process scopes differ, where LCA compare environmental performance from cradle to grave while LCC mainly aim at determining the cost-effectiveness of alternative investments, including the use phase. In this paper, we explore the challenges faced when considering life cycle aspects when making subsequent decisions in the process from planning, design and procurement to management of road infrastructure. These challenges are related to system boundaries, scope as well as decision definition and delimitations. Examples are given of how decision makers can find support in our developed tools, such as Geokalkyl and climate calculations (GHG emissions).

1 INTRODUCTION

1.1 *Background*

A vast number of factors need to be included during the planning, design, procurement, and production of investments and reinvestments in infrastructure assets. Even though practices differ to some extent between countries, the principles are more or less similar. However, on a more detailed level there are differences in legislations and requirements that influence the process.

The investment process adopted by the Swedish Transport Administration, STA, is described in more detail below, but includes the following steps (with the main purpose or goals in parentheses):

- Project formulation and selection (to define and select the project with the best value)
- Planning and preliminary design (to meet project goals and legal requirements)
- Detailed design and tender document (to ensure that solutions are feasible and requirements meet expectations)
- Production and guarantee or contract period (to ensure fulfillment of requirements)

In all of these stages, environmental considerations are essential along with costs, but the decision-making process involve much more than that. Traditionally, investment costs and time aspects have had a strong influence throughout the whole process. For more than a decade there have been many efforts made to achieve a better balance between these aspects, and especially to increase the emphasis on environmental effects and total costs (not only investment costs). Of special interest to the scope of this paper are the requirements implemented in the framework for procurement of consultants and their deliveries (plans, preliminary designs, permit applications, and tender documents). Between year 2013 and 2015 The STA's templates for procurement of consultants was revised, and tasks included separate deliveries of life cycle cost investigations and climate impact (as greenhouse gas estimations). These tasks have since evolved and become an integral part of the planning and design process. The STA is also committed to deliver on the UN Agenda 2030 having 17 global development goals that include an economic dimension as the driving force to reach social sustainability within the range of the planet's boundaries. In the long term, all decisions taken must be sustainable, which means that they are based on:

- Well-founded selection of measures
- Established working methods
- Measurable indicators for sustainability

1.2 *Aims and purpose*

This paper seeks to explore the needs and challenges in the decision-making process during infrastructure planning, design, and production when including different aspects of sustainability, especially climate goals and future costs. The current measures taken are also described, and examples of practices and tools are given. Emphasis is on road construction projects and specifically on pavement design aspects, even though the same principles may apply to other infrastructure assets.

2 CURRENT PRACTICE IN SWEDEN

2.1 *Project process*

In order to prepare for an examination of decision making related to sustainability in investment projects, the process is summarized below. The process adopted by the STA starts in the planning department where strategic planning is performed, including public and stakeholder consultation, cost-benefit analysis, and a thorough investigation to decide on the type of action. If the action involves some kind of construction and gets funded, then a project specification is established and the investment department is engaged to execute the project. Consultants and contractors then carry out most of the remaining planning, design, and construction, closely monitored by the STA.

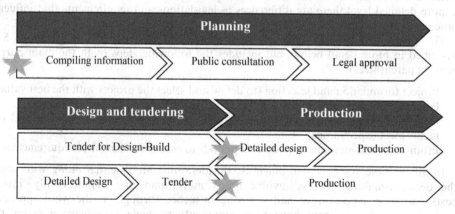

Figure 1. Project process for road and railway investments with the procurement of consultants and contractors indicated by stars (original figure by authors).

Because most of the work is performed by procured resources, it is essential that there exist clearly defined requirements and tasks for consultants and contractors to pursue sustainable solutions. A backbone to sustainable solutions is inherit in defined procedures, design manuals and regulations. Furthermore, life cycle aspects and sustainability should be considered and documented in the planning and design process, even though it is difficult to define these tasks and to estimate costs during bidding.

Consultants are required to deliver on:

- Climate calculations (GHG emissions, CO2ekv) along with proposed mitigation measures
- Life cycle aspect inventories and LCC estimations

Contractors are required to deliver on:

- Reduction of GHG emissions to meet requirements either set as a reduction in percent related to an estimated CO2ekv budget or set on specific materials (concrete, steel, and asphalt)
- Performance requirements that, among other things, consider sustainability
- General functional and technical requirements

The contractor's scope of action is limited by the delimitations made in earlier stages, as well as lack of knowledge of the user phase. The same goes for the consultant if it does not have an understanding of the entire life cycle. This is often reflected in the delivery of traditional measures of action or detailed solutions too early in the process. For example, in the planning stage propose detailed design or choice of material intended to reduce climate impact or meet stakeholder requirements. The early planning stages are not only important for creating the conditions for future planning and design stages, this stage is also where the greatest potentials are to reduce GHG and save resources. To produce sustainable roads under these circumstances is challenging. Constructed roads are often criticized for neglecting the use phase and not properly exploit opportunities or consider sustainability.

2.2 Defining systems and their boundaries

The project selection and execution process requires decisions to be made that take into account a vast number of different factors. Consequently, the above-mentioned project process is designed to first deal with overall strategic decisions and then step by step pass through (more or less formalized) tollgates where the project becomes better defined and targeted. At these tollgates, decisions are made based on the current level of detail and knowledge of project conditions and requirements. It is of utmost importance that all relevant aspects of the decisions are included and considered in one way or another. Calculations of investment costs have traditionally been an integral part of the process as well as an environmental impact assessment (EIA). The focus on investment costs has been criticized for not considering the future costs, and the environmental impact assessment has been criticized for only identifying issues but not being precise enough to decide on adequate measures as well as not considering future maintenance and traffic.

By experience, it has become clear that the definitions of systems and system boundaries used to form the basis for decisions at the tollgates are a key to successful implementation of life cycle aspects within the project process. A view on systems adopted by the STA follows, for example, SS-ISO 15686-5:2008 (Buildings and constructed assets – Service-life planning – Part 5: Life-cycle costing) where three levels of systems are defined together with an overall strategic level (socio-economic cost benefit analysis). These four levels can be defined in agreement with the project process and its tollgates:

1) Strategic level. Decisions are made regarding actions that should be taken in the road network to achieve the best value for a set budget level. This usually accounts for costs and benefits over a period of 40–60 years, or less depending on the period for which that the action is expected to be beneficial.

2) Strategic project level. Decisions are made regarding how to formulate a project in order to achieve the best value for the money while fulfilling the project goals. This usually considers a 40–60 year period.
3) System level. Decisions are made regarding how to fulfill the required function at the lowest overall cost while at the same time considering other (external) costs for road users and society. The time frame considered is the functional design life.
4) Detailed component level. Decisions are made regarding the design of each component with respect to its expected technical design life or replacement interval.

Each system level is made up of several subsystems (A, B,... below) or, in contrast, several subsystems are assembled into a new system with superior functionality.

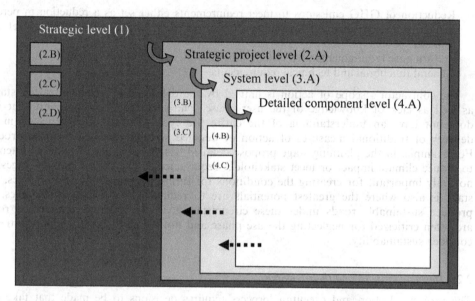

Figure 2. System levels used in the project process to further define sub-systems for decision-making. Dotted arrows indicating the need to aggregate more detailed models to be used on higher system levels. (original figure by authors)

Traditionally LCA is based on a material flow cradle-to-grave assessment of products being used on multiple occasions as described, for example, in SS-ISO 14040:2006 (Environmental management - Life cycle assessment - Principles and framework). As such, LCA is well suited for level 4 calculations on components. The same models can also be used for aggregation to levels 1–3 for the investment phase. Using this approach, climate impact calculations have been widely used in investment projects since the year 2015. However, the service life/use phase has not been included other than briefly so far in the analysis. The relationships are uncertain, which means that calculation models are seldom sufficiently mature to cover other than the construction stage. Performing LCA analyses of construction products or roads requires methodological considerations on several dimensions, leading to a variety of explanations for differences in the results.

For civil engineering works, a framework for sustainability assessments is published in the standard EN 15643-5:2017 (Sustainability of construction works - Sustainability assessment of buildings and civil engineering works - Part 5: Framework on specific principles and requirement for civil engineering works). ISO standards give rules for how to perform LCA.

On the product level, an environmental product declaration (EPD) should be based on the standard EN 15804. An EPD of a construction product gives "cradle-to-gate" information. Information modules A1-A3 (covering the production) are mandatory, while the

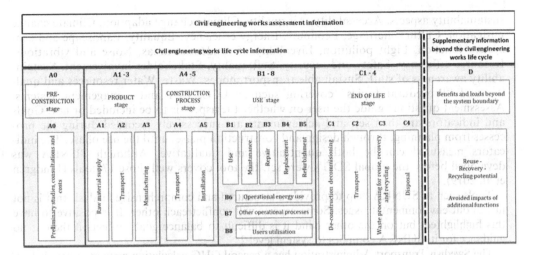

Figure 3. Information modules applied in the assessment of environmental, social and economic performance of a civil engineering works (EN 15643-5:2017).

construction stage and use stage modules, as for now, are voluntary. After ongoing revision of EN 15804, module C (end of life) and module D (reuse and recovery) will probably be made mandatory. EPDs should also be published in accordance with ISO standard 14025.

Challenges posed while combining decisions based on materials that do not consistently include or exclude the use phase are further discussed below.

2.3 *Current tools and practices*

Procurement of consultants for planning and design is done based on a set of templates in which tasks are specified. The current version of these templates includes climate calculations and life cycle cost investigations along with other related tasks such investigating traffic during construction, risk analysis, socio-economic cost benefit calculations, and environmental impact assessments. Because this practice is more or less similar in all projects commissioned by the Swedish Transport Administration, there is a growing ability for the consultants to consider life cycle aspects.

The two most frequently mentioned obstacles for further development of these practices are the lack of input into these analyses and the need for more clear-cut work descriptions. The latter will remain a challenge because the nature of the issues suitable for alternative solutions is usually project specific. Otherwise, regulations should guide the process to the most optimum solution if such a solution can be generalized to cover a range of conditions. More thorough investigations into life cycle aspects should either focus on regulations or on conditions or possibilities/challenges that are specific for the project. To fulfill the need for input, there are some tools developed such as "Klimatkalkyl" and "Geokalkyl" to cover specific decision-making situations as described below. In our experience, the calculation tools are not the major obstacles, and instead it is the challenge to extract data from the management and business systems where the data are collected.

There is a common understanding of some aspects of sustainability, but not as a whole, and thus it is difficult to improve overall performance. The SUNRA tool (Sustainability, National Road Administrations) was developed as part of a joint European research project (Sowerby et al. 2014). The aim was to develop a common understanding and means of measuring, benchmarking, and improving sustainability performance in road projects. This Excel tool includes 26 selectable sustainability topics, representing a variety of sustainability aspects; Accessibility, Air quality, Climate change adaption, Climate change mitigation, Cultural heritage, Economy, Energy efficiency, Equality, Landscape and ecosystem function, Light pollution, Liveability of residential areas, Noise and vibrations, Resource efficiency, Safety and security, Soil quality, Stakeholder involvement, Sustainability awareness of staff, Sustainable transport modes, Waste, Water resources and quality, and "Procedural topics" covering such as LCC, EIA and management systems. Assessment questions guide the user on whether the aspect is to be included or not. Goals and indicators are then set, and achievements are successively recorded during the processes from planning, through design and construction phase, until the use phase. The indicators measured can be both quantitative and qualitative. Although SUNRA was developed before the global SDG's were set, the tool covers well those 17 goals (Lindgren et al. 2019).

Tools such as SUNRA help to provide a structured and comprehensive analysis that is not just about environmental issues. Some aspects will conflict each other. It is positive to have this highlighted, but on the other hand it is difficult to balance goal conflicts if there is no guidance for it. SUNRA complies to system level 2.

The Swedish Transport Administration has a general GHG calculation method, Klimatkalkyl, that is used in larger infrastructure projects. It was originally used for follow up purposes but has been further developed and made more detailed (STA 2019). It is also to be used in procurement. A GHG budget is calculated as a base from which a reduction requirement in percent is set that is to be reached by the designer and contractor. On average, we need to achieve a 15% reduction by 2020 (from 2015) and 50 % by 2030, in order to meet our national goal to be fossil free by 2045. In addition to percentage reduction requirements, we also set carbon requirements on the production of some frequently used materials with a high climate impact. These materials are concrete, rebar (reinforcement steel), and construction steel. Allowed CO2ekv emission levels from production of concrete and rebar steel are set in line with the 2045 zero emission goal. Environmental product declarations (EPD) are used as verification documents that requirements are met. So far, for construction steel we require only an EPD for the steel itself and not for the steel product. No maximum CO2ekv-level from the production stage is set as it is for concrete and rebar. This is due to the lack of product specific LCA data for steel components. The GHG calculation method was originally used on system levels 1 and 2 but has been used on system levels that are more detailed.

Geokalkyl ("Geo calculation" in English) is a tool for early comparisons of costs and GHG emissions from building infrastructure projects (STA 2018). The tool compares alternative alignments within a selected road or railway corridor. It is constructed in GIS format, and the calculations are carried out in Excel. Requested input data for roads include elevation data, soil maps, traffic volume, and design parameters such as the number of lanes and the climate zone. The tool is specialized for earthworks, resulting in an optimized design of the road including any need for geotechnical reinforcements. With regard to soil quality, road profile, and surrounding terrain, the most appropriate reinforcement method is proposed. Where the planned route crosses another road, watercourse, etc., the tool proposes a suitable measure, such as a crossing, a roundabout, a pipe, a bridge, a tunnel, etc. The results are displayed both in tables and in diagrams or visualized graphically. This is illustrated in Figure 4 where the bars show the site-specific costs or GHG emissions for three routes in comparison. High costs or emissions are generally associated with weak ground conditions and advanced constructions such as tunnels and bridges. By zooming in on the image (Figure 5), one can see more details of what information each bar represent.

The aggregated results are also presented in more detail where costs and emissions are clustered into pavement, cut, fill, and reinforcement.

One of the main advantages with Geokalkyl is that costs and climate impacts can be handled more thoroughly in the early planning process than former standards, thus providing better decision-making support. Anticipated challenges are thus given more time to plan for and to supplement previous investigations. Cost-heavy items or sections can be compared to GHG-intensive ones, and thus it includes additional dimensions in the decision-making data.

Figure 4. Cost comparison of three railway routes visualized by the heights of the bars (original figure by authors).

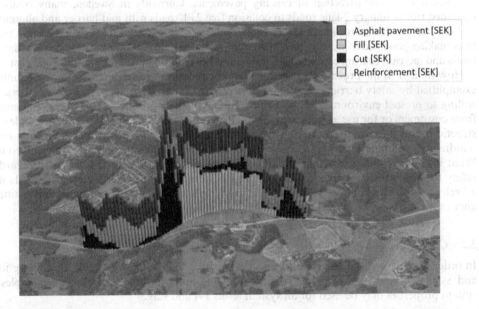

Figure 5. Costs for pavement, cut, fill, and reinforcement visualized in different colors (here in grey shades) and bar heights along a stretch of road (original figure by authors).

A multi-criteria analysis can be added, taking a number of perspectives other than construction costs or GHG into account. These include conservation areas such as nature and cultural heritages, real estates, terrain, soil, business interests, etc. Information sheets are added as layers in GIS, and by giving each the same value (1–10) or individual values, one can highlight potential conflicts or evaluate what impacts different interests have.

Geokalkyl provides many calculation possibilities, but the results should primarily be used as guidance for optimization purposes. Geokalkyl complies with system level 3.

Tools for overall LCC and pavement analysis on system level 3 and 4 regarding LCA and LCC have been developed in various spreadsheet applications. However, these tools have not been extensively developed and used for reasons described below.

3 DEFINING DECISION-MAKING CHALLENGES AND NEEDS

3.1 *Examples from practice*

Typical questions that arise early in the planning stage (system levels 2 and 3) where life cycle aspects need to be considered are:

– horizontal and vertical road alignment (geotechnics, water and drainage, external effects such as noise, volume distribution, and transport)
– cross-sectional geometry (type of mid-barrier, width of lanes, and other parts, widening)
– bearing capacity (condition of existing road, existing tar-containing asphalt, traffic during construction)

All of the above issues include many life cycle aspects and have long-term consequences for, not only, pavement performance and maintenance. The three above issues are important for several aspects and technical areas. Recent project experiences highlight the need to early identify life cycle aspects and weigh these different aspects against each other. Sometimes on a fairly technical level (system level 3). Before proceeding to the next stage in planning and design.

The design stage (system levels 3 and 4) involve the assessment of for example design of rehabilitation or reconstruction of existing pavements. Currently in Sweden, many roads are upgraded from ordinary 2-lane roads to collision free 2+1-roads with mid-barrier and alternating 1 and 2 lanes in each direction. In these projects there is a great need to assess different alternatives making good use of existing pavement and its material, and propose solutions to widen the roads and increase bearing capacity with a minimum of environmental impact and total cost.

How to later in the process set effective GHG or procurement requirements is a dilemma as exemplified by safety barriers. Manufacturers of road safety equipment, as well as others, are willing to present environmental footprints (EPDs) of their products. This is due to demand from customers or for use as sales arguments. Road safety barriers might look like simple constructions, but they are made up of a vast number of designs and components, in total many hundreds of combinations. Which system boundaries and functional units are relevant to use? What if GHG requirements conflict with safety requirements? Is there a risk with regard to safety that a low-GHG barrier per kilometer rules out a high-GHG concrete barrier? Is it at all relevant to compare "per kilometer"? Moreover, how can installation work and maintenance in the use stage be taken into account?

3.2 *Common system definitions*

In order to make well-balanced decisions, it is essential that comparable models, assumptions, and system boundaries be used. To illustrate this importance, the following examples of system properties may be used for all system levels 1-4 and stages:

– Definitions of alternative actions to analyze and compare in order to select the optimum alternative. The definitions must be suitable for each life cycle aspect and different steps in the project process and not cause unnecessary constraints later in the process or contradict earlier decisions in the project process.
– The time period to include in the calculation of life cycle-related costs and effects (such as climate impacts).
– Assumptions on maintenance activities and their results over time.
– Assumptions regarding important uncertainties and variations that will influence life cycle aspects, such as access to data and quality of data.

Shortcomings in our abilities to correctly predict future consequences for road managers, road users, and the environment need to be handled. In our experience, an LCA-based approach is easier to apply to systems with well-known products and designs (level 4), while it is more difficult to model on aggregated system levels (1–3). Costs, on the other hand, can often be estimated based on procurement data that include a package of different actions on a small road network, which can be used to model costs on levels 1–3, while more detailed costs (level 4) cannot be extracted from collected data. Other important effects are sometimes impossible to quantify or even describe. This create a challenge because there are limited benefits in estimating one consequence with high accuracy if the other consequences of importance cannot be estimated or can only be described in qualitative terms. Currently this is handled by careful separation with respect to accuracy and requirements in transparent documentation, which allows the decision maker to assess all aspects and to use their judgment to balance these aspects. A common view on system definitions is believed to facilitate this process. Another important aspect to future consequences is the fact that experience taught us that conditions and expectations will change in the future. The ability to handle future changes in traffic, climate, safety requirements, legislations, etc. will be of great importance to life cycle consequences.

3.3 Measureable parameters and indicators

When using parameters as requirements in procurement, it is obviously necessary that the requirements can be verified. However, it might not be as obvious that the adequacy of decisions made during the whole project process and the regulatory framework should also be verified. This is essential for creating feedback in the project process. In practice, when using the system-level view of the project process feedback in objective terms sometimes poses a challenge. An example is given to illustrate this. At the network level (level 1), system planners mainly deal with terms such as accessibility, reliability, safety, or environmental impact. At the component system level, the engineers and regulatory framework mainly deal with design issues to fulfill design criteria in order to achieve expected functional properties. To achieve a minimum environmental impact and low total costs, it is well known that activities in the production stage and having a long service life are key factors. This means that decisions in the planning and design stages (levels 2-4) need to be fixed by requirements that can be verified later on in the process, preferably by quantifiable parameters.

Verification of energy use and GHG emissions during construction and the entire service life is a challenging task. EPD:s handle some aspects during the construction phase but addressing transport and equipment as well as the entire service life will need further work involving both the use of Quality Management Systems and predictive models.

3.4 Empirical data and knowledge

Following the above discussion on measureable parameters, it can be concluded that pavement engineers foresee a very promising future with new abilities to make measurements in the field and to document actions made during production and maintenance. Consequently, the empirical evidence will increase significantly in the future. Empirical evidence alone may not be very useful, without the benefits of fundamental understanding gained by decades of research on pavement materials and design-related topics. The STA has a long tradition of collecting pavement surface data, pavement maintenance actions, and utilization of data in PMS (Pavement Management System) and procurement. Some examples of new abilities gained by the STA are (1) a system that collects information on all repair actions, (2) a system for asset management with information on all assets including the unbound pavement layers and geotechnical information, and (3) a drainage-related management system.

The benefits of an advancement in data collection and knowledge of pavement performance and deterioration might not be shown by a more precise prediction of asphalt thickness to reach a 20-year design life. It might be useful to avoid creating vulnerable sections, to create resource-efficient solutions, and to make better use of existing conditions along the road, including both existing new sections. In our experience, it is often a minor part of the pavement sections that triggers maintenance and the actual reason behind road deterioration is often poorly understood. In order to advance practice in this respect, a number of abilities need improvements:

– Prediction of adequate properties and performance of the subgrade layer during seasonal changes and drainage conditions (including topology).
– A closer link between design parameters, requirements, testing, and field measurements (including network and object surveys) that better reflect functional requirements or deterioration mechanisms.

As an example of the latter inability, FWD data and criteria related to asphalt and concrete cracking are used in the design phase, but no data are currently collected on a regular basis in the use phase. However, this will soon be changed since both deflection and crack data will be collected but still there is a long journey to create useful parameters and models based on these.

4 CONCLUSIONS

There are a number of more or less well-known aspects that influence life cycle assessments and their results. The goal and scope differ between different analyses and there is a variety in the functional units used. Currently, GHG impact estimations and total costs comprise different number of components and level of detail. In addition to this, we have a set of standards that are open for choices and interpretations. In practice, and for the true life cycle perspective, the system boundaries and choice of life cycle stages to be included are the most important. However, as long as the interpretation of the use phase is fraught with a number of weaknesses, life cycle assessments might give a distorted picture.

It is also essential that future projects and management systems are organized in such a manner that information from later stages in the project process/sub system levels can give feedback. For example, to validate previous decisions or agreements as well as to provide feedback to improve decisions or design models.

REFERENCES

EN 15643-5 Sustainability of construction works – Sustainability of buildings and civil engineering works – Part 5 Framework on specific lprinciples and requirement for civil engineering works. *CEN 2017.*

Lindgren, Å., Friberg, F. 2019. Management Methods of Infrastructure Projects Towards Sustainability Swedish Transport Administration, Dept of Technology and Environment, Sweden. *Paper presented at PIARC World Congress 2019.*

Sowerby, C. et al. 2014. SUNRA - a sustainability rating system framework for National Road Administrations. *Paper presented at the Transport Research Arena 2014, Paris, France.*

STA, Swedish Transport Administration 2018. Geokalkyl. Available online (in Swedish): https://www.trafikverket.se/tjanster/system-och-verktyg/Prognos–och-analysverktyg/geokalkyl/

STA, Swedish Transport Administration 2019. Klimatkalkyl. Available online (in Swedish): https://www.trafikverket.se/tjanster/system-och-verktyg/Prognos–och-analysverktyg/Klimatkalkyl/.

Combined life cycle assessment and life cycle cost analysis for the Illinois Tollway

Egemen Okte
Department of Civil and Environmental Engineering, University of Illinois at Urbana–Champaign, USA

Imad L. Al-Qadi
Department of Civil and Environmental Engineering, Bliss Professor of Engineering, Illinois Center for Transportation, University of Illinois at Urbana–Champaign, USA

Hasan Ozer
School of Sustainable Engineering and Built Environment, Arizona State University, USA

ABSTRACT: Life cycle assessment (LCA) and life cycle cost analysis (LCCA) are the main pillars of pavement sustainability. Even though they are usually considered separately, both LCA and LCCA are needed for decision making. This study uses an Illinois-developed LCA/LCCA tool to analyze two sections on the Illinois Tollway. The first scenario is a 12.8-mi-long section of the Tri-State Tollway in Chicago, Illinois. The initial year of the contract is 2001, and the analysis period is 60 years. The second scenario is a 4.7-mi-long section of the Ronald Reagan Tollway in Chicago. The initial year of the contract is 2004, and the analysis period is 78 years. Both emissions and costs (agency and user) during the analysis period are considered. A sensitivity analysis was conducted to assess the effect of changing traffic, international roughness index (IRI) progression rate and discount rate. Both scenarios showed that more than 70% of the emissions are due to fuel consumption; and pavement roughness is one of the key factors that controls costs and emissions.

1 INTRODUCTION

The Illinois Center for Transportation has recently developed a user-friendly LCA/LCCA tool. The LCA analysis module development in 2014. Several Illinois Tollway case studies were analyzed (Al-Qadi et al. 2015). In 2018, an LCCA add-on was developed for the tool, making the tool capable of conducting both LCA and LCCA (Okte et al. 2019). In this study, the LCA/LCCA tool is used to analyze two sections of the Illinois Tollway, reporting both LCA and LCCA results as well as a sensitivity analysis. Please note that the study is not meant to be comparative. It presents two different case studies.

2 METHODOLOGY

2.1 *System boundary*

The scope of the study covers the material, construction, maintenance and rehabilitation, use and end-of-life phases of the pavement life cycle. A cutoff approach is used to allocate environmental impacts associated with recycled and virgin materials. For example, the impacts of using recycled asphalt pavement (RAP) and recycled asphalt shingles (RAS) are derived only from the processing and handling of the recycled material after removal from the source. Therefore, the burdens from producing, removing and transporting the original material (i.e., RAP and RAS) are attributed solely to the virgin material. For end of life, it is assumed that

95% of asphalt is recycled offsite and 5% is transported to a landfill. Also, 45% of concrete is recycled offsite, 50% is reused onsite and 5% is transported to a landfill.

Agency costs, user-delay costs and vehicle-operating costs are considered for LCCA. Crash costs and emission costs are not considered in this study due to lack of data availability.

2.2 Functional unit

The functional unit is a reference unit to which the inputs and outputs of the LCA are normalized. The functional unit selected for this study is vehicle miles travelled (VMT). Therefore, VMT is reported alongside the results.

2.3 Development of LCI and cost databases

For materials and equipment operation, reported values in literature were used and modelled using SimaPro 7.3.3 (PRé Consultants 2012). The use-stage models were developed using MOVES software and regression analysis (Ziyadi et al. 2018). For life cycle costing, FHWA LCCA guidelines were followed (Walls and Smith 1998). The unit cost values and consumer price index were obtained from the Bureau of Labor Statistics website. Crash rate and emission cost calculations were adopted from Mallela and Sadasivam (2011). Details for development of databases and models used in the analysis are discussed somewhere else (Al-Qadi et al. 2015; Okte et al. 2019).

2.4 Development of the LCA/LCCA tool

Preliminary work on LCA has been documented in studies by Yang et al. (2014); Kang (2013); and Kang et al. (2014). After the LCCA add-on development, Illinois Tollway scenarios were analyzed and the results are presented herein.

3 PROJECT SCENARIO ANALYSES

3.1 Scenario 1: Tri-State tollway

The first scenario study is a 12.8-mi-long section of the Tri-State Tollway in Chicago, Illinois, between mileposts 12.3 to 30.1. There are four lanes in each direction; the southbound direction is analyzed in this study. It has an annual daily traffic of 70,968 vehicles with 17.1% trucks measured in 2013. The project description for this scenario is "Roadway partial resurfacing and rehabilitation (3-in HMA overlay over four lanes)."

The structure of the roadway is a 12 in stone matrix asphalt (SMA) overlaid jointed plain concrete pavement (JPCP) with 12-ft-long lanes. The construction year is 2001, and there are four overlays assumed on years 15, 28, 39 and 48. The analysis period is 58 years. The details of the work-zone configurations are gathered from contractor documents. The Illinois Tollway has reported that it takes 20 hrs to complete one mile of maintenance activity (Ghosh et al. 2018). Table 1 reports the results of the analysis. The functional unit is 21,032 million VMT.

Figure 1 illustrates the contribution percentage of each stage to global warming potential (GWP) and total primary energy for LCA. As expected with high-volume sections, around 72% of GWP contributions are coming from the use stage. The materials and construction stage is second with 26.2%.

Figure 2-A reports that the LCCA user cost, which is 71% vehicle-operating cost and 29% vehicle-delay cost. As Figure 2-B illustrates, the user cost of the project follows the same trend as the predicted international roughness index (IRI) progression of the section. There are small deviations at the year of maintenance because work-zone activities induce additional vehicle-operating costs. For example, at years 39 and 48, the vehicle operating costs are higher than years 40 and 49 respectively. This is due to additional VOC during work zone activities. VOC is composed of fuel consumption, tire wear-tear and vehicle repair cost. Also, reported VOC is additional or excess VOC compared to a perfect pavement with IRI of 40 in/mile.

Table 1. Base results of scenario 1.

Tabulated Results (per project-analysis period)	Entire Project	Materials & Construction	Maintenance & Rehabilitation	Use	End of Life
Present Agency Cost ($1000)	46,102	$3,352	$42,750	$0	$0
Present User Cost ($1000)	16,502	$0	$0	$16,502	$0
Global Warming Potential (tonnes CO2e)	1.79E+05	3.02E+03	4.69E+04	1.28E+05	6.11E+02
Total Primary Energy (GJ)	2.92E+06	1.16E+05	9.86E+05	1.81E+06	7.50E+03
Primary Energy as Fuel (GJ)	2.39E+06	4.30E+04	5.24E+05	1.81E+06	7.50E+03

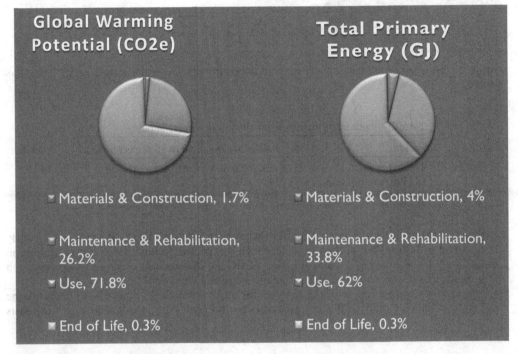

Figure 1. Percentages of global warming potential (A) and total primary energy (B).

For each scenario, sensitivity analysis is conducted by changing the average daily traffic, IRI progression rate and discount rate. The values for the sensitivity analysis are chosen to illustrate the sensitivity of VOC and delay cost to different parameters. Table 2 summarizes the cases considered.

3.2 Scenario 1: Sensitivity analysis

Figure 3-A presents the impact of average daily traffic on GWP. As would be expected, there is a semi-linear proportional relationship between traffic level and GWP for the use stage; vehicle emission during the use stage has a linear function with traffic. This relationship suggests that the impact of additional work-zone emissions is not significant enough to disrupt the linear trend. Additionally, because use stage is 72% of the GWP emissions, a change in traffic would impact the total emissions of the project.

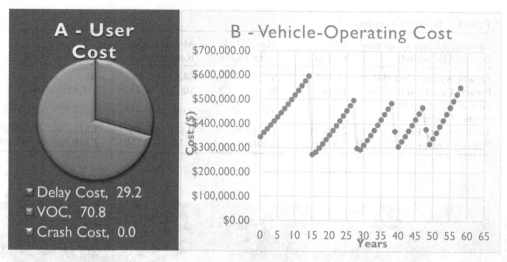

Figure 2.   Distribution of user costs (A) and vehicle-operating costs over the analysis period in real dollars (B).

Table 2.   Parameters for sensitivity analysis (base case in bold).

Factors	Scenarios Compared (with respect to Base)
Traffic Levels	115%, 100%, 85%, 70%
IRI Progression Rate	125%, 100%, 75%
Discount Rate	3%, 4%, 5%

Figure 3-B illustrates the impact of IRI progression rate on GWP. IRI rate has a nearly linear trend with the GWP of the use stage, which is expected because the energy consumption model used in this study is a linear function of IRI. A 25% reduction in the IRI progression rate results in a 7% decrease in GWP of the use stage and vice versa.

Figure 4 presents the results of the analyses for life cycle costs. As Figure 4-A illustrates, the relationship between daily traffic and user costs is not linear. Although vehicle-operating costs

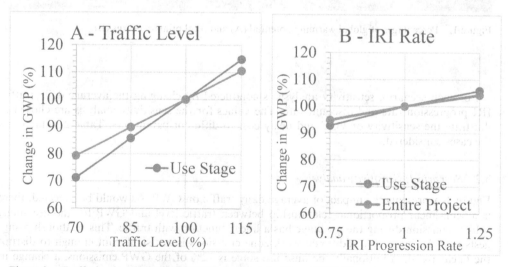

Figure 3.   Traffic level vs GWP (A) and IRI progression rate vs GWP (B).

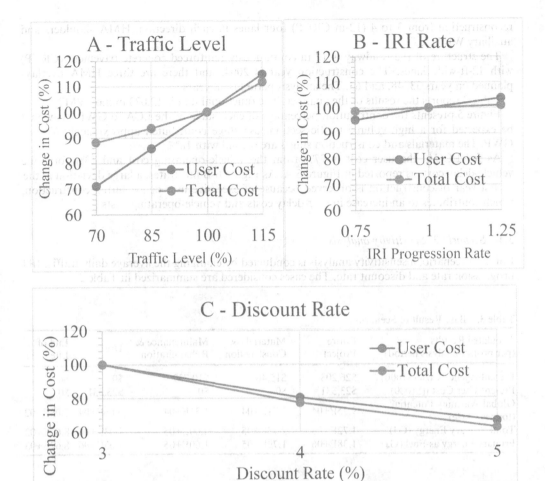

Figure 4. Traffic level vs cost (A), IRI progression rate vs cost (B) and discount rate vs cost (C).

are changing linearly with daily traffic, delay costs are dependent on traffic conditions and queue formation. In this case, a 15% increase in traffic results in an almost 45% increase in user cost, and a 15% decrease in daily traffic decreases the user cost by 25%.

A similar trend is observed with the total project cost. Because agency cost is not affected by change in traffic level, the impact of traffic on overall project cost is lower. As shown in Figure 4-B, there is a linear relationship between user costs and IRI progression rate because only

vehicle-operating costs are affected by the progression rate. The IRI progression rate has a similar trend to cost as that to GWP. Finally, Figure 4-C shows the impact of discount rate on life cycle costs. As would be expected, an increase in discount rate will decrease life cycle costs. Total project costs are slightly more affected by the discount rate because agency costs only incur every approximately 15 years; while user costs incur each year.

3.3 Scenario 2: Ronald Reagan Memorial Tollway

The second scenario is a 4.6-mi-long section of Ronald Reagan Memorial Tollway in Chicago, Illinois, between mileposts 122.9 to 127.5. There are four lanes in each direction, and the east-bound direction is analyzed for this study. It had an annual daily traffic of 67,005 vehicles with 9.1% trucks measured in 2013. The project description is "Roadway widening and reconstruction from 3 to 4 (12-in CRCP) four lanes in each direction, HMA shoulders and auxiliary WB lane."

The structure of the roadway is 12 in continuously reinforced concrete pavement (CRCP) with 12-ft-wide lanes. The construction year is 2004, and there are three HMA overlays planned on years 33, 48, and 63. The analysis period is 78 years.

Table 3 reports the results of the analyses. The functional unit is 12,073 million VMT.

Figure 5 presents the contribution percentage of each stage of the LCA to GWP. As would be expected for a high-volume traffic section, use stage contributed approximately 68% of GWP. The materials and construction stages are second with 18%.

As for LCCA, the user cost is 77% from the vehicle-operating cost and 23% form the vehicle-delay cost, as reported in Figure 6-A. As Figure 6-B illustrates, a large deviation in the first year of construction is observed because of excessive user delay during construction, which contributes to an increase in user-delay costs and vehicle-operating costs.

3.4 Scenario 2: Sensitivity analysis

For each scenario, a sensitivity analysis is conducted by changing the average daily traffic, IRI progression rate and discount rate. The cases considered are summarized in Table 2.

Table 3.   Base Result of Scenario 2.

Tabulated Results (per project-analysis period)	Entire Project	Materials & Construction	Maintenance & Rehabilitation	Use	End of Life
Present Agency Cost ($1000)	$26,203	$15,463	$10,739	$0	$0
Present User Cost (($1000)	$23,331	$0	$0	$23,331	$0
Global Warming Potential (tonnes CO2e)	9.75E+04	1.71E+04	1.31E+04	6.66E+04	7.08E+02
Total Primary Energy (GJ)	1.72E+06	2.75E+05	4.00E+05	1.03E+06	8.67E+03
Primary Energy as Fuel (GJ)	1.38E+06	1.79E+05	1.60E+05	1.03E+06	8.67E+03

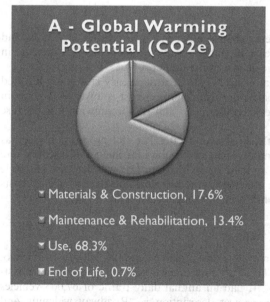

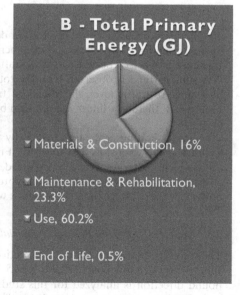

Figure 5.   Percentages of global warming potential (A) and total primary energy (B).

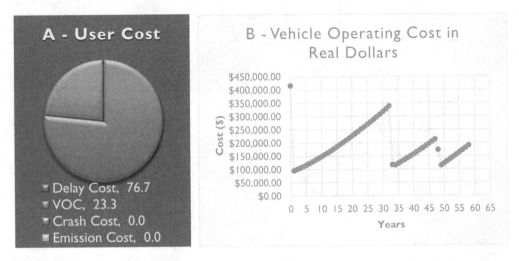

Figure 6. Distribution of user costs (A) and vehicle-operating costs over the analysis period in real dollars (B).

Figure 7-A presents the impact of average daily traffic on GWP, which showed a linear relationship because vehicle emission during the use stage has a linear function with traffic. A change in traffic would impact the total emissions of the project because use stage is 68% of the GWP emissions.

Figure 7-B illustrates the linear relationship impact of IRI progression rate on GWP; the energy consumption model used in this study is a linear function of IRI. A 25% reduction in the IRI progression rate results in a 7% decrease in GWP of the use stage and vice versa.

Figure 8 presents the results of the analyses for the life cycle costs. Figure 8-A illustrates that the relationship between daily traffic and user costs, which shows the attribution of queue formation and traffic delay. However, because there is already queue formation in the base traffic level, the 15% increase in traffic does not increase the delay cost exponentially. However, a 15%

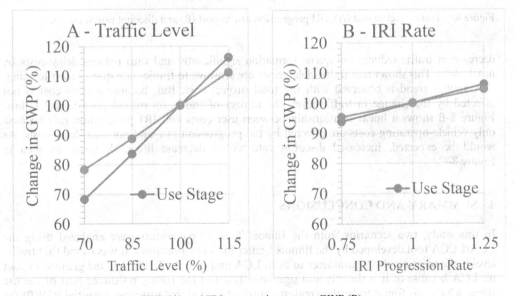

Figure 7. Traffic level vs GWP (A) and IRI progression rate vs GWP (B).

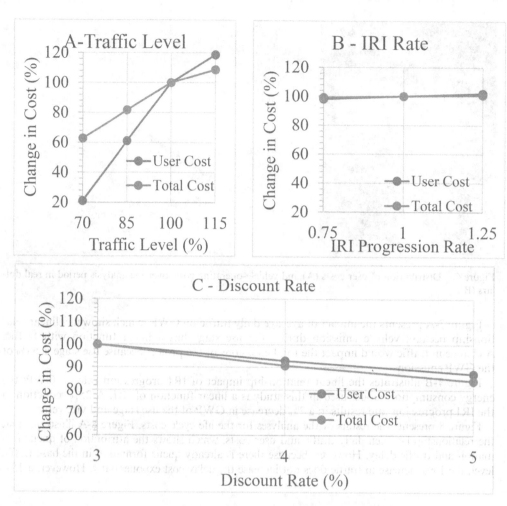

Figure 8.   Traffic level vs cost (A), IRI progression rate vs cost (B) and discount rate vs cost (C).

decrease in traffic reduces the queue formation significantly and thus reduces delay costs by nearly 40%. This shows that traffic-delay costs are sensitive to traffic once queues start forming.

A similar trend is observed with the total project cost. But, because agency cost is not affected by the change in traffic level, the impact of traffic on overall project cost is low. Figure 8-B shows a linear relationship between user costs and IRI progression rate because only vehicle-operating costs are affected by the progression rate and not user-delay costs. As would be expected, increased discount rate would decrease life cycle costs, as seen in Figure 8-C.

4   SUMMARY AND CONCLUSIONS

In this study, two scenarios from the Illinois Tollway pavements were analyzed using the LCA/LCCA tool developed by the Illinois Center for Transportation. It was found that traffic level is the most sensitive parameter to both LCA and LCCA. Traffic has the greatest impact on LCA because of it is directly and significantly affect the rolling resistance part of the use stage, which was found to be the greatest contributor to global warming potential (GWP) for highly trafficked highways. In addition, traffic conditions highly affect the user-delay costs once a queue starts forming; hence, impacting the LCCA.

REFERENCES

Al-Qadi, I.L., Yang, R., Kang, S., Ozer, H., Ferrebee, E., Roesler, J.R., Salinas, A., Meijer, J., Vavrik, W.R., & Gillen, S.L. 2015. Scenarios developed for improved sustainability of Illinois Tollway: Life-cycle assessment approach. *Transportation Research Record: Journal of the Transportation Research Board* 2523 (1): 11–18.

Ghosh, L., Abdelmohsen, A., El-Rayes, K., & Ouyang, Y. 2018. Temporary traffic control strategy optimization for urban freeways. *Transportation Research Record: Journal of the Transportation Research Board* 2672 (16): 68–78.

Kang, S.G. 2013. The development of a regional inventory database for the material phase of the pavement life-cycle with updated vehicle emission factors using MOVES. Master's Thesis.

Kang, S., Yang, R., Ozer, H., & Al-Qadi, I.L. 2014. Life-cycle greenhouse gases and energy consumption for material and construction phases of pavement with traffic delay. *Transportation Research Record: Journal of the Transportation Research Board* 2428 (1): 27–34.

Mallela, J., & Sadasivam, S. 2011. *Work zone road user costs, concepts and applications*. Publication No. FHWA-HOP-12-005. Washington, DC: Federal Highway Administration.

Okte, E., Al-Qadi, I.L., & Ozer, H. 2019. Effects of pavement condition on LCCA user costs. *Transportation Research Record: Journal of the Transportation Research Board* 2673 (5): 339–350.

PRé Consultants. 2012. SimaPro, Version 7.3.3. Amersfoort, Netherlands.

Walls, J., & Smith, M. 1998a. *Life-cycle cost analysis in pavement design*. Publication No. FHWA-SA-98-079. Washington, DC: Federal Highway Administration.

Walls, J., & Smith, M. 1998b. *Life-cycle cost analysis in pavement design—Interim technical bulletin*. Publication No. FHWA-SA-98-079. Washington, DC: Federal Highway Administration.

Yang, R., Ozer, H., Kang, S., & Al-Qadi, I.L. 2014. Life cycle impact of producing bituminous mixes with varying degrees of recycled asphalt materials. In *Proc., International Symposium on Pavement LCA*: 14–16.

Ziyadi, M., Ozer, H., Kang, S., & Al-Qadi, I.L. 2018. Vehicle energy consumption and an environmental impact calculation model for the transportation infrastructure systems. *Journal of Cleaner Production* 174: 424–436.

eLCAP: A web application for environmental life cycle assessment of pavements focused on California

John T. Harvey*, Jon Lea, Arash Saboori, Maryam Ostovar & Ali Azhar Butt
Department of Civil and Environmental Engineering, University of California Pavement Research Center, University of California Davis, USA

ABSTRACT: This paper introduces eLCAP, a standalone LCA software for quantification of the environmental impacts of pavement projects. eLCAP is a web-based tool for conducting life cycle assessment for pavement using the UCPRC life cycle inventory database and aims to serve a wide variety of users: academia, government agencies, and contractors. It is focused on project level decision making and comparison among design alternatives. eLCAP is integrated with PaveM (Caltrans pavement management system), the Caltrans version of the RealCost life cycle cost analysis tool, CA4PRS (to simulate construction, congestion, and user delays), CalME (pavement design tool for asphalt pavement), and Caltrans' implementation of PavementME (pavement design tool for concrete pavement) through use of same traffic databases, pavement performance models (where applicable), definitions, and other data. This provides Caltrans with a comprehensive suite of tools for improved design and management of their network.

1 INTRODUCTION

Caltrans has a growing need to be able to quantify its greenhouse gas (GHG) emissions and other environmental impacts of pavement operations, and to consider GHG and other impacts in pavement management, conceptual design, design, materials selection, and construction decisions. Caltrans also faces a number of policy issues that require similar types of analyses, all of which can be performed using LCA.

Caltrans has a need for an LCA tool that models the details of the construction and maintenance life cycle of a pavement project when a user needs environmental impact results and has sufficient input data for the project. These data are typically available at approximately the 30 percent design completed stage, where lengths and widths are known and pavement structures, materials and construction logistics, and traffic handling, are being considered and designed. eLCAP can also be used to compare "typical" structures from past projects earlier in the process when alternatives choices are being considered and first-order estimates of infrastructure emissions are being considered along with other sources of emissions. eLCAP is also intended for use when considering new specifications and policies such as for construction and materials quality and properties specifications, recycling and reclamation policies and specifications, or evaluating new technologies. It is also intended for use in developing typical emissions factors for materials, construction and transportation of pavement preservation, maintenance, and reconstruction treatments and new pavements for use in the Caltrans pavement management system.

All of these uses point to the need for a project-level LCA tool that uses LCIs specific to the materials and equipment typically used in California and by Caltrans, and that can be easily updated as LCIs are improved and new treatments, materials, and practices are added; eLCAP fills this need. eLCAP is being designed so that it can also produce concept-level evaluations with California-specific data, and include roadway and drainage features, landscaping, and standard highway overcrossing bridges in the future.

A separate interface is being developed for eLCAP for use by cities and counties that does not rely on the Caltrans location reference system and traffic databases.

2 SCOPE OF ELCAP

eLCAP models the life cycle history of a pavement project by allowing a user to specify any number of construction-type events, occurring at a user-specified date, followed by an automatically generated Use Stage event that begins immediately afterward and lasts until the next construction-type event or the End-of-Life (EOL) date. It covers all treatments and processes used by Caltrans.

Many agencies use Caltrans specifications, particularly in the northern part of the state. LCIs will be developed for materials uniquely included in the Southern California Greenbook (Public Works, 2018) in the future.

Figure 1 shows the scope of eLCAP software.

eLCAP computes 18 different impact category values, among them, Global Warming Potential (GWP), Human Health Particulate Air, Acidification, Primary Renewable Energy, etc.

Users interact with eLCAP via a web browser that accesses its user interface (UI). The main UI web page contains the controls necessary to define the life cycle of a pavement project: Construction, Maintenance/Rehab, Materials, Transport and Equipment. Data for a pavement project is grouped into a project trial; there can be an unlimited number of project trials for a project, and a user can have an unlimited number of projects. All user data is stored in a database, currently SQL Server.

In addition, a user can save the data for a project trial to a local hard disk in a "json"-formatted file. These downloaded files can act as a backup to the user database or as project documentation; they can also be uploaded to eLCAP for processing.

3 ELCAP LCA DATABASE GENERATION

3.1 Processes for creating LCI

The plan for eLCAP is to use the best available data at reasonable cost. At this time, for materials and other input flows publicly developed datasets are sparse for some important pavement flows, and industry-produced data are just becoming more available. To fill gaps in these data sets, UCPRC has negotiated the purchase of selected curated and documented data from thinkstep's GaBi™ models (thinkstep 2016). These data are fixed at the time of purchase with no further updating from thinkstep until the next purchased download. There are also cases in which specific models for the state of California has been developed that are used in the eLCAP model, such as the GREET model (Wang 1999) for the impacts of the fuel cycle and vehicle cycle of different vehicle categories. The documentation report of the UCPRC LCI database covers all the details in this regard (Saboori et al. in press).

For LCI generated from GaBi, the left side of Figure 3 shows the processing procedure carried out to generate the eLCAP LCA database. The steps in the procedure are:

1. An LCA analyst builds a California-specific model in *GaBi* and exports flows for the main process in the model.
2. A separate software tool, DB Gen, is used to process the *GaBi*-exported CSV files; this software tool reads these files and generates *process and flow definition* files.
3. DB Gen is next used again to process the generated *process and flow definition* text files and also the manually generated *definition* and *flow* files.
4. The above steps result in an XML database file that eLCAP loads when the application starts up. It has a structure shown at the center of Figure 1.
5. *Model definition* files are created. All sources of LCA data are now available for eLCAP.
6. A user accesses eLCAP via a web browser. eLCAP reads the LCA XML database file into memory and also it reads and processes the model definition files into memory.

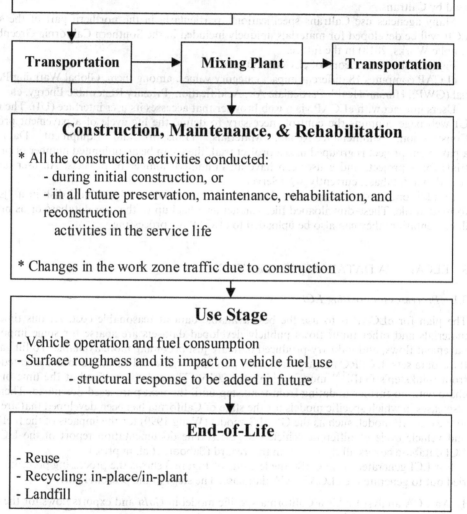

Material Production

- raw materials extraction
- transportation to plant
- plant processes

for a comprehensive list of all the construction materials used in pavement projects in California, including materials not currently widely used (such as reflective coatings)

| Transportation | → | Mixing Plant | → | Transportation |

Construction, Maintenance, & Rehabilitation

* All the construction activities conducted:
    - during initial construction, or
    - in all future preservation, maintenance, rehabilitation, and reconstruction
        activities in the service life

* Changes in the work zone traffic due to construction

Use Stage

- Vehicle operation and fuel consumption
- Surface roughness and its impact on vehicle fuel use
    - structural response to be added in future

End-of-Life

- Reuse
- Recycling: in-place/in-plant
- Landfill

Figure 1.   Scope of the eLCAP software.

7. The structure for the in-memory version of the XML database file mimics the structure of the XML file
8. The user requests that an analysis be performed. eLCAP builds the *balance model*, balances it, computes the LCI for the construction-type event, and then the impact factors for it.

The significant efforts by the pavement industry to produce national-average LCAs, product category rules (PCR), and the pilot program of Caltrans requiring environmental product

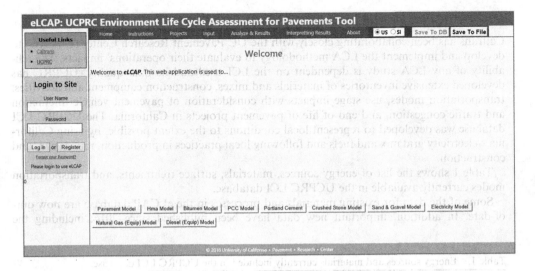

Figure 2.　The eLCAP homepage.

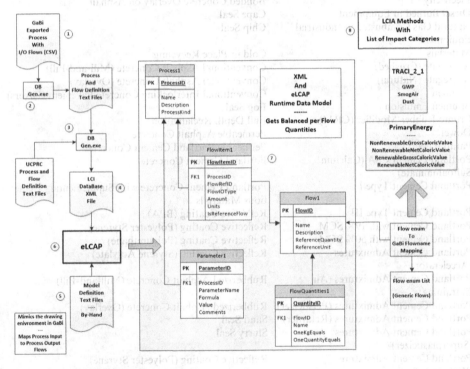

Figure 3.　Overall eLCAP data processing and operation.

declarations (EPD) for pavement materials (Caltrans website) are expected to provide timely and region-specific data for use in future versions of eLCAP.

267

Caltrans has been collaborating closely with the UC Pavement Research Center (UCPRC) to develop and implement the LCA methodology to evaluate their operations' impacts. The reliability of any LCA study is dependent on the LCI used and its data quality. UCPRC has developed extensive inventories of materials and mixes, construction equipment and activities, transportation modes, use stage impacts with consideration of pavement vehicle interaction and traffic congestion, and end of life of pavement projects in California. The UCPRC LCI database was developed to represent local conditions to the extent possible, by using California's electricity grid mix and fuels and following local practices in production, mix design, and construction.

Table 1 shows the list of energy sources, materials, surface treatments, and transportation modes currently available in the UCPRC LCI database.

Some of the data for existing materials and processes in the eLCAP database are now out-of-date. In addition, important new data have been published since 2016, including the

Table 1. Energy sources and materials currently included in the UCPRC LCI database.

Energy Sources	Surface Treatments
Electricity	Bonded Concrete Overlay on Asphalt
Diesel Burned in Equipment	Cape Seal
Natural Gas Combusted in Industrial Equipment	Chip Seal
Materials	Cold in-Place Recycling
Aggregate (Crushed)	Conventional Asphalt Concrete (Mill-and-Fill)
Aggregate (Natural)	Conventional Asphalt Concrete (Overlay)
Bitumen	Conventional Interlocking Concrete Pavement (Pavers)
Bitumen Emulsion	Fog Seal
Crumb Rubber Modifier (CRM)	Full Depth Reclamation
Dowel	Permeable Asphalt Concrete
Paraffin (Wax)	Permeable Portland Cement Concrete
Portland Cement CSA (Calcium Sulfoaluminate)	Portland Cement Concrete
Portland Cement Type I	Portland Cement Concrete with Supplementary Cementitious Materials
Portland Cement Type III	Reflective Coating (BPA)
Portland Cement with 19% SCM	Reflective Coating (Polyester Styrene)
Portland Cement with 50% SCM	Reflective Coating (Polyurethane)
Portland Cement Admixtures (Accelerator)	Reflective Coating (Styrene Acrylate)
Portland Cement Admixtures (Air Entraining)	Rubberized Asphalt Concrete (Mill-and-Fill)
Portland Cement Admixtures (Plasticizer)	Rubberized Asphalt Concrete (Overlay)
Portland Cement Admixtures (Retarder)	Sand Seal
Portland Cement Admixtures (Superplasticizer)	Slurry Seal
Portland Cement Admixtures (Waterproofing)	Reflective Coating (Polyester Styrene)
Quicklime	Portland Cement Concrete with Supplementary Cementitious Materials
Reclaimed Asphalt Pavement (RAP)	Transportation
Reflective Coating (BPA)	Barge Transport
Reflective Coating (Polyester Styrene)	Heavy Truck (24 Tonne)
Reflective Coating (Polyurethane)	Ocean Freighter
Reflective Coating (Styrene Acrylate)	
Styrene Butadiene Rubber (SBR)	
Tie Bar	

Asphalt Institute LCA (AI 2019), and new materials and treatments are being used in California, such as better inventories for thin (bonded) concrete overlays on asphalt (BCOA), mix designs reflecting the change from Hveem to Superpave mix design methods, new types of supplementary cementitious materials (SCM), and cold central plant recycling (CCPR). Other new materials and processes currently being evaluated by Caltrans and some in use by cities for which LCI have been developed since 2016 include asphalt mixes with reclaimed asphalt pavement (RAP) contents greater than 15 percent, dense graded asphalt mixes with five to 10 percent recycled tire rubber. The database is currently being updated with latest available data, changes in Caltrans materials specifications, new materials and processes, and updates to the electricity grid mix.

3.3 Critical review

The eLCAP database (Saboori et al. in press) was submitted for third-party verification in 2016 according to ISO requirements and was passed.

The updated inventory will be submitted for a new critical review in late 2019.

3.4 XML database file

eLCAP's LCA database is an XML file that is read into memory when the application starts up. This is done for performance reasons since access to it occurs frequently. The file is constructed as discussed in the previous section. The schema for the file, and the in-memory version of it, is shown in Figure 4. Data is accessed using the DAL, as usual, for any data source used by eLCAP.

3.5 Use stage modeling

Use Stage-type events, which are automatically generated for each user-defined construction-type event, have a start date immediately after the end of the construction event and an end date specified by the user. For the Use Stage the current version of eLCAP only models well-

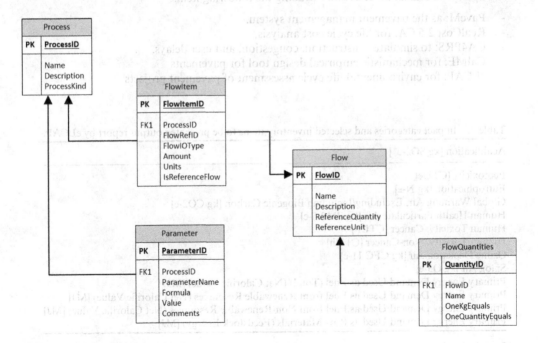

Figure 4. XML file schema.

269

to-wheel impacts of additional fuel use caused by pavement roughness as measured by the International Roughness Index (IRI). The tool models the environmental effects of "using" the pavement project by computing the greenhouse gases (GHG) from traffic (cars and trucks) driving over it during the time span of the Use Stage and including the effects of increasing IRI and traffic with time. Default IRI models from the Caltrans pavement management system are assigned to the treatment, and can be changed by the user, or user defined IRI model parameters can be input and saved. The IRI models will be updated as they are updated for the Caltrans PMS, approximately every three years.

Current research will be included in the next versions of eLCAP, starting in 2020. The impacts of pavement structural response will be included in a future version of eLCAP as current research results are finalized. Recently completed models for changes in traffic drive cycles from construction work zones will also be added to a future version. The full well-to-wheel impacts from vehicle use of the pavement, rather than just changes in fuel use from pavement-vehicle interaction, will be added in a future version of eLCAP to allow planners and policy-makers to compare infrastructure and vehicle use impacts as was done in the software PE-2 (Mukherjee & Cass 2012). This will include consideration of relative impacts of changes in vehicle fleet propulsion energy types and consumption.

3.6 *Life Cycle Impact Assessment (LCIA)*

eLCAP uses EPA TRACI 2.1 impact assessment methodology for reporting the LCIA results. Table 2 lists the main impact categories and life cycle inventory items that are presented in the reports generated by eLCAP.

4 INTEGRATION WITH OTHER PAVEMENT TOOLS

eLCAP is the latest tool in a suite of software that enables an integrated and comprehensive pavement management system that incorporates full life cycle thinking in terms of costs and environmental impacts (Figure 5 & Figure 6). All the software in this suite are web based so the updates are instantaneous and including the following items:

- PaveM: as the pavement management system,
- RealCost 2.5 CA: for life cycle cost analysis,
- CA4PRS: to simulate construction, congestion, and user delays,
- CalME: for mechanistic empirical design tool for pavements
- eLCAP: for environmental life cycle assessment of pavement projects

Table 2. Impact categories and selected inventory items to be presented output report by eLCAP.

Acidification [kg SO2-e*]
Ecotoxicity [CTUe]
Eutrophication [kg N-e]
Global Warming Air, Excluding/Including Biogenic Carbon [kg CO2-e]
Human Health Particulate Air [kg PM2.5-e]
Human Toxicity, Cancer [CTUh]
Human Toxicity, Non-Cancer [CTUh]
Ozone Depletion Air [kg CFC 11-e]
Smog Air [kg O3-e]
Primary Energy Demand Used as Fuel (Total) (Net Calorific Value) [MJ]
Primary Energy Demand Used as Fuel from Renewable Resources (Net Calorific Value) [MJ]
Primary Energy Demand Used as Fuel from Non-Renewable Resources (Net Calorific Value) [MJ]
Primary Energy Demand Used as Raw Materials (Feedstock Energy) [MJ]

* equivalent

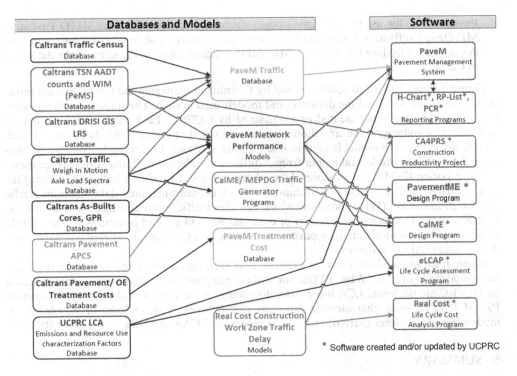

Figure 5.   Caltrans pavement related software and common databases and models.

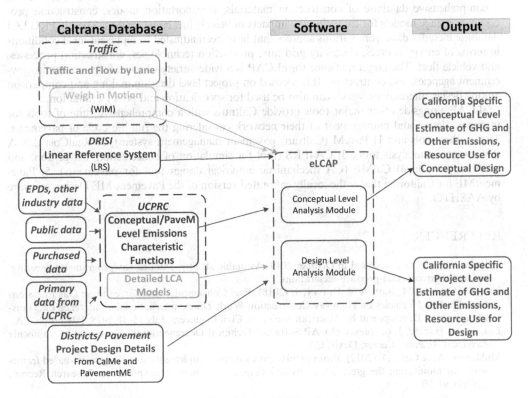

Figure 6.   Data for caltrans network and project level LCA software.

- PavementME for use in California (locally calibrated version of the AASHTO Pavement ME Design software). California traffic data are used as input to a traffic generator tool developed for feeding the PavementME and the climate data are the same as the data used in CalME.

Figure 5 shows the list of software used by Caltrans in managing the transportation infrastructure in California and the database and models used in each. The software with an asterisk in their title are either created or maintained by UCPRC. Each model can have one or multiple data sources which are shown by the arrows that connect the databases to each model. Similarly, each model is used in one or more software. For example, the PaveM Network Performance Models take data from 5 different databases: Caltrans performance measurement system (PeMS) which stores data related pavement condition indices, Caltrans GIS and linear referencing system (LRS) database which contains mapping data, Caltrans traffic and weigh in motion (WIM) database which contains traffic count and truck axle load data, Caltrans as-built maps and test results of core samples and ground penetrating radar (GPR) which provide structural details of pavement sections, and Caltrans automated pavement condition survey (APCS) database.

Figure 6 shows the data needs for network level (conceptual level) and project level (design level) analysis in eLCAP. While EPDs, public data, purchased data, and primary data collected by UCPRC are the initial LCA input data, depending on the type of analysis either conceptual PaveM level emission characteristic functions (for network level analysis) or detailed LCA models (for project-level analysis) are used as inputs to eLCAP alongside traffic and LRS data.

5 SUMMARY

This paper introduces eLCAP, a web-based tool for conducting environmental life cycle assessment focused on transportation infrastructure. eLCAP uses the UCPRC LCI database, which is a comprehensive database of construction materials, transportation modes, construction processes, use stage models for capturing the impacts of vehicle fuel consumption. The UCPRC LCI database includes data from various sources that have been adjusted to represent local conditions in terms of energy sources, electricity grid mix, production technologies, construction processes, and vehicle fleet. The target audience for eLCAP is a wide variety of users from academia, government agencies, and contractors. It is focused on project level decision making and comparison among design alternatives, which can also be used for specification and policy evaluation.

eLCAP alongside other major tools provide Caltrans with a comprehensive suite of tools for improved design and management of their network considering the full life cycle of pavements. These other tools are: 1) PaveM (Caltrans pavement management system), 2) RealCost 2.5CA (life cycle cost analysis tool), 3) CA4PRS (tool for simulation of construction, congestion, and user delays), and 4) CalME (CA mechanistic empirical design tool for pavements). 5) PavementME for California that is the locally calibrated version of the Pavement ME Design software by AASHTO.

REFERENCES

Caltrans Website. Last accessed in October 2019. Available at: https://dot.ca.gov/programs/engineering-services/environmental-product-declarations.

Kim, C., Ostovar, M. Butt, A. & Harvey, J. (2018). Fuel Consumption and Greenhouse Gas Emissions from On-Road Vehicles on Highway Construction Work Zones. International Conference on Transportation and Development by American Society of Civil Engineers. July 15–18,2018. Pittsburgh, PA.

Lea, J. and Harvey, J. (in press). eLCAP Software Technical Documentation. University of California Pavement Research Center, Davis, CA.

Mukherjee, A. & Cass, D. (2012). Project emissions estimator: implementation of a project-based framework for monitoring the greenhouse gas emissions of pavement. Transportation Research Record, 2282(1), 91–99.

Public Works Standards (2018). *The Greenbook: Standard Specifications for Public Works Construction*: BNi Building News, Vista, CA.

Saboori, A., Harvey, J., Li, J., Lea, J. Kendall, A., Wang, T. in press. Documentation of UCPRC Life Cycle Inventory used in CARB/Caltrans LBNL Heat Island Project and other LCA Studies, University of California Pavement Research Center, Technical Report, Davis, CA. thinkstep (2016). GaBi Software version 6.3. thinkstep International, Stuttgart, Germany. thinkstep (2019). Life Cycle Assessment of Asphalt Binder. Asphalt Institute, Lexington, KY. Available at: http://www.asphaltinstitute. org/engineering/lca-study-on-asphalt-binders/

Wang, M. Q. (1999). GREET 1.5-transportation fuel-cycle model-Vol. 1: methodology, development, use, and results (No. ANL/ESD-39 VOL. 1). Argonne National Lab., IL (US).

Pavement, Roadway, and Bridge Life Cycle Assessment 2020 – Harvey et al (eds)
© 2020 Taylor & Francis Group, London, ISBN 978-0-367-55166-7

Green Up pavement rehabilitation decision tool

D. Andrei & R.E. Kochan
California State Polytechnic University, Pomona, California, USA

ABSTRACT: Green Up is a pavement rehabilitation decision tool that uses symbols and colors to illustrate the sustainable aspects of proposed pavement rehabilitation strategies, primarily for the cradle-to-gate phase of the life cycle. In addition, Green Up provides designers with succinct educational content as well as links to outside resources about sustainable materials and pavement rehabilitation strategies.

1 INTRODUCTION

1.1 Background

When comparing pavement rehabilitation alternatives, decisions are most often based on cost, the corresponding life extension, and the experience or familiarity of the designer with the proposed rehabilitation solutions. Sustainability considerations are rarely included unless specifically required by the client or owner agency. Green Up was developed to address the need for a rapid and easy to use online tool to illustrate and compare the sustainable features of pavement rehabilitation alternatives. An initial version of Green Up was proposed by Andrei (2014). This paper describes a subsequent version of the methodology which was finalized in 2019 with support from the State of California's Senate Bill 1 "The Road Repair and Accountability Act of 2017" and administered by the California State University Transportation Consortium. A macro-enabled Excel spreadsheet was developed in this study and was used to generate the graphics and results presented in this paper. An online tool based on the methodology described here will be developed in the future.

1.2 The concept

Unlike other sustainability rating systems that award points for sustainable practices to calculate a final score – the higher the better - Green Up uses colored areas and symbols to convey the complex combination of factors that contribute to the overall sustainability of a solution. No total score is calculated, instead designers can visualize the differences between the proposed alternatives and identify potential areas of improvement. The end goal is to improve or "green up" the proposed solutions by becoming aware of their positive and negative aspects and by learning how the negative aspects can be reduced or eliminated.

Two Green Up graphics are presented in Figure 1, side by side. Both depict key sustainability aspects for one square foot of pavement – hence the "1 SF" label on the central white rhombus which symbolizes the unit area for comparison. The graphic on the left illustrates a mill-and-fill strategy where the top 4 inches of an existing asphalt concrete pavement are milled and then replaced with the same thickness of new hot mix asphalt concrete (HMA). The graphic on the right shows the alternative which consists of milling 2 inches of the surface, cold recycling in place the next 2 inches of pavement, and then adding a 2-inch HMA overlay.

For both scenarios, the estimated life extension is 10 years which dictates the color of the heart symbol. The other three symbols in the center white rhombus of the graphic represent surface permeability (the water drop), solar reflectivity (the sun), and noise (the speaker).

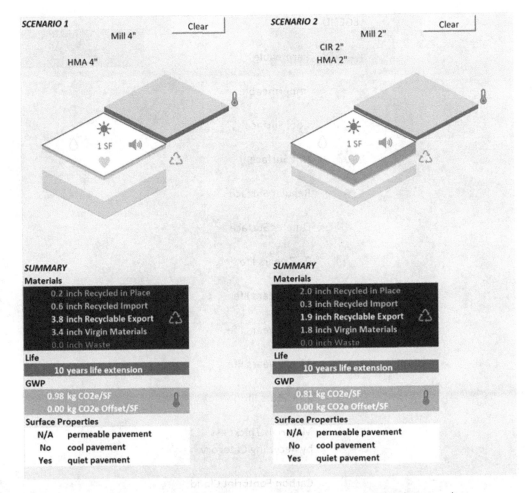

Figure 1.   Comparison of Green Up graphics and summaries for two rehabilitation scenarios.

Colors are also used to identify the relative quantities of virgin materials, recycled materials, and waste which are shown in the area below the surface, according to the following criteria:

- recycled in place, such as cold in place recycling: green
- recycled import, such as reclaimed asphalt pavement (RAP) or fly ash: light green
- recyclable export, such as RAP or recycled concrete aggregate (RCA): yellow
- virgin materials, such as virgin asphalt, portland cement, or virgin aggregate: orange
- waste, or materials that will be disposed of in a landfill: red

To produce the colored areas below the surface, quantities of materials in each category are estimated and then the corresponding thickness of material is calculated and shown in the form of colored layers below the surface, next to a green recycling symbol. A scale factor of 1:1 is used to determine the thickness of these layers relative to the dimensions of the white rhombus that represents the surface (12 by 12 inches). The full legend of symbols and colors is shown in Figure 2.

To the right of the white rhombus representing the surface, the 100-year global warming potential (GWP) of the strategy is depicted as a gray prism next to a red thermometer symbol. The height of the prism represents the quantity of CO_2 equivalent in kg/ft^2. A scale of 1 inch: 1 kg/ft^2 is used to determine the height of the GWP prism in the Green Up graphic.

Note that the cloud is representative only of the cradle-to-gate phase of the life cycle. The information used in Green Up was retrieved from currently available US-based Environmental Product Declarations (EPD) or life cycle assessments for portland cement (Portland Cement

Permeable

Impermeable

Cool Surface

Hot Surface

Regular Surface

Quieter Surface

0 - 7 years life

8 - 14 years life

15 - 22 years life

23 - 29 years life

30 + years life

Materials Thickness
by Recycling Category

Carbon Footprint Cloud

Not Applicable

Figure 2.   Legend for the Green Up graphic.

Association, 2016), asphalt cement (Asphalt Institute & thinkstep AG, 2019), portland cement concrete (National Ready Mixed Concrete Association, 2016), and hot mix asphalt concrete (Emerald Eco Label, 2019). These EPD's only account for the life cycle of the product up to the stage where the asphalt leaves the refinery, the cement leaves the cement plant, and the portland cement concrete (PCC) or HMA leave the mixing plant. Transportation from the plant to the job site, construction, service, future maintenance and rehabilitation, and ultimately disposal or recycling at the end of life are not included or reflected in the GWP cloud at this time.

GWP is easiest to measure and document over the cradle-to-gate stage of the life cycle. This may explain why most EPD's available are limited to the cradle-to-gate scenario. There is however growing evidence that increased vehicle emissions over the service life of a pavement - caused by surface roughness and rolling resistance - may in some cases exceed the cradle-to-gate GWP by more than one order of magnitude (Santero & Horvath, 2009). Accurately predicting traffic, roughness, and rolling resistance over the service life of a pavement structure remains a challenging task. As more data becomes available and more accurate predictive methodologies will be developed, this important contribution to GWP could be included in future versions of Green Up. Although roughness and rolling resistance are not included at this time, Green Up does consider some important measures of sustainability relevant to the service stage of the life cycle:

– life extension (heart symbol)
– surface permeability (water drop symbol)
– surface solar reflectivity (sun symbol)
– tire-pavement noise (speaker symbol)

As illustrated in Figure 2, the heart symbol can take five different colors, from red to green, the longer the life extension the closer to green. Surface properties on the other hand can only have three values: true (green), false (red), or not applicable (white). For example, a permeable pavement surface would not be recommended on a high traffic roadway where the permeable surface would not survive the loading. For the two scenarios illustrated in Figure 1, surface permeability was considered not applicable. The HMA surface was not solar reflective but it was quieter which explains the colors of the sun and speaker symbols.

2 GREEN UP INPUTS

2.1 *General project inputs*

The following general inputs are needed to produce the Green Up graphic:

– Life extension in years: users can directly input life extension in years. The information is used to give a different color to the heart symbol, as described earlier and in Figure 2. Long-life designs will generally result in lower maintenance needs and traffic closures over the life of the structure. Therefore the longer the life extension, the closer to green the corresponding color. A link to an online information page on long-life design is also provided. On the information page additional links are provided to resources such as Pavement Designer (PD Collective, 2019) or PaveXpress (Pavia Systems, 2019) which users can use to more accurately estimate time till the next major rehabilitation for a given strategy.
– Project area, in square feet.
– Carbon offsets: users can input carbon offsets in kg CO_2e (100-year global warming potential). Offsets are divided by the entire project area and the corresponding value in kg/ft^2 is shown as a thickness of green-colored GWP cloud (1 inch: 1 kg/ft^2). A link to an online information page on carbon offsets is provided for users interested in learning more about options to offset emissions.
– Permeable pavement: users can indicate whether permeable pavement would be beneficial and applicable for their specific project. When the entire pavement structure is not permeable or when the pavement is subject to heavy traffic, using a permeable surface may not be recommended. An online information page was developed to help users learn more about permeable pavements and solutions such as pervious concrete and porous asphalt.
– Heat island effect: users can indicate whether a cool, or solar-reflective surface would be beneficial for their project. The urban heat island effect may not be an issue in rural areas or for pavements that are shaded by trees or other structures. An online information page on cool pavements was developed to provide users with more information and links to other relevant online resources.
– Tire-pavement noise: users can indicate whether a quiet or quieter surface may be beneficial to their project. Noise may not be an issue on pavements that are far from residential areas or other situations where sound pollution is not an issue. An online information page was developed to provide users with more information on tire-pavement noise and links to other online resources.

2.2 Rehabilitation strategies

A limited number of rehabilitation treatments are currently available in Green Up:

– Jointed plain concrete pavement (JPCP)
– Hot mix asphalt (HMA)
– Removal/demolition of existing JPCP or HMA surface
– Rhombus grinding JPCP
– Cold planning/milling HMA
– Crack and seat JPCP
– Cold in-place recycling (CIR)
– Full depth reclamation (FDR) with cement or with asphalt emulsion

In addition, the following sustainable pavement strategies were also included:

– Permeable pavement (pervious concrete or porous asphalt)
– Cool pavement
– Quiet pavement
– Long-life pavement

More rehabilitation treatments and strategies are available and could be added to future versions of the Green Up tool. Due to the inherent time and budget constraints for this project, only the above treatments/strategies were included at this time. Detailed inputs for JPCP and HMA are described in the following sections. Inputs for the other strategies were not included due to the page limit for this paper.

2.3 Jointed plain concrete pavement inputs

Inputs and default values for JPCP are shown in Table 1. Note that all values can be adjusted by users, but these are the values that will be included by default.

Table 1. JPCP inputs and default values.

Input	Description	Default Value
Thickness	Average layer thickness, in inches	User Input
Surface or Base	Users select whether the concrete will serve as the surface of the pavement	Yes/No
Pervious	If a surface, users select whether the surface is permeable	Yes/No/NA
Cool	If a surface, users select whether the surface is cool	Yes/No/NA
Quiet	If a surface, users select whether the surface is quiet	Yes/No/NA
Portland Cement	Amount of portland cement, in lb/yd^3	445
Fly Ash	Amount of fly ash, in lb/yd^3	85
Slag Cement	Amount of granulated blast furnace slag (GGBFS), in lb/yd^3	0
Silica Fume	Amount of silica fume, in lb/yd^3	0
Virgin Aggregate	Amount of coarse and fine virgin aggregate, in lb/yd^3	3186
Recycled Aggregate	Amount of coarse and fine recycled aggregate, in lb/yd^3	0
EPD	Reference environmental product declaration	NRMCAEPD: 10003
Mix ID	Reference mix in the EPD	1412570
Declared Unit	Declared unit in the EPD	1m^3
GWP	Cradle-to-gate global warming potential in kg CO2e (100 years) per declared unit	462

In Green Up, JPCP will be assumed to be a cool surface by default due to the lighter color and higher albedo of regular concrete. Users that want to include pervious concrete will have the option to indicate that the surface is permeable, and they can adjust the mix design to reflect a pervious concrete mix. Virgin aggregates and cement will contribute to the total amount of virgin materials. Recycled aggregate, fly ash, slag, and silica fume will contribute to the recycled import category.

Figure 3 illustrates the contribution of the JPCP layer alone or in combination with rhombus grinding to the Green Up graphic. Life extension (represented by the color of the heart symbol) was not included in these scenarios since this would depend on the existing pavement structure, future traffic, etc. The first scenario is a 4-inch permeable and cool concrete surface for a parking lot application. The second scenario is for a 9-inch cool surface with 0.25-inch rhombus grinding. The last of the three scenarios is a repeat of the second scenario but it illustrates the effect of carbon offsets.

2.4 *Hot mix asphalt inputs*

Inputs and default values for HMA are shown in Table 2. Note that all values can be adjusted by users, but these are the values that will be included by default.

In Green Up, asphalt cement and virgin aggregate will contribute to the total amount of virgin materials reported. Crumb rubber and RAP will contribute to recycled import. A checkbox is included for warm mix technology - although checking the box will not influence the appearance of the Green Up graphic, users will be provided with a link to Warmmix-asphalt.org, a web page with information and resources about warm mix and its advantages created and maintained by the National Asphalt Pavement Association (NAPA, 2019).

Figure 4 Illustrates the contribution of the HMA layer alone to the Green Up graphic. The first scenario is a 3-inch porous asphalt surface for a parking lot application. The second scenario is a 6-inch dense graded surface. The last scenario is a quieter, 9-inch rubberized asphalt surface.

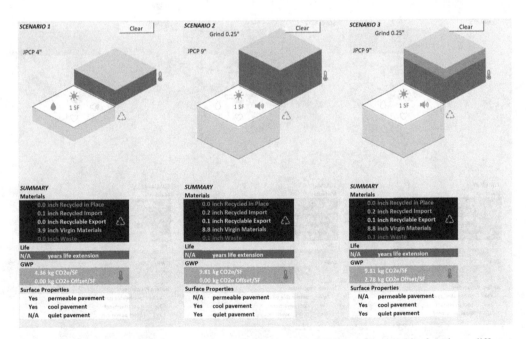

Figure 3. Contribution of the concrete surface alone to the green up graphic for three different scenarios.

Table 2. HMA inputs and default values.

Input	Description	Default Value
Thickness	Average layer thickness, in inches	User Input
Surface or Base	Users select whether the concrete will serve as the surface of the pavement	Yes/No
Pervious	If a surface, users select whether the surface is permeable	Yes/No/NA
Cool	If a surface, users select whether the surface is cool	Yes/No/NA
Quiet	If a surface, users select whether the surface is quiet	Yes/No/NA
Warm Mix	Users select whether warm mix technology will be used	Yes/No
Virgin Aggregate	Amount of virgin aggregate, % by weigh of total dry aggregate	85
RAP	Amount of recycled aggregate, % by weight of total dry aggregate	15
Asphalt Cement	Amount of asphalt cement, % by weight of HMA mix	5.5
Rubber	Amount of crumb rubber modifier, % by weight of asphalt binder	8
G_{mb}	Bulk specific gravity of the compacted HMA mix	2.35
EPD	Reference environmental product declaration	Emerald 19.55.52.9
Mix ID	Reference mix in the EPD	2018-TM-004D-VSS-64-10-R0
Declared Unit	Declared unit in the EPD	1 US Short Ton
GWP	Cradle-to-gate global warming potential in kg CO2 eq (100 years) per declared unit	38.5

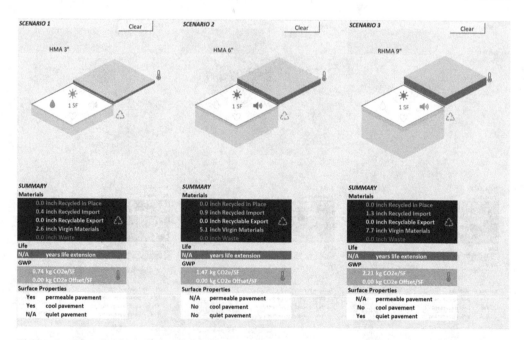

Figure 4. Contribution of the HMA surface alone to the Green Up graphic for three different scenarios.

3 INFORMATION PAGES

To assist users in entering the correct inputs, quick help balloons and links to other online resources are included in the user interface. Information pages have been developed for the following key terms and inputs in Green Up:

– Carbon offsets
– Urban heat island
– Long-life pavements
– Permeable pavements (pervious concrete and porous asphalt)
– Cool pavements
– Quiet pavements

These web pages can be accessed through links embedded in the user interface and are hosted on the website of the Pavement Recycling and Reclaiming Center at Cal Poly Pomona (PRRCenter.org, 2019). It is the intention of the authors to maintain and update the content of these pages in the future to keep the information current and relevant to Green Up users.

4 CASE STUDY

The City of Palmdale, California, considered two candidate strategies for the rehabilitation of Pearblossom Higway, a 4-lane major arterial in need of major rehabilitation or reconstruction. The existing pavement structure consists of asphalt concrete over granular base over sandy subgrade soils. A traffic index (TI) value of 13 or about 20 million equivalent single axle load (ESAL) was estimated for current and future traffic loading. The two design alternatives considered are shown in Figure 5.

While at this time Green Up does not include lean concrete base and geogrid in the list of available pavement rehabilitation strategies, a comparison can still be made between the two alternatives. The geogrid was ignored, and the lean concrete base was modeled as JPCP, since the amount of cement and aggregate in lean concrete base is comparable to concrete. The result is shown in Figure 6.

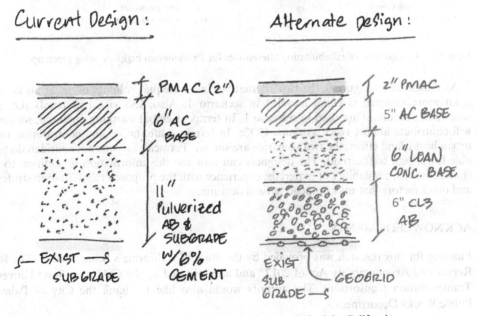

Figure 5.   Rehabilitation alternatives for Pearblossom highway, Palmdale, California.

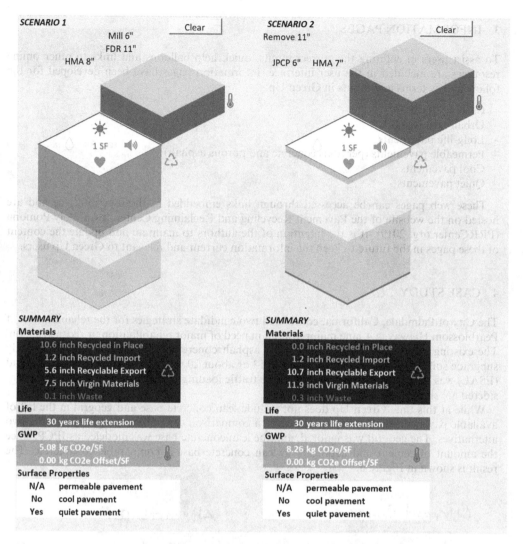

Figure 6. Comparison of rehabilitation alternatives for Pearblossom highway using green up.

As illustrated in Figure 6, the two alternatives involve equal volumes of materials however a lot more material is being recycled in scenario 1. Also, less virgin materials (i.e. non-renewable resources) are used in scenario 1. In terms of global warming potential, scenario 2 will contribute almost twice as much CO2e. In terms of life, noise, and contribution to the urban heat island effect, the two scenarios are similar. Permeability was not considered applicable for a high traffic road. The designers can now use this information in addition to cost considerations, availability of materials, experience with the proposed rehabilitation strategies, and other factors that may influence a final decision.

ACKNOWLEDGMENTS

Funding for this research was provided by the State of California's Senate Bill 1 "The Road Repair and Accountability Act of 2017" and administered by the California State University Transportation Consortium. The authors would also like to thank the City of Palmdale Public Works Department.

REFERENCES

Andrei, D. 2014. Design Alternative Comparison System for Pavements. 12th ISAP Conference Proceedings, International Society for Asphalt Pavements, Raleigh, North Carolina, June 2014. <https://www.taylorfrancis.com/books/e/9780429226779/chapters/10.1201/b17219-48>

Asphalt Institute & thinkstep AG, 2019. Life Cycle Assessment of Asphalt Binder. <http://www.asphaltinstitute.org/engineering/lca-study-on-asphalt-binders/>

Emerald Eco Label, 2019. An Environmental Product Declaration for Asphalt Mixtures. Declaration Number 1955.52.9 issued March 27, 2019. <https://asphaltepd.org/published/epd/52/>

National Asphalt Pavement Association, 2019. <http://www.warmmixasphalt.org/>

National Ready Mixed Concrete Association, 2016. NRMCA Member Industry-Wide EPD for Ready Mixed Concrete. Declaration Number: EPD 10080 issued October 20, 2016. <https://www.nrmca.org/sustainability/EPDProgram/Downloads/EPD10080.pdf>

Pavement Recycling and Reclaiming Center, 2019. <http://prrcenter.org/?page_id=59>

Pavia Systems, 2019. PAVEXpress <https://pavexpress.com/> accessed 3/7/2019

PD Collective, 2019. Pavement Designer <https://www.pavementdesigner.org/> accessed 3/7/2019

Portland Cement Association, 2016. Environmental Product Declaration. Portland Cements (per ASTM C150, ASTM C1157, AASHTO M 85 or CSA A3001). Declaration Number EPD 035 issued June 1, 2016. <https://www.cement.org/docs/default-source/sustainabilty2/pca-portland-cement-epd-062716.pdf?sfvrsn=2>

Santero, N.J. & Horvath, A. 2009. Global warming potential of pavements. *Environ. Res. Lett.* 4 (2009) 034011 (7pp) doi:10.1088/1748-9326/4/3/034011 <https://iopscience.iop.org/article/10.1088/1748-9326/4/3/034011>

Fuel and non-fuel vehicle operating costs comparison of select vehicle types and fuel sources: A parametric study

Rami Chkaiban
*Department of Civil & Environmental Engineering, Graduate Research Assistant, Pavement Engineering &
Science Program, University of Nevada, Reno, NV, USA*

Elie Y. Hajj
*Department of Civil & Environmental Engineering, Pavement Engineering & Science Program, University
of Nevada, Reno, NV, USA*

Gary Bailey
Senior Modeling and Simulation Engineer, Nevada Automotive Test Center, Carson City, NV, USA

Muluneh Sime
Director of Engineering, Nevada Automotive Test Center, Carson City, NV, USA

Hao Xu
*Department of Civil & Environmental Engineering, Transportation Program, University of Nevada, Reno,
NV, USA*

Peter E. Sebaaly
*Department of Civil & Environmental Engineering, Pavement Engineering & Science Program, University
of Nevada, Reno, NV, USA*

ABSTRACT: Greenhouse emissions, energy consumption, roadway use, preservation/main-
tenance/rehabilitation, and user costs are some of the inputs required to conduct an LCA
study. User costs are based on three main inputs: time delay, fuel vehicle operating costs
(VOCs), and non-fuel VOCs. Generally, time delay and fuel VOCs have been considered in
user costs determination. Non-fuel VOCs have been mostly overlooked due to a lack of reliable cost models adaptive to recent vehicle technologies changes. This paper presents
a comparison of fuel and non-fuel VOCs for three vehicle types on representative roadway
properties (e.g., grade, curvature, roughness) and vehicle speeds. Three fuel sources were considered. Newly developed FHWA fuel and non-fuel VOC predictive models were used.
A comparison between the estimated fuel and non-fuel VOCs for the different scenarios was
conducted. Comparable magnitudes were observed for fuel and non-fuel VOCs. Thus, highlighting the importance of non-fuel VOCs consideration in user costs determination.

1 INTRODUCTION

Life Cycle Assessment (LCA) of pavements and roadways is a technique for evaluating the
impact of a specific infrastructure on the environment over the entire life cycle. Conducting
a pavement LCA requires an examination of all the inputs and outputs over the pavement's
life-cycle stages; from material production, to design, construction, use phase, preservation
maintenance/rehabilitation, and finally end of life (Harvey et al., 2016). For each stage, the
energy and materials consumed are input data, and the emissions and waste created are
output data.

Modeling of the various stages requires consideration of several processes and their impacts during the pavement life cycle (Harvey et al., 2016). For instance, modeling the material production stage requires, for each material input to the pavement system, a life-cycle inventory (LCI) of the foreground and background processes used to produce a raw or finished material. Modeling of construction, preservation, maintenance, and rehabilitation stages requires the consideration of processes and impacts related to equipment mobilization and demobilization, equipment use at the site, transport of materials to and from the site, energy used on site, and changes to traffic flow (e.g., including work zone speed changes and delay). Modeling of the use phase requires the consideration of pavement characteristics (structural responsiveness, macrotexture, roughness, etc.) and their impact on pavement use phase (vehicle fuel consumption and emission, noise, safety, etc.). Similar to other stages, considerations at the end of life stage need to be given for equipment use and related fuel consumption, the reuse of materials, the "production" of recycled materials like reclaimed asphalt pavement (RAP) or recycled concrete aggregate (RCA), etc.

LCA outputs can be applied in product development and improvement, strategic planning, public policy making, and marketing (Harvey et al., 2016). The selection of pavement material or structural design is one of the applications that agencies in the U.S. use to identify opportunities to improve the environmental performance of products at the various LCA stages (Kendall, 2012). This application is conducted in conjunction with Life Cycle Cost Analysis (LCCA). Pavement LCCA is the economic analysis of the pavement by evaluating the pavement total cost, initial cost, and discounted future costs over an analysis period that is usually the pavement design life (Walls III and Smith, 1998). LCCA is primary composed of agency and user costs. Agency costs are composed of initial construction costs and preservation/maintenance/rehabilitation costs. User costs are mainly composed of fuel and non-fuel vehicle operating costs (VOCs) and travel time delay. Numerous factors influence agency and user costs. For example, agency costs are influenced by the construction material, design decisions, and preservation/maintenance/rehabilitation strategies. On the other hand, user costs are influenced by roadway characteristics, roadway design, and agencies preservation/maintenance/rehabilitation strategy, in addition to road congestion.

Several VOC models exist for estimating the various components of user costs (Chatti and Zaabar, 2012). However, vehicle and highway technologies have considerably changed over the past several decades. Thus, the need for new VOC models that reflect the influence of changes in technologies. Accordingly, a recent Federal Highway Administration (FHWA) study was completed to improve the fuel and non-fuel VOCs estimation by taking into consideration changes in vehicle technology and extended transportation factors (e.g., facility type, roadway characteristics, driving cycles) for use in benefit-cost analysis (Hajj Y. E. et al., 2017, Hajj E.Y. et al., 2018). The new FHWA models for VOCs estimation were developed for an array of vehicle types, traffic conditions, and highway design scenarios as defined by roadway conditions such as speed limit, curvature, grade, and pavement roughness. The non-fuel VOC models consisted of four separate models for oil consumption including labor cost, tire wear, mileage-related vehicle depreciation, and repair and maintenance costs.

In this paper, the new FHWA models were used to estimate user costs in terms of fuel and non-fuel VOCs for select vehicles and highway scenarios. A decrease in VOCs is an important component of the benefits derived from transportation network improvements. The effect of roadway characteristics (i.e., curvature, grade, and pavement roughness) on user costs were examined. Three vehicle types were considered in this exercise: small light duty vehicle (SLDV), large light duty vehicle (LLDV), and combination truck. Two fuel sources were considered for the SLDV vehicle: gasoline and hybrid-electric (HE). On the other hand, a gasoline and a diesel fuel source were considered for the LLDV and combination truck, respectively. The excess fuel and non-fuel VOCs were estimated, and their relative magnitudes were compared for the different highway scenarios considered in this effort. Thus, highlighting the significance of non-fuel VOCs and the need for consideration as part of user costs for LCA.

2 METHODOLOGY

Physics-based full vehicle simulation models are the basis for the new FHWA models for fuel and non-fuel VOCs (Bailey G. et al., 2017b, Bailey G. et al., 2017a). Thus, allowing for simulating the vehicle response to roadway characteristics while taking into consideration vehicle dynamics. Synthetically optimized (SO) driving cycles, extracted from actual real driving cycles data, were used, providing the ability to simulate different congestion levels and speed variation on different facility types such as full access control (FAC) and partial or no access control (PNAC). The new FHWA models were developed for 30 different vehicle and fuel types. In this paper, the following four vehicles were considered: a gasoline-powered and an HE-powered midsize car, a gasoline-powered large sport utility vehicle (SUV), and a diesel-powered fully loaded tractor-trailer—80,000 gross vehicle weight (GVW). Because of page limitations, the models and equations adopted in this paper are explicitly described in Hajj E.Y. et al., 2017, Hajj E. Y. et al., 2018.

It should be noted that the vehicle age of the midsize car and large SUV was assumed equal to the average age of passenger vehicles (i.e., 11 years) and light trucks (11.6 years) in the U.S., respectively (U.S. Department of Transportation, last accessed on April 19, 2017, Williams et al., 2017). The vehicle age of the tractor trailer was assumed equal to the average age of combination trucks in the U.S. (i.e., 5.5 years) (Hooper and Murray, 2018). The average vehicle age is an input for the estimation of mileage-related vehicle depreciation and repair and maintenance costs using the new FHWA non-fuel VOC models.

Different roadway scenarios were used for every vehicle, illustrating the relative magnitudes of the excess VOCs under multiple conditions. Each evaluated scenario had a different vehicle average speed, posted speed limit, curvature, grade, and pavement roughness.

3 STUDIED HIGHWAY SCENARIOS

3.1 *SLDV scenarios*

Three scenarios were assigned to the two midsize car vehicles. These scenarios highlight the influence of road curvature and grade on VOCs for an FAC facility. Scenario 1 consisted of a straight and flat roadway condition (i.e., grade of 0% and curvature of 0°). Scenario 2 consisted of a straight roadway with an upslope grade of 3.45%, thus highlighting the influence of road grade on VOCs. Scenario 3 consisted of flat roadway with a curvature of 7.4°, thus highlighting the influence of road curvature on VOCs. A speed limit and an average vehicle speed of 65 and 50 mph, respectively, was considered in all three scenarios. The pavement surface was considered to be in good condition with an international roughness index (IRI) less than 95 inch/mile. Table 1 summarizes the characteristics of the three different scenarios.

3.2 *Lldv scenarios*

Three scenarios were assigned to the large SUV vehicle. These scenarios highlight the influence of average vehicle speed and congestion on VOCs for PNAC. The roadway characteristics between all three scenarios were similar except for the average vehicle speed. Thus, resulting in

Table 1. SLDV category scenarios.

Roadway characteristics	Scenario 1(baseline)	Scenario 2	Scenario 3
Facility type	FAC	FAC	FAC
Speed limit (mph)	65	65	65
Grade (%)	0.00	3.45	0.00
Curvature (°)	0.0	0.0	7.4
IRI (inch/mile)	≤ 95	≤ 95	≤ 95
Average vehicle speed (mph)	50	50	50

different ratios between average vehicle speed and speed limit, with a lower ratio representing a more congested condition. Scenario 1 consisted of a free flow condition as reflected with a ratio of 1 for the average vehicle speed to speed limit. Scenario 2 represented a congested scenario (average speed to speed limit ratio equals to 0.75), followed by a more congested scenario 3 (average speed to speed limit ratio equals to 0.5). Table 2 summarizes the characteristics of the three different scenarios.

3.3 Combination truck scenarios

Two different scenarios were assigned for the combination truck. These scenarios highlight the influence of pavement roughness on VOCs for FAC. The roadway characteristics between the two scenarios were similar except for IRI. Scenario 1 has an IRI less than 95 inch/mile while scenario 2 had an IRI between 171 and 200 inch/mile. Table 3 summarizes the characteristics of the two scenarios.

4 RESULTS

The following cost figures were assumed in order to calculate fuel and non-fuel VOCs (all costs are in 2018 U.S. dollars):

- The average gasoline and diesel fuels were assumed equal to the U.S. national average price of $2.562 and $2.926, respectively (AAA, last accessed on September 8, 2019).
- The cost to replace one tire was assumed equal to $170.00 for the selected vehicles.
- The cost of an oil quart was assumed equal to $8.00.
- The labor cost of an oil change was assumed equal to $67.68, $66.10, and $142.33 for mid-size car, large SUV, and tractor trailer, respectively (Hajj E. Y. et al., 2018).

Two separate analyses are presented in this paper for each of the selected vehicles. The first analysis illustrates fuel and non-fuel VOC proportions relative to the total VOCs. The second analysis illustrates the excess fuel and non-fuel VOCs relative to the VOCs estimated for the baseline scenario (i.e., scenario 1).

Table 2. LLDV category scenarios.

Roadway characteristics	Scenario 1 (baseline)	Scenario 2	Scenario 3
Facility type	PNAC	PNAC	PNAC
Speed limit (mph)	40	40	40
Average Number of Traffic Control Devices per mile	4	4	4
Grade (%)	3.45	3.45	3.45
Curvature (°)	0.0	0.0	0.0
IRI (inch/mile)	≤ 95	≤ 95	≤ 95
Average vehicle speed (mph)	40	30	20
Ratio of average vehicle speed to speed limit	1	0.75	0.5

Table 3. Combination trucks category scenarios.

Roadway characteristics	Scenario 1 (baseline)	Scenario 2
Facility type	FAC	FAC
Speed limit (mph)	65	65
Grade (%)	0	0
Curvature (°)	0	0
IRI (inch/mile)	≤ 95	171-200
Average speed (mph)	55	55

287

4.1 Gasoline-powered midsize car

Figure 1 shows the fuel and non-fuel VOC proportions relative to the total VOCs. Based on the data presented in Figure 1 for the studied scenarios the following observations can be made:

- A 2% and 23% decrease in fuel consumption portion was observed for scenario 2 (grade effect) and scenario 3 (curvature effect), respectively, when compared to scenario 1.
- A 23% increase in tire wear portion was observed for scenario 3 (curvature effect) when compared to scenario 1.
- Repair and maintenance portion was the largest within the non-fuel VOC components. An increase in the repair and maintenance portion of 3% and 7% was observed with road grade and curvature, respectively.

On the other hand, Figure 2 presents the excess fuel and non-fuel VOCs in scenarios 2 and 3 relative to the baseline scenario 1. It can be seen that the excess of each of the non-fuel VOC components is either equal or exceeding the excess of fuel VOC. In general, the influence of road grade on all fuel and non-fuel VOCs was similar except being slightly higher for repair and maintenance. However, the influence of road curvature varied, with the maximum excess of 97% and 63% being for tire wear and repair and maintenance, respectively.

4.2 He-powered midsize car

Similar to gasoline-powered midsize car, Figure 3 shows the fuel and non-fuel VOC proportions relative to the total VOCs. Based on the data presented in Figure 3 for the studied scenarios the following observations can be made:

- A 16% and 33% decrease in fuel consumption portion was observed for scenario 2 (grade effect) and scenario 3 (curvature effect), respectively, when compared to scenario 1.
- A 10% increase in tire wear portion was observed for scenario 3 (curvature effect) when compared to scenario 1.
- Repair and maintenance portion was the largest within the non-fuel VOC components. An increase in the repair and maintenance portion of 22% and 37% was observed with road grade and curvature, respectively.

On the other hand, Figure 4 presents the excess fuel and non-fuel VOCs in scenarios 2 and 3 relative to the baseline scenario 1. It can be seen that the excess of each of the non-fuel VOC components is either equal or exceeding the excess of fuel VOC. The influence of road grade on

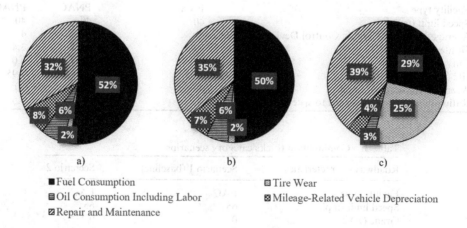

Figure 1. Fuel and non-fuel VOC proportions relative to total VOCs for gasoline-powered midsize car: a) scenario 1, b) scenario 2, and c) scenario 3.

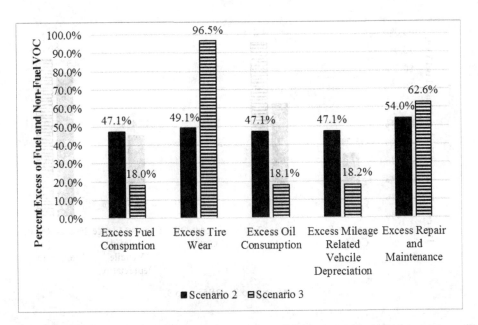

Figure 2. Excess fuel and non-fuel VOCs in scenario 2 (grade effect) and scenario 3 (curvature effect) relative to scenario 1 for gasoline-powered midsize car.

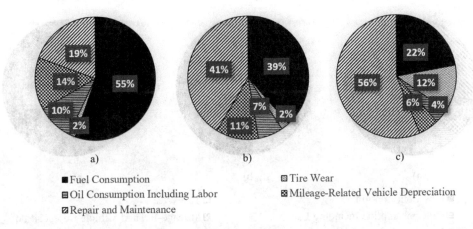

Figure 3. Fuel and non-fuel VOC proportions relative to total VOCs for HE-powered midsize car: a) scenario 1, b) scenario 2, and c) scenario 3.

all fuel and non-fuel VOCs varied between 34% for fuel and oil consumptions and 78% for repair and maintenance. The influence of road curvature was similar on fuel and oil consumptions, and slightly higher for mileage-related vehicle depreciation. However, road curvature resulted in 94% and 86% increase in tire wear, and repair and maintenance, respectively

4.3 Gasoline-powered large SUV

Figure 5 shows the effect of congestion on fuel and non-fuel VOC proportions relative to the total VOCs. Based on the data presented in Figure 5 for the studied scenarios the following observations can be made:

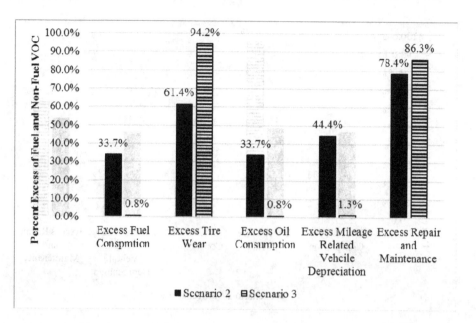

Figure 4. Excess fuel and non-fuel VOCs in scenario 2 (increase in road grade) and scenario 3 (increase in road curvature) relative to scenario 1 for HE-powered midsize car.

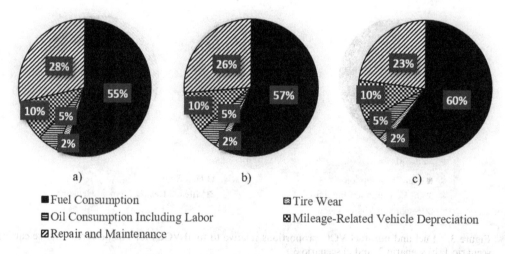

Figure 5. Fuel and non-fuel VOC proportions relative to total VOCs for gasoline-powered large SUV: a) scenario 1, b) scenario 2, and c) scenario 3.

- A 2% and 5% increase in fuel consumption portion was observed for scenario 2 (average vehicle speed over speed limit ratio equals to 0.75) and scenario 3 (average vehicle speed over speed limit ratio equals to 0.5), respectively, when compared to scenario 1.
- Repair and maintenance portion was the largest within the non-fuel VOC components. A decrease in the repair and maintenance portion of 2% and 5% was observed with the decrease in average vehicle speed over speed limit ratio to 0.75 and 0.5, respectively.

290

On the other hand, Figure 6 presents the excess fuel and non-fuel VOCs in scenarios 2 and 3 relative to the baseline scenario 1. It can be seen that the excess of each of the non-fuel VOC components is either equal or less than the excess of fuel VOC. The influence of congestion was similar on fuel and oil consumptions, and slightly lower for mileage-related vehicle depreciation. However, the excess in tire wear and repair and maintenance was minimal or negative with the increase in congestion level that is reflected with an excess ranging between -1.5% and +1%. This is mainly due to both, the tire wear and repair and maintenance models for gasoline-powered large SUV exhibiting minimum values within 20 to 30 mph average vehicle speeds. (Sime M. et al., 2018, Hajj E. Y. et al., 2018).

4.4 *Diesel-powered tractor trailer*

Figure 7 shows the fuel and non-fuel VOC proportions relative to the total VOCs. Based on the data presented in Figure 7 for the studied scenarios the following observations can be made:

- Fuel consumption portion was the largest and constituted 87-88% of the total VOCs.
- A 1% decrease in fuel consumption portion was observed for scenario 2 (higher pavement roughness) when compared to scenario 1.
- Repair and maintenance portion was the largest within the non-fuel VOC components. A slight increase in the repair and maintenance portion of 1% was observed with the increase in pavement roughness.

On the other hand, Figure 8 presents the excess fuel and non-fuel VOCs in scenarios 2 relative to the baseline scenario 1. It can be seen that the excess of each of the non-fuel VOC components is either equal or exceeding the excess of fuel VOC. The influence of pavement roughness on all fuel and non-fuel VOCs varied, with the highest being at 19% and 7% increase in repair and maintenance, and tire wear, respectively.

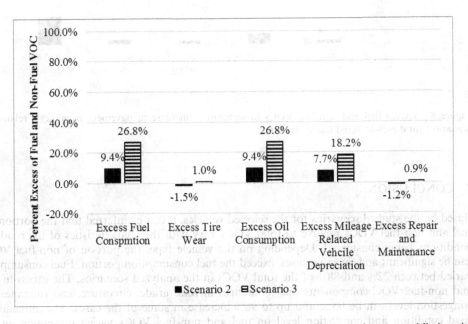

Figure 6.   Excess fuel and non-fuel VOCs in scenario 2 (average vehicle speed over speed limit ratio of 0.75) and scenario 3 (average vehicle speed over speed limit ratio of 0.5) relative to scenario 1 ((average vehicle speed over speed limit ratio of 1) for gasoline-powered large SUV.

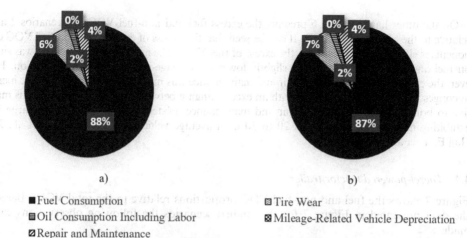

a)   b)

■ Fuel Consumption   ⊠ Tire Wear
⊟ Oil Consumption Including Labor   ⊠ Mileage-Related Vehicle Depreciation
☒ Repair and Maintenance

Figure 7.   Fuel and non-fuel VOC proportions relative to total VOCs for diesel-powered tractor trailer: a) scenario 1, and b) scenario 2.

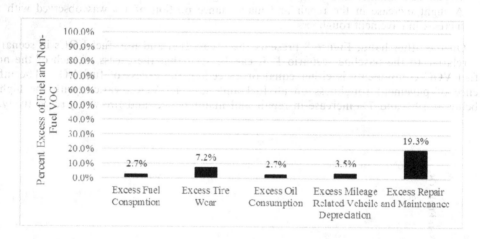

Figure 8.   Excess fuel and non-fuel VOCs in scenario 2 (increase in pavement roughness) relative to scenario 1 for diesel-powered tractor trailer.

5   CONCLUSION

Based on the studied scenarios for the selected vehicles, it can be inferred that the portions of fuel and non-fuel VOCs relative to total VOCs vary with the characteristics of the roadway conditions being considered. Depending on the vehicle type, the portion of non-fuel VOCs can be significant and, in some cases, exceed the fuel consumption portion. Fuel consumption varied between 22% and 88% of the total VOCs in the analyzed scenarios. The excess in fuel and non-fuel VOC components due to changes in road grade, curvature, and roughness or congestion level can be large and up to 96% increase in some of the cases. The influence of road condition and congestion level on fuel and non-fuel VOCs varied depending on the vehicle type, fuel source, and roadway scenarios considered. In general, the paper highlights through the presented examples the importance and significance for considering non-fuel VOCs as part of user costs estimation in an LCCA or an LCA.

REFERENCES

AAA last accessed on September 8, 2019. Gas prices. https://gasprices.aaa.com/.

Bailey G., Sime M., Hajj E. Y., P. E. Sebaaly, Chkaiban R. & A., H. 2017a. Enhanced Prediction of Truck Fuel Economy and Other Truck Operating Costs. Task 4a.-b: Fuel Consumption Estimates for Roadways with Partial or No Access Control. prepared for FHWA, Report No WRSC-UNR-201603.

Bailey G., Sime M., Hajj E. Y., P. E. Sebaaly, Chkaiban R., Kazemi S.F. & A., H. 2017b. Enhanced Prediction of Truck Fuel Economy and Other Truck Operating Costs. Task 3a.-b: Fuel Consumption for Controlled Access Facilities. prepared for FHWA, Report No WRSC-UNR-201612.

Chatti, K. & Zaabar, I. 2012. Estimating the effects of pavement condition on vehicle operating costs. NCHRP Report 720. *Transportation Research Board of the National Academies, Washington, DC.*

Hajj E. Y., Xu. H., Bailey G., Sime M., Chkaiban, R., Kazemi S.F. & Sebaaly, P. E. 2017. Enhanced Prediction of Truck Fuel Economy and Other Truck Operating Costs. Phase I: Modeling The Relationship Between Vehicle Speed and Fuel Consumption. prepared for FHWA, Report No WRSC-UNR-201705.

Hajj E. Y., Sime M., Chkaiban R., Bailey G., H., XU. & Sebaaly P. E. 2018. Enhanced Prediction of Truck Fuel Economy and Other Truck Operating Costs. Phase II: Modeling The Relationship Between Pavement Roughness, Speed, Roadway Characteristics and Vehicle Operating Costs. prepared for FHWA, Report No WRSC-UNR-201809-01.

Harvey, J., Meijer, J., Ozer, H., Al-Qadi, I. L., Saboori, A. & Kendall, A. 2016. Pavement Life Cycle Assessment Framework. United States. Federal Highway Administration.

Hooper, A. & Murray, D. 2018. An Analysis of the Operational Costs of Trucking: 2018 Update.

Kendall, A. 2012. Life Cycle Assessment for Pavement: Introduction. *Presentation, Federal Highway Administration Sustainable Pavements Task Group,*Davis, CA, April.

Sime M., Bailey G., Hajj E. Y., Chkaiban R., H., XU. & Sebaaly, P.E. 2018. Enhanced Prediction of Truck Fuel Economy and Other Truck Operating Costs. Task 7a. – b: Effects of Infrastructure Physical and Operating Characteristics on Non-Fuel Vehicle Operating Costs. prepared for FHWA, Report No WRSC-UNR-201806-01.

U. S. Department of Transportation, last accessed on April 19, 2017. Table 1-26: Average Age of Automobiles and Trucks in Operation in the United States. www.rita.dot.gov/bts/sites/rita.dot.gov.bts/files/publications/national_transportation_statistics/html/table_01_26.html_mfd: Bureau of Transportation Statistics.

Walls III, J. & Smith, M. R. 1998. Life-cycle cost analysis in pavement design-interim technical bulletin.

Williams, S. E., Davis, S. C. & Boundy, R. G. 2017. Transportation Energy Data Book: Edition 36. Oak Ridge National Lab. (ORNL), Oak Ridge, TN (United States).

293

Bayesian economic analyses of including reclaimed asphalt pavements in flexible pavement rehabilitation

Hongren Gong, Miaomiao Zhang, Wei Hu & Baoshan Huang*
The University of Tennessee, Knoxville, TN, USA

ABSTRACT: Sustainable infrastructure requires pavement maintenance and rehabilitation to be environmentally beneficial and economically efficient. With the ever-depleting natural resources and growing demands for sustainable infrastructure, the use of reclaimed asphalt pavement (RAP) materials has seen a steady increase. This study used Bayesian-based life cycle cost analysis (LCCA) to determine the long-term economic effectiveness of using RAP in rehabilitation. Field observed performance data retrieved from the database of Specific Pavement Study 5 (SPS-5) of the Long-Term Pavement Performance (LTPP) project were used to develop the performance model, which was then employed to determine the service life of the competing alternatives. According to the posterior distributions identified through the Markov chain Monte Carlo (MCMC) simulations, the overlay thickness was best fitted by the log-normal distribution. The LCCA indicated that the use of RAP in rehabilitative projects reduced the life cycle cost around 20%.

1 INTRODUCTION

Recycling used asphalt pavement materials into new or rehabilitative pavement construction activities have been gaining ever-growing efforts for both of its economic and environmental benefits. Enormous laboratory studies have been performed to investigate the short-term and long-term performance of adding reclaimed asphalt pavement (RAP) into new materials, developing new mixture design methods in considering the impacts of the aged binder in the RAP on the resulting mixtures, examine the blending and mixing status of the aged and virgin binder using both macro and micro techniques (Gong et al., 2018; Huang et al., 2005; Shu et al., 2012). Extensive studies have been focusing on evaluating the field performance of the asphalt mixtures containing RAP along with other techniques in promoting more environmental advantages (Aurangzeb & Al-Qadi, 2014; Shu et al., 2012).

Due to the budget constraint, state highway agencies have been seeking for optimal maintenance and rehabilitation (M&R) timing and M&R strategies to maximize the service life span and serviceability limited with minimized costs (Gong et al., 2018; Harvey et al., 2012). To these ends, the life cycle cost analysis (LCCA) has been frequently employed to identify the most economical option from a set of mutually competing alternatives (Harvey et al., 2012; Ozbay et al., 2004; Swei et al., 2013; Tighe, 2001; Walls & Smith, 1998). LCCA incorporates initial and various discounted future costs over the life of the alternatives to determine the economically optimal investment strategy (Reigle & Zaniewski, 2002).

Many methods can be used for LCCA. Some of the more common economic analysis strategies include net present value (NPV), equivalent uniform annual cost (EUAC), rate of return (ROR), benefit-cost (B/C) ratios, and break-even analysis (Rangaraju et al., 2008; Swei et al., 2013; Walls & Smith, 1998). The FHWA recommends the NPV being the method of choice (Walls & Smith, 1998). Tighe (2001) advocates the use of constant dollars and real discount

---

* Corresponding author

rate over nominal dollars, which is compliance with the recommendation of the FHWA (Walls & Smith, 1998).

Conventionally, LCCA is carried out for grids of discrete input variables, and then generates point estimates of the target economic indicators, such as the net present value (NPV) or equivalent uniform annual cost (EUAC) (Reigle & Zaniewski, 2002; Swei et al., 2013). However, due to the variability inherent in different sources of the input in computing the economic indicators, deterministic estimation of the life cycle cost is unable to capture the uncertainty of the inputs and characterize the reliability of the resulting LCC (Tighe, 2001). To investigate the effect of a particular input variable, sensitivity analyses are frequently performed (Batouli et al., 2017). However, the use of deterministic methods has long been blamed for its inability to test the significance between alternatives (Harvey et al., 2012; Xu et al., 2014) and heavily relies on the subjective experience of experts and engineers.

To address these quandaries with deterministic LCCA, alternatives that can account for the uncertainty in both the sources of input and insufficient performance modeling have been proposed for decades (Harvey et al., 2012; Swei et al., 2013; Tighe, 2001; Walls & Smith, 1998). To incorporate the input under risk consideration, Li & Madanu (2009) employed the Beta distribution that allows for any degree of skewness and kurtosis. The overlay thickness in rehabilitative projects can be best considered by a log-normal distribution (Swei et al., 2013; Tighe, 2001). FHWA recommends using the triangle distribution to model the impacts of varied discount rate (Walls & Smith, 1998). The Weibull distribution has been applied to characterize the pavement performance (Harvey et al., 2012; Osman, 2005). In the final step of an LCC analysis, the LCCs were frequently determined through sampling techniques, including Markov chain Monte Carlo (MCMC) (Swei et al., 2013; Wang & Wang, 2019), Latin Hypercube sampling (Li & Madanu, 2009; Walls & Smith, 1998).

Although many studies on LCCA have attempted to incorporate diversified sources of uncertainty, very few have approached the probabilistic LCC by obtaining a potential distribution that governs the data generation process of the target of interest. To this end, a full Bayesian analysis-based method was proposed that not only inherently considers the uncertainty of the inputs and models; it also generates comprehensive summary and diagnostic statistics for the reliability of the analysis. To learn a realistic Bayesian model, the data retrieved from the database of the Long-Term pavement performance (LTPP) project were used.

2 PROCEDURES OF LCCA

According to the Life-Cycle Cost Analysis in Pavement Design: Interim Technical Bulletin (Walls & Smith, 1998), an LCCA consists of the eight major steps. As SPS-5 experiments have already specified the experimental factors, this study only focused on determining the agency costs, computing the NPV, and interpreting the final results.

This study focused on the maintenance and rehabilitation construction costs of overlays with HMA containing RAP against those using conventional HMA with virgin binder. A design period of 20 years was used to potentially cover two cycles of maintenance and capture the economic effects of alternatives with shorter service life. The salvage value was ignored at the end of the relatively long analysis period.

The life-cycle cost (LCC) analysis is an economic evaluation tool for comparing and selecting pavement scheme of optimal long-term performance and economic benefits. The benefits can be considered in many ways; two of the most frequently used ones are the net present value (NPV) and the equivalent uniform annual cost (EUAC). The NPV, as shown in Equation (1), is calculated by assigning monetary values to benefits and costs, discounting future benefits (PV benefits) and costs (PV costs) using a proper discount rate and subtracting the total discounted costs from the total discounted benefits. Numerous existing studies considered the NPV to be the economic efficiency indicator of choice (Aurangzeb & Al-Qadi, 2014; Okte et al., 2019; Swei et al., 2013; Walls & Smith, 1998).

$$NPV = Initial\ Cost + \sum\nolimits_{k=1}^{N} Rehab\ Costs_k/(1+k)^n \qquad (1)$$

where NPV is the net present value; Costs initial is the initial construction, maintenance, and rehabilitation costs; Cost rehab$_k$ is the cost incurred at the k-th rehabilitation; i is the discount rate; n is the number of years into future; N is the total number of maintenances. In this study, the NPV in constant currency recommended by the FHWA (Walls & Smith, 1998) was adopted.

3 DATA SOURCES

The present analysis was based on data collected from the Specific Pavement Studies 5 (SPS-5) of the Long-Term Pavement Performance (LTPP) program for all the analyses. The LTPP database is currently managed by the Federal Highway Administration (FHWA) and has recorded performance data up to 30 years (Gong et al., 2018). The SPS-5 experiment was designed to investigate the effectiveness of various rehabilitative strategies, including: 1) surfacing materials: RAP versus virgin hot mix asphalt; 2) pre-overlay preparation intensity: minimal treatment of existing defected surface versus mill existing surface and 3) overlay thickness (thin (2 inches) vs. thick (4 inches)).The SPS-5 comprised of a total of 18 sites, with a purpose to consider a diversity of conditions such as temperature, precipitation, materials, preparation intensity before the placement of the new surface layers. In general, each LTPP section is 152.5 m (500 ft) long and 7.3 m (24 ft) wide.

4 METHODOLOGY

4.1 *Bayesian based uncertainty quantification*

Full Bayesian formulations of the different inputs were adopted to quantify the uncertainty in the LCCA. In this study, the uncertainty in agency costs regarding materials (overlay thickness), service life (pavement performance) were incorporated. The user costs that are common to the mutually competing alternatives canceled out each other and were ignored. There are other studies explored in this aspect (Aurangzeb & Al-Qadi, 2014; Chatti & Zaabar, 2012; Walls & Smith, 1998).

$$P(\theta|D) = \frac{P(D|\theta)P(\theta)}{P(D)} \qquad (2)$$

$$N(x|\mu,\sigma^2) = \frac{1}{\sqrt{2\pi\sigma^2}}\exp\left(-\frac{(x-\mu)^2}{2\sigma^2}\right) \qquad (3)$$

Equation (2) is the Bayesian equation, the denominator $P(D)$ is the evidence that can be determined through marginalizing over the prior distribution of the parameters $P(\theta)$, the numerator of the equation is the likelihood. In a Bayesian regression model, instead of obtaining point estimates of the model as did in the ordinary least square regression, the posterior distribution of the parameters was obtained given the data (D). In estimating the posterior distribution, closed-form solutions are only available to a limited number of prior distributions, such as the Gaussian (normal) distribution, see Equation (3). In situations where no closed-form solution available, the Markov chain Monte Carlo (MCMC) simulation can be used (Gelman et al., 2014). This study used an implementation of the MCMC called Hamiltonian Monte Carlo sampling or hybrid Monte Carlo sampling (Gelman et al., 2014). The stan language was used to create the Bayesian models (Carpenter et al., 2017).

In a Bayesian context, many metrics are available for model comparison, including those derived from maximum likelihood estimation, such as the Akaike information criterion (AIC) and Bayesian information criterion (BIC); those based on Bayesian theory such as the Deviance information criterion (DIC) and the widely available information criterion (WAIC), see Equation (5). The WAIC has a desirable property of averaging over the posterior distribution rather than conditions on a point estimate and is more relevant in a predictive context (Gelman et al., 2014). This study adopted the WAIC as the metric for determining the optimal distribution to the data.

$$lppd(y, \theta) = \sum_i \log \frac{1}{S} \sum_s^S p(y_i|\theta_s) \tag{4}$$

$$WAIC(y, \theta) = -2\left(llpd - \sum_s^S var_\theta \log p(y_i|\theta)\right) \tag{5}$$

where $lppd$ is the log-pointwise-predictive-density; y is the observations; S is the number of samples in the MCMC simulation; θ_s is the s-th set of sampled parameter values in the posterior distribution, $var_\theta \log p(y_i|\theta)$ is the variance in log-probabilities for each observation y_i.

4.2 Source of uncertainty

Four primary sources of uncertainty in computing the LCC were incorporated in this study: overlay thickness, material cost, the service life of the rehabilitation, and discount rate.

Uncertainty in overlay thickness. In the SPS-5 experiments, two nominal overlay thickness values were used to consider the effects of the thickness, 51 mm (2 in) and 102 mm (4 in). However, existing literature has shown that the overlay thickness can sometimes deviate from the original design value significantly. To consider the uncertainty in the overlay thickness, studies have shown that the log-normal distribution fits the overlay thickness best (Swei et al., 2013; Tighe, 2001).

The log-normal distribution can be understood as the variable after a logarithmic transformation follows a normal distribution. Due to the constraint of the logarithmic transformation on the concerning variable, the variable concerned must be positive ($x > 0$). For comparison purposes, several other probability distributions enjoyed wide application were also considered, including the Gamma distribution, skewed normal distribution, Weibull distribution. Detailed definitions of these probability distributions can be found in Gelman et al. (2014). According to the histogram given in Figure 1, the in-situ overlay thickness varies significantly, and both scenarios present a duo-mode pattern (two peaks).

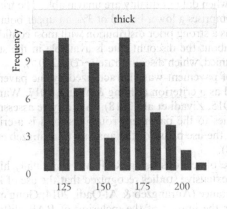 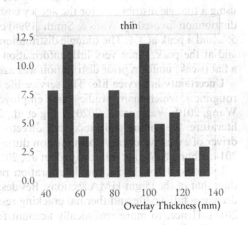

Figure 1. Histograms of overlay thickness.

Table 1. Comparison of different distribution for overlay thickness.

Overlay Type	Distribution	WAIC	Standard Error
Thicker	Log-Normal	0	0
	Skewed Normal	717.45	5.22
	Gamma	721.22	2.24
	Normal	723.31	1.99
	Weibull	726.8	3.15
Thinner	Log-Normal	0	0
	Skewed Normal	1.07	1.03
	Gamma	4.28	4.43
	Normal	4.80	4.66
	Weibull	630.3	5.06

To quantify the uncertainty in the overlay thickness, a simple Bayesian model with an intercept term was used to model the data and obtain the underlying posterior distribution. As shown in Table 1, according to the WAIC criterion, the distribution that best fits the thickness of thinner overlays is the log-normal distribution, whilst other distributions have similar WAIC values. For the thicker overlays, the skewed normal distribution showed a marginally smaller WAIC value than that of the Weibull and Log-Normal distributions. Given this negligible difference and to be consistent, the log-normal distribution was adopted as the underlying distribution for the overlay thickness. Figure 2 gives the posterior distribution of the thickness of the overlays for both the thin and thicker scenarios. For the MCMC simulation, four chains were used, each with 4000 iterations and 2000 iterations were used for initiation (warm-up or burn-in). Figure 2 gives the posterior distributions of the thickness of the overlays for both the thin and thicker scenarios.

Uncertainty in material cost. Although existing literature largely agrees on the overall benefits of adding RAP for rehabilitation projects, the rate of saving compared to the virgin HMA varies vastly and strongly depends on the percentage of RAP and recycling technologies. For a RAP recycling rate of 20% to 40%, the rate of savings ranges from 10% to 30%, with the most frequent number centered around 20% (Alkins et al., 2008; Aurangzeb & Al-Qadi, 2014; Horvath, 2003; Robinette & Epps, 2010). As the SPS-5 experiments used a RAP dosage of 30%, the uncertainty in the rate of saving in using RAP was model by a uniform distribution with a lower bound of 10% and upper bound of 30% (savings rate $\mho(0.1, 0.3)$). A more complicated distribution that places more probability mass around 20% can be beneficial, such as the log-normal distribution or triangle distribution. However, the use of log-normal distribution necessitates significant efforts in identifying the model parameters.

Uncertainty in discount rate. Regarding the discount rate, Walls & Smith (1998) recommend using a triangle distribution for the agency cost when data of quality are unavailable. The triangle distribution favored by Walls & Smith (1998) comprises a lower bound of 3%, an upper bound of 5%, and a peak at 4%. The triangle distribution is a strong prior distribution with most confidence laid at the peak. Since very little information about the discount rate is available in the study, a flat (weak) uniform prior distribution was assumed, which discount rate $i \sim \mho(0.2, 0.7)$.

Uncertainty in service life. The service life of pavement was characterized by the pavement roughness, which many studies have employed as a criterion (Chong & Wang, 2017; Wang & Wang, 2019; Wang et al., 2020; Yang et al., 2015; Ziyadi et al., 2018). In life cycle assessment literature, the rolling resistance that directly ties to the pavement roughness and is a critical driver of excessive energy consumption during the use phase of the analysis (Aurangzeb et al., 2014; Chong & Wang, 2017; Ziyadi et al., 2018).

In terms of roughness, the deterioration rate of sections using RAP was marginally higher than that of the virgin HMA sections. Besides, extensive studies recognized that the use of RAP decreases fatigue life and thermal cracking resistance (Aurangzeb & Al-Qadi, 2014; Gong et al., 2018). Hence, to more realistically account for the impact of the inclusion of RAP, different prior distributions for the service life of these two types of pavement were used.

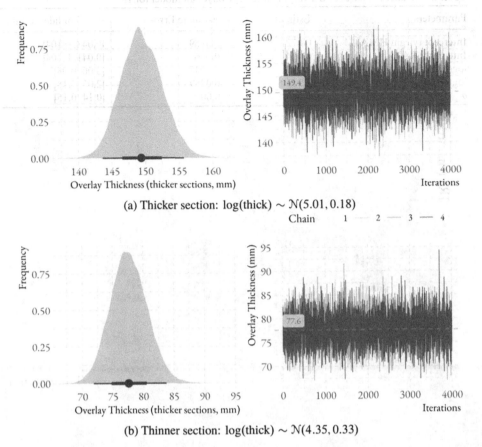

(a) Thicker section: log(thick) $\sim N(5.01, 0.18)$

(b) Thinner section: log(thick) $\sim N(4.35, 0.33)$

Figure 2. Bayesian estimates of thickness of overlays. The numbers annotated on the right panels are the median of the respective posterior samples, in mm.

$$\log(IRI) \sim N\left(\mu, \sigma^2\right) \tag{6}$$

$$\mu = b_0 + b_1 time_{log} + b_2 h_{log} + b_3 mat[i] \tag{7}$$

$$\sigma \sim \exp(1)$$

The model represented by Equation (7) can be formulated in arithmetic scale, which can be considered a non-linear function of the time, overlay thickness, and the use of RAP or not, as shown in Equation (8). Figure 3 gives the goodness-of-fit for the constructed model.

$$IRI = \exp(b_0) time^{b_1} thickness^{b_2} \exp(b_3\ mat[i]) \tag{8}$$

5   LIFE CYCLE COST ANALYSIS

A full probabilistic formulation of the LCCA requires to specify the prior distributions for all the components involved in the computation process. In this study, based on the analysis of the uncertainty of the different types of inputs, the following distributions were used for the materials cost, the overlay layer thickness, the pavement service life, and the discounting rate.

Table 2. Model estimates of the varying intercept Bayesian model for IRI.

Parameters	Estimates	Estimate Error	95% Confidence Interval
Intercept (virgin HMA)	0.072	0.0159	[0.043, 0.104]
Intercept (RAP)	0.074	0.016	[0.041, 0.104]
log(time)	0.0796	0.0043	[0.06, 0.08]
log(IRI_0)	2.116	0.0289	[2.05, 2.18]
σ	0.14	0.00	[0.14, 0.15]

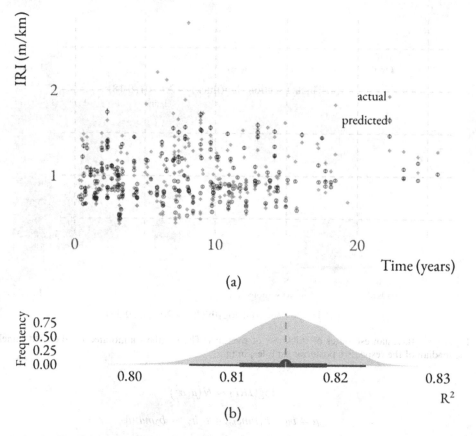

(a)

(b)

Figure 3. Verification of goodness of the IRI model. Shown only 300 randomly draw points out of a total of 1401 observations. σ: standard error; open circle: predicted values; solid circle: actual observations. A median R^2 of 0.815 was obtained.

- This study assumed an analysis period of 20 years, so as to cover two M&R actions.
- Overlay layer thickness: log-normal distribution. The unit cost of the virgin HMA is 70 $/ton (2017 value), the unit cost of the RAP mixture varies for different dosage and different recycling technology (in-place or in-place). Existing literature has reported that the rate of savings in using RAP ranges from 10% to 30% for a dosage of 25%-40% (Newcomb et al., 2016; Robinette & Epps, 2010; Williams et al., 2017).
- Material cost: uniform distribution $U(0.1, 0.3)$ for savings rate due to RAP inclusion;
- Service life: log-normal distribution, the RAP overlays were assumed to be last slightly shorter than the virgin HMA counterparts. The service life was determined by using 1.7 m/km as a cutting of value for both types of overlays.
- Discounting rate: uniform distribution $U(0.2, 0.6)$.

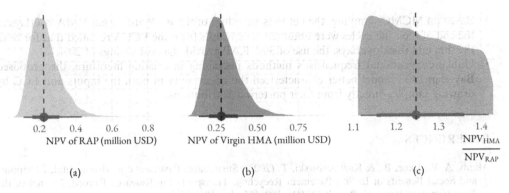

NPV of RAP (million USD)   NPV of Virgin HMA (million USD) $\dfrac{NPV_{HMA}}{NPV_{RAP}}$

(a)   (b)   (c)

Figure 4.   Simulated LCC using MCMC sampling for thinner overlays.

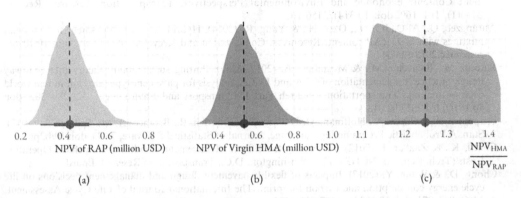

NPV of RAP (million USD)   NPV of Virgin HMA (million USD) $\dfrac{NPV_{HMA}}{NPV_{RAP}}$

(a)   (b)   (c)

Figure 5.   Simulated LCC using MCMC sampling for thicker overlays.

Based on the specified distributions of the input factors, the simulated LCC was given in Figure 4. Given the factors for thinner overlays, the median NPVs of the RAP and virgin HMA are 0.28 million $/km and 0.225 million $/km, respectively. The ratio of these two, as presented in Figure 4(c), has a median of 1.247.

For the thicker overlays, as shown in Figure 5, the median NPVs of the RAP and virgin HMA are 0.539 and 0.430 million $. The median ratio of the two (HMA over RAP) is 1.250.

6 CONCLUSIONS

The use of reclaimed asphalt pavement (RAP) into new and rehabilitation pavement projects has yielded significant economic and environmental benefits. This paper presented a Bayesian treatment of LCCA that considered the uncertainty inherent in different sources of inputs, including the material cost, overlay layer thickness, pavement service life, and the LCC itself. Although the presented analysis was not primarily focused on the environmental effects of using RAP for pavement construction and rehabilitation, which is the main target of a pavement life cycle assessment analysis, the methodology proposed can be extended to quantify the uncertainty related to that as well. Based on the analyses performed, the following observations were obtained:

- The thickness of the overlays can be best described with a log-normal distribution, which concurred with previous studies (Swei et al., 2015; Tighe, 2001);
- Given the initial level of roughness, the evolution of IRI can be estimated via a linear model after a logarithmic transformation with relatively high accuracy (mean R^2 D 0:821) and reliability;

- Based on MCMC sampling, the net present values of the RAP and virgin HMA overlays at the end of their life cycles were obtained. The results from the LCCA revealed that for both the thin and thick overlays, the use of 30% RAP provided a cost savings of 20%.
- Unlike conventional frequentist's methods prevailing in existing literature, the proposed Bayesian LCC model better characterized the uncertainty in both the inputs and LCC by drawing samples directly from their posterior distributions.

REFERENCES

Alkins, A. E., Lane, B., & Kazmierowski, T. (2008). Sustainable Pavements: Environmental, Economic, and Social Benefits of In Situ Pavement Recycling. Transportation Research Record: Journal of the Transportation Research Board, 2084(1), 100–103. doi: 10/dzb3zk

Aurangzeb, Q., & Al-Qadi, I. L. (2014). Asphalt Pavements with High Reclaimed Asphalt Pavement Content: Economic and Environmental Perspectives. Transportation Research Record, 2456(1), 161–169. doi: 10.3141/2456-16

Aurangzeb, Q., Al-Qadi, I. L., Ozer, H., & Yang, R. (2014). Hybrid life cycle assessment for as- phalt mixtures with high RAP content. Resources, Conservation and Recycling, 83, 77–86. doi: 10.1016/j.resconrec.2013.12.004

Batouli, M., Bienvenu, M., & Mostafavi, A. (2017, May). Putting sustainability theory into roadway design practice: Implementation of LCA and LCCA analysis for pavement type selection in real world decision making. Transportation Research Part D: Transport and Environment, 52, 289–302. doi: 10.1016/j.trd.2017.02.018

Carpenter, B., Gelman, A., Hoffman, M. D., Lee, D., Goodrich, B., Betancourt, M., Riddell, A. (2017). Stan: A Probabilistic Programming Language. Journal of Statistical Software, 76(1). doi: 10/b2pm

Chatti, K., & Zaabar, I. (2012). Estimating the Effects of Pavement Condition on Vehicle Operating Costs (Tech. Rep. No. NCHRP 720). Washington, D.C.: Transportation Research Board.

Chong, D., & Wang, Y. (2017). Impacts of flexible pavement design and management decisions on life cycle energy consumption and carbon footprint. The International Journal of Life Cycle Assessment, 22(6), 952–971. doi: 10.1007/s11367-016-1202-x

Gelman, A., Carlin, J. B., Stern, H. S., Dunson, D. B., Vehtari, A., & Rubin, D. B. (2014). Bayesian Data Analysis (3rd ed.). Boca Raton, FL: CRC Press.

Gong, H., Huang, B., & Shu, X. (2018). Field performance evaluation of asphalt mixtures containing high percentage of RAP using LTPP data. Construction and Building Materials, 176, 118–128.

Harvey, J. T., Rezaei, A., & Lee, C. (2012). Probabilistic Approach to Life-Cycle Cost Analysis of Preventive Maintenance Strategies on Flexible Pavements. Transportation Research Record: Journal of the Transportation Research Board, 2292(1), 61–72. doi: 10.3141/2292-08

Horvath, A. (2003, September). Life-Cycle Environmental and Economic Assessment of Using Recycled Materials for Asphalt Pavements.

Huang, B., Li, G., Vukosavljevic, D., Shu, X., & Egan, B. K. (2005). Laboratory Investigation of Mixing Hot-Mix Asphalt with Reclaimed Asphalt Pavement. Transportation Research Record (1929), 37–45.

Li, Z., & Madanu, S. (2009). Highway Project Level Life-Cycle Benefit/Cost Analysis under-Certainty, Risk, and Uncertainty: Methodology with Case Study. Journal of Transportation Engineering, 135(8), 516–526. (00083) doi: 10/dvnxg9

Newcomb, D., Epps, J., & Zhou, F. (2016). Use of RAP & RAS in high binder replacement asphalt mixtures: A synthesis (Tech. Rep. No. Special Report 213). Lanham, MD: National Asphalt Pavement Association.

Okte, E., Al-Qadi, I. L., & Ozer, H. (2019). Effects of Pavement Condition on LCCA User Costs. Transportation Research Record, 2673(5), 339–350. doi: 10.1177/0361198119836776

Osman, H. (2005). Risk-Based Life-Cycle Cost Analysis of Privatized Infrastructure. Transportation Research Record (1924), 192–196.

Ozbay, K., Jawad, D., Parker, N. A., & Hussain, S. (2004). Life-Cycle Cost Analysis: State of the Practice Versus State of the Art. Transportation Research Record, 1864(1), 62–70. doi: 10.3141/1864-09

Rangaraju, P. R., Amirkhanian, S., & Guven, Z. (2008). Life Cycle Cost Analysis for Pavement Type Selection (Final Report No. FHWA-SC-08-01). Clemson, SC: Clemson University.

Reigle, J. A., & Zaniewski, J. P. (2002, January). Risk-Based Life-Cycle Cost Analysis for Project-Level Pavement Management. Transportation Research Record, 1816(1), 34–42. doi: 10.3141/1816-05

Robinette, C., & Epps, J. (2010). Energy, Emissions, Material Conservation, and Prices Associated with Construction, Rehabilitation, and Material Alternatives for Flexible Pavement. Transportation Research Record: Journal of the Transportation Research Board, 2179(1), 10–22. doi: 10/ffhb96

Shu, X., Huang, B., Shrum, E. D., & Jia, X. (2012, October). Laboratory evaluation of moisture susceptibility of foamed warm mix asphalt containing high percentages of RAP. Construction and Building Materials, 35, 125–130. doi: 10.1016/j.conbuildmat.2012.02.095

Swei, O., Gregory, J., & Kirchain, R. (2013). Probabilistic Characterization of Uncertain Inputs in the Life-Cycle Cost Analysis of Pavements. Transportation Research Record, 2366(1), 71–77. (00025) doi: 10.3141/2366-09

Swei, O., Gregory, J., & Kirchain, R. (2015). Probabilistic Life-Cycle Cost Analysis of Pavements: Drivers of Variation and Implications of Context. Transportation Research Record: Journal of the Transportation Research Board, 2523(1), 47–55. (00000) doi: 10/gf3thq

Tighe, S. (2001). Guidelines for Probabilistic Pavement Life Cycle Cost Analysis. Transportation Research Record, 1769(1), 28–38. doi: 10.3141/1769-04

Walls, I. J., & Smith, M. R. (1998). Life-Cycle Cost Analysis in Pavement Design (Tech. Rep. No. FHWA-SA-98-079).

Wang, H., & Wang, Z. (2019). Deterministic and probabilistic life-cycle cost analysis of pavement overlays with different pre-overlay conditions. Road Materials and Pavement Design, 20(1), 58–73. doi: 10.1080/14680629.2017.1374996

Wang, H., Wang, Z., Zhao, J., & Qian, J. (2020). Life-Cycle Cost Analysis of Pay Adjustment for Initial Smoothness of Asphalt Pavement Overlay. Journal of Testing and Evaluation, 48(2), 20170529. doi: 10.1520/JTE20170529

Williams, B. A., Copeland, A., & Ross, C. T. (2017). Asphalt Pavement Industry Survey on Recycled Materials and Warm-Mix Asphalt Usage (Tech. Rep. No. Information Series 138). Washington, D. C.: Federal Highway Administration.

Xu, X., Noshadravan, A., Gregory, J., & Kirchain, R. (2014). Scenario analysis of comparative pavement life cycle assessment using a probabilistic approach. In International symposium on pavement LCA.

Yang, R., Kang, S., Ozer, H., & Al-Qadi, I. L. (2015). Environmental and economic analyses of recycled asphalt concrete mixtures based on material production and potential performance. Resources, Conservation and Recycling, 104, 141–151. doi: 10.1016/j.resconrec.2015.08.014

Yang, R., Ozer, H., Kang, S., & Al-Qadi, I. L. (2014). Environmental impacts of producing asphalt mixtures with varying degrees of recycled asphalt materials. In Proceedings of the international symposium on pavement life-cycle assessment, davis, CA, USA (p. 14–16).

Ziyadi, M., Ozer, H., Kang, S., & Al-Qadi, I. L. (2018). Vehicle energy consumption and an environmental impact calculation model for the transportation infrastructure systems. Journal of Cleaner Production, 174, 424–436. doi: 10/gc25vq

Pavement, Roadway, and Bridge Life Cycle Assessment 2020 – Harvey et al (eds)
© 2020 Taylor & Francis Group, London, ISBN 978-0-367-55166-7

Global warming potential and fossil depletion of enhanced rubber modified asphalt

A. Farina, A. Anctil & M.E. Kutay
Michigan State University, East Lansing, Michigan, USA

ABSTRACT: Crumb Rubber (CR) from scrap tires is used worldwide in road pavement applications. The objective of this study was to quantify the environmental impacts due to the production of hot mix asphalts (HMAs) modified with two types of enhanced CR. A cradle-to-gate life cycle assessment was performed to compare 1 ton of HMA modified with i) engineered CR (ECR), ii) devulcanized rubber (DVR), and iii) ECR and DVR together. All mixtures were compared with a reference styrene-butadiene-styrene (SBS) mix. Results revealed that the CR modified mixes had lower Global Warming Potential and Fossil Depletion than the SBS mixture. The use of ECR and DVR increased the electricity consumption during the manufacturing process with respect to the SBS, but it was beneficial due to the environmental credits obtained by recycling the rubber from scrap tires.

1 INTRODUCTION

In the USA, more than 250 million tires are disposed every year. Of this amount, 80% is reused as fuel for energy recovery, aggregates for construction, ground rubber for asphalt mixtures and other road surfaces (U.S. Tire Manufacturers Association, 2017). Only 16% of scrap tires are still disposed in landfills. To produce ground rubber, scrap tires are shredded and chipped in dedicated plants using machinery that cuts the tires into granules of different sizes. During this process, steel and textile fibers are removed and recycled (Feraldi et al., 2013). The crumb rubber (CR) obtained using this process has been used as an additive in asphalt pavements since the 1960s in the USA (United States Environmental Protection Agency, 1991), due to its chemical composition, the resistance to high temperature, biodegradation, and chemical reagents (Zanetti et al., 2015). The other advantage of ground rubber from scrap tires is that it can improve the mechanical performance of the base material (Kocak and Kutay, 2016; Santagata et al., 2018). However, the asphalt paving application is not the primary market for ground rubber and in 2017, the pavement application accounted for only 12% of the market while the rest was used for playgrounds and sports turfs surfaces.

The wet and dry processes are the two well-accepted techniques to incorporate CR in the hot mix asphalt (HMAs) (Epps, 1994). The wet process uses CR particles as a polymeric modifying agent of the base bitumen in which they can be partially or completely digested. The dry process uses CR particles as a partial replacement for the aggregate structure, and their interaction with the asphalt binder is minimal. Improved CR technologies such as devulcanized CR (DVR) and engineered CR (ECR) have lower rubber degradation and higher economic value. The devulcanization process is used to improve the processability and plasticity of the rubber from tires, which then facilitates its use in pavement material.

To break the sulfur-to-sulfur bonds of the rubber, chemical, thermal, mechanical, and thermomechanical techniques can be used (Garcia et al., 2015). The DVR is used to modify the binder at the asphalt plant in the wet process. The main advantage of this technology is that the DVR particles can be dissolved almost completely when mixed with the asphalt binder at high temperatures. This way, the modified binder is fluid with practically no suspended particles. However, using only the DVR is not sufficient since it does not improve the asphalt binder characteristics as the performance grade. Thus, it can be used to replace the SBS modification partially. One other emerging material is a chemically enhanced ECR, also named polymer-coated crumb rubber. In the ECR, a polymeric emulsion partially covers the surface area of the rubber particles. The polymer promotes and controls the reaction with the asphalt and allows the development of strong bonds between the materials during the mixture production process. The advantage of the ECR is that it can be used as a dry technology.

Life Cycle Assessment (LCA) is a comprehensive methodology used to evaluate and quantify the environmental impacts of products and services (EC JRC, 2010). When applied to road pavements, this methodology can help support paving solutions decisions by adding environmental factors to economic and mechanical performance. LCA is particularly useful to evaluate the benefits of asphalt mixtures containing recycled materials such as CR, plastic, fly ash, and glass to estimate the resource depletion. LCA has been used by many practitioners to identify the asphalt mixture with the lowest environmental impacts. Multiple studies have found using LCA, that the Global Warming Potential (GWP), Cumulative Energy Demand (CED), and all impacts from the Recipe and Eco-indicator 99 methods are lower for reclaimed asphalt pavement (RAP) compared to a traditional mix (Chiu et al., 2008; Farina et al., 2017; Bressi et al., 2019). RAP is the recycled material obtained from milling operations of old pavements and it can partially substitute the natural aggregates. The carbon footprint and embodied energy are lowered by using 30-40% by weight of RAP in the new mix and lower mixing temperature (150°C instead of 160-170°C) (Giunta et al., 2020). The environmental benefit of using CR in current LCA literature is not so clear. The outcomes of one study reported a 16%increase in Eco points from the Eco-indicator 99 method by using CR compared to the unmodified mix (Chiu et al., 2008). The higher results were associated with a higher asphalt binder content and heat usage to produce the rubberized mix. However, the same thickness of the surface layer and equal service life were used for all mixes considered. Other studies showed that these two factors, thickness and service life, play an important role since CR improves the mechanical performance of road pavements. The GWP and CED reduction varied from 36 to 45% when using CR instead of the traditional mix, due to higher durability and reduced thickness of the surface layer (Farina et al., 2017; Bressi et al., 2019).

However, the existing LCA literature on rubberized asphalts focuses on the use of wet and dry technologies referred to as regular CR in HMAs. The LCA of asphalt mixes modified with enhanced CR like the ECR or HMAs modified with a combination of different CR types has not been done. The objective of this study was to quantify the global warming potential (GWP) and the fossil depletion (FD) due to the production phase of asphalt mixes modified with i) ECR, ii) DVR, and iii) ECR and DVR together.

2 MATERIALS AND METHODS

2.1 Scenarios

This study included the HMAs modified with CR from scrap tires and the reference scenario:

- SBS mix: reference scenario. The bitumen was modified using 3.5 % of SBS by the weight of the asphalt binder used. This amount corresponds to 0.18% by HMA total weight. A sulfur catalyst solution was also used as a cross-linker with a ratio of 0.4% by weight of the total binder. The bitumen with a Performance Grade (PG) of 70-28 was 5.83% by weight of the total mixture.
- ECR 0.5% and 1% mixes: the ECR was added using the dry technology with a ratio of 0.5% and 1%, respectively, by weight of the asphalt mixture. The bitumen with a PG of 58-28 was 5.67% by weight of the total mix in both cases.
- DVR mix: the DVR was added in the mix using the wet technology at a rate of 7% by weight of total bitumen (0.31% by HMA total weight), with 2% of SBS content and 0.4% of cross-linker. The bitumen with a PG of 70-28 was 5.45% by weight of the total mixture.
- ECR DVR mix: the ECR and DVR were added in the blend utilizing the wet technology at a rate of 8% and 7% respectively, by weight of total bitumen (0.86% of scrap rubber by

HMA total weight), with 0.8% of cross-linker. The bitumen with a PG of 70-28 was 6.5% by weight of the total mixture.

Mixes were designed and prepared in the laboratory as the first step of this research project, investigating the mechanical performance of mixes through asphalt performance testing. Mixes were designed for a surface layer and an expected traffic level of 3 million equivalent single axle load (ESALs), for SBS, ECR, and ECR DVR mixes, and 1 million ESALs for the DVR mixture. The mix design for all HMAs followed the Michigan Department of Transportation's (MDOT) Superpave specifications (Michigan Department of Transportation, 2012). The mixtures compared in this study were prepared in the laboratory and had similar aggregate gradation with a nominal maximum size of 12.5 mm and a target air voids of 3±0.5%. The RAP content was 15% by weight of the total mixture for all scenarios except for the mix modified with DVR, where the RAP was 20% of the total weight. The binder content in the RAP was measured to be 5.25% by weight, and it was considered fully blended in the mix, as standard practice in asphalt plants in Michigan. Consequently, the virgin bitumen was reduced by 12 to 19% by weight of the total asphalt binder needed in the mixes. Table 1 shows the materials consumption for each blend based on the mix design.

2.2 LCA goal and scope

The goal of this study was to evaluate the environmental impact of asphalt mixtures modified with different CR technologies. Attributional LCA was used to compare CR from scrap tires and the synthetic polymer (SBS) in asphalt mixtures and focused on the manufacturing phase (cradle-to-gate analysis). The functional unit (FU) was one metric ton of hot mix asphalt produced in a dedicated plant. The system boundaries included raw materials sourcing, composite materials production, and transportation. The explanation of all processes and data collection are available in 2.3. Allocation methods and virgin materials used in tires and RAP were not considered (Harvey et al., 2016). The two impact categories considered were global warming potential (GWP) and fossil depletion (FD), calculated using the TRACI 2.1 method (U.S. Environmental Protection Agency (EPA), 2012). The GWP, expressed in kg of CO_2 equivalent, is an indicator comparing the global warming impacts of different gases (greenhouse gases, methane, nitrous oxide, etc.). The GWP measures how much energy the emissions of 1 kg of gas can absorb over a period of time (usually, 100 years), relative to the emissions of 1 kg of carbon dioxide (U.S. Environmental Protection Agency (EPA), 2017). The fossil depletion is expressed in a surplus of energy (Surplus MJ). Surplus energy is a concept based on the assumption that as more of a resource is extracted over time, the quality of deposits still available tends to decrease. Each extraction in the present will lead to more energy-intensive mining from lower-quality and less accessible deposits in the future (Klinglmair et al., 2014).

Table 1. Material input for 1 ton of hot mix asphalt.

Material	SBS kg	ECR0.5% kg	ECR1% kg	DVR kg	ECR DVR kg
Virgin Aggregates	800.45	797.56	793.31	756.40	794.75
Aggregates from RAP	141.26	140.75	140.00	199.58	148.02
Neat Bitumen	48.50	48.90	48.94	39.88	48.19
Bitumen from RAP	7.83	7.80	7.76	10.48	7.77
Synthetic polymer	1.77	-	-	0.88	-
Polymer coated CR	-	5.00	10.00	-	4.58
Devulcanized CR	-	-	-	3.08	4.01
Cross-linker	0.20	-	-	0.18	0.46

Notes: SBS= Hot Mix Asphalt modified with Styrene-Butadiene-Styrene; ECR0.5% and ECR1% = Hot Mix Asphalt modified with 0.5% and 1% of Engineered Crumb Rubber, respectively; RAP= Reclaimed Asphalt Pavement.

2.3 Life cycle inventory

In this project, material quantities for each mix were determined by the mix design obtained from our asphalt laboratory. Additional information regarding the manufacturing of DVR and ECR was provided by the manufacturers. Data for aggregates sourcing, bitumen and CR production were collected from the available literature.

Distances for the materials' transportation were included based on the commodity flow survey for the USA (US Census Bureau, 2012). To perform the LCA, SimaPro 8.5 (PRé Sustainability, 2018) and existing databases, Ecoinvent 3.0 (Wernet et al., 2016), USLCI (Torcellini and Deru, 2004), and US-EI 2.2 (LTS, 2018) were used. Specifically, the processes for electricity, chemicals for SBS and DVR processes, fuel, and transportation were taken from the databases and adjusted to represent the average US conditions.

2.3.1 Aggregates and RAP

The aggregates from natural rocks are classified based on their size as crushed stones, gravels, and sands and define the structure of asphalt mixtures. The energy consumption needed in quarries activities was included as recommended by the Product Category Rules for asphalt mixtures (National Asphalt Pavement Association (NAPA), 2016) based on data from the Portland Concrete Association (Marceau, Nisbet and Vangeem, 2007). The activities included were blasting, wet drilling in unfragmented rock, product loading in open truck, unloading primary and crushing, screening, conveyor, transfer, and storage piles. As summarized in Table 2, depending on the size of the aggregates, either gravel and sand or coarse aggregates data was used.

After the old pavements are milled, the RAP is stored in asphalt mixing plants, ready to be used in new mixtures. The fuel consumption of a milling machine, producing 350 tons/hour, was assumed to be 56,397 Btu per ton of RAP produced (Mukherjee, 2016). The avoided RAP disposal in a landfill was included as an environmental credit. A distance of 50 km was assumed for transporting the RAP to the asphalt plant by truck.

2.3.2 Bitumen

Bitumen is a petroleum product derived from the portioned distillation of crude oil in a refinery. It is used in asphalt mixtures because of its water-resistant, thermoplastic and visco-elastic properties. LCA of bitumen was modeled using the method proposed by Yang (2014) using updated data up to 2016 for the United States. The system boundaries included domestic crude oil extraction, flaring, domestic and foreign crude oil transportation, refinery, and storage. The economic allocation was used since bitumen has many valuable coproducts such as kerosene, fuel oil, and petroleum coke (International Organization for Standardization (ISO), 2012). Allocation coefficient, mass residue yield, and allocation factor were calculated. The allocation coefficient is calculated using the ratio between the price of the bitumen multiplied by its mass over the summation of all coproduct's prices multiplied by their mass. The mass residue yield is obtained by dividing the mass of the bitumen produced by the total mass of all coproducts. The allocation factor is the ratio between the allocation coefficient and the

Table 2. Energy consumptions in quarries operations to produce 1 ton of aggregates for asphalt mixes (Marceau, Nisbet and Vangeem, 2007).

Unit Process	Gravel and sand (0.075 – 6.30 mm)	Coarse aggregates (6.30 – 12.50 mm)
Gasoline, combusted in equipment/US (gal)	5.43E-3	9.4E-3
Diesel, burned in building machine/GLO US (Btu)	9681	15,087
Natural gas, combusted in industrial boiler NREL/US (ft^3)	1.33	3.45
Electricity, medium voltage, at grid, 2015/US US-EI U (kWh)	2.41	2.96
Electricity, hard coal, at power plant/US US-EI U (kWh)	-	4E-6
Transport, lorry 16-32t, EURO4/US- US-EI (tkm)	50	50

mass residue yield (Yang, 2014). Based on the mass (EIA, 2019b) and the price (EIA, 2019a) of all coproducts produced from domestic and imported crude oil in 2016, the allocation coefficient was 0.0132, while the mass residue yield was 0.0125. Therefore, 1.32% of the total economic output of the refinery and 2.15% of total mass output was bitumen. The amount of crude oil, materials, and energy summarized in Table 3 were multiplied by the allocation factor equal to 0.615 to allocate the bitumen.

According to the Eurobitumen report (European Bitumen Association, 2012), the energy consumption for the asphalt modification with rubber and synthetic polymers, such as SBS, DVR, and ECR DVR mixes scenarios, is 72 MJ per ton of modified bitumen.

2.3.3 Crumb rubber from scrap tires

At the end of their life span, tires are collected and then sorted based on different second lives applications. An average distance of 300 km was assumed for the transportation of scrap tires. Based on the U.S. Tire Manufacturers Association (2017), in 2017 the use of tires as fuel in cement kilns was the primary recycling method (43%), followed by the production of ground rubber (25%), landfill disposal (16%), and other applications (16%). In general, other applications include export, the use as tire-derived aggregates, and other unknown purposes. Since scrap tires are a valuable waste, we used economic allocation. In the future, the mass allocation coefficient for ground rubber could be higher than other options because the trend is to favor the production of ground rubber instead of the incineration of scrap tires. Consequently, the economic allocation coefficient could be lower due to the decreasing price for producing ground rubber. Lower CR market price will result in higher environmental credits associated with the use of scrap tire in second life. Table 4 summarizes the average costs for scrap tires used to calculate the economic allocation factors. Prices are based on publicly available information (BCA Industries, 2018).

After the collection phase, tires components are separated to produce ground rubber, and part of the rubber is lost in the textile fraction (Feraldi et al., 2013). For this reason, on average only 0.69 tons of ground rubber is obtained from 1 ton of scrap tires. Table 5 reports the

Table 3. Materials and energy used to refine 1 ton of crude oil (Yang, 2014).

Unit Process (units)	
Natural gas, sweet, burned in production flare (m³)	0.013
Natural gas, sour, burned in production flare (m³)	0.002
Electricity, medium voltage, at grid, 2015/US (kWh)	0.105
Steam, in chemical industry, production (kg)	0.124
Natural gas, burned in boiler modulating >100kW/US (MJ)	2.200
Diesel, burned in diesel-electric generating set (MJ)	0.028
Bituminous coal, combusted in industrial boiler NREL/US (kg)	2.4E-5
Residual fuel oil, combusted in industrial boiler NREL/US (l)	0.001
Refinery gas, burned in furnace (MJ)	0.360
Liquefied petroleum gas, combusted in industrial boiler NREL/US (l)	0.009

Table 4. Price and economic allocation coefficient for scrap tires.

	Price	Allocation coefficient
	$/t	
Tire-derived fuel	40	0.08
Ground rubber	300	0.59
Landfill	150	0.3
Other	15	0.03

Table 5. Materials and energy used to produce 1 ton of crumb rubber (Feraldi *et al.*, 2013).

Unit Process	
Input	
Scrap tires for ground rubber (ton)	1.45
Steel blades (kg)	0.9
Tap water, tap water production, conventional treatment (kg)	67
Electricity, medium voltage, at grid, 2015/US US-EI U (kWh)	513
Avoided products	
Cast iron {RoW}| production (ton)	0.23
Petroleum coke {RoW}| petroleum refinery operation (ton)	0.16
Emissions to air	
Particulates, unspecified (kg)	0.09

avoided products referred to those virgin materials whose production can be saved by recycling the coproducts of the rubber. Recycled steel can avoid the production of virgin steel and the fibers, used (Feraldi *et al.*, 2013) as fuel in kilns, prevents the production of petroleum coke. Based on the calorific value of the petroleum coke and the fibers, it has been calculated that 0.8 tons of textile fibers can substitute 1 ton of traditional fossil fuel. A distance of 600 km was assumed to transport the CR from the production plant to the next facilities.

2.3.4 *ECR and DVR*

The ECR is a combination of regular crumb rubber and polymer emulsion. The styrene-butadiene-rubber (SBR) is usually used as polymer emulsion to create a film that partially covers the surface area of the rubber particles (Kurgan and Dongre, 2015). In this study, the ECR composition was assumed to be 5% of SBR and 95% of CR. The mixing energy was 72 MJ per ton of ECR produced, and the transportation was assumed to be 1000 km by truck because the only producer is located in Washington D.C

The DVR is a pellet obtained after applying a chemical and mechanical process to a regular 0.595 mm CR. The chemical process uses compatibilizers, plasticizers, and reagents, which are needed to break the sulfur-to-sulfur bonds of the rubber. The mechanical process is used to create pellets. According to the manufacturer (Full Circle Technologies, LLC, Cleveland Ohio) and based on the chemicals available in the databases, Table 6 shows the unit processes used to model the. The transportation distance was assumed to be 500 km.

2.3.5 *Hot mix asphalts (HMAs)*

The HMAs investigated in this study were hot mix asphalts prepared in the laboratory at temperatures between 150 and 175°C. The energy consumption for the mixes production was from an LCA study developed in 2016 to support the Product Category Rule for asphalt mixtures for the Environmental Product Declaration (Mukherjee, 2016), based on the average of 40 plants in the USA. The energy consumption was assumed to be 317 MJ per ton of HMA produced even though the temperature slightly changed for the various mixes.

Table 6. Materials and energy used to produce 1 ton of DVR.

Unit Process	
Crumb rubber (kg)	1000
Petroleum refining coproduct, unspecified, at refinery/kg NREL/US (kg)	20
2,5-dimethylhexane-2,5-dihydroperoxide production (kg)	20
SBS (kg)	20
Electricity, medium voltage, at grid, 2015/US US-EI (kWh)	429

3 RESULTS AND DISCUSSION

Figure 1 shows the total GWP and FD results for each asphalt mixture and the percentage reduction of the CR mixes compared to the SBS HMA. Overall, all mixes modified with CR performed better than the SBS mix.

Figure 2 depicts the difference between each scenario that includes CR and the reference mix modified with SBS, which is represented by the zero lines. The positive and negative contributions indicate respectively the environmental burdens and credits due to each material included in the mix. The figure shows only the DVR, ECR, SBS, and bitumen contribution because the other impacts due to the aggregates production, the use of RAP, and the energy consumption in the asphalt plant, assumed the same value for all mixes. The DVR mix is the scenario with the highest environmental credits due to the lowest virgin bitumen content.

The bitumen and the energy spent in the asphalt plant have the highest contribution with respect to the total impacts. For the GWP, based on the different scenarios, the contribution of the bitumen was 47 to 55% of the impacts, while the asphalt plant operations burdens were 26 to 29%. For the fossil depletion, 79 to 83% of the impacts were associated with bitumen usage, while 10 to 12% was due to the asphalt plant energy consumption. Figure 3 provides additional information on the impact of bitumen content in HMA to fossil depletion.

The use of ECR and DVR in dry and wet mixes increased the electricity consumption during the manufacturing process compared to the production of SBS, but overall, it was beneficial due to the environmental credits associated with the recycling of the rubber from scrap tires. In the specific case of the dry HMAs with two contents of ECR (0.5% and 1%) results revealed the need for a comprehensive LCA, besides the cradle-to-gate analysis, since including more scrap tires in HMAs doesn't necessarily increase the environmental benefits.

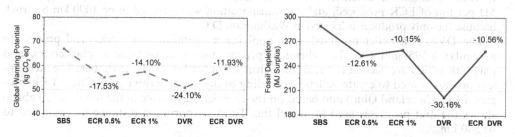

Figure 1. Global warming potential (kg CO_2 eq) and fossil depletion (MJ Surplus) per ton of mix produced.

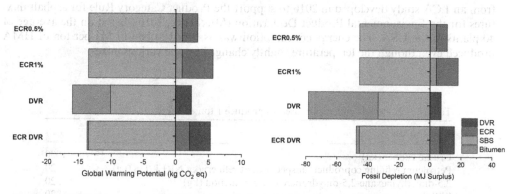

Figure 2. Global warming potential and fossil depletion comparison between CR asphalt mixes and SBS reference scenario.

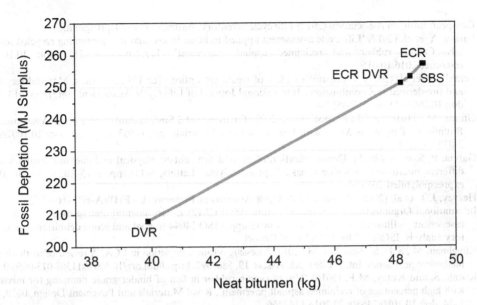

Figure 3. Fossil depletion (MJ Surplus) associated with the virgin bitumen content in the asphalt mix.

4 CONCLUSIONS AND FUTURE WORK

The goal of this study was to evaluate the environmental impacts of enhanced CR modified asphalt mixes by performing a cradle-to-gate LCA. Results of the assessment based on the production of 1 ton of each CR HMA compared to the SBS reference scenario, showed that overall, the mixes containing scrap tires performed better than the blend modified with the synthetic polymer. A reduction between 11.93 to 24.10% for GWP, and 10.15 to 30.16% for fossil depletion was calculated. Although the use of ECR and DVR in dry and wet mixes increased the electricity consumption during the manufacturing process, their usage was still beneficial due to the environmental credits obtained by the recycling of the rubber from scrap tires.

Future work should include sensitivity analysis on the system boundaries to consider the impacts associated with virgin resources used in the first life of recycled materials. The next step of this study will also consider construction, maintenance and reconstruction phases over the lifespan of a road pavement. Results of the life cycle assessment will be affected by the mechanical performance of each HMA, based on the results of the asphalt performance testing.

REFERENCES

BCA Industries (2018) Scrap Tire Processing. Available at: http://www.profitablerecycling.com/tirerecy cling.htm.

Bressi, S., Santos, J., Orešković, M., Losa, M., 2019. A comparative environmental impact analysis of asphalt mixtures containing crumb rubber and reclaimed asphalt pavement using life cycle assessment. Int. J. Pavement Eng. 0, 1–15. https://doi.org/10.1080/10298436.2019.1623404

Chiu, C.-T., Hsu, T.-H. and Yang, W.-F. (2008) 'Life cycle assessment on using recycled materials for rehabilitating asphalt pavements', Resources, Conservation and Recycling. doi: 10.1016/j .resconrec.2007.07.001.

EC JRC (2010) ILCD Handbook: General Guide for Life Cycle Assessment – Detailed Guidance. European Commission, Joint Research Centre, Institute for Environment and Sustainability, Luxembourg.

EIA (2019a) Open data. Available at: https://www.eia.gov/opendata/qb.php?category=714757.

EIA (2019b) Refinery&Blender Net Production. Available at: https://www.eia.gov/dnav/pet/pet_pn p_refp_dc_nus_mbbl_a.htm.

Epps, J. A. (1994) NCHRP 198. Uses of Recycled Rubber Tires in Highways.

European Bitumen Association (2012) Life cycle inventory: Bitumen. Brussels, Belgium.

Farina, A. et al. (2017) 'Life cycle assessment applied to bituminous mixtures containing recycled materials: Crumb rubber and reclaimed asphalt pavement', 117, pp. 204–212. doi: 10.1016/j.resconrec.2016.10.015.

Feraldi, R. et al. (2013) 'Comparative LCA of treatment options for US scrap tires: Material recycling and tire-derived fuel combustion', International Journal of Life Cycle Assessment, 18(3), pp. 613–625. doi: 10.1007/s11367-012-0514-8.

Giunta, M., Mistretta, M., Pratico', F.G., 2020. Environmental Sustainability and Energy Assessment of Bituminous Pavements Made with Unconventional Materials. pp. 1–503. https://doi.org/10.1007/978-3-030-29779-4

Garcia, P. S. et al. (2015) 'Devulcanization of ground tire rubber: Physical and chemical changes after different microwave exposure times', Express Polymer Letters, 9(11), pp. 1015–1026. doi: 10.3144/expresspolymlett.2015.91.

Harvey, J. T. et al. (2016) 'Pavement Life Cycle Assessment Framework - FHWA-HIF-16-014'.

International Organization for Standardization (ISO) (2012) Environmental management — Life cycle assessment — Illustrative examples on how to apply ISO 14044 to goal and scope definition and inventory analysis. ISO/TR 14049. Technical Report.

Klinglmair, M., Sala, S., Brandão, M., 2014. Assessing resource depletion in LCA: A review of methods and methodological issues. Int. J. Life Cycle Assess. 19, 580–592. https://doi.org/10.1007/s11367-013-0650-9

Kocak, S. and Kutay, M.E. (2016) 'Use of crumb rubber in lieu of binder grade bumping for mixtures with high percentage of reclaimed asphalt pavement', Road Materials and Pavement Design, 0629, pp. 1–14. doi: 10.1080/14680629.2016.1142466.

Kurgan G. and Dongre R. (2015) 'International Trends in Low-Carbon/Low-Energy Pavement Construction. Workshop on the Adoption of Innovative Technologies and Materials for Road Construction in India'.

Lee, J.C., Edil, T.B., Tinjum, J.M., Benson, C.H., 2010. Quantitative assessment of environmental and economic benefits of recycled materials in highway construction. Transp. Res. Rec. 138–142. https://doi.org/10.3141/2158-17

LTS (2018) DATASMART LCI Package. Available at: http://ltsexperts.com/services/software/datasmart-life-cycle-inventory/.

Marceau, M. L., Nisbet, M. A. and Vangeem, M. G. (2007) 'Life Cycle Inventory of Portland Cement Concrete'.

Michigan Department of Trasnportation (2012) 'Standard specifications for construction'.

Mukherjee, A. (2016) 'Life Cycle Assessment of Asphalt Mixtures in Support of an Environmental Product Declaration', (June).

National Asphalt Pavement Association (NAPA) (2016) Product Category Rules (PCR) for Asphalt Mixtures.

PRé Sustainability (2018) About SimaPro, https://simapro.com/about/.

Santagata, E. et al. (2018) 'Influence of lateral confining pressure on flow number tests', Bearing Capacity of Roads, Railways and Airfields - Proceedings of the 10th International Conference on the Bearing Capacity of Roads, Railways and Airfields, BCRRA 2017, pp. 237–242. doi: 10.1201/9781315100333-35.

Torcellini, P. and Deru, M. (2004) 'U.S. LCI Database Project Development Guidelines'. AthenaTM Sustainable Materials Institute Merrickville, Ontario, Canada. Available at: http://www.nrel.gov/lci/docs/dataguidelinesfinalrpt1-13-04.doc.

U.S. Environmental Protection Agency (EPA) (2012) Tool for the Reduction and Assessment of Chemical and other Environmental Impacts (TRACI) - USER ' S MANUAL.

U.S. Environmental Protection Agency (EPA), 2017. Greenhouse Gas Emissions - Understanding Global Warming Potentials [WWW Document]. URL https://www.epa.gov/ghgemissions/understanding-global-warming-potentials

U.S. Tire Manufacturers Association (2017) 2015 U.S. Scrap Tire Management Summary. Available at: https://www.ustires.org/scrap-tire-markets.

United States Environmental Protection Agency (1991) 'Markets for Scrap Tires -EPA/530-SW-90-074A October 1991'.

US Census Bureau (2012) '2007 Commodity Flow Survey', (February).

Wernet, G. et al. (2016) 'The ecoinvent database version 3 (part I): overview and methodology. The International Journal of Life Cycle Assessment'. Available at: http://link.springer.com/10.1007/s11367-016-1087-8.

Yang, R. Y. (2014) Development of a pavement life cycle assessment tool utilizing regional data and introducing an asphalt binder model. University of Illinois at Urbana-Champaign.

Zanetti, M. C. et al. (2015) 'Characterization of crumb rubber from end-of-life tyres for paving applications', Waste Management. Elsevier Ltd, 45, pp. 161–170. doi: 10.1016/j.wasman.2015.05.003.

Interpreting life cycle assessment results of bio-recycled asphalt pavements for more informed decision-making

K. Mantalovas
University of Palermo, Palermo, Italy

A. Jiménez del Barco Carrión
University of Granada, Granada, Spain

J. Blanc, E. Chailleux & P. Hornych
IFSTTAR, Bouguenais, France

J.P. Planche
Western Research Institute, Laramie, USA

L. Porot
KRATON, Almere, The Netherlands

S. Pouget
EIFFAGE, Corbas, France

C. Williams
Iowa state university, Ames, USA

D. Lo Presti
University of Palermo, Palermo, Italy; University of Nottingham, Nottingham, UK

ABSTRACT: Due to emerging climate challenges, engineers are looking to replace the conventional asphalt mixtures by utilizing bio-materials combined with Reclaimed Asphalt (RA). However, there is insufficient record in the literature assessing their environmental impacts. This study addresses the analysis of the results obtained from a life cycle assessment exercise of bio-recycled asphalt pavements, containing bio-materials and RA within their binder courses, developed within the BioRePavation project. The aim is to analyze the environmental benefits achieved, by means of hotspot and sensitivity analyses of the most impactful factors of their lifecycles. Two alternatives are compared to a conventional asphalt pavement and recommendations on how to improve the current practices are provided. Results rank the raw material acquisition and replacement as the most impactful stages, while through sensitivity analysis, linear relationships between the impact category indicator values and the RA% used, are detected.

1 INTRODUCTION

For the requirements of the road engineering industry to be satisfied, significant amounts of raw materials are extracted, and high quantities of asphalt mixtures are produced in a global scale. Approximately 5.2 million kilometers of roads are constructed within Europe according to the European Asphalt Pavement Association (2016). Approximately 280 million tonnes of asphalt mixtures are produced in a European scale annually, in order to construct and maintain these roads. Noteworthy effects are forced on the natural resources and the nature itself.

Since sustainable development and the conservation of the ecosystems have recently been in the focal point, the road engineering industry has been driven towards a more environmentally friendly framework of operating, in order to diminish its ecological effect as described by Rodríguez-Alloza (2015). This led to the development of technologies able to reduce the environmental impacts of the asphalt mixture production. Some of these technologies include the utilisation of alternative materials in order to enhance the environmental performance of asphalt pavements, while conserving or improving the natural environment (Aurangzeb et al., 2014; Dinis-Almeida and Afonso, 2015; Mohd Hasan and You, 2016; Sun et al., 2016; Wang et al., 2016; Zhang et al., 2017). The utilization of Reclaimed Asphalt (RA) for the production of asphalt mixtures, being one of these technologies, can have environmental benefits while maintaining an equivalently acceptable technical performance for the asphalt pavements; as it has already been proved (Giani et al., 2015; Presti et al., 2017). In fact, an asphalt mixture with RA, incorporates less bituminous binder and virgin aggregates when compared to a conventional one. This is thus the reason why the environmental impacts are reduced, and the energy demand can be decreased in certain cases by 23%, in comparison to conventional Hot Mix Asphalt (HMA) mixtures (Chiu et al., 2008; Newcomb et al., 2007; Song et al., 2018; Winkle, 2014). In addition, lately, the use of the so-called "by-products" is surprisingly increasing (Giani et al., 2015). Due to the depletion of the natural reserves of crude oil, engineers have been driven towards the utilization of alternative paving materials, while adopting approaches that can be less harmful for the ecosystems (Bre, 2013; Chowdhury et al., 2010; Dinis-Almeida and Afonso, 2015; Fini et al., 2012; Klabunde and Ph, 2014). A considerably feasible approach that is both effective and possibly less harmful for the ecosystems is the utilization of bio-materials for the production of asphalt mixtures (Zhang et al., 2017). Materials that originate from biomass can be characterized as a more sustainable alternative to the traditional, petroleum-based bituminous binders (Podolsky et al., 2016; Yang et al., 2017). They could come from rapeseed and soybeans (Onay and Koçkar, 2006; Şensöz and Kaynar, 2006) animal waste (Fini et al., 2012; Mills-Beale et al., 2014), green waste (Klabunde and Ph, 2014), microalgue (Yang et al., 2013), tea and coffee residue (Uzun et al., 2010). It seems that the technical performance of the asphalt mixtures containing materials derived from bio-mass are comparable with the performance of the conventional asphalt mixtures (Rodríguez-Alloza et al., 2015), but their environmental footprint has not appropriately been surveyed yet.

Life Cycle Assessment (LCA) is a tool that can be utilized to quantify the environmental impacts of a service, a product or a process over its whole life cycle. It is widely recommended for the assessment and the comparison of the environmental impacts of different types of asphalt pavements (Anastasiou et al., 2015; Chowdhury et al., 2010; Wolf et al., 2012). Moreover, the framework of LCA has been internationally standardized via ISO 14040 and 14044 (International Organization for Standardization, 2006; The International Standards Organisation, 2006). This study thus, addresses the interpretative analysis of the results of an LCA study conducted within the BioRePavation project. Asphalt pavements incorporating bio-materials have been constructed and validated within the aforementioned project and the results obtained from their environmental performance evaluation are furtherly analysed in this study. A hotspot analysis is performed in order to identify the most impactful life cycle stages, along with a sensitivity analysis on different RA percentages in order to tackle uncertainty aspects and reinforce the positive impacts of the RA utilization.

2 MATERIALS AND METHODS

2.1 Scope of the study

In this study, the focus is not oriented towards the implementation of the LCA itself. It goes a step further by attempting to analyze and interpret the results provided by the LCA case study, in order to provide practical recommendations to the involved stakeholders towards more sustainable and informed decision-making. Thus, a hotspot analysis was performed identifying the most impactful stages of the life cycle of the asphalt pavements, along with a sensitivity analysis on the re-recycling efficiency of the asphalt pavements during their life cycles.

2.1.1 *System boundaries and functional unit*

The functional unit is defined as the material included in $1m^2$ of asphalted surface and 14cm depth (wearing course and binder course) of a specific chainage of a French highway, meeting the performance requirements during the reference service life of the construction, which is set to be 30 years. The system boundaries are defined for a "cradle-to-grave" analysis as shown in Figure 1. This includes four stages of the pavement: product stage, construction stage, replacement and end-of-life (EoL). The full use stage has not been considered in this study but instead, only the replacement (B4) process has been taken under consideration, due to data limitations. The processes to be considered in each life cycle stage have been defined according to EN 15804 and Product Category Rules for Asphalt from the Norwegian EPD Foundation (2017).

The detailed life cycle inventory for the production of the asphalt mixtures that contain the bio-materials along with the detailed natural gas consumption, the machinery utilized, the transport distances of the materials and the electricity consumed are being published elsewhere. The final results are presented here in terms of values of MidPoint and EndPoint indicators, following the ReCiPe 2016 (H) impact assessment methodology (Huijbregts et al., 2016). This method translates emissions and resource extractions into a limited number of environmental impact scores by means of characterization factors, at midpoint and endpoint level. The former focuses on single environmental problems, for example climate change. The latter shows the environmental impact on three higher aggregation levels; effect on human health, biodiversity, and resource scarcity. The LCA was conducted according to ISO 14040, 14044, EN 15804 (2013), Product Category Rules and Environmental Product Declarations by EAPA (2016) and the Norwegian EPD Foundation (2009).

2.1.2 *Assumptions and considerations for the completion of the study*

For the completion of the LCA, some assumptions and considerations were made:

- Stack emissions data from the asphalt plant was available for the production for a specific asphalt mixture (binder, RA%, etc). The representativeness of this data was not judged adequate and thus this data was omitted.
- No transport distances were considered for the RA, since the stockpiling and processing infrastructure was located at the mixing plant.

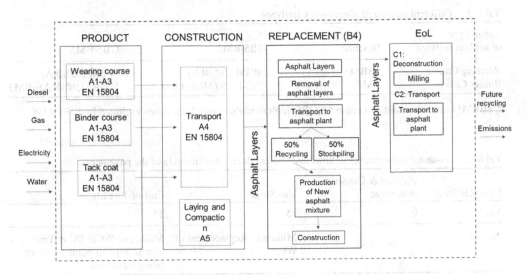

Figure 1. Definition of the system boundaries.

- Primary data for the life cycle inventory of the bio-materials and asphalt pavements were collected by the BioRePavation partners and will soon be available online.
- All the processes necessary to produce the bio-materials were taken under consideration; and thus, the bio-materials are not derived from end of life products.

2.2 *Materials and alternatives*

Three asphalt pavements have been environmentally assessed by utilizing the framework of life cycle assessment. A conventional asphalt pavement (baseline), an asphalt pavement incorporating a bio-based binder designed to completely replace virgin bitumen (BM2) and an asphalt pavement incorporating a bio-additive derived from epoxidized soya bean oil, utilized as a rejuvenator for asphalt mixtures with RA (BM3). The investigated pavement alternatives, following the same maintenance strategies can be seen in Table 1.

In the LCA of the described asphalt pavements, the stages of production, construction, replacement (maintenance) and end-of-life were all taken under consideration. However, during the end of life stage of the pavements, there has been the consideration that no waste is created; instead, the milled layers are transported to a mixing/storage plant for future recycling. Specifically, in the stage of replacement, since the case studies are taking place in the area of France, it is worth mentioning that the maintenance strategies were tailored for the French context. According to typical maintenance strategies in France wearing courses are replaced after 15 years of service life and binder courses are replaced after 30 years of service (Jullien et al., 2014; Laurent, 2004; Organisation de Cooperation et de developpement Économiques, 2005). In this regard, the End-of-Life of the system studied in this LCA is at year 30 from the construction, while at year 15 the wearing course is replaced. The processes involved in the replacement stage are the milling of the wearing course and its transport to the asphalt plant for HMA recycling, along with the RA processing and the asphalt mixture manufacturing for a new wearing course. Furthermore, the transport of the asphalt mixture to the construction site, the tack coat application and the paving process were also considered. Finally, the processes considered in the End-of-Life stage are the milling of the wearing and binder courses and their transport to the asphalt plant for stockpiling and/or screening and processing. Table 2 visualizes the assumed maintenance actions over the lifecycle of the pavements.

Table 1. Different pavement alternatives analyzed.

Layer/Type of asphalt mixture	Baseline	GB5-BM2	GB5-BM3
Wearing Course	*BBM (20%RA)	*BBM (20%RA)	*BBM (20%RA)
Binder Course	EME2 (20%RA)	**GB5 (50%RA)+BM2	**GB5 (50%RA)+BM3

* (BBM) bitumen bound macadam **(GB5) high modulus asphalt mixture developed by EIFFAGE

Table 2. Assumed maintenance actions undertaken during the lifecycle of the pavements.

Lifecycle Stage	Product & Construction Stage	Use Stage	End-of-Life Stage
Year	0	15	30
Action	-	Milling & Replacement of WC	Milling of WC& BC & Transport to asphalt plant for stockpiling/treatment

3 RESULTS OF THE LIFE CYCLE ASSESSMENT

3.1 *Life cycle assessment of asphalt pavements results*

In this chapter the results of the LCA exercise of the pavements is presented. The values of the impact category indicators are presented in Table 3 in the Supplemental Material. As ReCiPe 2016 has been utilized as the impact assessment methodology, the analysis has been made for MidPoint and EndPoint indicators.

As can be seen for the asphalt pavement with BM2 most of the values of the impact category indicators are increased compared to the baseline, but on the other hand climate change and damage to human health indicators when biogenic carbon is included, and metal and fossil depletion and damage to resources availability are significantly reduced. This is presumably occurring because all the virgin bitumen is replaced by BM2. Thus, less resources are utilized, while more water and land are exploited for the production of the bio-material into the asphalt mixtures. In other words, land and water related indicators are increased due to the necessary resources to produce this bio-based binder. Where BM3 was incorporated, a clear reduction in most of the values of the impact category indicators compared to the baseline is detected. However, it increases freshwater consumption and eutrophication, land use, marine eutrophication, stratospheric ozone depletion and terrestrial ecotoxicity. For the MidPoint indicators the bio-recycled pavements seem to reduce the values of the climate change indicator when biogenic carbon is included, the value of the metal depletion indicator but on the contrary, increase the value of the freshwater consumption, freshwater eutrophication, ionizing radiation, land use and terrestrial ecotoxicity indicators. This is happening because biomaterials seem to be strongly dependent upon water and land in to be produced. The most critical impact category indicator compared to the baseline is land use. As for the EndPoint indicator level, the values of the damage to human health including biogenic carbon and damage to resources availability indicators are reduced. It is worth reiterating that indicators related to land and water are significantly affected when BM2 is utilized because a full binder replacement is taking place by a bio-material, which utilizes significant amounts of water and land to be produced.

3.2 *Identification of key issues: Hotspot analysis*

As mentioned before, the LCA study has been conducted following a cradle to grave approach. These stages can also be grouped into "cradle-to-gate" (product stage) and "gate-to-grave" (construction + replacement + EoL). To identify hotspots, a first division in the life cycle stages has been made considering these groups to differentiate between the life stages in which different stakeholders make decisions; namely in "cradle-to-gate", where material producers and asphalt plant contractors are involved, and in "gate-to-grave", where road authorities make decisions about asset management. Figure 2 shows the relative contribution of "cradle-to-gate" and "gate-to-grave" stages for the baseline.

Results show that for an analysis period of 30 years the "cradle-to-gate" stage is the hotspot in terms of environmental impacts. Every indicator is above the threshold of 68.26% contribution for the cradle-to-gate, reaching the highest value of 85.27% for the Ionizing Radiation. This clearly indicates the most relevant stage of the product system and thus, implies that main actions should be taken at this stage. The same results were found for all the mixtures and only the baseline is shown as example. On the other hand, the "gate-to-grave" stage has been further broken down into construction, replacement and EoL stages to further understand the contribution of each process to the environmental impacts, as can be seen in Table 4. From this disaggregation, it becomes apparent that the most impactful phase within the gate-to-grave system is the replacement of the wearing courses. For the midpoint indicators, the lowest contribution can be seen at 52.68% (freshwater eutrophication) while the highest reaches 99.95% for the metal depletion. The same can be observed for the endpoint indicators with all the indicators contributing more than 78.02%. This is because significant amounts of resources, materials and energy are

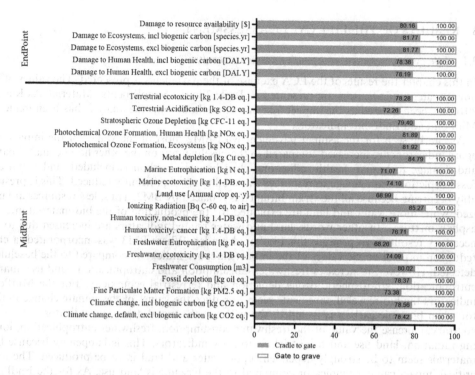

Figure 2. Contribution of cradle-to-gate and gate-to-grave stages to the impact category indicator values.

utilized during the replacement of the wearing courses, compared to their transport & construction and EoL.

Thus, Table 3 clearly shows that the main contributor in this stage is the replacement (although including 50% asphalt recycling) for all the impact category indicators. Hence, if road authorities want to environmentally improve the management of their assets, they should focus on improving the process of replacing the wearing courses of asphalt pavements when incorporating the investigated bio-materials.

4 SENSITIVITY ANALYSIS

4.1 Re-recycling efficiency of the wearing courses

In order to tackle the uncertainty originating from the RA percentage utilized in the replacement stage, a sensitivity analysis has been carried out, identifying and quantifying the different levels of influence for each impact category indicator when different percentages of RA are utilised. The relationship exhibited between the re-recycling percentage and the impact category indicators, in the stage of replacement, were linear. The uncertainty thus, can be minimised and actual impacts can be calibrated for the desirable RA percentage. This sensitivity analysis is meant to assess the influence of the percentage of old wearing course that is re-recycled in the production of the new wearing course in the replacement stage, studying in this way the re-recycling efficiency. In the system studied, the percentage of re-recycling was assumed to be 50%. Scenarios have been created for 20%, which would be the minimum used since the initial wearing course already included 20% RA, and 100%, to consider full re-recycling. This parameter affects only the replacement stage, producing therefore the same increase/decrease in all case studies for each impact category indicator.

Table 3. Values of Mid & EndPoint impact category indicators for all the alternatives studied.

	Baseline	GB5® - BM2	GB5® - BM3
Midpoint Indicators			
Climate change, excl biogenic carbon [kg CO2 eq.]	18.21	18.80	14.78
Climate change, incl biogenic carbon [kg CO2 eq.]	18.14	2.04	13.93
Fine Particulate Matter Formation [kg PM2.5 eq.]	0.01	0.01	0.01
Fossil depletion [kg oil eq.]	22.59	13.57	17.33
Freshwater Consumption [m3]	0.05	0.31	0.10
Freshwater ecotoxicity [kg 1,4 DB eq.]	0.01	0.01	0.01
Freshwater Eutrophication [kg P eq.]	3.47E-05	5.07E-04	1.89E-04
Human toxicity, cancer [kg 1,4-DB eq.]	0.02	0.01	0.01
Human toxicity, non-cancer [kg 1,4-DB eq.]	4.05	4.34	2.53
Ionizing Radiation [Bq C-60 eq. to air]	0.30	0.38	0.26
Land use [Annual crop eq.·y]	0.50	13.06	3.22
Marine ecotoxicity [kg 1,4-DB eq.]	0.03	0.02	0.02
Marine Eutrophication [kg N eq.]	2.30E-04	1.34E-03	1.22E-03
Metal depletion [kg Cu eq.]	6.49	5.06	5.04
Photochemical Ozone Formation, E [kg NOx eq.]	14.74	17.16	12.26
Photochemical Ozone Formation, [kg NOx eq.]	9.16	10.67	7.62
Stratospheric Ozone Depletion [kg CFC-11 eq.]	6.03E-06	9.49E-06	7.03E-06
Terrestrial Acidification [kg SO2 eq.]	0.04	0.05	0.03
Terrestrial ecotoxicity [kg 1,4-DB eq.]	2.58	3.75	2.62
Endpoint Indicators			
Dmg to Human Health excl biogenic carbon [DALY]	3.49E-05	3.81E-05	2.86E-05
Dmg to Human Health incl biogenic carbon [DALY]	3.48E-05	2.26E-05	2.78E-05
Dmg to Ecosystems, excl biogenic carbon [species.yr]	1.96E-06	2.40E-06	1.66E-06
Dmg to Ecosystems, incl biogenic carbon [species.yr]	1.96E-06	2.35E-06	1.66E-06
Dmg to resource availability [$]	10.71	6.10	8.16

Table 4. Percentage contribution of each stage within the gate to grave system.

	Transport & Construction	Replacement	E-o-L
Midpoint Indicators			
Climate change, default, excl biogenic carbon [kg CO2 eq.]	11.65%	77.66%	10.68%
Climate change, incl biogenic carbon [kg CO2 eq.]	11.46%	78.22%	10.32%
Fine Particulate Matter Formation [kg PM2.5 eq.]	19.34%	63.20%	17.45%
Fossil depletion [kg oil eq.]	9.00%	80.94%	10.06%
Freshwater Consumption [m3]	5.65%	88.04%	6.31%
Freshwater ecotoxicity [kg 1,4 DB eq.]	14.56%	69.17%	16.27%
Freshwater Eutrophication [kg P eq.]	22.34%	52.68%	24.97%
Human toxicity, cancer [kg 1,4-DB eq.]	11.45%	75.79%	12.76%
Human toxicity, non-cancer [kg 1,4-DB eq.]	18.05%	61.78%	20.17%
Ionizing Radiation [Bq C-60 eq. to air]	0.40%	99.15%	0.45%
Land use [Annual crop eq·y]	21.34%	54.80%	23.86%
Marine ecotoxicity [kg 1,4-DB eq.]	14.17%	69.99%	15.84%
Marine Eutrophication [kg N eq.]	19.77%	58.14%	22.10%
Metal depletion [kg Cu eq.]	0.02%	99.95%	0.03%
Photochemical Ozone Formation, Ecosystems [kg NOx eq.]	1.24%	97.41%	1.35%
Photochemical Ozone Formation, Human Health [kg NOx eq.]	1.32%	97.25%	1.42%
Stratospheric Ozone Depletion [kg CFC-11 eq.]	9.74%	79.60%	10.66%

(Continued)

Table 4. *(Continued)*

	Transport & Construction	Replacement	E-o-L
Terrestrial Acidification [kg SO2 eq.]	20.11%	61.62%	18.28%
Terrestrial ecotoxicity [kg 1,4-DB eq.]	5.69%	87.94%	6.36%
Endpoint Indicators			
Climate change, default, excl biogenic carbon [kg CO2 eq.]	11.58%	78.02	10.41%
Damage to human health, excl biogenic carbon [year]	10.86%	78.84%	10.30%
Damage to human health, incl biogenic carbon [year]	1.60%	96.63%	1.77%
Damage to ecosystems, excl biogenic carbon [species x yr]	1.59%	96.65%	1.76%
Damage to ecosystems, incl biogenic carbon [species x yr]	7.36%	83.75%	8.89%

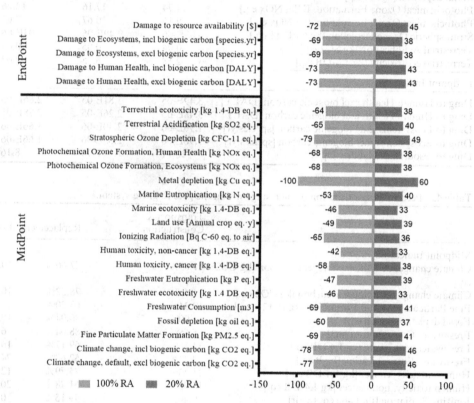

Figure 3. Environmental impacts per impact category indicator for different %RA compared to 50% re-recycling (blue line).

Figure 3 Shows the comparison between the relevant percentage variation of the values of the indicators originating from the different alternative RA% scenarios, along with the assumed recycling rate of 50% (Blue line). Increasing the percentage of RA from 50% to 100% utilised in the replacement of the wearing course, the environmental impacts follow a decreasing trend, proving the efficiency of increased re-recycling. In MidPoint level, all the impact category indicators are decreasing with the same magnitude, with the most affected indicators being metal depletion, stratospheric ozone depletion and climate change (including and excluding biogenic carbon). The least improvement is exhibited at human, freshwater toxicity and marine ecotoxicity. At the endpoint level, again all the indicators have improved. The improvement ranges between 73% for damage to human health (including and excluding biogenic carbon) and 69% for damage to ecosystems (including and excluding biogenic carbon). When 20% instead of 50%RA is utilized in the replacement stage of the asphalt pavements' lifecycle, from Figure 3, an increase in the percentage variation of the impact category indicators in both mid and endpoint can be seen. This supports the fact that the higher is the percentage of RA in the mixtures, the higher are the environmental benefits for the investigated cases. At midpoint level, the most affected indicator is again metal depletion that increases by 60%, followed by the stratospheric ozone depletion indicator, accounting for 49% increase. At endpoint level, the trend keeps being the same. Decreasing the RA% from 50% to 20%, the percentage variation of the impact category indicators increased. The most affected indicator is damage to resource availability reaching at 45% decrease and the less affected is the damage to ecosystems indicator (including and excluding biogenic carbon) that increased by 38%.

It can hence be concluded that increasing the percentage of RA into the replacement stage of the asphalt pavements with or without bio-materials, leads to environmental benefits. These benefits are strongly related to the percentage of RA utilized. Noteworthily, the rate of increase of the environmental benefits follows a linear relationship with the incorporated RA%.

5 CONCLUSIONS AND RECOMMENDATIONS

Within the perspective of a post-fossil fuel society based on renewable resources, pavement recycling techniques using bio-materials to completely or partially substitute bituminous based materials are being promoted. In this study the results of a full cradle-to-grave process-based LCA of a conventional asphalt pavement and two bio-recycled asphalt pavements are presented and furtherly analysed through the means of hotspot and sensitivity analyses. ReCiPe 2016 was chosen as the LCIA methodology, which allows performing the analysis with Midpoint and Endpoint impact category indicators. Focus was given within the interpretation phase of the LCA exercise. A hotspot analysis was undertaken to identify the most impactful stage of the cradle-to-gate and gate-to-grave life cycle phases of the asphalt pavements, while a sensitivity analysis of the re-recycling efficiency of the wearing courses was furthermore undertaken, to produce recommendations for stakeholders.

The comparison of each bio-recycled pavement in terms of impact category indicators is different. However, some common points can be detected.

- The bio-recycled pavements reduce the climate change (global warming potential) indicator when biogenic carbon is included, reduce the metal depletion indicator, and increase freshwater consumption, freshwater eutrophication, ionizing radiation, land use and terrestrial ecotoxicity indicator values.
- The most critical impact category indicator compared to the baseline is land use. This is presumably due to the fact that bio-materials require land and water for their production.
- Under the analysis period of 30 years, the "cradle-to-gate" stage is the most relevant in terms of environmental impacts. This means that in the short-term, actions to reduce the potential environmental impacts have to be focused in the processes included in "cradle-to-gate", which are raw material acquisition, transport to plant and plant production of the asphalt mixture. In these processes, materials producers, asphalt mixture designers and asphalt plant contractors are involved.
- The analysis of the "gate-to-grave", shows that the replacement stage is the most influential in terms of potential environmental impacts. Hence, if road authorities want to environmentally improve the management of their assets, they should focus on improving the process of replacing the wearing courses of asphalt pavements incorporating the specific bio-materials.
- A strong relationship between the percentage of recycling and the impact category indicators has been found, leading to the conclusion that this parameter may be used to optimise environmental impacts and be used to target desirable thresholds for specific indicators set by environmental agencies.

- The advantages of increasing recycling are clear, however, in order to move towards higher RA contents, asphalt plants have to be willing to adopt to challenges concerning the RA% that they are able to re-recycle within asphalt mixtures and road authorities have to accept the similar performance of asphalt mixture with and without RA.

This study is an assessment of two bio-recycled pavements, which include the use of high RA contents and bio-materials in pavements, produced under certain conditions. Therefore, the results are limited by the data collected, the choices made, uncertainties and the system assessed. In order to overcome these limitations, some of the conditions of the study, such as unknown durability and re-recycling content, were subjected to sensitivity analysis to evaluate the influence in the results. In any case, the conclusions of this study may not be extended to other technologies using high RA contents and bio-materials.

FUNDING

The results presented in this paper were obtained with the support of the BioRePavation project, as well as the SMARTI ETN project. BioRePavation has received funding from INFRA-VATION - ERA-NET Plus under the grant n.618109 and SMARTI ETN from the European Union's Horizon 2020 Program under the Marie Curie-Skłodowska actions for research, technological development and demonstration, under grant n.721493.

REFERENCES

Anastasiou, E.K., Liapis, A., Papayianni, I., 2015. Comparative life cycle assessment of concrete road pavements using industrial by-products as alternative materials. Resources, Conservation and Recycling 101, 1–8. https://doi.org/10.1016/j.resconrec.2015.05.009

Aurangzeb, Q., Al-Qadi, I.L., Ozer, H., Yang, R., 2014. Hybrid life cycle assessment for asphalt mixtures with high RAP content. Resources, Conservation and Recycling 83, 77–86. https://doi.org/10.1016/j.resconrec.2013.12.004

Bre, B.R.E., 2013. Product Category Rules for Type III environmental product declaration of construction products to EN 15804:2012 43.

Chiu, C.-T., Hsu, T.-H., Yang, W.-F., 2008. Life cycle assessment on using recycled materials for rehabilitating asphalt pavements. Resources, Conservation and Recycling 52, 545–556. https://doi.org/10.1016/j.resconrec.2007.07.001

Chowdhury, R., Apul, D., Fry, T., 2010. A life cycle based environmental impacts assessment of construction materials used in road construction. Resources, Conservation and Recycling 54, 250–255. https://doi.org/10.1016/j.resconrec.2009.08.007

Dinis-Almeida, M., Afonso, M.L., 2015. Warm Mix Recycled Asphalt - A sustainable solution. Journal of Cleaner Production 107, 310–316. https://doi.org/10.1016/j.jclepro.2015.04.065

EAPA, 2016. Guidance Document for preparing Product Category Rules (PCR) and Environmental Product Declarations (EPD) for Asphalt Mixtures by the European Asphalt Pavement Association (EAPA) 2 nd November 2016.

European Asphalt Pavement Association, 2016. Asphalt in Figures 2016.

Fini, E.H., Al-Qadi, I.L., You, Z., Zada, B., Mills-Beale, J., 2012. Partial replacement of asphalt binder with bio-binder: Characterisation and modification. International Journal of Pavement Engineering 13, 515–522. https://doi.org/10.1080/10298436.2011.596937

Giani, M.I., Dotelli, G., Brandini, N., Zampori, L., 2015. Comparative life cycle assessment of asphalt pavements using reclaimed asphalt, warm mix technology and cold in-place recycling. Resources, Conservation and Recycling 104, 224–238. https://doi.org/10.1016/j.resconrec.2015.08.006

Huijbregts, M.A.J., Steinmann, Z.J.N., Elshout, P.M.F.M., Stam, G., Verones, F., Vieira, M.D.M., Zijp, M., van Zelm, R., 2016. ReCiPe 2016: A harmonized life cycle impact assessment method at midpoint and enpoint level - Report 1 : characterization. National Institute for Public Health and the Environment 194. https://doi.org/10.1007/s11367-016-1246-y

International Organization for Standardization, 2006. ISO 14040-Environmental management - Life Cycle Assessment - Principles and Framework. International Organization for Standardization 3, 20. https://doi.org/10.1016/j.ecolind.2011.01.007

Jullien, A., Dauvergne, M., Cerezo, V., 2014. Environmental assessment of road construction and maintenance policies using LCA. Transportation Research Part D: Transport and Environment 29, 56–65. https://doi.org/10.1016/j.trd.2014.03.006

Klabunde, K.J., Ph, D., 2014. Sustainable Asphalt Pavements Using Bio- Binders from Bio-Fuel Waste.

Laurent, G., 2004. Evaluation économique des chaussées en béton et classiques sur le réseau routier national français. Paris.

Mills-Beale, J., You, Z., Fini, E., Zada, B., Lee, C.H., Yap, Y.K., 2014. Aging Influence on Rheology Properties of Petroleum-Based Asphalt Modified with Biobinder. Journal of Materials in Civil Engineering 26, 358–366. https://doi.org/10.1061/(ASCE)MT.1943-5533.0000712

Mohd Hasan, M.R., You, Z., 2016. Ethanol based foamed asphalt as potential alternative for low emission asphalt technology. Journal of Traffic and Transportation Engineering (English Edition) 3, 116–126. https://doi.org/10.1016/j.jtte.2016.03.001

Newcomb, D.E., Ray brown, E., Epps, J.A., 2007. Designing HMA mixtures with High RAP content: A practical guide41..

Onay, O., Koçkar, O.M., 2006. Pyrolysis of rapeseed in a free fall reactor for production of bio-oil. Fuel 85, 1921–1928. https://doi.org/10.1016/j.fuel.2006.03.009

Organisation de Cooperation et de developpement Économiques, 2005. Évaluation économique des chaussées à longue durée de vie.

Podolsky, J.H., Buss, A., Williams, R.C., Cochran, E., 2016. Comparative performance of bio-derived/chemical additives in warm mix asphalt at low temperature. Materials and Structures/Materiaux et Constructions 49, 563–575. https://doi.org/10.1617/s11527-014-0520-3

Presti, D., Airey, G., Liberto, M., Noto, S., Mino, G., Blasl, A., Falla, G., Wellner, F., 2017. AllBack2-Pave: Transport Infrastructure and Systems 109–118. https://doi.org/10.1201/9781315281896-17.

Rodríguez-Alloza, A.M., Malik, A., Lenzen, M., Gallego, J., 2015. Hybrid input-output life cycle assessment of warm mix asphalt mixtures. Journal of Cleaner Production 90, 171–182. https://doi.org/10.1016/j.jclepro.2014.11.035

Şensöz, S., Kaynar, I., 2006. Bio-oil production from soybean (Glycine max L.); Fuel properties of Bio-oil. Industrial Crops and Products 23, 99–105. https://doi.org/10.1016/j.indcrop.2005.04.005

Song, W., Huang, B., Shu, X., 2018. Influence of warm-mix asphalt technology and rejuvenator on performance of asphalt mixtures containing 50% reclaimed asphalt pavement. Journal of Cleaner Production 192, 191–198. https://doi.org/10.1016/j.jclepro.2018.04.269

Sun, Z., Yi, J., Huang, Y., Feng, D., Guo, C., 2016. Properties of asphalt binder modified by bio-oil derived from waste cooking oil. Construction and Building Materials 102, 496–504. https://doi.org/10.1016/j.conbuildmat.2015.10.173

The International Standards Organisation, 2006. INTERNATIONAL STANDARD assessment — Requirements and guilelines. The International Journal of Life Cycle Assessment 2006, 652–668. https://doi.org/10.1007/s11367-011-0297-3

The Norwegian EPD Foundation, 2009. Product-Category Rules (PCR) for preparing an Environmental Product Declaration (EPD) for Product Group Asphalt and crushed stone1–22..

Uzun, B.B., Apaydin-Varol, E., Ateş, F., Özbay, N., Pütün, A.E., 2010. Synthetic fuel production from tea waste: Characterisation of bio-oil and bio-char. Fuel 89, 176–184. https://doi.org/10.1016/j.fuel.2009.08.040

Wang, H., Chen, C., Li, J., Yang, X., Zhang, H., Wang, Z., 2016. Modified first dorsal metacarpal artery island flap for sensory reconstruction of thumb pulp defects. Journal of Hand Surgery: European Volume 41, 177–184. https://doi.org/10.1177/1753193415610529

Winkle, C.I. Van, 2014. Laboratory and field evaluation of hot mix asphalt with high contents of reclaimed asphalt pavement.

Wolf, M.-A., Pant, R., Chomkhamsri, K., Sala, S., Pennington, D., 2012. The International Reference Life Cycle Data System (ILCD) Handbook, European Commission. JRC references reports. https://doi.org/10.2788/85727

Yang, X., Mills-Beale, J., You, Z., 2017. Chemical characterization and oxidative aging of bio-asphalt and its compatibility with petroleum asphalt. Journal of Cleaner Production 142, 1837–1847. https://doi.org/10.1016/j.jclepro.2016.11.100

Yang, X., You, Z., Dai, Q., 2013. Performance evaluation of asphalt binder modified by bio-oil generated from waste wood resources. International Journal of Pavement Research and Technology 6, 431–439. https://doi.org/10.6135/ijprt.org.tw/2013.6(4).431

Zhang, R., Wang, H., You, Z., Jiang, X., Yang, X., 2017. Optimization of bio-asphalt using bio-oil and distilled water. Journal of Cleaner Production 165, 281–289. https://doi.org/10.1016/j.jclepro.2017.07.154

Use of recycled aggregates in concrete pavement: Pavement design and life cycle assessment

Xijun Shi
Center for Infrastructure Renewal, Texas A&M University, USA

Zachary Grasley
Zachry Department of Civil and Environmental Engineering, Center for Infrastructural Renewal, Texas A&M University, USA

Anol Mukhopadhyay
Texas A&M Transportation Institute, USA

Dan Zollinger
Zachry Department of Civil and Environmental Engineering, Texas A&M University, USA

ABSTRACT: Pavement recycling has become a common practice for many states in the U.S. While material properties of concrete containing recycled aggregates (reclaimed asphalt pavement and recycled concrete aggregate) as virgin aggregate replacements have been extensively characterized, very little effort has been made to address concrete pavement design concerns and assess potential sustainability benefits of this application. In this study, recommendations from a pavement design perspective for recycled aggregates based concrete pavement were reviewed and summarized. Two life cycle assessment case studies were presented. Based on the results, it is concluded that the use of recycled aggregates in concrete pavement not only is a technically sound strategy, but also can lead to significant sustainability benefits.

1 INTRODUCTION

Aggregate is an indispensable ingredient of concrete 0materials, accounting for nearly 70-80% of the total volume of concrete. Due to the enormous demand for concrete in new construction, many of good aggregate sources have been quickly depleted, leading to a continuous increase in aggregate cost. Recycled aggregates from different sources are available substitutes for natural aggregate. The utilization of recycled aggregates in pavement construction is considered one of the most effective strategies to reduce the rate of natural recourse depletion and dispose of construction debris in a sustainable manner.

Recycled aggregates derived from pavement debris include reclaimed asphalt pavement (RAP) and recycled concrete aggregate (RCA). While RAP is dominantly used in hot mix asphalt and RCA is largely regarded a base material for pavement applications, a considerable amount of effort has been made to explore using these recycled aggregates in concrete pavement slab (Shi et al., 2019a, Shi, 2018, Debbarma et al., 2019, Singh et al., 2018, Brand and Roesler, 2015, Gress et al., 2009). Among the existing studies, most of the investigations focused on testing material properties of concrete containing the recycled aggregates, and only very limited work touched upon the aspects of pavement design and life cycle assessment (LCA). This paper presents an overview of the existing findings on pavement design and life cycle assessment for concrete pavement containing recycled aggregates based on the authors' recent studies.

2 EFFECT OF RECYCLED AGGREGATES ON CONCRETE PROPERTIES

Aggregate properties have profound effects on concrete properties. Due to the presence of aged asphalt in RAP and reclaimed mortar (RM) in RCA, the produced recycled aggregates based concrete could have significantly different properties compared to plain concrete. From the mechanical perspective, the aged asphalt layers and reclaimed mortar are considered weak zones in the RAP based concrete and RCA based concrete, respectively. Based on thin section observations using optical microscope, the crack propagation through the asphalt layers and reclaimed mortar is clearly evident (Figure 1).

Other than being regarded as the major weak zones, the asphalt layers and reclaimed mortar are major reasons for many other changes of concrete properties caused by the addition of RAP and RCA. A summary of the altered concrete properties relevant to concrete pavement performance is presented in Table 1.

3 PAVEMENT DESIGN CONSIDERATIONS

The incorporation of the recycled aggregates causes considerable changes on concrete properties, consequently leading to significant impacts on pavement behavior. One immediate concern of using the recycled aggregates in concrete pavement is the reduced strengths. More specifically, concrete tensile strength is one of the most important inputs for pavement slab

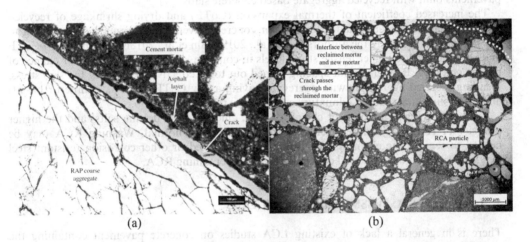

(a) (b)

Figure 1. Weak zone in recycled aggregate based concrete: (a) The asphalt layer is considered the weak zone in RAP based concrete; cracks tend to pass through the asphalt layer (Mukhopadhyay and Shi, 2019) (b) The reclaimed mortar is considered the weak zone in RCA based concrete; cracks tend to pass through the reclaimed mortar (Shi et al., 2019b).

Table 1. Effect of the recycled aggregates (RAP or RCA) on concrete properties.

Concrete properties	Changes due to the recycled aggregates incorporation
Compressive strength	Lower
Flexural strength	Lower
Splitting tensile strength	Lower
Coefficient of thermal expansion	Higher
Modulus of elasticity	Lower
Shrinkage	Higher
Aggregate interlock	Worse

thickness design, and the use of RAP or RCA leads to significant detrimental effects on the flexural strength and splitting tensile strength. Fortunately, it has been reported that the tensile strength reduction can be controlled within 20 % if not more than 40% coarse RAP is used to replace the same volume of virgin coarse aggregate in concrete (Shi et al., 2017). For the RCA based concrete, the tensile strength reduction is not as significant as that of RAP based concrete, and it is feasible to build concrete slab with 100% coarse RCA (Mukhopadhyay et al., 2019).

The asphalt in RAP (even though aged) and reclaimed mortar in RCA are less stiff than cement matrix and virgin aggregates. Therefore, the produced recycled aggregates based concrete invariably has reduced modulus of elasticity (MOE). Concrete pavement slab with reduced MOE is anticipated to have a better cracking resistance due to the lower stress level (Shi et al., 2019d). In addition, the higher viscoelasticity of RAP based concrete could potentially lead to higher creep in the concrete structures, which further relaxes stress. On the other hand, the low MOE could cause higher slab deflections, though. The higher differential energy caused by the deflection differential between the unloaded and loaded slabs results in a higher amount of base erosion, which eventually leads to higher slab faulting (Shi et al., 2019d). This finding has been validated in an RCA concrete pavement field study in Oklahoma. Falling weight deflectometer results from the field study showed that slabs containing 100% coarse RCA exhibited higher slab deflection differential compared to control slabs. Distress survey data confirmed that there existed higher joint faulting in the RCA concrete pavement section relative to the control section (Shi et al., 2019b). To account for the higher base erosion caused by the softer slab, using stronger base materials is highly recommended for pavements built with recycled aggregate based concrete slabs.

The increased coefficient of thermal expansion (CoTE) and drying shrinkage of recycled aggregate based concrete could lead to lower concrete pavement performance. According to the simulation studies by the authors (Shi et al., 2019d, Shi et al., 2018b), the increased CoTE and shrinkage cause higher tensile stress levels in concrete labs, leading to a higher chance of fatigue cracking. Reducing joint spacing turned out to be effective in reducing the slab stress; blending recycled aggregates with aggregate having a low CoTE (such as granite or basalt) and using shrinking reduced admixture can mitigate the problems, too.

Further, the higher amount of chemicals (e.g., chloride from deicing salts) and the higher porosity of RCA may cause steel to corrode faster than normal. Washing RCA may be needed to remove the chemicals. The use of epoxy-coated or other corrosion resistant steel may be necessary for reinforced concrete pavement containing RCA.

4 LIFE CYCLE ASSESSMENT CASE STUDIES

There is in general a lack of existing LCA studies on concrete pavement containing the recycled aggregates. Especially, no previous LCA investigations were found on RAP based concrete pavement. Only a few life cycle cost analysis (LCCA) or LCA studies on RCA based concrete pavement are reviewed in this section.

Verian et al. (2013) conducted a cost-benefit analysis on using RCA in new concrete pavement. A hypothetical 3-lane mile-long pavement built with 50% coarse RCA in the concrete layer and 100% coarse RCA in the base layer was evaluated. They found that around $3M saving could be realized by considering the cost savings due to the natural aggregate replacement and landfilling a less amount of pavement debris. Reza and Wilde (2017) presented a more comprehensive LCCA on using RCA in concrete pavement covering different hypothetical pavement construction scenarios. A total of eight alternatives with varying RCA replacement level, w/cm, slab thickness, and pavement service life to the first major concrete pavement rehabilitation were considered. Table 2 summarizes the pavement scenarios.

The LCCA results indicate that RCA based concrete pavement construction can yield significant cost savings due to avoiding the high cost of purchasing and hauling natural aggregate for concrete. These cost savings could compensate the higher expense related to additional sawcuts and dowel bars and higher construction cost of a two-lift pavement

Table 2. Different pavement scenarios for the life cycle cost analysis by Reza and Wilde (2017).

Scenarios	%RCA	w/cm	Slab thickness (in.)	Analysis life period (year)
1	0	0.37	9.0	59
2	50	0.37	9.0	50
3	50	0.37	10.1	59
4	50	0.36	9.0	59
5	100	0.37	9.0	46
6	100	0.37	11.2	59
7	100	0.35	9.0	59
8	50	0.39	6-in. RCA-PCC as lower lift, 3-in. exposed aggregate concrete as upper lift	59

construction. Additionally, the study concluded that it is more economical to put RCA in concrete slab than to use it in the base. To compensate the reduced strength for RCA based concrete, either strengthening the mix (e.g., increase the cementitious material and decrease the w/cm) or increasing the slab thickness was used in the case study. It was shown that strengthening the mix is more cost-effective than increasing pavement thickness.

Ram et al. (2011) conducted a life cycle assessment for a selected number of Michigan DOT concrete pavement sections to evaluate the sustainability benefits of using recycled and industrial by-product materials in concrete pavement. Among the various recycled materials, the use of RCA was found to lead to significant positive environmental benefits. The environmental benefits can be maximized by using on-site recycling instead of regional recycling, which is primarily due to the reduction of pollution associated with transportation. Reza and Wilde (2017) also conducted a life cycle analysis to compare Scenario 7 against Scenario 1 in Table 2 with the PaLATE tool; they found that the RCA-PCC pavement appeared to be more environmentally friendly compared to the conventional pavement.

To evaluate potential sustainability benefits of using recycled aggregates in concrete pavement, life cycle assessment case studies to compare recycled aggregates based concrete pavement with plain concrete pavement covering all three pillars of sustainability categories (i.e., economic, social, and environmental) were performed using an economic input-output life cycle assessment (EIO-LCA) approach. The EIO theory was proposed by Nobel Prize winner Wassily Leontief. It uses a general interdependent model that quantifies the interrelationships among sectors of an economic system while identifying the direct and indirect economic inputs. An online EIO-LCA software developed by Carnegie Mellon University (CMU) was used for the case studies (CMU, 2018). The method has produced numerous LCA publications in different fields (Rew et al., 2018, Shi et al., 2019c, Shi et al., 2018a).

4.1 RAP based concrete pavement EIO-LCA case study

A single-lift pavement made of plain concrete slab, a single-lift pavement made with RAP based concrete slab, and a two-lift concrete pavement using RAP based concrete as the bottom lift material were directly compared in the case study. The determination of pavement thickness for the plain concrete pavement and RAP based concrete slab was based on commonly used pavement design tools (i.e., AASHTO 1993, TxCRCP-ME, and Pavement ME) by considering the concrete property changes caused by the RAP addition. The results indicated that one needs to slightly increase the slab thickness if RAP aggregate is used. The material selection and slab thickness determination of the two-lift concrete pavement was challenging since no design procedures had yet been established. Therefore, the pavement structure referred to the existing two-lift concrete pavement field sections in the US. A stronger concrete mix with a higher cementitious material content was assigned as the top layer material to balance the low strength of the bottom lift where RAP aggregate was used.

Figure 2 shows the pavement structures that were justified to behave equivalently under same loading and environmental conditions.

Three hypothetical pavement sections with the structures in Figure 2 were studied using the EIO-LCA. The pavements were assumed to be 4.8 km long and 15 m wide. The quantity of each raw material was calculated subsequently, and the total material cost was computed based on the 2017 RS Means database (Gordian, 2017). It was assumed that the RAP material was obtained from excess stockpiles and was considered as waste in the analysis, so the cost of RAP was set to 0 for the RAP based concrete cases. All the extra RAP material was assumed to be disposed of in a landfill site that is 80 km away. The haul distance between the natural aggregate sources and the ready-mix concrete plant was assumed to be 48 km. The RAP land-fill cost was $2/ton and the aggregate haul cost was $0.22/ton/km. The paving cost for the single-lift pavement and the two-lift concrete pavement adopted $3.12/m^2 and $5.12/m^2 based upon a previous study (Hu et al., 2014).

Using the aforementioned information, the economic activity containing materials, transportation, and construction costs was readily calculated and input to the software. The EIO-LCA was subsequently performed, and the economic activity, transportation, land use, energy consumption, conventional air pollutants, greenhouse gasses emissions, toxic releases, and water withdrawals were output. Based on the results (Table 3), the use of RAP based concrete for pavement applications was found to yield considerable economic, social, and environmental benefits compared to the plain concrete pavement. The case study also showed that, among all the pavement types, the single-lift RAP based concrete pavement achieved the

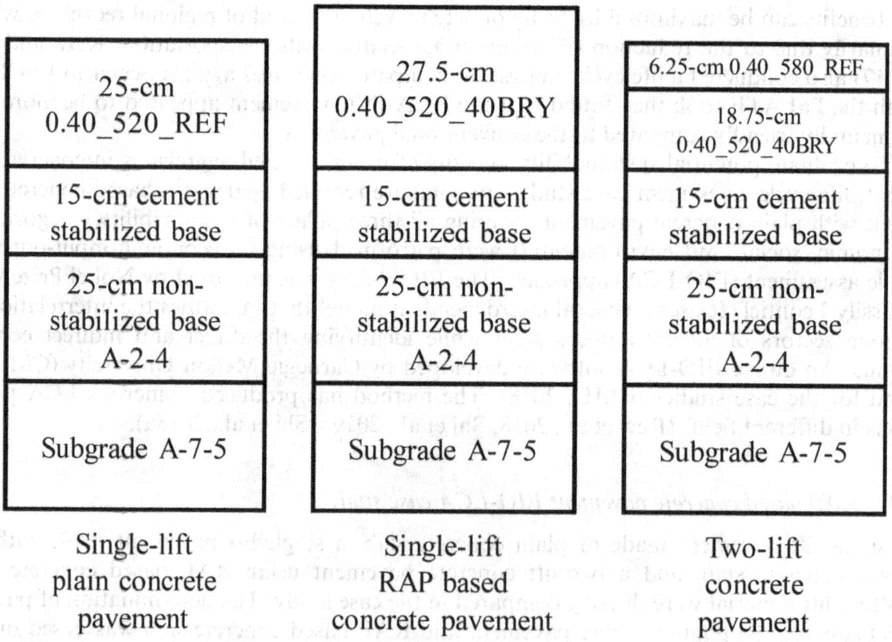

25-cm 0.40_520_REF	27.5-cm 0.40_520_40BRY	6.25-cm 0.40_580_REF
		18.75-cm 0.40_520_40BRY
15-cm cement stabilized base	15-cm cement stabilized base	15-cm cement stabilized base
25-cm non- stabilized base A-2-4	25-cm non- stabilized base A-2-4	25-cm non- stabilized base A-2-4
Subgrade A-7-5	Subgrade A-7-5	Subgrade A-7-5
Single-lift plain concrete pavement	Single-lift RAP based concrete pavement	Two-lift concrete pavement

e.g: 0.40_520_REF: a plain concrete mixture with 0.40 w/cm ratio, 520 lb/cy cementitious content

0.40_520_40BRY: a RAP based concrete mixture with 0.40 w/cm ratio, 520 lb/cy cementitious content, 40% virgin coarse aggregate replaced by RAP from the BRY source

Figure 2. Pavement structures evaluated in the EIO-LCA case study for RAP based concrete pavements.

Table 3. RAP based concrete pavement EIO-LCA case study results.

Category	Stressor	Change in stressors for pavements containing RAP concrete relative to the single-lift plain concrete pavement (%)	
		Single-lift RAP concrete pavement	Two-lift concrete pavement
Economic	Total economic	-10.90	-6.92
Social	Transportation movement	1.05	-3.38
	Land use	-8.14	1.16
	Energy use	-1.25	-4.13
Environmental	Conventional air pollutants	-3.81	-2.42
	Greenhouse gases emissions	3.00	-1.30
	Toxic releases	2.60	-0.23
	Water withdrawal	-6.70	-9.20

highest economic benefits, while the two-lift construction had highest positive impacts from social and environmental perspectives.

4.2 RCA based concrete pavement EIO-LCA case study

Similarly, an RCA based concrete pavement case study was carried out using the EIO-LCA tool. The RCA based concrete case study, however, included most of pavement life cycles (i.e., material production and construction, use, and end-of-life phases), while the RAP based concrete pavement case study detailed in the previous section only focused the material production and construction phase. An RCA based concrete pavement with virgin coarse aggregate fully replaced by coarse RCA was compared with a plain concrete pavement. Both the pavements were assumed to be 12.8 km long and 14.4 m wide with a same concrete layer thickness of 25 cm. Due to the detrimental effect of the RCA incorporation on concrete properties (see Table 1), the RCA based concrete pavement is likely to have a shorter service life compared to the control pavement with the same thickness. The reduced service life of the RCA based concrete pavement was well considered in the LCA study based on a survey of approximately 341 km of existing RCA based concrete pavement sections in the US conducted by Reza et al. (2018). In the survey, the average time to reach the condition of the first major concrete pavement rehabilitation was found to be 27 years for RCA based concrete pavement and 32 years for plain concrete pavement. Accordingly, the pavement service life for the RCA based concrete pavement and the plain concrete pavement in this LCA study was set as 27 years and 32 years, respectively. The analysis period was chosen as 27 years, and the 5-year remaining service life of the plain concrete pavement was included in the calculation of the end-of-life salvage value.

The economic input for the material production and construction phase was prepared similarly as the RAP based concrete pavement LCA study. The major components considered in the use phase are vehicle operation costs (VOC), which are the costs relevant to vehicle repair and maintenance, fuel consumption, and tire wear. The vehicle operation costs for the two concrete pavement cases were estimated using the VOC model developed under NCHRP 720 (Chatti and Zaabar, 2012). In the VOC model, the input that is different between the plain concrete pavement and the RCA concrete pavement is pavement international roughness index (IRI), and the IRI was collected from the RCA based concrete pavement survey by Reza et al. (2018). The IRI deterioration models were provided by Reza et al. (2018) using historical condition data. The economic activity input for the entire pavement life cycle was determined by summing up the costs during the material production and construction phase and the use phase and subtracting the end-of-life salvage value. An annual discount value of 1.5% was use in the study.

The EIO-LCA outputs included the economic activity, transportation, land use, energy consumption, conventional air pollutants, greenhouse gasses emissions, toxic releases, and water withdrawals. Additionally, impact analysis by the tool for reduction and assessment of chemicals and other environmental impacts (TRACI) was also reported (Table 4).

The major findings are:

- The inventory of stressors for the materials production and construction phase indicates that the RCA based concrete pavement yields significantly less economic, environmental, and social burden compared to the plain concrete pavement. The sustainability benefits of the RCA based concrete pavement in this phase of life cycle is attributed to less consumption of virgin aggregate, less virgin aggregate transported to the ready-mix plant, and less concrete debris transported and disposed of at the landfill site.
- The RCA based concrete pavement is slightly less sustainable compared to the plain concrete pavement during the use phase. The rougher pavement surface of the RCA based concrete pavement causes higher tire wear and fuel consumption for vehicles, which poses higher negative impacts on economy, environment, and society.
- The results of the total stressors and the characterization factors in the TRACI for the entire pavement life cycle are slightly mixed. Although the benefits of using RCA in concrete pavement during the materials production and construction phase are invariably achieved for all the categories, the higher amount of negative impacts by using RCA in pavement during the use phase can be more dominating throughout the entire life cycle for a few categories. As a result, the unfavorable impacts from the use phase cancel out the benefits achieved during the materials production and construction for the RCA based concrete pavement to some extent. Still, the pavement made with RCA based concrete is

Table 4. RCA based concrete pavement EIO-LCA case study results.

| TRACI category | Impacts | | Change in impacts for the pavement containing RCA concrete relative to the plain concrete pavement (%) |
	Plain concrete pavement	RCA concrete pavement	
Global warming potential (ton CO_2 eq)	1,319,708	1,318,834	-0.07%
Acidification air (ton SO_2 eq)	4,538	4,536	-0.05%
Human health particulate air (ton PM_{10} eq)	1,198	1,174	-2.05%
Eutrophication air (ton N eq)	136	136	0.33%
Eutrophication water (ton N eq)	0.470	0.471	0.14%
Ozone depletion air (ton CFC-11 eq)	0.401	0.402	0.14%
Smog air (ton O_3 eq)	80,385	80,280	-0.13%
Ecotoxicity (low) (ton 2,4D)	67	52	-22.30%
Ecotoxicity (high) (ton 2,4D)	68	53	-21.90%
Human health cancer (low) (ton benzene eq)	99	90	-9.93%
Human health cancer (high) (ton benzene eq)	506	458	-9.51%
Human health non-cancer (low) (ton toluene eq)	63,266	56,300	-11.01%
Human health non-cancer (high) (ton benzene eq)	1,345,530	1,061,000	-21.15%

generally more environmentally and socially friendly compared to the pavement made with virgin aggregates; such sustainability benefits are especially significant for the categories of ecotoxicity (both low and high), human health cancer (both low and high) and human health non-cancer (both low and high).

5 CONCLUSION

Pavement recycling has become a common practice for many states in the U.S. While material properties of concrete containing recycled aggregates (RAP and RCA) as virgin aggregate replacements were extensively characterized, very little effort has been made to address concrete pavement design concerns and assess potential sustainability benefits of this application. In this study, recommendations from a pavement design perspective for recycled aggregates based concrete pavement were reviewed and summarized. Two life cycle assessment case studies were presented. Based on the results, it is concluded that the use of recycled aggregates in concrete pavement not only is a technically sound strategy, but also can lead to significant sustainability benefits.

REFERENCES

Brand, A. S. & Roesler, J. R. 2015. Ternary concrete with fractionated reclaimed asphalt pavement. *ACI Materials Journal*, 112.
Chatti, K. & Zaabar, I. 2012. Estimating the effects of pavement condition on vehicle operating costs, *Report NCHRP 720*, Transportation Research Board.
Cmu. 2018. *Economic input-output life cycle assessment* [Online]. Available: http://www.eiolca.net/ [Accessed:10 October 2018].
Debbarma, S., Singh, S. & Ransinchung Rn, G. 2019. Laboratory investigation on the fresh, mechanical, and durability properties of roller compacted concrete pavement containing reclaimed asphalt pavement aggregates. *Transportation Research Record: Journal of the Transportation Research Board*, 0361198119849585.
Gordian. 2017. *RSMeans data* [Online]. Available: https://www.rsmeans.com/ [Accessed:1 March 2017].
Gress, D., Snyder, M. & Sturtevant, J. 2009. Performance of rigid pavements containing recycled concrete aggregate: update for 2006. *Transportation Research Record: Journal of the Transportation Research Board*, 99–107.
Hu, J., Siddiqui, M. S. & David Whitney, P. Two-lift concrete paving–case studies and reviews from sustainability, cost effectiveness and construction perspectives. *Proceedings of the TRB 93rd Annual Meeting*, 2014.
Mukhopadhyay, A. & Shi, X. 2019. Microstructural characterization of portland cement concrete containing reclaimed asphalt pavement aggregates using conventional and advanced petrographic techniques. *ASTM International Selected Technical Papers*, 1613.
Mukhopadhyay, A., Shi, X. & Zollinger, D. 2019. Recycling and reuse of materials in transportation projects—current status and potential opportunities including evaluation of RCA concrete pavements along an Oklahoma interstate highway. *Report FHWA-OK-18/04*, US. Dep. of Transportation, Oklahoma.
Ram, P., Van Dam, T., Meijer, J. & Smith, K. 2011. Sustainable recycled materials for concrete pavements. *Report RC-1550*, US. Dep. of Transportation, Michigan.
Rew, Y., Shi, X., Choi, K. & Park, P. 2018. Structural design and lifecycle assessment of heated pavement using conductive asphalt. *Journal of Infrastructure Systems*, 24, 04018019.
Reza, F., Wilde, W. & Izevbekhai, B. 2018. Performance of recycled concrete aggregate pavements based on historical condition data. *International Journal of Pavement Engineering*, DOI: 10.1080/ 10298436.2018.1503272
Reza, F. & Wilde, W. J. 2017. Evaluation of recycled aggregates test section performance. *Report MN/ RC 2017-06M*, US. Dep. of Transportation, Minnesota.
Shi, X. 2018. *Evaluation of portland cement concrete containing reclaimed asphalt pavement for pavement applications*. Ph.D Dissertation, Texas A&M University.

Shi, X., Mirsayar, M., Mukhopadhyay, A. K. & Zollinger, D. G. 2019a. Characterization of two-parameter fracture properties of portland cement concrete containing reclaimed asphalt pavement aggregates by semicircular bending specimens. *Cement & Concrete Composites*, 95, 56–69.

Shi, X., Mukhopadhyay, A. & Liu, K.-W. 2017. Mix design formulation and evaluation of portland cement concrete paving mixtures containing reclaimed asphalt pavement. *Construction and Building Materials*, 152, 756–768.

Shi, X., Mukhopadhyay, A. & Zollinger, D. 2018a. Sustainability assessment for portland cement concrete pavement containing reclaimed asphalt pavement aggregates. *Journal of Cleaner Production*, 192, 569–581.

Shi, X., Mukhopadhyay, A. & Zollinger, D. 2019b. Long-term performance evaluation of concrete pavements containing recycled concrete aggregate in oklahoma. *Transportation Research Record: Journal of the Transportation Research Board*, 5-2673, 429–442.

Shi, X., Mukhopadhyay, A., Zollinger, D. & Grasley, Z. 2019c. Economic input-output life cycle assessment of concrete pavement containing recycled concrete aggregate. *Journal of Cleaner Production*, 225, 414–425.

Shi, X., Mukhopadhyay, A., Zollinger, D. G. & Huang, K. 2019d. Performance evaluation of jointed plain concrete pavement made with portland cement concrete containing reclaimed asphalt pavement. *Road Materials and Pavement Design*. DOI: 10.1080/14680629.2019.1616604

Shi, X., Zollinger, D. G. & Mukhopadhyay, A. K. 2018b. Punchout study for continuously reinforced concrete pavement containing reclaimed asphalt pavement using pavement ME models. *International Journal of Pavement Engineering*, 1–14. DOI:10.1080/10298436.2018.1533134

Singh, S., Ransinchung Rn, G. & Kumar, P. 2018. Performance evaluation of RAP concrete in aggressive environment. *Journal of Materials in Civil Engineering*, 30-10, 04018231.

Verian, K. P., Whiting, N. M., Olek, J., Jain, J. & Snyder, M. B. 2013. Using recycled concrete as aggregate in concrete pavements to reduce materials cost. *Report FHWA/IN/JTRP-2018/18*, US. Dep. of Transportation, Indiana.

Carbon footprint of asphalt road pavements using warm mix asphalt with recycled concrete aggregates: A Colombian case study

D. Vega Araujo
Universidad del Norte, Barranquilla, Colombia

J. Santos
University of Twente, The Netherlands

G. Martinez-Arguelles
Universidad del Norte, Barranquilla, Colombia

ABSTRACT: The construction of road pavements is an important consumer of natural resources and a major contributor to the human-driven emission of greenhouse gases (GHGs). As increasing attention has been paid to sustainability and natural resources preservation, highway agencies are looking for alternative materials and innovative asphalt mix technologies to reduce the carbon footprint of their pavement construction practices. In this study, a comparative carbon footprint assessment was conducted to evaluate the potential environmental benefits related to the use of recycled concrete aggregates (RCA) as a partial replacement of coarse natural aggregates in the production of Warm Mix Asphalt (WMA). In order to estimate the GHGs emissions associated with the use of these alternative resources in the construction of road pavements in Barranquilla, Colombia, primary data was gathered from companies in the region. The SimaPro software version 8.4.0 was used for modelling the processes analyzed in the case study according to the methodologies proposed in the ISO 14067 standards "Greenhouse gases- Carbon footprint of products- Requirements and guidelines for quantification and communication" and ISO 14040 standards "Environment management- Life cycle assessment- Principles and framework". The pavement life cycle phases and processes included within the system boundaries were the following: (1) materials production and transportation to the mixing plant; (2) materials processing and mixtures production at the mixing plant; (3) mixtures transportation to the construction site; and (4) pavement construction. Finally, three percentages of RCA replacements were considered in the production of WMA.

1 INTRODUCTION

With the ever-increasing awareness of climate change, practitioners in transportation infrastructure have been striving for innovations to save natural resources and to reduce energy consumption and emissions. The asphalt pavement industry, in particular, has increased the efforts to develop and evaluate new and innovative methods and technologies that can potentially be more sustainable than the conventional ones. Among the new technologies, the use of asphalt mixtures requiring lower manufacturing temperatures, such as the Warm Mix Asphalt (WMA), have received particular attention. Usually these techniques allow for a reduction in the mixing temperature in a range of 20 to 40 °C (D'Angelo et al. 2008, Vega et al. 2019, Vidal et al. 2013). Furthermore, they are also associated with savings in emissions of greenhouse gases (GHGs) such as CO_2 of up to 40% (EAPA 2010, Hassan 2009, Vaitkus et al. 2009) comparatively to those generated by the production of conventional Hot Mix Asphalt (HMA).

Another practice in activities of construction and/or rehabilitation of asphalt pavements that can be considered as eco-friendly is the use of recycled materials as partial replacement of the natural aggregates (NA) in the mixtures. The use of different types of recycled materials has been in the spotlight during the last two decades (Martinez-Arguelles et al. 2019, Vidal et al. 2013, Wang et al. 2018). Reclaimed Asphalt Pavement (RAP) and Recycled Concrete Aggregates (RCA) have been two of the most used materials when trying to reduce the use of NA in asphalt mixtures. Moreover, the literature reports their application in different pavement layers (Cho et al. 2011, Ding et al. 2016, Farooq et al. 2018, Mills-Beale & You 2010). Although RCA have been widely studied and showed promising results as a replacement for coarse NA in HMA (Pasandín & Pérez 2015, Perez et al. 2012, Zulkati et al. 2013) the research studies related to its application in WMA are still limited.

Motivated by the growing use of both RCA and WMA in road pavement construction and maintenance practices, this study aims to assess and compare the potential carbon footprint savings associated with the use of different percentages of RCA as a replacement of coarse natural aggregates in WMA mixtures.

2 METHODS

The study presented in this paper was performed taking into account as much as possible the ISO 14067 standards "Greenhouse gases- Carbon footprint of products- Requirements and guidelines for quantification and communication" (ISO 2013), the ISO 14040 standards "Environment management- Life cycle assessment- Principles and framework" (ISO 2006) and the Federal Highway Administration's (FHWA's) Pavement Life Cycle Assessment (LCA) Framework (Harvey et al. 2016).

2.1 *Goal and scope definition*

2.1.1 *Goal*
The main goal of this study was to estimate the potential environmental benefits, expressed as CO_2 equivalent emissions, related to the production of WMA with several RCA contents, namely 15, 30 and 45%. The results were compared with those associated with the use of conventional WMA, and are intended to provide insights in the extent to which recycling based materials and WMA technologies with similar features to those considered in this case study and employed in similar circumstances, are efficient in fostering environmental sustainability in the road pavement sector.

2.1.2 *System boundaries description*
The study was performed according to a cradle-to-laid approach and the system boundaries included four pavement life cycle phases (Figure 1): (1) materials production and transportation to the mixing plant; (2) materials processing and mixtures production at the mixing plant;

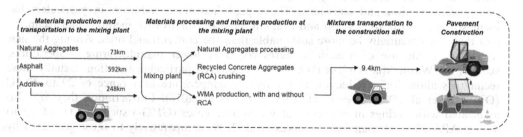

Figure 1. System boundaries considered in the case study.

334

(3) mixtures transportation to the construction site; and (4) pavement construction. Furthermore, the boundaries for the pavement structure were limited to the binder course (BC).

Finally, a 'cut-off' allocation approach was considered for dealing with the RCA. Thus, only the impacts associated with the processing of the recycling materials were included in the system boundaries (Santos et al. 2018, Schrijvers et al. 2016). That means that the environmental burdens related to the pavement demolition and the transportation of the recycled materials to the recycling facility were disregarded from the system boundaries.

2.1.3 Functional unit

The functional unit is the basis for comparisons between different products with the same utility for the same function. In the pavement sector this means a unit of pavement that can safely and efficiently support the same volume of traffic over the same project analysis period. Thus, it is defined by its geometry, service life, and levels of traffic supported. In this case study it was defined as a typical Colombian highway section with 1km in length and 1 lane 3.5 m wide.

The pavement structures were designed according to the conventional characteristics of traffic and subgrade support in Barranquilla, Colombia. More specifically, they were designed for a traffic load equal to 5×10^6 Equivalent Single Axle Load (ESAL) of 80kN, a CBR of 7.5% and a service life of 10 years. The geometric characteristics of a pavement structure designed with a conventional WMA (i.e., 0% RCA content) in the BC are illustrated in Figure 2.

In order to ascertain the potential environmental advantages related to the use of WMA with and without RCA content in the BC, the reference pavement structure (Figure 2) was compared to three alternative structures with equal geometry, but in which the BC of the initial structure was made of WMA with three RCA contents. Those alternatives represent structures with equivalent structural capacity, where the only design parameter that changed was the thickness of the asphalt BC. Tests carried out in the laboratory were performed with the purpose of determining the components proportions and mixtures performance (i.e., resilient modulus, fatigue and rutting resistance).

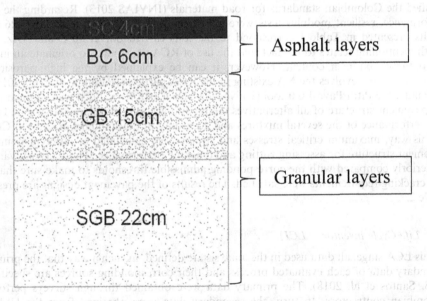

Figure 2. Geometric characteristics of a pavement structure designed with a conventional WMA. Acronyms: SC- Surface Course; BC- Binder Course; GB- Granular Base; SGB- Sub Granular Base.

Table 1. Composition and characteristics of the mixtures.

Item	Mixture			
	WMA0	WMA15	WMA30	WMA45
Natural Aggregate				
Quantity (%)[1]	95.6	88.3	80.9	73.5
Absorption (%)	3			
Recycled Concrete Aggregate				
Quantity (%)[2]	-	15	30	45
Bitumen				
Quantity (%)[1]	4.4	4.5	4.8	5.2
Additive				
Type	Chemical			
Quantity (%)[3]	0.3			
Properties				
Density (kg/m³)	2366	2310	2305	2289
Air voids (%)	4.3	4.8	4.6	4.8
Voids filled with asphalt (%)	66.6	66.5	67.2	66.0
Voids in the mineral aggregates (%)	12.7	14.2	13.9	14.2
Stability (kN)	17.2	14.8	16.7	20.1
Flows (mm)	2.9	2.7	3.0	3.4
Resilient modulus (MPa)	1633	1501	1372	1374

[1] Percentage of total mixture weight; [2]Percentage of coarse aggregates; [3]Percentage of asphalt weight.

Table 1 presents the composition and characteristics of the mixtures analyzed in the case study. They are identified according to the key "XY", where "X" stands for the type of mixture (i.e., WMA) and "Y" represents the percentage of RCA (i.e., 0, 15, 30 or 45%). In addition, all the mixtures contain 50% of coarse aggregates and 50% of fine aggregates. In this way, the RCA replacements were made in 50% of the total mass of the aggregates. The mixtures were designed according to Marshall design specifications (INVIAS 2014) and all samples satisfied the Colombian standards for road materials (INVIAS 2013). Regarding the mixtures performance, resilient modulus tests were performed according to the EN 12697-26 (C). The results presented in Table 1 correspond to the tests carried out at 40°C and 4Hz. A result worth mentioning relates to the fact that the use of RCA was found to originate an increase in the optimum bitumen content. However, it can be explained by the high porosity of the mortar layer that evolves the NA existing in the RCA particles (Pasandín & Pérez 2015).

Finally, the Pitra Pave 1.0.0 tool (Universidad De Costa Rica 2015) was adopted to design the pavement structure of all alternatives taking into account the characteristics and mechanical performance of the several mixtures and according to the standard practice in Colombia. In this way, maximum critical stresses and strains were calculated at different points of the pavement structure for assessing rutting and cracking performance. Those critical values were posteriorly compared with the corresponding admissible thresholds to make sure that rutting and cracking requirements were satisfied. The results of the pavement designs are presented in Table 2.

2.2 Life Cycle Inventory (LCI)

In this LCA stage, all data used in the analysis are defined, in such a way that the primary and secondary data of each evaluated process and their corresponding sources are specified (ISO 2013, Santos et al. 2018). The primary data were obtained through surveys performed in Colombian contractors. In turn, the secondary data were obtained from the USLCI and Ecoinvent v.3 databases and literature. Table 3 presents the processes evaluated per pavement LCA phase and the corresponding type of input data considered. Table 4 presents the values of the input data.

Table 2.   Pavement design for each type of mixture.

| Type of mixture | Thickness (cm) | | | | |
| | Asphalt layers | | Granular layers | | Total |
	SC	BC	GB	SGB	
WMA0	4.0	6.0	15.0	22.0	47.0
WMA15	4.0	6.5	15.0	22.0	47.5
WMA30	4.0	7.0	15.0	22.0	48.0
WMA45	4.0	7.0	15.0	22.0	48.0

Acronyms: WMA- warm mix asphalt; SC- surface course; BC- binder course; GB- granular base; SGB- sub granular base.

Table 3.   Summary of the type of data and corresponding sources.

Pavement LCA phase	Process	LCI type	Source
Materials production and transportation to the mixing plant	Natural Aggregates (NA) extraction	Primary	Previous investigation (Martinez-Arguelleset al. 2019)
	NA load movements and transportation	Primary	Previous investigation (Martinez-Arguelleset al. 2019)
	Asphalt production	Secondary	*"bitumen, at refinery/kg/ US"*- USLCI database
	Asphalt transportation	Primary	Survey data
	Additive production	Secondary	*"fatty acid/market for/ Alloc Def, U"*- Ecoinvent database
	Additive transportation	Primary	Survey data
Materials processing and mixtures production at the mixing plant	NA processing	Primary	Survey data
	RCA crushing	Primary	Survey data
	Mixture production (binder course layer), with and without RCA replacements	Primary	Survey data
Mixture transportation to the construction site	Mixture transportation	Primary	Survey data
Pavement construction	Finisher operation	Secondary	Literature data (Thenoux & Dowling 2007)
	Vibratory roller operation	Secondary	Literature data (Thenoux & Dowling 2007)
	Pneumatic roller operation	Secondary	Literature data (Thenoux & Dowling 2007)

In the mixture production phase, the thermal energy (TE) provided by the combustion of heavy fuel oil (HFO) was determined according to the energy balance proposed by Santos et al. (2018). It is represented by Equation 1.

Table 4. Input data considered in the case study.

Item	Diesel (gal/ton)	Lubricant (gr/ton)	Electricity (kWh/ton)	Water (kg/ton)
Materials production and transportation to the mixing plant phase				
Natural Aggregates				
Extraction (Martinez-Arguelles et al. 2019)	1.85	20	-	-
Load to the dump truck (Martinez-Arguelles et al. 2019)	1.85	20	-	-
Transportation to the mixing plant	0.56	9.42	-	-
Asphalt				
Transportation to the mixing plant	4.17	70.66		-
Additive				
Transportation to the mixing plant	1.94	32.95	-	-
Materials processing and mixtures production at the mixing plant phase				
Natural Aggregates				
Processing (Martinez-Arguelles et al. 2019)	0.075	0.69	2.33	100
Recycled Concrete Aggregates				
Crushing	0.075	0.69	2.33	100
Mixtures transportation to the construction site phase				
			Capacity [m³]	
Dumper	0.072	1.21	18	-
Pavement construction phase				

Construction Equipment	Performance (L/h)	Capacity (m³/h)	Diesel (L/m³)	Energy (MJ/m³)
Finisher (Thenoux & Dowling 2007)	13	60	0.22	8.39
Vibratory roller (Thenoux & Dowling 2007)	18	65	0.28	10.72
Pneumatic roller (Thenoux & Dowling 2007)	16	65	0.25	9.53

$$TE = [\sum_{i=1}^{M} m_i \times C_i \times (t_{min} - t_o) + m_{asph} \times C_{asph} \times (t_{mix} - t_o)$$

$$+ \sum_{i=1}^{M} m_i \times W_i \times C_{water} \times (100 - t_o) \qquad (1)$$

$$+ L_v \times \sum_{i=1}^{M} m_i \times W_i + \sum_{i=1}^{M} m_i \times W_i \times C_{vap} \times (t_{mix} - 100)] \times (1 + CL)$$

where TE = thermal energy in MJ/ton mixture, required to produce an asphalt mixture; M = number of aggregates fractions; m_i = mass of aggregates of fraction i; C_i = specific heat of aggregates of fraction i. The name of the remaining parameters and the values of all parameters considered in Equation 1 are presented in Table 5. Table 6 presents the TE and corresponding HFO required to produced 1 ton of the mixtures considered in the case study.

2.3 Life Cycle Inventory Assessment (LCIA)

The LCIA, which assesses the potential carbon footprint, was performed according to the Climate Change impact category specified by the TRACI v.2.1. impact methodology.

Table 5. Values of the parameters considered in Equation 1.

Parameter		Value	Unit
t_o	Ambient temperature	25	°C
$t_{mix-WMA}$	Mixing temperature of WMA with 0, 15, 30 and 45% RCA replacements	120	°C
C_{agg}	Natural aggregates specific heat[1]	0.74	KJ/Kg/°C
W_{agg}	Natural aggregates water content	3	% by mass of aggregates
W_{RCA}	RCA water content	3	% by mass of RCA
C_{RCA}	Recycled concrete aggregates specific heat[1]	0.74	KJ/Kg/°C
C_{water}	Water at 15°C specific heat	4.19	KJ/Kg/°C
L_v	Water latent heat of vaporization	2256	kJ/kg
C_{vap}	Water vapor specific heat	1.83	kJ/kg
C_{asph}	Asphalt specific heat	2.09	KJ/Kg/°C
CL	Casing loses factor[2]	27	%

[1] Value for granitic aggregates (Santos et al. 2018); [2]Value taken from the literature (Santos et al. 2018, West et al. 2014).

Table 6. Thermal energy (TE) and heavy fuel oil (HFO) consumed for producing 1 ton of each type of mixture.

Mixture	TE (MJ/ton mixture)	Fuel consumption (Kg HFO/ton mixture)
WMA0	202.75	4.71
WMA15	202.19	4.79
WMA30	201.96	4.79
WMA45	201.90	4.79

Acronyms: TE- Thermal energy; HFO- Heavy fuel oil.

3 RESULTS AND DISCUSSION

The SimaPro software version 8.4.0 was used for modelling the processes analyzed in this case study.

For each asphalt mixture implemented in the BC, Figure 3 presents the contribution of the various material production processes and construction equipment usage to the carbon footprint, expressed in kilograms of CO_2-equivalent.

According to the results presented in Figure 3, it can be concluded that the use of recycled materials (RCA in this case) in the production of WMA does not translate into savings of CO_2-eq emissions. Indeed, under the conditions of this case study, WMA with RCA were found to have a lower performance than that of the conventional WMA, and therefore, require thicker BC layers. Specifically, the mixtures WMA15, WMA30 and WMA45 were found to be 8, 17 and 17% thicker that the mixture WMA0, respectively. That represents an increase in the CO_2-eq emissions equal to 5, 13 and 14%, respectively, in relation to those generated with the use of the mixture WMA0 in the BC.

As far as the breakdown of the total emissions is concerned, Figure 3 shows that the major processes contributing to the total CO_2-eq emissions are those related to the bitumen production, mixture production and NA extraction. Specifically, the production of bitumen was found to contribute up to 40% in WMA45, while for the other alternative mixtures this value dropped to approximately 35%. The mixture production was found to contribute to approximately 30% for all alternatives evaluated. In turn, the NA extraction can contribute up to 24% of the total CO_2-eq emissions in the case of the WMA0. In accordance with the results described previously, it can be concluded that the hypothetical savings generated by the use of

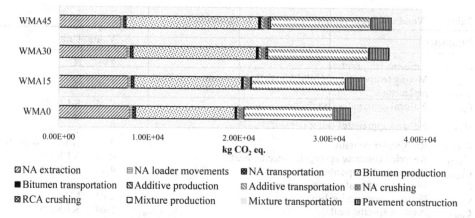

Figure 3. Contribution of the various processes to the total carbon footprint associated with the several asphalt mixtures implemented in the BC.

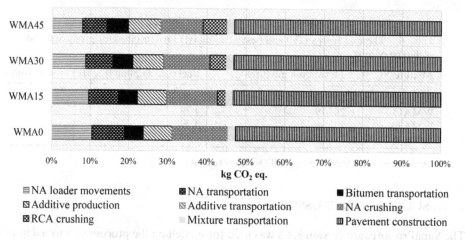

Figure 4. Breakdown of the total CO2-eq emissions without the contribution of the processes bitumen production, mixture production and NA extraction.

RCA in WMA are offset by the additional emissions generated by the increase in bitumen contents.

Finally, in order to understand the contribution of the remaining processes, Figure 4 presents the results without the contribution of the bitumen production, mixture production and NA extraction. From the analysis of this Figure, it can be noted that pavement construction and NA crushing are the processes that contribute most to the total CO_2-eq emissions generated for each alternative mixture studied. In the case of the process pavement construction, its contribution can represent around 53% of the total CO_2-eq emissions for all WMA mixtures, while the process NA crushing was found to contribute between 11% (WMA45) and 14% (WMA0) of the total carbon footprint.

4 CONCLUSIONS

A comparative carbon footprint assessment was performed to estimate the potential savings related to the use of different levels of RCA replacements in coarse natural aggregates of

WMA. Specifically, three contents of RCA were considered (i.e. 15, 30 and 45%) and the results compared with a control mixture. The processes analyzed were modeled in SimaPro software version 8.4.0, using primary data collected by surveys of representative companies in the region of Barranquilla as well as secondary data from the USLCI and Ecoinvent v.3 databases and literature. Both types of data were combined with laboratory characterization tests of the mixtures assessed. Finally, the LCIA was conducted according to TRACI v.2.1 impact methodology.

Based on the results obtained, the following points are worth highlighting:

- The use of RCA is likely to increase the optimum bitumen content and, therefore, the additive content up to 18% (for WMA45);
- Mixtures with RCA contents were found to show a lower performance than that of the control mixture, which translates into an increase in the thickness of the layer obtained from the pavement design;
- According to the conditions of the study, the bitumen production, mixture production and NA extraction were found to be the processes that contributed most to the carbon footprint of the analyzed mixtures;
- When the results are analyzed without the consideration of the processes mentioned above, the pavement construction and NA crushing can represent approximately 65% of the total carbon footprint when considered together.
- Finally, it can be concluded that the use of RCA in WMA is likely not to be an environmentally friendly solution, if the mixtures with RCA require greater bitumen content and show a lower performance than the control mixture.

The results of this study can be used to elucidate decision makers within the asphalt pavement industry interested in lowering the carbon footprint of their practices.

REFERENCES

Cho, Y. H., Yun, T., Kim, I. T., & Choi, N. R. 2011. The application of recycled concrete aggregate (RCA) for hot mix asphalt (HMA) base layer aggregate. *KSCE Journal of Civil Engineering*, 15(3), 473–478.

D'Angelo, J., Harm, E., Bartoszek, J., Baumgardner, G., Corrigan, M., Cowsert, J., ... & Prowell, B. 2008. Warm-mix asphalt: European practice (No. FHWA-PL-08-007). *United States. Federal Highway Administration. Office of International Programs*.

Ding, T., Xiao, J., & Tam, V. W. 2016. A closed-loop life cycle assessment of recycled aggregate concrete utilization in China. *Waste management*, 56, 367–375.

EAPA. 2010 (January). The Use of Warm Mix Asphalt. *Masterbuilder. Co.in*. 1–13. http://www.master builder.co.in/the-use-of-warm-mix-asphalt/.

Farooq, M. A., Mir, M. S., & Sharma, A. 2018. Laboratory study on use of RAP in WMA pavements using rejuvenator. *Construction and Building Materials*, 168, 61–72.

Harvey, J., Meijer, J., Ozer, H., Al-Qadi, I. L., Saboori, A., & Kendall, A. 2016. Pavement Life Cycle Assessment Framework (No. FHWA-HIF-16-014). *United States. Federal Highway Administration*.

Hassan, M. 2009. Life-cycle assessment of warm-mix asphalt: an environmental and economic perspective. *Lousiana Univerty*, *Civil Engineering Class*.

INVIAS. 2013. Secciones 700 y 800 - Materiales y Mezclas Asfálticas y Prospección de Pavimentos (Primera Parte). *Instituto Nacional de Vías*.

INVIAS 2014. Capítulo 4 – Pavimentos Asfálticos. *Instituto Nacional de Vías*.

ISO 14040. 2006. Environmental management - Life cycle assessment - Principles and framework. *International Organization for Standardization, Geneva, Switzerland*.

ISO 14067. 2013. Greenhouse Gases - Carbon Footprint of Products - Requirements and Guidelines for Quantification and Communication. *International Organization for Standardization, Geneva, Switzerland*.

Martinez-Arguelles, G., Acosta, M. P., Dugarte, M., & Fuentes, L. 2019. Life Cycle Assessment of Natural and Recycled Concrete Aggregate Production for Road Pavements Applications in the Northern Region of Colombia: Case Study. *Transportation Research Record*, 0361198119839955.

Mills-Beale, J., & You, Z. 2010. The mechanical properties of asphalt mixtures with recycled concrete aggregates. *Construction and Building Materials*, 24(3), 230–235.

Pasandin, A. R., & Pérez, I. 2015. Characterization of recycled concrete aggregates when used in asphalt concrete: a technical literature review. *European journal of environmental and civil engineering, 19*(8), 917–930.

Pérez, I., Pasandín, A. R., & Gallego, J. 2012. Stripping in hot mix asphalt produced by aggregates from construction and demolition waste. *Waste Management & Research, 30*(1), 3–11.

Santos, J., Bressi, S., Cerezo, V., Presti, D. L., & Dauvergne, M. 2018. Life cycle assessment of low temperature asphalt mixtures for road pavement surfaces: A comparative analysis. *Resources, Conservation and Recycling, 138*, 283–297.

Schrijvers, D. L., Loubet, P., & Sonnemann, G. 2016. Critical review of guidelines against a systematic framework with regard to consistency on allocation procedures for recycling in LCA. *The International Journal of Life Cycle Assessment, 21*(7), 994–1008.

Thenoux, G., Gonzalez, A., & Dowling, R. 2007. Energy consumption comparison for different asphalt pavements rehabilitation techniques used in Chile. *Resources, Conservation and Recycling, 49*(4), 325–339.

Universidad De Costa Rica. 2015. PITRA PAVE 1.0.0 - Software de Multicapa Elástica.

Vaitkus, A., Čygas, D., Laurinavičius, A., & Perveneckas, Z. 2009. Analysis and evaluation of possibilities for the use of Warm Mix Asphalt in Lithuania. *Baltic Journal of Road & Bridge Engineering (Baltic Journal of Road & Bridge Engineering), 4*(2).

Vega, D. L., Martinez-Arguelles, G & Santos, J. 2019. Life Cycle Assessment of Warm Mix Asphalt with Recycled Concrete Aggregate. In *IOP Conference Series: Materials Science and Engineering* (Vol. 603, No. 5, p. 052016). IOP Publishing.

Vidal, R., Moliner, E., Martínez, G., & Rubio, M. C. 2013. Life cycle assessment of hot mix asphalt and zeolite-based warm mix asphalt with reclaimed asphalt pavement. *Resources, Conservation and Recycling, 74*, 101–114.

Wang, H., Liu, X., Apostolidis, P., & Scarpas, T. 2018. Review of warm mix rubberized asphalt concrete: Towards a sustainable paving technology. *Journal of cleaner production, 177*, 302–314.

West, R., Rodezno, C., Julian, G., Prowell, B., Frank, B., Osborn, L. V., & Kriech, T. 2014. Field performance of warm mix asphalt technologies. (No. Project 9–47A).

Zulkati, A., Wong, Y. D., & Sun, D. D. 2012. Mechanistic performance of asphalt-concrete mixture incorporating coarse recycled concrete aggregate. *Journal of Materials in Civil Engineering, 25*(9), 1299–1305.

Sustainable flexible pavement overlay policy for reduced life-cycle cost and environmental impact

Chunfu Xin & Qing Lu

Department of Civil and Environment Engineering, University of South Florida, USA

ABSTRACT: Pavement is a critical part of the highway infrastructure, which deteriorates over time due to the combined effects of material aging, traffic loadings, and climatic factors. As pavement condition deteriorates, vehicle operating costs and adverse environmental impact would increase. As a typical pavement maintenance and rehabilitation (M&R) technique, asphalt overlay is commonly used to restore pavement performance. To make a cost-effective and eco-friendly decision in scheduling asphalt overlays, the effect of asphalt overlay strategy on life-cycle costs and environmental impact needs to be evaluated. In this study, an integrated life cycle assessment (LCA) - life cycle cost analysis (LCCA) approach is proposed to evaluate the life-cycle environmental and economic impacts of different asphalt overlay strategies. A post-overlay roughness progression model is incorporated in the integrated LCA-LCCA analysis. Based on the case study, for 2-in asphalt overlay projects, the application of 30% reclaimed asphalt pavement (RAP) materials would reduce the life-cycle energy consumption, greenhouse gas (GHG) emissions, criteria air pollutants, and economic costs by 1.6%, 1.8%, 13.7%, and 5.8%, respectively. Due to main impact of the use phase, the optimum overlay strategy for achieving the minimum life-cycle energy consumption and GHG emissions is 4-in milling and asphalt overlay with 30% RAP materials. While, due to main impact of production and construction phase, the optimum overlay strategy for achieving the minimum life-cycle criteria air pollutants and costs is 2-in asphalt overlay with 30% RAP materials.

1 INTRODUCTION

1.1 Background

An effective highway transportation infrastructure is a key factor in economic and social development [1]. Highway pavement, as a critical component of the highway transportation infrastructure, supports more than nine trillion tonne-kilometers of freight and transports passengers more than fifteen trillion kilometers around the world every year [2,3]. After the construction of a pavement system, pavement condition will deteriorate over time due to a combination effect of material aging, traffic loading, and environmental factors. As pavement condition deteriorates, vehicle operating costs and their corresponding environmental impacts would increase significantly [4]. As a typical pavement maintenance and rehabilitation (M&R) technique, asphalt overlay is commonly used to restore pavement performance. Meanwhile, asphalt overlay itself consumes a huge amount of energy and natural resources. To make a cost-effective and eco-friendly decision in scheduling asphalt overlays, the effect of asphalt overlay strategy on life-cycle costs and environmental impact needs to be evaluated. However, in practice, asphalt overlay strategies are primarily determined based on lowest life-cycle agency costs. The environmental impacts of asphalt overlay strategies over the analysis period are ignored.

1.2 Literature review

In recent years, several researchers have compared the environmental impacts of different pavement overlay systems with the life cycle assessment (LCA) approach. In 2009, Zhang et al. estimated the environmental impacts for concrete overlay, asphalt overlay, and engineered cementitious composites [ECC] overlay using a pavement LCA model. They found that, compared to a conventional concrete overlay system, the ECC overlay system can reduce the life-cycle energy consumption, greenhouse gas (GHG) emissions, and costs by 15%, 32%, and 40%, respectively [5]. In 2010, Zhang et al. optimized pavement overlay system by integrating dynamic LCA and LCCA with an autoregressive pavement overlay deterioration model. They found that the optimal overlay strategies would reduce the total life-cycle energy consumption, GHG emissions, and costs by 5-30%, 4-40%, and 0.4-12%, respectively [6]. In 2012, Yu and Lu compared the environmental impacts of three pavement overlay systems (concrete overlay, asphalt overlay, and crack, seat and asphalt overlay) with the LCA approach. They found that, relative to asphalt overlay, concrete overlay may reduce GHG emissions by about 35.7% [7]. In 2013, Lidicker et al. minimized the effects of asphalt overlay policies on life-cycle costs and GHG emissions by solving a bi-objective optimization problem. They found that there was a tradeoff between costs and emissions when developing an asphalt overlay policy [8]. In 2014, Wang and Gangaram quantified the impact of four pavement preservation treatments on life-cycle energy consumption and GHG emissions. They found that the thin overlay would achieve the largest benefits in reducing energy consumption and GHG emissions [9].

However, current studies only considered the effect of overlay type on life-cycle energy consumption and GHG emissions. The impacts of detailed asphalt overlay design on post-overlay pavement roughness progression and its corresponding life-cycle air pollutants were not considered.

1.3 Research objective

The objective of this study is to evaluate the effects of asphalt overlay strategies on economic costs and environmental impact over the analysis period. The remainder of this paper is structured as follows. The data collection for roughness model development and life-cycle analysis are illustrated in Section 2. The post-overlay roughness models are illustrated in Section 3. The research methodology for evaluating the life-cycle economic and environmental effects of asphalt overlay strategy is illustrated in Section 4. A case study for comparing the effect of different asphalt overlay strategies on life-cycle environmental and economic impacts is provided in Section 5. Finally, Section 6 summarizes the major findings and provides suggestions for identifying eco-friendly or cost-effective asphalt overlay strategy.

2 DATA COLLECTION

2.1 Long-Term Pavement Performance (LTPP)

In this study, to develop the post-overlay roughness model, asphalt overlay projects extracted from the LTPP SPS-3, SPS-5, and GPS-6 programs were incorporated [10]. For SPS-3 program, only pavement sections treated with thin asphalt overlay were selected. For SPS-5 program, the control sections were excluded since they provide no information related to asphalt overlay. In addition, roadway sections whose pre-overlay pavement roughness were not available were excluded. After being extracted from the LTPP database, 15-year post-overlay roughness data were evaluated for completeness and reasonableness. For missing roughness data, the linear interpolation approach was used by calculating a mean between the values before the missing data and the value after. Since the missing data account for less than 1% of total sequential observations, the impact of missing data on overall quantitative analysis for continuous pavement deterioration process can be ignored. In summary, as shown in Figure 1, 271 asphalt overlay projects were extracted from the LTPP database.

The potential contributing factors of post-overlay roughness were categorized as asphalt overlay design, original pavement performance, as-built pavement structural characteristics, traffic loadings, and climatic factors. These contributing factors can be collected from the

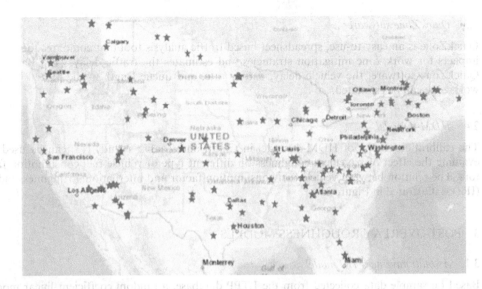

Figure 1. Spatial distribution of LTPP asphalt overlay projects.

LTPP maintenance module, rehabilitation module, test module, monitoring module, traffic module, and climate module. The overlay thickness can be extracted from the LTPP rehabilitation module or the LTPP test module. Since the actual overlay thickness deviated from the experimental design data, they were determined with pavement core testing data from the LTPP test module.

2.2 Athena pavement LCA

Athena pavement LCA is an LCA-based software package that makes life cycle assessment data easily accessible to transportation engineers [11]. The life cycle inventory (LCI) databases for building materials and products can be applied in the study. The fuel consumption factor and production rate for a list of overlay-related equipment are summarized in Table 1.

Table 1. A list of equipment during overlay activity [8].

Equipment	Fuel Consumption		Production Rates	
	Value	Unit	Value	Unit
Asphalt Paver	0.0620	l/tonne	1215	tonne/day
Asphalt Remixer	3.6409	l/tonne	8.30	tonne/hour
Black Topper	0.0009	l/m^2	10000	m^2/hour
Cold In-Place Recycler	0.0438	l/tonne	1713	tonne/hour
Compactor	0.0237	l/tonne	2,726	tonne/day
Heating Machine	1.1307	l/tonne	8.30	tonne/hour
HMA Transfer	0.0935	l/(tonne·km)	1,215	tonne/day
Roller	0.0533	l/tonne	1,215	tonne/day
Concrete Truck	3.7854	l/m^2	60	m^3/day
Dump Truck	0.2271	l/tonne	1,000	tonne/day
Water Truck	0.0114	l/m^2	20,00	m^2/day
Pavement Breaker	0.1345	l/m^2	1,000	m^2/day
Diamond Grinder	1.0759	l/m^2	125	m^2/day
Milling Machine	0.4203	l/m^3	40	m^3/hour

2.3 QuickZone software

QuickZone is an easy-to-use, spreadsheet-based traffic analysis tool that compares the traffic impacts for work zone mitigation strategies and estimates the traffic delays [12]. Based on QuickZone software, the vehicle delay, detour rate, and queue length in the partial closure work zone can be estimated.

2.4 HDM-4 model

The calibration results of HDM-4 model in Chatti and Zaabar's study [13] can be used to evaluate the effect of pavement roughness on different type of vehicle fuel consumption factors. The relation between vehicle fuel consumption factor and international roughness index (IRI) is illustrated in Figure 2.

3 POST-OVERLAY ROUGHNESS MODELS

3.1 As-built pavement IRI model

Based on sample data collected from the LTPP database, a random coefficient linear model was proposed to evaluate the difference between the original pavement IRI and as-built pavement IRI. Then, the as-built IRI model can be calculated as follows.

$$IRI_i^0 = 40.596 + \left[0.253 - 0.071ML - EN \cdot f\left(\beta|0.054, 0.06^2\right) - 0.075NT\right]IRI_i^B \\ - 1.348TK - 0.199FAT + 3.912RUT + \varepsilon_i \tag{1}$$

where, IRI_i^B and IRI_i^0 are original IRI (inches/mile) and as-built IRI (inches/mile), respectively. ML is the milling indicator (1 if milling is involved, 0 otherwise), EN is endogenous overlay design indicator (1 if overlay design is dependent on current pavement condition and traffic, 0 otherwise), NT is north central region indicator (1 if in LTPP north central region,

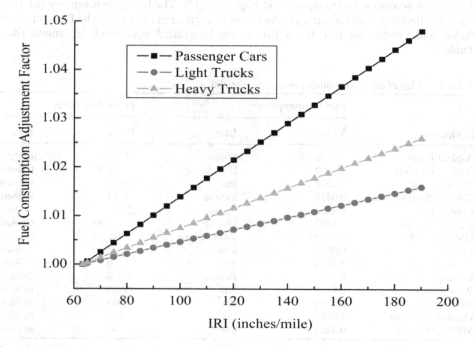

Figure 2.   Effect of pavement roughness on vehicle fuel consumption [13].

346

0 otherwise), TK is asphalt overlay thickness (inches), FAT is pre-overlay pavement relative fatigue cracking area (%), RUT is pre-overlay severe pavement rutting indicator (1 if rut depth is over 10 mm, 0 otherwise), $f(\beta|0.054, 0.06^2)$ is the probability density of the normal distribution with a mean of 0.054 and a standard deviation of 0.06, and ε_i is random error.

3.2 Post-overlay pavement IRI model

Based on sample data collected from the LTPP database, a correlated random coefficient linear model was proposed to evaluate the post-overlay IRI progression rate. Then, the post-overlay IRI progression model can be predicted as follows.

$$\Upsilon_i^t = [1 - 1.91TK + 0.24(OVR \cdot FAT) + 0.082FWD - 0.96BST$$
$$+2.73FSG + 7.767ESAL + 0.008FI]\,t \times 10^{-3} + f(\sigma_i) + \mu_i^t \tag{2}$$

$$\mu_i^t = 0.9857\mu_i^{t-1} + \eta_{it} \tag{3}$$

$$IRI_i^t = IRI_i^0 e^{\Upsilon_i^t} \tag{4}$$

where, IRI_i^t is post-overlay IRI (inches/mile) of pavement section i at overlay age t, IRI_i^0 is as-built IRI (inches/mile), Υ_i^t is the average natural logarithmic value of post-overlay IRI progression rate of pavement section i over t years, TK is asphalt overlay thickness (inches), OVR is asphalt overlay indicator (1 if no milling operation is involved, 0 otherwise), FAT is pre-overlay pavement relative fatigue cracking area (%), FWD is maximum FWD deflection on as-built pavement (μm), BST is bounded subbase layer thickness (inches), FSG is fine-grained subgrade indicator (1 if subgrade material is fine-grained soil, 0 otherwise), $ESAL$ is annual average daily 18-kip ESAL ($\times 10^6$ ESAL), FI is annual average freezing index (°C · days), $f(\sigma_i)$ is the probability density of the normal distribution with a mean of 0 and a standard deviation of 0.000018, μ_i^t is a time-series correlated error, and η_{it} is a random error.

4 RESEARCH METHODOLOGY

4.1 An integrated LCA-LCCA approach

As shown in the authors' previous study [4], the life-cycle environmental and economic impacts of different pavement overlay strategies can be evaluated with an integrated LCA-LCCA approach.

4.2 Life Cycle Assessment (LCA)

The material module of a pavement LCA includes raw material acquisition and material processing in the process of pavement overlay activities. The type and the amount of material consumed in construction may vary from one asphalt overlay strategy to another. The construction module includes equipment use and energy use at the construction site. As listed in Table 1, the fuel types of all construction equipment are assumed to be diesel. The transportation module accounts for transport of materials and equipment to and from the construction site.

The congestion module accounts for the environmental impacts created by construction-related traffic congestion, traffic delay, and traffic detour. Based on the case study assumptions, the vehicle delay, detour rate, and queue length in the partial closure work zone can be estimated with QuickZone software. When construction-related vehicle delay and congestion are identified, they can be coupled with fuel consumption and vehicle emissions to quantify their environmental impacts.

The usage module quantifies the environmental impacts of vehicle operations within the analysis period. Different asphalt overlay strategies would change vehicle fuel economy by affecting pavement roughness progression. In practice, pavement sections are most likely to remain in place at the end of the analysis period, a "cut-off" allocation method is used to assign no environmental impacts to the end-of-life module for all pavement overlay strategies.

4.3 Life Cycle Cost Analysis (LCCA)

The LCCA procedure consists of selecting an analysis period, selecting a discount rate, selecting a measure of economic worth, and determining monetary agency costs and user costs [14, 15]. The selected analysis period is 40 years, which is the same as the analysis period for life cycle assessment. The discount rate is assumed to be 3%. The net present value (NPV) expressing all costs and benefits over the analysis period in terms of their equivalent values in the initial year of the analysis period is selected as the measure of economic worth. Agency costs include all costs (e.g., material costs, equipment use fee, labor costs, temporary traffic control, and mobilization cost) incurred directly by highway agencies over the analysis period. The residual value of the recent overlay pavement structure at the end of the analysis period is deducted from agency costs.

The user costs include vehicle operating costs (VOC), user delay costs and vehicle crash costs. The vehicle operating costs are estimated as the monetary value of extra fuel consumption of vehicle traveling on an overlaid pavement relative to that on an ideally smooth pavement (i.e., IRI = 63 inches/mile). Based on the FHWA report [16], the rates of delay cost for passenger cars, light-duty trucks, and heavy-duty trucks are 11.58 \$/veh-hr (vehicle hour), 18.54 \$/veh-hr, and 22.31\$/veh-hr, respectively. The delay cost rates are in 1996 dollars and updated to 2020 dollars in the LCCA model using the rate of inflation (2%). The vehicle crash costs are estimated with the increased crash risk due to overlay construction activities. The increased crash risk costs for construction-related work-zone traffic and detour traffic are estimated as 0.22\$/VMT (vehicle-miles-traveled) and 0.15\$/VMT, respectively [17].

5 CASE STUDY

5.1 Goal and scope definition

The research objective is to evaluate the environmental and economic impacts of different overlay strategies over a 40-year analysis period. In the case study, sixteen asphalt overlay strategies were analyzed and summarized in Table 2.

The general roadway segment information, existing pavement structure, existing pavement performance, traffic characteristics, climatic factors, and construction project information for these overlay projects are summarized in Table 3.

The functional unit is a 10-km long, 3.7-m wide overlay system over the outer lane of the existing asphalt pavement. The asphalt overlay strategies are scheduled based on the post-overlay roughness progression model and overlay trigger value (IRIc = 170 inches/mile). The specific construction schedules for different asphalt overlay strategies are shown in Figure 3. The effects of different overlay strategies on future pavement roughness progression are illustrated in Figure 4.

5.2 Results and discussions

The environmental impact performance indicators include global warming potential (GWP), acidification potential (AP), human health (HH) particulate, smog potential (SP), and total primary energy consumption (TPE) [17]. The GWP is expressed on an equivalency basis relative to CO_2, where GWP is 1 for CO_2, 25 for CH_4, and 298 for N_2O [19]. The AP of air or water emission is calculated by its SO_2 equivalent effect. The HH particulate includes the

Table 2. Design of different asphalt overlay strategies.

Scheme	Overlay Thickness (inches)	Asphalt Milling (1-yes/0-no)	30% RAP (1 or 0)
1	2	1	0
2	2	1	1
3	2	0	0
4	2	0	1
5	4	1	0
6	4	1	1
7	4	0	0
8	4	0	1
9	6	1	0
10	6	1	1
11	6	0	0
12	6	0	1
13	8	1	0
14	8	1	1
15	8	0	0
16	8	0	1

Table 3. System definition of asphalt overlay projects.

Category	Item Description	Value
General information	Interstate highways (1-yes, 0-no)	1
	Number of lanes in each traffic direction	2
	Speed limit (km/h)	120
	Segment length (km)	10
	Main lane width (m)	3.7
	Inside shoulder width (m)	1.5
	Outside shoulder width (m)	2.5
Existing pavement structure	Structural course SP-12.5 thickness (inches)	4
	Structural course SP-19.0 thickness (inches)	6
	Lime-rock (LR) base course thickness (inches)	10
	Subgrade type (1-course-grained, 0-fine-grined)	0
	Subsurface drainage condition (1-good, 0-poor)	1
	IRI threshold for asphalt overlay	170
Existing pavement performance	Area of fatigue cracking in 10-km lane (%)	4
	Average rut depth in 10-km lane (mm)	8
Traffic information	Annual average daily traffic (AADT) (vehicles/day)	17,000
	Percentage of trucks in AADT (%)	12
	Average truck factor: an equivalent number of 80-kN ESAL	1.3
	Annual traffic growth rate (%)	0
Climatic factors	Climate zone (1-wet freeze zone, 0-otherwise)	0
	Annual average rainfall (mm)	1300
	Annual average freeze index (°C*days)	0
	Annual average daily temperature (°C)	24
	Average daily maximum temperature in July (°C)	34
	Average daily minimum temperature in January (°C)	10
Construction project information	Average distance from plant to site (km)	100
	Average distance from site to stockpile (km)	100
	Average distance from equipment depot to site (km)	100

particulate matter of various sizes (PM_{10} and $PM_{2.5}$). The smog potential is expressed on a mass of equivalent O_3, which is a product of interactions of volatile organic compounds (VOCs) and nitrogen oxides (NO_x). Since sulfur dioxide (SO_2), particulate matter (PM), and

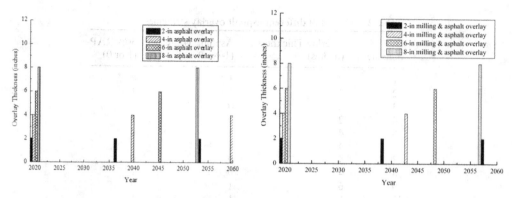

Figure 3. Construction schedule of different overlay strategies.

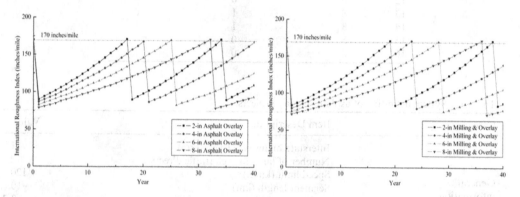

Figure 4. Effect of different overlay strategies on pavement roughness.

nitrogen dioxide (NO_2) are criteria air pollutants (CAP), the CAP can be calculated as the sum value of AP, HH, and SP. In addition, since feedstock energy stored in asphalt mixture can be harvested later during the recycling process, it is ignored in the total primary energy consumption. Thus, the TPE, GWP, and CAP are identified as three environmental impact indicators [18].

5.2.1 Energy consumption

The energy consumption and GHG emissions in material module, construction module, transportation module, congestion module, and usage module of different asphalt overlay strategies is illustrated in Figure 5. As shown in Figure 6, the three major LCA modules dominating energy consumption of asphalt overlay projects are usage module, construction module, and material module. The life-cycle energy consumption for 2-in asphalt overlay (scheme 3), 4-in asphalt overlay (scheme 7), 6-in asphalt overlay (scheme 11), and 8-in asphalt overlay (scheme 15) are 2.37×10^5 GJ, 2.53×10^5 GJ, 2.54×10^5 GJ, and 2.77×10^5 GJ, respectively. While, the life energy consumption for 2-in milling and asphalt overlay (scheme 1), 4-in milling and asphalt overlay (scheme 5), 6-in milling and asphalt overlay (scheme 9), and 8-in milling and asphalt overlay (scheme 13) are 2.46×10^5 GJ, 2.32×10^5 GJ, 2.44×10^5 GJ, and 2.82×10^5 GJ, respectively. The comparison results for the above fuel consumptions indicate that the milling operation is suggested when the asphalt overlay thickness is between 4 inches and 6 inches.

Relative to the traditional asphalt overlay, the inclusion of 30% RAP materials for 2-in and 8-in asphalt overlays reduces the life-cycle energy consumption by 1.6% and 3.7%, respectively. For 4-in asphalt overlay in the study, inclusion of 30% RAP material can reduce the total energy consumption by 5.19×10^3 GJ. This is because the inclusion of 30% RAP materials does not affect pavement roughness progression. Meanwhile, they can reduce the use of

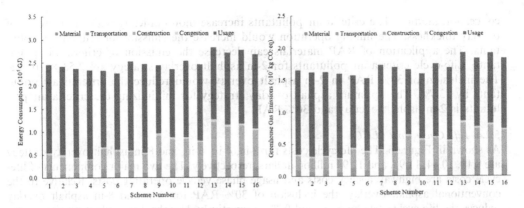

Figure 5.   Life-cycle energy consumption and GHG emissions of different overlay strategies.

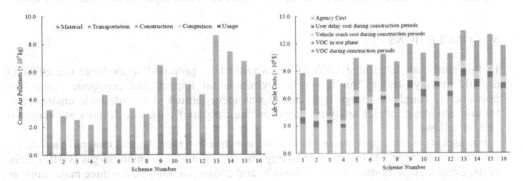

Figure 6.   Life-cycle criteria air pollutants and costs of different overlay strategies.

virgin aggregates and virgin asphalt binders in the production of asphalt mixtures. In the case study, the optimum overlay strategy consuming the least amount of energy in the analysis period is 4-in milling and asphalt overlay, and 30% RAP materials.

5.2.2 Greenhouse Gas (GHG) emissions

As shown in Figure 5, the three major LCA components for GHG emissions are usage module, construction module, and material module. Based on the calculation results, life-cycle GHG emissions for 2-in, 4-in, 6-in, and 8-in asphalt overlay strategies are 1.63×10^7 kg CO2 eq, 1.73×10^7 kg CO2 eq, 1.75×10^7 kg CO2 eq, and 1.92×10^7 kg CO2 eq, respectively. While, life-cycle GHG emissions for 2-in, 4-in, 6-in, and 8-in milling & asphalt overlay strategies are 1.65×10^7 kg CO2 eq, 1.57×10^7 kg CO2 eq, 1.66×10^7 kg CO2 eq, and 1.93×10^7 kg CO2 eq, respectively. The comparison results on these four pairs of fuel consumption indicate that the milling operation is suggested when the asphalt overlay thickness is between 4 inches and 6 inches. Relative to the traditional asphalt overlay, the inclusion of 30% RAP for 2-in and 8-in asphalt overlay reduces the life-cycle GHG emissions by 1.8% and 4.1%, respectively. For 4-in asphalt overlay in the study, inclusion of 30% RAP materials can reduce the total GHG emissions by 3.93×10^5 kg CO_2 eq. In this study, the optimum overlay strategy emitting the least amount of greenhouse gases in the analysis period is 4-in milling and asphalt overlay, and 30% RAP materials.

5.2.3 Criteria air pollutants

The criteria air pollutants and life cycle costs in different life cycle stages for different overlay strategies are illustrated in Figure 6. As can be seen, the three major LCA components for criteria air pollutants are construction module, material module, and congestion module. The criteria air pollutants increase monotonically with the increase of overlay thickness. The milling operation would increase the emission of criteria air pollutants. The application of RAP materials can decrease the emission of criteria air pollutants. Life-cycle criteria air pollutants for 2-in asphalt overlay strategy are 2.53×10^5 kg. The inclusion of 30% RAP in 2-in asphalt overlay would reduce the criterial air pollutants by 13.7%. The optimum asphalt overlay strategy for minimizing the criteria air pollutants is 2-in asphalt overlay and 30% RAP.

5.2.4 *Life Cycle Cost (LCC)*

As shown in Figure 6, The life-cycle costs for 2-in, 4-in, 6-in, and 8-in asphalt overlay strategy are 8.11, 10.84, 11.93, and 12.94 million dollars, respectively. The two major components of life-cycle costs are highway agency costs and usage phase vehicle operating costs. Relative to the conventional asphalt overlay, the inclusion of 30% RAP for 2-in and 8-in asphalt overlay reduces the life-cycle costs by 5.8% and 9.3%, respectively. For the 4-in asphalt overlay, inclusion of 30% RAP materials can reduce the total life-cycle cost by 848,500 dollars. The optimum asphalt overlay strategy with the minimum life-cycle cost is 2-in asphalt overlay and 30% RAP.

6 CONCLUSIONS

In this study, an integrated LCA-LCCA approach is proposed to evaluate the effect of asphalt overlay strategy on life-cycle environmental impact and economic costs. The post-overlay roughness progression model was incorporated in the life-cycle environmental and economic impact analysis. Based on the case study, the conclusions are summarized as follows.

For asphalt overlay projects, usage module, construction module, and material module are three major components of life-cycle energy consumption and greenhouse gas (GHG) emissions. Construction module, material module, and congestion module are three major components of life-cycle criteria air pollutants. Highway agency costs and usage phase vehicle operating costs are two dominant components of life-cycle costs.

The life-cycle energy consumption, GHG emissions, criteria air pollutants, and costs for 2-in asphalt overlay projects on 10-km long and 3.7-m wide pavement section over 40 years is 2.37×10^5 GJ, 1.63×10^7 kg CO_2 eq, 2.53×10^5 kg, and 8.11×10^6 dollars, respectively. The application of 30% RAP material in 2-in asphalt overlay projects would reduce the life-cycle energy consumption, GHG emissions, criteria air pollutants, and costs by 1.6%, 1.8%, 13.7%, and 5.8%, respectively. In practice, the application of RAP materials is strongly suggested in pavement rehabilitation.

Based on the sustainability goal of life-cycle energy consumption and GHG emissions, the optimum asphalt overlay strategy for the case study is 4-in milling and asphalt overlay with 30% recycled material. While, based on the sustainability goal of life-cycle criteria air pollutants and costs, the optimum overlay strategy is 2-in asphalt overlay with 30% recycled material. The research findings in the study can serve as a reference for highway agencies to schedule asphalt overlay in an eco-friendly way. It is worth mentioning that, since 30% RAP is not a typical RAP content, highway agencies may need to consider the uncertainty of future pavement performance by adjusting the RAP content in their overlay projects.

ACKNOWLEDGMENT

The authors acknowledge the support from the U.S. Department of Transportation (DOT) Center for Transportation, Environment, and Community Health (CTECH).

REFERENCES

[1] Paterson, W. D. O. Road deterioration and maintenance effects: models for planning and management. 1987.

[2] IRF. World Road Statistics 2010. Geneva, Switzerland: International Road Federation (IRF). 2010.

[3] Bureau of Transportation Statistics (BTS). National transportation statistics. Washington, DC, 2010.

[4] Lu, Q., Mannering, F.L., and Xin, C. An integrated life cycle assessment (LCA) – life cycle cost analysis (LCCA) framework for pavement maintenance and rehabilitation. CTECH report (I). 2018.

[5] Zhang, H., Lepech, M. D., Keoleian, G. A., Qian, S. and Li, V.C. Dynamic life-cycle modeling of pavement overlay systems: capturing the impacts of users, construction, and roadway deterioration. Journal of Infrastructure Systems 16, no. 4 (2009): 299–309.

[6] Zhang, H., Keoleian, G. A., Lepech, M. D., & Kendall, A. (2010). Life-cycle optimization of pavement overlay systems. *Journal of infrastructure systems*, *16*(4), 310–322.

[7] Yu, B., and Lu, Q. Life cycle assessment of pavement: Methodology and case study. Transportation Research Part D: Transport and Environment 17, no. 5 (2012): 380–388.

[8] Lidicker, J., Sathaye, N., Madanat, S., & Horvath, A. (2012). Pavement resurfacing policy for minimization of life-cycle costs and greenhouse gas emissions. *Journal of Infrastructure Systems*, *19*(2), 129–137.

[9] Wang, H., & Gangaram, R. (2014). *Life Cycle Assessment of Asphalt Pavement Maintenance* (No. CAIT-UTC-013). Rutgers University. Center for Advanced Infrastructure and Transportation.

[10] Lu, Q. and Xin, C. Pavement rehabilitation policy for reduced life-cycle cost and environmental impact based on multiple pavement performance measures. CTECH Report (II). 2018.

[11] Athena Sustainable Materials Institute (ASMI). User manual and transparency document: pavement LCA v3.1-web application. 2018. https://calculatelca.com/software/pavement-lca/.

[12] Federal Highway Administration (FHWA). Work zone and traffic analysis tools. https://ops.fhwa.dot.gov/wz/traffic_analysis/quickzone/Access in 2019.

[13] Chatti, K., and Zaabar, I. Estimating the effects of pavement condition on vehicle operating costs. Vol. 720. Transportation Research Board. 2012.

[14] Lu, Q., Li, M., Gunaratne, M., Xin, C., Hoque, M. and Rajalingola, M. Best practices for construction and repair of bridge approaches and departures. 2018.

[15] Shi, X., Mukhopadhyay, A., & Zollinger, D. (2018). Sustainability assessment for portland cement concrete pavement containing reclaimed asphalt pavement aggregates. *Journal of Cleaner Production*, *192*, 569–581.

[16] James, W., and Smith, M. Life-cycle cost analysis in pavement design-interim technical bulletin. Report No. FHWA-SA-98-079. 1998.

[17] MDOT. Standard specifications for construction. Michigan Department of Transportation, Ann Arbor, Michigan. 2002.

[18] Bare, J. C. Tool for the Reduction and Assessment of Chemical and Other Environmental Impacts (TRACI), Version 2.1 - User's Manual; EPA/600/R-12/554 2012.

[19] Watson, R. T., Zinyowera, M.C., and Moss, R. H. Climate change 1995. Impacts, adaptations and mitigation of climate change: scientific-technical analyses. 1996.

353

LCA and cost comparative analysis of half-warm mix asphalts with varying degrees of RAP

T. Mattinzioli, F. Moreno, M. Rubio & G. Martínez
Faculty of Civil Engineering, LabIC.UGR, Campus Fuentenueva, University of Granada, Spain

ABSTRACT: This paper describes the environmental impacts and associated direct costs of two half-warm asphalt mixtures (HWMA) at 90°C, with 70% and 100% RAP, respectively, compared to a reference AC 16 hot mix asphalt (HMA). In order to evaluate the environmental impacts of the mixtures, local data was studied and its quality reviewed. The process used highlights the benefits for HWMA recycled mixtures and the limitations of the current inventory data available in the local region. Findings include: 1) the combination of RAP and HWMA processes provide significant environmental savings and economic benefits when compared to conventional HMA mixtures, in a cradle-to-gate assessment, 2) the 70% and 100% RAP mixtures were found to provide an average energy saving of 31% and 35%, respectively, 3) while there was a limited amount of data found for Spain, sufficient open-access data availability was found at a European level for conducting a LCA.

1 INTRODUCTION

In response to the burgeoning need to reduce energy use and carbon footprints in all operations worldwide and to adopt a circular economy, decision making support tools are required in order to ensure the choices now made in all sectors are sustainable and responsible. In the European Commission's (EC) communication on Integrated Product Policy (COM (2003)302) (EC, 2019a), the EC concluded that life-cycle assessment (LCA) provides the best framework for assessing the environmental impacts of products, however it was underlined that more consistent data and consensus is required. Hand in hand with LCA, is life-cycle cost analysis (LCCA), which quantifies the economic viability of projects, and throughout Europe there have been various initiatives for LCCA development, in the format of guides by EUPAVE (EUPAVE, 2018) and projects such as LCC-DATA (initially for buildings (EC, 2019b)).

These same assessment methods can be implemented in order to tackle the large footprint the transportation sector imposes on the global environment. The main advancements made with regards to successfully reducing the environmental footprint of road pavements are the use of reduced temperature asphalt mixtures, and the recycling of aggregates (or reclaimed asphalt pavement, RAP). However, road authorities require justified solutions in order for them to be implemented. The use of LCA and cost analysis for new technologies can further reinforce their viability to road authorities, transporting them out of the laboratory and into real-life practice.

The use of both lower production mixture temperatures and RAP has been proven to reduce the adverse environmental impacts of standard asphalt mixtures throughout various studies (all with varying project assumptions, system boundaries and locations (Gulotta et al., 2019; Rubio et al., 2013; Vidal et al., 2013), while also being structurally viable alternatives to conventional processes and materials (Lizárraga et al., 2018, 2017). However, no studies were found for the LCA of HWMA combined with RAP.

As it stands, pavement LCA can be considered in its infancy. Having been around for close to 20 years (considering Stripple (2001) to have provided the first study), it is still considered to have various issues such as high purchase prices or limited inventory quality for tools and databases (Arteaga, 2018). Furthermore, given that sustainability is a context-based solution (Van Dam et al., 2015), a generic data inventory would not be valid for all regions, but rather local databases would be required (Santos et al., 2017). As it stands there is no local database available for Spain. European projects have considered various locations in their projects (All-Back2Pave, 2015; Lo Presti and D'Angelo, 2017) (Italy, Germany & UK), and some European countries have developed their own LCI (Jullien et al., 2015; Stripple, 2001). However, in Spain neither have taken place. Moral Quiza (2016) defined the environmental impacts of all the standardized pavement structures in Spain, according to the Spanish design standard (Ministerio de Fomento, 2003), with some local data and the Ecoinvent database in SimaPro. Santos et al. (2017) applied various LCA tools to a reconstruction project in Spain and Acciona (2013) carried out an Environmental Product Declaration of the N-340 road in Alicante, but again no local database was utilized.

1.1 Scope

This study aims to evaluate the viability of innovative, laboratory tested, asphalt mixtures, via the use of local, openly-available data. The mixtures will both contain varying quantities of RAP (70% and 100%) and will be produced at half-warm temperatures (90°C), and compared to a HMA reference mixture (160°C). The objectives to achieve this aim are to 1) explore the environmental impacts and 2) the economic impacts of using RAP along with HWMA processes, and 3) to review current local data available.

2 METHODOLOGY

In order to achieve the aforementioned aim of this study, the ISO 14040 framework will be adopted for LCA, where the following section will describe the functional unit, system boundary conditions, inventory and impact analysis indicators used for this study. The discussion will then review and interpret the results for the use of HWMA mixtures with a high content of RAP.

2.1 Functional unit and system boundaries

The functional unit (FU) considered is 1km of a typical Spanish highway surface course (T00 traffic class). Therefore, the dimensions of the FU are 1000m in length, 9m wide (two 3.5m lanes, with two 1m shoulders) and with a depth of 0.05m. This study considers the environmental impacts and economical costs of two different surface course layers. Therefore, impacts and costs related to the binder coarse, base course, sub-base, road markings, fences and railings, road signs and lighting are not within the scope of this study. Furthermore, maintenance operations are not included either. The same type of truck was used for transporting the materials to the plant (40km) and mixes to site (30km), and the same paving and compaction machinery was also used on-site for the projects. The assessment carried out will cover the cradle-to-gate (CEN, 2014) of the two mixtures, in order to establish the viability of the HWMA with varying amounts of RAP, where use phase energy and emissions will also be excluded in order to not overshadow the impacts of the two mixtures of interest. End-of-life considerations will also be excluded. The impacts used for RAP refer to its processing. Only the stockpile-plant and plant-site transport is considered.

The purpose of an LCCA is to estimate the total costs for project alternatives, in order to identify the most economically viable solution. This assessment does not consider a full LCCA, but would contribute to it. It can be stated that emulsion reduced-temperature RAP mixtures are as adequate as virgin mixtures (EAPA, 2008; Ojum, 2015), always when correct construction techniques are used, and therefore for all of the cases considered it is assumed that the three mixtures will all provide an equal service life before a first rehabilitation activity is required. Both through LabIC fatigue testing (UGR-FACT) and in Lizárraga et al. (2018, 2017), 100% RAP mixtures were found to offer similar resistance to fatigue, water sensitivity, and permanent deformations to HMA. Therefore, in this study, the boundary conditions for the cost analysis were not be influenced by durability. Furthermore, given this study is carrying out a cradle-to-gate analysis, only the direct costs for material, production and machinery fuel will be considered.

2.2 Materials

Firstly, the volume and quantity of the two mixtures was determined. The mixture design for the HMA (AC 16 S) following local standards (Ministerio de Fomento, 2014) and the HWMA was designed according to that established by ATEB (ATEB, 2014, chap. 5). From these calculations, the asphalt quantities are provided in Table 1. The HMA consists of fine and coarse aggregate, cement filler (<0.063mm) and bitumen, whereas the two HMWA consist of a bituminous emulsion (considered to already include rejuvenators for mixture workability) and 70% and 100% RAP aggregate. The RAP is assumed to originate from a stockpile, and therefore milling is outside of the project boundary.

With the mixtures designed, the HMA was produced in a fuel oil conventional discontinuous plant (batch asphalt plant), where the aggregates are weighted and measured, as specified, and heated to 180°C. The aggregates were then mixed with the bitumen and the filler at roughly 160°C in the pugmill. The production process for the HWMA was very similar to the manufacture of HMA in a conventional plant, but mixed at a lower production temperature (60-100°C) (Rubio et al., 2012); aggregate weighted and heated to 100°C, and mixed with the filler and bituminous emulsion at 60-70°C (Rubio et al., 2013). Because of the water in the HWMA bituminous emulsion, compactors require more passes to achieve optimal density (Rubio et al., 2013), therefore the compactors for the HWMA were assumed to pass 50% more than for the HMA mixture (HMA: 2-3 passes, HWMA: 5-6).

2.3 LCA and cost inventory

The secondary goal of this study is to assess at what stage LCI data is available in Spain. With that said, the most up-to-date, openly available and local sources were searched for. In the event that no local data could be found, the next most applicable source was selected. In general, it is accepted that data is not readily available, and its quality is usually low (Muench et al., 2014). Available data sources were identified and reviewed and are defined in Table 2, according to their geographical and temporal representativeness with the scale provided by

Table 1. Material inventory for mixtures.

Mixture	Material	Density (t/m^3)[a]	Quantity (%)	Weight (t)
HMA 1	Bitumen	2.50 t/m^3	4.75%	49.16 t
	Filler		5.72%	59.15 t
	Fine Aggregate		28.58%	295.75 t
	Coarse Aggregate		60.95%	630.94 t
HWMA 70% RAP	Emulsion	2.26 t/m^3	2.75%[c]	27.97 t
	RAP		68.08%	692.37 t
	Filler		1.75%	17.80 t
	Fine Aggregate		8.75%	88.99 t
	Coarse Aggregate		18.67%	189.87 t
HWMA 2 100% RAP	Emulsion	2.26 t/m^3	2.50%[c]	25.43 t
	RAP[b]		97.5%	991.58 t

[a] Density values are taken from real specimens.
[b] Assume 100% reclamation of aggregates.
[c] Reduced binder content as RAP already contains bitumen.

Table 2. Energy and emission inventory.

Item	Source	Source Proximity[a]	Source Date[b]
Bitumen	(eurobitume, 2012)	Good-Very Good	Fair
Emulsion	(eurobitume, 2012)	Good-Very Good	Fair
Aggregates (Sand & Coarse)	(Stripple, 2001)	Good	Poor
Filler	(MITECO, 2017)	Very Good	Very Good
HMA Plant	(Rubio et al., 2013)	Very Good (primary data)	Fair
HWMA Plant	(Rubio et al., 2013)	Very Good (primary data)	Fair
RAP	(EAPA, 2016; UNPG, 2011)	Good	Good-Fair
Paver	(Caterpillar, 2019)	Good	Very Good
Compactor	(Caterpillar, 2019)	Good	Very Good
Dumper Truck	(Brock, 2017; Stripple, 2001)	Good	Very Good-Poor

[a] Source Proximity key: very good = local source; good = European source; fair = non-European source.
[b] Source Date key: very good = <3 years, good = <6 years, fair = <10 years, poor = >10 years.

(Callahan et al., 2011, chap. 8). Feedstock energy not included, despite being required by ISO 14044, as is common practice and is problematic in pavement LCA (Harvey et al., 2016, fig. 6–12; Muench et al., 2014).

It is found, that when unavailable, data from other European countries can be applied to the country of interest; as done by Jullien, Dauvergne & Proust (2015), with Stripple (2001).

The life-cycle cost assessment was carried out via a document published by the Spanish Department of Transportation establishing project baseline costs (Ministerio de Fomento, 2016), and another local Spanish study (Tauste, 2014) for the costs related to HWMA and RAP (less available data); the data for this study was obtained via the collaboration of the Andalusian asphalt industry.

2.4 Impact assessment

From the inventory sources established, the outputs were collected. The impact categories (mid-point) selected to compare the mixtures are: global warming potential (GWP, $kgCO_{2e}$) (IPCC, 2013), acidification ($kgSO_{2e}$) (Hauschild and Wenzel, 1998), eutrophication ($kgPO_{4e}$) (Heijungs et al., 1992), photochemical oxidation (kg of Non-Metallic Volatile Organic Compounds) (van Zelm et al., 2008) and particulate matter 10 (kgPM). The energy (MJ), virgin material use (ton), recycled material use, and finally, the direct cost (€) shall be reported for the mixtures too.

3 RESULTS

3.1 LCA results

In order to establish the viability of the layers assessed, their environmental benefits will be first reviewed. The final results of the impact analysis for each mixture is displayed in Table 3, according to the pre-established environmental indicators. A comparison of the mixtures' emissions between the LCA stages is provided in Figure 1, and comparison of material use in Figure 2.

Overall, both HWMA mixtures offer energy and environmental impact savings, as seen in Table 3. The savings are due to the RAP and lower production temperatures of HWMA plants. In accordance with other studies (Muench et al., 2014; Wu and Qian, 2014), the production of the materials has the highest environmental impact, across all mixtures, meaning

Table 3. Impact analysis results, per 1km of a typical Spanish highway surface course.

		Total Primary Energy[a] (MJ)	Global Warming Potential (kgCO2e)	Acidification (kgSO2e)	Eutrophication (kgPO4e)	Photochemical Oxidation (kg NMVOC)	PM (kgPM)
HMA AC16	Material Production	4.29E+5	7.52E+4	8.28E+1	2.44E+1	1.73E+2	9.67E+0
	Mixture Production	3.47E+5	2.93E+4	1.52E+2	1.00E+2	1.72E+2	4.26E+0
	Construction	3.78E+3	2.76E+2	6.39E-2	1.57E-1	1.22E+0	9.03E-3
	Transport[a]	5.12E+4	3.69E+3	1.77E+0	4.31E+0	3.33E+1	1.32E+0
	Total	8.31E+5	1.08E+5	2.37E+2	1.29E+2	3.80E+2	1.53E+1
HWMA 70% RAP	Material Production	1.99E+5	2.61E+4	5.56E+1	1.80E+1	6.08E+1	5.48E+0
	Mixture Production	1.16E+5	1.10E+4	1.47E-1	3.02E+1	5.24E+1	1.64E+0
	Construction	3.93E+3	2.84E+2	6.48E-2	1.59E-1	1.23E+0	9.57E-3
	Transport[a]	4.64E+4	3.35E+3	1.60E+0	3.91E+0	3.02E+1	1.20E+0
	Total	3.65E+5	4.07E+4	5.74E+1	5.23E+1	1.45E+2	8.32E+0
HWMA 100% RAP	Material Production	1.28E+5	8.32E+3	5.34E+1	1.39E+1	2.27E+1	4.72E+0
	Mixture Production	1.16E+5	1.10E+4	1.47E-1	3.02E+1	5.24E+1	1.64E+0
	Construction	3.93E+3	2.84E+2	6.48E-2	1.59E-1	1.23E+0	9.57E-3
	Transport[a]	4.64E+4	3.35E+3	1.60E+0	3.91E+0	3.02E+1	1.20E+0
	Total	2.94E+5	2.29E+4	5.52E+1	4.82E+1	1.07E+2	7.56E+0

[a] Feedstock energy not included.
[b] Stockpile-Plant and Plant-Site transport distances combined.

that if the goal is to reduce the environmental impact of asphalt mixtures then the production materials should be a primary focus.

It can be seen more specifically from Figure 1, the half-warm process for mixture production provides a saving of 67% in energy consumption compared to hot mix plants. However, the highest savings due to half-warm plants is the saving of SO_{2e} emissions. Significant overall environmental benefits were seen for HWMA. The 70% RAP mixture provided a 54% energy saving for material production, and the 100% RAP mixture a 70% saving, compared to conventional virgin aggregates. From these results, it can be seen that the use of either RAP or half-warm plants would environmentally benefit a pavement construction project, with the benefits increasing even further with their combined use. The only increase in environmental impacts found can be associated to the construction machinery, where more on-site roller compaction is required for cooler mixtures, compared to HMA. Overall, an average energy saving of 31% and 36%, respectively, was found for the 70% and 100% RAP-containing HWMAs through the cradle-to-gate assessment.

With regards to the savings in terms of natural materials (Figure 2), it is apparent that the mixture with 100% RAP aggregate offers the most benefit with a 97.5% saving of virgin material (assuming 100% aggregate reclamation), thanks also to the residual binder found on the RAP (reducing the amount of virgin binder needed). The mixture with 70% RAP also offers a high saving of natural materials (67.6% recycled materials). With regards to the impacts per material, despite the binders and filler having the lowest quantity per weight, they offer the highest environmental impact in the mixtures. Despite the 100% RAP HWMA being constituted of 97.5% RAP, the aggregates only account for 36.7% of the energy usage.

Overall, both the 70% and the 100% RAP HWMA mixtures can offer large environmental savings (emissions and virgin materials), in comparison with a conventional HMA AC 16 surface layer.

Environmental Impacts of RAP-based HWMA versus conventional HMA

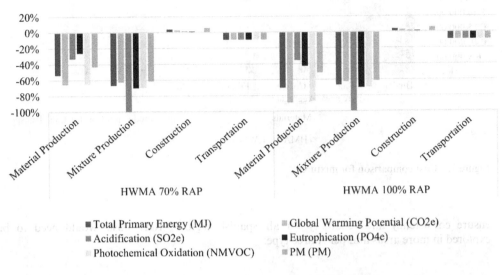

Figure 1. Impact assessment for HWMA mixtures compared to HMA.

Comparison between Material Quantities and their Energy Usage

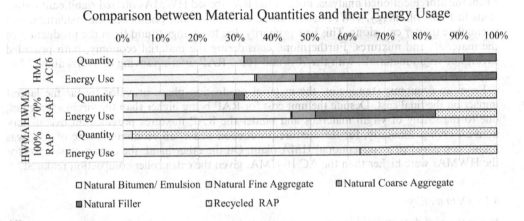

Figure 2. Material comparison for mixtures, with their corresponding energy usage.

3.2 Cost comparison

The direct costs for the cradle-to-gate evaluation are demonstrated in Figure 3. From the results displayed, it is apparent that the cost of the materials severely outweighs the production, construction and transportation costs. The highest material cost is found to be due to the binders, followed by the cement filler. With regards to economic savings for the production, roughly a 25% saving was made from HMA to HWMA processes, per FU. Despite the HWMA's need for more roller passes to reach the desired density, not much difference in paving cost was found. The 70% RAP HWMA mixture was found to offer an overall saving of around 33%, while the 100% RAP HWMA mixture around 35% saving, per FU. Therefore, by saving the amount of binder required due to RAP residual binder and the amount of filler (both most expensive, per unit weight), a significant cost reduction may be achieved. To

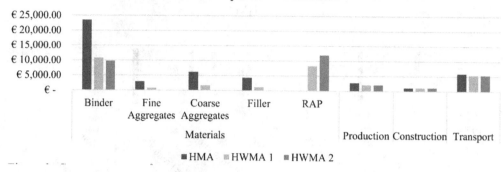

Figure 3. Cost comparison for mixtures.

ensure consistency and reliability for all Spanish regions, RAP costs would need to be explored in more areas and per source type.

4 DISCUSSION

From the aforementioned analysis, both of the RAP-based HWMAs offered significant reductions in terms of energy, GWP, acidification, eutrophication, photochemical oxidation, and particulate matter emissions. This was primarily due to savings found from the production of the materials and mixtures. Furthermore, considering the material economy, both provided savings of virgin materials, with the 70% and 100% RAP offering saving of 67.6% and 97.5%, respectively.

From an economic standpoint, the material production phase provided by far the largest impact on the final cost. Despite the unit cost for RAP being higher than the virgin aggregates, due to the saving of virgin materials and binder the RAP mixtures provided an overall economical saving. Similarly, for the mixture production phase, an economic saving of 25% was found, compared to a conventional HMA plant. On the other hand, the construction costs for the HWMAs were higher than the AC 16 HMA, given the extra roller compaction required.

4.1 Data quality

While the cost data utilized is purely from the target region, emission factors were proven difficult to obtain. The virgin aggregate data reviewed for this study was found to provide the largest variations in impact results, as seen in Table 4. In this study, the Stripple (2001) inventory was chosen as it has been used in various previous studies (Balieu et al., 2019; Muench et al., 2014), and as aggregate data for Spain was not found (except for filler data (MITECO, 2017)). While general emission inventories are published by the Spanish government (Gobierno de Aragón, n.d.; Junta de Andalucía, 2005), these documents only provide generic emission data for whole sectors.

Table 4. Variance in energy demands per source.

	Sand	Coarse Aggregate		HMA Plant
Stripple (2001)	6.25 MJ/t	65.94 MJ/t	Stripple (2001)	730.32 MJ/t
Marceau et al. (2007)	23.19 MJ/t	35.44 MJ/t	Rubio et al. (2013)	295.39 MJ/t
Jullien et al. (2015)	48.50 MJ/t	48.50 MJ/t	Jullien et al. (2015)	303.82 MJ/t

With regards to mixture production, also in Table 4, further disparities between plant operations include the energy value for a HMA plant with fuel oil, where for Swedish sources (Stripple, 2001) 730 MJ/t would be required, however French (Jullien et al., 2015) and Spanish sources (Rubio et al., 2013) require 304 MJ/t and 295 MJ/t, respectively. This difference can be due to the French and Spanish sources considering the amount of fuel required for production, while the Swedish source was found to also consider the electricity, natural gas and uranium for the plant. These extra factors, plus the 12-14 year difference between the sources can explain the great difference between these inventory values. No previous comparable data for HWMA was found.

This study acknowledges that quantifying RAP is a difficult exercise due to its novelty within the sector and the variability in its production (Gulotta et al., 2019), therefore the values for this process could vary considerably in other assessments with different data. For example, some studies model RAP as a free process (Muench et al., 2014), although in Lu et al. (2017) similar GWP values were used, but with lower energy values. The price of RAP is acknowledged to vary depending on process type and source, a report by the Spanish Public Works Research Centre states a 5-30% saving can be made (CEDEX, 2011). This is relative to the location, as RAP costs depend on availability and the cost of primary aggregates, landfill and transport (EAPA, 2008).

Machinery inventory values were obtained from their productivity information (Caterpillar, 2019), and then applied to emission values determined by the EPA (2014, 2016). A similar methodology was also adopted for the Sacramento Road Emissions Model (SMAQMD and Ramboll, 2016). Although, the SO_2 values found in the EPA documents (2014, 2016) are much higher than the target values imposed by legislations (Ntziachristos and Samaras, 2018), showing that if old machinery is used in projects, it does have an impact; though minimal compared to material and mixture production.

Despite eurobitume being used in various studies to provide bitumen inventory information, it is shown to provide some of the lowest energy and CO_{2e} results when compared to various other inventories (Yang et al., 2014).

The quality of the data obtained has been reviewed (Table 5) using a similar approach to Table 2, using a three-tier approach, according to its geographical and temporal representativeness. The plant and filler data were all deemed applicable for local use, whereas the majority of the sources are not from the local region. This study has provided a strong baseline for understanding what data is available in Spain, and it is therefore now possible to establish

Table 5. Review of LCI data used, according to geographical and temporal representativeness.

Item	Quality Comment	Item	Quality Comment
Bitumen	Medium: applicable to Spain, but not local primary data. Source should be updated.	HWMA Plant	High: local source; source could be updated.
Emulsion	Medium: applicable to Spain, but not local primary data; source should be updated.	RAP	Medium: relatively local, but due to importance local recent primary data should be found.
Aggregates (Sand & Coarse)	Low: high aggregate data variation found, local recent sources need to be explored.	Paver	Medium: reliable data source, but not local primary data & assumes modern equipment used.
Filler	High: local recent source.	Compactor	Medium: reliable data source, but not local primary data & assumes modern equipment used.
HMA Plant	High: local source; source could be updated.	Dumper Truck	Medium: reliable data source, but not local primary data & assumes modern equipment used.

Quality key: high = OK for local LCI, medium = sufficient, but primary data should be explored; low = not suitable for local LCI, primary data needs to be explored.

priorities for working towards a local inventory; the lowest data quality was found for aggregates. Finally, the data values used describe ideal construction conditions, therefore the energy consumption, emissions, and costs for construction and mixture production could increase under different conditions.

5 CONCLUSIONS

Different asphalt mixtures have been compared through the comparison of their cradle-to-gate environmental impacts and economical cost. In this paper, the impacts of two RAP-containing mixtures produced at a lower than conventional temperature (90°C compared to 160°C) were compared through quantitative LCA and cost comparison, using an AC 16 HMA as a reference mix. The two specific mixtures evaluated contained 70% and 100% RAP. As a secondary objective, aside of comparing the two mixtures, the quality of available data for the Spanish region was assessed. From the results of the LCA and cost comparison carried out, several results were found:

- The use of RAP combined with lower temperature mix technologies can offer large energy and carbon savings, when considering a cradle-to-gate approach. The 70% and 100% RAP mixtures were found to provide a 31% and 36% energy use saving, respectively. Similarly, from the environmental indicators, an average of 31% and 35% savings were found.
- Similarly, due to the saving of virgin filler, the need for less binder in the RAP mixtures (due to residual bitumen on aggregate), and the lower production temperatures, the half-warm RAP mixtures provided a 33% and 35% cost saving compared to a conventional HMA.
- Limited data is available for the Spanish region, where only plant and filler data was found. Aggregate data was found to be the least reliable, due to large disparities between the sources found and the age of the source used. However, at a European level, data can be considered as available for carrying out LCA. Local cost data is readily available for asphalt mixtures, except for new technologies (RAP and HWMA plants).

Due to the lack of primary data found for many of the items in the inventory, future work will involve sending out a survey to local constructors in order to obtain more reliable local data.

REFERENCES

Acciona Infraestructuras, 2013. Environmental Product Declaration: N-340 road.
AllBack2Pave, 2015. AllBack2Pave [WWW Document]. CEDR Transnatl. Road Res. Proj.
Arteaga, E.L., 2018. Life Cycle Assessment (LCA) of Pavements.
ATEB, 2014. Mezclas Templadas con Emulsión Bituminosa. Asociación Técnica de Emulsiones Bituminosas.
Balieu, R., Chen, F., Kringos, N., 2019. Life cycle sustainability assessment of electrified road systems. Road Mater. Pavement Des. 20, S19–S33.
Brock, J.D., 2017. Technical Paper T-135: Hot Mix Asphalt Trucking.
Callahan, W., James Fava, S.A., Wickwire, S., Sottong, J., Stanway, J., Ballentine, M., 2011. Product Life Cycle Accounting and Reporting Standard GHG Protocol Team.
Caterpillar, 2019. Caterpillar Performance Handbook 49.
CEDEX, 2011. Report: Reclaimed Asphalt Pavements.
CEN, 2014. BS EN 15804:2012 - Sustainability of construction works — Environmental product declarations — Core rules for the product category of construction products.
dos Santos, J.M.O., Thyagarajan, S., Keijzer, E., Flores, R.F., Flintsch, G., 2017. Comparison of Life-Cycle Assessment Tools for Road Pavement Infrastructure. Transp. Res. Board 2646, 28–38.
EAPA, 2016. Guidance Document for preparing Product Category Rules (PCR) and Environmental Product Declarations (EPD) for Asphalt Mixtures 1–22.
EAPA, 2008. Arguments to stimulate the government to promote asphalt reuse and recycling: EAPA-Position Paper.
EC, 2019a. European Platform on Life Cycle Assessment (LCA).

EC, 2019b. LCC DATA: Life-Cycle-Cost in the Planning Process. Constructing Energy Efficient Buildings taking running costs into account.

EPA, 2016. Nonroad Compression-Ignition Engines: Exhaust Emission Standards (EPA-420-B-16-022, March 2016).

EPA, 2014. Emission Factors for Greenhouse Gas Inventories.

EUPAVE, 2018. A guide on the basic principles of Life-Cycle Cost Analysis (LCCA) of pavements. European Concrete Paving Association. eurobitume, 2012. Life Cycle Inventory: Bitumen. Brussels, Belgium.

Gobierno de Aragón, n.d. Inventario de emisiones a la atmosfera en la comunidad de Aragón.

Gulotta, T.M., Mistretta, M., Praticò, F.G., 2019. A life cycle scenario analysis of different pavement technologies for urban roads. Sci. Total Environ. 673, 585–593.

Harvey, J.T., Meijer, J., Ozer, H., Al-Qadi, I.L., Saboori, A., Kendall, A., 2016. Pavement Life Cycle Assessment Framework. Washington DC, USA.

Hauschild, M., Wenzel, H., 1998. Environmental Assessment of Products, Volume 2: ed.

Heijungs, R., Guinée, J.B., Huppes, G., Lankreijer, R.M., Udo de Haes, H.A., Wegener Sleeswijk, A., Ansems, A.M.M., Eggels, P.G., Duin, R. van, Goede, H.P. de, 1992. Environmental life cycle assessment of products: guide and backgrounds (Part 1). CML, Leiden.

IPCC, 2013. AR5 Climate Change 2013: The Physical Science Basis.

Jullien, A., Dauvergne, M., Proust, C., 2015. Road LCA: the dedicated ECORCE tool and database Road LCA : the dedicated ECORCE tool and database. Int. J. Life Cycle Assess.

Junta de Andalucía, 2005. Capítulo 2. Emisiones de las Plantas Industriales.

Lizárraga, J.M., Jimenez Del Barco-Carrion, A., Ramírez, A., Díaz, P., Moreno-Navarro, F., Rubio, M. C., 2017. Mechanical performance assessment of half warm recycled asphalt mixes containing up to 100% RAP. Mater. Construcción 67.

Lizárraga, J.M., Ramírez, A., Díaz, P., Marcobal, J.R., Gallego, J., 2018. Short-term performance appraisal of half-warm mix asphalt mixtures containing high (70%) and total RAP contents (100%): From laboratory mix design to its full-scale implementation. Constr. Build. Mater. 170, 433–445.

Lo Presti, D., D'Angelo, G., 2017. Review and comparison of freely-available tools for pavement carbon footprinting in Europe, in: Pavement LCA 2017 Symposium. Illinois.

Lu, Y., Wu, H., Liu, A., Ding, W., Zhu, H., 2017. Energy Consumption and Greenhouse Gas Emissions of High RAP Central Plant Hot Recycling Technology using Life Cycle Assessment: Case Study, in: Pavement LCA 2017 Symposium. Illinois.

Marceau, M.L., Nisbet, M.A., Vangeem, M.G., 2007. Life Cycle Inventory of Portland Cement Concrete. Illinois.

Ministerio de Fomento, 2016. Orden Circular 37/2016 Base de Precios de Referencia de la Dirección General de Carreteras. Dirección General de Carreteras. Ministerio de Fomento.

Ministerio de Fomento, 2014. Pliego de Prescripciones Técnicas Generales para Obras de Carreteras y Puentes (PG-3) - Parte 2-Materiales Básicos, Parte 5-Firmes y Pavimentos, Parte 7-Señalización, Balizamiento y Sistemas de Contención de Vehículos.

Ministerio de Fomento, 2003. Norma 6.1 IC.

MITECO, 2017. Fabricación de Cemento (Combustión).

Moral Quiza, A., 2016. La herramienta ambiental análisis de ciclo de vida en el estudio de secciones de firme - Evaluación ambiental de varias secciones de firme de categoría de tráfico T00 a T2 conforme a la norma 6.1-1C. Universidad Alfonso X Sabio.

Muench, S.T., Lin, Y.Y., Katara, S., Armstrong, A., 2014. Roadprint: Practical Pavement Life Cycle Assessment (LCA) using generally available data, in: International Symposium on Pavement LCA 2014. Davis, California, USA.

Ntziachristos, L., Samaras, Z., 2018. EMEP/EEA air pollutant emission inventory guidebook.

Ojum, C.K., 2015. The design and optimisation of cold asphalt emulsion mixtures. University of Nottingham, UK.

Rubio, M. del C., Moreno, F., Martínez-Echevarría, M.J., Martínez, G., Vázquez, J.M., 2013. Comparative analysis of emissions from the manufacture and use of hot and half-warm mix asphalt. J. Clean. Prod. 41, 1–6.

Rubio, M.C., Martínez, G., Baena, L., Moreno, F., 2012. Warm mix asphalt: an overview. J. Clean. Prod. 24, 76–84.

Santos, J., Thyagarajan, S., Keijzer, E., Flores, R., Flintsch, G., 2017. Comparison of life cycle assessment tools for road pavement infrastructure, in: Pavement LCA 2017 Symposium. Illinois.

SMAQMD, Ramboll, 2016. Road Construction Emissions Model, Version 8.1.0.

Stripple, H., 2001. IVL Report: Life Cycle Assessment of Road - A Pilot Study for Inventory Analysis, 2nd ed. IVL Swedish Environmental Research Institute, Gothenburg, Sweden.

Tauste, R., 2014. Desarrollo de una Herramienta para el Análisis Económico y Ambiental de los Procesos de Fabricación y Puesta en Obra de Mezclas Bituminosas. University of Granada, Spain.

UNPG, 2011. Module d'informations environnementales de la production de granulats recyclés. Données sous format FDES conformes à la norme NF P 01-010. Union Nationale des Producteurs de Granulats.

Van Dam, T., Harvey, J., Muench, S., Smith, K., Snyder, M., Al-Qadi, I., Ozer, H., Meijer, J., Ram, P., Roesler, J., Kendall, A., 2015. Towards Sustainable Pavement Systems: A Reference Document. Washington D.C.

van Zelm, R., Huijbregts, M.A.J., den Hollander, H.A., van Jaarsveld, H.A., Sauter, F.J., Struijs, J., van Wijnen, H.J., van de Meent, D., 2008. European characterization factors for human health damage of PM10 and ozone in life cycle impact assessment. Atmos. Environ. 42, 441–453.

Vidal, R., Moliner, E., Martínez, G., Rubio, M.C., 2013. Life cycle assessment of hot mix asphalt and zeolite-based warm mix asphalt with reclaimed asphalt pavement. Resour. Conserv. Recycl. 74, 101–114.

Wu, S., Qian, S., 2014. Comparison of Warm Mix Asphalt and Hot Mix Asphalt Pavement Based on Life Cycle Assessment, in: International Symposium on Pavement LCA 2014. Sacramento, California, USA.

Yang, R., Ozer, H., Kang, S., Al-Qadi, I.L., 2014. Environmental Impacts of Producing Asphalt Mixtures with Varying Degrees of Recycled Asphalt Materials, in: International Symposium on Pavement LCA 2014. Sacramento, California, USA.

Pavement, Roadway, and Bridge Life Cycle Assessment 2020 – Harvey et al (eds)
© 2020 Taylor & Francis Group, London, ISBN 978-0-367-55166-7

Life cycle assessment of a thin bonded concrete overlay of asphalt project in Woodland, California

M. Ostovar, A.A. Butt, A. Mateos & J.T. Harvey
Department of Civil and Environmental Engineering, University of California Pavement Research Center (UCPRC), Davis, USA

ABSTRACT: Thin bonded concrete overlay of asphalt (BCOA), formerly known as thin whitetopping, can be defined as a rehabilitation alternative consisting of a 100 to 175 mm thick concrete overlay on an existing flexible or composite pavement. Thin BCOA has been used as a 20-year design life rehabilitation alternative for asphalt pavements in fair to good condition under low and intermediate traffic levels. While the technology for thin BCOA has been common on highways and conventional roads in several US states and other countries, its use has been very limited in California. As with any pavement rehabilitation, the materials and construction stages of thin BCOA result in significant environmental impacts, in terms of energy use, material resource consumption, waste generation, and emissions during the life of the BCOA pavement. This paper presents a life cycle assessment (LCA) that quantifies the potential environmental impacts due to the material and construction stages of a BCOA pilot project that has been implemented in Woodland, California. The scope of the study is from cradle-to-laid, including material and construction stages and transportation of the materials. However, sensitivity analysis has been performed in this paper to compare different BCOA design alternatives. The scope of the comparison selected is from cradle-to-gate. The impact categories included in this study are primary energy demand (fuel and non-fuel), greenhouse gas emissions, particulate matter, and smog formation. HVS PCC Type III mix had the highest energy consumption and environmental impacts followed by PCC Type II/V mix designs.

1 INTRODUCTION

BCOA or bonded concrete overlay on asphalt is a rehabilitation alternative that consists of placing a hydraulic cement concrete overlay on an existing asphalt pavement (Harrington & Fick, 2014). This study is mainly focused on thin BCOA, where overlay thickness is 100 to 175 mm (4 to 7 in). BCOA with an overlay thickness of 50 to 100 mm (2 to 4 in), typically referred to as ultrathin, is primarily used in urban areas with light traffic. While the technology for thin BCOA has been used on highways and conventional roads in several US states as well as in other countries for at least 20 years, use of thin BCOA has been very limited in California. After the successful evaluation of thin BCOA under accelerated trafficking conducted with the Heavy Vehicle Simulator (HVS) (Mateos et al., 2019), Caltrans decided to move forward and built a pilot thin BCOA project in State Route 113 (SR 113) in Woodland, California. The experimental data for the study presented in this paper come from the Woodland thin BCOA construction.

BCOA construction typically includes milling the existing asphalt layer to remove surface distresses and/or because of geometry constraints (e.g. to maintain road surface elevation). Asphalt surface pre-overlay repairs like localized patching and cracks sealing may be included as well. Sweeping multiple times and air blasting the asphalt surface to remove dust and debris are always required. Wetting the asphalt surface is also required. BCOA construction includes placing the concrete overlay and sawing joints to form 5×5 to 8×8 slabs. An alternative where the asphalt surface has more deterioration is to place a thin gap-graded rubberized hot mix asphalt (RHMA-G) overlay before placement of the concrete overlay. This approach showed the best bonding of the concrete overlay to the underlying layers and performance in the HVS test sections; however it is not certain that this performance will also occur on the Woodland SR 113 pilot sections. Thin BCOA joints are not always sealed because of the high cost and doubtful outcome (Mateos et al., 2019). All these activities were considered in modeling the construction stage except for sawing.

The main conclusion from recently concluded research on thin BCOA (Mateos et al., 2019) is that a well-designed, well-built 6×6 thin bonded concrete overlay placed on top of an asphalt base that is in fair to good condition can potentially provide 20 years of good serviceability on most of California's non-interstate roadways. Still, LCCA and LCA analyses of this rehabilitation alternative and comparison to other rehabilitation options (e.g. asphalt overlay) are pending. This study is expected to help bridge this gap by presenting a methodology that can be applied to conduct the LCA analysis of thin BCOA.

This paper presents the development of the life cycle inventories and some initial impact analysis of the BCOA technology as is being piloted in California. As a sensitivity, this paper compares several alternative BCOA cross-sections with the pilot project BCOA design, and the concrete mix used in the pilot project with alternative mixes for faster and slower strength gain. The paper does not include comparisons with other technologies because any comparisons will be highly context-driven and cannot be comprehensively evaluated yet. Instead, the intent is to place these new inventories in the software program eLCAP (under development) which will allow users to evaluate their own scenarios.

1.1 *Life cycle assessment*

Life cycle assessment is a technique that can be used for analyzing and quantifying the environmental impacts of a product, system, or process. LCA provides a comprehensive approach to evaluating the total environmental burden of a product or process by examining all of the in-puts and outputs over the life cycle, from raw material production to end-of-life. This systematic approach identifies where the most relevant impacts occur and where the most significant improvements can be made while identifying potential trade-offs.

As shown in Figure 1, the life cycle of pavements can be divided into four main phases including 1) Material production, 2) Construction, maintenance, and rehabilitation (M&R), 3) Use phase, and 4) End-of-Life (EOL) (Harvey et al. 2016). According to the goal and scope of this study, LCA does not consider the whole life cycle. This study is a cradle-to-laid or cradle-to-end of construction study which considers material stage, transportation stage, and construction stage. Therefore, this initial LCA study does not consider M&R, use stage and EOL. (Figure 1)

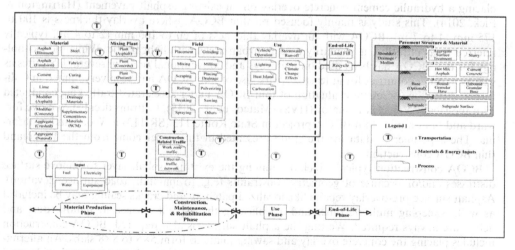

Figure 1. UCPRC LCA framework showing major life-cycle stages of a pavement (Harvey et al. 2016).

2 GOAL AND SCOPE OF THE STUDY

2.1 *Goal of the study*

Goal and scope define the system boundary and scope of the study, duration of the study and a suitable functional unit (Harvey et al., 2016). The goal of this study was to quantify the potential environmental impacts due to material and construction stages of thin BCOA. The scope of the study is from cradle-to-laid, including material and construction stages and transportation of the materials. The standalone LCA analysis is focused on a thin BCOA pilot project built in Woodland, California, in 2018-2019. The considered layer includes 150 mm thick PCC overlay on top of a new RHMA pavement, and 150 mm thick PCC overlay on top of a milled old asphalt. These configurations are referred to as 2B and 2A, respectively, in Table 1.

Additionally, a sensitivity analysis scoped at cradle–to-gate is performed comparing ten other BCOA design alternatives with the two Woodland pilot project alternatives. Each design alternative consists of a PCC overlay on top of either a new RHMA overlay or the milled asphalt pavement. Three concrete mix design alternatives are evaluated: (1) a rapid strength concrete, 4-hour opening time (OT), made with PCC Type III (1A and 1B in Table 1), (2) PCC Type II/V designed to be open to traffic in 24 hours constructed in Woodland (2A and 2B in Table 1), and (3) a normal strength concrete, 10-day design OT, made with PC Type II/V (3A and 3B in Table 1). The first one was used to build one of the sections that was tested with the HVS for a former research project on thin BCOA (Mateos et al., 2019). The third mix represents the typical concrete mix used in Caltrans standard jointed plain concrete pavements. For each of the three BCOA design alternatives, additional three designs with a thickness of 125mm are also considered in the sensitivity analysis (4A to 6B in Table 1). Table 1 shows the twelve different BCOA alternatives considered in this paper.

2.2 *Scope of the study*

The scope of this study was limited to "cradle-to-laid" for Woodland pilot project in which the materials and construction stages as well as transportation of materials in the life cycle of the pavements were considered. The use stage and end-of-life were not included in this study's

Table 1. Different considered BCOA cases.

Case Number	Material	Concrete Thickness mm (inch)	RHMA Thickness mm (inch)
1-A	HVS PCC Type III (4-hr OT)+ Tie Bar	150 (6)	—
1-B	HVS PCC Type III (4-hr OT)+ Tie Bar+ RHMA	150 (6)	30 (1.2)
2-A	Woodland PCC Type II/V (24-hr OT)+ Tie Bar	150 (6)	—
2-B	Woodland PCC Type II/V (24-hr OT)+ Tie Bar+ RHMA	150 (6)	30 (1.2)
3-A	Caltrans normal strength PCC Type II/V (10-day OT)+ Tie Bar	150 (6)	—
3-B	Caltrans normal strength PCC Type II/V (10-day OT) +Tie Bar+ RHMA	150 (6)	30 (1.2)
4-A	HVS PCC Type III (4-hr OT)+ Tie Bar	125 (5)	—
4-B	HVS PCC Type III (4-hr OT)+ Tie Bar+ RHMA	125 (5)	30 (1.2)
5-A	Woodland PCC Type II/V (24-hr OT)+ Tie Bar	125 (5)	—
5-B	Woodland PCC Type II/V (24-hr OT)+ Tie Bar+ RHMA	125 (5)	30 (1.2)
6-A	Caltrans normal strength PCC Type II/V (10-day OT) + Tie Bar	125 (5)	—
6-B	Caltrans normal strength PCC Type II/V (10-day OT)+Tie Bar+ RHMA	125 (5)	30 (1.2)

scope. The functional unit defined for this study is construction of 1 lane-km of pavement surface including the materials life cycle stage and transportation of materials. The Woodland pilot project cross-sections consisted of 150 mm (6 inches) of new concrete on varying thicknesses of old hot mix asphalt (HMA) with its surface milled, and an alternative with the same thickness of PCC, but with a 30 mm (1.2 inches) layer of RHMA-G placed between the old HMA and the new PCC.

The material stage included extraction of raw materials from the ground, transportation to processing plants, and the plant processing. Transportation of the materials from the plant to the site was also included. The intended audience of the study includes local governments, pavement researchers and practitioners, and pavement designers.

3 LIFE CYCLE INVENTORY (LCI) AND LIFE CYCLE IMPACT ASSESSMENT (LCIA)

In the inventory phase, all the inputs and outputs to the system boundary within the life cycle are quantified. The inputs are normally in the form of input flows of raw materials and energy and output flows of waste and pollution (depending on the system boundary), emissions to air, water, and soil as well as the flow of product output. Life cycle data for all inputs to and outputs from the system are assessed and assembled at this step (Harvey et al., 2016).

During an impact assessment (LCIA) phase, the LCI results are classified and categorized into several environmental impact categories such as global warming, primary energy consumption, ozone layer depletion, human health and more. The purpose of impact assessment is to better understand the environmental significance of the LCI by translating environmental flows into environmental impacts. (Butt et al., 2019).

Global warming, Photochemical Ozone Creation Potential, and particulate matters smaller than or equal to 2.5 micrometers were the life cycle impact assessment categories selected to be reported in this study. Renewable Primary Energy Demand and Non-renewable Primary Energy Demand, and Feedstock Energy were used for reporting energy consumption. Feedstock energy is a primary energy demand stored in the construction materials (such as asphalt) that is not consumed and can be recovered later:

➢ Global Warming Potential (GWP): in kg of CO2e.
➢ Photochemical Ozone Creation Potential (POCP): in kg of O3e (a measure of smog formation).
➢ Human Health (Particulate): in kg of PM2.5 (particulate matters smaller than or equal to 2.5 micrometers in diameter).
➢ Renewable Primary Energy Demand (PED-R) used as fuel from renewable resources (net calorific value excluding feedstock energy): in MJ.
➢ Non-renewable Primary Energy Demand (PED-NR) used as fuel from nonrenewable resources (net calorific value excluding feedstock energy): in MJ.
➢ Feedstock Energy (PED-FS) is Primary Energy Demand used as a material from nonrenewable resources (also called PED (non-fuel)): in MJ.

3.1 *LCI data*

The LCI database created by the UCPRC (Saboori et al., 2018) was used in this study including the details of model development, data sources, and the assumptions.

The PCC mix designs, including PCC Type III used for the HVS test sections with 4 hours OT, PCC Type II/V used for the Woodland project with 24 hours OT, and the normal strength PCC Type II/V use by Caltrans with the 10 days OT, as well as RHMA mix design used in the pavement layers of the project can be seen in Table 4. The PCC with 4 hours and 24 hours OT were designed to provide 3 MPa (450 psi) flexural strength (Caltrans requirement for opening the lane to traffic) after 24 hours while the PCC with 10 day10 days OT was designed to provide 4.5 MPa (650 psi) flexural strength at 10 day10 days.

Table 2. Energy input for 1 kg of PCC.			Table 3. Energy input for 1 kg of RHMA.	
Electricity	0.00618 MJ		Electricity	0.0076319 MJ
Natural Gas	0.000122 m³		Natural Gas	0.0103261 m³
Diesel	2.54E-007 m³			

The 2017 electricity grid mix for California that was used to calculate the impacts of materials and construction for this case study is shown in Table 2 and Table 3 (Saboori, 2018).

3.2 LCI data and report results

3.2.1 Material production stage
Table 5 and Table 6 show the environmental impacts of BCOA during the material stage.

3.2.2 Transportation and construction stages
Table 7 shows the transportation impacts for a functional unit of 1,000 kg-km of materials being transported. Table 8 and Table 9 depict transportation information and the impacts from transportation stage, respectively, for PCC Type II/V with 24-hour OT used in Woodland project.

Table 10 shows the fuel LCIAs and PEDs that were used to prepare impact assessments for the material and construction stages. The impact of construction activities for each pavement layer is calculated by estimating total fuel consumption for 1 ln-mile of the road by considering the equipment used, engine horsepower and fuel efficiency, and number of passes needed. The construction information can be seen in Table 11. Table 12 shows the impacts results due to construction stage for PCC Type II/V with 24-hour OT used in Woodland project.

4 INTERPRETATION

Interpretation might occur during all stages, but may be most important after impact assessment, because it will guide the development of conclusions and recommendations based on a study's outcome. (Harvey et al., 2016).

In this analysis, Figure 2 and Figure 3 depict the impacts for the different stages of the life cycle for the Woodland PCC Type II/V with 150 mm thickness and 24 hours OT, RHMA layer, and the whole BCOA.

In this analysis, surprisingly, the transportation stage can be considered as the hot spot with more environmental impacts (global warming and air pollution) and non-renewable consumed

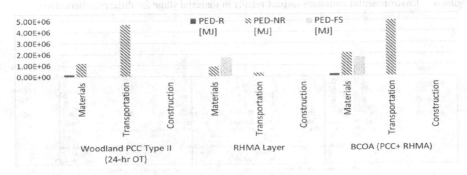

Figure 2. Consumed Energy per life cycle stage per pavement layer (Woodland case study).

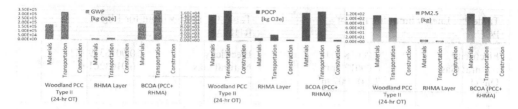

Figure 3. Environmental emissions impacts per life cycle stage per pavement layer (Woodland case study).

energy resource use than the material and construction stages (Table 13). It should be mentioned that the distances between quarries to plants is considered in the scope of this study which can be the most important factor leading to higher environmental impacts in the transportation stage. As shown in the sensitivity analyses of several papers, where transportation of materials is the most energy-intense process in the life cycle of a pavement, long transportation distances using trucks, leads to the increase in the environmental impacts (Butt et al., 2019, Butt et al., 2015, Pacheco-Torgal and Ding 2013). Different types of concrete mixes are evaluated in the sensitivity analysis of this paper.

5 SENSITIVITY ANALYSIS (CRADLE-TO-GATE)

As can be seen in Figure 4 and Figure 5, thickness of surface layer is an important factor leading to major changes in environmental impacts and energy consumptions in the material stage. The second influential criterion is the new RHMA layer under the surface rigid layer which results in significant increases in the environmental emission impacts and primary energy demands. The results show the increase of 8%-13% in GWP, POCP, PM$_{2.5}$, and PED-R. The sharp increase in PED-NR (75%) can also be seen.

The difference in the concrete mix designs is another notable factor which causes changes in emission impacts. and energy consumptions. HVS PCC Type III mix with 4 hours OT has the

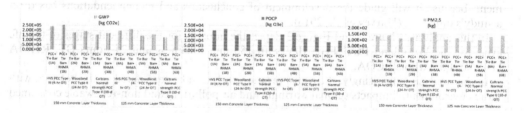

Figure 4. Environmental emissions impact results in material stage for different alternatives.

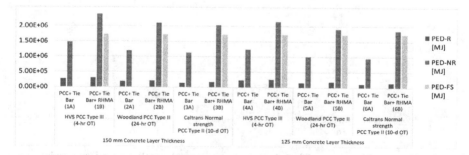

Figure 5. Energy consumptions results in material stage for different alternatives.

Table 4. PCC and RHMA mix designs, and number of tie bars in BCOA layers.

HVS PCC Mix Design Type III (4-hour OT)			Woodland PCC Type II/V Mix Design (24-hour OT)			Normal Strength PCC Type II/V Mix design (10-day OT)			RHMA Mix Design	
Material	Mass per Volume (lbl yd3)	% by mass	Material	Mass per Volume (lb/yd3)	% by mass	Material	Mass per Volume (lbl yd3)	Percentage by mass	Material	% by mass
Accelerator	37.436	0.89	Accelerator	0.00	0.00	Accelerator	76	1.62	Crushed	92.50
Flyash	0.00	0.00	Flyash	101	2.55	Flyash	704.153	15.00	Natural	0
Crushed Aggregate	1787	31.86	Crushed Aggregate	1200	30.34	Crushed Aggregate	1350	28.76	Bitumen	6.00
Natural Aggregate	1348	42.23	Natural Aggregate	1787	45.18	Natural Aggregate	1875	39.94	Extender oil	0.15
Type III Portland Cement	799	18.88	Type II/V Portland Cement	574	14.51	Type II/V Portland Cement	429	9.14	Crumb Rubber Modifier (CRM)	1.35
Retarder	4	0.095	Retarder	0.897	0.023	Retarder	0.2	0.004	Polymer	0
Water Reducing Admixture	6.25	0.15	Water Reducing Admixture	1.614	0.041	Water Reducing Admixture	2	0.040	RAP	0
Water	250	5.91	Water	291	7.36	Water	258	5.50		

Number of Tie bars
Number of tie bars per slab (slabs are 6 ft long) 2
Number of tie bars per 1 km 1094

Table 5. Impacts of material functional unit (1 kg) during production.

Material	Unit	GWP (kg CO2e)	POCP (kg O₃e)	PM2.5 (kg)	PED-R (MJ)	PED-NR (MJ)	PED-FS (MJ)
HVS PCC Type III (4-hr OT)	1kg	1.78E-01	1.50E-02	9.72E-05	2.08E-01	1.08E+00	0.000E+00
Woodland PCC Type II/V (24-hr OT)	1kg	1.296E-01	1.120E-02	8.502E-05	1.418E-01	8.652E-01	0.000E+00
Caltrans Normal Strength PCC Type II/V (10-d OT)	1kg	1.169E-01	8.228E-03	1.183E-04	1.076E-01	8.150E-01	0.000E+00
RHMA	1kg	5.628E-02	5.977E-03	4.036E-05	9.329E-02	3.408E+00	6.487E+00
Tie Bar	Each	3.343E+00	1.667E-01	1.616E-03	1.443E+00	4.147E+01	0.000E+00

Table 6. Alternative cases of BCOA material production impacts in 1 ln-km.

Case	Material	Concrete Thickness mm (inch)	RHMA Thickness mm (inch)	GWP (kg CO₂e)	POCP (kg O₃e)	PM2.5 (kg)	PED-R (MJ)	P-ED-NR (MJ)	PED-FS (MJ)
1-A	HVS PCC Type III (4-hr OT)+ Tie Bar	150 (6)	30 (1.18)	2.41E+05	2.02E+04	1.31E+02	2.79E+05	1.48E+06	0.00E+00
1-B	HVS PCC Type III (4-hr OT)+ Tie Bar+ RHMA	150 (6)	30 (1.18)	2.56E+05	2.18E+04	1.42E+02	3.03E+05	2.39E+06	1.73E+06
2-A	Woodland PCC Type II (24-hr OT)+ Tie Bar	150 (6)	30 (1.18)	1.763E+05	1.510E+04	1.150E+02	1.904E+05	1.198E+06	0.00E+00
2-B	Woodland PCC Type II (24-hr OT)+ Tie Bar+ RHMA	150 (6)	30 (1.18)	1.913E+05	1.669E+04	1.258E+02	2.153E+05	2.106E+06	1.73E+06
3-A	Caltrans Normal Strength PCC Type II (10-d OT)+ Tie Bar	150 (6)	30 (1.18)	1.594E+05	1.114E+04	1.593E+02	1.449E+05	1.131E+06	0.00E+00
3-B	Caltrans Normal Strength PCC Type II (10-d OT)+ Tie Bar+ RHMA	150 (6)	30 (1.18)	1.744E+05	1.273E+04	1.701E+02	1.697E+05	2.039E+06	1.73E+06
4-A	HVS PCC Type III (4-hr OT)+ Tie Bar	125 (5)	30 (1.18)	2.01E+05	1.68E+04	1.10E+02	2.32E+05	1.24E+06	0.00E+00
4-B	HVS PCC Type III (4-hr OT)+ Tie Bar+ RHMA	125 (5)	30 (1.18)	2.16E+05	1.84E+04	1.20E+02	2.57E+05	2.15E+06	1.73E+06
5-A	Woodland PCC Type II (24-hr OT)+ Tie Bar	125 (5)	30 (1.18)	1.475E+05	1.261E+04	9.614E+01	1.590E+05	1.006E+06	0.00E+00
5-B	Woodland PCC Type II (24-hr OT)+ Tie Bar+ RHMA	125 (5)	30 (1.18)	1.625E+05	1.420E+04	1.069E+02	1.838E+05	1.914E+06	1.73E+06
6-A	Caltrans Normal Strength PCC Type II (10-d OT)+ Tie Bar	125 (5)	30 (1.18)	1.334E+05	9.316E+03	1.331E+02	1.210E+05	9.500E+05	0.00E+00
6-B	Caltrans Normal Strength PCC Type II (10-d OT)+ Tie Bar+ RHMA	125 (5)	30 (1.18)	1.484E+05	1.091E+04	1.438E+02	1.458E+05	1.858E+06	1.73E+06

Table 7. Transportation impacts for a functional unit of 1,000 kg-km of materials.

	Functional Unit	GWP (kg CO2e)	POCP (kg O3e)	PM2.5 (kg)	PED-R (MJ)	PED-NR (MJ)	PED-FS (MJ)
Heavy Truck	1000 kg-km	7.798E-02	1.243E-02	2.492E-05	0.000E+00	1.116E+00	0.000E+00

Table 8. Transportation information.

Material	Transportation	Material in 1lane-km (kg)	No. of trips	1000 kg-km (1 Lane-km)
PCC Type II	1-way 40 km (25 mile) from plan to the construction field	1,332,000	56	2,983,680
Cement	1-way 692km (430 mile) from cement plant to the mixing plant	193,292	9	1,203,822
RHMA	1-way 56km (35 mile) from plan to the construction field	266,400	12	179,021
Bitumen	1-way 435km (270 mile) from refinery to the plant	15,974	1	6,949
Crushed Agg.	1-way 40 km (25 mile) from quarry to the plant	246,420	11	108,425

Table 9. Transport impacts.

Material	GWP (kg CO2e)	POCP (kg O3e)	PM2.5 (kg)	PED-R (MJ)	PED-NR (MJ)	PED-FS (MJ)
Woodland PCC Type II	3.265E+05	5.206E+04	1.044E+02	0.000E+00	4.673E+06	0.000E+00
RHMA	2.296E+04	3.660E+03	7.337E+00	0.000E+00	3.286E+05	0.000E+00
Total Transport. Impact	*3.495E+05*	*5.572E+04*	*1.117E+02*	*0.000E+00*	*5.002E+06*	*0.000E+00*

Table 10. Impacts of non-electricity energy source.

Diesel Burned (1 gallon)	GWP (kg CO2e)	POCP (kg O3e)	PM2.5 (kg)	PED-R (MJ)	PED-NR (MJ)	PED-FS (MJ)
	1.194E+01	5.273E+00	9.369E-03	0.000E+00	1.645E+02	0.000E+00

Table 11. Construction information.

Layer	Equipment/ Activity	Engine Power kw(hp)	Hourly Fuel Use m3/hr (gal/hr)	Speed km/h (ft/ min)	Time for 1Pass over 1lane-km (hr)	No. of Passes	Fuel Used m3(gal)	Total Fuel Used for 1lane-km m3 (gal)
Woodland PCC Type II	Milling for 25 mm (1 in)	522 (700)	0.076 (20)	0.183 (10)	5.47	1	0.41 (109.36)	0.49 (129.05)

(Continued)

Table 11. (*Continued*)

Layer	Equipment/ Activity	Engine Power	Hourly Fuel Use	Speed	Time for 1Pass over 1lane-km	No. of Passes	Fuel Used	Total Fuel Used for 1lane-km
		kw(hp)	m3/hr (gal/hr)	km/h (ft/min)	(hr)		m3(gal)	m3 (gal)
	Sweeping (multiple times)	59.66 (80)	0.008 (2)	1.83 (100)	0.55	2	0.01 (2.19)	
	Wetting	59.66 (80)	0.008 (2)	1.83 (100)	0.55	1	0.004 (1.09)	
	Concrete Placement	67.11 (90)	0.011 (3)	0.183 (10)	5.47	1	0.06 (16.40)	
RHMA	Prime coat application	260.995(350)	0.027 (7.2)	0.457 (25)	2.19	1	0.06 (16.40)	0.54 (143.15)
	RHMA placement	186.43(250)	0.040 (10.6)	0.274 (15)	3.65	1	0.15 (39.62)	
	Rolling (vibratory)	111.86(150)	0.031 (8.1)	0.457 (25)	2.19	2	0.13 (34.34)	
	Rolling (static)	111.86(150)	0.031 (8.1)	0.457 (25)	2.19	3	0.2 (52.83)	

Table 12. Construction impacts.

Material	Activity	GWP (kg CO2e)	POCP (kg O3e)	PM2.5 (kg)	PED-R (MJ)	PED-NR (MJ)	PED-FS (MJ)
Woodland	Milling for 25mm (1in)	1.306E+03	8.304E+01	1.475E-01	0.000E+00	2.591E+03	0.000E+00
PCC Type II	Sweeping(multiple-times)	2.612E+01	1.153E+01	2.049E-02	0.000E+00	3.599E+02	0.000E+00
	Wetting	1.306E+01	5.767E+00	1.025E-02	0.000E+00	1.799E+02	0.000E+00
	Concrete placement	1.959E+02	8.650E+01	1.537E-01	0.000E+00	2.699E+03	0.000E+00
	Total	*1.541E+03*	*1.868E+02*	*3.320E-01*	*0.000E+00*	*5.830E+03*	*0.000E+00*
RHMA	Prime coat application	1.883E+02	8.315E+01	1.477E-01	0.000E+00	2.594E+03	0.000E+00
	RHMA placement	4.620E+02	2.040E+02	3.625E-01	0.000E+00	6.366E+03	0.000E+00
	Rolling (vibratory)	4.236E+02	1.871E+02	3.324E-01	0.000E+00	5.838E+03	0.000E+00
	Rolling (static)	6.355E+02	2.806E+02	4.986E-01	0.000E+00	8.756E+03	0.000E+00
	Total	*1.709E+03*	*7.549E+02*	*1.341E+00*	*0.000E+00*	*2.355E+04*	*0.000E+00*
Total Construction Impact		***3.250E+03***	***9.417E+02***	***1.673E+00***	***0.000E+00***	***2.938E+04***	***0.000E+00***

Table 13. Final impacts of BCOA in different stages (Woodland case study).

Layer	Life Cycle Stage	Percent of total					
		GWP	POCP	PM2.5	PED-R	PED-NR	PED-FS
Woodland PCC Type II	Materials	32.41%	20.59%	48.09%	88.45%	16.79%	0.00%
(24-hr OT)	Transportation	60.01%	70.97%	43.66%	0.00%	65.47%	0.00%
	Construction	0.28%	0.25%	0.14%	0.00%	0.08%	0.00%
	Total	92.70%	91.81%	91.89%	88.45%	82.34%	0.00%
RHMA	Materials	2.76%	2.17%	4.50%	11.55%	12.72%	100.00%
	Transportation	4.22%	4.99%	3.07%	0.00%	4.60%	0.00%
	Construction	0.31%	1.03%	0.56%	0.00%	0.33%	0.00%

(*Continued*)

Table 13. *(Continued)*

| Layer | Life Cycle Stage | Percent of total | | | | | |
		GWP	POCP	PM2.5	PED-R	PED-NR	PED-FS
BCOA (PCC+ RHMA)	*Total*	7.29%	8.19%	8.13%	11.55%	17.65%	100.00%
	Materials	35.16%	22.76%	52.59%	100.00%	29.50%	100.00%
	Transportation	64.24%	75.96%	46.71%	0.00%	70.08%	0.00%
	Construction	0.60%	1.28%	0.70%	0.00%	0.41%	0.00%
TOTAL for the Functional Unit		5.44E+05	7.34E+04	2.39E+02	2.15E+05	7.14E+06	*1.73E+06*
		[kg CO$_2$e]	[kg O$_3$e]	[kg]	[MJ]	[MJ]	[MJ]

highest environmental impacts and energy consumption followed by PCC Type II/V mix designs. It should be mentioned that finer grinding of Type III PC as well as the higher amount of cement compared to Type II/V PC leads to higher environmental impacts.

Caltrans normal strength PCC Type II/V with 10-day OT has higher percentage of cement compared to Woodland PCC Type II/V with 24 hours OT (14% vs. 9%, respectively). According to Figure 4 and Figure 5, Caltrans normal strength mix has a slightly lower impacts in terms of GWP, POCP, and energy consumption compared to the Woodland mix. However, the Caltrans normal strength mix has the highest impacts in terms of PM$_{2.5}$. This might be because of the higher amount of fly ash in this mix compared to the other mixes.

This paper demonstrates the use of LCA to quantify and evaluate the environmental impacts of alternative materials, construction and designs for a pavement structure. This analysis should consider the relative performance of the different designs if it is expected to be different. Performance data are not yet available for these alternative designs.

ACKNOWLEDGMENT

The authors would like to thank Nick Burmas and Joe Holland from the Caltrans Division of Research, Innovation, and System Information for support of this project.

DISCLAIMER

REFERENCES

Butt, A.A., Harvey, J.T., Saboori, A., Ostovar, M., and Reger, D. (2019). *Airfield Life Cycle Assessment: Benchmark Study of a Project at JFK International Airport*. Airfield and Highway Pavements 2019: Innovation and Sustainability in Highway and Airfield Pavement Technology, ASCE, P 456–464.

Butt, A.A., Harvey, J.T., Reger, D., Saboori, A., Ostovar, M., and Bejarano, M. (2019). *Life-Cycle Assessment of Airfield Pavements and Other Airside Features: Framework, Guidelines and Case Studies*. Research Report, Federal Aviation Administration, Performing Organization No.: DOT/FAA/TC-19/2.

Butt, A.A. and Birgisson, B. (2016). *Assessment of the attributes based life cycle assessment framework for road projects*. Structure and Infrastructure Engineering Journal. Vol. 12, No. 9. P 1177–1184.

Butt, A.A., Toller, S., and Birgisson, B. (2015). *Life Cycle Assessment for the Green Procurement of Roads: A Way Forward*. Journal of Cleaner Production, 90: 163–170. DOI: 10.1016/j. jclepro.2014.11.068.

Harvey, J. T., Meijer, J., Ozer, H., Al-Qadi, I. L., Saboori, and A., Kendall, A. (2016). *Pavement Life Cycle Assessment Framework*. U.S. Department of Transportation Federal Highway Administration, FHWA-HIF-16-014, July 2016, Washington, DC.

Harrington, D. S., and G. Fick. (2014). *Guide to concrete overlays: Sustainable solutions for re-surfacing and rehabilitation existing pavements*. National Concrete Pavement Technology Center.

ISO (2006). *Environmental Management – Life Cycle Assessment – Principles and Framework*. International Organization for Standardization, ISO 14040, Geneva, Switzerland.

Kendall, A. (2012). *Life Cycle Assessment for Pavement: Introduction*. Presentation in Minutes, FHWA Sustainable Pavement Technical Working Group Meeting, April 25-26, 2012, Davis, CA.

Mateos, A., J. Harvey, F. Paniagua, J. Paniagua, and R. Wu. (2019). *Development of Improved Guidelines and Designs for Thin BCOA: Summary, Conclusions, and Recommendations*. University of California Pavement Research Center, Summary Report: UCPRC-SR-2018-01.

Pacheco-Torgal, F. and Ding, Y. eds. (2013). Handbook of recycled concrete and demolition waste. Elsevier.

Saboori, A., John Harvey, Ting Wang, and Hui Li. (2018). *Documentation of UCPRC Life Cycle Inventory used in CARB/Caltrans LBNL Heat Island Project and other LCA Studies*. University of California Pavement Research Center, Davis, CA.

Sensitivity analysis of the benefits of replacing virgin materials with RAP considering rejuvenator type and hauling distance

Mohamed Elkashef, Arash Saboori*, Maryam Ostovar, John T. Harvey, Ali Azhar Butt & David Jones
Department of Civil and Environmental Engineering, University of California Pavement Research Center, University of California, Davis, USA

ABSTRACT: Using reclaimed asphalt pavement (RAP) in pavement projects reduces the amount of virgin materials needed, averts the need to land-fill old materials, and reduces environmental impacts. The environmental and cost benefits of using RAP in asphalt mixes are dependent on multiple factors: the percent RAP content, the properties of the binder and aggregate in the RAP and content in RAP, the amount of virgin binder replacement with RAP binder, hauling distances from the original location to the mixing plant/site, and the amount and type of the rejuvenator that needs to be added. This paper conducted a sensitivity analysis of the changes in life cycle environmental impacts of hot mix asphalt due to different combinations of RAP content, rejuvenator type, and hauling distance. The analysis showed that long hauling distances reduces the environmental benefit of using RAP as measured by the reduction in global warming potential (GWP). For mixes containing 25 percent RAP with no rejuvenator and mixes containing 50 percent RAP with a soybean-derived rejuvenator, it was found out that a hauling distance of 114 miles and 159 miles, respectively, cancels out the reduction in GWP.

1 INTRODUCTION

Reclaimed Asphalt Pavement (RAP) is old Hot Mix Asphalt (HMA) that is milled off existing pavement and can be used to partially replace virgin asphalt binder and aggregate in new HMA. Caltrans has allowed contractors to use up to 15 percent RAP (by weight) in HMA without any additional engineering for a number of years (Caltrans, 2018), which is considered as the baseline for this strategy. Up to 25 percent RAP was allowed in the past several years, however, tight specifications and complex laboratory procedures essentially eliminated the use of more than 15 percent RAP. In 2018, the specifications were changed to allow up to 25 percent RAP without the need for the testing that was considered by industry to be onerous. Caltrans is working on developing approaches to include approximately 40 percent RAP in HMA in the future.

The RAP binder is much less expensive than virgin binder and the replacement of part of the virgin binder in a new mix with aged asphalt binder left on RAP particles is intended to obtain similar performance with a mix that has only virgin binder. The properties of RAP binder vary considerably depending on the original properties it when it was placed, the amount of aging, and later processing.

RAP binder is stiffer and more brittle than virgin binder. The use of RAP leads to a reduction in the cracking and fatigue resistance of asphalt mixtures (Li et al., 2008). The impact on cracking and fatigue resistance increases with increasing amounts of RAP in the mix (West et al., 2009). Therefore, higher percentages of RAP often require the use of a softer

---

*Corresponding author

virgin binder for the portion of the total binder not replaced by the RAP binder and the addition of softening additives, called rejuvenators, to facilitate blending of the aged and the virgin binders and creation of the desired blended properties. Rejuvenators help to restore the properties of aged RAP binders and hence improve their thermal cracking and fatigue resistance (Elkashef et al., 2018). It is important that mixes containing RAP have similar performance to mixes with virgin binder with respect to fatigue, low-temperature cracking, and rutting or else any cost and/or environmental benefits are in jeopardy because reduced performance will require replacement that is more frequent.

Three common types of rejuvenating agents (RA) used in the USA are aromatic extracts made from petroleum, bio-based RAs made from soy oil, and bio-based RAs made from tall oil that comes from coniferous trees (Zaumanis et al., 2014, Elkashef et al., 2017). Another method for handling the issue of aged binder is to use softer virgin binder, which eliminates the need of rejuvenators, however, this method is only applicable for RAP contents up to about 25 percent.

2 GOAL AND SCOPE DEFINITION

A first stage of this study developed life cycle assessment (LCA) models needed to quantify the changes in the global warming potential (GWP) of increasing RAP content in the total HMA consumed annually in Caltrans projects. This paper uses that model to investigate the impact of the rejuvenator type and content on the benefits realized through increased use of RAP in HMA. The goal of this study is to compare changes in greenhouse gas (GHG) emissions reductions by using rejuvenators of different types and contents while increasing the maximum RAP content in HMA mixes, going from 15 to 25, 40, and 50 percent binder replacement, and scale the use of HMA on the state network in California. Three cases were considered for rejuvenators: aromatic, bio-based, and no-rejuvenator. The study scope was further expanded to analyze the impact of hauling distance. Four hauling distances were considered: 0, 25, 50, and 100 miles.

The scope of this study is cradle to gate and considers all the impacts from materials extraction to transportation to plants, and all the processes conducted in the plant to prepare the final mix. It was assumed that the construction process and field performance of the mixes with higher RAP content is the same as the base scenario. The functional unit for this study is defined to be the California highway network. The analysis period considered for this study is 33 years, from 2018 to 2050. Figure 1 shows the system diagram considered in this study.

3 MODELING ASSUMPTIONS

The amount of dense-graded HMA to be used in Caltrans projects was taken from Caltrans' Pavement Management System software, PaveM (Figure 2). The current run provides data up to 2046. Due to lack of a better alternative to estimate material needed from 2046 to 2050, it was assumed that the average masses of HMA used in the 10 years prior will be applied during that time period. It should be noted that a large percentage of Caltrans asphalt paving uses rubberized hot mix asphalt (RHMA-G) and open graded mixes with both conventional and rubberized binders, however no RAP is allowed in RHMA materials and only low RAP is allowed in conventional open-graded mixes at this time. The large increase in the expected amount of HMA that will be used that can be seen in Figure 2 to occur in 2021 is the result of the new fuel tax passed by the state legislature in California in 2017 and sustained by popular vote in late 2018.

Four mixes with increasing maximum RAP content for HMA were considered for this study as shown in Table 1. Mixes with RAP content above 25 percent require a rejuvenating agent to be added. Due to very limited information available regarding the materials in rejuvenating agents, developing LCA models in GaBi was not an option and the life cycle inventory (LCI) of Aromatic BTX was used as a placeholder for aromatic extracts and the LCI of

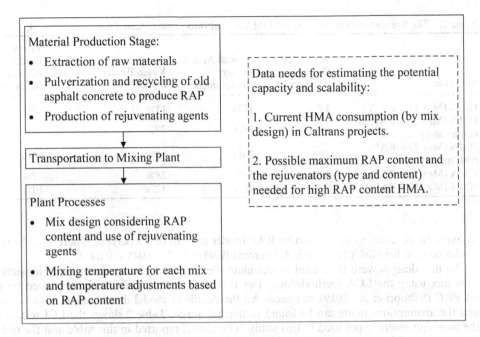

Figure 1. System diagram for increased use of RAP in HMA used in Caltrans projects.

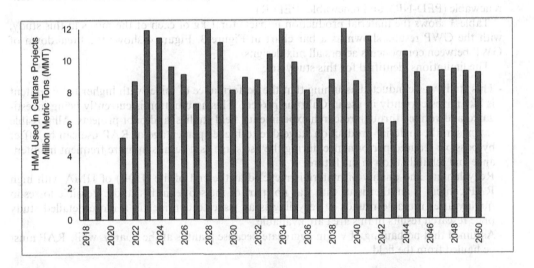

Figure 2. Materials needed per year between 2018-2050 based on PaveM outputs.

soy oil was used for bio-based RAs. Assumptions of this type must be made unless environmental product declarations (EPD) become available for these materials. Their proprietary nature precludes publication of their chemical formulae. This sensitivity analysis reports the results for HMA with maximum of up to 25, 40, and 50 percent RAP with aromatic rejuvenating agent (RA), bio-based RA, and with no RA (only for the scenario of up to 25 percent RAP).

The baseline mix design for HMA was taken from the UCPRC Case Studies report (Wang et al., 2012) and assumed a 4.7 percent virgin binder content. Furthermore, it was assumed that the RAP materials had a binder content of 5 percent by mass with a 90 percent binder

Table 1. The five scenarios considered for HMA for caltrans projects across the entire network.

Mix Title	Max RAP Content Allowed	Typical Actual Binder Replacement	Virgin Binder Replaced by RAP	Virgin Binder Replaced by Rejuvenator
HMA (Max 15% RAP)	15%	11%	11%	0%
HMA (Max 25% RAP, no rejuvenator)	25%	20%	20%	0%
HMA (Max 25% RAP, rejuvenator)	25%	20%	15%	5%
HMA (Max 40% RAP)	40%	35%	28%	7%
HMA (Max 50% RAP)	50%	42%	32%	10%

recovery ratio, resulting in an effective RAP binder content of 4.5 percent. Therefore, the total binder content for HMA baseline is 4.7 percent (0.04 + 0.15 * 0.045 = 0.047).

The mix designs were then used to calculate the cradle to gate environmental impacts of each mix using the LCA methodology. For this purpose, the LCI database created by the UCPRC (Saboori et al., 2019) was used. All the details of model development, data sources, and the assumptions made can be found in that reference. Table 2 shows the LCI results for the two rejuvenator types used in this study. The results reported in this table and the rest of this paper are selected impact categories, namely: global warming potential (GWP), smog formation potential, PM2.5 which indicates the amount of particulate matters less than 2.5 micrometer in diameter, and primary energy demand (PED) in three categories: total, non-renewable (PED-NR), and renewable (PED-R)

Table 3 shows the material production impacts for 1 kg of each of the mixes in this study, with the GWP results shown as a bar chart in Figure 3. Figure 4 shows the breakdown of GWP between components across all mix designs.

The limitations identified for this study are:

- This study was conducted assuming that the performance of mixes with higher RAP content is like mixes currently in use in Caltrans project. This assumption is currently being investigated and verified through research experiments, field studies, and pilot projects. All possible savings in the material production stage due to higher percentage of RAP use can be offset by possible reduced performance during the use stage as it results in more frequent maintenance and rehabilitation in the future.
- Recyclability and quality of materials recycled at the end of life (EOL) of HMA with high RAP content is another issue that was not part of the scope in this study. Possible losses in quality after multiple rounds of recycling is an issue to consider in a more detailed study once research results in this area are available.
- Assumed that no hauling of virgin aggregate because plant is at the quarry, while RAP must be hauled from the field

Table 2. LCI of the items used in this study.

Item	Unit	GWP [kg CO^{2}e]	Smog [kg O^{3}e]	PM 2.5 [kg]	PED-Total [MJ]	PED-NR [MJ]	PED-R [MJ]
Rejuvenator, Bio-based (Soy Oil)	1 kg	3.00E-1	2.60E-2	1.73E-4	3.48E+0	3.48E+0	0.00E+0
Rejuvenator, Aromatic BTX	1 kg	6.44E-1	1.57E-4	3.20E-2	4.78E+1	4.76E+1	2.18E-1

Table 3. Environmental impacts of material production stage for 1 kg of each of the mixes.

Mix Title	Rejuvenator Type	GWP [kg CO2e]	Smog [kg O3e]	PM 2.5 [kg]	PED-Total [MJ]	PED-NR [MJ]	PED-R [MJ]
HMA (Max 15% RAP)	N/A	4.95E-2	4.68E-3	3.25E-5	2.56E+0	2.54E+0	2.23E-2
HMA (Max 25% RAP)	Aromatic BTX	4.91E-2	4.37E-3	1.06E-4	2.46E+0	2.44E+0	2.10E-2
HMA (Max 25% RAP)	Bio-Based (Soy Oil)	4.83E-2	4.43E-3	3.12E-5	2.36E+0	2.34E+0	2.05E-2
HMA (Max 25% RAP)	No Rejuvenator	4.78E-2	4.41E-3	3.09E-5	2.35E+0	2.33E+0	2.01E-2
HMA (Max 40% RAP)	Aromatic BTX	4.69E-2	3.90E-3	1.33E-4	2.16E+0	2.15E+0	1.78E-2
HMA (Max 40% RAP)	Bio-Based (Soy Oil)	4.58E-2	3.99E-3	2.87E-5	2.02E+0	2.00E+0	1.71E-2
HMA (Max 50% RAP)	Aromatic BTX	4.64E-2	3.67E-3	1.77E-4	2.07E+0	2.05E+0	1.67E-2
HMA (Max 50% RAP)	Bio-Based (Soy Oil)	4.48E-2	3.79E-3	2.76E-5	1.86E+0	1.85E+0	1.57E-2

Table 4. Total changes in GHG emissions compared to the baseline for the analysis period (2018 to 2050).

Mix	Total GHGs (Tonne CO2e)	CO2e Reductions (Tonne CO2e)	Percent Reduction in GHGs vs Baseline (%)
Max 15%, no Rejuv	14,125,517	0	0.0%
Max 25% RAP, BTX	14,029,843	-95,674	-0.7%
Max 25% RAP, Soy Oil	13,799,359	-326,158	-2.3%
Max 25% RAP, no Rejuv	13,655,723	-469,794	-3.3%
Max 40% RAP, BTX	13,396,501	-729,016	-5.2%
Max 40% RAP, Soy Oil	13,073,824	-1,051,693	-7.4%
Max 50% RAP, BTX	13,255,578	-869,939	-6.2%
Max 50% RAP, Soy Oil	12,794,611	-1,330,906	-9.4%

4 RESULTS & DISCUSSION

The results of the analysis considering only changes in RA type and a hauling distance of 50 miles are shown in Table 3. The previous stage of this study showed that increasing the RAP content from the original 11.5 percent (for the max 15 percent RAP baseline) by 8.5, 23.5, and 30.5 percent can result in 96, 729, and 870 thousand metric tonnes of CO_2e savings compared to the baseline, respectively, when using aromatic BTX RAs during the 33-year analysis period. These numbers change to 326, 1,052, and 1,331 thousand metric tonnes of CO_2e when bio-based RA is utilized.

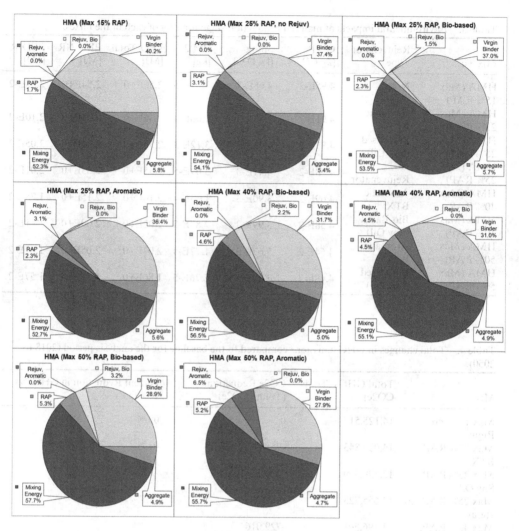

Figure 3. Breakdown of GWP between components across all mix designs considered in this study.

In terms of percent decrease in GWP, use of aromatic BTX RAs result in 0.7, 5.2, 6.2 percent reductions in GHG emissions compared to the baseline while bio-based RAs result in 2.3, 7.4, and 9.4 percent reductions compared to the baseline. HMA mixes with bio-based RAs result in more savings in GHG emissions compared to HMA mixes with BTX RAs as was expected due to GWP of bio-based RAs being 53 percent lower compared to BTX RAs

For the hauling distance, four different values were considered: 0, 25, 50, and 100 miles. Table 4 shows the GWP of RAP under different hauling distance and Figure 5 shows the percent reduction in GWP compared to the baseline (max 15% RAP, no rejuvenator) due to changes in hauling distance for all the combinations of RAP content and rejuvenator type considered in this study. The results show -29 to +3 percent change in GWP compared to baseline across all hauling distances and mixes. Only two cases had increase in GWP compared to the baseline, HMA with maximum 15 percent RAP, no rejuvenator, and a 100-mile hauling distance (3.1 percent increase) and HMA with max 25 percent RAP and BTX rejuvenator, and 100-mile hauling distance (2.8 percent increase).

Increasing RAP content and use of bio-based rejuvenator results in reduction of GWP, while increasing hauling distance will reduce the savings possible through use of RAP in

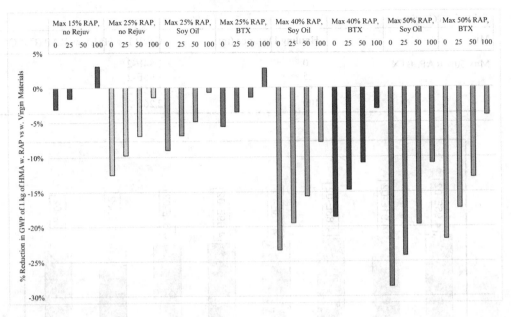

Figure 4. Percent reduction in GWP due to changes in hauling distance (0, 25, 50, 100 miles) for all the combinations of RAP content and rejuvenator type considered in this study.

Table 5. GWP ratio (HMA with RAP over HMA with virgin materials).

Mix	Hauling Distance (mi)	GWP 1kg HMA w. RAP (Cut-Off)
Max 15% RAP, no Rejuv	0	2.29E-2
	25	2.32E-2
	50	2.36E-2
	100	2.43E-2
Max 25% RAP, no Rejuv	0	2.06E-2
	25	2.13E-2
	50	2.19E-2
	100	2.33E-2
Max 25% RAP, Soy Oil	0	2.15E-2
	25	2.20E-2
	50	2.24E-2
	100	2.34E-2
Max 25% RAP, BTX	0	2.23E-2
	25	2.28E-2
	50	2.33E-2
	100	2.42E-2
Max 40% RAP, Soy Oil	0	1.81E-2
	25	1.90E-2
	50	1.99E-2
	100	2.17E-2
Max 40% RAP, BTX	0	1.92E-2
	25	2.01E-2
	50	2.10E-2
	100	2.29E-2
Max 50% RAP, Soy Oil	0	1.68E-2
	25	1.79E-2
	50	1.89E-2
	100	2.10E-2

(*Continued*)

Table 5. *(Continued)*

Mix	Hauling Distance (mi)	GWP 1kg HMA w. RAP (Cut-Off)
Max 50% RAP, BTX	0	1.84E-2
	25	1.95E-2
	50	2.05E-2
	100	2.26E-2

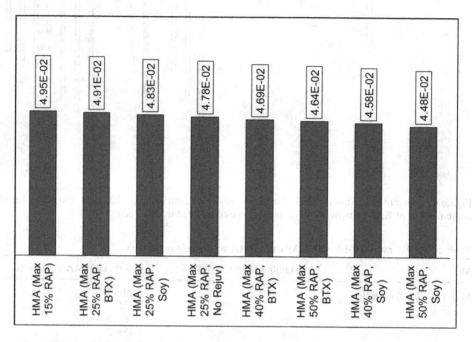

Figure 5. GWP (kg CO2e) of material production stage for 1 kg of each of the mixes.

HMA. Increasing RAP hauling distances counterbalance the effects of increased use of RAP in reducing the GWP of HMA mixes, both in terms of cost and environmental impacts. Depending on the rejuvenator type and the increase in RAP content, the reduction in GWP compared to the baseline resulted by increased RAP use is cancelled after specific transportation distance. For example, this distance for "Max 25% RAP, no Rejuv" is 114 miles and 159 miles for "Max 50% RAP, Soy Oil".

5 SUMMARY

This study investigated the impact of rejuvenator type and hauling distance on the LCI of 1 kg of HMA produced with different RAP contents. Bio-based rejuvenators have less GWP compared to Aromatic RAs, therefore, HMA with RAP mixes that use bio-based rejuvenator show 8 to 34 percent reduction in GWP on average across all RAP contents while this range for mixes with Aromatic rejuvenators is 5 to 19 percent. Increasing RAP hauling distances counterbalance the effects of increased use of RAP in reducing the GWP of HMA mixes. Changing hauling distance between 0 to 100 miles results in -29 to +3 percent change in GWP compared to baseline across all hauling distances and mixes.

REFERENCES

Caltrans (2018). Caltrans Fact Booklet 2018. California Department of Transportation, Sacramento, CA. Last accessed in March 2019 at: http://www.dot.ca.gov/drisi/library/cfb/2018-CFB.pdf

Elkashef, M., Williams, R.C. and Cochran, E., 2018. Investigation of fatigue and thermal cracking behavior of rejuvenated reclaimed asphalt pavement binders and mixtures. International Journal of Fatigue, 108, 90–95.

Elkashef, M., Podolsky, J., Williams, R.C. and Cochran, E., 2017. Preliminary examination of soybean oil derived material as a potential rejuvenator through Superpave criteria and asphalt bitumen rheology. Construction and Building Materials, 149, 826–836

Li, X., Marasteanu, M.O., Williams, R.C. and Clyne, T.R., 2008. Effect of reclaimed asphalt pavement (proportion and type) and binder grade on asphalt mixtures. Transportation Research Record, 2051(1), 90–97.

Saboori, A., John Harvey, Hui Li, Jon Lea, Alissa Kendall, and Ting Wang. Documentation of UCPRC Life Cycle Inventory used in CARB/Caltrans LBNL Heat Island Project and other LCA Studies University of California Pavement Research Center, Technical Report, Davis, CA 2019.

Wang, T., Lee, I. S., Kendall, A., Harvey, J., Lee, E. B., & Kim, C. (2012). Life cycle energy consumption and GHG emission from pavement rehabilitation with different rolling resistance. Journal of Cleaner Production, 33, 86–96.

West, R., Kvasnak, A., Tran, N., Powell, B. and Turner, P., 2009. Testing of moderate and high reclaimed asphalt pavement content mixes: laboratory and accelerated field performance testing at the national center for asphalt technology test track. Transportation research record, 2126(1),100–108.

Zaumanis, M., Mallick, R.B., Poulikakos, L. and Frank, R., 2014. Influence of six rejuvenators on the performance properties of Reclaimed Asphalt Pavement (RAP) binder and 100% recycled asphalt mixtures. Construction and Building Materials, 71, 538–550.

A framework for selection between end-of-life alternatives at the project-level considering full life cycle environmental impacts

Arash Saboori*, John T. Harvey, Jeremy Lea & David Jones
*University of California Pavement Research Center, Department of Civil and Environmental Engineering,
University of California, Davis, USA*

ABSTRACT: In-place recycling is gaining more popularity as an end-of-life (EOL) treatment as it results in less consumption of virgin aggregates and less environmental impacts in
hauling of materials to and from the site. However, there are challenges that needed to be
addressed so that the benefits of in-place recycling can be objectively quantified, and the contexts in which recycling provides benefits compared to conventional options can be identified.
This paper addresses this issue by defining a framework for comparison across all available
alternatives by quantification of the impacts throughout the full life cycle of each option. The
framework provides guidance on what parameters to considers, how to collect data, and how
to create models at each life cycle stage for quantification of the impacts.

1 INTRODUCTION

Sustainability is an integral part of the mission statement of the California Department of Transportation (Caltrans): "Provide a safe, sustainable, integrated, and efficient transportation system
to enhance California's economy and livability" which has resulted in Caltrans pursuing innovative materials and construction processes in their projects for many years. As pavements reach
their end-of-life (EOL), there are multiple options available and recycling has become increasingly popular due to the general perception of recycling as a more sustainable alternative than
other alternatives. This has also been driven by cost considerations for virgin versus reclaimed
asphalt binder and virgin versus recycled aggregate. Virgin aggregate has become scarcer in California near locations of greatest use which is another contributing factor to the recent trends
observed. Furthermore, it is widely believed that recycling will always lead to significant environmental benefits, to the point that many pavement engineers exclusively associate the term "sustainability for pavements" with recycling. Objective quantification of the environmental impacts
of recycling alternatives throughout their full life cycle and in different contexts is needed to have
a better understanding of the performance of different alternatives in terms of sustainability.

There has been significant progress in the field of transportation infrastructure sustainability and its assessment in the past 10 years in the U.S., primarily through the use of life cycle
assessment (LCA) to support decision-making. The early studies were conducted without utilizing a pavement specific LCA standard or framework, which resulted in similar studies
having different assumptions, scope, and system boundaries. Therefore, the results of initial
studies often produced conflicting conclusions, making it difficult for users of the conclusions
to understand and compare studies.

The scope of early LCA studies of pavement recycling was mostly at the project level and they
were mostly focused on benchmarking the impacts of design alternatives. The life cycle inventory
(LCI) databases used in early studies often did not represent local conditions in terms of technology and energy sources, transportation details, and the manufacturing or construction processes.

---

*corresponding author

Since then, pavement LCA has matured considerably, leading to the current situation in which national guidelines and frameworks are already available (Van Dam et al., 2015; Harvey et al., 2016), tools and databases with much better data quality, documentation, and inventory and reporting features are mainstream (both commercial like GaBi, SimaPro, and developed by federal and state agencies such as GREET and MOVES). More importantly, the extent of LCA application has expanded to analyze topics such as policy analysis (Gilbert et al., 2017; Lidicker et al., 2012), pavement network management, (Torres-Machi et al., 2014; Lee et al., 2017; and Wang, 2014) optimization (Chan, 2012; Kucukvar et al., 2014), evaluation of new materials (Glass et al., 2013; Li et al, 2016), quantification of uncertainty and understanding the sensitivity of the results to data quality and model assumptions (Milachowskl et al., 2011; Li et al., 2017), and the LCAs, production of product category rules (PCRs) for pavement materials (NAPA, 2017; NSF International, 2019) and environmental product declarations (EPDs.)

2 PROBLEM STATEMENT

However, there are still important questions and challenges that need to be addressed to continue the development of LCA for pavement. In particular for the EOL stage of pavements, these are the current gaps in the knowledge identified in the previous section:

- There are LCI datasets available for materials, surface treatments, and construction activities, but these datasets generally: a) do not include a comprehensive list of all available EOL options, b) are outdated, and c) are not representative of the local conditions in terms of processes, mix designs, and energy sources.
- There are limited data for quantifying the environmental impacts of EOL strategies for flexible pavements; information is needed for conventional methods such as reconstruction, mill-and-fill, and overlays, as well as in-place and plant-recycling recycling strategies such as full depth reclamation (FDR) and cold in place recycling (CIR).
- There are no well documented and calibrated performance prediction models for pavements built from recycled materials. Therefore, it is unclear whether such sections perform better, equally, or worse compared to conventional strategies. If the performance is worse, it is possible that the expected reductions in environmental impacts from recycling are offset by the need for more frequent maintenance and rehabilitation over the life cycle compared to conventional strategies.
- There is a scarcity of information and models regarding surface roughness from recycling treatments that is needed to characterize the impacts on vehicle fuel consumption during the use stage.
- There is no consensus on the methodology for modeling or allocation of the environmental impacts and potential benefits of recycling between the upstream project that is at its EOL and the downstream project that is going to use the recycled materials.

In this paper, research objectives to fully address the gaps identified above are defined and framework is developed for achieving them.

3 PENDING QUESTIONS ABOUT RECYCLING VERSUS CONVENTIONAL METHODS

The gaps identified in the previous section result in pending unanswered questions related to different life cycle stages of pavement sections. Some of the questions related to the material production, transportation to site, and construction are:

- What are the environmental impacts of the EOL strategies for flexible pavements in California considering only the material production, construction and EOL stages and without considering the use stage performance?
- How much does each life cycle stage contribute to the total impacts (without considering the use stage)?

– How do current alternative EOL strategies for flexible pavements in California compare with each other? Does in-place recycling (IPR) offer any reduction in environmental impacts versus conventional strategies such as mill-and-fill or overlay? If yes, how significant are they?
– How much does the hauling distance of materials contribute to the differences observed? what is the transport distance for new materials (for FDR, CIR, and overlays) at which the different environmental impacts are the same?

Regarding the use stage performance of different alternatives, answers to the following questions are important:

– How do IPR options compare with conventional strategies in terms of field performance during the use stage? Is the difference significant enough to warrant more frequent maintenance and rehabilitation, resulting in more environmental impacts and costs? Another important factor to consider during the use stage is surface roughness which directly impact vehicle fuel efficiency. How do IPR sections and conventional ones compare considering changes in surface roughness with time?
– Combining the results of the two phases and considering the full life cycle of a pavement section, what are the benefits and pitfalls of IPR and conventional strategies?

4 DEFINING RESEARCH OBJECTIVES TO ADDRESS CURRENT GAPS

In order to address the gaps identified before and answer the questions that were raised in the previous section, a research project was defined with the following objectives:

– Complete the LCI database for the materials and surface treatments used by Caltrans and local governments on their pavement projects.
– Develop models to quantify the environmental impacts of EOL alternatives for flexible pavements, using the practice in California as the case study.
– Develop performance prediction models for recycled sections to understand how their roughness and structural deterioration as manifested by cracking change with time and therefore be able to consider the full life cycle of maintenance and rehabilitation for recycling and conventional methods.
– Use the developed performance models to conduct a full life cycle assessment of in-place recycling and comparison with conventional methods.
– Use the results and findings of the full LCA to investigate and recommend a methodology for handling allocation of the EOL impacts and provide understanding on how it can affect decision making in terms of selecting the treatments based on their environmental impacts.

5 FRAMEWORK FOR FULL LIFE CYCLE COMPARISON OF PAVEMENTS EOL ALTERNATIVES

Figure 1 shows the framework defined and used for developing the models and databases needed for full life cycle comparison of available pavement EOL alternatives in this study.

The material production stage inventories were cradle-to-gate for all the materials used in the construction. This means that the LCIs included the energy consumption and emissions of all the production processes: raw material acquisition from the ground, transportation to the plant, and further processing of the raw materials in the plant until the final materials were ready to be shipped at the plant gate. The models represented the local conditions, technologies, and practices used in local plants and the construction processes in the local region, in this case California.

To model the construction activities, local practice and specifications were closely simulated. Figure 2 shows the approach used in a recently completed study (Saboori, in press) for

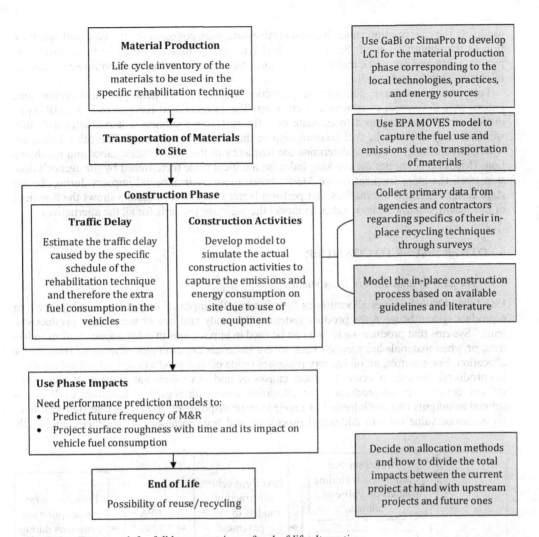

Figure 1. Framework for full lca comparison of end-of-life alternatives.

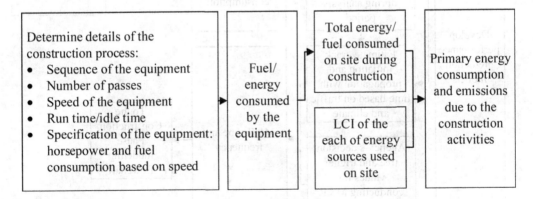

Figure 2. Modeling approach for the construction stage.

modeling the construction stage. Representative data were collected for the type and specification of the equipment used (horsepower, fuel type, mpg), materials used in the construction process (type and amount), and the actual construction process (order of equipment, number of passes, speeds, and more).

To model the use stage, performance prediction models for two main pavement performance indices were developed, roughness and cracking. The International Roughness Index (IRI) prediction model was developed to estimate how the surface smoothness would change with time which directly affected the fuel consumption in the vehicles on the road. The other index, the wheelpath cracking, is used to determine the frequency of the future maintenance and rehabilitations (M&R). The sooner the cracking index hit a critical value (determined by the agency's decision tree), the more frequent future M&R, hence more environmental impacts during the use stage compared to the alternatives that perform better in this area. Figure 3 shows the flowchart of the activities that were undertaken to model the use stage properly for all the alternatives.

6 OTHER ISSUES TO CONSIDER

6.1 *The issue of allocation in LCA studies*

ISO 14040 (2006a) defines allocation as "partitioning the input or output flows of a process or a product system between the product system under study and one or more other product systems." Systems that produce waste that can be used in applications in other applications or industries, or when materials are recycled back to use materials are examples of areas of concern for allocation. For example, an oil refinery processes crude oil and produces several refined products (co-products), how to allocate the total emissions and environmental impacts of the whole refinery between the co-products is an allocation issue with regard to co-products. Waste is defined as outputs that are intended or required to be disposed of, indicating that they that have no economic value without additional processing and transportation. As another example, fly

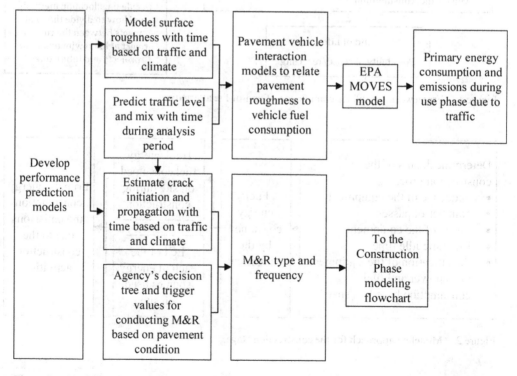

Figure 3. Modeling the use stage for comparing various EOL alternatives.

ash is a waste of combustion of coal in power plants and can be used as a substitute for portland cement. Dividing the impacts of coal combustion between the electricity generated by the plant (the main product) and the fly ash (the waste if it has no economic value, and the co-product if it has economic value beyond the cost of additional processing and transportation) is an allocation issue depending on whether the fly ash is treated as a waste or a co-product (by-product and co-product have the same meaning in LCA (Harvey et al., 2016).

Recycled materials are another area of concern. Recycling can be open-loop or closed-loop. In closed-loop recycling, the material is recycled back into the original product system, such as the case for aluminum cans or reclaimed asphalt pavement (RAP) back into hot mix asphalt (HMA), and theoretically can be recycled infinitely. In open-looped recycling, the material is recycled into other product systems with a substantial change in the inherent properties and possible losses of quality that will eventually lead to the disposal of the material as waste (ISO, 2006b). ISO 14040 and 14044 recommends avoiding allocation, whenever possible, by: (1) Dividing the unit process into sub-processes and collecting inputs and outputs related to each, or (2) Expanding the system boundaries to include the additional functionalities related to the co-products. When the need for allocation is unavoidable, ISO recommends partitioning the impacts based on the underlying physical relationships between them such as mass proportions or energy content ratios. ISO also allows allocation based on economic value where appropriate.

The issue of allocation is present in many aspects of pavement LCA studies such as environmental impacts of asphalt binder as a co-product of an oil refinery, or supplemental cementitious materials (SCMs) such as fly ash, ground blast furnace slag, silica fume, and waste of other industries. Another major area of concern is recycled materials such as RAP and reclaimed concrete aggregate (RCA.) While allocation of co-products and by-products in pavements have been considered by multiple studies (Sayagh, 2010; Huang, 2013; and Kang 2014), best practice for allocation of recycled materials, either in-place or at plant is still an area of debate.

Recycling pavement materials, either in-place or at a plant, will displace the use of some or all the virgin aggregates and sometimes the binders and therefore reduce or eliminate the impacts of producing virgin materials and neat binders. This can be considered as an environmental credit that can be allocated between the upstream project that provides the recycled materials and the downstream project that uses it, or just be applied to the downstream project as done in the cut-off method of allocating the impacts. There are also emissions and energy consumption, referred here as environmental burdens, from processing and transportation (where required) of the recycled materials. These burdens for in-place recycling include pulverization, processing, transportation and addition of virgin stabilizing agents and of virgin aggregate where needed to obtain the desired gradation. For in-plant recycling the burdens are caused by: demolition, transportation to the plant, processing done at the plant, and transportation of the recycled materials to the new construction site. The allocation of these environmental burdens between the upstream and downstream projects is not straight forward.

Three different approaches have been suggested for allocation of recycled materials (Harvey, 2016), as explained in Figure 4:

– Cut-off method: burdens of production of P1 all go to pavement 1, the burdens of all processes of recycling go to the downstream project (pavement 2 is responsible for R1, no credits to pavement 1 for producing recycled materials) and pavement 2 has reduced impacts from using the recycled materials
– 50/50 method: half the impacts of recycling processes and the reduced impacts of using recycled materials in pavement 2 go to the second pavement and half to the first pavement
– Substitution method: The burdens of recycling processes go to pavement 1, and the first pavement is also given the full benefits of reduced impacts by substituting recycled materials for virgin materials in pavement 2)

Currently, there is no consensus on a framework or methodology for handling the allocation. Furthermore, some other questions that need to be addressed are:

– Is there a difference in recyclability of a new pavement at its EOL versus a section that is built using any of the conventional rehabilitation techniques?

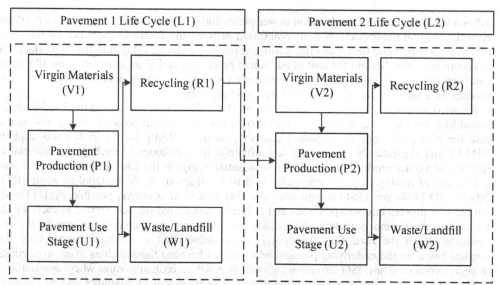

| Pavement 1 Life Cycle (L1) | Pavement 2 Life Cycle (L2) |

V1: Environmental impacts of material production for pavement 1
V2: Environmental impacts of material production for pavement 2 (V1<V2 due to using recycled materials that re-place virgin aggregate)
R1: Environmental impacts of the recycling processes
P1, P2: Environmental impacts of pavement construction
U1, U2: Environmental impacts of the use stage
W1, W2: Environmental impacts of waste management

Figure 4.   EOL allocation rules potentially applicable for pavements (Van Dam et al., 2015).

– How many times can a recycling strategy be repeated for a given material and do the con-struction activities change with subsequent recycling? If it is limited, which means the qual-ity deteriorates with each recycling, how will the allocation of the impacts between the current and future recycling (assuming the system expansion includes all future recycling)
– Does consequent recycling have a detrimental impact on the pavement performance? Is the same performance model applicable to a section recycled once and a section that that has been recycled multiple times?
– Should the LCA consider repeated use of the same treatment or are there paths in the ana-lysis period in which different alternatives should be considered?

6.2   Cost effectiveness

It is generally believed that in-place recycling practice costs less compared to conventional EOL of treatments such as mill-and-fill or total reconstruction as in-place recycling involves less virgin aggregate consumption and less hauling of the materials to the site. However, this assumption is usually based only on initial cost. Complete comparison should include all life cycle stages and requires life cycle cost analysis, as more frequent future M&R may offset the savings in the initial construction.

7   APPLICATION OF THE FRAMEWORK FOR EOL ALTERNATIVES FOR FLEXIBLE PAVEMENTS IN CALIFORNIA

The framework developed in this research was used to compare alternative EOL options for flex-ible pavements in the state of California in a two-part study (Saboori, in press). The first part of the study was conducted to benchmark the environmental impacts of few EOL treatments used in California for flexible pavements at their end of service life. The system boundary consisted of material production, transportation to the site and construction activities that together make up the EOL treatment. The system boundary did not include other life cycle stages such as use stage, future maintenance and rehabilitation, and traffic delays during construction activities. Twelve treatments were studied, consisting of ten in-place recycling alternatives, CIR and FDR with different stabilization methods (percent of foamed asphalt and/or portland cement added, and the thickness of wearing courses on top, and two conventional treatments (HMA overlay and HMA mill-and-fill). The results of the study showed that (Saboori, in press):

– There are large differences between the initial impacts of the treatments in each impact category
– Differences of an order of magnitude were observed between the lowest and highest values across all the impact categories
– The material production stage is dominant in all impact categories for all treatments
– The results were in line with the initial expectation that the additives used (type and amount) plus overlay thickness are the main contributing factors.

In the second part of the study performance prediction models for IRI and wheelpath cracking (WPC) were developed for sections built using CIR and FDR in the state of California. Caltrans' pavement management system (PMS) condition survey database consists of data from previously used visual pavement condition survey (PCS) and automated pavement condition survey (APCS) since 2011. This database was used to obtain data for sections that have had CIR or FDR at any point during the service-life.

The modeling processes for IRI and WPC were empirical-mechanistic, meaning that the variables included in the model and the equation form of the model were predetermined based on the behavior expected from mechanistic analysis, but they were empirical because no mechanics calculations were performed. Other variables were treated as category variables and separate equations with time as the explanatory variable were developed for each permutation of the other explanatory variables. The other explanatory variables included in these models were regional climate, truck traffic expressed in equivalent single axle loads (ESAL) per year, and the type and thickness, if HMA overlay on top of the CIR or FDR. WPC model consisted of two parts, crack initiation and crack progression.

The results (Saboori, in press) showed that, in terms of crack initiation, sections with CIR have a similar service life as the sections with FDR. However, in crack progression, CIR sections deteriorate at a much higher rate compared to FDR sections after cracks appear on top. Therefore, all CIR savings in GHG emissions and energy consumption during the construction stage compared to FDR may be offset by more frequent M&R in the future. Roughness models for FDR and CIR sections were also developed that showed that FDR sections with no stabilization had the worst performance and highest rate of increase in roughness with time, while FDR section stabilized with foamed asphalt performed better than the CIR sections and consistently maintained lower IRI values with time.

8 CONCLUSIONS

This paper identifies the current gaps in knowledge for proper selection among available alternatives for EOL of flexible pavements in the state of California. Then the paper proceeds to define clear research objectives to address the identified gaps and presents a general framework for developing the models and databases needed for such purpose.

DISCLAIMER

This study was funded, partially or entirely, by a grant from the National Center for Sustainable Transportation (NCST), supported by U.S. Department of Transportation's University Transportation Centers Program. The contents of this report reflect the views of the authors, who are responsible for the facts and the accuracy of the information presented herein. This document is disseminated in the interest of information exchange. The U.S. Government assumes no liability for the contents or use thereof. The contents of this report reflect the views of the author(s), who is/are responsible for the facts and the accuracy of the information presented. This document is disseminated in the interest of information exchange.

REFERENCES

Glass, Jacqueline, Tom Dyer, Costas Georgopoulos, Chris I. Goodier, Kevin Paine, Tony Parry, Henrikke Baumann, and Pernilla Gluch 2013. Future use of life-cycle assessment in civil engineering.

Gilbert, Haley E., Pablo J. Rosado, George Ban-Weiss, John T. Harvey, Hui Li, Benjamin H. Mandel, Dev Millstein, Arash Mohegh, Arash Saboori, and Ronnen M. Levinson 2017. Energy and environmental consequences of a cool pavement campaign. *Energy and Buildings* 157 (2017): 53–77. Available at: https://www.sciencedirect.com/science/article/pii/S0378778817309908?via%3Dihub

Chan, P., and S. L. Tighe. 2012. Quantifying Pavement Sustainability in Economic and Environmental Perspective. 89th Annual TRB Meeting. Transportation Research Board, Washington, DC.

Harvey, J. T., Meijer,J., Ozer, H., Al-Qadi, I. L., Saboori, and A., Kendall, A. Pavement Life Cycle Assessment Framework 2016. *U.S. Department of Transportation Federal Highway Administration*, FHWA-HIF-16-014, Washington, DC.

Huang, Y., A. Spray, and T. Parry 2013. Sensitivity Analysis of Methodological Choices in Road Pavement LCA. *The International Journal of Life Cycle Assessment*. Volume 18, Issue 1. Springer.

International Organization for Standardization (ISO) 2006a. *Environmental Management — Life Cycle Assessment — Principles and Framework. ISO 14040*. International Organization for Standardization, Geneva, Switzerland.

International Organization for Standardization (ISO) 2006b. *Environmental Management, Life Cycle Assessment, Requirements and Guidelines. ISO 14044*. International Organization for Standardization, Geneva, Switzerland.

Kang, S., Yang, R., Ozer, H., & Al-Qadi, I. 2014. Life-cycle greenhouse gases and energy consumption for material and construction phases of pavement with traffic delay. *Transportation Research Record: Journal of the Transportation Research Board*, (2428), 27–34.

Kucukvar, M., S. Gumus, G. Egilmez, and O. Tatari. 2014. Ranking the Sustainability Performance of Pavements: An Intuitionistic Fuzzy Decision-Making Method. *Automation in Construction*. Volume 40. Elsevier, Philadelphia, PA.

Lee, Jinwoo, and Samer Madanat 2017. Optimal policies for greenhouse gas emission minimization under multiple agency budget constraints in pavement management. *Transportation Research Part D: Transport and Environment* 55 (2017): 39–50.

Li, H., J. Harvey, A. Saboori, J. Lea, N. Santero, A. Kendall, and X. Cao 2017. Cool Pavement LCA Tool: Inputs and Recommendations for Integration", in press, University of California Research Center, Davis, CA.

Li, H., Saboori, A. and Cao, X., 2016. Information synthesis and preliminary case study for life cycle assessment of reflective coatings for cool pavements. *International Journal of Transportation Science and Technology*, 5(1), pp.38–46.

Lidicker, J., Sathaye, N., Madanat, S., & Horvath, A 2012. Pavement resurfacing policy for minimization of life-cycle costs and greenhouse gas emissions. *Journal of Infrastructure Systems*, 19(2), 129–137.

Milachowskl, C., T. Stengel, and C. Gehlen 2011. Life Cycle Assessment for Road Construction and Use. European Concrete Paving Association, Germany. http://www.eupave.eu/documents/technical-information/inventory-of-documents/inventory-of-documents/eupave_life_cycle_assessment.pdf

NAPA 2017. Product Category Rule for Asphalt Mixtures (version 1.0). National Association of Pavement Association, Greenland, MD. Available at: https://www.asphaltpavement.org/PDFs/EPD_Program/NAPA_Product_Category_Rules_%20final.pdf

NSF International 2019. Product Category Rule for Environmental Product Declarations of Concrete. National Center for Sustainability Standards, Washington, D.C. Available at: https://www.nsf.org/newsroom_pdf/concrete_pcr_2019.pdf

Sayagh, S., Ventura, A., Hoang, T., François, D., & Jullien, A. 2010. Sensitivity of the LCA allocation procedure for BFS recycled into pavement structures. *Resources, Conservation and Recycling*, 54(6), 348–358.

Saboori, Arash. (in press). Integrated and Data-Driven Transportation Infrastructure Management through Consideration of Life Cycle Costs and Environmental Impacts. PhD Dissertation, University of California Davis.

Torres-Machi, C., Chamorro, A., Yepes, V., & Pellicer, E. 2014. Current models and practices of economic and environmental evaluation for sustainable network-level pavement management. *Revista de la Construcción*, 13(2).

Van Dam, Thomas John, John T. Harvey, Stephen T. Muench, Kurt D. Smith, Mark B. Snyder, Imad Al-Qadi, I., Hasan Ozer et al. Towards sustainable pavement systems: a reference document. US Department of Transportation, *Federal Highway Administration*. Washington, D.C.

Wang, T., Harvey, J., & Kendall, A. 2014. Reducing greenhouse gas emissions through strategic management of highway pavement roughness. *Environmental Research Letters*, 9(3), 034007.

Quantification of potential reductions in greenhouse gas emissions by allowing increased use of reclaimed asphalt pavement in Caltrans projects

Arash Saboori, John T. Harvey, Ali Azhar Butt & Mohamed Elkashef
Department of Civil and Environmental Engineering, University of California Pavement Research Center, University of California, Davis, USA

ABSTRACT: Hot mix asphalt (HMA) is the surface type for approximately 75 percent of the California state highway network and a widely used structural material in a number of different pavement applications. Reclaimed asphalt pavement (RAP) is HMA that is milled off the existing surface and can be used to partially replace virgin asphalt binder and aggregate in new HMA. Currently, a maximum of 25 percent RAP by weight of mix is allowed in HMA by Caltrans. The goal of this study is to use the life cycle assessment (LCA) methodology and quantify the changes in greenhouse gas emissions (GHG) and other environmental impacts by allowing RAP contents of up to 25 percent, as allowed under a recent change in specification that will facilitate this, and then up to 40 and 50 percent in HMA, considering virgin binder replacement, versus the recent practice of 15 percent. Sensitivity analysis considers alternative rejuvenating agents based on assumed inventories.

1 INTRODUCTION

The California Department of Transportation (Caltrans) is currently working with industry and academia developing technical approaches for increasing the percentage of reclaimed asphalt pavement (RAP) in hot mix asphalt (HMA) and rubberized hot mix asphalt (RHMA) without reducing the performance of these materials. Caltrans has allowed contractors to use up to 15 percent RAP (by weight) in HMA without any additional engineering for a number of years (Caltrans, 2018a), which is considered as the baseline in this study. Up to 25 percent RAP was allowed in the past several years, however, the specifications called for a conservative approach to the engineering of the blended RAP/virgin binder and the use of expensive and time consuming testing that also required the use of highly regulated solvents, all of which essentially eliminated the use of more than 15 percent RAP. In 2018, the specifications were changed to allow up to 25 percent RAP without the need for the testing that was considered by industry to be onerous. Caltrans is working on developing approaches to include approximately 40 percent RAP in HMA in the future, and to begin to use up to 10 percent coarse RAP in RHMA. Coarse RAP consists of larger particle sizes in the material and has low binder content, with the result that this strategy would be to replace virgin aggregate with little or no replacement of virgin binder. The use of RAP in RHMA was not considered in this study.

2 ABATEMENT STRATEGY

An important portion of Caltrans environmental impacts are due to projects awarded each year to contractors for maintaining nearly 80,000 lane-km (50,000 lane-miles) of state-owned highway pavement infrastructure in California. Additional pavement infrastructure includes ramps, parking lots, turnouts, shoulders, rest areas, gore areas, drainage facilities, dikes, and curbs.

There are many different types of materials used in pavement projects; however, this study only focuses on the increased use of RAP in flexible pavements as a starting point. Similar evaluations should also be conducted for other transportation infrastructure materials such as portland cement concrete, metals, plastic polymers, and additives. The purpose of this case study was to assess how much Caltrans can reduce the environmental impacts due to the HMA materials used in these projects, specifically by increasing the amount of RAP to replace virgin materials. The scale of the study was the entire state network, as opposed to a project-level analysis.

As noted, HMA and RHMA are used for the majority of the pavement surfaces in California (Caltrans, 2015). The use of up to 15 percent RAP in asphalt is a mature and common practice across the nation. Asphalt surface layers can be milled at the end of their service life and used as RAP in new construction or maintenance and rehabilitation (M&R) activities by blending it with virgin asphalt binder and aggregate to create new HMA, hence reducing virgin materials (aggregate and binder) in the mix through replacement. The use of RAP provides cost savings to materials producers. This is particularly true for the RAP binder which is much less expensive than virgin binder. The replacement of part of the virgin binder in a new mix with aged asphalt binder left on RAP particles can produce similar performance with a mix that has only virgin binder if the residual binder in RAP is able to blend with the virgin binder in the new mix. To achieve the same performance, the properties of blended binder need to be similar to those the 100 percent virgin binder that would have been used otherwise.

RAP binder is more oxidized and therefore stiffer than the specified virgin binder for a given climate region and application. The properties of RAP binder vary considerably depending on the original properties it had before it was placed, the amount of aging, and later processing. RAP piles at asphalt plants also have RAP from multiple locations. Even if the RAP is milled from one location, it often has a mix of multiple asphalt layers placed over the years. RAP should be processed to create greater uniformity before being used in new mixes. How to measure and engineer the resultant properties of the blended binder, and also determine the degree of blending that occurs during mixing, are some of the technological challenges to using higher percentages of RAP (Epps Martin et al., 2017; Jones et al., 2017). Higher percentages of RAP often require the use of a softer virgin binder for the portion of the total binder not replaced by the RAP binder and the addition of softening additives, called rejuvenators, to facilitate blending of the aged and the virgin binders. It is important that the mix containing the RAP have similar performance to a mix with virgin binder with respect to fatigue and low-temperature cracking and rutting or else any cost and/or environmental benefits are in jeopardy.

3 GOAL & SCOPE DEFINITION

The goal of this study is to calculate how much reduction in GHG emissions can be achieved by increasing the maximum RAP content in HMA mixes, going from 15, to 25, 40, and 50 percent binder replacement, and scale those results to the use of HMA on the state network in California. The scope of this study considers all the impacts from materials extraction to transportation to plants, and all the processes conducted in the plant to prepare the final mix. This is an example of an LCA with a *cradle to gate* scope. It was assumed that the construction process and field performance of the mixes with higher RAP content is the same as the base scenario. Therefore, the construction stage, use stage, and end-of-life were excluded from the scope of this study. This assumption is not consistently valid and depends on the ability of the asphalt technology to adjust for the RAP properties, which is considered in this study, but it is sufficiently valid for this first-order analysis.

The functional unit for this study is defined to be the California highway network. The analysis period considered for this study is 33 years, from 2018 to 2050.

A major part of the effort in this study was spent on estimating the amount of materials used each year on the state highway network during the analysis period. For this purpose, two options were considered:

- Material quantity estimates projected over the analysis period based on programmed work in the Caltrans pavement management system (PaveM), or
- Projection of material quantities from historical data of construction projects published annually in the Construction Cost Data Book (Caltrans, 2018b).

PaveM is an asset management tool used for project prioritization, timing of future maintenance and rehabilitation, and budget allocation. User inputs into PaveM include a multitude of decision-making factors such as available budget, network characteristics (climate, traffic, dimensions), and agency decision trees that trigger treatment based on current and predicted values of key performance indices such as cracking and surface roughness for each segment in the network.

The output of PaveM is the type of treatment applied to each network segment, or "do nothing" for each year during the analysis period within the defined budget limits. The treatments for asphalt concrete are defined as thin, medium, and thick overlays, and provide an indication of the thickness of the asphalt concrete treatment, which can be multiplied by the lane width and length of the segment to get a volume. The volume can be converted to mass units typically used for asphalt materials, using a typical density. The PaveM estimates will tend to be lower than the actual total amount of asphalt concrete used by Caltrans because it only considers the lanes in the travelled way. It does not consider any paving on shoulders, ramps, parking lots, gore areas, use as a base under concrete pavement, and other places where Caltrans uses this material.

A PaveM run was conducted under the current default budgeting scenario which projected an expenditure of 267 million dollars in 2018 for asphalt paving materials. However, the data in the 2018 Construction Cost Data Book (items 390132, 390135, 390136, 390137, 390401, 390402, 395020, and 395040)[1] shows 545 million dollars in the same year for the same items. To address this discrepancy and calculate material consumption in each year during the analysis period, the tonnages of materials from the CCDB 2018 were multiplied in all years after 2018 by the ratio of CCDB purchases in 2018, divided by the PaveM projected purchases in 2018.

This study assumed that the current projected work plans to 2050 will not change considerably, and that current costs are representative of future costs. The study also assumes that current recycling strategies will not show much improvement. All of these assumptions are highly unlikely; however, they are reasonable for at least the next 5 to 10 years. It can be seen in Figure 1 that there is a very large increase in projected amount of HMA to be used after 2018 according to current master work plans for the state highway network. This reflects a large new influx of funding for state highway maintenance and rehabilitation coming from sustainment in November 2018 through direct popular vote of increases in the state taxes on gasoline and diesel.

It should also be noted that northern California local agencies often follow Caltrans specifications, and any effects of Caltrans specification changes would be amplified by agencies in the northern counties following those specifications. Changes in local government practices were not considered in this study.

Figure 1 shows the amount of HMA and RHMA needed each year between 2018 and 2050 in Caltrans projects. The amount of materials was taken from PaveM. The current run provides data up to 2046. Due to lack of better alternative to estimate material needs in 2046 to 2050, it was assumed that the average masses of HMA and RHMA used in the 10 years prior will be applied during that time period.

---

1 http://ppmoe..ca.gov/hq/esc/oe/awards/2018CCDB/2018ccdb.pdf

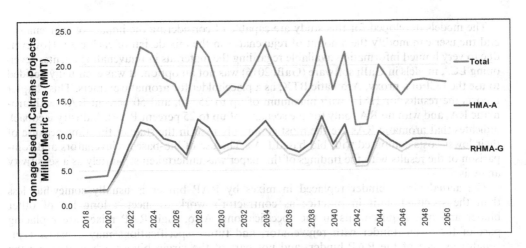

Figure 1.   Materials needed per year between 2018-2050 based on PaveM outputs.

Four HMA mixes with increasing maximum RAP content for HMA were considered as shown in Table 1. To avoid heating RAP to high temperatures, which can damage the residual binder in RAP, the heating temperature for RAP was limited to 350° C (177°F) while virgin aggregate were heated up to 500°C (260°F) to compensate and reach the required mixing temperature needed for the blended materials. These differences in temperature do not have any implications for the life cycle inventory of energy use for each mix because setting a fixed final mixing temperature for all mixes requires the same amount of heat for 1 kg mix of blended RAP and virgin materials, independent of the mass ratios of the two components. Increasing the RAP content will only result in higher temperatures needed for virgin aggregate materials to achieve the same mixing temperature for the blend. As noted previously, the use of RAP in RHMA is not currently allowed in Caltrans projects. Results for RHMA-G (gap-graded) without RAP were included for comparison.

Mixes with RAP content above 25 percent require a rejuvenating agent to be added to ensure reliable performance due to the higher percentage of aged binder recovered from RAP. Three common types of rejuvenating agents (RA) are aromatic extracts made from petroleum, bio-based RAs from made soy oil, and bio-based RAs made from tall oil that comes from trees. Another method for handling the issue of aged binder is to use softer virgin binder, which eliminates the need of rejuvenators, however, this method was assumed to be only applicable for RAP contents up to 25 percent based on recent research (Epps Martin et al., 2019).

Table 1.   The five scenarios considered for HMA for Caltrans projects across the entire network.

Mix Title	Max RAP Content	Actual Binder Replacement	Virgin Binder Replaced by RAP	Virgin Binder Replaced by Rejuvenator
HMA (Max 15% RAP)	15%	11%	11%	0%
HMA (Max 25% RAP)	25%	20%	15%	5%
HMA (Max 25% RAP)	25%	20%	20%	0%
HMA (Max 40% RAP)	40%	35%	28%	7%
HMA (Max 50% RAP)	50%	42%	32%	10%

The models developed for this study are capable of considering the impact of rejuvenators and the user can modify the amount of rejuvenator in the mix design of each case. However, due to very limited information available regarding the materials in rejuvenating agents, developing LCA models in GaBi software (GaBi 2016) was not an option. It was eventually decided to use the LCI of a proxy, Aromatic BTX, as a placeholder for aromatic extracts. This chapter reports the results for HMA with maximum of up to 25, 40, and 50 percent RAP with aromatic RA, and with no RA (only for the scenario of up to 25 percent RAP). Industry feedback indicates that aromatic RAs are the most commonly used in the state at this time because of their lower costs compared with bio-based RAs. The use of bio-based rejuvenators and comparison of the results with the findings of this paper was undertaken separately as a sensitivity analysis.

The actual virgin binder replaced in mixes by RAP binder is usually somewhat less than the specified limit in practice as contractor's work to meet a long list of other binder and mix requirements in the specifications. Also, high RAP mixes are replacing part of the virgin binder with rejuvenator, and future specifications may consider rejuvenator as part of the RAP binder, and not part of the virgin binder, when allowing the maximum RAP content.

The baseline mix designs for HMA and RHMA gap graded (RHMA-G) were taken from the UCPRC Case Studies report (Wang et al., 2012) and are presented in Table 2. It was assumed that the RAP materials had a binder content of 5 percent by mass with a 90 percent binder recovery ratio, resulting in an effective RAP binder content of 4.5 percent. Therefore, the total binder content for HMA baseline is 4.7 percent (0.04 + 0.15 * 0.045 = 0.047). The RHMA total binder content is 7.5 percent. These data combined with the information in Table 1 were used to develop the mix designs for new HMA scenarios which are shown in Table 2.

5 RESULTS

The mix designs were then used to calculate the cradle to gate environmental impacts of for each mix using the LCA methodology. For this purpose, the LCI database created by the UCPRC (Saboori et al., in press) was used. All the details of model development, data sources, and the assumptions made can be found in that reference.

Table 3 shows the LCI results for the main construction materials used in this study taken from the UCPRC LCI database. The electricity grid mix used for modeling the material production stage was based on the 2012 grid mix in California (CEC Website). Only the following impact categories and inventory items are reported in this study: Global Warming Potential (GWP), Smog Formation Potential, Particulate Matter 2.5 (PM 2.5), Primary Energy Demand (PED) which is reported as Total, Non-Renewable (NR) and Renewable (R).

Table 4 shows the material production impacts for 1 kg of each of the mixes in this study.

Table 2. Mix design component quantities by mass of mix for HMA scenarios and RHMA used in this study.

Mix	RAP	Rejuvenator	Virgin Binder	Virgin Aggregate	CRM	Extender Oil	Total Binder
HMA (Max 15% RAP)	11.5%	0.00%	4.18%	84.3%	0.0%	0.0%	4.7%
HMA (Max 25% RAP)	15.7%	0.24%	3.76%	80.3%	0.0%	0.0%	4.7%
HMA (Max 25% RAP, no Rejuv.)	20.9%	0.00%	3.76%	75.4%	0.0%	0.0%	4.7%
HMA (Max 40% RAP)	29.2%	0.33%	3.06%	67.4%	0.0%	0.0%	4.7%
HMA (Max 50% RAP)	33.4%	0.47%	2.73%	63.4%	0.0%	0.0%	4.7%
RHMA-G	0.0%	0.00%	5.81%	92.5%	1.5%	0.2%	7.5%

Table 3. LCI of the items used in this study.

Item	Unit	GWP [kg CO_2e]	Smog [kg O_3e]	PM 2.5 [kg]	PED-Total [MJ]	PED-NR [MJ]	PED-R [MJ]
Electricity (CA Grid Mix)	1 MJ	1.32E-1	4.28E-3	2.54E-5	3.09E+0	2.92E+0	1.70E-1
Natural Gas (Combusted)	1 m3	2.41E+0	5.30E-2	1.31E-3	3.84E+1	3.84E+1	0.00E+0
Asphalt Binder	1 kg	4.75E-1	8.09E-2	4.10E-4	4.97E+1	4.93E+1	3.40E-1
Aggregate (Crushed)	1 kg	3.43E-3	6.53E-4	1.59E-6	6.05E-2	5.24E-2	8.03E-3
Crumb Rubber Modifier	1 kg	2.13E-1	6.90E-3	1.05E-4	4.70E+0	3.60E+0	1.10E+0
RAP	1 kg	7.16E-3	1.39E-3	2.70E-6	1.02E-1	1.02E-1	0.00E+0
Rejuvenator, Aromatic BTX	1 kg	6.44E-1	1.57E-4	3.20E-2	4.78E+1	4.76E+1	2.18E-1

Table 4. Environmental impacts of material production stage for 1 kg of each of the mixes.

Mix Title	Rejuvenator Type	GWP [kg CO_2e]	Smog [kg O_3e]	PM 2.5 [kg]	PED-Total [MJ]	PED-NR [MJ]	PED-R [MJ]
HMA (Max 15% RAP)	N/A	4.95E-2	4.68E-3	3.25E-5	2.56E+0	2.54E+0	2.23E-2
HMA (Max 25% RAP)	Aromatic	4.91E-2	4.37E-3	1.06E-4	2.46E+0	2.44E+0	2.10E-2
HMA (Max 25% RAP)	No Rejuv	4.78E-2	4.41E-3	3.09E-5	2.35E+0	2.33E+0	2.01E-2
HMA (Max 40% RAP)	Aromatic	4.69E-2	3.90E-3	1.33E-4	2.16E+0	2.15E+0	1.78E-2
HMA (Max 50% RAP)	Aromatic	4.64E-2	3.67E-3	1.77E-4	2.07E+0	2.05E+0	1.67E-2
RHMA-G	N/A	6.00E-2	5.97E-3	1.00E-4	3.50E+0	3.46E+0	4.53E-2

The total GHG emissions due to material production stage of HMA and RHMA mixes in Caltrans projects can be quantified by combining the amount of materials used each year and data in Table 4 (LCA results for unit mass of each mix).

The material production impacts of HMA in the entire analysis period of 33 years (2018 to 2050) results in close to 14.1 million metric tonnes (MMT) of CO_2e for the baseline scenario. RHMA production impacts within the same time period are about 15.52 MMT CO_2 e. RHMA is responsible for about 52 percent of the combined GHG emissions of HMA and RHMA and since use of RAP is currently not permitted in RHMA mixes, use of RAP in RHMA is a significant untapped area for cutting emissions if it becomes technically possible to obtain same performance.

Increasing the RAP content from the original 11.5 percent (for the max 15 percent RAP baseline) by 8.5, 23.5, and 30.5 percent can result in 96, 729, and 870 thousand metric tonnes of CO_2e savings compared to the baseline, respectively, during the 33-year analysis period, when using aromatic BTX RAs. These reductions are equivalent to 0.7, 5.2, and 6.2 percent reductions in GHG emissions compared to the baseline. These results are presented in Table 5 as well.

It can be seen that percent reductions in CO_2e are not commensurate with percent reductions in replacement of virgin binder because of the high impacts of the assumed aromatic RA material and the percentages of RA that are needed to achieve blending. The use of the BTX inventory as an assumed surrogate for aromatic RA may be over- or under-estimating the impacts of these materials. This points out the need for better information regarding these materials, particularly if they are to be used on a more widespread basis to help increase the use of RAP in HMA.

Reductions in CO_2e emissions shown in Table 5 can be erased, or CO_2e emissions can potentially be increased, if the mixes with high RAP contents have worse performance than current HMA. Worse performance would increase the replacement frequency for HMA

Table 5. Total changes in GHG emissions compared to the baseline for the analysis period (2018 to 2050).

Metric	% Virgin Binder	% Binder from RAP	% RA	Total GHGs (Tonne CO2e)	CO2e Reductions (Tonne CO2e)	Percent Reduction in GHGs vs Baseline (%)
Max 15%, no Rejuv	4.18%	0.52%	0.00%	14,125,517	0	0.0%
Max 25% RAP, BTX	3.76%	0.71%	0.24%	14,029,843	-95,674	-0.7%
Max 25% RAP, no Rejuv	3.76%	0.94%	0.00%	13,655,723	-469,794	-3.3%
Max 40% RAP, BTX	3.06%	1.32%	0.33%	13,396,501	-729,016	-5.2%
Max 50% RAP, BTX	2.73%	1.50%	0.47%	13,255,578	-869,939	-6.2%

surfaces, and faster increases in roughness would increase vehicle emissions, both of which would increase CO_2e emissions if equivalent performance is not achieved.

6 SUMMARY

Increasing the amount of RAP in HMA mixes used by Caltrans in their projects can result in reductions in GHG emissions and cost savings. As shown in previous sections, RHMA production is also as significant as HMA production in terms of environmental impacts (annual RHMA impacts are about 67 percent of HMA impacts). However, Caltrans currently does not allow RAP to be used in RHMA which signifies a major untapped area for further reducing the GHG emissions of material production in Caltrans projects.

There are existing concerns, however, regarding the performance of HMA with higher RAP content. This study was conducted assuming similar performance during the use stage across all the scenarios. Decreases in performance can result in more frequent maintenance and rehabilitation needs in the future. Higher surface roughness due to poor performance can cause in an increase in vehicle fuel consumption. These issues can result in not only offsetting the original savings due to use of higher RAP content, but also causing higher environmental impacts compared to the base scenario.

Therefore, further research is needed to investigate the performance of HMA with higher than 15 percent RAP content, and also RHMA with RAP. The research findings would allow design guidelines to be developed and unintended consequences, that can arise from good intentions, to be avoided.

Assumed inventories had to be used for the aromatic RA modeled in this study. There is currently very little information regarding RAs used for increased RAP contents in HMA. Better information regarding these materials, particularly if they are to be used on a more widespread basis to help increase the use of RAP in HMA.

REFERENCES

Caltrans (2015). State of Pavement Report. California Department of Transportation, Sacramento, CA. Available at: http://dot.ca.gov/hq/maint/Pavement/Offices/Pavement_Management/PDF/SOP-2015.pdf

Caltrans (2018a). Caltrans Fact Booklet 2018. California Department of Transportation, Sacramento, CA. Last accessed in March 2019 at: http://www.dot.ca.gov/drisi/library/cfb/2018-CFB.pdf

Caltrans (2018b). Contract Cost Data Book. California Department of Transportation, Sacramento, CA. Available at: http://ppmoe.dot.ca.gov/des/oe/contract-cost-data.html

CEC (California Energy Commission) website, last accessed Sep 2017: http://www.energy.ca.gov/almanac/electricity_data/total_system_power.html

Epps Martin, A., E. Arámbula-Mercado, F. Kaseer, L. Garcia Cucalon, F. Yin, A. Chowdhury, J. Epps, C. Glover, E. Hajj, N. Morian, S. Pournoman, J. Daniel, C. Ogo, R. Rahbar-Rastegar, G. King

(2017). The Effects of Recycling Agents on Asphalt Mixtures with High RAS and RAP Binder Ratios. National Cooperative Highway Research Program Transportation Research Board. Washington, D.C. Available at: https://apps.trb.org/cmsfeed/TRBNetProjectDisplay.asp?ProjectID=3645

GaBi Software (2016), version 6.3. thinkstep International, Stuttgart, Germany, 2016.

Jones, D., M. Elkashef, L. Jiao, J. Buscheck, and J. Harvey (2017). "Pavement Recycling: Workplan for Continued Development of Guidelines for Determining Binder Replacement in High RAP/RAS Mixes." UCPRC-WP-2017-09.2, University of California Pavement Research Center.

Saboori, A., J. Harvey, H. Li, J. Lea, A. Kendall, and T. Wang (in press). Documentation of UCPRC Life Cycle Inventory used in CARB/Caltrans LBNL Heat Island Project and other LCA Studies University of California Pavement Research Center, Technical Report, Davis, CA.

Wang, T., I. S. Lee, A. Kendall, J. Harvey, E. B. Lee, and C. Kim (2012). Life cycle energy consumption and GHG emission from pavement rehabilitation with different rolling resistance. Journal of Cleaner Production, 33, pp.86–96.

Pavement recycling: A case study of life-cycle assessment and life-cycle cost analysis

Qingwen Zhou, Imad L. Al-Qadi & Hasan Ozer
Illinois Center for Transportation, University of Illinois at Urbana-Champaign, USA

Brian K. Diefenderfer
Virginia Transportation Research Council, USA

ABSTRACT: Transportation agencies and pavement industries are challenged with identifying cost-effective pavement rehabilitation techniques to extend pavement service lives. To meet this challenge, pavement recycling techniques have been introduced several decades ago. Although these techniques significantly reduce the cost and energy required to produce pavement layers, the assessment of their life cycle is not well quantified. To address this, a user-friendly life cycle assessment (LCA) and life-cycle cost analysis (LCCA) tool was developed specifically for asphalt pavement preservation and rehabilitation techniques. The tool employs life-cycle inventory data to calculate the material- and construction-stage impacts. The use-stage impact is calculated considering pavement related rolling resistance. The inventory data compiled in the tool are consistent with the ISO 14044:2006 standards. After a brief introduction of the tool, an agency-based case study is presented to demonstrate the environmental benefits of cold in-place recycling (CIR), cold central-plant recycling techniques (CCPR), and full-depth reclamation (FDR) compared with conventional reconstruction by milling and applying new overlays. Project specific data from two interstate construction projects in Virginia were collected and analyzed and the system boundary of the LCA and LCCA considered in this study includes materials and construction.

1 INTRODUCTION

Pavement recycling is a widely acknowledged technique to enhance the structural capacity and extend the service life of an existing pavement by recycling existing pavement and mixing it with recycling/stabilizing agent. The associated benefits of using pavement recycling techniques include reducing virgin materials use, utilizing stockpiled recycled materials, such as reclaimed asphalt pavement (RAP) and recycled asphalt shingles (RAS), reducing fuel consumption and resultant emissions during construction, and minimizing the time of lane closures. Pavement recycling techniques include hot in-place recycling (HIR), cold in-place recycling (CIR), cold central-plant recycling (CCPR) and full-depth reclamation (FDR). However, HIR is relatively less commonly applied because of the perceived higher fuel consumption during construction compared to cold recycling approaches (Horvath, 2004).

Cold in-place recycling involves recycling and mixing the upper 2 to 6 in of the existing asphalt concrete (AC) pavement; asphalt emulsion or foamed asphalt may be used as recycling agents and cement or lime are included as an active filler (ARRA, 2015). An AC overlay or non-structural treatment, such as chip seals, is typically applied to provide desired ride quality while controlling moisture infiltration. In 2004, the Nevada Department of Transportation (NDOT) described CIR with AC overlay on low- and medium-volume roads. The laboratory testing and long-term field performance of CIR projects indicated reduction in rutting, reflective cracking, and thermal cracking (Sebaaly et al., 2004). Lane and Kazmierowski (2005) implemented CIR with foamed asphalt and applied this technology on a highway section.

Testing using falling weight deflectometer (FWD) to backcalculate layer modulus, and measurements of pavement roughness and rutting using automatic road analyzer after one year of service showed no discernible distortion, rutting, or cracking.

Cold central-plant recycling (CCPR) is a process in which the recycled material is milled from a roadway and brought to a centrally located recycling plant that incorporates the recycling agents into the material (Apeagyei and Diefenderfer, 2012). Similar to CIR, asphalt emulsion or foamed asphalt may be used as recycling agents while cement or lime are sometimes included as an active filler. CCPR is beneficial in that it can reduce the use of virgin materials and also reuse existing stockpiled RAP as a stabilized base layer. In addition, CCPR operation reduces the impacts of construction to traffic. However, only few studies investigated the benefits of CCPR technique (Diefenderfer et al., 2015 and 2016).

Full-depth reclamation (FDR) typically pulverizes 6 to 12 in of existing pavement and remixes it with stabilizing agents such as cement, asphalt emulsion, or foamed asphalt (ARRA, 2015). If asphalt-based stabilizing agents are used, an active filler such as lime or cement may also be included. The recycled layer then performs as a stabilized base layer. Mallick et al. (2002a) conducted FWD testing on FDR pavement and analyzed the modulus to evaluate the improvement in pavement life and structural capacity. In another study by Mallick et al. (2002b), it was demonstrated that the addition of lime and cement with asphalt emulsion would increase the rate of strength gaining.

To optimize available funding and to utilize limited resources, public agencies gradually adopted CIR and FDR as rehabilitation strategies. The implementation of CIR and FDR in Nevada pavement network saved more that $600 million over 20 years (Bemanian et al., 2006). However, there is no systematic tool or approach conduct cost and environmental impacts evaluation on different recycling techniques in different projects. Hence, a user-friendly LCA and LCCA tool is needed for agencies to quantify both economic and environmental benefits of utilizing different pavement recycling techniques.

2 OBJECTIVE

The objective of this study was to evaluate pavement recycling techniques using a LCA and LCCA tool. This user-friendly LCA and LCCA tool was specifically developed for AC pavement preservation and rehabilitation techniques. Case studies of pavement recycling projects from Virginia Department of Transportation (VDOT) were used to demonstrate the economic and environmental benefits of CIR, CCPR and FDR on the premise of their good performance.

3 SUMMARY OF TOOL DEVELOPMENT

Pavement LCCA is defined as a process to compute total economic worth of a project segment by analyzing both initial costs and discounted future cost over the life of the project. Pavement LCA is a sustainability assessment methodology to quantify the environmental impact throughout the pavement life cycle (Van Dam et al., 2015). A pavement's life cycle begins with raw material acquisition and moves through production, use, rehabilitation, and end-of-life recycling, including final disposal. In this tool, the analysis includes the agency cost and environmental impacts related to the materials acquisition/production and construction of the pavement recycling techniques, as well as the user cost and environmental impacts resulting from the interaction of pavement with the traffic and environment in the use stage.

To simplify the computation of costs and environmental impacts within materials and construction stages, pay items were integrated in this tool. A pay item represents a unit of work for which a price or environmental impact is provided. The agency cost of materials and construction stage was computed based on the unit cost of each pay item and the total units of the corresponding pay item included. For the environmental impact evaluation, each pay item compiles corresponding unit processes regarding the materials production and construction, and unit processes summarize the material, construction, fuel, and/or hauling processes, including upstream and downstream data and models. For example, the unit process of asphalt binder includes crude oil extraction, flaring, transportation, refining, and blending terminal storage. The emission resulting from the unit process was represented by summing up the upstream and downstream emissions of those subprocesses.

In the use stage, this tool simulates the user cost which includes the extra fuel cost because of pavement roughness changes and traffic delay cost due to work zone. This tool also summarizes the environmental impacts resulting from the extra fuel consumption due to pavement roughness and texture change, as well as the global warming potential (GWP) emission as a result of changes in the pavement surface radiative forcing with respect to time.

In this study, the analyzed cases are still in operation, thus their analysis periods were not settled to determine their use stage impacts and user cost. Only the agency cost and environmental impacts from materials and construction stages were estimated for the following two cases using this tool.

4 CASE STUDIES

Two projects were investigated to evaluate the economic and environmental benefits of IPR and CCPR as rehabilitation methods. The first project was constructed in 2011 and the performance data demonstrated the effectiveness of pavement recycling. Due to the limited available cost data, only LCA evaluation was performed. The second project was started in 2015 and both LCA and LCCA were conducted.

4.1 Case I: Interstate-81

On the western side of Virginia, Interstate 81 was constructed in the late 1950s to early 1960s and is mostly comprised of approximately 12 in of asphalt concrete (AC) over 10 to 12 in of aggregate base. One-way corridor traffic averages approximately 26,000 vehicles per day, with peaks up to 34,000, and truck percentages range from 16 to 33 percent. The project was located at a 3.66-mi pavement section of southbound I-81 between Mileposts 217.66 and 214.00. The interstate in this section is a four-lane divided highway having lane widths of 12 ft. Traffic data in 2008 showed that this section carried 23,000 vehicles per day with 28 percent of five- and six-axle trucks, which is approximately equal to 1.7 million 18-kip equivalent single-axle load (ESAL) repetitions per year in the right lane and 0.3 million 18-kip ESAL repetitions in the left lane (Diefenderfer et al., 2015).

Prior to construction, VDOT's annual network-level condition survey identified this section as having frequently recurring structural distresses, despite regular periodic maintenance. Three alternative designs were developed for this reconstruction project, utilizing the aforementioned techniques including FDR, CCPR and CIR in accordance with ARRA (2015), see Figure 1. The conventional reconstruction is also presented in the left of Figure 1. In the left lane, a new 4-in AC layer was placed on a 5-in CIR. In the right lane, a new 4- and 6-in AC layer were placed on 8- and 6-in CCPR layer, respectively which was in turn placed on a 12-in FDR.

The dense graded AC mix was designed with 5 percent asphalt binder content, 4 percent air voids and 10 percent RAP; the maximum specific gravity Gmm was 2.53. While for the stone mastic asphalt (SMA) surface, it was designed with 7 percent asphalt binder content and 4 percent air voids. No RAP was used and the Gmm value was 2.53 as well. The CCPR and CIR materials were produced with a cement content of 1 percent and a foamed asphalt content of 2 percent. The FDR layer was stabilized using a 3 percent combination of cement and a proprietary lime kiln dust product.

VDOT has conducted periodic measurements of international roughness index (IRI) and rut depth of the I-81 sections after reconstruction. Measurements show that left lane had a greater rut depth and IRI than the right lane. After the first 34 months, the left lane rut

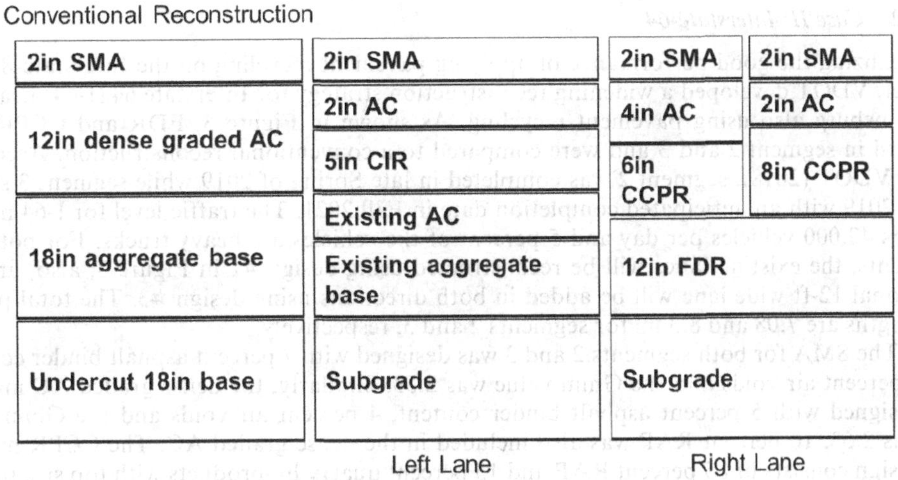

Conventional Reconstruction

2in SMA	2in SMA	2in SMA	2in SMA
12in dense graded AC	2in AC	4in AC	2in AC
	5in CIR	6in CCPR	8in CCPR
	Existing AC		
18in aggregate base	Existing aggregate base	12in FDR	
Undercut 18in base	Subgrade	Subgrade	

Left Lane   Right Lane

Figure 1. Pavement cross sections on I-81 reconstruction section.

depth and IRI were 0.04 in and 61 in/mi, while for the right lane were 0.03 in and 46 in/mi, respectively (Diefenderfer et al., 2015). The VDOT classified the ride quality for both lanes as excellent demonstrated good performance of CCPR, CIR and FDR techniques.

The environmental impacts including GWP and total primary energy consumption of conventional reconstruction and three alternatives are shown in Figure 2. The three recycled alternatives produce less GWP and consume less energy com-pared to the conventional reconstruction. As would be expected, the left lane, passing lane with fewer truck traffic volume, compared to the right lane, resulted in lower energy consumptions and GWP. Because of the dominative role of binder production in energy consumption and GWP, the section with 8-in CCPR consumed less energy and produced less GWP compared to the section with 6-in CCPR, although it consumed more cement which is also critical in GWP contribution.

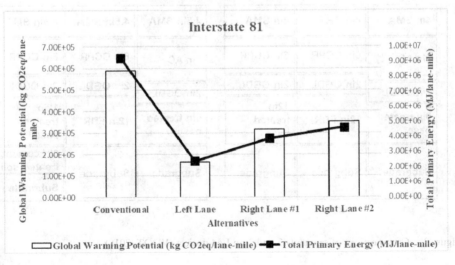

Figure 2. Comparative environmental impact analysis of I-81 reconstruction project (using environmental inventory data from Illinois Tollway [Al-Qadi et al. 2015] and FHWA [Al-Qadi et al. 2017]).

Realizing the good performance of applying pavement recycling on the Interstate-81 project, VDOT developed a widening reconstruction strategy for Interstate 64 (I-64) near Williamsburg also using pavement recycling. As shown in Figure 3, FDR and CCPR were used in segment 2 and 3 and were compared to a conventional reconstruction. According to VDOT (2018), segment 2 was completed in late Spring of 2019 while segment 3 started in 2019 with an anticipated completion date in Fall 2021. The traffic level for I-64 in 2017 was 42,000 vehicles per day and 5 percent of the vehicles are heavy trucks. For both segments, the existing lanes will be reconstructed using design #2 in Figure 3; also, an additional 12-ft wide lane will be added in both directions using design #3. The total project lengths are 7.08 and 8.3 mi for segments 2 and 3, respectively.

The SMA for both segments 2 and 3 was designed with 7 percent asphalt binder content, 4 percent air voids and the Gmm value was 2.53. Similarly, the dense graded AC mix was designed with 5 percent asphalt binder content, 4 percent air voids and the Gmm value was 2.53. 10 percent RAP was also included in the dense graded AC. The CCPR mixture design consists of 85 percent RAP and 15 percent quarry by-products with top size passing No. 10 sieve stabilized with 2.5 percent foamed asphalt binder and 1 percent cement. The FDR layer and the cement treated aggregate base layer in the new lanes were comprised of (RAP) stabilized with 5 percent cement.

In this case, costs for the reconstruction and widening activities are shown in Table 1 and Table 2. The cost reductions were realized with the use of recycled materials, especially replacing new AC with CCPR mix and replacement of natural aggregate base layer with in-place recycling asphalt and treated layers. Overall, the agency was able to save up to $5.4 million and $4.8 million in the mainline construction cost for segments 2 and 3, respectively.

In both segments 2 and 3, energy consumptions and GWP of designs #2 and #3 are lower than the design #1 which is the conventional reconstruction (see Figure 4). The savings of energy consumption may reach up to 40 percent by utilizing FDR and CCPR. It is noted that the savings are not as significant as that in I-81 project because more cement is used in I-64 project. Similarly, design #3 of segments 2 and 3 are less environmental-friendly due to the extra cement use.

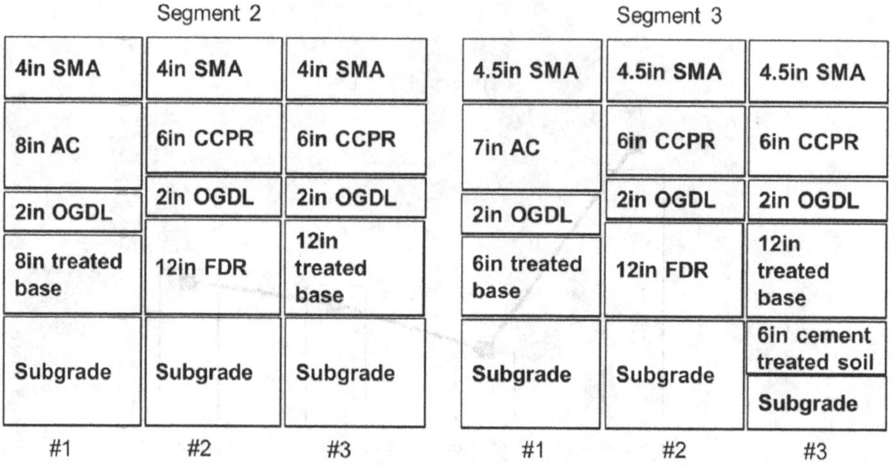

Figure 3. Cross sections of segments 2 and 3 on I-64 project (VDOT, 2018).

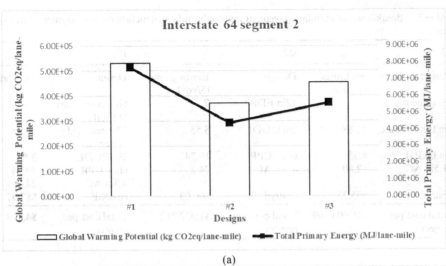

(a)

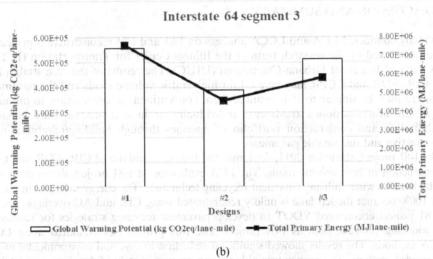

(b)

Figure 4. Comparative environmental impact analysis of I-64 reconstruction project (a) segment 2; (b) segment 3. (using environmental inventory data from Illinois Tollway [Al-Qadi et al., 2015] and FHWA [Al-Qadi et al., 2017]).

Table 1. Breakdown of mainline pavement materials and construction costs of segment 2 in I-64 project.

#1		#2		#3	
Design	All Lanes ($/yd^2)	Design	Existing lane ($/yd^2)	Design	New lane ($/yd^2)
8in treated base	20.25	2in OGDL	4.05	12in treated base	27.34
2in OGDL	6.75	12in FDR	6	2in OGDL	4.05
12in AC	55.68	6in CCPR	14.58	6in CCPR	14.58
		4in AC	14.9	4in AC	14.9
cost/yd^2	$82.68	cost/yd^2	$39.53	cost/yd^2	$60.87
Total cost per project	$12,363,107	Total cost per project	$3,940,479	Total cost per project	$3,033,769

409

Table 2. Breakdown of mainline pavement materials and construction costs of segment 3 in I-64 project.

#1		#2		#3	
Design	All Lanes ($/yd^2)	Design	Existing lane ($/yd^2)	Design	New lane ($/yd^2)
6in cement slab	7	12in FDR	7	6in cement trea-ted soil	7
6in treated base	16.88	2in OGDL	8.55	12in recycled materials	22.48
2in OGDL	8.55	6in CCPR	19.74	2in OGDL	8.55
11.5in AC	62.80	4.5in AC	24.8	6in CCPR	19.74
				4.5in AC	24.8
cost/yd^2	$95.23	cost/yd^2	$60.09	cost/yd^2	$82.57
Total cost per project	$16,693,109	Total cost per project	$7,022,212	Total cost per project	$4,824,511

5 DISCUSSION AND SUMMARY

This study conducted LCA and LCCA analysis on I-81 and I-64 reconstruction projects using a tool developed by the research team at the Illinois Center for Transportation (ICT) of the University of Illinois at Urbana-Champaign (UIUC). The results of the case studies indicate that CIR, FDR and CCPR may be used for high traffic volume roads and the pavement performance may be similar to that resulted from conventional reconstruction techniques. The use of the aforementioned treatments individually or in combination may enhance the rehabilitation and construction portfolio of agencies through building/rehabilitating more cost-effective and sustainable pavements.

The I-81 project, started in 2011, demonstrated the successful use of CIR, FDR and CCPR in reconstruction of high-volume roads. The LCA evaluation of I-81 project shows great environmental benefits when utilizing pavement recycling techniques. The energy savings can reach more than 100% because the left lane is mildly rehabilitated using CIR and AC overlays. The success of I-81 project encouraged VDOT to develop pavement recycling strategies for reconstruction and widening projects, such as I-64 project. The I-64 project was evaluated using LCA and LCCA methods. The results showed significant reduction in cost and environmental impacts in the recycled sections. The agency was able to save ap-proximately $5.4 million and $4.8 million for 7.08- and 8.3-mi segments, respectively, in the materials production and construction.

The cost savings and environmental impact reductions of applying recycling techniques are mainly due to reduction of virgin materials usage, especially the usage of asphalt binder. Asphalt binder is one of the most influential factors to the overall environmental impacts and also one of the costliest materials of pavement construction. In addition, pavement recycling can also significantly re-duce the hauling distance, thus, the fuel consumption and cost related to materials hauling are decreased.

Pavement recycling project success also depends on the guided material selection, suitable mixture design for AC and recycled layers, as well as quality control. There is an opportunity to set up comprehensive pavement recycling design guidelines to assist agencies in applying those recycling techniques considering both performance, cost, and environmental impacts using the ICT tool.

ACKNOWLEDGEMENTS

This publication is based on the results conducted in cooperation with the University of Transportation Center (UTC) on Preservation, Illinois Center for Transportation (ICT) of the University of Illinois at Urbana-Champaign (UIUC), The Sustainable Pavement Program of the Federal Highway Administration, and Virginia Department of Transportation (VDOT). The authors would like to acknowledge the assistance provided by many individuals including Kurt Smith, Karim Chatti, and Heather Dylla. The contents of this document reflect the views of the authors, who are responsible for the facts and the accuracy of the information presented herein. The aforementioned organizations assume no liability for the contents or use thereof. This report does not constitute a standard, specification, or regulation.

REFERENCES

Al-Qadi, I. L., H. Ozer, M. Krami Senhaji, Q. Zhou, R. Yang, S. Kang, M. Thompson, J. Harvey, A. Saboori, A. Butt, H. Wang, and X. Chen. 2017. A Life-Cycle Methodology for Energy Use by In-Place Pavement Recycle Techniques. Report No. ICT-17-023, FHWA Contract No. DTFH6114C00046. Federal Highway Administration, McLean, VA.

Al-Qadi, I. L., R. Yang, S. K. Kang, H. Ozer, E. Ferrebee, J. R. Roesler, A. Salinas, J. Meijer, W. V. Vavrik, and S. L. Gillen. 2015. Scenarios Developed for Improved Sustainability of Illinois Tollway. *Transportation Research Record: Journal of the Transportation Research Board*. Transportation Research Board, Washington, DC. 2523:11–18.

Apeagyei, A. K., and Diefenderfer, B. K. 2012. Evaluation of Cold in-place and Cold Central-plant Recycling Methods Using Laboratory Testing of Field-cored Specimens. *Journal of Materials in Civil Engineering*, 25: 1712–1720.

Asphalt Recycling and Reclaiming Association (ARRA). 2015. *Basic Asphalt Recycling Manual (BARM)*. FHWA-HIF-14-001. Asphalt Recycling and Reclaiming Association, Annapolis, MD.

Diefenderfer, B. K., B. F. Bowers, and A. K. Apeagyei. 2015. Initial Performance of Virginia's Interstate 81 In-Place Pavement Recycling Project. *In Transportation Research Record: Journal of the Transportation Research Board*. Transportation Research Board, Washington, DC. 2524: 152–159.

Diefenderfer, B. K., Diaz-Sanchez, M., Timm, D. H., and Bowers, B. F. 2016. *Structural Study of Cold Central Plant Recycling Sections at the National Center for Asphalt Technology (NCAT) Test Track* (No. VTRC 17-R9). Virginia Transportation Research Council.

Horvath, A. 2004. *A life-cycle analysis model and decision-support tool for selecting recycled versus virgin materials for highway applications*. Final Report for RMRC Research Project 23: 35.

Lane, B., and Kazmierowski, T. 2005. Implementation of Cold In-Place Recycling with Expanded Asphalt Technology in Canada. In Transportation Research Record: Journal of the Transportation Research Board. Transportation Research Board of the National Academies, Washington, DC, 1905: 17–24.

Mallick, R.B., Bonner, D.S., Bradbury, R.L., Andrews, J.O., Kandhal, P.S., and Kearney, E.J. 2002a. Evaluation of Performance of Full-Depth Reclamation Mixes. In Transportation Research Record: Journal of the Transportation Research Board. Transportation Research Board of the National Academies, Washington, DC. 1809: 199–208.

Mallick, R.B., Teto, M.R., Kandhal, P.S., Brown, E.R., Bradbury, R.L., and Kearney, E.J. 2002b. Laboratory Study of Full-Depth Reclamation Mixes. In Transportation Research Record: Journal of the Transportation Research Board. Transportation Research Board of the National Academies, Washington, DC. 1813: 103–110.

Sebaaly, P. E., Bazi, G., Hitti, E., Weitzel, D., & Bemanian, S. 2004. Performance of Cold In-place Recycling in Nevada. Transportation Research Record, 1896: 162–169.

Van Dam, T.J., Harvey, J., Muench, S.T., Smith, K.D., Snyder, M.B., Al-Qadi, I.L., Ozer, H., Meijer, J., Ram, P., Roesler, J.R. and Kendall, A. 2015. Towards Sustainable Pavement Systems: A Reference Document (No. FHWA-HIF-15-002). United States. Federal Highway Administration.

Virginia Department of Transportation (VDOT). 2018. Construction Documentation of I-64 and I-81 Projects. Virginia Department of Transportation, Richmond, VA.

ACCIONA's expertise in the use of LCA in construction sector

Andrea Casas Ocampo & Edith Guedella Bustamante
ACCIONA Construction, Madrid, Spain

ABSTRACT: The construction sector is aware of the need of measuring, avoiding and redu-cing environmental damage caused by construction materials, processes, services, construction methods and sites, under a life cycle perspective. ACCIONA has the strategy of considering Life Cycle Assessment (LCA) as a tool for both environmental and business management as decision-making tool for their projects. The last years, the company has increased the use of innovated materials like Fiber Reinforced Polymer (FRP) instead of concrete and steel in civil infrastructures with the aim to reduce environmental impacts and improving sustainability of infrastructures. The application of LCA methodology in these cases has contributed to meas-ure impacts reduced and improving construction processes. Thanks to this expertise, ACCIONA has been the first company in the world in obtaining an Environmental Product Declaration (EPD) of a whole road project (N-320 road).

1 INTRODUCTION

1.1 *Life cycle assessment*

The term "Life Cycle" refers to the major activities in the course of the product's life span from its manufacture, use and maintenance, to its final disposal. LCA is a "cradle-to-grave" approach for assessing products, processes or systems. "Cradle-to-grave" approach begins with gathering raw materials from the earth to create the product and it ends when all the gathered materials are returned to the earth. LCA assesses all stages of a product's life from the perspective that they are interdependent, meaning that one stage leads to the next. LCA enables the estimation of the cumulative environmental impacts resulting from all stages in the product life cycle, including impacts not considered in more traditional analyses (e.g., raw material extraction, material trans-port, ultimate product disposal, etc.). Including the impacts throughout the product life cycle, LCA provides a comprehensive view of the environmental aspects of the product or process and a more accurate picture of the true environmental trade-offs in product and process selection.

Specifically, LCA is a technique to assess the environmental aspects and potential impacts associated with a product, process or service, by: compiling an inventory of relevant energy and material inputs and environmental releases; evaluating the potential environmental impacts associated with identified inputs and releases and interpreting the results to help the decision-makers to make a more outstanding decision.

LCA has been used commonly in other industries but not in construction industry for the evaluation of environmental performance of processes. ACCIONA uses LCA methodology as a selection criterion in order to make the optimum decision in regard construction technolo-gies, processes and materials with better environmental performance (i.e. tunnels, roads, bridges, etc.). In this context, environmental assessment methodologies provide a valuable tool for helping decision makers and engineers to identify and select the best alternative design regarding environmental issues.

ACCIONA has committed in reducing environmental impacts in its construction sites betting on innovative materials and construction processes by replacing traditional ones by them. Some LCA examples highlight encouraging results for innovative materials in comparison with traditional ones. GaBi ts software (Professional version) is the tool used for calculating environmental impacts in all examples of this paper. This software is provided by SPHERA, in collaboration with IKP University of Stuttgart. It incorporates its own database with information of many processes and includes Ecoinvent and ELCD databases updated in 2019.

Some examples of ACCIONA's know how are: environmental assessment of Fiber Reinforced Polymer (FRP) bridge and its comparison with its analogue structure in concrete and steel; railway bridge (Arroyo Valchano, Orense, Spain), excavation and sustaining of a tunnel in a 40MPa rock considering two different procedures, drilling and blasting and construction of road (Cieza-Fuente de la Higuera Road, Valencia, Spain) and structural reinforcement (with concrete, steel or FRP) of beams and columns. All of them are described below.

Fiber Reinforced Polymer (FRP) bridge and its comparison with its analogue structure in concrete and steel (Madrid, Spain). During the last years, different FRP structures have been designed and built in Spain by ACCIONA (including bridges, footbridges, and a lighthouse). These kind of structures represent a different way for building road infrastructures and LCA was performed in order to evaluate the environmental damage. The analysis was carried out for a 30 m span and 12 m width bridge and it was divided in 3 stages:

- *Production stage:* the beams in composite materials were manufactured as prefabricated elements that were ready to be installed on site. This stage took into account the quantity of each raw material, energy consumed and emissions produced in the extraction, processing and transportation of raw materials to the factory.
- *Construction stage:* At this stage, energy emission made inside the production line to transform raw materials into a composite bridge was considered. This phase also included the use of energy and material resources, water use, and waste management due to infrastructure exploitation.
- *End of life:* It the demolition of the structure was assumed, considering possible recycling of demolition waste.

The evaluation of environmental impacts caused by the construction of a FRP bridge and its comparison with an analogue structure in concrete were done taking into account general emissions (Figure 1); emissions to air (Figure 2); Carbon Dioxide (CO_2) emissions (Figure 3) and CML 2001 methodology (Figure 4). CML 2001 is an impact assessment method that restricts quantitative modelling to early stages in the cause-effect chain to limit uncertainties. Results are grouped in midpoint categories according to common mechanisms or commonly accepted groupings. In this case, the categories analyzed were: Terrestric Ecotoxicity Potential (TETP) (kg DCB-Equiv); Photochemical Ozone Creation Potential (POCP) (kg Ethene-Equiv); Ozone Layer Depletion Potential (ODP) (kg R11-Equiv); Human Toxicity Potential (HTP) (kg DCB-Equiv); Global Warming Potential 100 years (GWP) (kg CO2-Equiv) and Freshwater Aquatic Ecotoxicity Potential (FAETO) (kg DCB-Equiv)[1]

The obtained results show that, in general, the FRP bridge has a better environmental performance than the bridge whose girders are made of concrete. This fact is due to the lightness of the FRP. In this way, considering the entire system, from cradle to grave, the transport and installation stages become crucial, since the cranes needed are less powerful and, therefore require less energy consumption, thus reducing the environmental impacts. Once considered all the stages of the life cycle of the two types of bridges analyzed (FRP and concrete), it is clear that global warming potential is the greatest impact derived from the emissions produced during their life cycle. One of the hotspots where the environmental impacts are higher is the production stage due to the raw materials, thus ACCIONA is working in implementing

---

1 https://www.universiteitleiden.nl/en/research/research-output/science/cml-ia-characterisation-factors

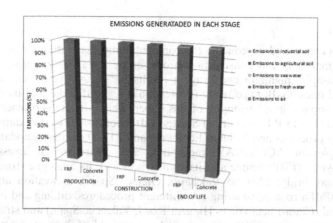

Figure 1. Emissions generated in each stage.

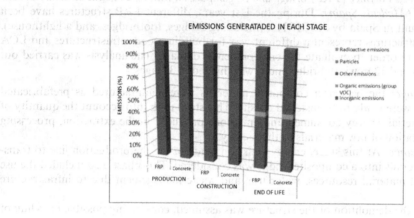

Figure 2. Emissions generated in each stage.

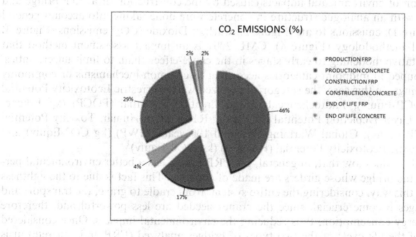

Figure 3. CO_2 emissions generated.

alternatives to substitute some of them by others with less environmental impact, working closely with resin and fiber manufactorers.However, the computation of global emissions and impacts taking into account the entire life cycle offers an advantage to the FRP structures

414

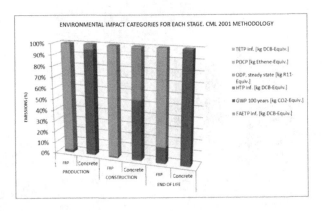

Figure 4.   Environmental impact categories for each stage. CML 2001 methodology.

versus concrete or steel. This fact is mainly due to the great advantage that FRP materials have in terms of their lightness, requiring less energy and fuel consumption during the construction phase (including installation and transport) and their good behavior in terms of durability and maintenance.

Excavation and sustaining of a tunnel construction considering two different construction processes (drilling and blasting): The goal of this study was the comparison of the environmental impacts attributable to the excavation and maintenance of 1m³ of tunnel in a rock of 40MPa built by two different processes: drilling and blasting. Environmental impacts were taken into account throughout the entire life cycle of the process; it means, from the origins of the raw materials used, to the end of the waste generated. In this Life Cycle Analysis knowledge in environmental matters related to the excavation of tunnels were generated and allowed improving different stages of construction processes. Not only the adverse environmental effects derived from the excavation process were considered, but other aspects such as environmental damage caused by the production and transport of necessary raw materials, maintenance of tunnel and impacts caused by the management of waste. The analysis was divided in 3 stages:

- *Production stage*: The quantity of each raw material used in the factory was determined, as well as the energy consumed and the emissions produced in the extraction, processing and transport to factory of raw materials. Furthermore, it was determined quantities of raw material used in the manufacture of the necessary elements for the excavation of the rock and support of the tunnel (concrete and steel), as well as the energy consumed and the emissions generated in these steps of the processes.
- *Construction stage:* Raw materials and energy required for the excavation-support of 1 m³ of 40MPa rock tunnel were considered for each process (drilling and blasting).
- *End of life:* the main waste considered were the excavated materials as well as the tungsten carbides from the widia pikes (in the case of drilling). All this waste was disposed in an authorized landfill

The evaluation of environmental impacts were analyzed following CML 2001 methodology (Figures 5, 6 Y 7) explained before and Eco-Indicator 99 (Figure 8). Eco-Indicator 99[2] is a life cycle impact assessment tool developed by PRé Consultants B.V that helps designers to make an environmental assessment of a product by calculating eco-indicator scores for materials and processes used. The resulting scores provide an indication of areas for product

2 Eco-Indicator 99.Manual for designers. A damage oriented method for Life Cycle Impact Assessment. Publication of Ministry of Housing, Spatial planning and the Environment Communication Directorate. The Netherlands.

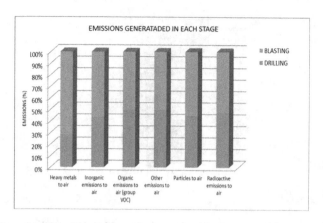

Figure 5. Emissions generated in each stage.

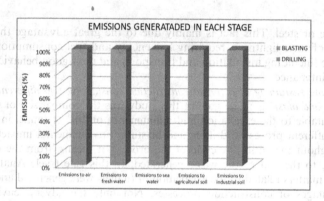

Figure 6. Emissions generated in each stage.

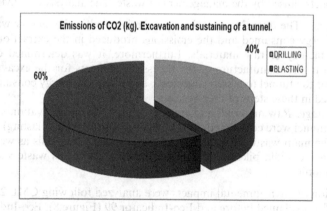

Figure 7. Emissions of CO_2 in two processes analyzed.

improvements. Eco-indicator 99 offers a way to measure various environmental impacts, and shows the final result in a single score. The results obtained are shown in the next figures.

The results obtained conclude that the excavation-sustaining of a m^3 of tunnel in 40MPa rock using drilling, in general, causes less environmental and human impacts than the excavation-sustaining of the same rock by the procedure of blasting. Taking into account CO_2

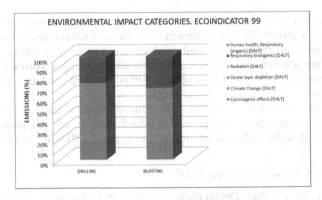

Figure 8. Environmental impact categories. Ecoindicator 99.

emissions into the air, it can be ensured that the excavation with drilling produces 33% less CO_2 emissions into the air that the excavation using blasting technique, due in part to the amount of emissions generated during the manufacturing stage of raw materials used in the blasting process, mainly of explosives and steel.

2 ENVIRONMENTAL PRODUCT DECLARATION (EPD)

Environmental labelling approaches are currently used tools to enhance the evaluation of environmental performance. These approaches can be classified, according to different ISO Standards as[3]:

- *Environmental Labels Type I*: The EU Ecolabel is an environmental Label Type I (third party organizations establish criteria for judging the environmental-friendliness of products and grants a Type I environmental label to them based on said criteria) identifies products and services that contribute to sustainability because they have demonstrated a reduced environmental impact throughout their life cycle. There are already more than 17,000 EU Ecolabelled products on the market, but there are no references for road products and infrastructures.
- *Environmental Labels Type II*: Environmental labels Type II are self-declared environmental labels (often a single attribute, sometimes a company's own environmental logo). There are already a huge number of self-declared labels of several products on the market, but there are no references for road products and infrastructures.
- *Environmental Labels Type III*: The Environmental Product Declaration (EPD) is a Label Type III and it is a registered trademark. An environmental declaration is defined, in ISO 14025, as quantified environmental data for a product with pre-set categories of parameters based on the ISO 14040 series of standards, but not excluding additional environmental information. Environmental Product Declarations (EPD) add several new market dimensions to inform about environmental performance of products and services. EPDs are based on principles inherent in the ISO standard for Ecolabel Type III environmental declarations (ISO 14025) giving them a wide-spread international acceptance. The EPD methodology, including a structured and well-defined procedure based on LCA for mapping all relevant environmental aspects of goods and services in a life-cycle perspective.

EPDs have become a trend for being an excellent way to communicate transparently the environmental impacts assessed through LCA. EPDs are also useful not only to obtain environmental information on the products but also as verification on environmental requirements in the tendering documents. They are considered to be a useful tool in green purchasing and

---

3 https://www.environdec.com/

procurement within both the public and business sector. EPDs used for that purpose may include all sorts of information, e.g. content of hazardous substances, information about disassembly, recovery and recycling of used products and waste. As a source of information for these applications, a number of EPDs can be outlined for people involved in green purchasing and procurement, and assessment of suppliers. EPDs might be used as a strategic tool for different types of in-company environmental work.

ACCIONA has obtained six EPDs, with four of them dedicated to construction self-performed products and processes, providing added value to the company given its concern and excellence both technically and environmentally. Some EPDs are mentioned below:

2.1 *Arroyo valchano railway bridge, 2013*[4]

The Environmental Product Declaration of "Arroyo Valchano" railway bridge (Madrid-Galicia North-Northwest high-speed line. stretch: Zamora-Lubián; sub-stretch: Otero de Bodas-Cernadilla) was the first EPD of an infrastructure project presented by a private company in the world. ACCIONA proved its commitment to sustainability by becoming the first construction company worldwide to register this EPD. It enabled the company to identify the hot spots or points to be improved in order to achieve a better environmental performance of the company. ACCIONA also considers EPDs as an essential contribution towards gaining benefits in tendering processes as well as in green public procurement.

This EPD covers bridge "structure" only. Within the International EPD system based on ISO standard 14025, this EPD was drawn up in accordance with the Product Category Rules (PCR) 2013:23 *bridges and elevated highways* and with CEN standard 15804 (Sustainability of construction works). The aim of this EPD was providing experts and scientists (in the construction and infrastructure sectors) objective and reliable information on the environmental impact of constructing a railway bridge. The railway bridge over the Arroyo de Valchano was built to save the step of the way on this river. The railway bridge is a double track railway and it was built only for transportation of passengers. The environmental performance section of the declaration is based on a life cycle assessment (LCA) carried out by ACCIONA in 2013. A full set of impact categories were calculated and the results are presented in Figures 9 and 10. The results are given per m of railway bridge and include construction, maintenance and operation per year during 60 years.

The different resource uses have been broken down into single material uses. It must be noted that natural aggregates and water are the most relevant nonrenewable and renewable materials, respectively. In Figure 11, the different impact categories have been split into overview activity areas (construction stages) in order to show the main sources of the emissions and impacts. According to the dominance analysis, it can be highlighted that execution of foundation and shaft of piers and execution of abutments are the most pollutant phases during the construction of "Arroyo Valchano railway bridge". Furthermore, construction phase was separated in upstream and core contribution, detailing each contribution to the global environmental performance. As shown from the figures, the production of different materials (upstream) for the bridge is the most important factor for the overall environmental performance. The construction work and transports (core) play some role while operation activities are small.

2.2 *Road construction (Cieza-Fuente de la Higuera Road, Valencia, Spain and N-340 Road, Elche, Alicante, Spain). Sector E-40, 201.*[5]

The main goals of the road were improving the road network taking advantage of existing infrastructure and improving access to the industrial park. The section developed for the main

---

4 http://www.environdec.com/en/Detail/?Epd=9342
5 http://www.environdec.com/en/Detail/?Epd=9697

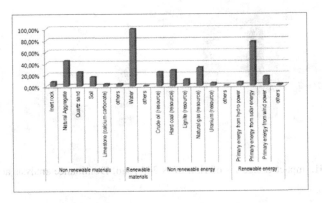

Figure 9. Specification of resources (materials and energy) for 1m of bridge.

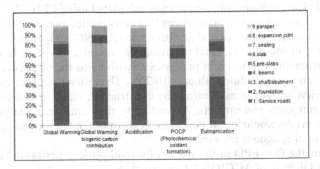

Figure 10. CML impact categories for construction of 1m bridge. Dominance analysis according to different construction stages.

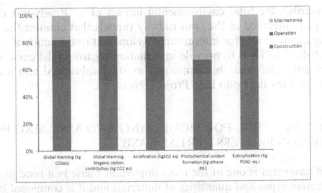

Figure 11. CML impact categories of 1km of N-340 road. Dominance analysis according to different life cycle stages.

road has four lanes, divided into two roadways. Total length of the axes was 943 m and 625 m on service roads. The goal of this analysis was to evaluate the environmental impact of 1Km of road. It was assumed that the lifespan of road was 20 years and maintenance of the top layer of the road was required at half of the lifespan. It must be highlighted that raw materials, transport, energy consumption during installation and construction processes were included in LCA calculations, following the PCR 213:20 (Highways (except elevated highways), streets, roads) recommendation of included upstream and core modules. Due to lack of credible and robust data, CO_2 sequestration due to concrete carbonatation were not taken

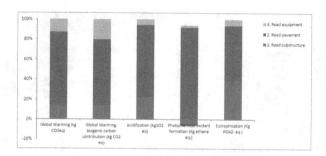

Figure 12. CML impact categories of 1km of N-340 road. Dominance analysis according to construction stage.

into account. CO_2 emissions occurred during land deforestation needed for road construction were accounted. The environmental impact was calculated using CML 2001 methodology.

According to figures, it can be pointed out that construction processes have the highest contribution to the global environmental performance (50 - 60%). Operation activities represent around 10 – 30 % of total impacts and maintenance around 15% - 30%. In general terms, as it is shown, upstream and downstream processes contribute to the total environmental impacts with 80% - 90% and core module with only 10-20%. The different impact categories have been split into overview activity areas during the construction stage in order to show the main sources of the emissions and impacts. According to the dominance analysis, it can be highlighted that the production of road pavement is the most pollutant phase, contributing with 60 -70% of the total emissions released.

This EPD was the first EPD in the world of a road project, performed by the Technology and Innovation Division of ACCIONA Construction.

2.3 Follo line project, 2017[6]

These EPDs describes the total environmental impact of a "Ready-mix concrete" and "bi-component mortar" included in the twin railway tunnels that connect the cities of Oslo and Ski, establishing a central axis for interurban development to the south of the Norwegian capital. The goal of these EPDs is to provide information regarding the environmental impact of the "Ready-mix concrete" and "bi-component mortar" fabricated and used by ACCIONA Construction in 2017 for the Follo Line Project (Norway).

3 USE OF LCA AS A TOOL FOR MODELLING AND ASSESSING PAVEMENT CONSTRUCTION PROCESS IN REAL CASE

Construction of pavement is one of the most important phase in a road project. This process needs many different types and quantities of materials and it is composed by different stages. ACCIONA has started a research that aims to evaluate environmental impacts of pavement construction of a real site in Norway (E6 Project: Trondheim-Vaernes). It will be created a model that includes all processes carried out for construction of the pavement based on real data. The aim is not only to determine such critical processes and to integrate different alternatives for reducing environmental impacts but also to generate a complete model that can be used in any pavement construction.

---

6 https://www.environdec.com/Detail/?Epd=12614
  https://www.environdec.com/Detail/?Epd=12615

4 CONCLUSIONS

LCA can be used in construction as a powerful tool for the evaluation of environmental impacts in construction products and processes. The results obtained show the importance of considering the emissions and impacts produced in each of the stages of the life cycle of the structures (from the manufacturing of the raw materials till the end of life of the waste), in order to know which of these stages influences more on the complete life cycle of the system and to be able to make an effective decision making.

LCA results could be also used for improving the environmental behavior of the construction processes, identifying the hot spots where the environmental impacts are higher, and improving the energy consumption with the optimization of the production processes (thus, reducing the whole environmental impact).

It is essential to mention the importance of LCA methodology as a key tool for helping to make right decisions related to the development of any new technology, process or product considering environmental performances. ACCIONA is aware of the environmental impact of its construction methods and sites and uses LCA methodology for assessing, measuring and reducing them. This paper shows different examples of using LCA method to evaluate which stage of the life cycle of a product/process is susceptible to have changes or improvements for increasing the sustainability of the overall process, hence achieving the optimum technology from both technical and environmental point of view.

It is important to evaluate new processes and materials for construction processes not only from a technical point of view but also for their environmental behavior. A complete analysis allows companies to know and determining the most critical processes or/and materials that could be modified for improving the sustainability of construction processes. ACCIONA works in identifying such critical processes in different construction projects that need to be modified or improved and applies LCA methodology for determining the best solution integrating technical and environmental behavior. Results obtained allow the company to take decisions based on a real and validated information. All examples explained in this paper were carried out with the aim to improve construction processes where the company operates.

REFERENCES

Ding, Grace K.C. 2008. Sustainable construction- The role of environmental assessment tools. *Journal of environmental management*, 86, 3, 451–464.

EN 15804:2012 Sustainability of construction works. Environmental product declaration. *Core rules of the product category of construction products.*

EPA. 2006. Life cycle assessment. *Principles and practice. EPA/600/R-06/060.*

Erlandsson, M. & Borg, M. (2003). Generic LCA-methodology applicable for buildings, constructions and operation services—today practice and development needs. *Building and Environment*, 38,7, 919–938.

Fawer Matthias; 1997; *Life Cycle Inventories for the Production of Sodium Silicates.*

GaBi professional. GaBi Software-System and Database for Life Cycle Engineering. Thinkstep AG.

García, L. (2011). Life Cycle Assessment of railway bridges. Developing a LCA tool for evaluating Railway Bridges. *Master Thesis 323. Structural Design and Bridges.* ISSN 1103.4297. ISRN KTH/BKN/EX-323-SE.

ILCD: International Reference Life Cycle Data System.

EPA Publication 739 – 'Guidelines for the preparation of Environment Improvement Plans'.

Buyles, M., Braet, J. & Audenaert, A. 2013. Life cycle assessment in the construction sector: A review. *In Renewable and Sustainable Energy Reviews* 26. 379–388.

ISO 14020:2000 Environmental labels and declarations. General principles.

ISO 14025:2006 Environmental labels and declarations-Type III Environmental Declarations-Principles and procedures.

ISO 14040:2006 Environmental management-Life Cycle Assessment-Principles and framework.

ISO 14044:2006 Environmental management-Life Cycle Assessment-Requirements and guidelines.

ISO 21930:2007 Sustainability in building construction. Environmental declaration of building products.

Norcem AS; 16.10.2013; EPD® CEM II, Anlegg FA og Standard FA Sement (ISO 14025; ISO 21930; EN15804).

LCA of construction and demolition waste recycling: Case study of production phase

T. Desbois
Cerema, Direction Ouest, Laboratoire de Saint-Brieuc, Saint-Brieuc, France

O. Yazoghli-Marzouk
Cerema, Direction Centre Est, Laboratoire Autun, France

A. Feraille
Université Paris-Est, Laboratoire Navier CNRS, Ecole des Ponts ParisTech, IFSTTAR, Marne-la-Vallée, France

ABSTRACT: This paper presents the study of the environmental assessment of construction and demolition waste (C&DW) production using life cycle assessment (LCA). In the LCA framework, the phases of life cycle corresponding to primary production is in general not considered. Therefore, the process of C&DW primary production was investigated in this study. The data were collected from a demolition site of a building to make LCIs for this case study. A "data collection form" was defined as regards energy and water consumption, CO_2 emissions and wastes production, the main relevant flux for building owners. Then environmental impact indicators were determined using mid-point method. The results show that the major contributor to all the impacts categories calculated is due to "Materials from demolition and asbestos removal", followed by "building site machineries" and "incoming". Another important result is that the transportation impacts of materials from demolition are small compared to their treatment. At least a sensitivity study was carried out in order to determine the environmental gain that would represent the halving of unsorted rubbles, by making them recyclable (for example by better sorting). The results lead to a reduction of 3 to 13% of the total impact.

1 INTRODUCTION

Construction industries produce 850 million tonnes of construction and demolition waste (C&DW) per year in Europe (Sàrez & Osmani 2019). These alternative materials can be produced by construction process (CW), demolition process (DW) or both (C&DW) (Bovea & Powell 2016). They can result from building, bridge and road works. This life stage is generally followed by waste processing and stockpiling stages, before being reused or recycled in construction industry and closing the loop (circular economy). Indeed, the last two decades have shown a growing interest in the use of C&DW materials within the area of road construction (Ofrir 2010, Jullien et al. 2019, Silva et al. 2019, Tavira et al. 2018, Xuan et al. 2015, Simon et al. 2006, Eighmy & Chesner 2000). Their recycling allows saving non-renewable natural resources, reducing waste storage or landfilling and promoting local economy. To support the development of recycling this alternative material, laws, directives, guidelines and methods were published (BMUB 2013, Cerema 2011, Cerema 2016) and the opportunity of reuse alternative materials locally have become important.

This paper presents the study of environmental assessment of DW production using life cycle assessment (LCA). Generally, the environmental assessment of recycling waste materials is taken into account only through waste processing (sorting/crushing/grinding). Jullien et al.

(2019) recently studied the phases of stockpiling and use. The authors demonstrate the importance of including such impact to LCA for recycling assessment and propose a model to allocate all the leaching potential to the alternative resource. The phase of life cycle corresponding to primary production of waste from demolition process (Bovea & Powell 2016) has not been yet considered, due to lack of primary data. Thus, in this study experimental data were collected from a demolition site to build life cycle inventories (LCIs) for this case study. A "data collection form" was defined as regards energy and water consumption, CO_2 emissions and waste production, the main relevant flux for building owners. Then environmental impact indicators were determined using mid-point method.

2 IMPACT ASSESSMENT

LCA is the commonly used global assessment methodology. It aims to evaluate a project, a process, a product on its entire lifecycle, from its cradle-raw material extraction to its end-of-life or grave-disposal, through different impact indicators. The multi-criteria aspect of LCA allows avoiding a pollution transfer from one environment to another. LCA allows the comparison of the effects on the environment of various alternative materials resources for construction projects.

LCA is also a diagnostic tool that enables to improve the global environmental profile of any system considered. It can be decomposed into successive steps that depend upon the authors: 1/system description, 2/elementary process, 3/flux calculation, 4/build the appropriate model, 5/analyze and interpret the results and do the report. The system usually gathers the elementary processes that are defined as the smallest unit of the system with inputs and outputs related to the industrial operation of interest (AFNOR ISO 14040 2006).

2.1 Indicators

LCA requires the identification of relevant and reliable indicators for impact assessment applicable to a project, a process, a product, from its cradle-raw material extraction to its end-of-life or grave-disposal (SETAC 1993).

In this paper, LCA method has been undertaken complying with the standards AFNOR ISO 14040 and 14044 (2006) of the International Organization for Standardization (ISO). The indicators in Table 1 are assessed according to the European standard EN15804+A1 (2014) and its French complement NF EN 15804/CN. It is highlighted that the methods shown in Table 1 have been chosen for their international representativeness.

Table 1. Environmental impact indicators considered and calculation methods used.

Environmental indicator	Unit	Calculation method
Consumption of energy resources	MJ	Cumulative Energy Demand (CED)
Wastes	kg	Environmental Design of Industrial Products (EDIP)
Fresh water	m^3	In house method based on EN 15804
Water pollution	m^3	NF EN 15804/CN
Air pollution	m^3	NF EN 15804/CN
Depletion of abiotic resources (fossil)	MJ, net calorific value	Impact-oriented characterisation (CML 2001)
Depletion of abiotic resources (elements)	kg Sb equiv	Impact-oriented characterisation (CML 2001)
Acidification of soil and water	kg SO_2 equiv	Impact-oriented characterisation (CML 2001)
Ozone depletion	kg CFC^{-11} equiv	Impact-oriented characterisation (CML 2001)
Global warming	kg CO_2 equiv	Impact-oriented characterisation (CML 2001)
Eutrophication	kg $(PO_4)^{3-}$ equiv	Impact-oriented characterisation (CML 2001)
Photochemical ozone creation	kg Ethene equiv	Impact-oriented characterisation (CML 2001)

2.2 System boundaries

Performing materials LCA involves underlying objectives leading to compare products (or processes) or providing environmental information (for public and/or private organizations). In the first case, the system includes only processes and life cycle steps that may induce differences between the compared products. In the second case, we need to choose wide systems.

Figure 1 gives the overall life cycle of a demolition waste (DW) from its primary production phase to its recycling or use in road. In our study, we focus on DW primary production life span.

3 MATERIALS AND METHODS

3.1 Demolition waste

In our case study, demolition waste (DW) was produced by a demolition operation of a social building composed by 80 apartments, located in France. The total floor area was 5,487m². The demolition includes asbestos removal and demolition works.

The total quantity of DW produced was 7,115 tons. They were composed of hazardous waste and radioactive wastes from asbestos removal operations and of non-hazardous waste for recycling. Neither components for re-use nor energy for recovery were produced.

According to the recycling application type (example filling, embankment, etc.), the non-hazardous DW can be processed or not.

3.2 Life Cycle Inventory (LCI) for DW primary production

Experimental or primary data were collected from the demolition site (case study) to build LCI for DW primary production. Thus, a "data collection form" was defined to be completed by the company in charge of the demolition. The contract concluded between the public authority and the company drawn-up the control procedure and the financial penalties in case of non-communication of the data.

The data collection form includes several topics to facilitate its filling:
a/ Topic 1: description of the worksite (location, duration, companies involved)
b/ Topic 2: consumptions (electricity, water and fuel)
c/ Topic 3: wastes (designation, quantity, destination i.e. type of treatment, distance of transport from worksite to treatment centers, way of transport (vehicle or other). People can also add comments about this topic),
d/ Topic 4: designation, quantity and supply of materials used for demolition,
e/ Topic 5: building site machineries with definition, use (consumption fuel and immobilization) and transport,
f/ Topic 6: equipment: Description and amortization of equipment,

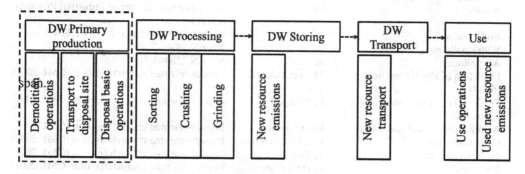

Figure 1. Life cycle phases of DW and system boundary in dashed lines (- - -).

g/ Topic 7: transport of people (distance and mean of transport or type of vehicle),

h/ Topic 8: site facilities (offices, refectory, sanitary, material container) with their use (electricity consumption and amortization), area/mass and transport.

The data collected has been verified and validated by the company as well as by the authors. We therefore consider these data reliable and precise, according to experts. However, they remain specific to these demolition works.

3.3 Life cycle assessment of DW primary production

The LCA of recyclable DW primary production was calculated. Recyclable DW primary production being the primary production of only the rubble able to be used as secondary raw material for the production of recycled aggregate.

The functional unit considered was "producing 1 ton of rubble to be use as secondary raw material for recycled aggregate production". So all the values coming from the worksite were divided by 6,737 (quantity of rubble produced by this demolition case in tons).

The flows taken into account were:

- the production and the transportation of materials used (backfill materials)
- the production and the transportation of consumables
- the transportation of limited reuse supplies
- the transport and the consumption of building site machineries
- the transport of people
- the materials from demolition and their transport
- the transport of site facilities
- the water consumption.

The flows not taken into account were:

- the production of limited reuse supplies
- the production of building site machineries
- the administrative department.

A diagram of the key processes for the demolition of buildings is presented in Figure 2.

In this study, no cut-off has been made. However, consumables were excluded from the modeling, due to the lack of data, especially their mass.

This LCA has been conducted by affecting environmental data at each deconstruction data. The environmental database used was Ecoinvent v.3.2 cut-off. The LCA was modeling by OpenLCA software.

3.4 Main values of the demolition worksite

This worksite included 3 to 4 storeys apartment buildings. Here are some values relating to their demolition (before division due to functional unit):

- Water consumption: $37m^3$
- Electricity consumption: 29,639kWh
- Backfill materials: 2,522 tons
- Transport of backfill materials: 2,368 km
- Asbestos waste: 146 tons
- Materials waste: 6,969 tons
- Fuel consumptions of building site machineries: 9,784 liters

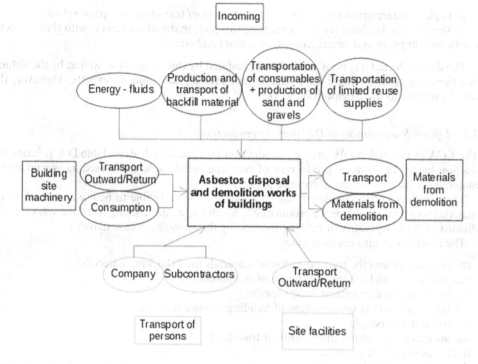

Figure 2. Diagram of the key processes for the demolition of buildings.

4 IMPACT RESULTS AND DISCUSSION

The results of the life cycle assessment are given in Table 2 and Figure 3 shows the distribution of site impacts according to the different items represented in Figure 2.

Figure 3 shows that "Materials from demolition and asbestos removal" is the major contributor to all the impacts categories calculated (more than 60%), followed by "building site machineries" (3 to 36%) and "incoming" (4 to 12%).

Table 2. Results of life cycle assessment.

Impact category	Reference unit	Result
Acidification of soil and water	kg SO_2 eq.	0.134
Ozone depletion	kg CFC^{-11} eq.	$4.68\ 10^{-6}$
Eutrophication	kg PO_4^{2-} eq.	$2.25\ 10^{-2}$
Photochemical ozone creation	kg ethylene eq.	$7.38\ 10^{-3}$
Air pollution	m^3	112.60
Water pollution	m^3	87,907.81
Global warming	kg CO_2 eq.	22.05
Depletion of abiotic resources (fossil)	MJ, net calorific value	387.68
Depletion of abiotic resources (elements)	kg Sb eq.	$4.09\ 10^{-9}$
Consumption of renewable energy resources	MJ	33.95
Consumption of non-renewable energy resources	MJ	447.45
Radioactive waste	kg	$2.91\ 10^{-3}$
Hazardous waste	kg	26.09
Non-hazardous waste	kg	1,032.71
Fresh water	m^3	$6.22\ 10^{-2}$

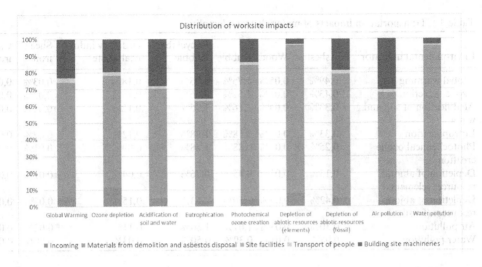

Figure 3. Distribution of worksite impacts (items are explained in Figure 2).

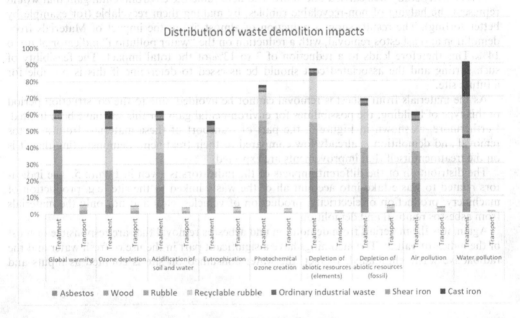

Figure 4. Distribution of "materials from demolition" impacts compared to overall impacts of demolition.

The detail of the distribution of the impacts of Materials from demolition and asbestos removal is given in Figure 4. This latter shows that the impacts come precisely from materials from asbestos removal and unsorted rubbles. The transportation impacts of materials from demolition are small compared to their treatment. Table 3 presents the share of transport of the various materials resulting from demolition on the overall impacts. The most important impact is below 3% on the overall impacts and attributed to recyclable rubble.

For asbestos materials, two treatment centers are involved: located at 670 km for the vitrification of materials and the landfill site allowing the bruying of contaminated tarmac is at 8km. For the other materials, the treatment centers are located between 8 and 19 km. So, the total distance covered for the 7,115 tons of materials from demolition is 6,535km.

Table 3. Transportation impacts of materials from demolition.

Environmental indicator	Asbestos	Wood	Rubble	Recyclable Rubble	Ordinary industrial waste	Shear iron	Cast iron
Global warming	0,49%	0,05%	1,75%	2,95%	0,18%	0,03%	0,02%
Ozone depletion	0,43%	0,05%	1,52%	2,56%	0,15%	0,02%	0,02%
Acidification of soil and water	0,33%	0,04%	1,16%	1,95%	0,12%	0,02%	0,02%
Eutrophication	0,33%	0,04%	1,18%	1,98%	0,12%	0,02%	0,02%
Photochemical ozone creation	0,25%	0,03%	0,88%	1,48%	0,09%	0,01%	0,01%
Depletion of abiotic resources (elements)	0,13%	0,01%	0,45%	0,76%	0,04%	0,01%	0,01%
Depletion of abiotic resources (fossil)	0,42%	0,05%	1,50%	2,53%	0,15%	0,02%	0,02%
Air pollution	0,30%	0,03%	1,07%	1,80%	0,11%	0,02%	0,01%
Water pollution	0,08%	0,01%	0,30%	0,50%	0,03%	0,00%	0,00%

A sensitivity study was carried out in order to determine the environmental gain that would represent the halving of non-recyclable rubbles, by making them recyclable (for example by better sorting). The results show a reduction of almost 5% in the impact of Materials from demolition and asbestos removal, with a reduction on the "water pollution" indicator of up to 14%. This therefore leads to a reduction of 3 to 13% of the total impact. The feasibility of such sorting and the associated cost should be assessed to determine if this is possible for a future site.

As the materials from asbestos removal cannot be avoided, due to the construction period of this type of building, the possibilities for environmental gain for this site have been limited. Furthermore, as shown in Figure 5, the part of transport of these materials from asbestos removal and demolition is already low compared to their treatment, demonstrating that it is on the treatment itself that improvements are expected.

The distribution of the different impacts of the indicators is given in Figure 5. The indicators related to waste take into account all of the waste linked to the site (e.g. production of machinery, production of electricity, production of vehicles, etc.) and not only the materials from asbestos removal and demolition.

Again, it is the materials from demolition and asbestos removal that are responsible of most of the indicator values. The incomings have a significant part in the use of fresh water and the radioactive waste eliminated. In fact, the water consumed on site is counted as inputs and

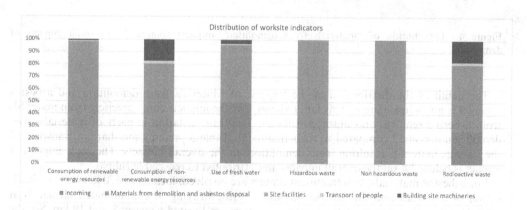

Figure 5. Distribution of the impacts of the indicators.

represents 48% of this indicator. Regarding materials from asbestos removal and demolition, this corresponds to the water necessary for their treatment.

5 CONCLUSIONS

This paper presents the study of the environmental assessment of recyclable DW primary production using life cycle assessment (LCA). Usually, the environmental assessment of recycling waste materials is taken into account only through waste processing (sorting/crushing/grinding). The phase of life cycle corresponding to primary production of recyclable waste from demolition process (Bovea & Powell 2016) has not been yet considered, due to lack of primary data.

A "data collection form" was defined for the collection of the primary or experimental data from the demolition site. Then these data were collected, by the company in charge of the demolition, to build LCI for recyclable DW primary production.

The results show that "Materials from demolition and asbestos removal" is the major contributor to all the impacts categories calculated. Results show that the transportation impacts of materials from demolition are small compared to their treatment. These result should be explored in detail in order to evaluate its sensitivity to the case studied. In an eco-conception approach, as the materials from asbestos removal cannot be avoided, due to the construction period of this type of building, the possibilities for environmental gain can be evaluated only on unsorted rubbles. Thus a sensitivity study was carried out in order to determine the environmental gain that would represent the halving of unsorted rubbles, by making them recyclable. The results show a reduction of 3 to 13% of the total impact. The feasibility of such sorting and the associated cost should be assessed to determine if this is possible for a future center.

Considering that these results were obtained from a unique demolition worksite, it would be important to extend this approach to other demolition worksites in order to succeed in discriminating the values by characteristics of demolished buildings or sizes of demolished buildings.

REFERENCES

AFNOR ISO 14040 2006. Environmental management – Life cycle assessment – Principles and framework, in: I.O.f. Standardization (Ed.):33–46.
AFNOR ISO 14044 2006. Management environnemental – Analyse du Cycle de Vie – Exigences et lignes directrices.
Bovea, M.D. & Powell, J.C. 2016. Developments in life cycle assessment apllied to evaluate the environmental performance of construction and demolition waste. *Waste Management* 50: 151–172.
Cerema (ed) 2011. French Environmental legislation. *Acceptabilité de matériaux alternatifs en technique routière, évaluation environnementale*. France: Cerema, pp.28
Cerema (ed) 2016. *Acceptabilité environnementale de matériaux alternatifs en technique routière - Les matériaux de déconstruction issus du BTP*. France: Collection Référence, pp. 39.
Eighmy, T. & Chesner, W. 2000. User guidelines for by products and secondary use materials in pavement construction. Federal highway administration R&D Center (FHWA-RD-148). The document is available on http://www.rmrc.unh.edu.
EN 15804:2012+A1:2013. Sustainability of construction works. Environmental product declarations. Core rules for the product category of construction products.
Federal Ministry of the Environment (BMUB) 2013. Arbeitsentwurf zur Verordnung zur Festlegung von Anforderungen für das Einbringen oder das Einleiten von Stoffen in das Grundwasser, an den Einbau von Ersatzstoffen und für die Verwendung von Boden und bodenähnlichem Material (Ordinance on Groundwater Protection, Mineral Waste Utilization and Federal Soil Protection and Contaminated Sites, draft mutual release).
Jullien, A., Proust, C. & Yazoghli-Marzouk, O. 2019. LCA of alternative materials – Assessment of ecotoxicity and toxicity for road case studies. *Construction and Building Materials* 227: 116737. https://doi.org/10.1016/j.conbuildmat.2019.116737

Ofrir 2010. Observatoire Français du Recyclage dans les Infrastructures Routières, http://ofrir2.ifsttar.fr, [accessed 6-08-2018].

Tavira, J., Jiménez, J.R., Ayuso, J., Sierra, M. J. & Ledesma, E. F. 2018. Functional and structural parameters of a paved road section constructed with mixed recycled aggregates from non-selected construction and demolition waste with excavation soil. *Construction and Building Materials* 164:57–69.

Sàrez, P. V. & Osmani, M. 2019. A dignosis of construction and demolition waste generation and recovery practice in the European Union. *Journal of Cleaner Production* 241:1118400. https://doi.org/10.1016/j.jclepro.2019.118400

SETAC 1993. *Guidelines for Life-Cycle Assessment: a "code of practice"*, Ed. SETAC Foundation for Environmental Education, Florida.

Silva, R.V., De Brito, J. & Dhir, R. K. 2019. Use of recycled aggregates arising from construction and demolition waste in new construction applications. *Journal of Cleaner Production* 236: 117629.

Simon, M.J., Chesner, W., Eighmy, T. & Jongedyk, H. 2006. National Research Projects on Recycling in Highway Construction. Vol. 64. N°1, United States department of transportation. Federal highway administration.

Pavement, Roadway, and Bridge Life Cycle Assessment 2020 – Harvey et al (eds)
© 2020 Taylor & Francis Group, London, ISBN 978-0-367-55166-7

Life cycle assessment of pervious concrete pavements reinforced by recycled carbon fiber composite elements

M. Rangelov
National Research Council (NRC), Washington, USA

S. Nassiri & K. Englund
Washington State University, Pullman, USA

ABSTRACT: Mass production of fiber-reinforced polymer composite materials in the last decades resulted in increased amounts of accompanying excess materials from manufacturing lines and end-of-life products. The negative environmental impact of disposal in landfills or incineration necessitates the development of scalable reuse applications for these highly durable materials. In a series of studies, the authors investigated the reuse of excess carbon fiber reinforced polymers (CFRP) from the aerospace industry in pervious concrete (PC). By a mechanical refining, CFRP scrap pieces were processed into fibrous elements and used as a discrete reinforcement in PC. Results indicated that reinforced PC (rPC) demonstrated comparable or enhanced hydraulic, mechanical, and durability performance compared to a plain PC. This study extends previous efforts by evaluating the environmental footprint of rPC versus plain PC pavement using life cycle assessment (LCA). The chosen scope of LCA was cradle-to-site, assuming that the construction, use, maintenance, and end-of-life phases would produce comparable impacts for both options. PC pavement thickness was determined per a mechanistic design procedure, using mechanical properties established experimentally in previous studies. The results of LCA indicate that rPC sections have comparable or lower environmental impacts relative to the corresponding PC slabs. The improvement of environmental profile of rPC can be achieved primarily through slab thickness reduction, which is the effect of enhanced mechanical properties due to CFRP reinforcement addition. Increasing CFRP processing efficiency can also result in environmental benefits, however to a lesser extent than the slab thickness reduction.

1 INTRODUCTION

In recent years, the production and use of fiber-reinforced polymer materials have markedly increased in automotive, sport, energy, and aerospace industry. Demand for carbon fiber composites, for instance, has experienced a steady annual increase of approximately 10 percent from 2009 to 2013 (Witten *et al.*, 2014). As a result, the amount of excess material on manufacturing lines is also increasing (Ye *et al.*, 2013). As environmental regulations become restrictive and increased costs of landfilling, development of feasible and scalable recycling and reuse methods of fiber-reinforced polymer composites is becoming an urgent concern (Oliveux *et al.*, 2015). (Li *et al.*, 2016) investigated different end of life (EOL) scenarios for the treatment of carbon fiber reinforced polymer (CFRP) composites by evaluating costs and carbon dioxide emissions. Results of their study showed that the incineration has detrimental effects in terms of emissions and that landfilling may become economically unfeasible because of increased landfilling taxes (Li *et al.*, 2016). Accordingly, mechanical recycling was found to be beneficial from environmental perspective; however it was emphasized that 'higher value uses' for the recycled products should be targeted to achieve economic feasibility (Li, *et al.*, 2016).

As fiber-reinforced polymer materials are durable and characterized by high mechanical properties, they have the potential for use after recycling as discrete reinforcement in pavements. In a series of studies, authors have investigated potential utilization of mechanically refined CFRP material from Boeing Company in pervious concrete (PC) (Rangelov et al., 2016; Rodin III et al., 2018; AlShareedah et al., 2019). In the initial phase, CFRP material was downscaled into coarse, fibrous, and powdered fraction, all of which were tested in PC (Rangelov et al., 2016). Results have indicated that the fibrous fraction has the highest potential to simultaneously improve mechanical and durability characteristics, without compromising hydraulic performance (Rangelov et al., 2016). In the subsequent studies, different dosages of a fibrous fraction of CFRP material was investigated in different PC mixtures (Rodin III et al., 2018; AlShareedah et al., 2019). Since PC performs a dual function of water infiltration and load-bearing, the objective of these studies was to identify mixture designs and CFRP doses that present satisfactory performance in both domains. Therefore, each mixture was tested for physical (porosity), hydraulic (infiltration), mechanical (compressive, tensile, flexural strength), and durability (resistance to impact and abrasion) properties (Rodin III et al., 2018; AlShareedah et al., 2019). Results indicated that reinforced PC mixtures (rPC) demonstrate comparable or improved performance relative to the corresponding plain PC mixture (Rodin III et al., 2018; AlShareedah et al., 2019). The studies to date, however, did not investigate environmental performance of rPC compared to PC. Environmental considerations are becoming increasingly important part of infrastructure planning and decision-making (Antunes et al., 2018). Therefore, environmental performance evaluation can meaningfully supplement the current research results and inform PC mixture design selection.

Environmental performance of pavements can be quantified using life cycle assessment (LCA) methodology. For conventional pavements, LCA framework has been established (Harvey et al., 2016) and multiple research studies were completed (Santero et al., 2010; Azar-iJafari et al., 2016). Conversely, a number of LCA studies on permeable pavements is limited (Antunes et al. 2018). (Wang et al., 2010) developed a framework for life cycle cost analysis (LCCA) and outlined some basic considerations for LCA of permeable pavements. (Wang et al., 2018) demonstrated that permeable asphalt pavement can provide savings in terms of energy consumption, GHG, and air pollutants, compared to conventional dense-graded asphalt pavement. Maiolo et al. (2017) developed a sustainability index that incorporates hydraulic and environmental performance for evaluation of permeable pavements and green roofs. LCA studies on PC were limited to the assessment of greenhouse gases (GHG) (Yap et al., 2018), or GHG and energy consumption (El-Hassan and Kianmehr, 2017). Conclusively, a comprehensive analysis of PC pavements from the life cycle perspective is much needed (Antunes et al., 2018).

2 OBJECTIVE

The objective of this study is to extend previous efforts in rPC research by evaluating the environmental performance of rPC versus plain PC pavement using LCA methodology through various implementation scenarios.

3 METHODOLOGY

3.1 *Mixtures*

Two groups of PC mixtures investigated in previous experimental work, with characteristics summarized in Table 1 were included in the analysis. The first group was prepared with crushed basalt aggregate, typical for Eastern Washington (mixtures denoted as Basalt_PC and Basalt_rPC hereafter). The second group was batched with natural pea gravel aggregate, typical for Western Washington (mixtures denoted as Pea_PC and Pea_rPC hereafter). Details on experimental procedures and results can be found in reference (Rodin III et al., 2018) for

Table 1. Summary of mixture designs, physical, hydraulic and mechanical properties based on previous experimental work (Rodin III et al., 2018; AlShareedah et al., 2019).

Material constituent [kg/m3]	Basalt_PC	Basalt_rPC	Pea_PC	Pea_rPC
Aggregate	1376		1640	
Cement	413		285	
Water	112		98.5	
CFRP	0	46.8	0	4.15
Admixtures (set retarder and hydration stabilizer)	2.86×10^{-3}		2.34×10^{-3}	
28-day modulus of rupture (MR) [MPa]	2.2	3	2.26	2.4
Infiltration [cm/h]	~3000		~1100	
Porosity [unitless]	~0.25		~0.19	

Basalt and reference (AlShareedah et al., 2019) for Pea mixtures. It is noteworthy that in the previous experimental work, multiple rPC mixtures were tested and compared against the corresponding PC mixture. In this study, one rPC mixture which maximized mechanical performance while keeping the same hydraulic performance compared to the PC mixture was selected from each group for LCA. As seen in Table 1, the two mixture groups differ markedly in terms of porosity and infiltration and this difference can mainly be attributed to different aggregates. Crushed basalt is an angular aggregate that provides for relatively high-porosity PC (~0.25), as well as relatively high CFRP dosages (46.8 kg/m^3 or 3 percent by volume) and corresponding improvements in mechanical properties (Rodin III et al., 2018). In the case of pea gravel, aggregate grains are rounded, particle packing is facilitated and the porosity of PC is lower than that of basalt PC mixtures (~0.19) (AlShareedah et al., 2019). Accordingly, the amount of CFRP that can be incorporated into the mixture without compromising infiltration and mechanical properties is relatively low (4.15 kg/m^3 or 0.27% by volume) (AlShareedah et al., 2019). Therefore, considering different performance of the two mixture groups, in this study, comparisons were made solely among PC with rPC prepared with the same aggregate.

3.2 Scenario analysis

To make a meaningful comparison between PC and rPC mixtures, implementation scenarios were developed and presented in Table 2. First, two scenarios, correspondent to PC pavement in Eastern and Western Washington (Basalt vs. Pea mixtures) were differentiated in Table 2. In all cases, implementation of PC in parking lots was assumed and pavement thickness was calculated using the mechanistic design procedure for PC thickness design developed by AlShareedah and Nassiri (2019). Inputs of this model include modulus of rupture (MR) of each mixture (previously specified in Table 1), composite modulus of subgrade reaction (k-value), and average

Table 2. PC pavement thicknesses calculated based on the mechanistic procedure and different traffic loading levels.

	Scenarios			
	Eastern Washington scenario		Western Washington scenario	
PC thickness based on the traffic loading	Basalt_PC	Basalt_rPC	Pea_PC	Pea_rPC
Thickness [mm] ADTT=25, category B	203	165	191	178
Thickness [mm] ADTT=100, category C	229	191	216	203
Thickness [mm] ADTT=700, category D	216	191	216	203

daily truck traffic (ADTT). For both implementation scenarios, three traffic loading types were investigated and k-value was assumed as 100 psi/in. Note that thicknesses pertinent to category D are equal or lower than that for category C despite the higher ADTT values because of the difference in the truck types. Category C presents the truck with higher weights which results in higher design thickness of the slab (AlShareedah and Nassiri, 2019).

4 LCA METHODOLOGY

4.1 *Goal and scope*

The goal of this study is to compare the environmental profile of PC and rPC pavements. Elements of the LCA scope are functional unit, system boundary, methodological choices, and details of the analysis (data sources and quality) (Simonen, 2014). In this study, functional unit is PC (or rPC) pavement section with dimensions 3.66 m by 4.57 m (12 by 15 feet) (AlShareedah and Nassiri, 2019), with thickness and ADTT load as specified in Table 2 (several use cases are analyzed correspondent to rows of Table 2) and 12-year design life (Wang et al., 2018). It was assumed that the pavement will not need reconstruction during 12-year service life.

Analyzed product system for PC and rPC pavements are shown in Figure 1. As seen in Figure 1, the primary focus of the analysis was the cradle-to-site portion of the life cycle, which includes materials production, transportation to ready-mix plant, mixing, and transportation to the site (phases A1-4 in Figure 1). The impacts associated with the remaining life cycle phases were assumed as equivalent between PC and rPC pavements, and, since the study is comparative, were not included in the analysis (Harvey et al., 2016). This assumption is elaborated for each phase as follows.

In terms of construction (phase A5), the results of previous research work indicate that the rPC mixtures selected for this study present the workability comparable to the corresponding PC mixtures (Rodin III et al., 2018; AlShareedah et al., 2019). It is therefore assumed that the

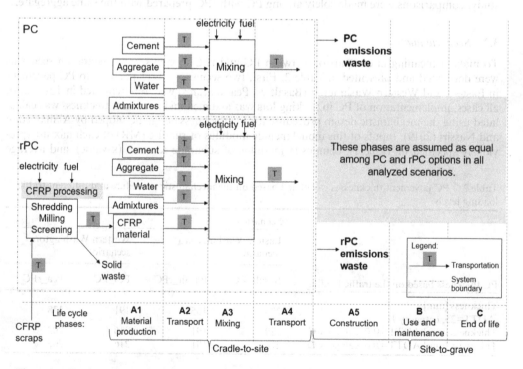

Figure 1. Product systems for PC and rPC pavements.

434

placement of both mixtures necessitates comparable compaction efforts. Additionally, chosen rPC mixtures also present comparable infiltration to their PC counterparts. Hence, it is assumed that PC and rPC pavements require the base reservoir of the same thickness and that the base layer can be excluded from the analysis. The most important maintenance procedure for PC is vacuuming to remove fines and debris that causes clogging. Because of comparable infiltration rates among PC and rPC, it is expected that the rates of clogging and maintenance schedule will be the same. Previous research studies have shown that introduction of CFRP into the previous concrete has no detrimental effect on the quality of the filtered stormwater runoff (Englund et al., 2016). Previous experimental results indicated that CFRP addition can improve abrasion resistance, which can potentially extend the service life of PC (Rodin III et al., 2018; AlShareedah et al., 2019). However, due to a lack of reliable models to correlate abrasion resistance to the service life of PC, that effect is disregarded in this study. Other aspects of use phase (albedo, carbonation, pavement-vehicle interaction) are beyond the scope of the analysis presented here. EOL phase for PC is associated with relatively high uncertainty. It is, therefore, assumed that both options would have comparable impacts associated to demolition and disposal, and EOL phase was disregarded. CFRP scraps are considered as a waste product with no economic value or feasible reuse option [cutoff allocation approach (Harvey et al., 2016)]. Therefore, only the processing and transport of CFRP scraps to obtain CFRP fibrous material was considered in the product system.

Modeling was performed using OpenLCA software, using TRACI 2.1. impact assessment methodology (Bare, 2011). Five categories of environmental impacts were considered: acidification potential (AP), eutrophication potential (EP), global warming potential (GWP), ozone depletion potential (ODP), and photochemical ozone creation potential (POCP).

4.2 Data sources

Data sources used in LCA are specified in Table 3. All transportation distances are estimated based on the correspondence with ready mix plants from Eastern and Western Washington and shown in Table 3.

Environmental emissions associated with CFRP material processing are estimated using the energy consumption. The production process consists of three steps: cutting of large scraps into 25-40 cm long pieces, shredding, and hammer milling through 25.4-mm screen. The hammer-milled CFRP consists of different particle sizes and shapes, which are further separated into finer fractions: Large (material passing 3.36-mm mesh and retained on 2-mm mesh), Medium (material passing 2-mm mesh and retained on 0.707-mm mesh), and Small (material passing 0.707-mm

Table 3. Data sources used in the study.

Life cycle phase	Flow	Emissions data	Transportation distance [km]	
			Eastern Washington scenario	Western Washington scenario
A1 & A2	Aggregate	(Marceau et al., 2007)*	3.2	
	Cement	(Marceau et al., 2007)	120	60
	Water	(NREL, 2012)	Local	
	CFRP material	Electricity (NREL, 2012)	500	30
	Admixtures	EFCA (2015)	500	60
A3	Mixing	(Marceau et al., 2007)	Not applicable	
A4	Transportation	(NREL, 2012)	30	

* Manufactured aggregate and natural aggregate from (Marceau et al., 2007) was used for Eastern and Western Washington scenario, respectively.

mesh). The Medium fraction is used in PC. The work is being done on researching feasible ways of repurposing Large, Small CFRP fraction, and for the purpose of this study, these materials are considered as solid waste. The outlined production process yields 18.3 percent Medium size CFRP. It is noteworthy that CFRP elements are not fibers, but rather bundles of carbon fibers bonded by the initial thermoset resin used in the original composite. To estimate the impacts of CFRP processing, energy consumption reported by (Shuaib and Mativenga, 2016) was used, since the processing methodology is similar to the one used in this study. Accordingly, it was estimated that total energy consumption needed to produce 1 kg of medium CFRP was 7 MJ and, since the entirety of used equipment was electrical, electricity for Northwestern United States was used from U.S. LCI database (NREL, 2012). It is assumed that the processing facility for CFRP is located in Western Washington (close to the original CFRP production) and transportation distance for CFRP material to the plant is estimated accordingly.

5 RESULTS AND DISCUSSION

5.1 Material constituents

Table 4 shows the impacts per 1 m^3 of all four mixture designs characterized in this study. As expected, rPC mixtures present higher impacts compared to plain counterparts and this difference is due to the effects of CFRP processing. Basalt mixtures present higher environmental impacts compared to the Pea gravel mixtures, which is primarily the effect of their higher cement content. It is noteworthy, however, that the Basalt mixtures analyzed in this study represent the old suite of PC mixtures and that mixture designs with lower cement content (typically 250-270 kg/m^3) are currently more common. Additionally, the direct comparison among mixtures using a declared unit of 1 m^3 undermines different functionality and performance of different mixtures. Therefore analysis of PC sections for cradle-to-site is provided in section 5.2, as a means of more appropriate comparison.

Percentage contribution of the constituents to 1 m^3 of each mixture type is provided in Figure 2. As the results in Figure 2 indicate, the component with the highest contribution is Portland cement in all cases, ranging from 77 to 99 percent for plain mixtures and between 64 to 99 percent for rPC mixtures. Transportation of materials adds up to 20 percent of environmental impacts in Eastern WA scenario and up to 10 percent for Western WA scenario, which is expected since the transportation distances were higher for Eastern Washington, as shown in Table 3. CFRP production is responsible for up to 11 percent of impacts for Basalt rPC mixture and up to 5 percent in Pea rPC mixture. Basalt mixture was prepared with higher CFRP content than the Pea mixture (3 versus 0.27 percent of CFRP by volume), resulting in higher contribution of CFRP. Aggregate constitutes up to 7 percent of the environmental footprint, while chemical admixtures and mixing make relatively small contribution (up to 2.5 combined).

5.2 Cradle-to-site analysis

Figure 3 Shows normalized environmental impacts for PC and rPC sections designed for different loading categories. Normalization was performed using US 2008 [year] impacts

Table 4. Environmental impacts of 1 m^3 of PC or rPC for all four mixture designs.

Environmental impacts	Declared unit: 1 m^3 of PC or rPC			
	PC_basalt	rPC_basalt	PC_pea	rPC_pea
AP [kg SO$_2$ eq.]	2.78E+00	3.22E+00	1.87E+00	1.90E+00
EP [kg N eq.]	8.38E-02	1.00E-01	5.41E-02	5.52E-02
GWP [kg CO$_2$ eq.]	6.46E+02	7.29E+02	4.15E+02	4.22E+02
ODP [kg CFC-11 eq.]	3.98E-06	3.98E-06	2.74E-06	2.74E-06
SCP [kg O$_3$ eq.]	3.95E+01	4.54E+01	2.62E+01	2.66E+01

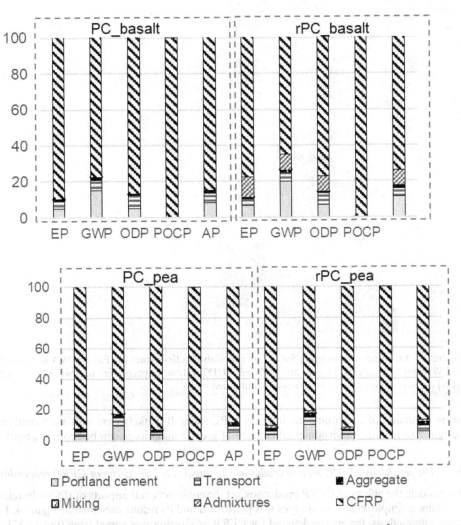

Figure 2.   Percentage contribution of different material components to impacts of 1m³ of PC and rPC.

(Ryberg *et al.*, 2014). As seen in Figure 3, in all cases AP, GWP, and POCP are the impacts with relatively high magnitude, while EP and ODP normalized values are low. Therefore, the reduction AP, GWP, and POCP has the highest potential to effectively improve the overall environmental profile of PC and rPC slabs. The comparison of PC and rPC slabs indicates that the impacts of the two are comparable at all ADTT levels. rPC slabs provide relatively small reduction of environmental impacts in all categories, however, the magnitude of this improvement is limited to approximately 4 percent. Relative to PC, environmental impacts of rPC slabs include additional effects of CFRP production which enables the increase of MR and reduction of pavement thickness. The results provided in Figure 3 indicate that the effects of CFRP processing and thickness reduction compensate one another, providing for a slab with equal or slightly lower environmental impacts. The thickness reductions were 25.4 mm for Basalt mixtures and 12.7 mm for Pea mixtures for all loading categories, as shown earlier in Table 2. Therefore, slabs designed for lower traffic loading have higher relative thickness reduction and experience somewhat higher reduction of environmental impacts due to CFRP usage compared to slabs with higher load (e.g. up to 4 percent reduction for Category B as opposed to up to 2 percent reduction for Category C, in the case of Basalt mixtures). Conclusively, the results indicate that CFRP does not have any detrimental effects and that it may

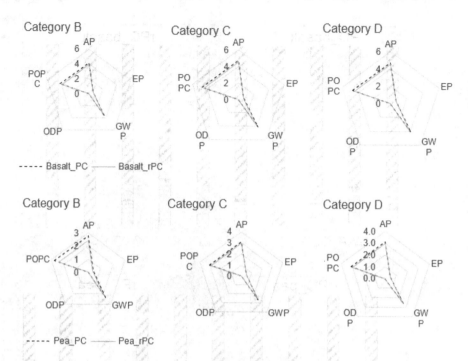

Figure 3. Environmental impacts for PC or rPC slab (cradle-to-site) for Eastern WA scenario (upper) and Western WA scenario (lower) and different ADDT values. Impacts were normalized using US 2008 [year] impacts (Ryberg *et al.*, 2014) and are multiplied by 10^{10}.

cause reduction of environmental impacts in PC slabs. It is likely that with additional mixture design optimization, environmental benefits of greater magnitude can be brought about.

5.3   *The sensitivity of rPC section cradle-to-site impacts to the energy of CFRP processing*

To evaluate the effects of CFRP production on the environmental impacts of rPC slabs relative to PC slabs, a simple sensitivity analysis was performed and its results presented in Figure 4. For the sensitivity analysis, the energy demand for CFRP production was varied from 0 to 12 MJ per kg

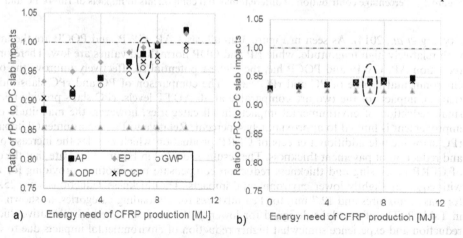

Figure 4.   The sensitivity of environmental impacts of rPC slab relative to that of PC slab as a function of CFRP production energy: a) Basalt mixture, ADTT=25 and b) Pea mixture, ADTT=25.

of Medium size CFRP fibrous material. Figure 4 shows the Category B traffic loading; comparable graphs were developed for other traffic loads but were not shown here due to limited space and resulting trends very similar to that in Figure 4. A hypothetical scenario with energy consumption set to 0 MJ represents the reduction of slab thickness with no additional impacts of CFRP production. This scenario is not realistic; however, it provides an indication of the maximum extent of impacts reduction if the processing energy is completely reduced. Data points confined with the dashed ellipse correspond to the processing energy of 7 MJ, which is used in the remainder of the study.

As expected, results in Figure 4 indicate that the reduction in CFRP processing energy results in the lowering of environmental impacts. For Basalt mixtures, this effect is more prominent than for Pea gravel mixtures because of their higher CFRP content. However, the magnitude of reduction is limited to up to 89 and 93 percent for the Basalt and Pea gravel mixtures, respectively. Accordingly, the improvements in CFRP processing may produce environmental savings of limited extent. Further reduction in design thickness would likely be the most efficient way to reduce footprint of rPC slabs. Figure 4 a) also shows 'a breakeven point' between PC and rPC slabs, which is positioned between 8 and 12 MJ, depending on the environmental impact of interest. For Pea gravel mixtures [Figure 4 b)], 'a breakeven point' is beyond the analyzed range of energy consumption since Pea gravel mixture exhibits lower sensitivity to impacts of CFRP and is batched with relatively low CFRP dosage. For ODP impact category, changes in CFRP processing do not yield impact reduction, since 99 percent of ODP is caused by the production of Portland cement. However, as seen in Figure 3, low normalized values of ODP indicate that this impact has low significance in the overall environmental footprint of PC and rPC slabs.

Altogether, considering analyzed environmental impact categories, the most effective way to reduce the environmental impacts of rPC is through slab thickness reduction, which is achieved through increased MR, if mechanistic design procedure is used. Additionally, if increase in MR and thickness reduction can be achieved with lower CFRP content, environmental impacts of rPC can be less contingent on CFRP processing energy.

5.4 *Study limitations and future work*

This study was limited to 5 environmental impact categories (AP, EP, GWP, ODP, and POCP). Since the key impetus for this study is the improvement of the waste management practices of CFRP scrap material, the further analysis of waste production, reuse, and EOL considerations would inform this study in a meaningful way. As stated in Section 4.2, CFRP processing yields 18 percent medium CFRP fraction, which is used in PC. At the moment, scalable CFRP processing methodologies for all material fractions are being developed. These new procedures will provide more reliable data source for LCA and waste inventory. Moreover, direct estimates of production energy instead of using a literature value would also inform and strengthen the conclusions of LCA. Authors recognize that the results of LCA are not deterministic but rather stochastic, and hence a more detailed sensitivity and reliability analysis will be included in the future work. Literature indicates that CFRP machine processing is associated with dust and particulate matter emissions (Miller, 2014), which may be assessed in the future work through Respiratory Effects impact category. As PC's key functionality is facilitating storm water filtration, assessment of ecotoxicity and eutrophication based on impact assessment models with high geographical resolution can be additional meaningful future development. Lastly, inclusion of the use phase into the scope of LCA may be beneficial, since rPC presents improved durability and abrasion resistance (Rangelov *et al.*, 2016; Rodin III *et al.*, 2018), which were not accounted for in this study.

6 CONCLUSIONS

This study expands on previous research efforts in the investigation of use of recycled carbon fiber reinforced polymer (CFRP) material from aerospace industry in pervious concrete (PC). PC mixtures reinforced with CFRP material (rPC) demonstrated comparable hydraulic performance and improved mechanical and durability properties relative to PC. The main objective of this study was the assessment and comparison of environmental impacts of PC and rPC pavement sections for cradle-to-site scope of analysis using life cycle assessment (LCA) methodology. Results demonstrate that rPC slabs present comparable or lower environmental impacts compared to corresponding PC slabs, up to 4 percent. Improvements are slightly higher for low traffic loading. Sensitivity analysis indicated that the thickness reduction due to improved mechanical properties of rPC is the most effective way to reduce impacts of rPC sections, while the reduction of CFRP processing energy has the limited repercussions on environmental impacts. Results of this study can be used to guide the future development of rPC mixtures in a manner that leverages environmental benefits and provides for satisfactory engineering performance.

ACKNOWLEDGEMENT

This research was performed while the author, M. Rangelov, held an NRC Research Associateship award at Federal Highway Administration (FHWA).

REFERENCES

AlShareedah et al. 2019. Field performance evaluation of pervious concrete pavement reinforced with novel discrete reinforcement. *Case Studies in Construction Materials*. Elsevier, 10, p. e00231.

AlShareedah, O. and Nassiri, S. 2019. Methodology for mechanistic design of pervious concrete pavements. *Journal of Transportation Engineering, Part B: Pavements*. American Society of Civil Engineers, 145(2), p. 4019012.

Antunes et al. 2018. Permeable pavements life cycle assessment: a literature review. *Water*. Multidisciplinary Digital Publishing Institute, 10(11), p. 1575.

AzariJafari et al. 2016. Life cycle assessment of pavements: reviewing research challenges and opportunities. *Journal of Cleaner Production*. Elsevier, 112, pp. 2187–2197.

Bare, J. 2011. TRACI 2.0: the tool for the reduction and assessment of chemical and other environmental impacts 2.0. *Clean Technologies and Environmental Policy*. 13(5), pp. 687–696.

El-Hassan, H. and Kianmehr, P. 2017. Sustainability assessment and physical characterization of pervious concrete pavement made with GGBS. *MATEC Web of Conferences*. EDP Sciences, p. 7001.

Harvey et al. 2016. Pavement Life Cycle Assessment Framework (FHWA-HIF-16-014). *FHWA-HIF-16-014*, pp. 1–246. doi: 10.1007/978-3-662-44719-2_1.

Englund et al. 2016. *Recycled Carbon Fiber Composites (rCFC) for Reinforcing Pervious Pavements*. Submitted to Boeing Company.

Li et al. 2016. Environmental and financial performance of mechanical recycling of carbon fibre reinforced polymers and comparison with conventional disposal routes. Journal of Cleaner Production, 127. pp. 451–460. ISSN 1879-1786'.

Maiolo et al. 2017. Synthetic sustainability index (SSI) based on life cycle assessment approach of low impact development in the Mediterranean area. *Cogent Engineering*. Taylor & Francis, 4(1), p. 1410272.

Marceau et al. (2007) *Life cycle inventory of portland cement concrete*. PCA.

Miller, J. 2014. Investigation of Machinability and Dust Emissions in Edge Trimming of Laminated Carbon Fiber Composites. Dissertation submitted to University of Washington.

National Renewable Energy Laboratory (NREL) 2012. *U.S. Life Cycle Inventory Database*. Available at: https://www.lcacommons.gov/nrel/search (Accessed: 20 September 2008).

Oliveux et al. 2015. Current status of recycling of fibre reinforced polymers. *Progress in Materials Science*. Elsevier, 72, pp. 61–99.

Rangelov, M. et al. 2016. Using carbon fiber composites for reinforcing pervious concrete. *Construction and Building Materials*. Elsevier, 126, pp. 875–885.

Rodin III et al. 2018. Enhancing mechanical properties of pervious concrete using carbon fiber composite reinforcement. *Journal of Materials in Civil Engineering*. ASCE, 30(3), p. 4018012.

Ryberg et al. 2014. Updated US and Canadian normalization factors for TRACI 2.1. *Clean Technologies and Environmental Policy*. Springer, 16(2), pp. 329–339.

Santero et al. 2010. *Life cycle assessment of pavements: a critical review of existing literature and research*. Lawrence Berkeley National Lab, Berkeley, CA, USA.

Shuaib, N. and Mativenga, P. 2016. Effect of process parameters on mechanical recycling of glass fibre thermoset composites. *Procedia CIRP*. Elsevier, 48, pp. 134–139.

Simonen, K. (2014) *Life cycle assessment*. Routledge.

Wang *et al.* (2010) *A Framework for Life-Cycle Cost Analyses and Environmental Life-Cycle Assessments for Fully Permeable Pavements*. Institute of Transportation Studies, UC Davis.

Wang, Y. *et al.* 2018. Initial evaluation methodology and case studies for life cycle impact of permeability of permeable pavements. *International Journal of Transportation Science and Technology*. Elsevier, 7 (3), pp. 169–178.

Witten *et al.* 2014. *Composites Market Report 2014. Market developments, trends, challenges and opportunities-The Global CRP Market*.

Yap *et al.* 2018. Characterization of pervious concrete with blended natural aggregate and recycled concrete aggregates. *Journal of Cleaner Production*. Elsevier, 181, pp. 155–165.

Ye *et al.* 2013. Parameter optimization of the steam thermolysis: a process to recover carbon fibers from polymer-matrix composites. *Waste and Biomass Valorization*. Springer, 4(1), pp. 73–86.

Effect of durability on fiber-reinforced asphalt mixtures sustainability

E. Lizasoain-Arteaga & D. Castro-Fresno
GITECO Research Group, Universidad de Cantabria, Santander, Spain

G.W. Flintsch
Center for Sustainable Transportation Infrastructure, Virginia Polytechnic Institute and State University, Blacksburg, USA

ABSTRACT: Fiber-reinforced asphalt overlays are a promising technology to rehabilitate rigid pavements. Based on laboratory results, they improve reflective cracking performance, thus increasing the service life of the road. However, research is needed to determine if their advantages overcome the drawbacks of having to produce a new additive. In this paper, a Life Cycle Assessment (LCA) and Life Cycle Cost Analysis (LCCA) were performed to determine the minimum service life extension needed for fiber-reinforced overlays to be considered sustainable. According to the results, the economic aspect is more restrictive, and fibers need to increase the service life of the layer by at least 8% to obtain an economic benefit.

1 INTRODUCTION

Asphalt concrete overlays are a common way to rehabilitate rigid pavements in the state of Virginia due to their inexpensive nature compared to most concrete rehabilitation alternatives and reduced impact on traffic (De León Izeppi et al., 2015). In addition to being more cost-effective, this practice reduces noise by improving the skid resistance of the road (Khazanovich et al., n.d.). However, the combination of concrete pavement joints together with the poor condition of the existing layer induces horizontal and vertical movements, which cause distresses that generate an early deterioration of the asphaltic overlay (Bennert, 2010).

Achieving sufficient durability of asphalt mixtures is crucial to constructing sustainable pavements (Santero et al., 2010). Increasing the service life of the road produces environmental, economic, and social benefits since it decreases natural resource depletion, energy consumption, and traffic affection, while increasing road safety at the same time. Therefore, developing innovative solutions that improve the mechanical performance and, consequently, the service life of the asphalt overlays is indispensable.

The addition of fibers into asphalt mixtures to try to increase the durability of roads has been studied for many years, and very promising results have been achieved. Button and Hunter (1984) analyzed the mechanical performance of eight types of synthetic fibers in the lab and field and, after monitoring the mixtures' behavior for 2 years, improvement was seen in the fatigue and cracking responses. However, they also concluded that more bitumen and compaction energy were necessary. Tapkın (2008) carried out similar research to study the effect of polypropylene fibers in the behavior of asphalt mixtures. Results showed an increase in the Marshall stability, rutting resistance, and fatigue life, and a decrease in reflection cracking. Alrajhi (2012) focused on testing the effect of combining polypropylene and aramid fibers as an additive in asphalt mixtures. Sixteen job-mix formulas were designed and tested. A direct relationship between the dynamic modulus and the amount of fibers added was observed, as well as an improvement of the stiffness in almost all the mixtures analyzed. Klinsky et al. (2018) corroborated these results when adding a mix of aramid and polypropylene fibers to a hot asphalt mixture, observing an improvement in the amount of permanent deformation accumulation and fatigue performance. However, the advantages achieved by adding fibers to a mixture should be checked against the drawbacks involved in their production and transportation.

This paper analyses the minimum service life extension that a Stone Matrix Asphalt reinforced with aramid and polyolefin fibers (FSMA) needs to be considered sustainable. To this aim, a Life Cycle Assessment (LCA) and Life Cycle Cost Analysis (LCCA) were performed to determine the environmental and economic impact this FSMA produces when different durabilities are assumed. The results were compared to the impact produced by a conventional overlay (CSMA) to determine the point at which both mixtures produce the same impact.

2 LIFE CYCLE ASSESSMENT (LCA)

An LCA is commonly structured in four steps: goal and scope, inventory analysis, impact assessment, and interpretation of the results (ISO, 2006a)(ISO, 2006b).

2.1 Goal and scope

As mentioned before, the goal of this analysis was to determine the minimum service life extension fiber-reinforced overlays need to be considered sustainable. For this study, a 1-km length of lane with a width of 3.75 m and a layer thickness of 0.076 m was selected as the functional unit.

In this rehabilitation method, in which an overlay is placed over a cracked rigid pavement, only the flexible layer is removed when it is deteriorated. Therefore, as the Federal Highway Administration (FHWA) recommends an analysis period long enough to include at least a major rehabilitation event within the system boundaries, a period equal to twice the overlay's durability was selected. The results were then annualized to achieve a fair comparison when different service lives were assumed for the FSMA.

The selection of the system boundaries was made based on the stages defined in the UNE-EN 15804:2012 standard regarding the environmental product declaration rules of construction products (UNE-EN, 2012). In this analysis, only four stages of the road life have been considered: material production, construction, maintenance, and end-of-life (see Figure 1), with

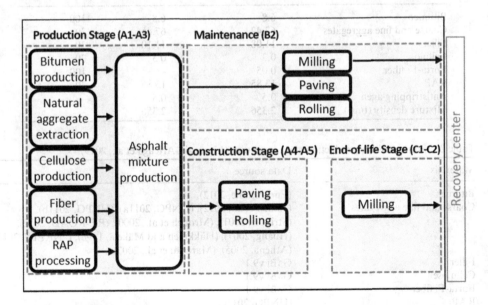

Figure 1. LCA boundaries.

construction and end of life being fictitious stages since the overlay is a maintenance technique itself. Furthermore, no leaching stage has been included in the analysis because, according to the provider (Forta-Fi), fibers are an inert material that neither produce leachate nor chemically react with the mixture's components and, consequently, no differences between the mixtures are expected in this regard.

2.2 *Inventory analysis*

The inventory analysis step of the LCA involves quantifying all the inputs and outputs associated with the functional unit regarding material and energy consumption, as well as the emissions produced. To this end, different databases were checked, including the United States database available in GaBi v9.1, for the production of energy and fuels.

2.2.1 *Production stage*

The production stage includes the extraction and production of the materials (bitumen, coarse and fine aggregates, cellulose, fibers, and RAP), their transportation to the asphalt plant, and the production of the asphalt mixture. The characteristics of the mixtures used for the analysis and the transport distances assumed can be seen in Table 1.

Forta-Fi fiber, the only difference between the FSMA and CSMA mixtures, is 13% aramid fibers and 87% polyolefin fiber. As the nature of the latter has been not revealed by the manufacturer, a polyethylene low density has been assumed based on its specific weight (0.91 g/cm^3) (Forta, 2013). Regarding the RAP, the recommendations made by the National Asphalt Pavement Association (NAPA, 2016) have been followed. In this sense, the emissions produced during the crushing and screening processes, as well as during its transportation to the asphalt plant, have been included in the analysis. Finally, no data were found regarding the chemical composition of the antistripping agent or the emissions generated during its production. Therefore, as the quantity is less than 1% and the same amount is used in both mixtures, it was decided to exclude it (UNE-EN, 2012). Table 2 shows the data sources used in this paper (Lizasoain-Arteaga et al., 2019)

Table 1. Definition of mixtures and transport distances assumed.

Materials	Asphalt mixture dosage (%wt)		Distance (km)
	FSMA	CSMA	
Bitumen	6.8	6.8	100
Coarse and fine aggregates	67.42	67.45	30
Filler	11.08	11.10	30
Cellulose	0.3	0.3	30
Forta-Fi fiber	0.05	-	100
RAP	13.85	13.85	30
Antistripping agen	0.5	0.5	-
Mixture density (t/m^3)	2.356	2.356	-

Table 2. Sources of the production stage inventory (Lizasoain-Arteaga et al., 2019).

Materials	Data source
Bitumen	(Eurobitume, 2012)
Coarse and fine aggregates	(Jullien et al., 2012), (UNPG, 2011a), (UNPG, 2011b), (Stripple, 2001), (Mroueh et al., 2000), (RE-ROAD, 2012), (Huang, 2007), (Häkkinen and Mäkelä, 1996), ("NREL," 2012) (Athena, 2005), (Marceau et al., 2007)
Filler	GaBi v9.1
Cellulose	GaBi v9.1
Forta-Fi fiber	GaBi v9.1
RAP	(UNPG, 2011c)

2.2.2 Construction stage

The construction stage considers the transportation of the asphalt mixture from the asphalt plant to the roadwork (30 km) and the paving and rolling of the mixture. As no differences in the workability of the mixtures has been detected during the mix design phase, the same consumption of diesel per ton of asphalt was assumed for both mixtures (1.56 l/tn) (Lizasoain-Arteaga et al., 2019).

2.2.3 Maintenance

Maintenance activities involve milling the deteriorated flexible overlay (0.41 l/tn), the transportation of the RAP produced from the roadwork to the recovery center, which has been fictitiously located 30 km away, and the production and construction of a new overlay. These activities are carried out when the asphalt mixture is deteriorated, which depends on the mixture's durability.

In this study, the CSMA has been considered to last 15 years, as indicated in VDOT (2018). On the other hand, no information regarding the durability of the designed FSMA is available at this moment. Thus, different durability hypotheses have been defined as a percentage of the durability of the CSMA (90%, 100%, 110%, 120%, 130%) (see Table 3).

2.2.4 End-of-life stage

This stage includes the milling of the asphalt overlay and the transportation of the RAP to the recovery center. However, the end of life of the overlay does not imply the end of the road's life since another layer would be constructed to replace the deteriorated one.

2.3 Impact assessment

Different characterization methods are available to transform the resources and emissions detected during the inventory stage into impacts. Each of these methods is characterized by using different impact categories and hypotheses to make the transformation, for example the number of substances covered, the weather conditions assumed, or the reference units used. Consequently, the results of the LCA can differ tremendously depending on the selections made. In this analysis, four characterization methods have been chosen to achieve reliable results (see Table 4): ReCiPe 1.08, ReCiPe 2016, CML 2001-January 2016 update, and TRACI.

ReCiPe enables the transformation of the emissions into 18 midpoint impacts, which are then transformed into endpoint impacts. In this way, the effect of the product life is calculated in three protection areas: damage to human health (HH), damage to ecosystem diversity (ED), and damage to resource availability (RA).

CML is a method commissioned by the Dutch ministries that is recommended for carrying out Environmental Product Declarations. It is used all over the word except for the United States, which relies on TRACI (see below). CML calculates 11 midpoint impacts, so if a single score is required to represent the environmental impact of a product, normalization and weighting steps are needed. In this work, the impact produced by the 28 member states of the European Union in 2000 and the weighting factors described in Lizasoain-Arteaga et al. (2019) as a combination of the EPA, BEES, NOGEPA, and BREE factors have been selected.

TRACI was developed to create an equivalent method to CML but particularized to the United States. TRACI calculates 10 midpoint impacts. The impact produced in the United States in 2008 and the weighting factors proposed in the software BEES have been used to achieve a single score.

Table 3. Maintenance schedule.

Mixture	Construction (year)	Maintenance (year)	End-of-life (year)
CSMA	0	15	30
FSMA (90%)	0	13.5	27
FSMA (100%)	0	15	30
FSMA (110%)	0	16.5	33
FSMA (120%)	0	18	36
FSMA (130%)	0	19.5	39

Table 4. Characterization methods and impact categories.

Impact	ReCiPe 1.08	ReCiPe 2016	CML 2001	TRACI
MIDPOINT				
Acidification (terrestrial)	X	X	X	X
Climate change	X	X	X	X
Fossil depletion	X	X	X	X
Ecotoxicity				X
Freshwater		X	X	
Marine	X	X	X	
Terrestrial	X	X	X	
Eutrophication			X	X
Freshwater	X	X		
Marine	X	X		
Human toxicity	X		X	
Cancer		X		X
No cancer		X		X
Ionizing radiation	X	X		
Land use		X		
Agricultural land occupation	X			
Urban land occupation	X			
Metal depletion	X	X	X	
Natural land transformation	X	X		
Ozone depletion	X	X	X	X
Particulate matter formation	X	X		X
Photochemical oxidant formation	X	X	X	X
Water depletion	X	X		
ENDPOINT				
Damage to human health	X	X		
Damage to ecosystem diversity	X	X		
Damage to resources availability	X	X		

3 LIFE CYCLE COST ANALYSIS (LCCA)

LCCA is a technique that uses economic analysis to evaluate the total cost of an investment option over an analysis period (Harvy et al., 2016). It normally involves two types of costs: agency costs and user costs. The former includes the expenditures that the owner of the road bears, like the initial construction, maintenance, or the end-of-life costs. On the other hand, the latter includes the costs that the road user incurs, like vehicle operating costs (which depend on the pavement conditions), delay costs produced during roadwork, and accident costs (Qiao et al., 2015). In this study, only the agency costs have been included since, as can be seen in Figure 1, neither the use phase nor the traffic delay produced during the maintenance activities are included within the system boundaries.

The data used in this analysis as well as the sources are shown in Table 5. To take into account the time value of money, a 4% discount rate was applied as recommended by the Virginia Transportation Research Council (VDOT, 2002) to discount the agency costs to year zero when calculating the Net Present Value.

4 RESULTS

The results achieved after performing the LCA and LCCA for both the CSMA and FSMA are shown in Table 6 and plotted in Figure 2. As expected, the greater the service life extension, the better the economic and environmental results.

When the environmental impacts are compared, only a 3% difference between the characterization methods is achieved. Results calculated using both versions of ReCiPe (1.08 and 2016)

Table 5. Costs database.

Material/process	Units	Costs	Source
Bitumen	$/tn	485.32	(M. de Fomento, 2016)
Coarse and fine aggregates	$/tn	11.29	(CYPE Ingenieros, 2019)
Filler	$/tn	45.62	(CYPE Ingenieros, 2019)
Forta-Fi fiber	$/tn	19,481.87	Provider
Cellulose	$/tn	1,330.00	Provider
RAP	$/tn	5.13	PaLaTe v2.0
Paving	$/tn	2.34	(CYPE Ingenieros, 2019)
Rolling	$/tn	2.89	(CYPE Ingenieros, 2019)
Milling	$/tn	32.32	(CYPE Ingenieros, 2019)
Transportation	$/(tn*km)	0.11	(CYPE Ingenieros, 2019)

Table 6. Annualized LCA and LCCA results.

| Mixtures/Impacts | CSMA | FSMA | | | | |
		90%	100%	110%	120%	130%
ReCiPe 1.08 HH [DALY]	4.95E-03	5.64E-03	5.07E-03	4.61E-03	4.23E-03	3.90E-03
ReCiPe 1.08 ED [species.yr]	2.19E-05	2.51E-05	2.26E-05	2.05E-05	1.88E-05	1.74E-05
ReCiPe 1.08 RA [$]	1.53E+02	1.79E+02	1.61E+02	1.46E+02	1.34E+02	1.24E+02
ReCiPe 2016 HH [DALY]	4.09E-03	4.66E-03	4.19E-03	3.81E-03	3.49E-03	3.23E-03
ReCiPe 2016 ED [species.yr]	1.13E-05	1.29E-05	1.16E-05	1.06E-05	9.67E-06	8.93E-06
ReCiPe 2016 RA [$]	3.29E+02	3.85E+02	3.47E+02	3.15E+02	2.89E+02	2.67E+02
CML [Points]	4.06E-08	4.70E-08	4.23E-08	3.85E-08	3.53E-08	3.25E-08
TRACI [Points]	9.03E+00	1.04E+01	9.33E+00	8.48E+00	7.77E+00	7.18E+00
LCCA [$]	3.06E+03	3.91E+03	3.40E+03	2.99E+03	2.66E+03	2.38E+03

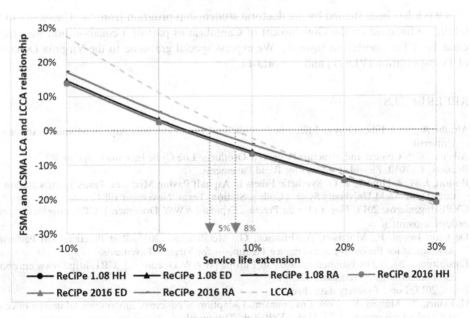

Figure 2. FSMA and CSMA LCA and LCCA relationship.

are very similar, with ReCiPe 1.08 predicting slightly more damage in the human and ecosystem protection areas than the updated version. Impacts calculated using CML and TRACI are also very similar, which make sense considering that TRACI was an adaptation of CML to

447

accommodate U.S. requirements. However, resource availability calculated with both ReCiPe characterization methods provides the most restrictive results, with the FSMA needing at least a 5% increase in service life to generate the same environmental impact as the conventional mixture.

Regarding the economic analysis, the steeper slope of the line plotted in Figure 2 shows the great benefit that extending the service life of the road has on the investment. However, this approach is more restrictive than the environmental one, with the FSMA needing at least an 8% increase in overlay durability to achieve economic benefits.

5 CONCLUSIONS

This study assessed the minimum life cycle extension that fiber-reinforced asphalt mixtures need to be considered as sustainable as conventional mixtures. An LCA and LCCA were performed for the modified and conventional mixtures to compare their impacts to determine the point at which both mixtures generate the same impact.

After analyzing four stages of the road life, no significant differences between characterization methods were seen, but resource availability had the more demanding impact. However, the economic approach was more restrictive and service life should increase by at least 8% to obtain a sustainable mixture.

For future work, the durability of both asphalt mixtures should be predicted to determine if the minimum 8% increase is reached. To do so, different lab tests are being performed to feed FlexPAVE, software developed by North Carolina State University and sponsored by the FHWA that simulates the behavior of the pavement. Furthermore, FSMA is being tested in the Heavy Vehicle Simulator at Virginia Tech, which will shed more light on this issue.

ACKNOWLEDGEMENTS

This work has been funded by the doctoral studentship program from the University of Cantabria, co-financed by the Government of Cantabria as part of a collaboration with the Virginia Tech Transportation Institute. We express special gratitude to the Virginia Department of Transportation (VDOT) and to Forta-Fi.

REFERENCES

Alrajhi, A., 2012. Fiber Dosage Effects in Asphalt Binders and Hot Mix Asphalt Mixtures. Arizona State University.

Athena, 2005. Cement and Structural Concrete Oroducts: Life Cycle Inventory Update #2,.

Bennert, T., 2010. Flexible Overlays for Rigid Pavements.

Button, J. W. & Hunter, T. G. Synthetic Fibers in Asphalt Paving Mixtures. Texas Transportation Institute, Texas A&M University System College Station, Texas, November 1984., n.d.

CYPE Ingenieros, 2019. Generador de Precios. España [WWW Document]. URL http://www.generad ordeprecios.info/

De León Izeppi, E., Morrison, A., Flintsch, G., McGhee, K., 2015. Best Practices and Performance Assessment for Preventive Maintenance Treatments for Virginia Pavements.

Eurobitume, 2012. The bitumen industry in Europe [WWW Document]. URL http://www.eurobitume. eu/bitumen/industry/ (accessed 12.30.17).

Forta, 2013. Forta-Fi safety data sheet.

Häkkinen, T., Mäkelä, K., 1996. Environmental adaption of concrete. Environmental impact of concrete and asphalt pavements. VTT Tied. - Valt. Tek. Tutkimusk.

Harvey, J., Meijer, J., Ozer, H., Al-Qadi, I., Saboori, A., Kendall, A., 2016. Pavement Life Cycle Assessment Framework.

Huang, Y., 2007. Life Cycle Assessment of Use of Recycled Materials in Asphalt Pavements.

ISO, 2006a. ISO 14040: Environmental Managemen - Life Cycle Assessment - Principles and Framework, 2 end. International Organization for Standarization.

ISO, 2006b. ISO 14044: Environmental Management - Life Cycle Assessment - Requirements and Guidelines, 1 edn. International Organization for Standarization.

Jullien, A., Proust, C., Martaud, T., Rayssac, E., Ropert, C., 2012. Variability in the environmental impacts of aggregate production. Resour. Conserv. Recycl. 62, 1–13. https://doi.org/https://doi.org/10.1016/j.resconrec.2012.02.002

Khazanovich, L., Lederle, R., Tompkins, D., Harvey, H., Signore, J., n.d. Guidelines for the Rehabilitation of Concrete Pavements Using Asphalt Overlays.

Klinsky, L.M.G., Kaloush, K.E., Faria, V.C., Bardini, V.S.S., 2018. Performance characteristics of fiber modified hot mix asphalt. Constr. Build. Mater. 176, 747–752. https://doi.org/https://doi.org/10.1016/j.conbuildmat.2018.04.221

Lizasoain-Arteaga, E., Indacoechea-Vega, I., Pascual-Muñoz, P., Castro-Fresno, D., 2019. Environmental impact assessment of induction-healed asphalt mixtures. J. Clean. Prod. 208, 1546–1556. https://doi.org/https://doi.org/10.1016/j.jclepro.2018.10.223

M. de Fomento, 2016. Ordin Circular 37/2016. Base de precios de referencia de la Dirección General de Carreteras.

Marceau, M.L., Nisbet, M. a, Vangeem, M.G., 2007. Life Cycle Inventory of Portland Cement Concrete. Cycle.

Mroueh, U.M., Eskola, P., Laine-Ylijoki, J., Wellman, K., Mäkelä, E., Juvankoski, M., 2000. Life cycle assessment of road construction. Finnra Reports 17/2000.

NAPA, 2016. Product Category Rules (PCR) For Asphalt Mixtures.

NREL [WWW Document], 2012. U.S. Life Cycle Invent. Database. URL https://www.lcacommons.gov/nrel/search (accessed 11.19.12).

Qiao, Y., Dawson, A.R., Parry, T., Flintsch, G.W., 2015. Evaluating the effects of climate change on road maintenance intervention strategies and Life-Cycle Costs. Transp. Res. Part D Transp. Environ. 41, 492–503. https://doi.org/https://doi.org/10.1016/j.trd.2015.09.019

RE-ROAD, 2012. Life Cycle Assessment of Reclaimed Asphalt.

Santero, N., Masanet, E., Horvath, A., 2010. Life Cycle Assessment of Pavements: A Critical Review of Existing Literature and Research. https://doi.org/10.2172/985846

Stripple, H., 2001. Life Cycle Assessment of Road. A Pilot Study for Inventory Analysis.

Tapkın, S., 2008. The effect of polypropylene fibers on asphalt performance. Build. Environ. 43, 1065–1071. https://doi.org/https://doi.org/10.1016/j.buildenv.2007.02.011

UNE-EN, 2012. UNE-EN 15804:2012+A1:2013. Sustainability of construction works. Environmental product declarations. Core rules for the product category of construction products.

UNPG, 2011a. Module d'informations environnementales de la production de granulats issus de roches meubles.

UNPG, 2011b. Module d'informations environnementales de la production de granulats issus de roches massives.

UNPG, 2011c. Module d'informations environnementales de la production de granulats recyclés.

VDOT, 2018. Chapter VI of Materials Division's Manual of Instruction.

VDOT, 2002. Life Cycle Cost Analysis Pavement Options.

Life Cycle Assessment (LCA) study on asphalt binders manufactured in North America

M. Buncher
Asphalt Institute, Lexington, KY, USA

R. Corun
Associated Asphalt, Ocean City, USA

Maggie Wildnauer & Erin Mulholland
thinkstep, Boston, USA

ABSTRACT: The Asphalt Institute (AI) provided an industry-average LCI dataset on asphalt binder representative of North American industry conditions through a recent LCA study.

Asphalt binder is one of many product streams from a refinery, so attributing material and energy inputs/outputs to just asphalt was a challenge. The main material and energy flows to be allocated were crude oil, thermal energy and associated emissions, electricity consumption, and emissions.

This LCA was based on information supplied by twelve AI member refineries and eleven terminals in the U.S. and Canada for 2015 and 2016. The scope of this cradle-to-gate study includes raw material sourcing and extraction, transportation to refineries, refining of crude oil into asphalt, transport to the terminal, and final blending of the asphalt binders at the terminal. Only those refinery processes associated with asphalt production were included in the assessment.

Results showed a large fraction of the potential environmental burden is due to the upstream crude oil extraction.

1 BACKGROUND

Driven by green building standards (e.g., LEED, Living Building Challenge, IgCC) and other initiatives, the demand for accurate life cycle inventory (LCI) and life cycle impact assessment (LCIA) data has increased for products used in the construction sector, including for pavement. Transportation agencies, such as the Illinois Toll Road Authority and the California Transportation Department, are starting to look at how they can use such information in their project plans and designs (Ozer 2017, California DOT 2018).

Therefore, the Asphalt Institute (AI) undertook this project to ensure the published LCI data on asphalt binder is as accurate as possible and representative of North American industry conditions.

Along with aggregate, asphalt binder is the major component of asphalt mixtures.

The AI is a US-based international trade association of petroleum asphalt producers, manufacturers and affiliated businesses. Founded in 1919, Institute members represent 90% of the asphalt binder produced in North America.

The AI hired thinkstep AG, an international consulting firm, to lead this life cycle assessment (LCA) due to their expertise in sustainability and life cycle assessment. In March 2019, the AI released its industry average life cycle assessment (LCA) report on asphalt binder produced in North America. This paper summarizes the LCA study and full report (Wildnauer, et al. 2019).

The National Asphalt Pavement Association (NAPA) has an Environmental Product Declaration (EPD) program for asphalt mixtures (http://www.asphaltpavement.org/EPD). The declared unit and reported impact categories in AI's study align with NAPA's EPD program and associated Product Category Rules (PCR). NAPA plans to incorporate the new AI LCI datasets into their EPD.

2 SCOPE OF THE STUDY

This study covers these four asphalt binders manufactured by AI members in North America:

- Asphalt binder without additives
- Asphalt binder with 3.5% styrene-butadiene-styrene (SBS)
- Asphalt binder with 8% ground tire rubber (GTR) (terminal blend)
- Asphalt binder with 0.5% polyphosphoric acid (PPA)

This industry-average assessment is based on information supplied by twelve AI member refineries (from nine companies) and eleven terminals (from four companies) in the U.S. and Canada. The declared unit of these four products is 1 kilogram of asphalt binder. The scope of this cradle-to-gate study includes raw material sourcing and extraction, transportation to refineries, refining of crude oil into asphalt, transport to the terminal, and final blending of the asphalt binders at the terminal. Only those refinery processes that are associated with asphalt production were included in the assessment.

3 REPRESENTATIVENESS

The data are intended to represent asphalt production during the 2015 and 2016 calendar years for the four types of asphalt binders as produced by AI members in North America.

Temporal. All primary data were collected for the years 2015-2016. All secondary data come from the GaBi 2017 databases and are representative of the years 2012-2016. As the study intended to compare the product systems for the reference years 2015-2016, temporal representativeness is considered to be very good.

Geographical. To assess geographical representativeness, the total asphalt and road oil capacity of the five US PADD regions and Canada were compared to the reported asphalt capacity of the participating facilities. In Table 5-2 of the full report, the weighted average deviation is 12%, thus the geographical representativeness is considered to be good per the method described by (Koffler, et al. 2016).

Technological. All primary and secondary data were modeled to be specific to the technologies or technology mixes under study. Where technology-specific data were unavailable, proxy data were used. Technological representativeness is considered to be good. The representative asphalt data provided accounts for 24% of asphalt capacity in the United States and Canada as of the beginning of 2017 and 27% of annual production for 2016 (EIA 2017b, Government of Canada 2017). In addition, the technological representativeness was evaluated based on facility size, as a proxy for overall efficiency, according to reported asphalt and road oil production capacity. All facility capacity data are publicly available (EIA 2017a, The Oil & Gas Journal 2017). The full report compared capacity of the population with that of the sample in Table 5-3 and found the weighted average deviation to be 14%. Thus, the technological representativeness is 86% and therefore considered to be good.

4 ALLOCATION

Since asphalt is only one product stream in a complex multi-product system (refinery), it is crucial that the allocation methodology appropriately captures only that share of the total impacts of the system that can be attributed to the asphalt binder. For this study, the main material and energy inputs that needed to be allocated were crude oil input, thermal energy consumption (including associated emissions), and electricity.

4.1 *Previous methodologies*

There have been a variety of allocation methods used for LCAs of petroleum refineries. Eurobitume's LCI study allocates between refinery products based on economic factors calculated from European market averages over 7 years, grouping the outputs of each tower as residues and distillates (Eurobitume 2012). The GaBi refinery model allocates crude oil inputs by energy content and electricity, thermal energy, and emissions by mass (thinkstep 2016). Both these assessments rely on theoretical refinery models supplemented by industry average values as opposed to primary data collected directly from refineries where each refinery is modeled separately.

PRELIM (the Petroleum Refinery Life Cycle Inventory Model) estimates energy use and GHG emissions for crude oil products. The tool, which models separate unit processes for different refinery setups and crude assays, allows the user to choose between mass, energy, market value, and hydrogen contents for allocation at the sub-process level. (Abella, et al. 2016)

The NAPA EPD Program is currently using the NREL USLCI data for crude oil at refinery, allocated based on a Master's thesis by Rebecca Yang (NREL 2003, Yang 2014). Yang's allocation considers both the economic values of the refinery co-products as well as the mass yield of those co-products, allocating at the refinery level. Alternatively, the user of the USLCI data could choose to allocate the NREL data based on mass or energy content.

With the exception of Eurobitume's LCI, all these studies use an average crude slate representative of crude oil consumption in the US for all refinery products, not just asphalt. In contrast, this study aimed to collect data only when asphalt was being produced at the refinery.

4.2 *Background data*

Allocation of background data (energy and materials) taken from the GaBi 2017 LCI database is documented online at http://www.gabi-software.com/support/gabi/gabi-database-2018-lci-documentation. A report published by the Joint Research Commission has additional information on the energy datasets found in GaBi (Garrain, de la Rua and Lechon 2013).

4.3 *Co-product and multi-input allocation*

Since asphalt is one product stream in a complex, multi-product system, it is crucial that the allocation methodology appropriately captures and allocates the total impacts of the system. For this study, the main material and energy inputs to be allocated are crude oil input, thermal energy consumption and associated emissions, and electricity. A scenario analysis was included as part of the study to validate the approach. Baseline and scenario allocation methodologies are shown in Table 1.

4.3.1 *Electricity*
Mass allocation was selected for electricity in both the baseline and scenario analyses because the density of products is directly related to the electrical demand for pumping the products. No other allocation method is deemed an appropriate scenario for electricity. This methodology also aligns with existing GaBi datasets of refinery products (European Commission 2013, thinkstep 2016).

4.3.2 *Crude oil input*
Energy content of the co-products (using the net calorific value) is the baseline allocation methodology selected for crude oil input. This is an appropriate methodology because lighter refinery products with a higher net calorific value are preferred due to their higher market value and demand. Also, since this methodology aligns with existing GaBi datasets of refinery products, we ensure a consistent methodology with fuel datasets (European Commission 2013,

Table 1. Proposed and scenario allocation/subdivision methodologies.

Input or Output	Proposed baseline Methodology	Allocation Scenario 1	Allocation Scenario 2	Allocation Scenario 3
Electricity	Mass allocation	*(no other method deemed applicable)*		
Crude oil	Energy content Allocation (net calorific value)	Mass allocation	*(Same as baseline)*	Mass allocation
Thermal energy	Subdivision calculated as Sensible heat of asphalt, accounting for inefficiencies	*(Same as baseline)*	Energy content of allocation (using net calorific value)	Energy content of allocation (using net calorific value)
Direct emissions	Allocated based on thermal energy use	*(Same as baseline)*	Energy content of allocation (using net calorific value)	Energy content of allocation (using net calorific value)

thinkstep 2016). Because asphalt is a construction material and not purchased for its energy content, the scenario analysis will investigate the mass allocation for the crude oil input.

4.3.3 *Thermal energy*

Thermal energy requirements for asphalt is based on the sensible heat[1] of asphalt in the system. It is calculated from the temperature differential between the crude tank to the asphalt rundown line and the specific heat capacity of the asphalt, while accounting for the heating system's inefficiencies.

Each oil refinery is different and complex, with heat integration between production processes and utilities. All refinery energy usage must be accounted for across the refinery co-products through a defined allocation method. One approach would be a rigorous heat and material balance around the many process units relevant to asphalt production using a process simulation software (e.g. Aspen HYSYS, Petro-SIM). This software can be used to create a complex model using thermodynamic equations to calculate the energy flow associated with each co-product at the process level thus tracking the energy required for each co-product. However, each refinery's model is proprietary and could not be utilized in the LCA due to the complexity and confidentiality. This would greatly reduce transparency and it could not be guaranteed that models were consistent across refineries.

While lighter products are vaporized and sometimes condensed as part of their production from crude oil, asphalt is never vaporized nor condensed, which made employing a simpler method possible. The energy input to produce asphalt in refining is therefore calculated as the net sensible heat input (i.e., the energy required to raise the temperature of a material) into the asphalt fraction of the crude oil based on the temperature difference between the crude tank outlet and the asphalt storage tank rundown line, adjusted for the efficiency of the heaters, piping losses, and storage energy losses. This will be referred to as the *sensible heat method* and is described in further detail below.

The basic asphalt production process starts with crude in a crude tank. The crude is partially heated and mixed with water to dissolve salts. The water is separated from the crude and removed, removing the salts. The very low salt crude is heated again and distilled, with all products lighter than heavy gas oil vaporizing in the crude tower, leaving atmospheric residue, also known as reduced crude, at the tower bottom. The energy required for vaporization is attributable to the lighter products and hence excluded from consideration. The atmospheric residue is then heated again and further distilled under a vacuum, vaporizing all gas oils and

---

1. Sensible heat is the energy required to change the temperature of a substance with no phase change (retrieved from: http://climate.ncsu.edu/edu/health/health.lsheat, July 17, 2017)

any remaining diesel, with asphalt remaining as a hot liquid in the bottom of the vacuum distillation tower. The hot asphalt is exchanged with other refinery feeds, mostly in the crude and vacuum distillation units, to return heat to the process, before going to the asphalt rundown line and to asphalt storage. Thus, the net energy consumed by the asphalt is the net sensible heat input into the asphalt molecules from the crude tank to the asphalt rundown line. The total energy consumed to produce the asphalt then is the net sensible heat in the asphalt plus any energy losses in the heating system (heater, piping, and vessel losses). All additional energy consumption is attributed to the other co-products.

The sensible heat method multiplies the change in temperature of the asphalt by its heat capacity, divided by the average efficiency of the heaters. Then, the estimated piping and vessel losses were added. The initial temperature is that of the crude entering the process and the final temperature is after asphalt run down where heat is recovered – therefore, the calculation accounts for the heat recovered from the asphalt.

$$\frac{C \times \Delta T}{\eta} + L = Thermal\ energy\ required$$

Where,
  C = heat capacity (J/K)
  ΔT = temperature difference between crude oil input and asphalt run down (K)
  η = efficiency of heating system (unitless)
  L = losses (J)

It should be noted that even with very conservative estimate for losses used for these calculations, loss still accounted for less than 1% of the energy required.

The described method was developed jointly with the Asphalt Institute and validated by two refineries operated by the same company. The validation process compared the results of the sensible heat method to a complex thermodynamic model from a process simulation software. To assess the accuracy of using the sensible heat method, the complex model was modified so that 10,000 barrels of asphalt product were removed from the total output while all the other co-products and their production volumes remained unaltered. Then the model was run to establish the difference between the energy requirements of the original model and the modified model, which is the energy associated with producing those 10,000 barrels of asphalt. In both cases, it was found that that the result calculated with the complex model was within one percent of the energy consumption calculated using the sensible heat method.

4.3.4 Direct emissions
Direct emissions were allocated based on the fraction of total thermal energy use (excluding recovered heat) calculated for asphalt. While process-specific emissions are generally preferred, in some cases only site-wide emissions from fuel combustion will be available from participating refineries. Process-specific emissions were allocated based on the fraction of total thermal energy for the process and site-wide emissions were allocated based on the fraction of total thermal energy for the site.

5 PRODUCT SYSTEM

5.1 Crude oil slate
The production stage starts with extraction of crude oil and delivery to the refinery. Crude oil is modeled based on thinkstep's proprietary crude oil supply model, which considers the whole supply chain of crude oil (i.e., extraction, production, processing, the long-distance transport and the regional distribution to the refinery) and forms the basis of all refinery product inventories in the GaBi databases (thinkstep 2016). Companies were asked to provide crude name, region of origin, extraction technology, and mode of transportation. In many

cases, primary information on the extraction technology was not available, in which case it was selected and modeled based on crude name. When the name alone did not provide enough information to select an extraction technology, it was modeled using the region of origin's average crude slate mix as a proxy.

The resulting average crude oil slate for North American asphalt binder used for the LCA study represents a mix of conventional (primary, secondary and tertiary production) and unconventional (oil sands, in-situ) extraction technologies, as shown in Table 2. All the AI member's asphalt binder products were manufactured in Canada and the United States, with 85% of crude oil sourced from those nations.

5.2 *Asphalt production*

Crude oil refinery activities begin with the input of crude oil. Crude is fed to the desalter where it is partially heated and mixed with water to dissolve salts. The water is separated and removed. Next, the crude oil enters the atmospheric distillation unit, where it is heated and distilled. All products lighter than heavy gas oil vaporize and the energy required for vaporization is fully attributable to those lighter products. The residue from the atmospheric distillation is introduced to the vacuum distillation unit. The atmospheric residue is heated and further distilled under a vacuum, vaporizing all gas oils and any remaining diesel, with asphalt remaining as a hot liquid in the bottom of the vacuum distillation tower. The hot asphalt passes through heat exchangers alongside other refinery feeds, mostly in the crude and vacuum distillation units, to return heat to the process, before going to asphalt storage.

While process-specific electricity, thermal energy, water usage, and emissions would have been preferred, these data points were not available. Therefore, refinery-level data were collected for site-wide consumption of electricity, thermal energy, and water as well as direct emissions and allocated to the asphalt product as described in section 4.

5.3 *Asphalt terminal*

The processes within each refinery were vertically aggregated first and then combined into one production-weighted average. This average asphalt production process then provided the input of asphalt to the average terminal process.

At the asphalt terminal, hot liquid asphalt is stored, additives (ground tire rubber, styrene-butadiene-styrene, or polyphosphoric acid) are mixed or milled into the asphalt, and hot liquid asphalt is further distributed. The terminals consume electricity (mainly used for milling) and thermal energy (used for storage). Terminals can be either co-located with the refinery or off-site. For this study, all participating companies were located off-site. Inbound transportation from the refinery to the terminal is a production weighted average of the distances and modes collected from the companies.

Table 2. Crude oil extraction method of representative AI asphalt binder used in LCA.

Category of extraction technology	Percentage (by mass)
Crude from oil sands	44%
Primary extraction	22%
Secondary extraction	16%
Tertiary extraction, steam injection	15%
Tertiary extraction, CO2 injection	1%
Tertiary extraction, nitrogen injection	1%
Tertiary extraction, natural gas injection	1%
Other (refinery products)	<1%

6 RESULTS

The reported impact categories represent impact potentials, i.e., they are approximations of environmental impacts that could occur if the emissions (a) followed the underlying impact pathway and (b) met certain conditions in the receiving environment while doing so. In addition, the inventory only captures that fraction of the total environmental load that corresponds to the chosen functional unit (relative approach). Results are therefore relative expressions only and do not predict actual impacts, the exceeding of thresholds, safety margins, or risks.

The declared unit and reported impact categories have been selected to align with the requirements of the NAPA EPD program for asphalt mixtures mentioned in Section 1 of this paper. These impact categories also aligns with EN15804 (CEN 2013).

Table 3 presents the total cradle-to-gate environmental impact results for all four products.

Figure 1 presents the relative results of asphalt without additives leaving the terminal, broken down by crude oil extraction and transport, refinery operations, and terminal operations (including transport to the terminal).

Figure 2 presents the relative impacts for crude oil extraction and refinery operations (excluding the terminal). The crude oil is the primary driver of impact for GWP100, PED (non-renew.), SFP, and FF. Crude transport represents 1% to 20% of the impacts for all reported impact categories, with a 5% contribution to GWP100 and a 20% contribution to AP and SFP. For EP, the highest impacts for refinery operations are due to waste water treatment which contribute 46%.

Table 3. Impact assessment, per kg (IPCC 2013, EPA 2012).

Impact Category	Unit	Asphalt binder, no additives	Asphalt binder, with 8% GTR	Asphalt binder, with 0.5% PPA	Asphalt binder, with 3.5% SBS
IPCC AR5					
Global warming potential [GWP100]	Kg CO_2 eq	0.637	0.621	0.654	0.765
Global warming potential [GWP20]	kg CO_2 eq	0.766	0.745	0.786	0.918
TRACI 2.1					
Acidification potential (AP)	kg SO_2 eq	1.78E-03	1.69E-03	1.96E-03	2.12E-03
Eutrophication potential (EP)	Kg N eq	1.66E-04	1.57E-04	1.69E-04	1.82E-04
Smog formation potential (SFP)	Kg O_3 eq	0.0360	0.0347	0.0365	0.0427
Fossil fuel consumption (FF)	MJ (NCV)	5.32	4.98	5.36	5.66
Total use of non-renewable primary energy resources (PED)	MJ (NCV)	53.2	52.2	53.5	55.2
Use of net fresh water (excl. rain water)	L	1.01	0.92	1.06	1.44
Use of net fresh water (incl. rain water)	L	1.68	1.57	1.76	2.40

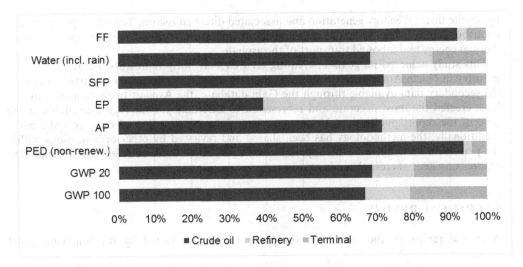

Figure 1.   Overall impacts of asphalt binder, no additives.

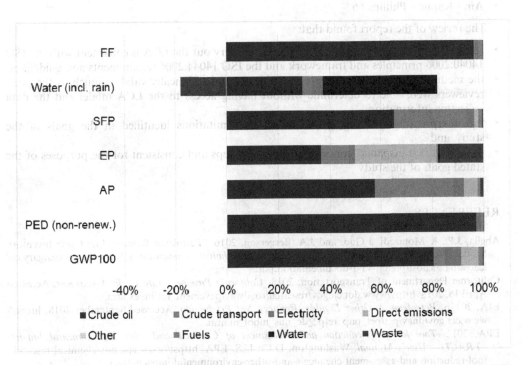

Figure 2.   Crude oil and refinery impacts of asphalt binder, no additives [TRACI 2.1, except PED (non-renew.) and Water (incl. rain)].

7   CONCLUSIONS

The extraction of the crude oil is the primary driver of all potential environmental impacts, due most significantly to the use of crude oil from oil sands or crudes extracted via a tertiary method. At the refinery itself, electricity is the most significant single driver of impact followed by on-site thermal energy generation and associated direct emissions. Terminal operations can contribute up to 20% of potential environmental impacts without additives, driven primary by thermal energy and inbound transport of the asphalt.

This study achieved its goals in creating an LCI that fairly represents the asphalt binder industry in North America. By combining the primary data collected from participants with the secondary data available through the Gabi database, the Asphalt Institute and thinkstep have created the most accurate and representative LCI data for the region available at the time of the report publication. Updates to these datasets are expected on a periodic basis. Additionally, the methodology has been shared and reviewed by other organizations within the petroleum industry, with the hope that they apply it to future petroleum LCI datasets to ensure consistency across the industry.

8 CRITICAL REVIEW

A critical review of the completed study report was conducted by the following panel members:

- Arpad Horvath – Consultant, Berkeley, California (Chair)
- Mike Southern – Eurobitume
- Amit Kapur – Phillips 66

The review of the report found that:

- the approach, described in the report, used to carry out the LCA is consistent with the ISO 14040:2006 principles and framework and the ISO 14044:2006 requirements and guidelines,
- the methods used in the LCA are scientifically and technically valid as much as the peer-reviewers were able to determine without having access to the LCA model and the data collection information,
- the interpretations of the results reflect the limitations identified in the goals of the study, and
- the report is transparent concerning the study steps and consistent for the purposes of the stated goals of the study

REFERENCES

Abella, J.P., K Motazedi, J Guo, and J.A. Bergerson. 2016. "Petroleum Refinery Life Cycle Inventory Model (PRELIM)." *PRELIM v1.1, User guide and technical documentation.* https://www.ucalgary.ca/lcaost/files/lcaost/prelim-v1-1-documentation.pdf.

California Department of Transportation. 2018. *California Pavement Life Cycle Assessment.* Accessed April 15, 2018. http://www.dot.ca.gov/research/roadway/pavement_lca/index.htm.

EIA. 2017b. *Petroleum & Other Liquids: Refinery Net Production.* Accessed September 2018. https://www.eia.gov/dnav/pet/pet_pnp_refp2_dc_nus_mbbl_m.htm.

EPA. 2012. *Tool for the Reduction and Assessment of Chemical and other Environmental Impacts (TRACI) – User's Manual.* Washington, D.C.: U.S. EPA. https://www.epa.gov/chemical-research/tool-reduction-and-assessment-chemicals-and-other-environmental-impacts-traci.

Eurobitume. 2012. "Life Cycle Inventory: Bitumen." http://www.eurobitume.eu/fileadmin/pdf-downloads/LCI%20Report-Website-2ndEdition-20120726.pdf.

European Commission. 2013. *Background analysis of the quality of the energy data to be considered for the European.* Editors: Simone Fazio, Marco Recchioni, Fabrice Mathieux. Authors: Daniel Garrain, Cristina de la Rùa, Yolanda Lechòn. European Commission, Joint Research Centre, Institute for Environment and Sustainability.

Garrain, Daniel, Cristina de la Rua, and Yoland Lechon. 2013. *Background analysis of the quality of energy data to be considered for the European Reference Life Cycle Database.* European Commission, Joint Research Centre, Institute for Environment and Sustainability.

Government of Canada. 2017. "Table 134-0004: Supply and disposition of refined petroleum products." *Statistics Canada.* Accessed September 2018. http://www5.statcan.gc.ca/cansim/a26?lang=eng&id=1340004.

IPCC. 2013. *Climate Change 2013: The Physical Science Basis.* Genf, Schweiz: IPCC.

ISO. 2006. *ISO 14040: Environmental management – Life cycle assessment – Principles and framework.* Geneva: International Organization for Standardization.

ISO. . 2006. *ISO 14044: Environmental management – Life cycle assessment – Requirements and guidelines.* Geneva: International Organization for Standardization.

Koffler, C., Shonfield, P., and Vickers. J. 2016. "Beyond pedigree—optimizing and measuring representativeness in large-scale LCAs." *Int J Life Cycle Assess.* doi:10.1007/s11367-016-1223-5.

Wildnauer, M., Mulholland, E. and Liddie, J., March 2019. *Life Cycle Assessment of Asphalt Binder, On behalf of the Asphalt Institute,* ISBN 978-1-934154-76-2, Copyright 2019.

NAPA. 2017. "Product Category Rules For Asphalt Mixtures." http://www.asphaltpavement.org/index.php?option=com_content&view=article&id=1119&Itemid=100361.

NREL. 2003. "U.S. Life Cycle Inventory Database; Crude oil, in refinery." https://uslci.lcacommons.gov/uslci/search.

Ozer, Hazan, and Iman L Al-Qadi. 2017. "Regional LCA Tool Development and Applications." *Pavement LCA Symposium.* Illinois Center for Transportation University of Illinois at Urbana Champaign.

thinkstep. 2016. "The GaBi Refinery Model." http://www.gabi-software.com/uploads/media/The_GaBi_Refinery_Model_2016.pdf.

Yang, R.Y. 2014. *Development of a Pavement Life Cycle Assessment Tool Utilizing Regional Data and Introducing an Asphalt Binder Model, MS Thesis.* Urbana-Champaign: University of Illinois at Urbana-Champaign. http://hdl.handle.net/2142/50651.

Preliminary evaluation of using intelligent compaction for life cycle assessment and life cycle cost analysis of pavement structures

S. Satani, S.F. Aval, J. Garrido & M. Mazari
Department of Civil Engineering, California State University Los Angeles, Los Angeles, CA, USA

ABSTRACT: Many factors can affect the variability of the pavement compaction quality, e.g., moisture content, surface temperature, and material quality. Such an approach may result in unnecessary passes, extended working hours, higher energy consumption, and environmental impacts. The use of intelligent compaction technology during construction can improve the performance and extend the life of pavement structure while reducing the risk for both the contractors and project owners. This study focused on a case study to evaluate the preliminary Life Cycle Assessment (LCA) and Life Cycle Cost Analysis (LCCA) of intelligent compaction compared to conventional compaction approach. The environmental impacts of the improved construction process were quantified based on limited data available from the case study. The analysis showed a reduction in energy consumption and the production of greenhouse gas (GHG) emissions as a result of utilizing intelligent compaction. However, a comprehensive review is required to quantify the benefits and establish more accurate performance indicators. The LCCA showed that the utilization of such technology potentially reduces the construction and maintenance costs in addition to the enhancement of the quality control/quality assurance (QC/QA) process. The LCCA performed in this study consisted of different scenarios in which the number of operating hours were used to evaluate the cost efficiency of the intelligent compaction technique during construction.

1 INTRODUCTION

Durability and service life of pavement structures depend on uniform compaction of different layers to achieve the maximum in-situ density. The advancement of compaction technologies over the past decades has helped to improve compaction quality and uniformity. However, the current state of practice for quality control and quality assurance (QC/QA) of compacted layers are still based on spot tests to estimate in-situ density. Even though these measurements are acceptable with some level of variability, they are only revealing information about a limited number of locations throughout the construction site. Such a limited number of tests cannot provide enough information to ensure the uniformity of the entire compaction process. With the recent advancements of construction techniques such as Intelligent Compaction (IC), a comprehensive database of the construction process can be collected. These systems usually include a vibration sensor (accelerometer) mounted inside the roller drum, a Global Positioning System (GPS) receiver mounted on the roller cabin, and a data acquisition system attached to a display that shows the real-time construction data to the operator. IC systems provide a comprehensive set of information during and after the construction process that could be used to improve and enhance the construction uniformity and quality (Mazari et al. 2016, Fathi et al. 2019, Tirado et al. 2019).

Several studies have been focusing on evaluating the benefits and improving the use of intelligent compaction systems (Mooney et al. 2007, White et al. 2007, Chang et al. 2011, Nazarian et al. 2015, and Kumar et al. 2016). The expected benefits of using IC systems during construction and compaction process can be listed as improved quality of compaction/uniformity,

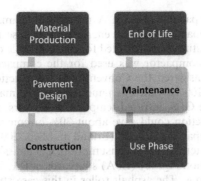

Figure 1.   Life-Cycle Assessment Stages for a pavement system (Harvey et al. 2016).

reduced over/under compaction costs, reduced time of compaction, identification of soft or weak spots, extended lifetime of the roller, and integration of design, construction, and performance.

Although these benefits have been included in different studies, there has been little information focusing on quantifying the benefits and improvement. Moreover, the focus on life cycle assessment (LCA) and life cycle cost analysis (LCCA) of pavement structures constructed with intelligent compaction technique has been overlooked in the literature. Although quantifying the added benefits in both the short and long term can be challenging, these measures help decision-makers, project owners, and contractors to evaluate the effectiveness of such technologies. The main goal of this study is to partially assess the impact of using intelligent compaction techniques on life cycle assessment and life cycle cost analysis of a pavement structure. The scope of this study is limited to the data available from a case study, and the analyses are limited in terms of factors that were considered for environmental impact analyses and cost savings over the life of the constructed pavement section.

The life cycle assessment stages for a pavement system are shown in Figure 1 (Harvey et al. 2016). Although a comprehensive LCA would involve all the illustrated stages from material production to end-of-life, the focus of this study is mainly on the construction phase. The maintenance phase is also predicted to be indirectly impacted by the use of intelligent compaction method. However, quantifying such impacts needs long term monitoring of the pavement section constructed with IC methods compared to the pavement sections constructed with conventional methods.

2   METHODOLOGY

2.1   Case study details

A hot mix asphalt (HMA) overlay was constructed using a double drum roller instrumented with the Intelligent Compaction equipment. The total length of this case study was 2.2 miles over a single lane. The roller operating durations, which is the basis for LCCA and LCA calculations, were estimated for both IC and conventional compaction processes. Complementary field data were also collected, which is beyond the scope of this paper.

2.2   Partial Life Cycle Assessment

This part of the study covers the comparison between environmental impacts of pavement compaction process using intelligent compaction and conventional compaction techniques. This assessment only includes the variation in energy/fuel consumption and gas emissions generated during the compaction process. A procedure summarized in the following tables shows how intelligent compaction can save energy/resource use and reduce emissions.

The calculations in this study are presented for one lane-mile of compacting hot mix asphalt overlay. A double drum compactor was used for the compaction process. The same compactor roller was also considered for Conventional Compaction calculations. The roller is equipped with a 75 kW (100 HP) diesel engine. One of the major assumptions in this study was that with an accurate GPS, infrared temperature sensors, and real-time data collection/display, intelligent compaction could save about 30% in compaction hours over a one-mile length of a pavement compaction process (Briaud and Seo, 2003). Tables 1 and 2 summarize LCA impact categories in terms of fuel consumption, CO emission, and NOx emission. Based on Environmental Protection Agency (EPA) standards (EPA, 2018), all engines must satisfy specific criteria for emissions. The asphalt roller in this case study is equipped with a Tier 3 non-road diesel engine with a power of 75 kW. Typical fuel consumption is 5.6 gallons per hour (Barrington Diesel Club, 2019). Based on EPA standards, the engine emits 5 grams of CO per hour and 4 grams of NOx per hour to generate 1 Kilowatt energy (Table 1). Hourly water consumption during the asphalt compaction process is 106 gallon-per-hour (gph), according to the field measurements. The project data indicates that the compaction time for 3.54 km (2.2 miles) of asphalt pavement overlay compaction is 5.16 and 6.71 hours for IC and Conventional Compaction (CC), respectively. Converting these values for 1.6 km (1 mile) results in 2.35 and 3.05 hours for IC and CC, respectively. Calculation of energy and resource use, shown in Table 2 and Table 3, demonstrate the standard comparison between Conventional Compaction and Intelligent Compaction over a one lane-mile and 10-hours work schedule. Based on the preliminary analysis, adopting Intelligent Compaction can reduce 3.4 gallons of fuel consumption, 263 grams of CO emission, 210 grams of NOx emission, and 74.2-gallon water consumption over only a 1 lane-mile length of a pavement overlay construction. Assuming a 10-hour compaction performance (see Table 3), using an IC roller can save about 23% in resource/energy use as well as a reduction in environmental emissions.

To further expand the preliminary analysis for the use of intelligent construction in this case study, a simplified LCA was performed using the Athena® pavement LCA tool. This tool includes a comprehensive database of construction materials as well as

Table 1. Life-Cycle Assessment impact categories for intelligent compaction[*].

Category	Unit	Number of units	Engine Power (kW)	Total
Fuel (Diesel)	Gallon/hr	5.6	75	5.6 gph
CO	Gram/kilowatt-hour	5	75	375 g/kWh
NOx	Gram/kilowatt-hour	4	75	300 g/kWh
Water	US Gallon/hr	106	N/A	106 gph

* Based on EPA Tier 3 non-road diesel engine (Engine Power = 75 kW)

Table 2. Comparison of energy use and environmental emissions for 1-mile compaction.

Item	Conventional Compaction			Intelligent Compaction		
	Unit	Number of hours	Total units	Unit	Number of hours	Total units
Fuel (Diesel)	5.6 gph	3.05	17 gal	5.6 gph	2.35	14 gal
CO	375 g/kWh	3.05	1144 g	375 g/kWh	2.35	881 g
NOx	300 g/kWh	3.05	915 g	300 g/kWh	2.35	705 g
Water	106 gph	3.05	323	106 gph	2.35	249 gal

Table 3. Comparison of energy use and environmental emissions for 10 hours compaction.

	Conventional Compaction			Intelligent Compaction		
Item	Unit	Number of hours	Total units	Unit	Number of hours	Total units
Fuel (Diesel)	5.6 gph	10	56 gal	5.6 gph	7.7	43 gal
CO	375 g/kWh	10	3750 g	375 g/kWh	7.7	2888 g
NOx	300 g/kWh	10	3000 g	300 g/kWh	7.7	2310 g
Water	106 gph	10	1060 gal	106 gph	7.7	816 gal

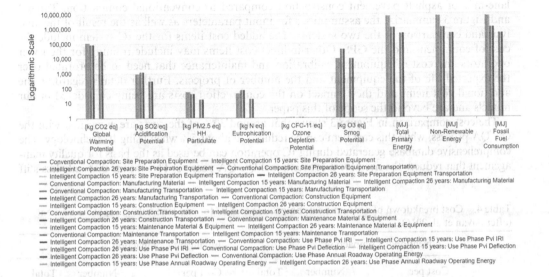

Figure 2. Summary measures of life cycle stage for intelligent compaction alternative.

incorporating the use-phase energy estimation and the pavement vehicle interaction (PVI) model. Although the main purpose of the software is to compare different design options to facilitate decision-making process, the modification of parameters for the construction phase of the pavement life cycle can give an estimate of the impact. However, the inputs are not quite configurable to show the compaction effort as one of the variables in the model. The overall construction equipment item may include the intelligent compaction equipment and the impact of their incorporation in the compaction process. Moreover, the long-term impact of improved compaction uniformity and quality can be quantified through improved smoothness, pavement vehicle interaction, and maintenance/rehabilitation costs over the life span of the pavement structure. Figure 2 illustrates summary measures of life cycle stages for an alternative that partially considers the use of intelligent compaction technology compared to conventional compaction for two scenarios assuming the extended life cycle for 15 and 26 years. Further analysis is required to compare the relevant construction and life cycle cost items.

2.3 Life Cycle Cost Analysis (LCCA)

LCCA process begins with the development of alternatives to accomplish the structural and performance objectives of the project, which is a subset of benefit-cost analysis (BCA). The lowest cost option (LCC) may not consider risks, budgets, and environmental concerns (Harvey et al. 2016). Contractors and agencies have accepted life-cycle cost analysis (LCCA) as one of the standards in determining the lowest-cost alternative for paving, maintenance, and rehabilitation. The Federal Highway Administration promotes the use of LCCA measurement analysis for state departments of transportation (DOTs) that allows transportation officials to determine alternative investment options. The current LCCA standard analysis allows state DOTs to quantify the differential costs of alternative investment options for a given project. The following sections include a summarize LCCA for the use of intelligent compaction compared to conventional compaction for a segment of a roadway.

2.4 Construction costs framework

Savan et al. (2016) reported that IC technology has the advantage of saving about $349 per lane-mile of asphalt pavement construction compared to conventional compaction. Table 4 and Figure 3 summarize the assumptions for input parameters as well as the result of the analysis and comparison of the two systems. The added cost items for the IC system include the cost of equipment and the GPS. Other indirect cost items may include training for the roller operators and cost of equipment calibration and maintenance that need to be prorated over the expected life of the equipment and the number of projects. Further details regarding the additional cost items and their impact on the construction costs are being considered in our models and are beyond the scope of this paper.

The cost comparison in Figure 3 shows that around 30% of the costs are associated with the QC/QA process, while this cost item can be reduced to about 5% when using IC technology. The comprehensive database generated during IC operation can be used as the basis for quality management that reduces the need for a conventional QC/QA process. This means that only less-stiff

Table 4. Cost breakdown of conventional compaction and IC for one lane mile of asphalt pavement (after Savan et al. 2016).

| | Conventional Compaction | | | | Intelligent Compaction | | | |
Item	Cost per unit ($)	Unit	Number of Units	Total Cost ($)	Cost per unit ($)	Unit	Number of Units	Total Cost ($)
Roller	36	hour	10	360	43	hour	7.7	328
Operator	30	hour	10	300	30	hour	7.7	231
GPS	-	-	-	-	0.9	hour	7.7	7
QC/QA	0.05	m²	5886	282	0.05	m²	557	27
Total	-	-	-	942	-	-	-	593

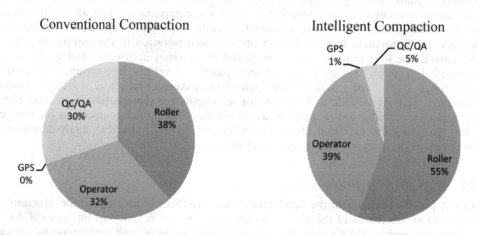

Figure 3. Cost of the asphalt pavement construction using conventional compaction and IC (after Savan et al. 2016).

Table 5. Compaction cost assuming IC has no improvement on the compaction speed.

| | Conventional Compaction | | | | Intelligent Compaction | | | |
Item	Cost per unit ($)	Unit	Number of Units	Total Cost ($)	Cost per unit ($)	Unit	Number of Units	Total Cost ($)
Roller	36	hour	10	360	43	hour	10	426
Operator	30	hour	10	300	30	hour	10	300
GPS	-	-	-	-	0.9	hour	10	9
QC/QA	0.05	m^2	5886	282	0.05	m^2	558	27
Total	-	-	-	942	-	-		762

areas, identified from IC stiffness map, can be selected as potential sub-lots for using spot tests for quality management. It is also observed that the higher hourly cost of IC equipment breaks even with QC/QA cost savings and compaction speed enhancement through the IC operation.

In a sensitivity analysis scenario performed for this case study, the 30% operating time savings in the IC process was ignored, and it was assumed that both conventional compaction and IC operators would operate for a similar duration (i.e., 10 hours). The results of the cost analysis for this scenario shows that IC technology reduces the costs as much as $180 per lane-mile per day (Table 5).

2.5 Roadway life cycle costs framework

Xu et al. (2011) showed that a heterogeneous pavement layer results in lower rutting and better fatigue performance. Based on that study, the utilization of intelligent compaction can improve the fatigue life of the asphalt pavement for up to 2.6 compared to a less uniform compaction scenario. However, other factors are influencing the life cycle of the pavement e.g., traffic parameters, environmental factors, and long-term properties. For the sensitivity analysis in this case study, both 26 years of lifetime improvement and a conservative assumption of 15 years lifetime improvement were considered. In this case, the average life span of the pavement structure was considered to be ten years.

According to Table 6, the annual cost of maintenance per lane-mile is estimated to be $25,000 (Savan et al. 2016). Since the life cycle of the pavement is assumed to be extended to 26 years, the cost associated with the maintenance is distributed over the 26 years. Table 6 shows that the annual, 10, and 26 years cost savings. In the next scenario (Table 7), the life cycle factor is adjusted to 1.5, which assumes an extended lifetime of 15 years when using intelligent compaction.

Table 6. Roadway lifecycle cost per mile for one year and 26 Years.

Compaction Type	Service Life (yrs)	Annual Cost ($)	Cost Over 10 Years ($)	Cost Over 26 Years ($)
Conventional	10	25,000	250,000	650,000
Intelligent	26	9,600	96,00	250,000
Savings		15,500	154,000	400,000

Table 7. Roadway lifecycle cost per mile for one year and 15 years.

Compaction Type	Service Life (yrs)	Cost Per Year ($)	Cost Over 10 Years ($)	Cost Over 15 Years ($)
Conventional	10	25,000	250,000	375,000
Intelligent	15	16,700	167,000	250,000
Savings		8,300	83,000	125,000

For this stage of the case study, two different scenarios have been performed and analyzed. The first scenario is assuming 30% operating time improvement during IC application, and the second scenario is assuming no operational improvement. The average roller speed was extracted for the entire length of the construction section to estimate the duration of compaction. The cost of the roller operator for conventional compaction was estimated from typical labor surcharge and equipment rental rates. The cost of the IC roller is conservatively assumed $100 per hour. The cost of QC/QA for conventional compaction and IC were estimated to be 30% and 5% of the total costs, respectively. The cost of GPS equipment was removed from the calculations due to the insignificant impact on the overall analysis.

2.6 Case study scenario (1): 30% operating time improvement with IC

According to the results for the construction site (2.2-miles-long) in this case study, the cost savings of using IC technology compared to the conventional compaction process was around 35 percent (see Table 8 for the breakdown of costs).

2.7 Case study scenario (2): No operating time improvement with IC

In the second scenario, it was conservatively assumed that using IC would not minimize the duration of the compaction process. This condition can happen when the operator has no prior experience with the IC equipment and follows the conventional compaction routine with an IC-equipped roller. Even in this scenario, the IC process can save about 17 percent of the compaction costs compared to the conventional compaction process (Table 9).

It should be noted that these preliminary analyses are based on limited information from the construction case study. Most of the improvements in the IC process are either difficult to quantify or need detailed information and long term monitoring of the constructed section. It is also noteworthy that the choice of GPS precision and the IC technology affects both the initial equipment costs and quality of the collected data.

Table 8. Cost analysis of conventional compaction and IC for asphalt pavement project.

| Item | Conventional Compaction | | | | Intelligent Compaction | | | |
	Cost per unit ($)	Unit	Number of Units	Total Cost ($)	Cost per unit ($)	Unit	Number of Units	Total Cost ($)
Roller	86	hour	6.7	580	100	hour	5.2	520
Operator	22	hour	6.7	150	22	hour	5.2	115
QC/QA		m^2		310		m^2		35
Total				1,000				660

Table 9. Cost analysis of conventional compaction vs. IC for asphalt pavement project with equal operating time.

| Item | Conventional Compaction | | | | Intelligent Compaction | | | |
	Cost per unit ($)	Unit	Number of Units	Total Cost ($)	Cost per unit ($)	Unit	Number of Units	Total Cost ($)
Roller	86	hour	6.7	576	100	hour	6.7	670
Operator	22	hour	6.7	148	22	hour	6.7	150
GPS	-	-	-	-		hour	6.7	-
QC/QA		m^2		310		m^2		45
Total	-	-	-	1,040	-	-		860

3 CONCLUSIONS

Intelligent Compaction (IC) technology can improve the compaction quality by providing complete coverage for the entire compacted section as well as a comprehensive construction quality management database. A case study was considered in this paper to perform a preliminary life cycle assessment, and life cycle cost analysis of a pavement construction project to understand the benefits of IC compared to the conventional compaction practice. The preliminary life cost assessment to quantify the environmental impacts showed that using the intelligent compaction method can reduce 3.5 gallons of fuel consumption, 260 grams of CO emission, 210 grams of NOx emission, and 74-gallon water consumption over a 1 lane-mile length of asphalt pavement overlay. Based on the partial life cycle cost analysis, intelligent compaction can provide a cost efficiency of up to $170 per lane-mile on only compaction costs. Although difficult to quantify, the compaction uniformity provided by using IC technology can potentially enhance the fatigue performance and permanent deformation during the life cycle of the compacted pavement section. Long term monitoring of the pavement performance during the life cycle is critical in quantifying the benefits of using intelligent compaction methods. In this case study, a number of default assumptions were made during the analyses that may not be applicable to other construction projects. Such preliminary analyses can help agencies and contractors to better understand the benefits of intelligent compaction technology in the construction of pavement sections.

ACKNOWLEDGMENT

This study was carried out with partial support from the California State University Transportation Consortium (CSUTC). The contents of this study reflect the views of authors and not necessarily those of the sponsors.

REFERENCES

Briaud, J-L, Seo, J. (2003). Intelligent compaction: overview and research needs. Texas A&M University, College Station, TX;

Chang, G., Xu, Q., Rutledge, J., Horan, B., Michael, L., White, D., & Vennapusa, P. (2011). Accelerated implementation of intelligent compaction technology for embankment subgrade soils, aggregate base, and asphalt pavement materials. Federal Highway Administration Report No. IF-12–002.

Fathi, A., Tirado, C., Mazari, M., Rocha, S., & Nazarian, S. (2019, September). Correlating Continuous Compaction Control Measurements to In Situ Modulus-Based Testing for Quality Assessment of Compacted Geomaterials. International Conference on Information technology in Geo-Engineering (pp. 585–595).

Harvey, J. T., Meijer, J., Ozer, H., Al-Qadi, I. L., Saboori, A., Kendall, A. (2016). Pavement Life- Cycle Assessment Framework. Washington, DC: Federal Highway Administration.

Kumar, S. A., Mazari, M., Garibay, J., Al Douri, R. E., Nazarian, S., & Si, J. (2016). Compaction quality monitoring of lime-stabilized clayey subgrade using intelligent compaction technology. In 2016 International Conference on Transportation and Development, American Society of Civil Engineers.

Mazari, M., Beltran, J., Aldouri, R., Chang, G., Si, J., & Nazarian, S. (2016). Evaluation and Harmonization of Intelligent Compaction Systems. In International Conference on Transportation and Development 2016 (pp. 838–846).

Mooney, M. A., & Rinehart, R. V. (2007). Field monitoring of roller vibration during compaction of subgrade soil. Journal of Geotechnical and Geo-environmental Engineering, 133(3), 257–265.

Nazarian, S., Mazari, M., Abdallah, I. N., Puppala, A. J., Mohammad, L. N., & Abu-Farsakh, M. Y. (2015). Modulus-based construction specification for compaction of earthwork and un- bound aggregate. Washington, DC: Transportation Research Board.

Pasetto, M., Pasquini, E., Giacomello, G., & Baliello, A. (2017). Life-Cycle Assessment of road pavements containing marginal materials: comparative analysis based on a real case study. In Pavement Life-Cycle Assessment (pp. 199–208). CRC Press.

Santos, J., Bryce, J., Flintsch, G., Ferreira, A., & Diefenderfer, B. (2015). A life cycle assessment of in-place recycling and conventional pavement construction and maintenance practices. Structure and Infrastructure Engineering, 11(9), 1199–1217.

Savan, C. M., Ng, K. W., & Ksaibati, K. (2016). Benefit-cost analysis and application of intelligent compaction for transportation. Transportation Geotechnics, 9, 57–68.

Tirado, C., Fathi, A., Mazari, M., & Nazarian, S. (2019). Design verification of earthwork construction by integrating intelligent compaction technology and modulus-based testing, Transportation Research Board, (No. 19–03542).

United States Environmental Protection Agency (2017). Greenhouse Gas Inventory Explorer, Retrieved Dec 2019.

White, D. J., Vennapusa, P., & Thompson, M. J. (2007). Field validation of intelligent compaction monitoring technology for unbound materials. Center for Transportation Research and Education, Iowa State University. Report No. MN/RC–2007–10.

Xu, X., Akbarian, M., Gregory, J., & Kirchain, R. (2019). Role of the use phase and pavement- vehicle interaction in comparative pavement life cycle assessment as a function of context. Journal of Cleaner Production, 230, 1156–1164.

Impact of allocation method on carbon footprint of pervious concrete with industry byproducts

X.D. Chen
Graduate Research Assistant, Rutgers University

H. Wang
Associate Professor, Rutgers University

ABSTRACT: This study evaluated the impact of allocation method on carbon footprint of pervious concrete with industry byproducts. Three pervious concrete mixtures were considered in the analysis, including the mixtures with regular Portland cement, fly ash, and blast furnace slag. Life-cycle assessment (LCA) was used to quantify greenhouse gas (GHG) emission of three pervious concrete mixtures. The study applied allocation methods based on physical mass and economic values to quantify the effects of allocation on impact assessment of fly ash and slag. The upstream proportion and sensitivity analysis of allocating coefficients were considered uniquely. The analysis results showed that the choice of allocation methods had significant effect on LCA assessment results. For the scenario of no allocation, the replacement of cement with industry byproducts substantially reduced carbon footprint of pervious concrete mixtures by 15%. Using economic allocation method, the mixtures with fly ash and slag caused 4-5% greater GHG emission as compared to the scenario of no allocation. However, pervious concrete mixture with industry byproducts could cause greater environment burden when mass allocation method was used in LCA.

1   INTRODUCTION

Pervious pavement system as one of green infrastructures can mitigate the negative environmental impacts associated with storm water runoff in urban areas. Pervious pavement system allows storm water runoff to flow through and get filtered by gravel and soil. This process mimics the natural rainfall infiltration with earth. The applications of pervious concrete are commonly limited to light traffic area such as parking lots and sidewalk because concrete has high porosity and relatively lower strength and durability than conventional concrete (Shu et al. 2011).

The replacement of Portland cement with supplementary cementitious materials (SCMs) such as slag and fly ash gains popularity in the recent years to reduce the carbon footprint and for a more sustainable development. Portland cement was well documented by previous studies that it is the primary source of CO_2 emission for commercially produced concrete mixtures (Chen et al. 2010). To quantify the environmental impacts of the mixtures with varying content of supplementary materials, it is important to decide which allocation methods should be used in life cycle assessment study. Most of the previous studies neglected the allocation process and assign no environmental impact to the industrial by-products. For example, some studies assumed by-products such as fly ash from coal fired power plants and steel slag from steel production had no "embodied impacts" (Tait and Cheung 2016), since electricity and steel productions are responsible for all environmental impacts from production processes which are not aiming at producing waste products (Marceau et al. 2006; McLellan, et al., 2011; Van den Heede and De Belie, 2010).

On the other hand, several studies have been conducted to calculate the impact of allocation methods of fly ash and steel slag in cement mixture (Chen et al., 2010; Sayagh, et al. 2010;

Habert, 2012). The ISO standard for LCA recommends testing the applicability of two allocation methods if the allocation cannot be avoided. The first method is to use principles of physical causality for allocation, such as mass or energetic value; and the second method is to use other principles of causality such as economic values. The basic procedure of environmental impact allocation is to apply an allocation coefficient to the primary process to distinguish the environmental impact associated with the by-product. The mass allocation coefficient is the ratio of the weight of by-product to the total weight of the primary product and by-products. The economic allocation coefficient gives higher weight to the primary product according to the benefits of selling primary product and industry by-product.

Chen et al. (2010) conducted a study to explore different allocation methods of SCMs in the Europe. The results show that the mass allocation method induced the greater environmental burden on by-products, while the economic allocation method appeared to be more reasonable in emphasizing the fact that SCMs are waste product and should not share the same environmental burden as the main products. Since the allocation methods of by-products are highly sensitive to the process and efficiency of production, the application of allocation methods in LCA should consider geographical difference and potential market fluctuations. Sayagh et al., (2010) studied the sensitivity of environmental impacts of slag recycled into various pavement structures in France. The authors used allocation coefficients based on the mass proportion contributing to a given impact category. Results showed that the supplementary material of slag contributes to the saving of natural resources of asphalt binder and cement materials. Habert (2012) proposed an economic allocation method for fly ash and slag based on the economic behavior of European Union Greenhouse Gas Emission Trading System. The method allocated environmental costs fairly over the industries within carbon trading system, which is suitable to apply in consequential life cycle assessment.

However, there are few studies to investigate the effect of allocation methods in LCA of pervious concrete mixture using fly ash and steel slag in the U.S. At the same time, there is absence of study on the industry by-products' upstream energy consumption and GHGs. Depending on the study boundary, when allocation methods are applied, the by-products are accountable for part of the upstream environmental impacts from the production processes.

Aiming to address the research gaps, life-cycle assessment was adopted in this study to quantify GHG emission and energy consumption of three pervious concrete mixtures. This study used allocation methods based physical mass and economic values in calculating the GHGs due to energy consumption of fly ash and steel slag. The upstream proportion and sensitivity analysis of allocating coefficients were considered uniquely. Finally, to support multi-criteria decision-making process in selecting pervious concrete mixtures, radar charts were presented with internal normalization of economic costs, environmental sustainability, and engineering properties.

2 QUANTIFICATION OF ENVIRONMENTAL IMPACTS

2.1 Mix design

Table 1 shows the mixtures design proportions of pervious concrete with and without SCMs. The three mixtures include pervious concrete with fly ash (PRC-FA), pervious concrete with blast furnace slag (PRC- GBFS), and conventional pervious concrete (PC-Regular). Portland Type I Cement was used in all pervious concrete mixtures. The replacement ratios of fly ash and slag were 15% and 25% in this study.

Table 1. Mixtures design proportions (lb./cubic yard).

Mix	Cement	3/8 in. Aggregate	Fly Ash	Slag	Water	Water/Cement Ratio
PC-Regular	620	2700			168	0.27
PC-FA	525	2500	95		168	0.32
PC-BFS	465	2500		155	168	0.36

Life-Cycle Assessment provides a comprehensive method to quantify the environmental impacts of three pervious concrete mixtures with industrial by-products. The functional unit is defined as one cubic yard (0.765 cubic meter) of pervious concrete mixtures. One of the advantages of using LCA is that it can avoid double counting GHG emissions in various stages of pavement's life cycle. The impact assessments include greenhouse gas (GHG) emission and energy consumption from the upstream and direct combustion processes. The greenhouse gases considered in this study include CO_2, CH_4 and N_2O, which are converted to CO_2 equivalent using the Global Warming Potential (GWP) in 100-year time horizon (IPCC, 2007). The system boundary covers material acquisition and manufacturing stage of all materials used in the selected pervious concrete mixtures.

Table 2 shows the inventory data of energy consumption and GHGs for sand or gravel, crushed stone, and Portland cement. It is obvious that Portland cement has significantly higher environmental impacts than aggregate or sand.

The impact assessment considered both combustion and upstream processes. The upstream environmental impact of a product is defined as the environmental impacts of producing a certain amount of fuel. This amount of fuel was used to support the life-cycle energy consumption of the product being studied. At the same time, the combustion or direct environmental impacts of this product are from the GHGs generated by combusting the certain amount of fuel. Table 3 shows the energy profile of materials used in the three pervious concrete mixtures. The energy profile of fly ash and slag are for post-processing such as drying, grinding, and stocking. Equation 1 was used to calculate the upstream energy consumption and GHG emission.

$$UEE = \sum_{i=1}^{n} CE \cdot PE_i \cdot UEE_i \tag{1}$$

Where, CE = Combustion energy (MMBTU/ton); UEE_i = Upstream energy consumption (BTU/MMBTU) or emission (g/MMBTU) for the ith type of energy (calculated from GREET) as shown in Table 4; UEE = Upstream energy consumption (BTU/ton) or emission (g/ton); PE_i = Percent of the ith type of energy in the energy matrix; i = Type of energy including coal, diesel, gasoline, liquefied petroleum gas, natural gas, distillate oil, petroleum coke, residual oil, and electricity; and n = Total number of energy type.

There are mainly two processes for industrial by-products. The fist process is called "primary production", which means the production process of the main product and the "waste". The secondary process is the treatment of the by-production for further use. Take fly ash as an example, the primary production is the generation of electricity in a coal fired power plant; and the secondary process include drying, transportation and stocking of the fly ash. The system boundary for steel slag and fly ash as industrial by-products is shown in Figure 1.

There are two allocation methods in LCA: mass allocation and economic allocation. The mass allocation coefficient C_m is shown in Equation 2, and the coefficient for economic allocation approach is calculated using Equation 3.

Table 2.   LCI data for portland cement and aggregate (Marceau et al. 2006).

Raw material/Process	Energy consumption (MJ/ton)		GHG emission (g CO_2 eq. ton)	
	Combustion	Upstream	Combustion	Upstream
Portland Cement	4,340	1,522	927,988	133,579
Sand or Cement	23	26	73	1,094
Crushed Stone	32	30	1,420	1,323

Table 3. Energy profile of materials used in the mixtures.

	Cement	Aggregate	Coal-fired Power plant	Fly Ash	Steel	Slag
Coal	57%	2%	99.95%		1%	
Diesel						1%
Gasoline		4%				
Natural Gas	1%	12%		58%	33%	47%
Distillate Fuel Oil	3%	42%	0.046%	26%		6%
Petroleum Coke	18%					
Residual Oil		7%			2%	
Nuclear Power	9%					
Electricity	12%	33%		16%	18%	46%
Other					27%	
Total	100%	100%	100%	100%	100%	100%
Reference	PCA, 2006	PCA, 2006	Chen etal., 2010	Chen etal., 2010	U.S EIA, 2006	Dunlan, 2003

Table 4. Upstream energy consumption (BTU/MMBTU) and emission (g/MMBTU) (Wang, et al. 2016).

	Coal	Natural Gas	Distillate Fuel Oil	Gasoline	Diesel	Residual Oil	Petroleum Coke	Electricity U.S mix
Total Energy (Btu/MMBtu)	22,974	199,510	309,634	309,840	230,070	133,514	314,688	2,313,860
CO2 eq. (g/MMBtu)	5,548	18,401	20,127	20,144	21,761	13,768	26.368	176,769

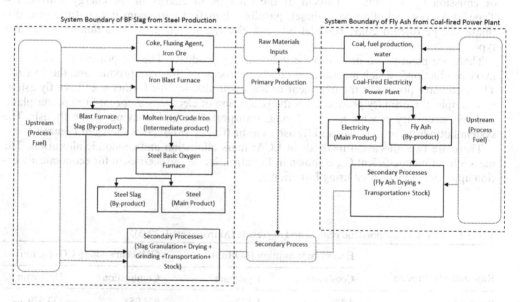

Figure 1. System boundary in LCA for steel slag and fly ash used in pervious concrete mixture.

$$C_m = \frac{m_{by-product}}{m_{primary\ product} + m_{by-product}} \qquad (2)$$

Where, m is the mass of product.

$$C_e = \frac{(p * m)_{by-product}}{(p * m)_{primary\ product} + (p * m)_{by-product}} \qquad (3)$$

Where, $(p * m)$ is the product of price per unit and the mass value of the material.

The upstream environmental impacts due to the production of the main products will be proportioned to fly ash and slag using economic and mass allocation coefficients. Equation 4 shows the calculation method.

$$I_{by-product} = C * \left[(I_{UEE} + I_{Direct})_{Primary\ process} \right] + (I_{UEE} + I_{Direct})_{Secondary\ process} \qquad (4)$$

Where, I_{UEE} refers to environmental impacts of the products associated with the upstream energy consumption or emission; I_{Direct} is the direct environmental impacts of the products; and C is the allocation coefficient.

Mass allocation coefficient C_m is the ratio of the by-product weight to the total weight of primary product and by-product. To calculate the mass allocation coefficient of slag, we need to know the production mass of slag, crude iron, and steel. The production of blast furnace slag was estimated to be 25% to 30% by the mass of crude iron output according to Mineral Commodity Summaries of U.S. 2017 with an average production rate of 27.5% (US Geological Survey, 2017). 98% of crude iron by mass could be used for steel production (World Steel Association, 2019). To calculate the mass allocation coefficient of fly ash, we need to know the production of electricity, coal, and fly ash. A study conducted by Babbitt and Lindner (2005) investigated four electricity power plants in Florida. The study showed that in general every 1000kg coal can generate 216kg coal combustion products (CCPs) and 9.68GJ electricity. It is noted that the production rates might vary across the U.S. since factors such as electricity production efficiency and CCPs rates are different. In a larger geographic scale of the whole country, the five-year (2012-2016) fly ash production ranges from 35% to 47% of total CCPs with an average is 41% according to the annual report of Coal Combustion Products (CCP) Production & Use Statistics published by American Coal Ash Association, (ACAA, 2016).

In the economic allocation method, the allocation coefficient C_e is estimated from both economic value and mass production of main products and byproducts. The average price of slag was $18.5 per metric ton from 2012 to 2016 ranging from $17 to $19.5 per metric ton (US Geological Survey, 2017). The five-year average price from 2012 to 2016 for crude steel (usually refers to hot-rolled steel) in the U.S. was $585.012 per metric ton, ranging from $462.25 to $656.57 per metric ton. The price of concrete quality fly ash ranges from $18.14 to $40.82 per metric ton (ACAA, 2017). For all sectors in 2012 to 2016, the price of electricity in the U.S. varied from $0.0984 to $0.1044 per kilowatt-hour with an average price of $0.10208 per kilowatt-hour (US EIA, 2017). The market price of by-products varies depending on many factors such as seasonal demands, supply amount, the market location, fuel price and transportation cost.

It is critical to get the inventory data of steel, slag, electricity, and fly ash before the allocation process. Table 5 and Table 6 shows the energy consumption and GHGs of per ton steel blast furnace slag, one-ton fly ash and one-kWh electricity generated from coal fired power plants.

Table 5. Inventory data for steel and slag.

Material	Energy Consumption (MJ/ton)		GHG Emission (g CO$_2$ eq./ton)		Data Reference
	Combustion	Upstream	Combustion	Upstream	
Primary Production per ton steel	43,924	24,355	2,488,326	1,784,212	(GREET, 2013) (EIA,2006)
Secondary Process per ton Slag	670	791	66,775	58,125	Dunlap (2003)

Table 6. Inventory data for electricity and fly ash.

Material	Energy Consumption (MJ/kWh)		GHG Emission (gCO$_2$ eq./kWh)		Data Reference
	Combustion	upstream	Combustion	upstream	
Primary Production per kWh (Coal-fired Power Plants)	11	0.254	1,001	58	(DOE, 2016
	(MJ/ton)		(gCO$_2$ eq./ton)		
Secondary Process per ton Fly Ash	140	79	10,055	5,839	(Chen et al., 2010)

3 RESULTS AND DISCUSSIONS

3.1 *Allocation coefficients*

Based on the collected data of production weights and prices, Table 7 shows the mass and economic allocation coefficients of slag and fly ash with average, maximum, and minimum values. In general, the coefficients of allocation by mass of slag and fly ash are considerably larger than coefficients of economic allocation. In different time periods, economic contexts, and geographic locations, the economic allocation coefficient might fluctuate (Chen et al. 2010). However, in the long run, mass allocation coefficient may be relatively stable because it is not likely to have a striking improvement of the production technologies of electricity and steel.

3.2 *Environmental impacts of slag and fly ash using allocation methods*

Figure 3 shows the environmental impacts of slag and fly ash in total of upstream and combustion processes for no allocation, economic allocation and mass allocation. The figures also show the GHG emission and energy consumption of one-ton cement as a comparison. The

Table 7. Allocation coefficients based on physical mass and economic values of fly ash and slag.

	Mass allocation coefficient		Econ allocation coefficient	
	Fly Ash	Slag	Fly Ash	Slag
Maximum	9.3%	22.2%	1.6%	1.2%
Average	8.0%	20.7%	1.0%	0.8%
Minimum	7.1%	19.2%	0.5%	0.6%

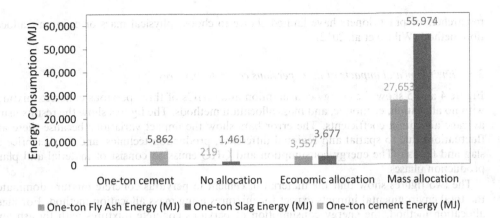

a)

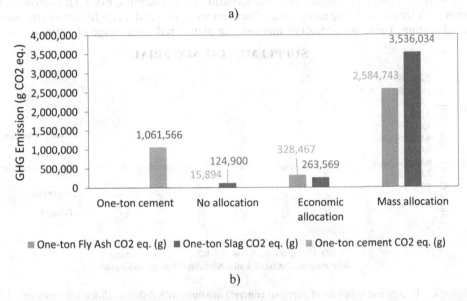

b)

Figure 3. a) Energy consumption, and b) GHG emission of fly ash and slag using no allocation, economic allocation, and mass allocation methods.

figures only present the environmental impacts using the average mass and economic allocation coefficients. It is noted that the scenario of no allocation did not allocate environmental burdens from primary production, instead, it only considered the secondary treatments of slag and fly ash such as drying and stocking. The results include raw material acquisition, plant production, and secondary production stages.

one-ton of slag and fly ash has substantial lower energy consumption and GHGs than one-ton cement for no allocation scenario and allocation based on economic values. With coefficient based on physical mass, the environmental impacts of both CSM materials sharply increase, which has more than 100% higher than the environmental impacts of cement with the same quantity, leaving no environmental advantages for using CSM materials in concrete mixture.

These results therefore suggest that the selection of allocation coefficients has a direct impact on the determining whether using CSM is more environmentally sustainable. It is also known that, driven by application considerations and data and information availability, researchers or practitioners have limited choice to choose physical mass or economic allocation method (Wiloso et al. 2012).

3.3 *Environmental impacts of three pervious concrete mixtures*

Figure 4 and 5 show the energy consumption and GHGs of three pervious concrete mixtures with no allocation, economic, and mass allocation methods. The figures show the results using average allocation coefficients. The error bars show the impact variations because there are fluctuations due to spatial and regional difference, production technics, and market price of slag and fly ash. The energy consumption and GHG emission consist of material and plant production phases.

The two figures show that the material of cement in pervious concrete mixture dominates the total environmental impacts expect for mixtures using mass allocation method. For mass allocation method, the energy consumption of pervious concrete mixtures with fly ash and slag were 77% and 262% higher than the scenario of no allocation. For GHG emission, mixtures with fly ash and slag using mass allocation were 43% and 101% higher as compared to no allocation. Using no allocation method, the plant production stage of pervious concrete

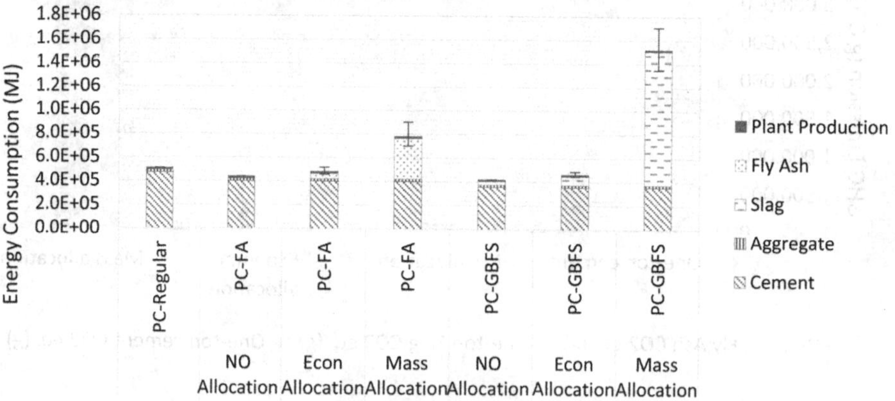

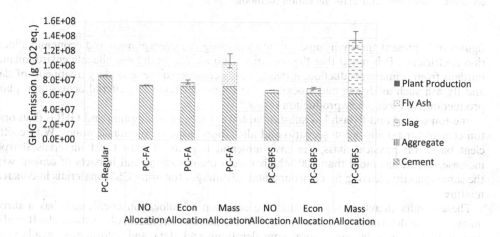

Figure 4. Energy consumption of pervious concrete mixtures with different allocation methods.

Figure 5. GHG emission of pervious concrete mixtures with different allocation methods.

mixture with fly ash and slag has 15% to 25% lower energy consumption than the regular pervious concrete mixtures. For allocation method based on economic values, the energy consumption of mixtures with fly ash and slag are 10% and 18% higher than the regular concrete mixture, and 6% and 9% higher in term of GHG emissions.

3.4 *Multiple performance characteristics of pervious concrete mixtures*

Although pavement LCA and LCCA are perfectly rational in itself, very limited thoughts have been given to the performance metrics as a whole. To assist decision making process in choosing the preferable pervious concrete mixture, we proposed a multiple performance characteristics analysis with internal normalization that collectively considers the engineering performance of pervious concrete mixtures, environmental sustainability, and economic justification.

Equation 5 shows the calculation of the internal normalization for cost and environmental sustainability; Equation 6 shows the internal normalization for engineering properties. There are two reasons for using two different internal normalization equations: the first purpose is to keep normalization results within 0 to 1; and the second purpose is using greater normalization values to reflect the desired trend, for example, the lower cost of a mixture is represented by the greater normalization value.

$$N_{ij}^{Int} = \frac{Min_i(a_{ij})}{a_{ij}} \tag{5}$$

$$N_{ij}^{Int} = \frac{a_{ij}}{Max_i(a_{ij})} \tag{6}$$

Where, N_{ij}^{Int} is internal normalization results; and a_{ij} are performance characteristics.

Table 8 presents the cost data in the U.S. of slag, fly ash, cement, water and aggregate for pervious concrete mixtures. As compared to the price per ton of cement, slag is 80% lower and fly ash is 70% lower. Thus, the economic cost of pervious concrete mixtures with supplementary material of fly ash and slag are expected to have more advantage than regular concrete mixtures.

The engineering properties, as shown in Table 9, are from laboratory testing including compressive strength tests, flexural strength tests, freeze-thaw resistance tests, and permeability tests. Testing details can be found from lab experiments conducted by Chen, et al., 2019.

The three radar charts in Figure 9 (supplemental material) present the internal normalization results of seven performance criteria as a metric of the performance of pervious concrete mixtures: the larger the area covered in the radar chart; the overall performance of that mixture is better. The criteria include sustainability of energy consumption and GHGs, economic cost in term of price, and engineering properties of freeze-thaw resistance, flexural strength, compressive strength, and porosity.

With no allocation (Figure 9 a) and with economic allocation method (Figure 9 b), using fly ash as supplementary material in the mixture has the largest area in the radar diagram, which means it has the best overall performance when the mixtures were evaluated with

Table 8. Cost data of pervious concrete mixtures (Chen, et al., 2019).

	Price (2015)	Unit	Reference
Slag	$19.50	metric ton	U.S. Geological Survey, 2017
Fly Ash	$29.48	metric ton	American Coal Ash Association 2017
Cement	$105.00	metric ton	U.S. Geological Survey, 2017
Water	$3.38	kgal	U.S. Department of Energy 2017
Aggregate	$11.31	metric ton	U.S. Geological Survey, 2015

Table 9. Engineering properties of the three pervious concrete mixtures (Chen, et al., 2019).

	Flexural Strength 28-day (Mpa)	Compressive Strength 28-day (Mpa)	Hydraulic Conductivity, k (cm/sec)	freeze thaw % of weight Remaining (100-cycle with salt)
PC-Regular	2.50	13.71	0.05	97.27
PC-FA	1.64	13.09	0.06	95.39
PC-GBFS	1.52	9.14	0.05	95.86

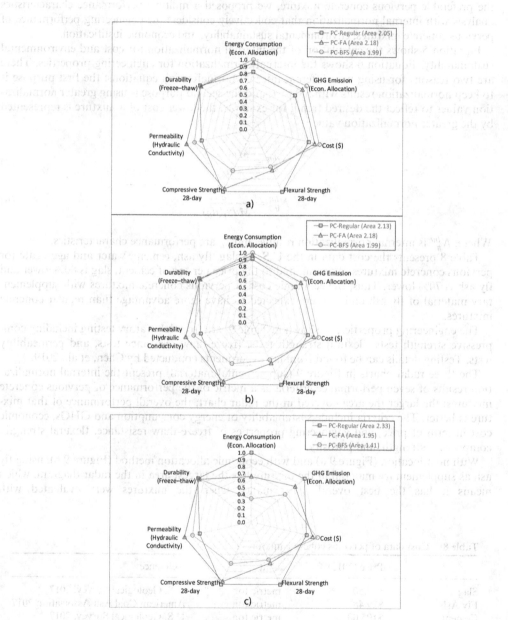

Figure 9. Multi-criteria assessment of pervious concrete mixtures a) no allocation; b) economic allocation method; c) physical mass allocation method.

environmental, economic, and engineering properties collectively. However, when mass allocation method was applied to fly ash and slag, Figure 9 (c) shows that the regular mixture with conventional cement has the largest covered area in the radar chart, because the supplementary materials of fly ash and slag has substantial higher energy consumption and GHGs than cement which considerably outranges their benefit of cost saving.

4 CONCLUSIONS

This study evaluated environmental impacts of three pervious concrete mixtures and explore the allocation method on industrial by-production of fly ash and slag as supplementary material to conventional Portland cement. The allocation methods include no allocation method, allocation based on economic values, and allocation based on physical mass of the products. To assist decision making process in choosing the preferable pervious concrete mixture, we proposed a multiple performance characteristics analysis with internal normalization that collectively considers the engineering performance of pervious concrete mixtures, environmental sustainability, and economic justification.

The environmental impacts of slag and fly ash with and without allocation methods were investigated. one-ton of slag and fly ash has substantial lower energy consumption and GHGs than one-ton cement for no allocation scenario and allocation based on economic values. With coefficient based on physical mass, the environmental impacts of both CSM materials sharply increase, which has more than 100% higher than the environmental impacts of cement with the same quantity, leaving no environmental advantages for using CSM materials in concrete mixture. For mass allocation method, the energy consumption of pervious concrete mixtures with fly ash and slag were 77% and 262% higher than the scenario of no allocation. For GHG emission, mixtures with fly ash and slag using mass allocation were 43% and 101% higher as compared to no allocation. Using no allocation method, the plant production stage of pervious concrete mixture with fly ash and slag has 15% to 25% lower energy consumption than the regular pervious concrete mixtures. For allocation method based on economic values, the energy consumption of mixtures with fly ash and slag are 10% and 18% higher than the regular concrete mixture, and 6% and 9% higher in term of GHG emissions. The choice of allocation method might be set by political decision. Economic allocation may be more appropriate than mass allocation if decision making process prefer to encourage the use of industry byproducts in concrete mixtures.

Using multi-criteria assessment of pervious concrete mixtures, the overall performance of each pervious concrete mixtures was quantified through the area covered in the radar diagram. With allocation method based on economic values and without using any allocation methods, mixture with fly ash as supplementary material has best overall performance when the it was evaluated with environmental, economic, and engineering properties collectively. With mass allocation, mixture with conventional Portland cement got the largest covered area in the radar chart since mixtures with supplementary materials of fly ash and slag has significantly higher environmental impacts.

REFERENCES

American Coal Ash Association (ACAA). (2017). About Coal Ash- How much are CCPs worth. Retrieved from https://www.acaa-usa.org/about-coal-ash.aspx.

Babbitt, C. W., & Lindner, A. S. (2005). A life cycle inventory of coal used for electricity production in Florida. *Journal of Cleaner Production, 13*(9), 903–912.

Chen, C., Habert, G., Bouzidi, Y., Jullien, A., & Ventura, A. (2010). LCA allocation procedure used as an incitative method for waste recycling: An application to mineral additions in concrete. *Resources, Conservation and Recycling, 54*(12), 1231–1240.

Chen, X. and Wang, H. (2018) Life cycle assessment of asphalt pavement recycling for greenhouse gas emission with temporal aspect, *Journal of Cleaner Production,* 187, 148–157.

Chen, X., Wang, H., Najm, H., Venkiteela, G., & Hencken, J. (2019). Evaluating engineering properties and environmental impact of pervious concrete with fly ash and slag. Journal of Cleaner Production, 237, 117714.

Chesner, W. H., Collins, R. J., & MacKay, M. H. (1998). User guidelines for waste and by-product materials in pavement construction, No. FHWA-RD-97-148.

Dunlap, R. (2003). Life cycle inventory of slag cement manufacturing process, Final report to Slag Cement Association.

Habert, G. (2013). A method for allocation according to the economic behaviour in the EU-ETS for by-products used in cement industry. The International Journal of Life Cycle Assessment, 18(1), 113–126.

Intergovernmental Panel on Climate Change (IPCC), Fourth Assessment Report (AR4), 2007.

Kevern, J. (2008). Advancement of pervious concrete durability. PhD Dissertation, Iowa State University.

Marceau, M., Nisbet, M. A., & Van Geem, M. G. (2006). Life cycle inventory of portland cement manufacture, PCA R&D Serial No. 2095b, Portland Cement Association.

McLellan, B. C., Williams, R. P., Lay, J., Van Riessen, A., & Corder, G. D. (2011). Costs and carbon emissions for geopolymer pastes in comparison to ordinary portland cement. Journal of Cleaner Production, 19(9), 1080–1090.

Sayagh, S., Ventura, A., Hoang, T., François, D., & Jullien, A. (2010). Sensitivity of the LCA allocation procedure for BFS recycled into pavement structures. Resources, Conservation and Recycling, 54(6), 348–358.

Shu, X., Huang, B., Wu, H., Dong, Q., & Burdette, E. G. (2011). Performance comparison of laboratory and field produced pervious concrete mixtures. Construction and Building Materials, 25(8), 3187–3192.

US Geological Survey (2001). Mineral Commodity Summaries, 2001: Government Printing Office.

US Geological Survey (2017). Mineral Commodity Summaries, 2017: Government Printing Office.

US Geological Survey (2015). 2015 Minerals Yearbook: Cement: Government Printing Office.

US Geological Survey (2015). 2015 Minerals Yearbook: Stone, curshed: Government Printing Office.

Tait, M. W., & Cheung, W. M. (2016). A comparative cradle-to-gate life cycle assessment of three concrete mix designs. The International Journal of Life Cycle Assessment, 21(6), 847–860.

Van den Heede, P., De Belie, N., Zachar, J., Claisse, P., Naik, T., & Ganjian, E. (2010). Durability related functional units for life cycle assessment of high-volume fly ash concrete. Paper presented at the Proceedings of the second international conference on sustainable construction materials and technologies. Volume three of three. Milwaukee: UWM Center for By-Products Utilization.

Wang, M. Q. (1999). GREET 1.5-transportation fuel-cycle model-Vol. 1: methodology, development, use, and results (No. ANL/ESD-39 VOL. 1). Argonne National Lab., IL.

Wang, H., Thakkar, C., Chen, X., & Murrel, S. (2016). Life-cycle assessment of airport pavement design alternatives for energy and environmental impacts. Journal of Cleaner Production, 133, 163–171.

Wiloso, E. I., Heijungs, R., & De Snoo, G. R. (2012). LCA of second generation bioethanol: a review and some issues to be resolved for good LCA practice. Renewable and Sustainable Energy Reviews, 16(7), 5295–5308.

World Steel Association. (2019, Feb). Raw Materials. Retrieved from World Steel Association: https://www.worldsteel.org/steel-by-topic/raw-materials.html

FocusEconomics. (2017). Steel Price History Data (USD per metric ton, aop). Retrieved from http://www.focus-economics.com/commodities/base-metals/steel-usa

U.S. Department of Energy (DOE). (2016). Executive Summary for the Environment Baseline, Volume 1: Greenhouse Gas Emissions from the U.S. Power Sector.

U.S. Energy Information Administration. (2017). Eletricity Data Browser. Retrieved from https://www.eia.gov/electricity/data/browser/

U.S. Energy Information Administration (EIA) (2006), Steel Industry Analysis, Accessed at http://www.eia.gov/consumption/manufacturing/briefs/steel/

Jin, N. (2010). Fly Ash Applicability in Pervious Concrete. (Electronic Thesis or Dissertation). Retrieved from https://etd.ohiolink.edu/

McCain, G., & Dewoolkar, M. (2010). Pervious concrete pavements: mechanical and hydraulic properties. Transportation Research Record: Journal of the Transportation Research Board, (2164), 66–75.

Argonne National Laboratory. (2012). GREET 2013 vehicle cycle model. Argonne National Laboratory.

Kalbar, P. P., Karmakar, S., & Asolekar, S. R. (2012). Selection of an appropriate wastewater treatment technology: A scenario-based multiple-attribute decision-making approach. Journal of environmental management, 113, 158–169.

Kim, H. H., & Park, C. G. (2016). Plant Growth and Water Purification of Porous Vegetation Concrete Formed of Blast Furnace Slag, Natural Jute Fiber and Styrene Butadiene Latex. Sustainability, 8 (4), 386.

Rahla, K. M., Mateus, R., & Bragança, L. (2019). Comparative sustainability assessment of binary blended concretes using Supplementary Cementitious Materials (SCMs) and Ordinary Portland Cement (OPC). Journal of Cleaner Production, 220, 445–459.

Ravindrarajah, R. S., & Yukari, A. (2010). Environmentally friendly pervious concrete for sustainable construction. In 35th Conference on Our World in Concrete & Structures, Singapore, 25–27.

Salonitis, K., & Stavropoulos, P. (2013). On the integration of the CAx systems towards sustainable production. Procedia CIRP, 9, 115–120.

U.S. Department of Energy (2017). Water and Wastewater Annual Price Escalation Rates for Selected Cities across the United States (No. DOE/EE-1670). Pacific Northwest National Lab. (PNNL), Richland, WA.

Development of pavement performance prediction models for in-situ recycled pavements in Virginia

E.A. Amarh
Virginia Tech Transportation Institute, Blacksburg, USA

J. Santos
University of Twente, Enschede, The Netherlands

G.W. Flintsch
Virginia Tech Transportation Institute, Blacksburg, USA

B.K. Diefenderfer
Virginia Transportation Research Center, Charlottesville, USA

ABSTRACT: The number of life cycle assessment studies of in-place recycled pavements is limited and the few existing barely considered the *Maintenance & Rehabilitation* (M&R) and *Use* stage. The main reason behind this limitation is the lack of information on how the performance of the pavement evolves over time after the application of a treatment, and how to determine the M&R frequencies and service life of each alternative treatment. Pavement performance prediction models (PPPMs) are therefore needed to fill this gap. However, the progression of deterioration in recycled pavement systems in the long-term is not clearly understood and there is limited data to support the development of the PPPMs. This paper presents the development of PPPMs for road pavement sections rehabilitated with in-place recycling treatments in the state of Virginia, United States. The developed models show that the deterioration rate of roughness in pavement sections rehabilitated with these treatments ranges from 0.7 to 5.2 in/mi/year depending on the type of treatment.

1 INTRODUCTION

Asphalt pavement recycling techniques, including hot in-place recycling (HIR), cold in-place recycling (CIR), cold central plant recycling (CCPR) and full-depth reclamation (FDR), have proven to be cost effective rehabilitation strategies that offer many advantages compared to traditional methods, such as milling and filling. Some of these advantages include reduction or complete elimination of the need for virgin materials, reduced traffic congestion and lower environmental impacts (Santos et al. 2017; Stroup-Gardiner 2011; Thenoux et al. 2007). Despite many successful experiences, some DOTs are still reluctant to use in-place pavement recycling treatments due to concerns about the performance of these treatments compared to more traditional pavement maintenance and rehabilitation (M&R) treatments.

The existence of criteria for selecting the right treatment to apply to the right candidate road section at the right time is one of the most commonly cited bottlenecks impeding the widespread use of asphalt pavement recycling treatments (Stroup-Gardiner 2011). The Federal Highway Administration's (FHWA) 2006 Recycled Materials Policy, revised in 2015 (FHWA 2018), aims to encourage the use of recycling techniques in pavement rehabilitation projects. An extract from the 2015 policy states, *"the determination of the use of recycled materials should include an initial review of engineering and environmental suitability"*. Several FHWA's publications in the form of technical guidelines and checklists provide information to support the review of the engineering suitability aspects of the policy statement, covering initial project level forensic examinations and the identification of the failure mechanism of candidate projects. However, there are no guidelines on how to assess the environmental suitability of these recycled materials.

Life cycle assessment (LCA), a standardized methodology intended to analyze and quantify the potential environmental impacts of a process, product or system, can be used to ascertain the suitability of the FHWA policy from the environmental perspective. Because pavement LCA is a relatively emerging field of study, and not many analysis involving recycling projects have been carried out, many important components are currently missing that need to be developed for the sake of a comprehensive analysis. Key among these are: (i) an inventory database that covers various unit processes related to the recycling of a road pavement using various techniques; (ii) PPPMs that predict how these recycled pavements will deteriorate in the future; (iii) a tool that conducts an inventory analysis and estimates the associated potential environmental impacts.

Some of the needs listed above are corroborated by the NCHRP Synthesis 421 that identified the lack of a well-designed experimental approach to assess the progression of pavement distresses and the overall decline in the pavement condition index that can provide information on life cycle cost and service life of in-place recycling techniques (Stroup-Gardiner 2011). Among the existing studies that have analyzed and documented the performance of in-place pavement recycling techniques (Jones et al. 2015; Lewis et al. 2006; Miller et al. 2006; Romanoschi et al. 2004) only a few have provided a time-evolution of project performance exceeding 5 years (Amarh et al. 2019; Lane and Kazmierowski 2012).

Wolters and Zimmerman (2010) reviewed state practices on performance modeling and developed three pavement performance modeling options for two groups of models – individual and family-type models – for PennDOT. Deterioration models were developed for Virginia DOT (VDOT) in a study that incorporated the structural capacity of the pavement in the form of a modified structural index, along with the pavement age (Ercisli 2015). Several model shapes were discussed in the author's research. However, models for in-place recycling projects were not specifically developed in any of these studies. A literature review identified three recent studies that discussed deterioration models for in-place recycling projects as part of a broader project (Saboori et al. 2017; Santos et al. 2015; Santos et al. 2017; Senhaji 2017). Senhaji (2017) discussed a two-pronged approach to estimating the performance and lifespan of in-place recycling treatments. Network-level condition data from the California DOT was used to develop deterministic models that predicted the progression of international roughness index (IRI) and wheelpath cracking from the treatment age. Linear models for conventional asphalt concrete overlays, CIR, and FDR were developed for this project. Santos et al (2015) conducted an LCA for in-place recycling project and compared the results to two other pavement maintenance alternatives, namely traditional reconstruction, and corrective maintenance. The authors developed an LCA model that included the use stage into the system boundaries. To determine the M&R strategies to be implemented in these projects, the authors used a quadratic model that predicted the pavement IRI progression from the treatment age. Data from an adjacent roadway section was used to calibrate the quadratic IRI model, as there was lack of long-term IRI measurements for the in-place recycling project under investigation. The IRI prediction was subsequently used to estimate the additional fuel consumption due to the rolling resistance (RR). Saboori et al (2017) estimated the potential environmental impacts of alternative end-of-life treatments, including pavement recycling treatments in California. The scope of the study did not include the M&R and use stages due to the lack of information on (i) how pavement roughness evolves over time (and thereby impacting vehicle fuel consumption), and (ii) how to determine the M&R frequencies and service life of each alternative treatment. The authors concluded the study emphasizing the need for PPPMs that can help to understand how pavement recycling-based treatments affect pavement performance.

Motivated by the needs discussed above, this paper presents the development of PPPMs for recycled asphalt pavements that will not only serve as critical inputs to enhance the comprehensiveness of pavement LCA modeling but also aid DOTs and other decision makers in (i) quantifying the service life of recycling projects and (ii) developing M&R strategies for better planning and allocation of funds in the future.

2 OBJECTIVES

The main goal of the research work presented in this paper is to develop PPPMs that VDOT and LCA practitioners in general can use to model the M&R and use stages of road pavement projects involving the application of pavement recycling treatments. The emphasis was placed on creating PPPMs for individual VDOT's in-place recycling projects to support an on-going comparative LCA study performed by the DOT.

3 METHODOLOGY

3.1 *Data processing*

Data from the VDOT's pavement management system (PMS) was used in this research work. Data related to IRI, rutting, fatigue cracking, transverse cracking, and summarized condition indices for 16 FDR, 4 CIR, 2 CCPR sections was collected. Data preparation steps involving the identification and removal of erroneous data (e.g., unreasonably high or low data points, negative values) were carried out to clean the raw data. For nonlinear modeling, a requirement of a minimum three time-series data points was set to clean the data further (Baladi et al. 2017). Therefore, all projects with less than three years of data were removed from the analysis. After the data processing/filtering steps a total of 8 FDR and 2 CIR projects detailed in Table 1 were considered suitable to proceed with the analysis.

3.2 *Exploratory data analysis*

The filtered VDOT's data was analyzed using the curve-fitting tools and visual graphs in JMP statistical software. The data comprised the pavement age computed from the project construction year, along with condition descriptors, such as fatigue cracking, rutting, IRI, transverse cracking, longitudinal cracking, patching and bleeding. Information on the overall condition of the projects in the form of summarized indices (i.e., critical condition index (CCI), load-related distress index and non-load related distress index) were also included. For flexible pavements, the load related distress rating (LDR) gives an indication of pavement condition concerning damage due to wheel loads applied to the pavement. It comprises distresses such as alligator (fatigue) cracking, wheel path patching and rutting. The LDR is a deduct-based index with a value of 100 when there are no discernible load related distresses on the pavement being evaluated. Deduct points are assigned for each of the distresses that are load-related depending on the type as well as severity and frequency of occurrence. Similar to the LDR, the non-load related distresses rating (NDR) represents the functional condition of the pavement, but the distresses assigned here are not load-related. Longitudinal and transverse cracking, non-wheel path patching and bleeding are examples of the distresses measured to calculate NDR. The CCI is the lower of the LDR and NDR and is used as an indicator to measure the overall pavement condition. Details on the development of these indices are discussed in (McGhee 2002). The pavement parameters of interest for conducting pavement LCA studies are the CCI, IRI, and age, with the latter being used as a predictor of future CCI and IRI values. A scatterplot matrix of the projects showing distributions, linear fits and correlation between these variables is presented in Figure 1.

The pooled data showed a strong negative correlation between the CCI and age, and a positive correlation between the IRI and age. These trends were expected, as they are characteristic of pavement deterioration with time. Summary statistics for the pavement parameters are presented in Table 2.

Since the underlying objective of the development of the PPPMs is to enable comparative LCA studies involving recycling projects, a decision was made to perform the analysis using "Route" as a grouping variable. This will result in analyzing the projects individually with the aim of developing PPPMs for each project instead of a single model to characterize a family of projects with similar characteristics. There was high variation in the CCI and IRI data per project. The age of the project ranged from a minimum of 5 years to a maximum of 9 years.

Table 1. Characteristics of in-place recycling projects used for the analysis.

Route (Length - mi)	Road class	Recycling methods[*]	AADTT[**]	Pavement structure (above subgrade)[***] Layer 1	Layer 2	Layer 3	Layer 4	Total thickness
IS00081SB (3.7)	Interstate	FDR Lime + CIR FA + CCPR	6943	2.0 in. SMA12.5D	4.0 in IM19.0D	6.0 in CCPR	12.0 in FDR	24.0 in.
SR00003EB (3.0)	Primary	FDR Cement	92	2.0-in SM12.5A	2.0-in IM19.0A	9.5-in FDR	-	13.5 in.
SR00003WB (3.0)	Primary	FDR Cement	85	2.0-in SM12.5A	2.0-in IM19.0A	9.5-in FDR	-	13.5 in.
SR00006EB (3.6)	Primary	FDR Cement	127	1.5-in SM12.5A	2.0-in IM19.0A	9.0-in FDR	-	12.5 in.
SR00013EB (3.6)	Primary	FDR Cement	172	1.5-in SM12.5A	2.0-in IM19.0A	9.0-in FDR	-	12.5 in.
SR00024EB (2.9)	Primary	FDR Cement	61	1.5-in SM9.5D	9.0-in FDR	-	-	10.5 in.
SR00040EB (0.25)	Primary	FDR FA	48	2.5-in SM9.5D	9.8-in FDR	-	-	12.3 in.
SR00040EBa (0.25)	Primary	FDR AE	48	2.5-in SM9.5D	9.8-in FDR	-	-	12.3 in.
US00017NB (9.8)	Primary	CIR AE	127	1.5-in SM12.5A	2.0-in IM19.0A	5.0 in. CIR	-	8.5 in.
US00017SB (9.8)	Primary	CIR FA	170	2.0-in SM12.5A	3.0-in IM19.0A	5.0 in. CIR	-	10.0 in.

*Recycling methods:
CCPR = Cold central plant recycling
FDR FA = Full-depth reclamation with foamed asphalt
CIR AE = Cold in-place recycling with asphalt emulsion
**AADTT = Average Annual Daily Truck Traffic
***Mix Type Definitions:
IM19.0D = Intermediate Mix with 19.0 mm maximum nominal aggregates, "D" stands for binder with performance grade 70-22
SMA12.5D = Stone Matrix Asphalt with 12.5 mm maximum nominal aggregates, "D" stands for binder with performance grade 70-22
SM9.5A = Surface Mix with 9.5 mm maximum nominal aggregates, "A" stands for binder with performance grade 64-22
SM9.5D = Surface Mix with 9.5 mm maximum nominal aggregates, "D" stands for binder with performance grade 70-22
SM12.5A = Surface Mix with 12.5 mm maximum nominal aggregates, "A" stands for binder with performance grade 64-22
SM19.0A = Surface Mix with 19.0 mm maximum nominal aggregates, "A" stands for binder with performance grade 64-22

3.3 *Pavement Performance Prediction Models (PPPMs) development*

Several approaches have been used to develop PPPMs. They include, among others, neural networks, fuzzy logic systems, genetic algorithms, neurofuzzy systems and regression methods (Flintsch and Chen 2004). Despite its simplicity compared to the other methods, the traditional regression analysis approach has the potential to satisfy the model validation criteria. Thus, it was adopted in this study. Other types of function such as exponential, sigmoidal and logistic that are commonly used outside the pavement domain were also evaluated. In addition, a negative binomial model capable of accommodating the over-dispersion usually observed in pavement data was examined (Byers et al. 2003). A brief description of the models is presented below, with the representative graphical forms and equations presented in Figure 2 and Table 3 respectively.

The exponential model describes a mathematical function whose growth rate value is proportional to the function's current value. Since the condition of a pavement section in a given year depends on the condition in the previous year, the inclusion of the exponential model is then justified. Specifically, the two-parameter and three-parameter forms were evaluated. The logistic model is an "S" shaped curve widely used to describe

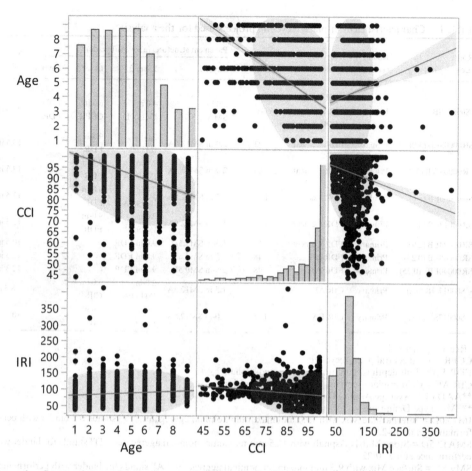

Figure 1. Scatterplot matrix showing correlations and distributions between CCI, IRI (in/mi) and pavement age (years).

Table 2. Summary statistics for pavement data.

Parameter	Mean	Standard deviation	Minimum	Maximum
Age (year)	6.60	2.25	5	9
CCI	93.21	10.51	44	100
IRI (in/mi)	86.14	30.81	33	412

growth and decay in several fields of study. For pavement deterioration modeling where the condition decreases with increasing age, the application of an inverted "S"-curve can be considered to describe the trends in observed condition data with respect to time. The Gompertz model is another "S"-curved function for time series analysis that describes growth as being slow at the beginning and end of the period under study. Though similar to the logistic model, the Gompertz curve approaches the lower asymptote value at a faster rate than the upper asymptote value (contrary to the logistic model where both asymptotes are approached symmetrically by the curve). The negative binomial model, derived from the Poisson distribution, permits the modelling of data with variance relatively different (larger or smaller) compared to the mean. Over-dispersion is commonly observed in pavement condition data (Ercisli 2015) and the negative binomial model presents a unique solution where linear models and Poisson distributions are

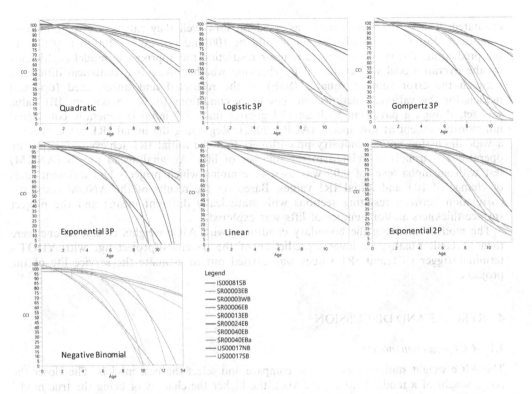

Figure 2. Potential CCI fitting curves as a function of pavement age.

Table 3. Test statistics for the comparison of CCI prediction models.

Function	General equation	AICc	AICc weight	SSE	RMSE
Negative Binomial	$a - Age^b \times Exp(c)$	270.7	54%	222.0	6.3
Quadratic	$a + b \times Age + c \times Age^2$	272.5	22%	228.2	6.5
Logistic 3P	$\dfrac{c}{1+Exp(-a \times (Age-b))}$	272.8	18%	229.5	6.6
Gompertz 3P	$a \times Exp(Exp((-b \times (Age - c)))$	275.1	6%	237.5	6.8
Exponential 3P	$a - b \times Exp(c \times Age)$	281.7	0%	262.9	6.6
Linear	$a - b \times Age$	338.6	0%	630.6	12.9
Exponential 2P	$a \times Exp(b \times Age)$	346.5	0%	713.1	14.6

Note: *a, b, c* = model coefficients

incapable of adequately describe the data and account for the over-dispersion respectively (Poch and Mannering 1996).

Regression analysis was performed to predict the CCI and IRI of the projects using the treatment age as the predictor variable. VDOT generally uses the CCI and other factors as a trigger to plan the type and frequency of pavement M&R schedules in the state. In pavement LCA models, IRI models are used to predict the evolution of surface roughness over time and subsequently assess the impacts on vehicle fuel consumption due to RR. A number of model-shapes from various functions (Figure 2) discussed in Ercisli (2015) were initially fitted to the data from individual projects to determine which models best fit the trends observed in scatter plots.

The most plausible models were then selected using the second-order Akaike information criterion (AICc) weight (calculated from the AICc). This estimator represents the relative likelihood of a model when comparing several models (where 1.0 being most likely). The nature of the curves, in terms of how sharply they evolve was visually evaluated. The curves were further assessed on how well they satisfy several boundary conditions. For CCI, the initial value should be 100 and not exceed 100 at any time in the prediction. The effect of constraining or restricting the appropriate model coefficients on the overall model was evaluated by checking whether there is a statistical difference between the error sum of squares (SSE) of the restricted and unrestricted (original) models for the various functions. No boundary conditions for the maximum IRI value was set, though a pavement with an IRI greater than 500 in/mi is generally considered not rideable except at low speeds (ACPA 2002). Regarding the initial IRI value, there is a wide discussion among industry practitioners that the initial IRI achieved after paving operations is usually correlated to the number of lifts. An analysis of means (ANOM) test with an alpha level of 0.05 was used to examine which projects had different rates of change of IRI and initial IRI values. Based on the results of the ANOM tests, the correlation between resulting sections with statistically different means and the project surface thickness and/or number of lifts was explored.

The models satisfying the boundary conditions with AICc weights closest to one were then selected. Finally, an inverse prediction of the pavement/project age with VDOT's terminal/trigger CCI and IRI values was carried out to estimate the service life of the projects.

4   RESULTS AND DISCUSSION

4.1   CCI prediction models

The AICc weight statistic was used to compare and select the best models. The closer the AICc weight of a model is to 1 (or 100%), the higher the chances of being the true model among those being compared.

Table 3 shows the test statistics from this step of the analysis. The negative binomial model was selected for further validation. As the CCI is an index, ranging from 0 to 100 the model was "refitted", this time with appropriate restrictions on the model coefficients. The effect of constraining or restricting the parameters (y-intercept) on the overall model was evaluated by checking whether there is a statistical difference between the error sum of squares (SSE) of the restricted and original model. The results of the test statistics are presented in Table 4. There is insufficient statistical evidence to conclude that the SSE for the compared models (original and fixed parameters) are statistically different as the p-value was found to be greater than 0.05. Thus, one can conclude that models with the intercept set to 100 can be used without any significant changes to the model. The coefficients and test statistics for the final model are presented in Table 5.

4.2   IRI prediction models

The IRI is used together with the CCI to make rehabilitation decisions. In pavement LCA modeling, IRI values are commonly used as a trigger for maintenance needs. Furthermore, they are also used to relate pavement roughness to vehicle fuel consumption during the use stage because of the RR. The model functions and corresponding statistics are presented Table 6.

Based on the AICc weight values the exponential 2P model was found to be the model that best predicts the IRI from the treatment age comparatively to the linear model. Table 7 shows

Table 4.   Test statistics for evaluating the effects of fixing parameters in the original model.

Model		SSE	DFE	MSE	Restrictions	F Ratio	Prob > F
Negative Binomial	original	237.5	35	6.79			
	a fixed	316.8	45	7.04	$a = 100$	0.4930	0.8828

Table 5. Parameter estimates and statistics for the CCI prediction models.

Project ID	Parameter estimates					
	a	Standard Error	b	Standard Error	c	Standard Error
IS00081SB	100	0	1.7	0.7	-0.7	1.3
SR00003EB	100	0	1.9	0.2	0.1	0.4
SR00003WB	100	0	1.8	0.2	0.3	0.3
SR00006EB	100	0	5.0	1.5	-8.1	3.3
SR00013EB	100	0	1.3	0.3	0.0	0.7
SR00024EB	100	0	2.1	1.7	-1.7	2.6
SR00040EB	100	0	3.9	0.7	-4.4	1.3
SR00040EBa	100	0	4.5	1.0	-5.8	1.9
US00017NB	100	0	2.0	2.3	-1.8	3.5
US00017SB	100	0	2.6	0.7	-1.3	1.1

Table 6. Test statistics for the comparison of IRI model.

Function	General equation	AICc	AICc weight	SSE	RMSE
Exponential 2P	$a \times Exp(b \times Age)$	495.4	73%	2927.9	8.1
Linear	$a + b \times Age$	497.4	27%	3019.0	8.2

Table 7. Parameter estimates and statistics for the IRI prediction models.

Project	Parameter estimates			
	a	p-value	b	p-value
IS00081SB	48.5	<.0001	0.008	0.79
SR00003EB	79.2	<.0001	0.034	0.11
SR00003WB	89.9	<.0001	0.021	0.30
SR00006EB	90.0	<.0001	0.007	0.52
SR00013EB	91.9	<.0001	0.014	0.19
SR00024EB	105.0	<.0001	0.004	0.87
SR00040EB	89.7	<.0001	0.035	0.02
SR00040EBa	100.8	<.0001	0.058	<.0001
US00017NB	74.2	<.0001	0.003	0.94
US00017SB	84.2	<.0001	0.013	0.65

the estimates and test statistics for the final IRI model selected. The IS00081SB project was found to have a significantly lower initial IRI value (48 in/mi) comparatively to the average value of all projects (i.e., 85 in/mi). The number of paved layers was higher for this project comparatively to that of the other projects analyzed. In turn, the SR00040EBa project was found to have a significantly high rate of IRI deterioration (7 in/mi) comparatively to the average value of all projects (i.e., 2 in/mi/year). The average rate of change of IRI for the cement-treated and bitumen-treated FDR treatments were found to be 1.5 and 5.2 in/mi/year, respectively, while the bitumen-treated CIR treatments were found to deteriorate at a rate value of 0.7 in/mi/year. Figure 3 presents the results in the ANOM graphs. Observations in red are statistically different from the mean.

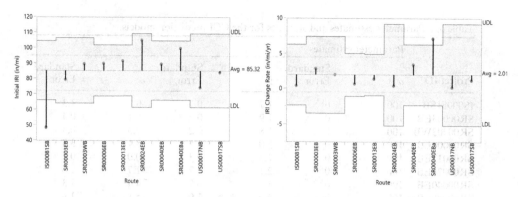

Figure 3. Comparison of the rates of change of IRI and initial IRI values across projects.

4.3 Estimation of treatment service life

Table 8 in the Supplemental Material shows the results of the estimated life of the various treatments from typical trigger values used by VDOT for rehabilitation decisions. Using a terminal CCI value of 40 (poor condition), the service life of the projects was estimated from an inverse prediction of the project age. It resulted in an average of 13 years with a standard deviation of 6 years. The average age of the cement-treated FDR and bitumen-treated FDR projects were estimated to be 14 and 9 years respectively. The bitumen-treated CIR sections averaged 13 years with a standard deviation of 7 years. The IS00081SB project, which combines FDR, CIR and CCPR, was estimated to last 18 years before a major rehabilitation

SUPPLEMENTAL MATERIAL

Table 8. Results of treatments service life estimation.

Model function	Project ID	Recycling method	ADTT	Parameter predicted	Trigger condition	Estimated life
	IS00081SB	FDR Lime + CIR FA + CCPR	6943		40	18
	SR00003EB	FDR Cement	92		40	8
	SR00003WB	FDR Cement	85		40	8
Negative Binomial	SR00006EB	FDR Cement	127		40	11
	SR00013EB	FDR Cement	172	CCI	40	26
	SR00024EB	FDR Cement	61		40	16
	SR00040EB	FDR FA	48		40	9
	SR00040EBa	FDR AE	48		40	9
	US00017NB	CIR AE	127		40	19
	US00017SB	CIR FA	170		40	8
	IS00081SB	FDR Lime + CIR FA + CCPR	6943		140	> 30
	SR00003EB	FDR Cement	92		140	17
	SR00003WB	FDR Cement	85		140	21
	SR00006EB	FDR Cement	127		140	> 30
Exponential 2 P	SR00013EB	FDR Cement	172	IRI (in/mi)	140	30
	SR00024EB	FDR Cement	61		140	> 30
	SR00040EB	FDR FA	48		140	13
	SR00040EBa	FDR AE	48		140	6
	US00017NB	CIR AE	127		140	> 30
	US00017SB	CIR FA	170		140	> 30

Note: FDR = Full-Depth Reclamation, CIR = Cold In-Place Recycling, CCPR = Cold Central Plant Recycling, FA = Foamed Asphalt, EA = Asphalt Emulsion treatment is required. Finally, it is worth mentioning that these results are aligned with the values found in literature (ARRA 2015; Peshkin et al. 2004) and with contractors' survey data reported by Senhaji (2017).

5 SUMMARY

In-service road pavements require maintenance and rehabilitation (M&R) treatments to keep them in compliance with structural and functional standards and requirements. With the increased focus on the sustainability of our roadway systems, it has become important to document the cost and potential environmental impacts of different M&R strategies over the pavements life cycle. In the last 12 years, the Virginia Department of Transportation (VDOT) has conducted various pavement recycling projects to evaluate the performance of emerging recycling-based M&R techniques, such as Cold Recycling (CR) and Full-Depth Reclamation (FDR). Life cycle assessment (LCA) studies involving in-situ recycled pavements are almost non-existent, and in general, the few existing studies did not comprehensively cover the M&R and use stages.

This paper presents (i) the development of pavement performance prediction models (PPPMs) to describe condition deterioration and (ii) the estimation of service life of end-of-life treatment projects in Virginia. The critical condition index (CCI) and pavement roughness, expressed in terms of IRI, were predicted through regression analysis by using as predictor the pavement age. The service life of the treatments was posteriorly estimated from the resulting models by using VDOT's terminal trigger values.

The initial IRI values and their rate of change for the treatments analyzed were found to range 48 - 85 in/mi and 0.7 - 5.2 in/mi/year, respectively, depending on the type of treatment. The age of the treatments applied in primary roads estimated from a trigger CCI value of 40 ranged from 8 - 26 years, with cement-treated projects generally lasting longer. The Interstate-81 project that combined FDR, CIR and CCPR is projected to last 18 years without the performance of any major rehabilitation.

6 FUTURE WORK

In the near future, this study will proceed in two main directions. First, PPPMs for other pavement condition parameters such as fatigue cracking, longitudinal cracking, and patching will be developed by using VDOT's and other States' data received from a survey carried out earlier. These models will be assessed against the current PPPMs used by VDOT's PMS with similar rehabilitation strategies. For example, the current VDOT's models for new reconstruction projects in interstate highways will be compared to the model developed for the FDR project implemented in the Interstate 81. Second, the developed PPPMs for cold recycling treatments will be included in the use stage model of a VDOT's tailored pavement LCA tool that is currently being developed by the authors.

REFERENCES

Amarh, E. A., Flintsch, G. W., Fernández-Gómez, W., Diefenderfer, B. K., and Bowers, B. F. (2019). "Eight-Year Field Performance of Portland Cement and Asphalt Stabilized Full-Depth Reclamation Projects." *Airfield and Highway Pavements 2019: Design, Construction, Condition Evaluation, and Management of Pavements*, American Society of Civil Engineers Reston, VA, 263–272.

American Concrete Pavement Association (2002). "The International Roughness Index (Iri): What Is It? How Is It Measured? What Do You Need to Know About It?" *American Concrete Pavement Association R&T Update, Concrete Pavement Research and Technology. Skokie, IL.*(3.07).

Asphalt Recycling and Reclaiming Association (2015). *Basic Asphalt Recycling Manual*, United States Federal Highway Administration, Annapolis, MD.

Baladi, G. Y., Dawson, T., Musunuru, G., Prohaska, M., and Thomas, K. (2017). "Pavement Performance Measures and Forecasting and the Effects of Maintenance and Rehabilitation Strategy on Treatment Effectiveness." United States. Federal Highway Administration.

Byers, A. L., Allore, H., Gill, T. M., and Peduzzi, P. N. (2003). "Application of Negative Binomial Modeling for Discrete Outcomes: A Case Study in Aging Research." *Journal of clinical epidemiology*, 56(6), 559–564.

Ercisli, S. (2015). "Development of Enhanced Pavement Deterioration Curves."Master's thesis, Virginia Tech, Blacksburg.

FHWA (2018). "Overview of Project Selection Guidelines for Cold in-Place and Cold Central Plant Pavement Recycling." *TechBrief*, Federal Highway Administration, Office of Asset Management, Pavements, and Construction, FHWA-HIF–17–042.

Flintsch, G. W., and Chen, C. (2004). "Soft Computing Applications in Infrastructure Management." *Journal of Infrastructure Systems*, 10(4), 157–166.

Jones, D., Wu, R., and Louw, S. (2015). "Comparison of Full-Depth Reclamation with Portland Cement and Full-Depth Reclamation with No Stabilizer in Accelerated Loading Test." *Transportation Research Record: Journal of the Transportation Research Board*(2524), 133–142.

Lane, B., and Kazmierowski, T. (2012). "Ten-Year Performance of Full-Depth Reclamation with Expanded Asphalt Stabilization on Trans-Canada Highway, Ontario, Canada." *Transportation Research Record*, 2306(1), 45–51.

Lewis, D., Jared, D., Torres, H., and Mathews, M. (2006). "Georgia's Use of Cement-Stabilized Reclaimed Base in Full-Depth Reclamation." *Transportation Research Record: Journal of the Transportation Research Board*(1952), 125–133.

McGhee, K. (2002). "Development and Implementation of Pavement Condition Indices for the Virginia Department of Transportation." *Phase I: Flexible Pavements*, Virginia Department of Transportation, Maintenance Division.

Miller, H. J., Guthrie, W. S., Crane, R. A., and Smith, B. (2006). "Evaluation of Cement-Stabilized Full-Depth-Recycled Base Materials for Frost and Early Traffic Conditions." Recycled Materials Resource Center, University of New Hampshire, Durham, Final Report, Project 28.

Peshkin, D. G., Hoerner, T. E., and Zimmerman, K. A. (2004). *Optimal Timing of Pavement Preventive Maintenance Treatment Applications*, 523, Transportation Research Board.

Poch, M., and Mannering, F. (1996). "Negative Binomial Analysis of Intersection-Accident Frequencies." *Journal of transportation engineering*, 122(2), 105–113.

Romanoschi, S., Hossain, M., Gisi, A., and Heitzman, M. (2004). "Accelerated Pavement Testing Evaluation of the Structural Contribution of Full-Depth Reclamation Material Stabilized with Foamed Asphalt." *Transportation Research Record: Journal of the Transportation Research Board*(1896), 199–207.

Saboori, A., Harvey, J., Butt, A., and Jones, D. (2017). "Life Cycle Assessment and Benchmarking of End of Life Treatments of Flexible Pavements in California." *Pavement Life-Cycle Assessment*, CRC Press, 241–250.

Santos, J., Bryce, J., Flintsch, G., Ferreira, A., and Diefenderfer, B. K. (2015). "A Life Cycle Assessment of in-Place Recycling and Conventional Pavement Construction and Maintenance Practices." *Structure Infrastructure Engineering*, 11(9), 1199–1217.

Santos, J., Flintsch, G., and Ferreira, A. (2017). "Environmental and Economic Assessment of Pavement Construction and Management Practices for Enhancing Pavement Sustainability." *Resources, Conservation and Recycling*, 116, 15–31.

Senhaji, M. K. (2017). "Development of a Life-Cycle Assessment Tool for Flexible Pavement in-Place Recycling Techniques and Conventional Methods."Master's thesis, University of Illinois at Urbana-Champaign, Urbana, Illinois.

Stroup-Gardiner, M. (2011). "Recycling and Reclamation of Asphalt Pavements Using in-Place Methods." NCHRP Synthesis of Highway Practice 421; Transportation Research Board.

Thenoux, G., Gonzalez, A., Dowling, R. J. R., Conservation, and Recycling (2007). "Energy Consumption Comparison for Different Asphalt Pavements Rehabilitation Techniques Used in Chile." 49(4), 325–339.

Wolters, A. S., and Zimmerman, K. A. (2010). "Current Practices in Pavement Performance Modeling Project 08-03 (C07): Task 4 Report Final Summary of Findings." Pennsylvania. Dept. of Transportation. Bureau of Planning and Research, No. FHWA-PA-2010-007-080307.

Mechanistic modeling of the effect of pavement surface mega-texture on vehicle rolling resistance

S. Rajaei
PSI-Intertek, TX, USA

K. Chatti
Michigan State University, MI, USA

ABSTRACT: Rolling resistance plays an important role in vehicle fuel consumption. Pavement surface texture and roughness are two of the pavement elements that affect the energy dissipation in the vehicle. The effect of roughness on rolling resistance has been the focus of several studies in the past, while a few studies investigated the effect of macro-texture. However, most of these studies either considered the mega-texture as a part of the roughness profile or did not include it in the texture spectrum at all. In this study, the effect of pavement surface mega-texture on rolling resistance was investigated mechanistically.

For this purpose, a finite element-based quarter-car model is developed. The model includes a full-3D finite element tire model combined with a suspension system. The effects of several profiles with different mega-texture levels on rolling resistance of the quarter-car and tire are evaluated and compared with the effect of roughness and macro-texture.

1 INTRODUCTION

Due to the increase in global warming and public awareness, energy efficiency has gained a lot of attention in the past few decades. The majority of the petroleum used in the United States is for transportation (71% in 2011 (EPA,2012)). The surface profile is one of the factors in the contact between the tire and pavement surface that influences the fuel consumption. The pavement profile includes different scales of roughness (unevenness), mega-, macro-, and micro-textures, from which mega-texture (profile with wavelengths between 5 cm and 50 cm and peak to peak amplitude of 0.1 to 50 mm.) is the focus of this study. The surface profile causes energy loss in the tire and the vehicle suspension by resisting the tire rolling. This energy loss is known as rolling resistance that involves different mechanisms such as; (i) vehicle dynamics (energy loss in the suspension system), (ii) tire vibration and deformation, and (iii) tire tread deformation (Bendtsen, 2004). The longer wavelengths of the profile affect the energy loss in the suspension system, while the shorter wavelength and surface texture influence the tire and tread deformation. Mega-texture of the profile includes a wide range of wavelengths from 5 cm to 50 cm; therefore, it affects rolling resistance by creating vibration in both the tire and the suspension system.

Several empirical and numerical studies have investigated the effect of surface profile on rolling resistance (specifically, effect of roughness and macro-texture). The studies on the effect of mega-texture are much more limited. The existing empirical studies are mostly related to the vehicle fuel consumption measurements which include built-in rolling resistance models within their fuel consumption models (Chatti & Zaabar, 2012, Hammarström et al, 2012). There are a few studies that investigated the effect of surface profile based on the actual rolling resistance measurements (Sandberg et al, 2018, Boere, 2009). Most of these studies do not differentiate between the roughness and mega-texture. One of the few studies that considered mega-texture separately, suggested that mega-texture can have a dominant effect on rolling resistance, and it can even be the main factor in rolling resistance which can affect fuel usage up to 9% (Descornet,1990). In another study by Sandberg (1990) the effect of different scales of the profile (ranging from 2 to 3500 mm) was investigated, and the mega-texture was reported to be influential on fuel consumption both at low and high speeds.

Similar to the empirical studies, the available numerical models mostly do not separate mega-texture from roughness and consider the influence of IRI which includes the profile beyond the contact patch of the tire (approximately 25 cm). Since mega-texture includes both texture within and beyond the contact patch, it affects both the energy dissipation within the tire and the suspension and therefore, all of the mechanisms involved in rolling resistance. Consequently, there is a lack of a numerical model that is able to capture the effect of the entire range. As a result, besides the experimental evidence, there are no mechanistic studies that specifically address the effect of mega-texture on rolling resistance.

From the common surface profile characterization parameters, the mean profile depth (MPD) parameter is usually used for texture characterization of the macro-texture. It considers the profiles in 10 cm lengths, so it only considers a small portion of the lower part of the mega-texture. On the other hand, the international roughness index (IRI) is usually used for characterization of the wavelengths larger than the contact patch, and therefore in addition to roughness, it only includes a portion of the higher part of the mega-texture scales. As a result, so far, the effect of mega-texture within the contact patch has been neglected. In addition, since the MPD and IRI do not consider the whole range of mega-texture, they are not adequate for defining mega-texture properly and another parameter should be considered for its characterization.

Therefore, to address these shortcomings, we develop in this study a mechanistic model that is able to capture the effect of the entire mega-texture range on the rolling resistance of the tire and vehicle suspension.

2 METHODOLOGY

Tire designs are very complicated, and they include many different parts with various material types. Therefore, the finite element method (FEM) can be a strong tool for modeling tires for different purposes, such as modeling the vehicle operation conditions, the effect of pavement structure or roughness on rolling resistance (Hernandez, 2015, Wei & Olatunbosun, 2014). To investigate the effect of the full mega-texture spectrum on rolling resistance, the developed model requires including both the effect of textures within and beyond the contact patch in one single model. For this purpose, a finite element-based quarter-car model is developed. This model consists of: (i) a detailed 3D FE tire model with mesh size that is small enough for capturing the effect of the smallest mega-texture wavelength on the tire deformation, and (ii) a quarter car suspension system that supports the vehicle mass with a parallel spring and dashpot system for capturing the effect of the mega-texture on the vehicle's suspension. This system is therefore able to include all of the mechanisms involved in the rolling resistance, from tread deformation and tire bending to the vehicle dynamics.

The measured surface profiles may include all or a few of the profile scales. Hence, for investigating the effect of mega-texture on rolling resistance, it should be isolated from the other scales, based on the definition from PIARC (Wambold et al., 1995). Then the effect of mega-texture on rolling resistance is compared to the effect of the macro-texture and roughness found previously by the authors (Rajaei, 2019), to show their importance on the rolling resistance of the vehicle.

3 MEGA-TEXTURE PROFILE CHARACTERIZATION

The Pavement International Association of Road Congresses (PIARC) defined 4 scales within a pavement surface profile as follows (Wambold et al., 1995):

- Micro-texture: wavelengths smaller than 0.5 mm,
- Macro-texture: wavelengths between 0.5 and 50 mm and peak to peak amplitude of 0.1 to 20 mm,
- Mega-texture: wavelengths between 50 and 500 mm and peak to peak amplitude of 0.1 to 50 mm.
- Roughness: wavelength larger than 500 mm.

In this study, pavement surface profiles from measurements conducted as part of NCHRP 1-45 study (Chatti & Zaabar, 2012) are used. The measurement device in this previous study had a longitudinal resolution of 7.5cm and the measurement length on different routes was up to 1000 m. So, the profiles contain both roughness and mega-texture, but do not include wavelengths between 5 and 7.5 cm. However, since they have the majority of the mega-texture spectrum, the dataset is considered suitable for this study.

For understanding the effect of the mega-texture on rolling resistance, first, the mega-texture should be isolated from the whole profile. For this purpose, filtering - a common method for decomposition of profiles and signals - is used. The filtering is performed in MATLAB software, using the built-in command BUTTER to eliminate the wavelengths larger than the highest limits of mega-texture (50 cm). In this command the normalized cutoff frequency (W_n) for upper and lower limits can be defined by:

$$W_n = \frac{\frac{\text{cutoff frequency}}{\text{sampling frequency}}}{2} \tag{1}$$

where the cutoff frequency is defined as the upper and lower limit frequency of each scale, while the sampling frequency is the frequency of the experimental data. As an example, Figure 1 depicts a sample of filtered surface from the NCHRP I-45 study.

Figure 1 (a) shows the full profile of 1000m before filtering. Figure 1, parts (b) and (c) show the filtered roughness and mega-texture, respectively. As it can be seen, any wavelength smaller 50cm and larger than 5cm, is excluded from the profile and considered as mega-texture.

After the mega-texture is separated from roughness, outlier or grade removal should be applied, if necessary, before profile characterization. It should be noted that the surface profiles can be transferred to the FE model directly; however, for comparison of the effect of different profiles they should be characterized. As it was mentioned before, since IRI does not include the textures within the contact patch (smaller than 25 cm) and the standard length for MPD measurements is limited to 10 cm, both of these measures are unable to define the entire mega-texture portion of the profile (wavelengths between 5 cm and 50 cm). Therefore, the root mean square (RMS) is selected for characterizing mega-texture, as defined below:

$$R_q^2 = \frac{1}{n} \sum_{i=1}^{n} (Z_i)^2 \tag{2}$$

where z_i is the height of each profile point when the mean of the profile is set to zero, and n is the number of points in the profile.

RMS is a statistical parameter and is therefore dependent on the sample size. The selected sample size should be large enough (i) to represent the whole spectrum of the mega-texture, and (ii) for the quarter-car FE model to reach the steady state condition. For the profile to represent the whole mega-texture spectrum, the length of the profile should be larger than twice of the maximum wavelength (>1 m). However, for the FE model to reach the steady state condition, the required length is related to some of the model characteristics, such as the size of the tire and the rolling velocity. Based on these criteria, the minimum length of 12 m is selected for the mega-texture profiles.

The profiles in the NCHRP1-45 database are then divided into 12 m profiles and the RMS values are calculated for each profile to find the range of the variation of the RMS values for mega-texture (see Figure 2).

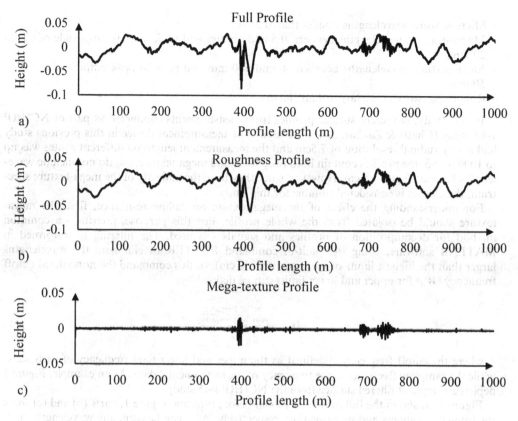

a)

b)

c)

Figure 1. Filtering mega-texture from the measured profile (data from NCHRP I-45 study (3)).

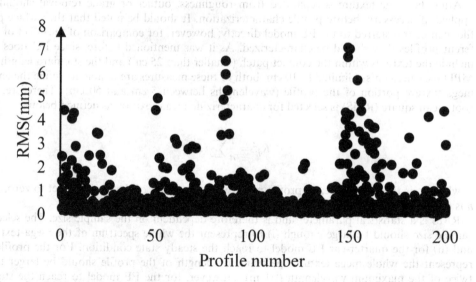

Figure 2. RMS variation for mega-texture.

As it can be seen, RMS values of mega-texture profiles can be as high as 7.5 mm, but the majority of the values are less than 4 mm. Based on the obtained range, the final surfaces are selected randomly, in order to properly cover the range of RMS values for each scale.

The FE quarter-car model developed in this study is a combination of a full 3D tire model and a parallel spring and dashpot suspension system. The commercial software ABAQUS is used for modeling. The tire model is a simplified to include only the main parts of the tires such as tread, sidewall, apex, two belt layers, cap layer, carcass layer, reinforcements, and rim. The geometry of the different parts and their material properties are taken from another study (Wei & Olatunbosun, 2014). The geometry of the cross section of the tire is depicted in (Figure 3(a)). Three different rubber materials are considered for the tire tread, sidewall, and apex. These materials are defined as hyper-viscoelastic material using Yeoh hyperelastic constitutive model and Prony series. The belts, cap layer, and reinforcements are defined as elastic materials and, similar to the construction process, they are embedded within the rubber materials. The rim of the tire and the pavement surface are defined as rigid. The vertical loads throughout the process are transferred to the tire through the rim.

To develop a rigorous quarter-car FE model that includes the dynamic effects of contact between the profile and the tire, three different models are developed. The first model is the base FE 3D tire model which is developed based on the information taken from a study by Wei & Olatunbosun (2014). The size of the tire is 235/60-R18. To build this model the following steps are followed: (i) Generation of the 2D model of half of the cross section of the tire (Figure 3(a)). In this step, the inflation pressure of 200 kPa is applied to the tire, (ii) generation of the 3D half tire model by revolving the 2D cross section axisymmetrically (Figure 3 (b)), (iii) development of the 3D full tire model, by mirroring the 3D half tire model and applying the vertical load on the centroid of the rim (Figure 3 (d)), (iv) steady state rolling analysis of the tire to reach the rolling velocity of the tire (10 km/h) in steady state condition, and (v) transient analysis of the tire rolling over a step (Figure 3 (d)). This model is then verified by comparing the longitudinal force applied to the rim at different velocities to the controlled test results. (see Figure 4).

For the second model, the size of the FE tire model and the boundary conditions are changed to mimic an actual rolling resistance measurement test at 80km/h speed. Then, in the third model, the FE quarter-car model is developed by adding the suspension system to the tire based on a standard quarter-car by Dixon (2008), similar to the study by Zabaar et al.

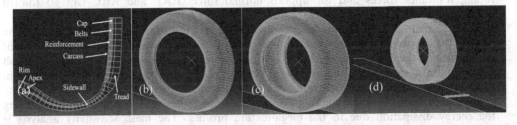

Figure 3. Tire model development, (a) 2D cross section, (b) 3D half tire, (c) 3D full tire, tire rolling over a step at 10 km/h velocity.

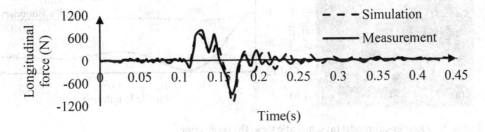

Figure 4. Tire model verification - comparison with measurements by (Wei & Olatunbosun, 2016).

(2018) (see Figure 5 (a)). The suspension system is a combination of a spring and a dashpot and divides the vertical load into sprung (axle load) and unsprung (vehicle load) masses. To do so, after defining the spring and the dashpot, the weight of the tire is calculated, and the remaining load of the unsprung mass is applied to the rim centroid. The sprung mass is then defined on top of the suspension system.

To verify the suspension model, the vertical displacement of the spring in the model is compared with the theoretical spring's displacement of $\Delta = mg/K$, where K and g are the stiffness of the spring and gravity, respectively. Due to the presence of the dashpot in the model, it is expected for the vertical displacement of the spring to be close to Δ (within a 5% boundary) (see Figure 5 (b)).

This model is then used for capturing the effect of different mega-texture profiles on the rolling resistance of tire and suspension system. It should be noted that before rolling the quarter-car model over the textured profile, it should be first rolled over a smooth surface to reach a steady condition. The length of the smooth surface (transition length) at 80 km/h is found to be 10m for the tire model and 30 m for the quarter-car model.

In this study, the rolling resistance force (RRF) and rolling resistance coefficient (RRC) are used for comparing the results of different profiles. The total rolling resistance force is the summation of the rolling resistance forces in the tire and suspension. RRF of the tire is defined as the average work per unit length:

$$RRF = \frac{\sum F.d}{d} \tag{3}$$

where F is the longitudinal force at the rim centroid and d is the driving distance.

While for the suspension, the average rolling resistance force is equal to the average energy dissipation (D) over the unit length, defined as:

$$D_s = \frac{C_s}{V} E[\dot{z}_s^2] \tag{4}$$

where V is the vehicle's velocity, and $\dot{z}_s$ is the relative velocity of the suspension.

The total rolling resistance coefficient (RRC) is then defined as the ratio between the total rolling resistance forces and the applied normal force. RRF and RRC can go toward a constant value (with small fluctuations) if the rolling distance is sufficiently long.

Since the main goal of the model is to capture the effect of the mega-texture profiles, the mesh in the longitudinal direction plays an important role on the results. As the wavelength of the profile and the distance between two consecutive points on the profile decreases, the mesh size should also decrease. This can also affect the mesh size within the tire cross section in order to keep a reasonable aspect ratio within the elements. For this reason, a mesh sensitivity analysis is performed for determining the minimum radial mesh size for reasonably capturing the energy dissipation due to the mega-texture profiles. The mesh sensitivity analysis is

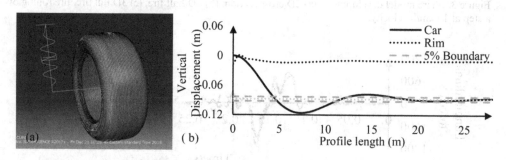

Figure 5.   Quarter-car model (a) schematic view, (b) verification.

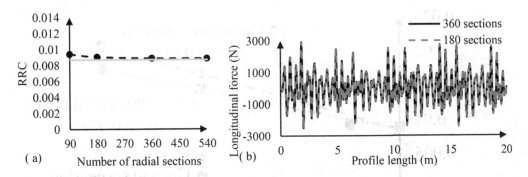

Figure 6. Mesh sensitivity analysis for (a) a smooth surface (b) mega-texture.

performed on both smooth and textured surfaces with radial section numbers of 90 (element size of 2.2 cm) to 540 (element size of 3.7 mm) (see Figure 6).

As it can be seen, for a smooth surface, the RRC variation for meshes with the number of sections higher than 180 is small, and the results tend toward a plateau value (Figure 6 (a)). For the effect of mega-texture, the difference between 360 and 180 radial sections is minimal (Figure 6 (d)). Therefore, the number of radial sections of 180 is chosen for the model.

5 RESULTS

For understanding the effect of mega-texture on the rolling resistance of the tire and quarter-car, different profiles were chosen from the database with different RMS values ranging from 0 to 6.5 mm. The main aim is to investigate the effect of mega-texture on rolling resistance for the tire and quarter-car models. In addition, the mega-texture effect on RRC is then compared with the effect of the rest of the profile spectrum; namely, macro-texture and roughness, obtained by the authors in a separate study (Rajaei, 2019). This result can indicate the importance of mega-texture on RRC.

Mega-texture includes wavelength within and beyond the tire contact patch. It is assumed that the textures within the contact patch affect the tire and the ones beyond the tire influence the suspension. This assumption has been the basis for many quarter-car models and is used in IRI calculation. In such cases, the profile enveloping is performed to exclude the texture within the contact patch from the profile, so that it does not affect the suspension. The developed quarter-car model in this study, however, does not have this limitation, since it is able to consider the effect of mega-texture on both tire and suspension simultaneously. Here, different mega-texture profiles with different RMS values are defined in the FE model. The tire is rolled over each profile for 12 m to reach a steady state condition (in addition to the transition length on the smooth surface mentioned before) and the corresponding RRC values are obtained (see Figure 7).

As it can be seen the effect of the mega-texture on the suspension of the vehicle is negligible for smaller RMS values and the effect of these profiles is limited to the tire energy dissipation caused by deformation and bending in the tire. However, for rougher surfaces, the effect of the profiles on the energy dissipation in the suspension can be quite noticeable. It should be noted that the RMS value of zero corresponds to the tire rolling on a smooth surface.

For a better comprehension of the effect of mega-texture on rolling resistance in the tire and suspension system, the parameter ΔRRC is introduced based on y_1 and y_2 relationships given in Figure 7:

$$\Delta RRC_{suspension} = RRC_{tire \ \& \ suspension, \ RMS=x} - RRC_{tire, \ RMS=x} = y, \ _{RMS=x} - y_{1, \ RMS=x} \quad (5)$$

$$\Delta RRC_{tire} = y_{tire, \ RMS=x} - RRC_{RMS=0} = y_{1, \ RMS=x} - y_{1, \ RMS=0} \quad (6)$$

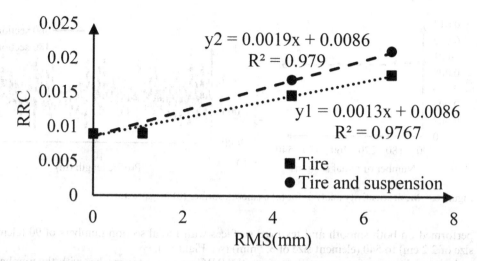

Figure 7.   Effect of mega-texture RMS on RRC.

Therefore, the ratio between the effect of mega-texture on rolling resistance from the suspension and that from the tire is $\Delta RRC_{suspension}/\Delta RRC_{tire} = 0.0004/0.0014 \approx 0.3$. This result shows that the effect of mega-texture on the tire is more than its effect on the suspension.

Figure 8 Shows the comparison of the effect of mega-texture with that of the other scales within the profile, namely roughness and macro-texture. Figure 8 (a) and (b) depicts the effect of the profile on the tire and quarter-car, respectively. As it can be seen, for RMS values lower than 2 mm, the RRCs for all of the scales are very similar to each other for both tire and quarter-car models. It should be mentioned that in all of the scales the effect of RMS values in this range on suspension is minimal. Therefore, whether the profile includes roughness or texture, for profiles with RMS values less than 2 mm, the rolling resistance is limited to the energy dissipation in the tire, which is expected.

For higher RMS values, the energy dissipation in the tire caused by the mega-texture profile is significantly higher than that from roughness. However, although the difference between the two scales is also significant in the quarter-car model, it should be noted that this difference is mainly due to effect of mega-texture on tire deformation (ΔRRC_{tire}), while the effect on the suspension ($\Delta RRC_{suspension}$) for both mega-texture and roughness is in the same range.

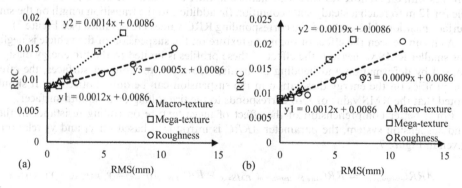

Figure 8.   Effect of RMS on RRC (a) comparison of different scales for tire (b) comparison of different scales for quarter-car model.

6 SUMMARY AND CONCLUSION

The effect of pavement mega-texture profile on tire and vehicle rolling resistance is investigated in this study. For capturing the effect of the mega-texture on tire deformation, including tire bending and tread deformation, a full 3D finite element tire model is developed. To add the influence of the profile on the energy dissipation in the vehicle suspension system, a suspension system is combined with the tire model to build a quarter-car FE model. The surface profile is decomposed in order to filter out the mega-texture profile. After characterization of the profile using RMS parameter, the effect of different profiles on the rolling resistance of the tire and suspension is evaluated.

A linear relationship is obtained between the RMS of mega-texture and RRC of the tire and the quarter-car model. In the profiles with low RMS, the effect is limited to the tire. In the profiles with higher RMS values, the effect on suspension is more evident, although, the effect of mega-texture on tire is still more dominant. In comparison to the other scales of the profile, namely roughness and macro-texture, mega-texture shows a higher RRC for both tire and quarter-car models. Although there is a difference between the effect of mega-texture and roughness on RRC, it is evident that this difference is more related to the effect of mega-texture on the tire rather than the suspension. Since the minimum wavelength for roughness is 50 cm, it is expected to have less of an effect on the tire and more on the suspension. A comparison between the effect of mega-texture and roughness on the suspension shows a similar result between the two scales. These obtained results are an indication of the importance of mega-texture on the rolling resistance of the vehicle.

REFERENCES

Bendtsen, H., 2004. Rolling resistance, fuel consumption-a literature review. Rolling resistance, Fuel consumption-a literature review, (23).

Boere, S., 2009. Prediction of road texture influence on rolling resistance. Eindhoven University of Technology.

Chatti, K. and Zaabar, I., 2012. Estimating the effects of pavement condition on vehicle operating costs (Vol. 720). Transportation Research Board.

Descornet, G., 1990. Road-surface influence on tire rolling resistance. In Surface characteristics of roadways: international research and technologies. ASTM International.

Sandberg, U.S., 1990. Road macro-and megatexture influence on fuel consumption. In Surface characteristics of roadways: International Research and Technologies. ASTM International.

Dixon, J.C., 2008. The shock absorber handbook. John Wiley & Sons.

EPA, 2012. "Greenhouse Gas Inventory Report".

Hammarström, U., Eriksson, J., Karlsson, R. and Yahya, M.R., 2012. Rolling resistance model, fuel consumption model and the traffic energy saving potential from changed road surface conditions. Statens väg-och transportforskningsinstitut.

Hernandez, J.A., 2015. Development of deformable tire-pavement interaction: contact stresses and rolling resistance prediction under various driving conditions (Doctoral dissertation, University of Illinois at Urbana-Champaign).

Rajaei, S. 2019, mechanistic modeling of the effect of pavement surface texture and roughness on vehicle rolling resistance, 2019, Michigan State University.

Sandberg, U., Bergiers, A., Ejsmont, J.A., Goubert, L., Karlsson, R. and Zöller, M., 2011. Road surface influence on tyre/road rolling resistance. Swedish Road and Transport Research Institute (VTI), Linköping, Sweden, accessed Sept, 23, p.2018.

Wambold JC, Antle CE, Henry JJ, Rado Z., 1995. PIARC (Permanent International Association of Road Congress) Report. International PIARC Experiment to Compare and Harmonize Texture and Skid Resistance Measurement, C-1 PIARC Technical Committee on Surface Characteristics, France.

Wei, C. and Olatunbosun, O.A., 2014. Transient dynamic behaviour of finite element tire traversing obstacles with different heights. Journal of Terramechanics, 56, pp.1–16.

Wei, C. and Olatunbosun, O.A., 2016. The effects of tyre material and structure properties on relaxation length using finite element method. Materials & Design, 102, pp.14–20.

Zaabar, I., Chatti, K. and Lajnef, N., 2018, July. Evaluation of fuel consumption models for pavement surface roughness effect. In Advances in Materials and Pavement Prediction: Papers from the International Conference on Advances in Materials and Pavement Performance Prediction (AM3P 2018), April 16-18, 2018, Doha, Qatar (p. 305). CRC Press.

Pavement, Roadway, and Bridge Life Cycle Assessment 2020 – Harvey et al (eds)
© 2020 Taylor & Francis Group, London, ISBN 978-0-367-55166-7

Effect of pavement structural response on vehicle fuel consumption: Lessons learned from data collection, processing and analysis

A.A. Butt, J.T. Harvey, D.T. Fitch & D. Reger
University of California, Davis, USA

D. Balzarini, I. Zaabar & K. Chatti
Michigan State University, East Lansing, USA

M. Estaji & E. Coleri
Oregon State University, Corvallis, USA

A. Louhghalam
University of Massachusetts Dartmouth, USA

ABSTRACT: A Phase 1 study that used modeling to estimate the effects of pavement non-elastic structural response on vehicle fuel economy concluded that there is an important effect on fuel consumption of vehicles, though small and highly dependent on the pavement structure, of roughness, texture and structural response. The goal of Phase II study was to measure the field vehicle fuel consumption on different pavement types over a range of air temperatures, and use the data to develop an empirical model and for calibration and validation of the mechanistic empirical models for vehicle energy consumption due to pavement structural response. The models are intended for use in pavement management and design, as well as policy development. This paper provides a brief discussion of design of the data collection experiment, data collection methods, test sections characterization, data cleaning, checking and analysis. Lessons learned during each step in this project are discussed further at the end of each section. The experiment included twenty-one sections in California with different pavement structure types. The vehicles selected and instrumented for the fuel economy measurements included a five axle semi-trailer tractor, a diesel truck, an SUV, a gasoline and a diesel car. Vehicles were run on cruise control and data recorded at 72 km/hr (45 mph) and 88.5 km/hr (55 mph), and 56 km/hr (35 mph) and 72 km/hr (45mph) on state and local roads, respectively. A linear fuel consumption empirical model was developed for each vehicle based on the collected data.

1 INTRODUCTION

Pavements have an influence on the fuel mileage of vehicles. The effect of pavement on the fuel mileage of individual cars and trucks is small relative to the effects of stop-and-start and "lead foot" driving, air resistance when driving at speeds over 75 km/hr (45 mph), and under-inflated tires—generally less than about 3 percent change in fuel mileage relative to a "perfect" pavement. However, the cumulative effect of pavement is magnified because it applies to every vehicle on the road. Fuel consumption is not only affected by the vehicle properties but also by the pavement itself as well. The pavement characteristics such as pavement roughness, texture and structural response can influence the fuel efficiency of vehicles, and related greenhouse gases (GHG) and air pollution emissions (Harvey et al. 2016). Models exist for the effects of roughness and texture that are calibrated using field data through the

NCHRP 1-45 project (Chatti and Zaabar 2012). Several structural response models do exist as well however, comprehensive field validation of the influence of structural response on vehicle fuel economy has not much been investigated. Therefore, the developed models have not been calibrated for a range of different pavement types, traffic and climatic conditions. It is hypothesized that a deflection bowl is formed on the pavement surface under vehicle loads and excess fuel is consumed by the vehicles to drive this bowl; more fuel is consumed if a vehicle is heavy or if a vehicle is driving over a less stiff pavement structure.

In Phase 1 of the study, researchers at Michigan State University (MSU), the Massachusetts Institute of Technology (MIT), and Oregon State University (OSU) developed models for the mechanics of pavement structural response on vehicle fuel consumption. University of California Pavement Research Center (UCPRC) researchers then used the models—along with existing models of roughness and texture—to run simulations of annual fuel economy in different climate regions, with urban and rural traffic speeds (congested, uncongested) and traffic flow (vehicle types) conditions (Harvey et al. 2016). The Phase I study recommended to investigate further the structural responsiveness of pavements and to validate and calibrate the models that were developed by the researchers at MIT, OSU and MSU. The report on the development of the models in Phase I is not published yet.

In order to isolate and determine the effects of structural response of different types of pavements, a field study was required in which the parameters affecting vehicle fuel consumption could be measured and separated out from the total fuel consumed by the different vehicle types. To do such an experiment, enough knowledge of the sections, vehicles, equipment being used, tests being performed and data type being collected was required. The steps followed to design the experiment, data collection process and data cleaning process are shown in Figure 1. This paper discusses the experimental design, the difficulties faced, and the lessons learned in this study. The details of data, analysis and results are reported in a comprehensive technical report expected to be published in early 2020 (Butt et al. 2020).

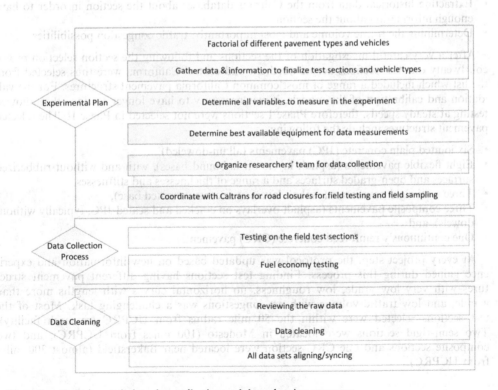

Figure 1. Experimental plan, data collection and data cleaning processes.

2 EXPERIMENTAL PLAN

2.1 *Factorial and selection*

2.1.1 *Test sections*

Test sections of different pavement structure types from all over California, including the ones in Phase I were included in the factorial. A section selection protocol was developed in order to find the sections with;

- Pavement surface grade to be as minimal as possible (<0.5%),
- No horizontal curves (straight sections),
- Lengths not to be lesser than 1 mile,
- Higher safety for traffic closures and fuel economy testing,
- No future pavement maintenance schedules during the time of field testing,
- Low traffic density routes, and
- Mostly low roughness (exception of a few high roughness sections for comparison purposes).

Other important section selection tasks include the following.

- Driving on the section and visually inspect the section and the area.
- Determining a safe location for installation of equipment for recording weather data during the testing.
- Determining safe location for research staff to monitor fuel economy testing.
- Determining locations on the section for extracting core samples and sub-surface temperature measurements.
- Determining location for sub-grade sampling.
- Determining points where the vehicles could be safely turned around such as on and off-ramps, roundabouts etc.
- Extracting historical data from the Caltrans databases about the section in order to have enough information about the section.
- Determining the traffic volume and most importantly, traffic congestion possibilities.

After a very careful investigation of the sections and following the section selection protocol, twenty one test sections across northern and central California were thus selected from the list which included a range of most common California pavement structures. For the validation and calibration of the models, it was necessary to have longer sections that allowed testing at steady speeds, therefore Phase I sections were not selected in Phase II. The selected pavement structures include the following.

- Six jointed plain concrete (JPC) pavements (all un-doweled),
- Eight flexible pavements (asphalt layers on unbound bases), with and without rubberized surfaces and open-graded surfaces and a range of thicknesses and stiffnesses,
- Three semi-rigid pavements (asphalt layers over cement treated base),
- Three composite pavements (asphalt overlays on cracked and seated JPC, typically without dowels), and
- One continuously reinforced concrete (CRC) pavement.

At every project step, the protocol was updated based on new information and experience gained during this process. Finding test sections having different pavement structures with very low grade, low roughness, no horizontal curves with lengths more than a mile, and low traffic volume with no congestions was a challenging task. Most of the test sections selected were within the 50 mile radius from UCPRC (research facility). Two semi-rigid sections were located in Modesto (100 miles from UCPRC), and two composite sections and one CRC section were located near Bakersfield (almost 300 miles from UCPRC).

2.1.2 *Vehicles for fuel economy testing*

A list of different vehicle types that included gasoline fueled, diesel fueled, hybrid and electric was created. A protocol was also developed to select the most appropriate vehicles for the study. The vehicle selection protocol stated:

- Vehicles to be no more than 5 years old so that the fuel consumption data collected is from todays' engine technology.
- Vehicles to have at least 20,000 miles so that the new engine wear variability is accounted for.
- Accident-free vehicles.
- Vehicles that are representative of the existing fleet.
- Same vehicles to be rented in winter and summer for the fuel economy testing.
- New tires allowed to be mounted to testing vehicles.

The University of California (UC) fleet services who own a large fleet were contacted for light vehicle rentals. An advantage of renting the vehicles from UC fleet services was the option of using the same fueling facility for re-fueling the vehicles from the same location for the entire fuel economy testing (exception was for sections more than 50 miles away); this way the fuel variability could be avoided. The renting of a truck was one of the most challenging task as the same truck was to be rented for two seasons (winter and summer) during the year. The trucking companies usually are over booked due to transportation commitments with goods companies. Moreover, convincing a driver (with 'class A' driver's license) to commit for the testing was not an easy task either as drivers consider such experiments unsafe because of not having any prior experience. Several trucking companies and truck drivers were contacted. Chavez trucking committed to provide an eighteen wheeler truck and one of its drivers showed interest in the study.

The vehicles selected for the fuel economy measurements included a five axle semi-trailer tractor truck (HHDT), a single axle dual tire diesel truck (Ford F450), a sport utility vehicle (SUV; Ford Explorer), a gasoline car (Chevrolet Impala) and a diesel car (Chevrolet Cruze).

2.2 *Variables and equipment*

As mentioned earlier, there were several variables that were to be determined and measured in order to be able to isolate the effects of structural response; the fuel consumption of vehicles depends on the vehicle engine, vehicle physical shape, climate, driving surface, tire-pavement interaction etc. An extensive review of several studies related to fuel economy and pavements (Smith 1970; Biggs 1988a; Biggs 1988b; Greenwood and Bennett 1995; Bennett and Greenwood 2001; Chatti and Zaabar 2012) was performed and several variables were identified. Then, the equipment and measuring methods were investigated. Some of the major variables identified includes;

- *Pavement surface gradient* – Gradient is one of the major variable that has the largest effect on the fuel consumption of the vehicles. It was important to have the gradient readings as precise as possible. Several attempts were made to measure the gradient including the hand-held Trimble^{TM} global positioning system (GPS) and Google Maps^{TM}. The measured altitude data from a previous ground penetrating radar (GPR) project covering the state network (which was available from accelerated pavement condition survey [APCS] database for 2011 and 2015) was found to have the least noise and is a much more accurate way of determining the slope. This is because the GPS unit used in the GPR has better precision as compared to the other sources used in this study. For non-state roads, GPR data was not available, therefore, accelerometers were used that had a built in GPS unit in them and capability to measure altitude.
- *Pavement roughness* – Roughness, as stated in the introduction, is another important variable that effects vehicle fuel consumption. Road profile was measured using the UCPRC profiler vehicle equipped with a Dynatest® inertial profiler, Mk-III. International roughness index (IRI) is the measure of roughness; high fuel consumption on rougher surfaces.

- *Air and pavement temperatures* – The weather stations were installed near the fuel economy testing cite during the testing in order to record the climate data that could explain the vehicles fuel consumption behavior. The air and pavement temperatures were measured by the sensors connected to the weather station. Solar powered weather stations with storage capability were used in the study as weather stations were to be used during the night time testing or during cloud covers as well.
- *Effective headwind* – Headwind and tailwind also affect fuel consumption. Although wind measuring device, anemometers, were installed on each vehicle to measure the wind speed, however it was found that the wind speed effects from head, tail or side winds couldn't be differentiated, therefore, wind speed and direction measured on the weather station was the most appropriate option selected for calculating effective headwind.
- *Air density* – It was determined that calculating the air density will give a better measure than the readings from the weather station. Weather station measured dew point and air temperature was used along with the altitude of the weather station location from the sea-level information to calculate the air density for each data point (per minute readings).
- *Fuel measurements* – A slight movement of the vehicle due to acceleration, braking, driving behavior, engine friction, or other external forces that are acting upon a vehicle effect fuel consumption. An On-board diagnostic (OBD) device has a capability of reading the performance of the vehicle's engine when connect to the vehicle. There are OBD devices that are equipped with storage capability. Several OBDs were studied and tested in order to determine the most appropriate for this project. An effort was also made to build and program an Arduino chip OBD device but due to lack of time; the effort halted. HEM Data OBD devices were acquired for the study. These devices were equipped with data logging capability and a slot for memory drive that records the vehicles electronics data. The OBD device is connected to the OBD connector in the vehicle (usually under the dashboard) that gets access to data from the engine control unit (ECU) or also referred to as the engine management system of the vehicle. Certain data could be collected from a vehicle as allowed by each vehicle manufacturer. There are though extended parameters that could be recorded as well but requires OBD device programming. The data received by the OBD is usually in series so more the parameters are to be collected, more time it will take to re-read the same parameters in the next cycle causing a delay between the two readings. Therefore, the OBD devices for each vehicle were programmed to record only the data that was required for the study purposes. In the plan to record fuel consumption data from the vehicles, it was also discussed to directly measure fuel injection by connecting sensors on the fuel pipeline of the vehicle at various locations or connect a device directly to the control area network (CAN) of the vehicle however, the vehicles were to be driven on state and county roads, and for safety purposes that was not pursued and only HEM OBD devices were used to collect the fuel consumption data from each vehicle.
- *Tire pressure and temperature* – new tires for all the vehicles were purchased and mounted to the vehicles. The vehicles were driven for few hours each day (fuel economy test trials were also done using new tires) for breaking in the new tires. Due to difficulty finding a cold storage unit, a cold storage trailer was acquired so that the tires could be wrapped (to avoid contact with air/oxygen) and stored in the cold unit when not being used (as the study was to have testing done in two seasons). The tire pressure and temperature changes due to the rotation of the tires, exposure to the external temperatures and road friction. The change in tire pressure means change of tire-pavement surface contact area hence effecting fuel consumption. A tire pressure monitoring system (TPMS) is an electronic system that can monitor the tires temperature and pressure. PressurePro® PULSE TPMS unit was used for this study. This single unit could monitor up to 80 wheel positions and 5 vehicles and had the ability to measure tire pressures and temperatures every minute. It also had data logging capability and could record data in a connected memory drive. Wireless dynamic sensors were used that have 1-2% pressure measurement accuracy up to 200 PSI and temperature resolution of 1.5 °C. Each sensor, having a unique ID, was installed on the tire valve and was synced to the PULSE monitoring unit, which was powered by the vehicle. It was found that when the vehicles were driven for a certain time (mainly 30 mins

on average), the air in the tires had reached an optimum tire pressure which did not change over time.

- *Density and energy density of fuels (diesel and gas)* – In order to convert fuel consumption (in MJ/kg) to energy; density (in kg/liter) and energy density (in MJ/liter) of the fuel is required. As the fuel quality effects vehicle's fuel consumption and performance; fuel samples were collected every time a vehicle was fueled and samples were tagged with a date, time and vehicle. Mostly, same fuel stations, as mentioned earlier UC Davis fleet services station for vehicles and Chavez trucking fuel tanker station for HHDT, were used. Bomb calorimeter test was performed on all the gasoline and diesel samples collected during winter and summer and no major variability was found in the samples. Therefore, only one density test for one of the samples of gasoline and diesel was performed.

3 DATA COLLECTION PROCESS

3.1 *Test section characterization*

As mentioned earlier, sections characterization was required to have knowledge and understanding of the surfaces and structures on which fuel economy testing was conducted. Four tests were mainly conducted by the UCPRC team on each section during the road closures organized by California Department of Transportation (Caltrans; for all state roads) and UCPRC (for county roads) which includes:

- Falling (Heavy) Weight Deflectometer (FWD) Tests - Three drop heights were setup in order to achieve an average load of 5,000, 8,000, and 12,000 lbs (25, 35 and 55 kN, respectively) and two repetitions for each weight drop height were applied at each station in each section. Full time history of the deflections was also collected in order to back calculate the viscoelastic properties of the non-elastic layers. Maximum number of readings were recorded based on the Caltrans allowable closure times per day per section. Closures were strategically planned so that; least traffic flow was affected, on a non-rainy day and on the availability of the Caltrans staff. For all pavement types, deflection readings were measured in the right wheel path in the outer (truck) lane equidistant along the length of the section. For JPCP, the deflections were measured on the un-cracked slabs in the mid-slab as well as on the slab joint in the outer wheel path. The FWD tests were conducted in winter and summer season in order to capture the effects of the coldest and hottest pavement temperatures.
- Laser Texture Scanner (LTS) Tests - LTS was used during the summer time only in order to measure the mean profile depth (MPD) on the surface of asphalt pavement, and mean texture depth (MTD) on the surface of concrete pavement sections. Macrotexture measurements for 15 minutes were conducted at two to three different locations on the sections during the FWD test closures.
- Temperature profiling - An eight inches long thermocouple was fabricated in such a way that when it is inserted in a drilled hole on the section, temperature readings could be recorded at every two inches depth. Air and pavement surface temperatures were also measured. The hole drilled in the pavement was vacuumed for debris and filled with motor oil before the insertion of the pre-fabricated thermocouple and first twenty minute readings were discarded. In cases when the temperature profile data was not collected during the experiments, due to mandatory moving road closures and such as for the JPCP and CRCP, temperature gradients were estimated using BELLS3 temperature formula (FHWA).
- Inertial Profiler: Road profiler was used to measure the roughness and texture of the test sections as described in Section 2.2.

Subgrade samples from all the sections were collected and wet sieve analysis was performed. 2-4 pavement section cores (6 inches diameter) between the wheel path for all the sections except CRCP and four (out of five) JPCP were also collected. A core was extracted from one of the JPCP section because the pavement cross section data for that section was unavailable.

For all the other cores collected, frequency sweep tests (shear test) were conducted at different temperatures to achieve the master curve.

3.2 *Fuel economy testing and data collection*

A protocol for the fuel economy testing was developed which was to be strictly followed in order to avoid variability in data collection procedures. The protocol was also necessary for safety reasons as the vehicle fleet was to be driven at comparatively slower speeds than the highway and county road posted speeds. The drivers for all the vehicles were the same for each testing cycle (winter and summer). SUV driver was only an exception however, in order to control the weight limits, sand bags were used in each vehicle that could be removed or replaced with heavier sand bags to control the weight of the vehicle during each testing season. Major reason to keep the same drivers during the entire fuel economy data collection experiment was to avoid driving behavior variability. The weight of the vehicles was also measured before each test cycle. Safety was the priority so amber lights were mounted on each vehicle and safety magnet signs (slow moving vehicle) were installed in the rear and driver side of all the vehicles. The windows of the vehicles were closed, headlights and radio turned on and ventilation/AC turned off at all times when on the test section.

The weather stations were set up (at least 20 minutes) on the test sites at pre-identified locations (safe location, free from obstructions) that recorded wind speed, wind direction, and pavement and air temperatures during the testing. The engines were warmed up and vehicles driven for at least 15 minutes before the start of the test runs on the sections. The vehicles were driven in a convoy; gas car followed by an SUV, an F450, an HHDT and lastly a diesel pilot car. Enough distance between the vehicles was ensured so that there are no tailgating effects captured in the measurements. On all state roads, the vehicles were run on cruise control at 72 km/hr (45 mph) and 88.5 km/hr (55 mph) speeds whereas on all county roads at 56 km/hr (35 mph) and 72 km/hr (45mph) speeds. Data from number of replicates on each section was recorded. More replicates were recorded on shorter sections. The first round of fuel economy measurements took place in winter 2015-2016 and second round in summer 2016.

One of the major difficulties was to select hot, warm and cold days as the weather was unpredictable; the testing days scheduled during the hottest and coldest months of the year had to be re-scheduled due to unforeseen change in weather conditions. The two major weather events that resulted in cancellation of testing were mainly rain and high winds. Non-rainy and un-windy days (wind speeds less than 5 mph) were the requirements of the testing. Furthermore, if a weather event occurred (rain or temperature change), the temperatures to reach back to higher ranges took time and testing had to be stopped. The testing schedule had major time limitation (to complete the testing within the short seasons, winter and summer); fuel economy data was to be collected on several sections for a range of temperatures. Availability of the research staff was another minor hurdle which did effect some of the schedules a few times.

During the testing, drivers often had to repeat the test runs in cases; if correct testing speeds were not reached in time on a section, brakes were applied due to traffic slowing down, weather station staff informed of high wind records, or if the drivers forgot to close the windows (especially in summer when the tests were performed without turning on the air conditions).

4 DATA PREPARATION AND ANALYSIS

Data processing was divided into two main steps:

1. Studying the raw data collected and checking for completeness
2. Data cleaning and aligning

4.1 Raw data checks

It was necessary to check the data for completeness after each test as an incomplete or missing dataset would result in the experiment un-analyzable. Each data file was checked for data completeness. First order graphs were developed and checked. The data was then loaded in Postgres database.

One of the major observations from the fuel economy dataset was the effect of gear and RPM on the fuel consumption of the vehicles. Fuel consumed by a vehicle on a given section on a certain testing day at the same speed was often different for different runs. This was mainly due to the vehicle running in two different gears. As the tests were performed on cruise control, the drivers had no control over the gear selection while driving and collecting the data. There are two scenarios possible when turning on cruise control at the same speed: 1) the vehicle will be in a lower gear if the vehicle accelerator paddle is hit hard and cruise control turned on. 2) vehicle will be in a higher gear if accelerated slowly before turning on the cruise control i.e. vehicles can drive at the same speed but in an upper limit lower gear (example gear 5) or in a lower limit higher gear (example gear 6). This led to further investigation of the gears and RPMs for all the vehicles used in the experiment. The vehicles were rented again and the fuel economy and RPM in each gear and speed recorded and matched with the fuel economy dataset which was collected during the fuel economy testing. The lesson learned in this context was to always collect gear information, RPM and speed in order to be able to correctly understand the fuel consumption of the vehicle.

4.2 Data cleaning and aligning

A first attempt was made to write a computer program in Python language that extracts the relevant/required data from the data files. Each dataset was divided into several sub-section lengths and the variability/noise in the datasets were observed. Representative volume element (RVE) statistical method was used to find out the smallest sub-section length that the section could be divided into. The datasets for gradient and roughness (two parameters that affect the fuel consumption the most) were studied for 9, 18, 27, 37, 73, 97, 152 m (30, 60, 90, 120, 240, 320, 500 feet, respectively) sub-section lengths for each section. It was determined that the variability/noise in the 9 m (30 feet) sub-section lengths was too high and 27 m (90 feet) sub-section lengths was low. Longer sub-section lengths could not be used as the most of the effects will not appear at all in mechanistic empirical analysis. Thus, 27 m (90 feet) sub-section length was decided for all the sections.

After several trials, the final process followed to clean the datasets is as follows.

1. Using C# language to extract the fuel economy data from the vehicles OBD .CSV files.
2. Using Matlab to translate .RSP file extension to understandable format .CSV file format and then using R code to extract roughness and MPD and MTD data for each of the section tested,
3. Using PostgreSQL to extract data for all section gradients.
4. Using python to extract data from weather station files.

The cleaning process included the removal of culverts which were present in some of the sections. This required removal of few sub-section before and after the culvert as the effect of roughness had to be removed from the dataset. Also, if brakes were applied or OBD recorded strange readings, the data for those runs had to be discarded. Any ambiguity is any dataset disqualified it from being used for the cleaned final dataset.

4.3 Empirical analysis

The empirical modeling framework included a broad conceptual model of the primary variables thought to influence the fuel consumed by a vehicle while set on conventional cruise control (see Figure 2 for a directed acyclic graph of the key variables and causal links). All detailed theories of pavement structural responses are avoided in this

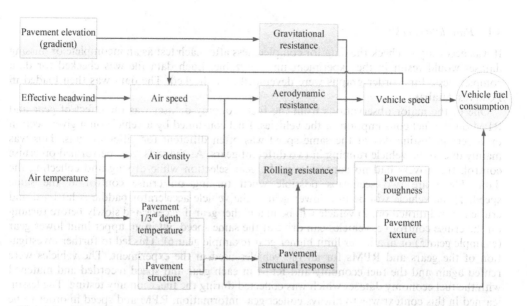

Figure 2.   Directed acyclic graph representing our conceptual model of vehicle fuel consumption where arrows indicate a causal effect direction. White boxes indicate measured predictor variables, colored boxes indicate un-measured constructs, and the white circle represents the outcome. Arrows In yellow leading to other arrows indicate moderating effects (interactions).

framework, as those theories are covered in the other mechanistic empirical modeling frameworks. The goal of this framework was to examine if pavement effects on fuel efficiency could be observed in the field on the basis of experimental design, and to explore the relative magnitude of those effects.

Translating the conceptual model in Figure 2 to statistical models proved challenging. First, as the conceptual model indicates, clear feedback loops in the model cannot be handled in most statistical modeling frameworks such as linear regression. While some more complex modeling frameworks (e.g. structural equation models) allow such feedbacks, the other complexities of the data (e.g. serial autocorrelation, large number of records, multi-level structure) forced the need for a parsimonious statistical framework.

Fuel consumed is modeled through linear regression because it is a continuous variable. While fuel consumed cannot be negative, the model is not constrained to positive reals because in some cases zero fuel was consumed in a sub-segment. Zero fuel was mainly due to vehicle speed adjustment by the cruise control unit while going downhill in order to maintain the test speed cruise control was set at. As mentioned above, all the complexities in Figure 3 could not be included in a linear regression. Because of this, the linear regression models are greatly simplified to remove feedbacks (see Figure 3 for the simplified directed acyclic graph). The most important simplification made was the removal of real-time speed of the vehicles. Because real-time speed is managed by the vehicle controller once cruise control is set, real-time speed acts as a mediating variable for all the upstream effects (e.g. gradient, tailwind, pavement structure, etc.) on fuel consumption. By removing real-time speed, the new model includes direct causal links from those variables to the response variable (fuel consumed). Other causal links in the model were simplified. For example, separate variables for air and pavement temperature in the statistical models could not be included because of extreme collinearity. This collinearity is because pavement temperature mediates the effect of air temperature on the structural response of the pavement (as is shown in Figure 2). Instead, only air temperature is used in the model which is interpreted as having an effect on fuel consumption through two pathways: (1) air density changes resulting in changes in aerodynamic resistance and perhaps slight changes to combustion efficiency, and (2) pavement temperature changes resulting in changes to the structural response of a pavement and thus varying losses of energy due to rolling resistance.

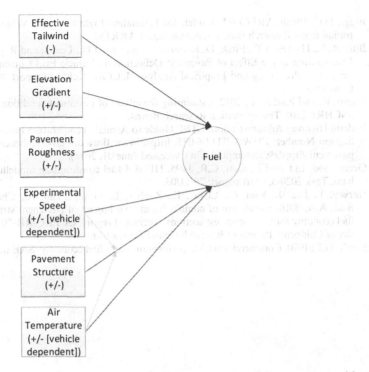

Figure 3. Simplified directed acyclic graph representing the conceptual model we represent with our linear regression. White boxes indicate measured predictor variables, and the white circle represents the outcome. Yellow arrows leading to other arrows indicate moderating effects (interactions).

Vehicle speed is included as a categorical variable (e.g. the speed that cruise control was set at in the experimental run) so that comparisons could be made for interactions of temperature, speed, and pavement type.

ACKNOWLEDGMENT

The authors would like to thank Nick Burmas and Joe Holland from the Caltrans Division of Research, Innovation, and System Information for support of this project.

DISCLAIMER

This document is disseminated in the interest of information exchange. The contents of this report reflect the views of the authors who are responsible for the facts and accuracy of the data presented herein. The contents do not necessarily reflect the official views or policies of the State of California or the Federal Highway Administration. This publication does not constitute a standard, specification or regulation. This report does not constitute an endorsement by the Department of any product described herein.

REFERENCES

Bennett, C.R. and Greenwood, I.D. 2001. Modelling road user and environmental effects in HDM-4. HDM-4 Highway Development and Management Series, Vol 7.
Biggs, D.C. 1988a. Fuel Consumption Estimation Using ARFCOM. Proceedings 14th Australian Road Research Board (ARRB) Conference, Part 3, pp. 280–293.

Biggs, D.C. 1988b. ARFCOM: Models for Estimating Light to Heavy Vehicle Fuel Consumption. Australian Road Research Board, research report ARR152.

Butt, A.A., Harvey, J.T., Fitch, D.,Kedarisetty, S., Lea, J.D., Lea, J. and Reger, D. expected early 2020. Investigation of the Effect of Pavement Deflection on Vehicle Fuel Consumption: Field Data Collection, Data Processing and Empirical Analysis. UCPRC technical report. Project funded by Caltrans. California.

Chatti, K. and Zaabar, I., 2012. Estimating the effects of pavement condition on vehicle operating costs. (NCHRP 720). Transportation Research Board.

Federal Highway Administration. LTPP Guide to Asphalt Temperature Prediction and Correction. Publication Number: FHWA-RD-98-085. https://www.fhwa.dot.gov/publications/research/infrastructure/pavements/ltpp/98085/tempred.cfm (Accessed June 01, 2018).

Greenwood, I.D. and Bennett, C.R. 1995. HDM-4 fuel consumption modelling. University of Birmingham, Task 2020-3, draft report: 29/11/03.

Harvey, J., Lea, J., Kim, C., Coleri, E., Zaabar, I., Louhghalam, A., Chatti, K., Buscheck, J. and Butt, A.A, 2016. Simulation of cumulative annual impact of pavement structural response on vehicle fuel economy for California test sections. Technical report. UCPRC-RR-2015-05. Prepared by University of California Pavement Research Center for Caltrans.

Smith, G.L. 1970. Commercial vehicle performance and fuel economy. SAE transactions, pp.729–751.

PVI related decision-making tools in use phase of LCA: A literature review

K. Mohanraj & D. Merritt
The Transtec Group Inc., Austin, USA

N. Sivaneswaran
Turner Fairbank Highway Research Center, USA

H. Dylla
Federal Highway Administration, USA

ABSTRACT: It is widely recognized that vehicle operating costs (VOC) and excess fuel consumption (EFC) are affected by various pavement characteristics, and pavement vehicle interaction (PVI). PVI related VOC and, more recently EFC with the increased emphasis on pavement life cycle assessment, are factors that agencies consider when making investment decisions or in evaluating strategies. Several studies have attempted to quantify the effects of PVI on the use-phase of pavement life cycle assessment (LCA) in the form of EFC and VOC. These studies in general investigate rolling resistance influencing factors such as pavement roughness, macrotexture, and structural responsiveness to generate models to predict VOC and EFC. While the impacts of roughness have been studied more extensively, agencies do not have enough guidance on quantifying all these aspects of PVI on use-phase LCA or Life-Cycle Costs Analysis (LCCA).

Applicability of EFC and VOC in pavement decision-making processes such as pavement type selection, timing of rehabilitation treatments etc. was studied along with challenges for implementation. A critical review of ten (10) most recent PVI models was performed by assessing their availability, identifying gaps and their compatibility and applicability for investment decisions. Some of the common challenge observed include lack of measurement methods for ground truth, data requirements and ease of use of the models, and lack of guidelines for implementation.

1 INTRODUCTION

As part of efforts towards more sustainable decision-making, state and federal highway agencies are utilizing pavement LCA to analyze and quantify the environmental impacts of systems such as roadway networks. Although extensive research has been performed on the cradle-to-gate portion of pavement LCA, the cradle-to-grave still remains incomplete due to certain data gaps in use phase. One component of the use-phase that is missing is related to the vehicle operation costs and the excess fuel consumption associated with pavement-vehicle interaction after construction of the pavement, and through its service life.

PVI is the effect of pavement characteristics and condition on vehicle performance and operations (1). VOC and EFC are affected by various pavement characteristics and PVI, and are factors that agencies must consider when making investment decisions or in evaluating strategies during pavement life cycle assessment. The effects of PVI on the use-phase of pavement LCA or LCCA in the form of EFC and VOC have been investigated primarily by studying rolling resistance. The pavement properties considered as influential on rolling resistance are pavement roughness/evenness, macrotexture, and structural responsiveness. While the impacts of roughness have been studied more extensively, state highway agencies are still in need of guidance to quantify all these aspects of PVI and implement them in use-phase LCA or Life-Cycle Costs Analysis (LCCA).

This paper discusses a study initiated by FHWA that looks at the applicability of EFC and VOC the decision-making processes such as pavement type selection, timing of rehabilitation treatments etc. The paper summarizes a critical review and assessment of the availability of, and gaps in existing PVI-related VOC and EFC models and their compatibility and applicability for investment decisions.

2 BACKGROUND

Rolling resistance is defined as the "energy consumed by a tire per unit distance covered (1)." It is one among five resistive forces a vehicle needs to overcome (others are aerodynamic, internal friction, gravitational, inertial). The total rolling resistance is aggregate of contribution from tire (vehicle) and from the riding surface (pavement).

The operating costs associated with a vehicle's fuel consumption (excess and instantaneous), oil consumption, tire wear, repair and maintenance, and mileage-related depreciation are lumped together and referred to as VOCs. These costs depend on various factors such as roadway geometry, climate, traffic characteristics, vehicle, and pavement characteristics, among other factors. The VOCs dependent on vehicle and pavement characteristics are realized through pavement-vehicle interaction.

EFC is the additional fuel needed for normal operation due to a change in conditions. The baseline fuel consumption may be referred to as Instantaneous Fuel Consumption (IFC), which is the fuel consumption of a vehicle at a given instant in time. Fuel consumption factors, which are a function of resistive forces are generally considered as: aerodynamic forces, internal frictional forces, gravitational forces, inertial forces, and rolling resistance forces.

3 HISTORICAL DEVELOPMENTS

The need to develop models to estimate VOCs for use in economic analysis has been recognized since the 1940s by several agencies and governments globally. In 1968, the World Bank in conjunction with Transport and Road Research Laboratory (TRRL) and the Laboratoire Central des Ponts et Chaussées (LCPC) developed a basic road project appraisal model (12, 26). At the same time, World Bank also commissioned the Massachusetts Institute of Technology (MIT) to develop a model that led to the then most advanced Highway Cost Model (HCM). The HCM study indicated the need for further research in developing countries (12, 26).

As a result of HCM, the World Bank initiated a study to develop a model to evaluate road user costs for low volume roads. In 1979, the World Bank produced the first version of Highway Design and Maintenance Standards Model (HDM) based on investigations performed in Kenya (12). This led to a further study that was intended to serve developing countries such as Kenya, India, Brazil, and the Caribbean. Field studies were conducted in these countries to develop models to estimate VOCs based on the deterioration observed. And in the developed countries of Europe and North America the emphasis was on road user costs associated with travel time. In 1987, HDM-III was published along with a computer program HDM-95 to estimate VOCs (12).

After seeing an extensive use of HDM-III across the globe, the World Bank extended the scope of the model and the Highway Development and Management Tool (HDM-4) was developed in year 2000. The emphasis in this study was given to applying the existing knowledge and reduce the use of empirical studies. HDM-4 model was developed with the intent to address traffic congestion effects in industrialized countries, effects of cold climates, different pavement types, road safety, and environmental effects (12).

514

In Australia and New Zealand several studies were performed to develop VOC models that led to NAASRA (National Association of Australian Road Authorities) Improved Model for Project Assessment & Costing (NIMPAC) and Australian Road Fuel Consumption Model (ARFCOM) models in Australia and (New Zealand Vehicle Operating Costs) NZVOC model in New Zealand. In Europe, Sweden spearheaded the study by developing the VETO model in 1991, and the British Department of Transportation followed this up with COBA VOC model in 1993 (4).

In the U.S., the Texas Research and Development Foundation (TRDF) produced a VOC model in year 1982. A decade later, through National Cooperative Highway Research Program (NCHRP) MicroBENCOST was developed in 1993 using the TRDF and HDM studies (4).

All studies mentioned so far developed empirical models. Several studies have been performed to in the past decade to estimate the PVI related VOCs and EFC resulting in both mechanistic and empirical models (summarized in next section). Most of the PVI related VOC models consider rolling resistance force or the energy dissipated due to pavement surface characteristics such as roughness and/or texture and the pavement structural response (deflection) due to vehicular loading. Some empirical models directly calculate fuel consumption based on field measurements and regression analysis.

4 LITERATURE REVIEW OF EXISTING MODELS

A comprehensive literature review to identify current information related to VOC and EFC associated with PVI was conducted as part of a current FHWA research study. The review criteria consisted of factors such as: consideration of different pavement type during model development, impact on vehicle types including heavy vehicles, the availability of these models for public use, assumptions, limitations and gaps associated with these models, and the data requirements for implementation of the models. The models identified were further categorized as surface characteristics models and structural response models.

While some of the models identified may not directly require pavement condition input data or may not produce an output in terms of fuel consumption or other VOCs (e.g. output in terms of energy consumed or dissipated), these models are intended to be used for comparison results from the other models.

Several other PVI related models for VOC and EFC were reviewed as part of the study, however they were not further considered due to limitations such as: use of broad developmental assumptions, model being independent of pavement related factors or condition, use of outdated or obsolete measurement methodologies, the model was further improved by the same or a different researcher and may be considered as outdated, or need for proprietary software or the execution of model requiring purchase of a special software. Some of the models not considered on this basis are listed below:

- ARRB Aggregate Models (2)
- NRC – Phase II Model (25)
- Novel Multiscale Numerical Model for Texture related Fuel Consumption (19)
- Roughness Model Describing Heavy Vehicle-Pavement Interaction (20)
- Mechanistic-probabilistic vehicle operating cost (PVOC) model (6)
- Microscopic fuel consumption model (Virginia Polytechnic Institute) (3)

4.1 Surface characteristics models

NCHRP 1-45 (HDM4/World Bank) VOC Models (4): This research, performed by Michigan State University, was used to develop models for estimating the effects of pavement condition on vehicle operating costs (further divided in to fuel consumption, tire wear, and repair and maintenance). The HDM-4 model developed by World Bank was adopted, and a large

515

amount of data and information was collected, which was reviewed and analyzed to identify the most relevant VOC models. The initial studies focused on calibrating the empirical models in HDM-4 and updating the coefficients and costs to U.S. conditions. Subsequent studies attempted to improve the models using mechanistic approach to study the effects of roughness, texture and deflection, and also pavement type.

The resulting models from this study provide means to calculate vehicle operating costs, or more specifically, instantaneous fuel consumption (mL/km), tire wear (%/km), and repair and maintenance (U.S. dollars and labor hours per 1,000 km). Although the models require several non-pavement-related input parameters, the pavement-related inputs for the model are: roughness (quantified as International Roughness Index, IRI, in m/km), and texture (quantified as mean texture depth using the "sand patch" method, in mm). And deflection (quantified through measurement using the Benkelman Beam rebound method, in mm) is not included in this model.

The only pavement-related input used by the repair and maintenance model is roughness, and the effect of texture and deflection is not considered. The study shows the model to work well, however there were some limitations noted during the literature review, which are listed below.

MIT - Roughness-Induced PVI Model (16,18): This research was performed by Massachusetts Institute of Technology and supported by Portland Cement Association and Ready Mixed Concrete Research and Education Foundation. The study focuses on roughness-induced energy dissipation during PVI, where the dissipated energy was estimated and converted to fuel consumption. A mechanistic model was developed that relates the vehicle suspension to the roughness profile of the pavement using the quarter car model.

The resulting model from this study provides means to calculate EFC in MJ/km. The only pavement-related input for this model is roughness, quantified as IRI, in m/km, but the pavement profile power spectral density (PSD) may be used to improve the model output.

National Research Council of Canada Phase III Model (NRC) (25): The Cement Association of Canada sponsored the Center for Surface Transportation Technology and National Research Council of Canada to investigate the effects of pavement structure on fuel consumption. This study is the third phase of testing and investigation of the effects of pavement structure on fuel consumption rate. Phase I (1998) considered only a single season to determine the fuel efficiency differences between concrete and asphalt pavements. Phase II (2000-02) considered not only multi-seasons, but also involved a variety of trucks and trailer combinations and a relatively wide spectrum of road roughness.

Phase III considered the lessons learned from Phases I and II and was complementary to them while it sought improvements in vehicle types used in the previous studies. A statistical model was developed to calculate fuel consumption, in L/100 km, for trucks. Additional models to calculate seasonal (winter and summer) fuel consumption in passenger cars were also developed. The pavement-related inputs for the model are: roughness (quantified as IRI, in m/km), pavement temperature in $^{\circ}$C, and pavement type (asphalt, concrete, composite).

MIRIAM/VETO Models (11, 22, 23): This study is part of the Models for rolling resistance In Road Infrastructure Asset Management Systems (MIRIAM) project led by the Swedish National Road and Transportation Institute (VTI). A rolling resistance model was developed based on empirical data obtained from coast-down measurements, using roughness and texture as pavement-related factors. The calculated rolling resistance was then incorporated in to a fuel consumption model.

The model in this study calculates fuel consumption, in terms of fuel consumed per unit time. Although the models require several input parameters that are non-pavement-related, the pavement-related inputs for the model are: roughness (quantified as IRI, in m/km) and texture (quantified as mean profile depth, MPD, in mm).

UIUC Roughness Model (13, 14): This study was performed through the eXtreme Science and Engineering Discovery Environment (XSEDE) program, which is supported by the National Science Foundation (NSF), and The Illinois State Toll Highway Authority Administration. The model was developed by the University of Illinois at Urbana-Champaign.

516

The model developed in this study presents a theoretical approach to calculate the vehicle fuel consumption induced by pavement's structural response. A quarter-car model was combined with a white noise model of road roughness. And the profile of pavement was modeled as stationary filtered white noise and an equation was proposed to accommodate a broad range of stochastic representations of the road roughness for analyzing vehicle response and tire damping. The Environmental Protection Agency's (EPA's) formula of energy equivalent to 1 liter of gasoline was used to convert the deflection-induced rolling resistance force to estimated fuel consumption, in l/km. The only pavement-related input for this model is roughness (quantified as IRI, in m/km); however the pavement profile power spectral density (PSD) may be used to improve the model output.

Updated FHWA HERS models (Enhanced Prediction of Vehicle Fuel Economy and Other VOC) (7, 8): FHWA's Highway Economic Requirements System (HERS) model to estimate VOCs was updated by University of Nevada at Reno and Nevada Automotive Test Center. This study was performed to improve the VOCs estimation for different vehicle types, traffic conditions, and highway design scenarios as defined by roadway properties. The study developed extensive driving cycles with new transportation datasets for fuel consumption and non-fuel VOCs simulations and modeling to address the need for a comprehensive database of driving cycles. Vehicle fuel economy was evaluated by developing full vehicle models to represent the vehicle fleet and a wide range of driving cycles.

The study was performed in two phases. Phase I focused on developing relationship between vehicle speed and fuel consumption based on driving cycles, and in Phase II pavement roughness, and roadway conditions such as curvature and grade were incorporated into the models. The models developed in the two phases provides means to calculate VOCs, more specifically: fuel consumption (in gallons/1,000 miles), tire wear (in inches/1,000 miles), oil consumption (in quarts/1,000 miles), mileage-related vehicle depreciation (% of MSRP), and repair and maintenance (in dollars/mile).

Although these models do not have direct pavement-related inputs, the simulations considered the effect of roughness (IRI in m/km), and also pavement profile PSD. The road roughness in the form of PSD was used to calculate total rolling resistance forces. Based on the simulations, pavement condition adjustment factors (PCAF) for all models were developed as a function of vehicle speed. These models will be used for comparing results from other models.

4.2 *Limitations of surface characteristics models*

In general, none of the models take into account the properties of pavement materials and their behavior under various loading and environmental conditions. Further, the effect of pavement deflection/stiffness on fuel consumption is not incorporated in any of the models. The impacts of aging of the pavement and distress such as cracking, rutting, faulting etc., which could affect the pavement performance, roughness and surface texture, have also not been considered. These models, with the exception of NCHRP1-45 and the updated FHWA HERS model, do not estimate other PVI related VOCs such as tire wear, and repair and maintenance.

As shown in Table 1, with the exception of NCHRP1-45 and the MIRIAM models, effect of surface texture has not been considered in model development. Although roughness is considered in the form of IRI, the effect of megatexture is not captured unless the models use the entire pavement profile in the form of PSD (MIT and UIUC models).

In the NCHRP1-45 model calibration, fuel consumption for heavy trucks is assumed to be the same for summer and winter conditions since winter data was not available. This assumption could affect the results obtained for flexible pavements. Also, the tire model as well as the repair and maintenance model consider passenger car and heavy trucks only, and not the entire fleet of vehicles.

Table 1. Surface characteristics models.

	NCHRP1-45 VOC Models	MIT Roughness-Induced PVI Model	NRC Phase III Model	MIRIAM/VETO	UIUC Roughness Model	Updated HERS Model
Considers	Roughness (IRI), Texture (MPD)	Roughness (IRI and PSD)	Roughness (IRI)	Roughness (IRI), Texture (MPD)	Roughness (IRI and PSD)	See footnote [a]
Calculates	Rolling Resistance Coefficient	Dissipated energy		Rolling Resistance Coefficient	Dissipated energy	IFC, Tire Wear, R&M, Oil consumption, Mileage related depreciation
VOC Outputs	IFC, Tire Wear, R&M	EFC	IFC	IFC	EFC	
Mechanistic/Empirical	Empirical	Mechanistic	Empirical	Empirical	Mechanistic	Mechanistic
Roughness	0.5 to 8.5 m/km	$1.1<w<3.8$ [b]$1<c<4$ [c]	0.6 to 1.5 m/km	0.44 to 9.32 mm/m	1 to 5 m/km $0.02\pi < w/V < 6\pi$ (rad/m) [d]	0.67 to 6.67 m/km (43 to 423 in/mi)
Texture	0.2 to 2.0 mm	Not used	Not used	0.34 to 2.29 mm	Not used	Not used
Gradient	-3.4% to 3.1%	Not specified	-3.4 to 3.3 %	-1.88 to 1.73 %	Not specified	+/- 0.2 to 10 %
Curvature	Not specified	Not specified	-0.16 to 0.18 %	Not specified	Not specified	0.0 to 55.9 degrees of curvature
Vehicle Type	Car to truck	Golden Car	Car & semi-trailer	Car to truck	Half car	Car to truck
Vehicle Weight	1,325 to 12,338 kg (1.46 to 13.6 ton)	Golden Car	1,756 to 49,400 kg	1,860 to 27,000 kg	Half car	1,474 to 36,287 kg (3,250 to 80,000 lb.)
Vehicle Speed	56 to 112 km/h	80 km/h	60 & 10 km/h	60 to 120 km/h	10 to 60 m/s	1 to 90 mph
Fuel Type	Gas and Diesel	Not specified	Gas, Diesel	Not specified	Not specified	Gas, Diesel, LNG, HE
Tire Pressure	35 to 110 psi	Not specified	Not specified	32.6 to 130.5 psi (2.25 to 9 bar)	Not specified	OEM
Tire Size	0.38 to 0.57 m	Not specified	Not specified	Not specified	Not specified	Wide Range

a No pavement related inputs; b waviness number; c unevenness number; d waviness number as a function of speed; R&M-Repair & Maintenance; OEM-Original Equipment Manufacturer

The NRC study was not intended as a comparison between all grades of concrete or asphalt, and this may be seen in the roads used to quantify fuel consumption. Also, IRI values were limited to a maximum of 2 m/km, which in turn limited the influence of roughness on the results.

Some additional limitations that may or may not be consequential are that these models do not take into account the effect of tire properties on PVI (except NCHRP1-45), including the change in tire behavior with speed, temperature, texture (and microtexture).

MIT Deflection-Induced PVI Model (1, 15, 17): This research was performed by Massachusetts Institute of Technology and supported by Portland Cement Association and Ready Mixed Concrete Research and Education Foundation. These studies focused on deflection-induced, dissipation-induced, and roughness-induced PVI models, where the dissipated energy was evaluated and converted to fuel consumption. The initial research was performed to address the fuel consumption due to deflection of pavements by modeling the pavement on viscoelastic foundation. Subsequently, research continued to improve the models, address gaps and limitations and included a separate model for the effect of roughness on fuel consumption, to result in current models. The study is based on the theory that the vehicle is perennially on an uphill slope, thus needing extra energy to overcome the additional rolling resistance.

The resulting model from this study provides means to calculate EFC, in MJ/km. The pavement-related inputs for the model are; Young's modulus of surface layer in MPa, subgrade modulus in MPa, and pavement surface layer thickness in m. The moduli measured using wave propagation from falling weight deflectometer (FWD) time history may be used.

Table 1 shows the properties along with the range considered either during model development or during validation or sensitivity analysis of the surface characteristics models.

UIUC - Structural Rolling Resistance Model (10, 24): This study was performed through the eXtreme Science and Engineering Discovery Environment (XSEDE) program, which is supported by the National Science Foundation (NSF), and The Illinois State Toll Highway Authority Administration. The model was developed by the University of Illinois at Urbana-Champaign.

The study primarily focuses on the effect of pavement structure on fuel consumption and how it affects LCA calculations. Generalized models were developed to assess the effect of pavement deflection on vehicle fuel consumption, using mechanistic-based approaches.

The model developed in this study presents a theoretical approach to calculate the vehicle fuel consumption induced by pavement's structural response. The EPA's formula of energy equivalent to 1 liter of gasoline was used to convert the deflection-induced rolling resistance force to estimated fuel consumption, in l/km.

Although these models do not have direct pavement-related inputs, the power of the contact force calculations considers the deflection of the pavement surface. The structural response of the pavement to a moving wheel load is calculated using rate- and history-dependent response of pavement, and the power dissipated in the process is determined.

Vehicle Fuel Economy Models (Caltrans) – OSU (9): The California Department of Transportation (Caltrans) tasked the research team led by University of California Pavement Research Center (UCPRC), with Massachusetts Institute of Technology (MIT), Michigan State University (MSU) and Oregon State University (OSU) as team members, to investigate the effect of pavement structural response (with the effects of roughness and texture included) on the vehicle energy consumption and emissions.

The OSU model uses a generalized Maxwell-type viscoelastic model, and the pavement structure was modeled in Abaqus using finite element method.

Vehicle Fuel Economy Models (Caltrans) – MSU (9): The MSU model assumes a slope due to the load while assuming an axisymmetric deflection basin. The slope calculated was added as rolling resistance to further estimate the dissipated energy.

The Caltrans study is currently on-going, and the review was performed on a preliminary report. However, the models that were developed and used in this study, specifically the OSU and MSU models will be considered to calculate fuel consumption. The model developed by MIT has already been considered as an independent model. The pavement-related inputs for the model are: OSU - pavement surface layer properties (asphalt only) and MSU - pavement layer properties.

Table 2 shows the properties along with the range considered either during model development or during validation or sensitivity analysis of the structural response models.

Table 2. Structural response models.

	Deflection-Induced PVI Model (MIT)	Structural Rolling Resistance Model (UIUC)	Vehicle Fuel Economy Model (Caltrans/OSU)	Vehicle Fuel Economy Model (Caltrans/MSU)
Considers	Deflection (FWD), Layer Thickness and material properties	Deflection Basin	Surface Layer Properties	Pavement Layer Thicknesses and material properties
Calculates	Dissipated Energy	Dissipated Energy	Dissipated Energy	Dissipated Energy
VOC Outputs	EFC	EFC	EFC	EFC
Mechanistic/ Empirical	Mechanistic-Empirical	Mechanistic	Mechanistic	Mechanistic
Pavement type	Flexible, Rigid, Composite	Flexible	Flexible, Rigid, Composite	Flexible, Rigid, Composite
Pavement layer properties	Temperature specific layer properties & thickness. Limited to 2 layers	Temperature specific layer properties & thickness.	HMA Master-curve. Limited to surface layer.	HMA Master-curve. Lower layers time independent.
Vehicle type	Medium car to heavy truck	Heavy truck	Medium car to heavy truck	Medium car to heavy truck
Vehicle Weight	1.46 to 13.6 ton and 36.29 kN/axle	35.5 kN/axle	1.46 to 13.6 ton	1.46 to 13.6 ton
Vehicle Speed	50 to 130 km/h	8 to 120 km/h	50 and 100 km/h	50 and 100 km/h
Tire pressure	242 to 749 kPa	758 kPa	242 to 749 kPa	242 to 749 kPa

4.4 *Limitations of structural response models*

None of the structural response models consider the effects of roughness and surface texture (pavement surface characteristics). Also, the impacts of aging of the pavement and distress such as cracking, rutting, faulting etc., which could affect the pavement structural behavior have not been considered. These models do not estimate other PVI related VOCs such as tire wear, and repair and maintenance.

In the MIT model, a two-layer pavement system is assumed, which means the base and/or subbase layers for flexible pavements are not considered separately and are lumped together with the subgrade instead using an empirical relationship.

The tire mechanics used in the UIUC study assumes a non-dissipative tire, which does not represent a realistic tire-pavement system. Also, as shown in Table 2, the study was performed considering a flexible pavement structure only and cannot be applied to rigid pavements.

The effect of layers below the surface layers, including subgrade, have not been considered in the OSU model (see Table 2). At this stage the OSU model can be used for flexible pavements only.

Some additional limitations that may or may not be consequential are that these models do not take into account the effect of tire properties on PVI, including the change in tire behavior with speed, temperature, texture (and microtexture). The effect or influence of preceding axle load is not considered on the axles to follow.

5 CHALLENGES FOR IMPLEMENTATION

Practicality and ease of use of the models – A few of the models that are primarily theoretical will be faced with implementation challenges. This challenge is amplified when the model requires additional finite element analysis (FEA) prior to each use, or additional calculation using physics for each vehicle type, speed, tire etc. These challenges are primarily anticipated as availability of FEA software and qualified personnel to perform such analyses are uncommon at SHA and/or private consultant level. Another challenge presents itself when analysis is required for an entire network consisting of varying fleet of vehicles and varying pavement sections, which for some models, may result in performing several analysis runs and could be considered time consuming and impracticle.

Lack of measurement methods for ground truth – At this time there is no accurate method to measure fuel consumption due to PVI related rolling resistance factors only. Although devices and measurement methods to estimate rolling resistance are available, due to the number of variables encountered in practice it is not possible to accurately measure VOCs and EFC due to this rolling resistance. Therefore, verification and validation of these models can only be an estimate.

Compatibility of models – It may not be possible to develop a single universally accepted model to quantify all these aspects of PVI on use-phase LCA. However, the surface characteristics and structural response models will need to be evaluated for compatibility with each other to calculate total VOCs and EFC associated with all three rolling resistance factors. This will include evaluation of model coefficients and adjustment factors to ensure that the models are truly independent of the rolling resistance factors not considered at the time of development. Additionally, a reasonable methodology to combine the outputs from multiple models to calculate total VOCs and EFC will need to be established.

Data requirements of the models – The data required for the models over an analysis period may not always be available. This would be true especially in case of analysis of a new construction where the future deflection basins and pavement profiles cannot be predicted.

Value and usefulness of results – There is still some uncertainty on the usefulness of results and the order of magnitude in terms of impact on road user costs.

Guidelines for implementation – There are no current guidelines on the use of these models. The guidelines needed by SHA would include but not limited to identification of critical situations where one model is more applicable over the other, data collection methods, interpretation of results in decision making process. Also, one of the implementation challenges is the lack of a master tool or spreadsheet that can analyze all the models.

6 IMPLEMENTATION PATH – NEXT STEPS

The first step in implementation of the findings of this study was to critically review and assess the availability and gaps in PVI models for VOC and EFC, such that these costs can more accurately be accounted for in pavement investment decision-making.

The next step will be accomplished through critical evaluation of the identified PVI models, with the assistance of an expert Technical Review Panel (TRP), and development of case studies to further evaluate the models to determine the significance of PVI in comparison to other factors that affect VOC and EFC.

Through the case studies the feasibility and efficacy of SHAs incorporating VOC and EFC as part of their network and project level pavement investment decisions will be assessed. The various stages of design, such as preliminary design stage, environmental impact study, final design and life-cycle cost analysis, etc. will be considered to determine the best stage for incorporation of the rolling resistance factors and their impact on added VOC and overall EFC.

Also, consideration will be given to the current state-of-practice for routine pavement data collection and the suitability of the models to the data readily available.

As a final step, the PVI-based rolling resistance factors evaluated during the case studies will be evaluated against other influencing measures that are beyond the control of transportation agencies, such as tire type and inflation, vehicle and cargo mass, engine size and type, fuel type, driving behavior etc. The intent will be to provide perspective to decision makers on the order of magnitude of these factors for any potential policymaking.

The final recommendations will include approaches outlined for highway agencies and local governments on best practice methods for the use of PVI related VOC and EFC models, which in turn could help in improving pavement durability and sustainability to maximize vehicle fuel economy, improve safety, ride quality, and road conditions, and to minimize the need for road and vehicle repairs.

REFERENCES

1 Akbarian, M., Moeini-Ardakani, S., Ulm, F. J., & Nazzal, M. (2012). Mechanistic Approach to Pavement-Vehicle Interaction and Its Impact on Life-Cycle Assessment. Transportation Research Record: Journal of the Transportation Research Board, 2306(2306), 171–179. https://doi.org/10.3141/2306-20

2 Akcelik, R. (1983). Progress in Fuel Consumption Modelling for Urban Traffic Management. Australian Road Research Board (Vol. 1983–4). https://doi.org/0 86910 123 4

3 Ahn, K. (1998). Microscopic Fuel Consumption and Emission Modeling. Masters Thesis, Civil and Environmental Engineering, Virginia Polytechnic Institute and State University.

4 Chatti, K., & Zaabar, I. (2012). Estimating the Effects of Pavement Condition on Vehicle Operating Costs. NCHRP Report 720. National Academies Press.

5 Chupin, O., Piau, J. M., & Chabot, A. (2013). Evaluation of the structure-induced rolling resistance (SRR) for pavements including viscoelastic material layers. Materials and Structures/Materiaux et Constructions, 46(4), 683–696.

6 Curtis F. Berthelot, Sparks, G. A., Blomme, T., Kajner, L., & Nickeson, M. (1997). Mechanistic-probabilistic vehicle operating cost model. Transportation, 122(5), 337–341.

7 Hajj, E. Y., Xu, H., Bailey, G., Sime, M., Chkaiban, R., Kazemi, S. F., and Sebaaly, P. E. (2017). "Enhanced Prediction of Vehicle Fuel Economy and other Operating Costs, Phase I: Modeling the Relationship between Vehicle Speed and Fuel," Technical Report, Federal Highway Administration, Washington, DC.

8 Hajj, E. Y., Sime, M., Chkaiban, R., Bailey, G., Xu, H., and Sebaaly, P. E. (2018). "Enhanced Prediction of Vehicle Fuel Economy and other Operating Costs, Phase II: Modeling the Relationship between Pavement Roughness, Speed, Roadway Characteristics and Vehicle Operating Costs," Technical Report, Federal Highway Administration, Washington, DC.

9 Harvey, J. T., Zaabar, I., Louhghalam, A., Coleri, E., & Chatti, K. (2016). Model Development, Field Section Characterization, and Model Comparison for Excess Vehicle Fuel Use Attributed to Pavement Structural Response. Transportation Research Record: Journal of the Transportation Research Board, 2589(1), 40–50.

10 Hernandez, J. A., Al-Qadi, I. L., & Ozer, H. (2017). Baseline rolling resistance for tires' on-road fuel efficiency using finite element modeling. International Journal of Pavement Engineering, 18(5), 424–432. https://doi.org/10.1080/10298436.2015.1095298

11 Karlsson, R., Carlson, A., & Dolk, E. (2012). Energy use generated by traffic and pavement maintenance: decision support for optimization of low rolling resistance maintenance treatments. Vti Notat, pp 43.

12 Kerali, H. R. (2000). Overview of the HDM-4 system, Vol. 1, The Highway Development and Management Series, International Study of Highway Development and Management (ISOHDM), World Roads Association (PIARC), Paris.

13 Kim, R. E., Kang, S., Spencer, B. F., Al-Qadi, I. L., & Ozer, H. (2017). New Stochastic Approach of Vehicle Energy Dissipation on Nondeformable Rough Pavements. Journal of Engineering Mechanics, 137(July), 826–833. https://doi.org/10.1061/(ASCE)EM

14 Kim, R. E., Kang, S., Spencer, B. F., Ozer, H., & Al-Qadi, I. L. (2017). Stochastic Analysis of Energy Dissipation of a Half-Car Model on Nondeformable Rough Pavement. Journal of Transportation Engineering, Part B: Pavements, 143(4), 1–10.

15 Louhghalam, A., Akbarian, M., & Ulm, F. J. (2014). Scaling Relationships of Dissipation-Induced Pavement-Vehicle Interactions. Transportation Research Record: Journal of the Transportation Research Board, 2457(2457), 95–104.

16 Louhghalam, A., Tootkaboni, M., & Ulm, F. J. (2015). Roughness-Induced Vehicle Energy Dissipation: Statistical Analysis and Scaling. Journal of Engineering Mechanics, 137(July), 826–833. https://doi.org/10.1061/(ASCE)EM

17 Louhghalam, A., Akbarian, M., & Ulm, F. J. (2014). Flügge's Conjecture : Dissipation- versus Deflection-Induced Pavement – Vehicle Interactions, 140(8), 1–10.

18 Louhghalam, A., Tootkaboni, M., Igusa, T., & Ulm, F. J. (2018). Closed-Form Solution of Road Roughness-Induced Vehicle Energy Dissipation. Journal of Applied Mechanics, 86(1), 011003. https://doi.org/10.1115/1.4041500

19 Mansura, D. A., Thom, N. H., & Beckedahl, H. J. (2018). Numerical and Experimental Predictions of Texture-Related Influences on Rolling Resistance. Transportation Research Record. https://doi.org/10.1177/0361198118776114

20 Papagiannakis, A. T., & Gujarathi, M. (1995). A Roughness Model Describing Heavy Vehicle-Pavement Interaction. Final Research Report, (3), 58 p. Retrieved from.

21 Pouget, S., Sauzéat, C., Benedetto, H. Di, & Olard, F. (2011). Viscous Energy Dissipation in Asphalt Pavement Structures and Implication for Vehicle Fuel Consumption. Journal of Materials in Civil Engineering, 24(5), 568–576. https://doi.org/10.1061/(asce)mt.1943-5533.0000414

22 Sandberg, U. (2011). Rolling resistance-Basic Information and State-of- the-Art on Measurement methods. Report from Models for rolling resistance in Road Infrastructure Asset Management systems (MIRIAM).

23 Sandberg, U., Bergiers, A., Ejsmont, J. A., Goubert, L., & Karlsson, R. (2011). Road surface influence on tyre/road rolling resistance. Report from Models for rolling resistance In Road Infrastructure Asset Management systems (MIRIAM).

24 Shakiba, M., Ozer, H., Ziyadi, M., & Al-Qadi, I. L. (2016). Mechanics based model for predicting structure-induced rolling resistance (SRR) of the tire-pavement system. Mechanics of Time-Dependent Materials, 20(4), 579–600.

25 Taylor, G. W., & Patten, J. D. (2006). Effects of Pavement Structure on Vehicle Fuel Consumption - Phase III.

26 Watanatada, T, et al. The highway design and maintenance standards model: Volume 1: Description of the HDM-III model. The Highway Design and Maintenance Standards Series, Baltimore: John Hopkins for the World Bank, 1987.

Context-specific assessment of the life cycle environmental performance of pavements considering neighborhood heterogeneity

Hessam AzariJafari, Xin Xu, Jeremy Gregory & Randolph Kirchain
Massachusetts Institute of Technology, Cambridge, USA

ABSTRACT: The impacts of pavement use phase on life cycle environmental impacts were often quantified using generalized models, which can result in an oversimplification of the assessment. In this study, we estimate the pavement deterioration and the induced extra fuel consumption (EFC) considering the context-related data such as temporally dynamic traffic volume and ambient temperatures. To have a representative resolution of urban morphology and the highways constructed within the city boundary, we apply an adapted analytical model for radiative forcing (RF) and a hybrid framework combining two different models for simulating building energy demand (BED). This framework estimates the impacts of changing pavement albedo for different urban neighborhoods considering the realistic heterogeneity. The impact of several context-specific factors, including location, cloudiness, shading, and morphology, were taken into account in this model. Then, the model was applied to a case study of road network in the city of Phoenix, Arizona. Comparative analysis reveals that the EFC induced by pavement roughness and deflection results in significant environmental benefits to change in the pavement network system. In urban neighborhoods with low density, the RF effect was a key driver for impacts due to changes in albedo. In addition, reflective pavements create net global warming potential benefit, which is mainly due to the RF benefit being more significant than the change in cooling and heating demand of buildings in the urban neighborhood. The conclusions from this model can support more informed decisions on the life cycle environmental impacts of pavement design and maintenance decisions, particularly in urban areas.

1 INTRODUCTION

As a consequence of population growth and economic development in cities, a significant trend in urbanization has been observed. This trend delineates the importance of environmental impacts assessment in urban neighborhoods, where a growing community is exchanging resources and emissions with the ecosphere. As a holistic tool, life cycle assessment (LCA) has been implemented to evaluate the environmental burdens of different products or services being provided in these neighborhoods. The LCA tool has been incorporated into various studies to assess the environmental impacts of city infrastructures, such as pavements that provide traffic services and to facilitate suburb-to-city commuting. Owing to its long lifetime, the use phase has been identified as the environmental hotspot of the pavement life cycle (Loijos et al. 2013, Xu et al. 2019). However, efforts to improve the assessment quality of the use phase is limited. In fact, the impacts of use phase, such as albedo, were often quantified using simplified models, which are not able to capture and distinguish the context of the study and consequently, results in an uncertainty in the results (Noshadravan et al. 2013, AzariJafari et al. 2016).

Pavement albedo perturbs the quantity of radiative forcing (RF) reflected from the surface and simultaneously, alters the building energy demand (BED) through the urban heat island (UHI) effect. Therefore, in an urban context, the albedo effect is a highlighted component of pavement LCA studies. As stated in the literature, the UHI effect speeds up smog formation and increases A/C energy demand in cities (US EPA 2013). In addition, previous research works showed the contribution of cool pavements to lower outdoor air temperatures. Rosenfield et al (1998) reported that implementing a cool pavement strategy would reduce peak power demand by 100 MW. Gilbert et al. reported that increasing the pavement albedo by 0.2 in Los Angeles would result in an annual cooling energy savings of 0.22 kWh/m^2·yr (Gilbert et al. 2017). Following the momentum that was made regarding the greenhouse gas (GHG) saving of cool surfaces, several governments, such as California, adapted certain regulations for the implementation of cooling surface strategies.

While the majority of literature confirms the benefits of implementing cool pavements in cooling energy consumption, certain studies highlighted the importance of including both the induced heating and cooling BED. Kolokotroni et al (2012) reported that the decrease in heating demand in winter can offset the cooling BED saving induced by the UHI effect in the city of London, UK, leading to 90% and 50% reduction in electricity consumption in short and long run, respectively. This study clearly demonstrated the importance of incorporating context, specifically, seasonal temperatures and the related trade-off between heating and cooling energy changes in buildings. On the other hand, AzariJafari et al. (2019) showed that switching to a cool pavement system in cold regions, such as those prevalent in Canada, can result in a net increase in BED (42 ± 9 MWh/yr/km pavement). Their findings motivate further research to incorporate the geographically specific climate condition instead of attributing a single CO_2 saving factor to BED saving.

The same observation for incorporating location-specific data in other use phase components, such as EFC, is seen in the pavement LCA literature. Louhghalam et al. (2014) reported that there is a significant change between fuel consumption of heavy vehicles as the speed and temperature increases. Based on their findings, the deflection-induced EFC of vehicles on asphalt pavement can be doubled at an ambient temperature of 30 °C compared to the consumption at 10 °C. This sensitivity emphasizes the importance of including context-specific information, such as high-resolution temperatures to the pavement LCA study. As a result, a revision in consideration of EFC emission based on local conditions and vehicle speed seems to be necessary.

In light of incorporating the albedo effects of pavement into climate change impact, it is critical to understand the scale of the environmental changes associated with RF and BED considering the whole life cycle components, such as EFC, lighting and carbonation, materials and construction, maintenance and repair, and end of life. This study investigates location-specific modeling for the pavement life cycle modeling. The effects of context-specific parameters, such as location, urban morphology, shadings, and high-resolution temperature were considered in the models. This research provides a comprehensive evaluation of the effectiveness of implementing different pavement strategies at a neighborhood scale. Ultimately, the outputs of this model can help urban planners, transportation agencies, and pavement designers have a high-resolution evaluation of the pavement decision-making in order to meet the GWP mitigation goals.

2 METHODOLOGY

The scope of this study is limited to cradle to grave impacts. The characterization factors associated with GWP100 were incorporated to estimate the climate change impact of the pavement life cycle (IPCC 2006). For the background unit processes, ecoinvent v.3.4 (2017) was used while city-specific data was implemented for foreground modeling. The use phase of pavements comprises albedo, carbonation, lighting, roughness and deflection-induced pavement-vehicle interaction (PVI). The climate change impact of road illumination and concrete carbonation were included using developed models that can be found in (Lagerblad 2005, Santero and Horvath 2009) while the models developed and incorporated for other components are presented in the following sections.

2.1 Adapted models for estimating EFC induced by IRI and deflection

The LCA model of the use phase accounts for both roughness and deflection components. In addition, this model considers the effect of pavement maintenance and repair on the physical specifications of the pavement and subsequently, the PVI. The deflection PVI was estimated using the model developed at CSHub (Akbarian et al. 2012, Louhghalam et al. 2014). The so-called GEN II model calculates the deflection-induced EFC of the pavement system over its lifetime using a mechanistic approach. Roughness is characterized by the international roughness index (IRI). The prediction of IRI over time was predicted from the results of the Mechanistic-Empirical Pavement Design Guide (MEPDG) approach using Pavement-ME software. The difference between yearly IRI and its value at initial construction was calculated and translated into the EFC using the calibrated HDM-4 model of the World Bank (Chatti and Zaabar 2012).

2.2 Adapted models for location-specific albedo

Albedo (shown as α) is a scale that represents the fraction of solar shortwave radiation that is reflected when it reaches to any surface. When the radiation is fully absorbed by the surface, then a zero value is considered to the surface albedo and if the solar radiation is fully reflected, then the albedo of the surface equals 1. The albedo of pavement surface induces environmental impacts in two ways: first, directly through adjusting the radiative forcing (RF), and second, indirectly by changing the ambient temperature and consequently affecting the energy demand in the built environment adjacent to the pavement. In this work, a baseline value of 0.3, which roughly represents the average reflectivity of the earth, was considered for the calculation of the albedo effect. $\Delta\alpha$ is the change in albedo calculated with reference to a baseline albedo of 0.3.

When it comes to pavement albedo modifications, a context-specific quantification of the RF impacts is required to support more informed decisions on their effectiveness at a local scale. In previous works, an empirical model for clear-sky was adopted from the literature and the model was calibrated for cloudy-sky conditions with data obtained from climate simulations. Details about the model development can be found in Xu et al (2018). The following equation was considered for the calculation of the global warming potential of RF (GWP_{alb}):

$$GWP_{alb} = \frac{-\int_0^{TH} \frac{A}{A_{earth}} R_{TOA}\tau f_a\, \Delta a_s dt}{\int_0^{TH} RF_{co_2} dt} \tag{1}$$

Where R_{TOA} is the downward solar radiation at top of the atmosphere (TOA); τ is the urban canyon transmittance, which is defined as the ratio of the solar radiation reflected from the pavement surface at the bottom of an urban canyon, to the radiation reflected from the same road surface, not in a canyon; f_a is the two-way transmittance factor accounting for absorption and reflection of solar radiation throughout the atmosphere, which is calculated for the realistic cloudy sky of the city using a cloudy sky multiplier, which itself is a function of cloudiness, precipitable water, and solar zenith angle, and aerosol optical depth of the location; the pavement functional area (A) divided by the surface area of the earth (A_{earth}) is considered as the localization factor. For a time horizon (TH) of 50 years, the radiative forcing of CO_2 was calculated as 5.02×10^{-14}.

For the BED effect of albedo change, two mechanisms were incorporated into the simulation. First, as the pavement albedo increases, a larger amount of incident radiation is reflected onto the nearby buildings, which consequently alters the BED. Moreover, as opposed to the incident radiation effect, increasing pavement albedo reduces the canyon temperature in the neighborhood. To capture both effects, a hybrid framework was developed and incorporated using Rhino® and Grasshopper®. The procedure of simulating the BED change as a result of albedo changes is presented in Figure 1. The energy and radiation simulation plugin for Grasshopper® - Ladybug and Honeybee, and the urban climate simulator, called Urban

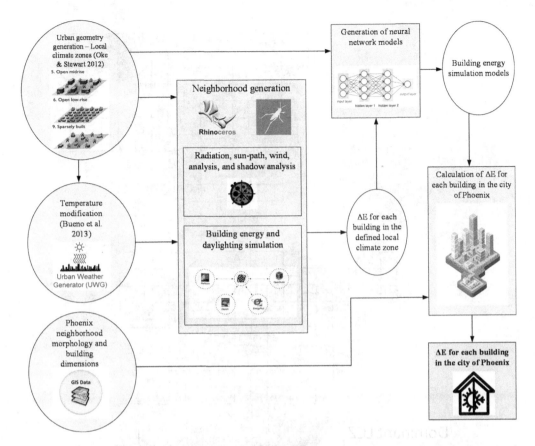

Figure 1. Simulation workflow for the hybrid modeling framework ($\Delta\alpha$ is pavement albedo change and ΔE is the associated change in BED).

Weather Generator (UWG) were incorporated to assess the effect of pavement albedo change on the local climate zones (LCZs) created based on the proposed neighborhood taxonomy by Stewart and Oke (2012). For more information about LCZ's definition and the framework details, readers are referred to method paper that explicitly presented the methodology of the BED modeling (Xu et al. 2020). Then, a machine learning-based method was used to simulate the BED of individual buildings for a realistic heterogeneous neighborhood obtained from geographic information system (GIS) data.

2.3 Case study of phoenix city

Previous efforts on pavement LCA incorporating albedo mainly focused on changing the business-as-usual scenario (i.e. asphalt in the majority of U.S. roads) with concrete as a common structural system that can support the traffic loads (Li et al. 2013, Gilbert et al. 2017). Hence, replacing a functionally equivalent flexible (asphalt) with a rigid (concrete) pavement was considered as a case study. The effect of maintenance and repair schedules (M&R) were included leveraging the MEPDG models associated with the Pavement-ME software. The functional unit in this analysis is the total urban road network of the city of Phoenix for an analysis period of 50 years. The system boundaries include the whole life cycle phases from the extraction of construction materials, construction, maintenance and repairs, use, and end of life. Traffic load for a local street in the city of Phoenix is obtained from government data (https://catalog.data.gov/dataset/phoenix-study). The average annual daily truck traffic (AADTT)

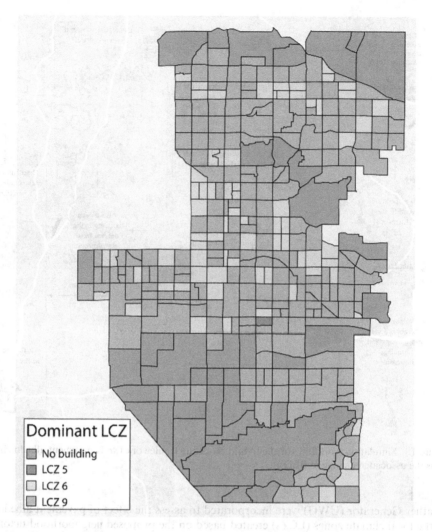

Figure 2. Dominant neighborhood morphology for census tracts of phoenix city according to the classi-fication of (stewart and oke 2012).

was 1600 for two directions with a 3% annual increase rate. Surface mix designs for asphalt and concrete were consistent with local agency practices and pavement design life. The dominant neighborhood morphology for each census tract (CT) of the city is presented in Figure 2.

3   RESULTS AND DISCUSSIONS

Results of the BED changes of the Phoenix buildings on a CT level are presented in Figure 3. As a result of replacing asphalt with concrete, the cooling and heating energy demand increases and decreases, respectively. Increasing pavement albedo results in more reflection of solar radiation from the pavement surface and at the same time, causes the ambient air temperature to drop. Therefore, both the longwave solar radiation and the convective flux from the environment decrease. The net balance of the increased shortwave and decreased longwave and convective flux affect the heat gain and energy demand indoor. As can be seen in Figure 3.a and 3.b, the incident radiation effect dominated the cooling effect that ultimately results in heat gain both in warm and cold seasons. Moreover, as shown in Figure 2, the majority of the

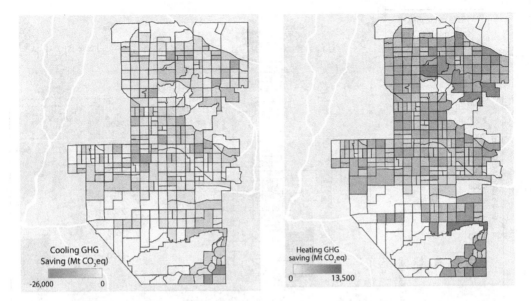

Figure 3.   Changes in a) cooling and b) heating energy demand of buildings in each census tract of Phoenix city (negative values imply excessive energy demand).

CTs in the city of Phoenix corresponds to sparsely built neighborhoods. When the aspect ratio of the neighborhood is small (e.g. in those CTs whose morphology are dominated by LCZ9), the context looks like a wide roadways, where buildings are far apart or the building height is not as large as those prevalent in downtown (In Figure 2, it is distinguished by red color). In this case, modifications to pavements induce a smaller impact on the BED. One should note that as aspect ratio in a neighborhood grows, there is either a grow in building height (and consequently, incident radiation received by the building facade grows) or a shrink in the canyon width. A narrower canyon width results in a less dissipation of incident radiation as opposed to faster dissipation when LCZ9 neighborhood is prevalent for a given census tract. Therefore, the cooling demand and heating saving increases as H/W grows. However, as the studied city is located in a warm and dry zone, the range of cooling demand is larger than that for heating and as a result, the net energy change tends to be a burden rather than a saving in various neighborhoods.

Figure 4 Shows the life cycle GHG saving of replacing asphalt (the predominant paving material in Phoenix) with concrete. The viscoelastic properties of asphalt pavements lead to higher deflection-induced EFC when exposed to elevated temperatures in hot and dry climates like Phoenix (Xu et al. 2019). This is the primary driver for the EFC saving of the urban roads. Our results also show that the GHG emissions of materials and equipment for the initial construction of concrete are higher than for asphalt. Nevertheless, the multiple overlays required for asphalt pavements ultimately result in net GHG savings in embodied impacts for the switch to concrete. In the northern and southern parts of the city, there are certain CTs that the neighborhood is too sparse to be considered in the LCZ characteristics. The large GHG savings in the northern CTs are attributed to the potential savings from The PVI effects. While in the southern CTs without buildings, the total mileage of the roads is extremely smaller than those in the northern part. Therefore, we observe negligible GHG savings in the south roads compared to those in the north.

The relative magnitude of GHG savings is a combined effect of the context-specific use phase impact and the total lane miles of pavements in each CT. Figure 4 shows the significance of context in impacting the eco-friendly alternative and the effectiveness of the GHG savings related to the pavement replacement. Ranges of the GHG savings reflect the variation associated with the urban morphology, especially due to BED changes in each CT. As shown

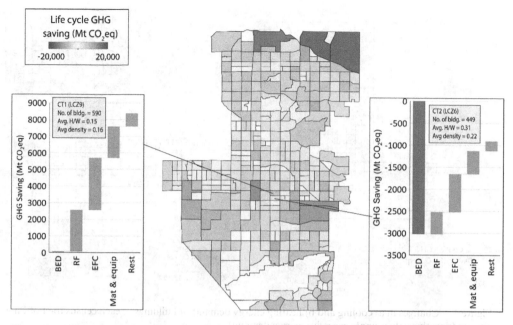

Figure 4.   Life cycle GHG savings of replacing asphalt with concrete incorporating materials and construction, maintenance and repair, albedo-radiative forcing (RF), albedo-building energy demand (BED), excess-fuel consumption (EFC) induced by surface roughness and deflection, and the rest of the life cycle components (carbonation, lighting, and end of life).

in Figure 4, one single strategy may not lead to a benefit, even in adjacent neighborhoods. In fact, owing to the denser neighborhood and higher aspect ratio, the 3,000 metric tons GHG burden induced by the BED effect in CT2 dominates the savings induced by other life cycle components. The overall life cycle saving for CT2 is negative, meaning that asphalt has lower life cycle GHG emissions than concrete in this scenario. Moreover, as shown in the left inset chart in Figure 4, the net BED effect of shifting from asphalt to concrete in CT1 is negligible compared to the GHG changes in other life cycle components. The small GHG saving associated with the BED change in CT1 can be explained by the fact that the increase in cooling demand is counterbalanced with the saving in heating BED in that neighborhood.

4   CONCLUSIONS AND OUTLOOK

This study developed and incorporated a context-specific pavement LCA model that accounts for all the life cycle components, starting from materials, construction, use, maintenance and repair, and end-of-life. Then, the model was applied to a case study of shifting from the business-as-usual scenario, i.e. asphalt, to a reflective rigid pavement, i.e. concrete. The main challenge of pavement LCA is to set the resolution of the impact assessment. Hence, in this study, to quantify the RF impact of albedo change, an empirical analytical model was adapted from the literature. This model incorporated the effects of climate and context conditions in the urban area. For BED, several simulation tools were incorporated to analyze the interactions between buildings and pavements. The context-specific models of EFC due to surface roughness and deflection were employed to complement the analysis of the use phase components. The carbonation and lighting were also considered in this study, taking into account the surface characteristics, mix design, and the number of rainy days. The presented case study comprises the urban road network in the city of Phoenix, which incorporates location-specific data extracted from climate simulations and GIS datasets. A comparison of the results across the different LCZs reflects the importance of having a high-resolution and context-specific model for albedo impact estimation. Urban-scale impact quantification is also necessary as albedo modifications can have different and even contradicting impact on BED depending on urban morphology. A certain area of opportunities still exists to improve the accuracy of the proposed model. The impact of building properties on BED as well as data collection on the evolution of pavement albedo due to aging and resurfacing would provide a clearer picture of the neighborhoods. The incorporation of real-time traffic volume representing the temporal dynamic of the city and its effect on the amount of radiation reflected from the surface can be considered for further research studies. Conducting these research studies in light of the associated uncertainty can result in a robust conclusion on the preferred scenarios.

REFERENCES

Louhghalam, M. A., FJ Ulm (2014). Pavement Infrastructures Footprint: The Impact of Pavement Properties on Vehicle Fuel Consumption. Computational Modelling of Concrete Structures.

Akbarian, M., S. S. Moeini-Ardakani, F.-J. Ulm and M. Nazzal (2012). "Mechanistic Approach to Pavement-Vehicle Interaction and Its Impact on Life-Cycle Assessment." Transportation Research Record: Journal of the Transportation Research Board 2306(-1): 171–179.

AzariJafari, H., A. Yahia and B. Amor (2019). "Removing Shadows from Consequential LCA through a Time-Dependent Modeling Approach: Policy-Making in the Road Pavement Sector." Environmental Science & Technology 53(3): 1087–1097.

AzariJafari, H., A. Yahia and M. Ben Amor (2016). "Life cycle assessment of pavements: reviewing research challenges and opportunities." Journal of Cleaner Production 112, Part 4: 2187–2197.

Chatti, K. and I. Zaabar (2012). Estimating the effects of pavement condition on vehicle operating costs, Transportation Research Board.

ecoinvent (2017). Ecoinvent v.3.4 database. Swiss Centre for Life Cycle Inventories. Zurich and Dubendorf, Switzerland.

Gilbert, H. E., P. J. Rosado, G. Ban-Weiss, J. T. Harvey, H. Li, B. H. Mandel, D. Millstein, A. Mohegh, A. Saboori and R. M. Levinson (2017). "Energy and environmental consequences of a cool pavement campaign." Energy and Buildings 157: 53–77.

IPCC. (2006). "IPCC Guidelines for National Greenhouse Gas Inventories." August 20th 2016, from http://www.ipcc-nggip.iges.or.jp/public/2006gl/pdf/0_Overview/V0_0_Cover.pdf.

Kolokotroni, M., X. Ren, M. Davies and A. Mavrogianni (2012). "London's urban heat island: Impact on current and future energy consumption in office buildings." Energy and Buildings 47: 302–311.

Lagerblad, B. (2005). "Carbon dioxide uptake during concrete life cycle–State of the art." Swedish Cement and Concrete Research Institute—CBI.

Li, H., J. T. Harvey, T. J. Holland and M. Kayhanian (2013). "The use of reflective and permeable pavements as a potential practice for heat island mitigation and stormwater management." Environmental Research Letters 8(1): 015023.

Loijos, A., N. Santero and J. Ochsendorf (2013). "Life cycle climate impacts of the US concrete pavement network." Resources, Conservation and Recycling 72(0): 76–83.

Louhghalam, A., M. Akbarian and F.-j. Ulm (2014). "Flügge's Conjecture: Dissipation- versus Deflection-Induced Pavement – Vehicle Interactions." 1–10.

Noshadravan, A., M. Wildnauer, J. Gregory and R. Kirchain (2013). "Comparative pavement life cycle assessment with parameter uncertainty." Transportation Research Part D: Transport and Environment 25(0): 131–138.

Rosenfeld, A. H., H. Akbari, J. J. Romm and M. Pomerantz (1998). "Cool communities: strategies for heat island mitigation and smog reduction." Energy and Buildings 28(1): 51–62.

Santero, N. J. and A. Horvath (2009). "Global warming potential of pavements." Environmental Research Letters 4(3): 034011–034011.

Stewart, I. D. and T. R. Oke (2012). "Local Climate Zones for Urban Temperature Studies." Bulletin of the American Meteorological Society 93(12): 1879–1900.

US EPA (2013). Heat Island Compendium | Heat Island Effect | US EPA, United States Environmental Protection Agency,.

Xu, X., M. Akbarian, J. Gregory and R. Kirchain (2019). "Role of the use phase and pavement-vehicle interaction in comparative pavement life cycle assessment as a function of context." Journal of Cleaner Production 230: 1156–1164.

Xu, X., H. Azarijafari, J. Gregory, L. Norford and R. Kirchain (2020). "An integrated model for quantifying the impacts of pavement albedo and urban morphology on building energy demand." Energy and Buildings: 109759.

Xu, X., J. Gregory and R. Kirchain (2018). The Impact of Pavement Albedo on Radiative Forcing and Building Energy Demand: Comparative Analysis of Urban Neighborhoods. Transportation Research Board 97th Annual Meeting. Washington D.C., Transportation Research Board.

Rapid ground-based measurement of pavement albedo

Sushobhan Sen
Department of Civil and Environmental Engineering, University of Pittsburgh, USA

Jeffery Roesler
Department of Civil and Environmental Engineering, University of Illinois at Urbana-Champaign, USA

ABSTRACT: The albedo of pavements is a key optical property, which affects the development of the Urban Heat Island (UHI) in cities. However, measuring albedo of in-service pavements is challenging, requiring lane closures, ideal weather conditions, and a significant time investment. A newly-developed, rapid ground-based instrument, D-SPARC, to measure pavement albedo was deployed in a survey of three city districts (University, Residential, and Commercial) in Urbana, Illinois. Each district was about 0.1 km^2 in area and took just 90 minutes to perform each albedo survey. The albedo survey was executed at various times during the day and the night as the D-SPARC did not require any sunlight to operate. The average albedo of each of the three districts was 0.22, 0.17, and 0.17 respectively, with the lowest and highest albedo measurement in each of district varying by approximately a factor of two. There was significant intra-district variation in albedo, even between adjoining roads, showing the need to rapidly measure albedo over large road networks in a city. Finally, the relationship between traffic volume and albedo in each of the three districts was examined. The correlation coefficient was low, indicating that albedo did not depending on the volume of traffic of the roads that were surveyed but depended largely on weathering and debris.

1 INTRODUCTION

The Urban Heat Island (UHI) effect, which leads to higher air temperatures in urban areas as compared to adjacent rural ones, has been observed for decades in hundreds of cities across the world (Arnfield, 2003, Peng et al., 2011, Tran et al., 2006). Kleerekoper et al. (2012) listed seven reasons for the development of UHI, one of them being the replacement of vegetated surfaces with low albedo artificial ones. These low albedo surfaces absorb more heat and retain it longer than vegetated surfaces, which leads to higher air temperatures. These artificial surfaces include pavements, roofs, and walls.

The effect of pavements on UHI has been widely examined in the literature. Santamouris (2013) and Qin (2015) reviewed hundreds of studies that investigated the impact of pavements and concluded that the most commonly-recommended measure to mitigate the UHI effect was implementing cool pavements, which have a higher albedo than typical pavements. Other studies, such as by Sen and Roesler (2017) and Gui et al. (2007), showed that the thermal properties of pavements could also be used to mitigate UHI. Qin et al. (2019) developed an analytical model for pavement surface temperature and concluded that albedo had the strongest effect on pavement surface temperature and hence the greatest potential for UHI mitigation.

Several studies (Li et al., 2013, Pisello et al., 2014, Richard et al., 2015, Sreedhar and Biligiri, 2016) have measured the albedo of pavements in the field using an albedometer (also called a double pyranometer). The principle of this is to measure the incident and reflected solar radiation over a small area of pavement, the ratio of which is the albedo. The typical albedo of new asphalt pavements was 0.05-0.10 and aged asphalt pavements was 0.15-0.20, while that of new and aged concrete pavements was 0.35-0.40 and 0.20-0.30 respectively.

These studies were based on the ASTM E1918 (2016) standard, which requires a clear, sunny day for measurement as well as a 4 m x 4 m area of pavement to either be cast in the lab or cordoned off in the field. The instrument also takes several minutes to setup. Some studies (Qin and He, 2017, Akbari et al., 2008, Sen et al., 2018) have developed ways to reduce the required size of the pavement, although the smallest possible size has been reported to be 1 m x 1 m, which is still a large size to cast in the lab or cordon off in the field. This, combined with the requirement to have clear, sunny days, has limited the speed at which pavement albedo can be measured using ground-based techniques, particularly over large road networks.

In a previous study by the authors (Sen, 2019) a new instrument, the Discrete SPectrAl RefleCtometer (D-SPARC), was developed to overcome these challenges. D-SPARC is a ground-based instrument capable of estimating pavement albedo at any time of the day or night from a 180 mm x 90 mm strip of pavement. It is much faster to use than an albedometer, enabling the rapid measurement of albedo of a road network. It has been shown to be quite reliable, with a root mean square error (RMSE) of 0.02 for measurements performed at night and 0.06 during the day. In the present study, the results from an albedo survey conducted in three different land use areas in Urbana, Illinois using D-SPARC are presented to show the variation in pavement albedo in those areas. The relationship between traffic volume and albedo are also investigated, which has not been considered in previous studies.

2 METHODOLOGY

2.1 Study area

For the pavement albedo field survey, three 3 x 3 urban blocks were chosen in Urbana, Illinois. Each of these blocks was about 0.1 km^2 in area. These three districts represent three different categories of land use:

1. **University District:** Several concrete roads along with several asphalt streets in the campus area with typical traffic patterns including a lot of buses, cars, cyclists, and a few trucks.
2. **Residential District:** Mostly asphalt streets as well as some brick roads surrounded by single-family housing units with traffic patterns almost entirely of cars, with a few buses and trucks.
3. **Commercial District:** This district was around a large shopping mall with several smaller businesses in the vicinity. Most roads were asphalt with a few concrete roads. The traffic pattern had a larger percentage of trucks than the other two districts, with buses and cars also forming a formidable part of the traffic as well.

A map of these three districts is shown in Figure 1. In addition to roads, the University District and Commercial District also had several large parking lots. The present study was limited to only the streets without including the albedo of the parking lots. The streets varied in age with several built as recently as two years ago, while others are much older without records indicating a date of construction.

2.2 D-SPARC instrument

To estimate the albedo of the roads in the three districts, an operator and an assistant visited each of the streets with the D-SPARC instrument shown in Figure 2. The instrument consists of three distinct units that are housed in two boxes. The Transmitter Unit (TU) contains a series of light sources at specific frequencies, which generates light to the pavement surface. The Receiver Unit (RU) has sensors that determine the intensity of light reflected by the surface. Finally, the Controller and Storage Unit (CSU) triggers the TU and RU measurements and stores the results. Figure 2(a) shows the instrument, where the TU and RU are stored in one box and the CSU in another box.

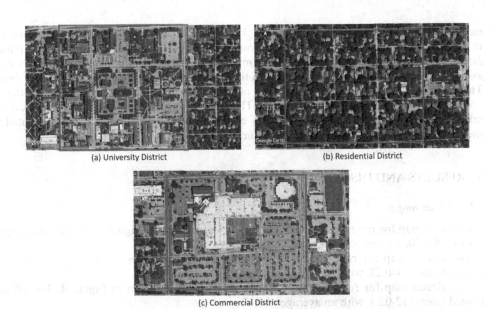

(a) University District

(b) Residential District

(c) Commercial District

Figure 1. Three districts chosen for pavement albedo survey: (a) University district (b) Residential district and (c) Commercial district.

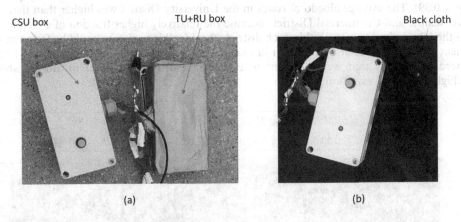

CSU box

TU+RU box

Black cloth

(a)

(b)

Figure 2. Photo of the D-SPARC components (a) the controller and storage unit (CSU) and transmitter and receiver unit (TU+RU) boxes shown next to each other with some connecting wires and (b) a black cloth placed on top of the TU+RU box to cut off solar radiation, with the CSU box placed on top of it.

To operate the instrument, it is important to cut off any sunlight from the box containing the TU and RU, which is accomplished by covering it with a black cloth as shown in Figure 2 (b). The CSU box is then placed on top of the black cloth, and measurement is initiated when the user presses the green button. One measurement takes about 15 seconds including the time required to write the results to storage, after which a new set of measurements can be taken immediately by pressing the button again.

2.3 *Survey procedure*

The albedo of each pavement measured between every intersection representing a segment of that street of approximate length 100 m. For each segment, three independent albedo measurements were taken along the pavement edges and then averaged to give a single albedo value for that segment. In this case, 'pavement edge' refers to the part of the pavement immediately adjacent to the curb and gutter. Many of the streets had minor patching, which were avoided, and testing was only performed on the representative pavement surface of the street. The survey was completed at various times of the day and night given the D-SPARC device does not require external light to operate. The survey in one of the districts took about 90 minutes to complete. Finally, the albedo of each of the pavement segment was presented in the form of an albedo map, as shown in the next section.

3 RESULTS AND DISCUSSION

3.1 Albedo maps

The albedo map for roads in the University District is shown in Figure 3. The albedo ranged from 0.15-0.30, with an average of 0.22.

The albedo map for roads in the Residential District is shown in Figure 4. The albedo ranged from 0.11-0.23, with an average of 0.17.

The albedo map for roads in the Commercial District is shown in Figure 4. The albedo ranged from 0.12-0.25, with an average of 0.17.

A box-and-whisker plot shown in Figure 6 provides the distribution of albedo in each district. The average albedo of the University District was statistically different from that of the Residential and Commercial districts ($p = 3.4 \times 10^{-4}$ and 3.7×10^{-5} respectively). However, the average albedo of the Residential and Commercial districts was statistically similar ($p = 0.39$). The average albedo of roads in the University District was higher than that of the Residential and Commercial Districts because the relatively higher fraction of concrete roads in the University District. Within each district, considerable variation could be found in pavement albedo as a result of variation in materials, age, weathering, and dirt/debris. Furthermore, pavements that were adjacent to each other also had a different albedo, showing a highly localized variation in albedo.

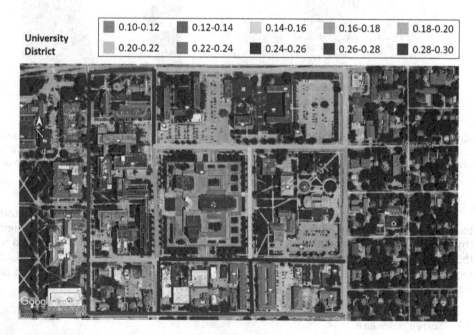

Figure 3.   Albedo map of streets in the university district.

	0.10-0.12	0.12-0.14	0.14-0.16	0.16-0.18	0.18-0.20
Residential District	0.20-0.22	0.22-0.24	0.24-0.26	0.26-0.28	0.28-0.30

Figure 4.   Albedo map of streets in the residential district.

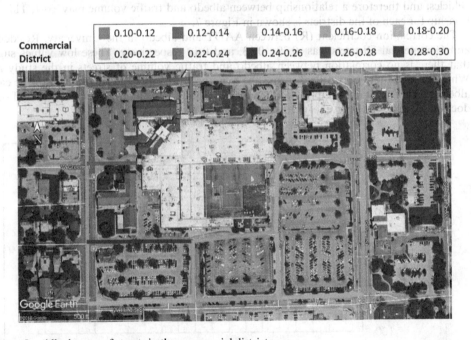

	0.10-0.12	0.12-0.14	0.14-0.16	0.16-0.18	0.18-0.20
Commercial District	0.20-0.22	0.22-0.24	0.24-0.26	0.26-0.28	0.28-0.30

Figure 5.   Albedo map of streets in the commercial district.

3.2   *Variation with traffic volume*

The traffic volume in terms of Average Annual Daily Traffic (AADT) for the roads in the three districts in 2018 was obtained from the Illinois Department of Transportation (IDOT, 2019) and compared to their measured albedo. Several roads in the Residential District did not have any traffic data and were omitted from the analysis. The objective of this comparison

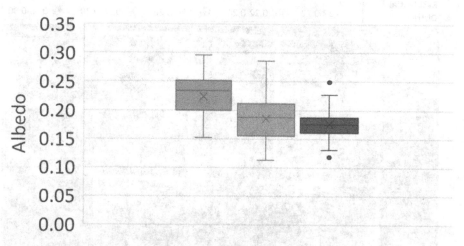

Figure 6. Albedo distribution for the three districts.

was to determine if traffic volume affected the optical properties of the road. It has been suggested that the albedo of pavements changes over time because of dirt and debris from vehicles, and therefore a relationship between albedo and traffic volume may exist. This relationship for each of the districts is shown in Figure 7.

The correlation coefficient (R) between AADT and albedo in the University, Residential, and Commercial Districts was -0.39, -0.13, and 0.12 respectively. These low values suggest that there is no correlation between albedo and traffic volume of streets in the study area. While this does not necessarily disprove the hypothesis that dirt and debris from traffic causes albedo to change, it does imply that the albedo of pavements after a significant period of use does not depend on the level of traffic.

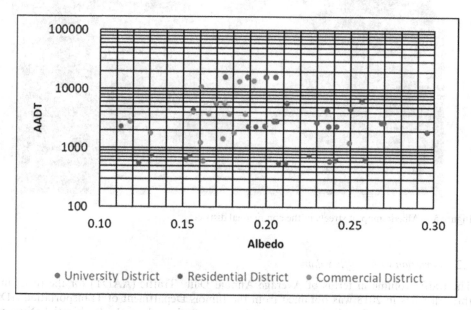

Figure 7. Relationship between AADT and albedo for each of the districts.

4 CONCLUSIONS

Albedo is one important property of pavements that contributes to the development of the UHI effect in cities. The measurement of pavement albedo in the field can be challenging, with current standard requiring clear, sunny days with a minimum of 4 m x 4 m section of the road cordoned off. Although efforts have been made to reduce the required sample size of the pavement area, it is still difficult to rapidly measure the albedo of a large street network. In this study, an albedo survey with a newly-developed instrument to rapidly measure pavement albedo, called D-SPARC, was presented.

The albedo of roads in three urban land use districts in Urbana, Illinois was measured using the D-SPARC device. These districts, called University, Residential, and Commercial Districts, differed in terms of their land use and traffic composition. Each district was about 0.1 km^2 in size and took only 90 minutes each to survey using D-SPARC. Albedo maps of the streets in these districts were then prepared from an average of three measurements per street segment (between consecutive intersections). In the University District, the albedo varied from 0.15-0.30, with an average of 0.22. In the Residential District, the corresponding values were 0.11-0.23 and 0.17 respectively, while albedo ranges were 0.12-0.25 and average 0.17 in the Commercial District. While the average albedo of the Residential and Commercial districts was found to be statistically similar, they were each found to be statistically different from that of the University District. The University District had a higher average albedo is attributed to the higher proportion of concrete pavements. Within each district, the smallest and largest measured albedo varied by approximately a factor of two, with differences even in the albedo for segments of roads that were adjacent to each other. This shows the localized variation of albedo that would be difficult to capture for large road networks using traditional albedo measurement techniques.

Finally, the variation of albedo with traffic volume (in terms of AADT) was examined. The correlation coefficient between the two was low, indicating no correlation between the traffic volume and albedo for the streets surveyed. The D-SPARC has been shown to produce reasonable albedo value for asphalt and concrete pavement in urban environment and could be used to build the input data for an urban heat island simulation.

REFERENCES

Akbari, H., Levinson, R. and Stern, S., 2008. Procedure for measuring the solar reflectance of flat or curved roofing assemblies. *Solar Energy, 82*(7), pp.648–655.

Arnfield, A.J., 2003. Two decades of urban climate research: a review of turbulence, exchanges of energy and water, and the urban heat island. *International Journal of Climatology: a Journal of the Royal Meteorological Society, 23*(1), pp.1–26.

ASTM E1918 (2016) Standard Test Method for Measuring Solar Reflectance of Horizontal and Low-Sloped Surfaces in the Field. *American Society for Testing and Materials (ASTM)*.

Gui, J., Phelan, P.E., Kaloush, K.E. and Golden, J.S., 2007. Impact of pavement thermophysical properties on surface temperatures. *Journal of materials in civil engineering, 19*(8), pp.683–690.

IDOT, 2019. https://www.gettingaroundillinois.com/gai.htm?mt=aadt. Accessed September 11, 2019.

Kleerekoper, L., Van Esch, M. and Salcedo, T.B., 2012. How to make a city climate-proof, addressing the urban heat island effect. *Resources, Conservation and Recycling, 64*, pp.30–38.

Li, H., Harvey, J. and Kendall, A., 2013. Field measurement of albedo for different land cover materials and effects on thermal performance. *Building and environment, 59*, pp.536–546.

Peng, S., Piao, S., Ciais, P., Friedlingstein, P., Ottle, C., Bréon, F.M., Nan, H., Zhou, L. and Myneni, R. B., 2011. Surface urban heat island across 419 global big cities. *Environmental science & technology, 46*(2), pp.696–703.

Pisello, A., Pignatta, G., Castaldo, V. and Cotana, F., 2014. Experimental analysis of natural gravel covering as cool roofing and cool pavement. *Sustainability, 6*(8), pp.4706–4722.

Qin, Y., 2015. A review on the development of cool pavements to mitigate urban heat island effect. *Renewable and sustainable energy reviews, 52*, pp.445–459.

Qin, Y. and He, H., 2017. A new simplified method for measuring the albedo of limited extent targets. *Solar Energy, 157*, pp.1047–1055.

Qin, Y., Hiller, J.E. and Meng, D., 2019. Linearity between pavement thermophysical properties and surface temperatures. *Journal of Materials in Civil Engineering, 31*(11), p.04019262

Richard, C., Dore, G., Lemieux, C., Bilodeau, J.P. and Haure-Touze, J., 2015. Albedo of pavement surfacing materials: in situ measurements. In *Cold Regions Engineering 2015* (pp. 181–192).

Santamouris, M., 2013. Using cool pavements as a mitigation strategy to fight urban heat island—A review of the actual developments. *Renewable and Sustainable Energy Reviews, 26*, pp.224–240.

Sen, S., 2019. Role of pavements in urban energetics. PhD Dissertation, University of Illinois at Urbana-Champaign, Urbana, IL.

Sen, S. and Roesler, J., 2017. Microscale heat island characterization of rigid pavements. *Transportation Research Record, 2639*(1), pp.73–83.

Sen, S., Roesler, J. and King, D., 2018. Albedo estimation of finite-sized concrete specimens. *Journal of Testing and Evaluation, 47*(2).

Sreedhar, S. and Biligiri, K.P., 2016. Comprehensive laboratory evaluation of thermophysical properties of pavement materials: Effects on urban heat island. *Journal of Materials in Civil Engineering, 28*(7), p.04016026.

Tran, H., Uchihama, D., Ochi, S. and Yasuoka, Y., 2006. Assessment with satellite data of the urban heat island effects in Asian mega cities. *International journal of applied Earth observation and Geoinformation, 8*(1), pp.34–48.

Author Index